Multimedia Systems, Standards, and Networks

edited by

Atul Puri
AT&T Labs
Red Bank, New Jersey

Tsuhan Chen
Carnegie Mellon University
Pittsburgh, Pennsylvania

MARCEL DEKKER, INC. NEW YORK · BASEL

ISBN: 0-8247-9303-X

This book is printed on acid-free paper.

Headquarters
Marcel Dekker, Inc.
270 Madison Avenue, New York, NY 10016
tel: 212-696-9000; fax: 212-685-4540

Eastern Hemisphere Distribution
Marcel Dekker AG
Hutgasse 4, Postfach 812, CH-4001 Basel, Switzerland
tel: 41-61-261-8482; fax: 41-61-261-8896

World Wide Web
http://www.dekker.com

The publisher offers discounts on this book when ordered in bulk quantities. For more information, write to Special Sales/Professional Marketing at the headquarters address above.

Current printing (last digit):
10 9 8 7 6 5 4 3 2 1

PRINTED IN THE UNITED STATES OF AMERICA

Multimedia Systems, Standards, and Networks

Signal Processing and Communications

1. Digital Signal Processing for Multimedia Systems, *edited by Keshab K. Parhi and Takao Nishitani*

2. Multimedia Systems, Standards, and Networks, *edited by Atul Puri and Tsuhan Chen*

Additional Volumes in Preparation

Embedded Multiprocessors: Scheduling and Synchronization, *Sundararajan Sriram and Shuvra S. Bhattacharyya*

Signal Processing for Intelligent Sensor Systems, *David C. Swanson*

Compressed Video Over Networks, *edited by Ming-Ting Sun and Amy Riebman*

Blind Equalization and Identification, *Zhi Ding and Ye (Geoffrey) Li*

Preface

We humans, being social creatures, have historically felt the need for increasingly sophisticated means to express ourselves through, for example, conversation, stories, pictures, entertainment, social interaction, and collaboration. Over time, our means of expression have included grunts of speech, storytelling, cave paintings, smoke signals, formal languages, stone tablets, printed newspapers and books, telegraphs, telephones, phonographs, radios, theaters and movies, television, personal computers (PCs), compact disc (CD) players, digital versatile disc (DVD) players, mobile phones and similar devices, and the Internet.

Presently, at the dawn of the new millennium, information technology is continuously evolving around us and influencing every aspect of our lives. Powered by high-speed processors, today's PCs, even inexpensive ones, have significant computational capabilities. These machines are capable of efficiently running even fairly complex applications, whereas not so long ago such tasks could often be handled only by expensive mainframe computers or dedicated, expensive hardware devices. Furthermore, PCs when networked offer a low-cost collaborative environment for business or consumer use (e.g., for access and management of corporate information over intranets or for any general information sharing over the Internet). Technological developments such as web servers, database systems, Hypertext Markup Language (HTML), and web browsers have considerably simplified our access to and interaction with information, even if the information resides in many computers over a network. Finally, because this information is intended for consumption by humans, it may be organized not only in textual but also in aural and/or visual forms.

Who Needs This Book?

Multimedia Systems, Standards, and Networks is about recent advances in multimedia systems, standards, and networking. This book is for you if you have ever been interested in efficient compression of images and video and want to find out what is coming next; if you have any interest in upcoming techniques for efficient compression of speech or music, or efficient representation of graphics and animation; if you have heard about existing or evolving ITU-T video standards as well as Moving Picture Experts Group (MPEG) video and audio standards and want to know more; if you have ever been curious about the space needed for storage of multimedia on a disc or bandwidth issues in transmission of multimedia over networks, and how these problems can be addressed by new coding standards; and finally (because it is not only about efficient compression but also about effective playback systems) if you want to learn more about flexible composition and user interactivity, over-the-network streaming, and search and retrieval.

What Is This Book About?

This is not to say that efficient compression is no longer important—in fact, this book pays a great deal of attention to that topic—but as compression technology undergoes standardization, matures, and is deployed in multimedia applications, many other issues are becoming increasingly relevant. For instance, issues in system design for synchronized playback of several simultaneous audio-visual streams are important. Also increasingly important is the capability for enhanced interaction of user with the content, and streaming of the same coded content over a variety of networks. This book addresses all these facets mainly by using the context of two recent MPEG standards. MPEG has a rich history of developing pioneering standards for digital video and audio coding, and its standards are currently used in digital cable TV, satellite TV, video on PCs, high-definition television, video on CD-ROMs, DVDs, the Internet, and much more. This book addresses two new standards, MPEG-4 and MPEG-7, that hold the potential of impacting many future applications, including interactive Internet multimedia, wireless videophones, multimedia search/browsing engines, multimedia-enhanced e-commerce, and networked computer video games. But before we get too far, it is time to briefly introduce a few basic terms.

So what is multimedia? Well, the term *multimedia* to some conjures images of cinematic wizardry or audiovisual special effects, whereas to others it simply means video with audio. Neither of the two views is totally accurate. We use the term *multimedia* in this book to mean digital multimedia, which implies the use of several digitized media simultaneously in a synchronized or related manner. Examples of various types of media include speech, images, text/graphics, audio, video, and computer animation. Furthermore, there is no strict requirement that all of these different media ought to be simultaneously used, just that more than one media type may be used and combined with others as needed to create an interesting multimedia presentation.

What do we mean by a *multimedia system*? Consider a typical multimedia presentation. As described, it may consist of a number of different streams that need to be continuously decoded and synchronized for presentation. A multimedia system is the entity that actually performs this task, among others. It ensures proper decoding of individual media streams. It ties the component media contained in the multimedia stream. It guarantees proper synchronization of individual media for playback of a presentation. A multimedia system may also check for and enforce intellectual property rights with respect to multimedia content.

Why do we need *multimedia standards*? Standards are needed to guarantee interoperability. For instance, a decoding device such as a DVD player can decode multimedia content of a DVD disc because the content is coded and formatted according to rules understood by the DVD player. In addition, having internationally uniform standards implies that a DVD disc bought anywhere in the world may be played on any DVD player. Standards have an important role not only in consumer electronics but also in multimedia communications. For example, a videotelephony system can work properly only if the two endpoints that want to communicate are compatible and each follows protocols that the other can understand. There are also other reasons for standards; e.g., because of economies of scale, establishment of multimedia standards allows devices, content, and services to be produced inexpensively.

What does *multimedia networking* mean? A multimedia application such as playing a DVD disc on a DVD player is a stand-alone application. However, an application requiring downloading of, for example, MP3 music content from a Web site to play on a hardware or software player uses networking. Yet another form of multimedia networking may involve playing streaming video where multimedia is chunked and transmitted to the decoder continuously instead of the decoder having to wait to download all of it. Multimedia communication applications such as videotelephony also use networking. Furthermore, a multiplayer video game application with remote players also uses networking. In fact, whether it relates to consumer electronics, wireless devices, or the Internet, multimedia networking is becoming increasingly important.

What Is in This Book?

Although an edited book, *Multimedia Systems, Standards, and Networks* has been painstakingly designed to have the flavor of an authored book. The contributors are the most knowledgeable about the topic they cover. They have made numerous technology contributions and chaired various groups in development of the ITU-T H.32x, H.263, or ISO MPEG-4 and MPEG-7 standards.

This book comprises 22 chapters. Chapters 1, 2, 3, and 4 contain background material including that on the ITU-T as well as ISO MPEG standards. Chapters 5 and 6 focus on MPEG-4 audio. Chapters 7, 8, 9, 10, and 11 describe various tools in the MPEG-4 Visual standard. Chapters 12, 13, 14, 15, and 16 describe important aspects of MPEG-4 Systems standard. Chapters 17, 18, and 19 discuss multimedia over networks. Chapters 20, 21, and 22 address multimedia search and retrieval as well as MPEG-7. We now elaborate on the contents of individual chapters.

Chapter 1 traces the history of technology and communication standards, along with recent developments and what can be expected in the future.

Chapter 2 presents a technical overview of the ITU-T H.323 and H.324 standards and discusses the various components of these standards.

Chapter 3 reviews the ITU-T H.263 (or version 1) standard as well as the H.263 version 2 standard. It also discusses the H.261 standard as the required background material for understanding the H.263 standards.

Chapter 4 presents a brief overview of the various MPEG standards to date. It thus addresses MPEG-1, MPEG-2, MPEG-4, and MPEG-7 standards.

Chapter 5 presents a review of the coding tools included in the MPEG-4 natural audio coding standard.

Chapter 6 reviews synthetic audio coding and synthetic natural hybrid coding (SNHC) of audio in the MPEG-4 standard.

Chapter 7 presents a high-level overview of the visual part of the MPEG-4 visual standard. It includes tools for coding of natural as well as synthetic video (animation).

Chapter 8 is the first of two chapters that deal with the details of coding natural video as per the MPEG-4 standard. It addresses rectangular video coding, scalability, and interlaced video coding.

Chapter 9 is the second chapter that discusses the details of coding of natural video as per the MPEG-4 standard. It also addresses coding of arbitrary-shape video objects, scalability, and sprites.

Chapter 10 discusses coding of still-image texture as specified in the visual part of the MPEG-4 standard. Both rectangular and arbitrary-shape image textures are supported.

Chapter 11 introduces synthetic visual coding as per the MPEG-4 standard. It includes 2D mesh representation of visual objects, as well as definition and animation of synthetic face and body.

Chapter 12 briefly reviews various tools and techniques included in the systems part of the MPEG-4 standard.

Chapter 13 introduces the basics of how, according to the systems part of the MPEG-4 standard, the elementary streams of coded audio or video objects are managed and delivered.

Chapter 14 discusses scene description and user interactivity according to the systems part of the MPEG-4 standard. Scene description describes the audiovisual scene with which users can interact.

Chapter 15 introduces a flexible MPEG-4 system based on Java programming language; this system exerts programmatic control on the underlying fixed MPEG-4 system.

Chapter 16 presents the work done within MPEG in software implementation of the MPEG-4 standard. A software framework for 2D and 3D players is discussed mainly for the Windows environment.

Chapter 17 discusses issues that arise in the transport of general coded multimedia over asynchronous transfer mode (ATM) networks and examines potential solutions.

Chapter 18 examines key issues in the delivery of coded MPEG-4 content over Internet Protocol (IP) networks. The MPEG and Internet Engineering Task Force (IETF) are jointly addressing these as well as other related issues.

Chapter 19 introduces the general topic of delivery of coded multimedia over wireless networks. With the increasing popularity of wireless devices, this research holds significant promise for the future.

Chapter 20 reviews the status of research in the general area of multimedia search and retrieval. This includes object-based as well as semantics-based search and filtering to retrieve images and video.

Chapter 21 reviews the progress made on the topic of image search and retrieval within the context of a digital library. Search may use a texture dictionary, localized descriptors, or regions.

Chapter 22 introduces progress in MPEG-7, the ongoing standard focusing on content description. MPEG-7, unlike previous MPEG standards, addresses search/retrieval and filtering applications, rather than compression.

Now that you have an idea of what each chapter covers, we hope you enjoy *Multimedia Systems, Standards, and Networks* and find it useful. We learned a great deal—and had a great time—putting this book together. Our heartfelt thanks to all the contributors for their enthusiasm and hard work. We are also thankful to our management, colleagues, and associates for their suggestions and advice throughout this project. We would like to thank Trista Chen, Fu Jie Huang, Howard Leung, and Deepak Turaga for their assistance in compiling the index. Last, but not least, we owe thanks to B. J. Clarke, J. Roh, and M. Russell along with others at Marcel Dekker, Inc.

Atul Puri
Tsuhan Chen

Contents

Contributors

Olivier Avaro Deutsche Telekom-Berkom GmbH, Darmstadt, Germany

Vahe Balabanian Nortel Networks, Nepean, Ontario, Canada

Mahesh Balakrishnan Philips Research, Briarcliff Manor, New York

Andrea Basso Broadband Communications Services Research, AT&T Labs, Red Bank, New Jersey

Stefano Battista bSoft, Macerata, Italy

Serge J. Belongie Computer Science Division, EECS Department, University of California at Berkeley, Berkeley, California

Chad Carson Computer Science Division, EECS Department, University of California at Berkeley, Berkeley, California

Shih-Fu Chang Department of Electrical Engineering, Columbia University, New York, New York

Tsuhan Chen Carnegie Mellon University, Pittsburgh, Pennsylvania

Xuemin Chen General Instrument, San Diego, California

Leonardo Chiariglione Television Technologies, CSELT, Torino, Italy

Mehmet Reha Civanlar Speech and Image Processing Research Laboratory, AT&T Labs, Red Bank, New Jersey

Yining Deng Electrical and Computer Engineering Department, University of California at Santa Barbara, Santa Barbara, California

Gianluca De Petris CSELT, Torino, Italy

Gianluca Di Cagno Services and Applications, CSELT, Torino, Italy

Peter K. Doenges Evans & Sutherland, Salt Lake City, Utah

F. Dufaux Compaq, Cambridge, Massachusetts

Touradj Ebrahimi Signal Processing Laboratory, Swiss Federal Institute of Technology (EPFL), Lausanne, Switzerland

Alexandros Eleftheriadis Department of Electrical Engineering, Columbia University, New York, New York

Gerard Fernando Sun Microsystems, Menlo Park, California

Yuval Fisher Institute for Nonlinear Science, University of California at San Diego, La Jolla, California

David A. Forsyth Computer Science Division, EECS Department, University of California at Berkeley, Berkeley, California

Guido Franceschini Services and Applications, CSELT, Torino, Italy

Jean Gelissen Nederlandse Phillips Bedrijven, Eindhoven, The Netherlands

Bernhard Grill Fraunhofer Geselshaft IIS, Erlangen, Germany

Barry G. Haskell AT&T Labs, Red Bank, New Jersey

Kaynam Hedayat Brix Networks, Billerica, Massachusetts

Carsten Herpel Thomson Multimedia, Hannover, Germany

Jürgen Herre Fraunhofer Geselshaft IIS, Erlangen, Germany

Caspar Horne Mediamatics, Inc., Fremont, California

Qian Huang AT&T Labs, Red Bank, New Jersey

Thomas Huang Department of Electrical and Computer Engineering, University of Illinois at Urbana-Champaign, Urbana, Illinois

Sergey Ioffe Computer Science Division, EECS Department, University of California at Berkeley, Berkeley, California

James D. Johnston AT&T Labs, Florham Park, New Jersey

Rob H. Koenen Multimedia Technology Group, KPN Research, Leidschendam, The Netherlands

Youngjik Lee ETRI Switching & Transmission Technology Laboratories, Taejon, Korea

Shipeng Li Microsoft Research China, Beijing, China

Weiping Li Optivision, Inc., Palo Alto, California

Jie Liang Texas Instruments, Dallas, Texas

Zvi Lifshitz Triton R&D Ltd., Jerusalem, Israel

Ajay Luthra General Instrument, San Diego, California

Wei-Ying Ma Hewlett-Packard Laboratories, Palo Alto, California

Jitendra Malik Computer Science Division, EECS Department, University of California at Berkeley, Berkeley, California

Bangalore S. Manjunath Electrical and Computer Engineering Department, University of California at Santa Barbara, Santa Barbara, California

Y. Nakaya Hitachi Ltd., Tokyo, Japan

Chiu Yeung Ngo Video Communications, Philips Research, Briarcliff Manor, New York

Joern Ostermann AT&T Labs, Red Bank, New Jersey

Fernando Pereira Instituto Superior Técnico/Instituto de Telecommunicações, Lisbon, Portugal

Eric Petajan Lucent Technologies, Murray Hill, New Jersey

G.D. Petris CSELT, Torino, Italy

Atul Puri AT&T Labs, Red Bank, New Jersey

Schuyler R. Quackenbush AT&T Labs, Florham Park, New Jersey

Hayder Radha Philips Research, Briarcliff Manor, New York

Ganesh Rajan General Instrument, San Diego, California

Dipankar Raychaudhuri C&C Research Laboratories, NEC USA, Inc., Princeton, New Jersey

Daniel J. Reininger C&C Research Laboratories, NEC USA, Inc., Princeton, New Jersey

Takashi Sato Philips Research, Briarcliff Manor, New York

Richard Schaphorst Delta Information Systems, Horsham, Pennsylvania

Eric D. Scheirer Machine Listening Group, MIT Media Laboratory, Cambridge, Massachusetts

Robert L. Schmidt AT&T Labs, Red Bank, New Jersey

Behzad Shahraray AT&T Labs, Red Bank, New Jersey

Julien Signès Research and Development, France Telecom Inc., Brisbane, California

Iraj Sodagar Sarnoff Corporation, Princeton, New Jersey

Gary J. Sullivan Picture Tel Corporation, Andover, Massachusetts

Viswanathan Swaminathan Sun Microsystems, Menlo Park, California

Raj Talluri Texas Instruments, Dallas, Texas

Peter van Beek Sharp Laboratories of America, Camas, Washington

Liam Ward Teltec Ireland, DCU, Dublin, Ireland

Jae-Woo Yang ETRI Switching & Transmission Technology Laboratories, Taejon, Korea

Ya-Qin Zhang Microsoft Research China, Beijing, China

Multimedia Systems, Standards, and Networks

1

Communication Standards: Götterdämmerung?

Leonardo Chiariglione
CSELT, Torino, Italy

I. INTRODUCTION

Communication standards are at the basis of civilized life. Human beings can achieve collective goals through sharing a common understanding that certain utterances are associated with certain objects, concepts, and all the way up to certain intellectual values. Civilization is preserved and enhanced from generation to generation because there is an agreed mapping between certain utterances and certain signs on paper that enable a human being to leave messages to posterity and posterity to revisit the experience of people who have long ago departed.

Over the centuries, the simplest communication means that have existed since the remotest antiquity have been supplemented by an endless series of new ones: printing, photography, telegraphy, telephony, television, and the new communication means such as electronic mail and the World Wide Web.

New inventions made possible new communication means, but before these could actually be deployed some agreements about the meaning of the "symbols" used by the communication means was necessary. Telegraphy is a working communication means only because there is an agreement on the correspondence between certain combinations of dots and dashes and characters, and so is television because there is an agreed procedure for converting certain waveforms into visible and audible information. The ratification and sometimes the development of these agreements—called standards—are what standards bodies are about. Standards bodies exist today at the international and national levels, industry specific or across industries, tightly overseen by governments or largely independent.

Many communication industries, among these the telecommunication and broadcasting industries, operate and prosper thanks to the existence of widely accepted standards. They have traditionally valued the role of standards bodies and have often provided their best personnel to help them achieve their goal of setting uniform standards on behalf of their industries. In doing so, they were driven by their role of "public service" providers,

Götterdämmerung: Twilight of the Gods. See, e.g., *http://walhall.com/*

a role legally sanctioned in most countries until very recently. Other industries, particularly the consumer electronics and computer industry, have taken a different attitude. They have "defined" communication standards either as individual companies or as groups of companies and then tried to impose their solution on the marketplace. In the case of a successful outcome, they (particularly the consumer electronics industry) eventually went to a standards body for ratification.

The two approaches have been in operation for enough time to allow some comparisons to be drawn. The former has given stability and constant growth to its industries and universal service to the general citizenship, at the price of a reduced ability to innovate: the telephone service is ubiquitous but has hardly changed in the past 100 years; television is enjoyed by billions of people around the world but is almost unchanged since its first deployment 60 years ago. The latter, instead, has provided a vibrant innovative industry. Two examples are provided by the personal computer (PC) and the compact disc. Both barely existed 15 years ago, and now the former is changing the world and the latter has brought spotless sound to hundreds of millions of homes. The other side of the coin is the fact that the costs of innovation have been borne by the end users, who have constantly struggled with incompatibilities between different pieces of equipment or software ("I cannot open your file") or have been forced to switch from one generation of equipment to the next simply because some dominant industry decreed that such a switch was necessary.

Privatization of telecommunication and media companies in many countries with renewed attention to the cost–benefit bottom line, the failure of some important standardization projects, the missing sense of direction in standards, and the lure that every company can become "the new Microsoft" in a business are changing the standardization landscape. Even old supporters of formal standardization are now questioning, if not the very existence of those bodies, at least the degree of commitment that was traditionally made to standards development.

The author of this chapter is a strong critic of the old ways of formal standardization that have led to the current diminished perception of its role. Having struggled for years with incompatibilities in computers and consumer electronics equipment, he is equally adverse to the development of communication standards in the marketplace. He thinks the time has come to blend the good sides of both approaches. He would like to bring his track record as evidence that a Darwinian process of selection of the fittest can and should be applied to standards *making* and that having standards is good to expand existing business as well as to create new ones. All this should be done not by favoring any particular industry, but working for all industries having a stake in the business.

This chapter revisits the foundations of communication standards, analyzes the reasons for the decadence of standards bodies, and proposes a framework within which a reconstruction of standardization on new foundations should be made.

II. COMMUNICATION SYSTEMS

Since the remotest antiquity, language has been a powerful communication system capable of conveying from one mind to another simple and straightforward as well as complex and abstract concepts. Language has not been the only communication means to have accompanied human evolution: body gesture, dance, sculpture, drawing, painting, etc. have all been invented to make communication a richer experience.

Writing evolved from the last two communication means. Originally used for point-to-point communication, it was transformed into a point-to-multipoint communication means by amanuenses. Libraries, starting with the Great Library of Alexandria in Egypt, were used to store books and enable access to written works.

The use of printing in ancient China and, in the West, Gutenberg's invention brought the advantage of making the reproduction of written works cheaper. The original simple system of book distribution eventually evolved to a two-tier distribution system: a network of shops where end users could buy books. The same distribution system was applied for newspapers and other periodicals.

Photography enabled the automatic reproduction of a natural scene, instead of hiring a painter. From the early times when photographers built everything from cameras to light-sensitive emulsions, this communication means has evolved to a system where films can be purchased at shops that also collect the exposed films, process them, and provide the printed photographs.

Postal systems existed for centuries, but their use was often restricted to kings or the higher classes. In the first half of the 19th century different systems developed in Europe that were for general correspondence use. The clumsy operational rules of these systems were harmonized in the second half of that century so that prepaid letters could be sent to all countries of the Universal Postal Union (UPU).

The exploitation of the telegraph (started in 1844) allowed the instant transmission of a message composed of Latin characters to a distant point. This communication system required the deployment of an infrastructure—again two-tier—consisting of a network of wires and of telegraph offices where people could send and receive messages. Of about the same time (1850) is the invention of facsimile, a device enabling the transmission of the information on a piece of paper to a distant point, even though its practical exploitation had to wait for another 100 years before effective scanning and reproduction techniques could be employed. The infrastructure needed by this communication system was the same as the telephony's.

Thomas A. Edison's phonograph (1877) was another communication means that enabled the recording of sound for later playback. Creation of the master and printing of disks required fairly sophisticated equipment, but the reproduction equipment was relatively inexpensive. Therefore the distribution channel developed in a very similar way as for books and magazines.

If the phonograph had allowed sound to cross the barriers of time and space, telephony enabled sound to overcome the barriers of space in virtually no time. The simple point-to-point model of the early years gave rise to an extremely complex hierarchical system. Today any point in the network can be connected with any other point.

Cinematography (1895) made it possible for the first time to capture not just a snapshot of the real world but a series of snapshots that, when displayed in rapid succession, appeared to reproduce something very similar to real movement to the eye. The original motion pictures were later supplemented by sound to give a complete reproduction to satisfy both the aural and visual senses.

The exploitation of the discovery that electromagnetic waves could propagate in the air over long distances produced wireless telegraphy (1896) and sound broadcasting (1920). The frequencies used at the beginning of sound broadcasting were such that a single transmitter could, in principle, reach every point on the globe by suitably exploiting propagation in the higher layers of atmosphere. Later, with the use of higher frequencies, only more geographically restricted areas, such as a continent, could be reached. Eventually, with the use of very high frequency (VHF), sound broadcasting became a more local business where again a two-tier distribution systems usually had to be put in place.

The discovery of the capability of some material to generate current if exposed to light, coupled with the older cathode ray tube (CRT), capable of generating light via electrons generated by some voltage, gave rise to the first communication system that enabled the real-time capture of a visual scene, simultaneous transmission to a distant point, and regeneration of a moving picture on a CRT screen. This technology, even though demonstrated in early times for person-to-person communication, found wide use in television broadcasting. From the late 1930s in the United Kingdom television provided a powerful communication means with which both the aural and visual information generated at some central point could reach distant places in no time. Because of the high frequencies involved, the VHF band implied that television was a national communication system based on a two-tier infrastructure. The erratic propagation characteristics of VHF in some areas prompted the development of alternative distribution systems: at first by cable, referred to as CATV (community antenna television), and later by satellite. The latter opened the television system from a national business to at least a continental scale.

The transformation of the aural and visual information into electric signals made possible by the microphone and the television pickup tube prompted the development of systems to record audio and video information in real time. Eventually, magnetic tapes contained in cassettes provided consumer-grade systems, first for audio and later for video.

Automatic Latin character transmission, either generated in real time or read from a perforated paper band, started at the beginning of this century with the teletypewriter. This evolved to become the telex machine, until 10 years ago a ubiquitous character-based communication tool for businesses.

The teletypewriter was also one of the first machines used by humans to communicate with a computer, originally via a perforated paper band and, later, via perforated cards. Communication was originally carried out using a sequence of coded instructions (machine language instructions) specific to the computer make that the machine would execute to carry out operations on some input data. Later, human-friendlier programming (i.e., communication) languages were introduced. Machine native code could be generated from the high-level language program by using a machine-specific converter called a compiler.

With the growing amount of information processed by computers, it became necessary to develop systems to store digital information. The preferred storage technology was magnetic, on tapes and disks. Whereas with audio and video recorders the information was already analog and a suitable transducer would convert a current or voltage into a magnetic field, information in digital form required systems called modulation schemes to store the data in an effective way. A basic requirement was that the information had to be "formatted."

The need to transmit digital data over telephone lines had to deal with a similar problem, with the added difficulty of the very variable characteristics of telephone lines. Information stored on a disk or tape was formatted, so the information sent across a telephone line was organized in packets.

In the 1960s the processing of information in digital form proper of the computer was introduced in the telephone and other networks. At the beginning this was for the purpose of processing signaling and operating switches to cope with the growing complexity of the telephone network and to provide interesting new services possible because of the flexibility of the electronic computing machines.

Far reaching was the exploitation of a discovery of the 1930s (so-called Nyquist sampling theorem) that a bandwidth-limited signal could be reproduced faithfully if sampled with a frequency greater than twice the bandwidth. At the transmitting side the signal was sampled, quantized, and the output represented by a set of bits. At the receiving side the opposite operation was performed. At the beginning this was applied only to telephone signals, but the progress in microelectronics, with its ability to perform sophisticated digital signal processing using silicon chips of increased complexity, later allowed the handling in digital form of such wideband signals as television.

As the number of bits needed to represent sampled and quantized signals was unnecessarily large, algorithms were devised to reduce the number of bits by removing redundancy without affecting too much, or not at all as in the case of facsimile, the quality of the signal. The conversion of heretofore analog signals into binary digits and the existence of a multiplicity of analog delivery media prompted the development of sophisticated modulation schemes. A design parameter for these schemes was the ability to pack as many bits per second as possible in a given frequency band without affecting the reliability of the transmitted information.

The conversion of different media in digital form triggered the development of receivers—called decoders—capable of understanding the sequences of bits and converting them into audible and/or visible information. A similar process also took place with "pages" of formatted character information. The receivers in this case were called browsers because they could also move across the network using addressing information embedded in the coded page.

The growing complexity of computer programs started breaking up what used to be monolithic software packages. It became necessary to define interfaces between layers of software so that software packages from different sources could interoperate. This need gave rise to the standardization of APIs (application programming interfaces) and the advent of "object-oriented" software technology.

III. COMMUNICATION STANDARDS

For any of the manifold ways of communication described in the preceding section, it is clear that there must be an agreement about the way information is represented at the point where information is exchanged between communicating systems. This is true for language, which is a communication means, because there exists an agreement by members of a group that certain sounds correspond to certain objects or concepts. For languages such as Chinese, writing can be defined as the agreement by members of a group that some graphic symbols, isolated or in groups, correspond to particular objects or concepts. For languages such as English, writing can be defined as the agreement by members of a group that some graphic symbols, in certain combinations and subject to certain dependences, correspond to certain basic sounds that can be assembled into compound sounds and traced back to particular objects or concepts. In all cases mentioned an agreement—a standard—about the meaning is needed if communication is to take place.

Printing offers another meaning of the word "standard." Originally, all pieces needed in a print shop were made by the people in the print shop itself or in some related shop. As the technology grew in complexity, however, it became convenient to agree—i.e., to set standards—on a set of character sizes so that one shop could produce the press while another could produce the characters. This was obviously beneficial because the print shop could concentrate on what it was supposed to do best, print books. This is the manufacturing-oriented definition of standardization that is found in the *Encyclopaedia Britannica*: "Standardisation, in industry: imposition of standards that permit large production runs of component parts that are readily fitted to other parts without adjustment." Of course, communication between the author of a book and the reader is usually not hampered if a print shop decides to use characters of a nonstandard size or a different font. However, the shop may have a hard time finding them or may even have to make them itself.

The same applies to photography. Cameras were originally produced by a single individual or shop and so were the films, but later it became convenient to standardize the film size so that different companies could specialize in either cameras or films. Again, communication between the person taking the picture and the person to whom the picture is sent is not hampered if pictures are taken with a camera using a nonstandard film size. However, it may be harder to find the film and get it processed.

Telegraphy was the first example of a new communication system, based on a new technology, that required agreement between the parties if the sequence of dots and dashes was to be understood by the recipient. Interestingly, this was also a communication standard imposed on users by its inventor. Samuel Morse himself developed what is now called the Morse alphabet and the use of the alphabet bearing his name continues to this day.

The phonograph also required standards, namely the amplitude corresponding to a given intensity and the speed of the disk, so that the sound could be reproduced without intensity and frequency distortions. As with telegraphy, the standard elements were basically imposed by the inventor. The analog nature of this standard makes the standard apparently less constraining, because small departures from the standard are not critical. The rotation speed of the turntable may increase but meaningful sound can still be obtained, even though the frequency spectrum of the reproduced signal is distorted.

Originally, telephony required only standardization of the amplitude and frequency characteristics of the carbon microphone. However, with the growing complexity of the telephone system, other elements of the system, such as the line impedance and the duration of the pulse generated by the rotary dial, required standardization. As with the phonograph, small departures from the standard values did not prevent the system from providing the ability to carry speech to distant places, with increasing distortions for increasing departures from the standard values.

Cinematography, basically a sequence of photographs each displayed for a brief moment—originally 16 and later 24 times a second—also required standards: the film size and the display rate. Today, visual rendition is improved by flashing 72 pictures per second on the screen by shuttering each still three times. This is one example of how it is possible to have different communication qualities while using the same communication standard. The addition of sound to the motion pictures, for a long time in the form of a trace on a side of the film, also required standards.

Sound broadcasting required standards: in addition to the baseband characteristics of the sound there was also a need to standardize the modulation scheme (amplitude and later frequency modulation), the frequency bands allocated to the different transmissions, etc.

Television broadcasting required a complex standard related to the way a television camera scans a given scene. The standard specifies how many times per second a picture is taken, how many scan lines per picture are taken, how the signal is normalized, how the beginning of a picture and of a scan line is signaled, how the sound information is multiplexed, etc. The modulation scheme utilized at radio frequency (vestigial sideband) was also standardized.

Magnetic recording of audio and video also requires standards, simpler for audio (magnetization intensity, compensation characteristics of the nonlinear frequency response of the inductive playback head, and tape speed), more complex for video because of the structure of the signal and its bandwidth.

Character coding standards were also needed for the teletypewriter. Starting with the Baudot code, a long series of character coding standards were produced that continue today with the 2- and 4-byte character coding of International Standardization Organization/International Electrotechnical Commission (ISO/IEC) 10646 (Unicode).

Character coding provides a link to a domain that was not originally considered to be strictly part of "communication": the electronic computer. This was originally a standalone machine that received some input data, processed them, and produced some output data. The first data input to a computer were digital numbers, but soon characters were used. Different manufacturers developed different ways to encode numbers and characters and the way operations on the data were carried out. This was done to suit the internal architecture of their computers. Therefore each type of computing machine required its own "communication standard." Later on, high-level programming languages such as COBOL, FORTRAN, C, and C++ were standardized in a machine-independent fashion. Perforations of paper cards and tapes as well as systems for storing binary data on tapes and disks also required standards.

With the introduction of digital technologies in the telecommunication sector in the 1960s, standards were required for different aspects such as the sampling frequency of telephone speech (8 kHz), the number of bits per sample (seven or eight for speech), the quantization characteristics (A-law, μ-law), etc. Other areas that required standardization were signaling between switches (several CCITT "alphabets"), the way different sequences of bits each representing a telephone speech could be assembled (multiplexed), etc. Another important area of standardization was the way to modulate transmission lines so that they could carry sequences of bits (bit/s) instead of analog signals (Hertz).

The transmission of digital data across a network required the standardization of addressing information, the packet length, the flow control, etc. Numerous standards were produced: X.25, I.311, and the most successful of all, the Internet Protocol (IP).

The compact disc, a system that stored sampled music in digital form, with a laser beam used to detect the value of a bit, was a notable example of standardization: the sampling frequency (44.1 kHz), the number of bits per sample (16), the quantization characteristics (linear), the distance between holes on the disc surface, the rotation speed, the packing of bits in frames, etc.

Systems to reduce the number of bits necessary to represent speech, facsimile, music, and video information utilized exceedingly complex algorithms, all requiring standardization. Some of them, e.g., the MPEG-1 and MPEG-2 coding algorithms of the Moving Picture Experts Group, have achieved wide fame even with the general public. The latter is used in digital television receivers (set-top boxes). Hypertext markup language (HTML), a standard to represent formatted pages, has given rise to the ubiquitous Web browser, actually a "decoder" of HTML pages.

The software world has produced a large number of software standards. In newspaper headlines today is Win32, a set of APIs providing high-level functionalities abstracted from the specifics of the hardware processing unit that programmers wishing to develop applications on top of the Windows operating system have to follow. This is the most extreme, albeit not unique, case of a standard, as its is fully owned by a single company. The Win32 APIs are constantly enriched with more and more functionalities. One such functionality, again in newspaper headlines these days, is the HTML decoder, alias Web browser. Another is the MPEG-1 software decoder.

IV. THE STANDARDS BODIES

It is likely that human languages developed in a spontaneous way, but in most societies the development of writing was probably driven by the priesthood. In modern times special bodies were established, often at the instigation of public authorities (PAs), with the goal of taking care of the precise definition and maintenance of language and writing. In Italy the Accademia della Crusca (established 1583) took on the goal of preserving the Florentine language of Dante. In France the Académie Française (established 1635) is to this day the official body in charge of the definition of the French language. Recently, the German Bundestag approved a law that amends the way the German language should be written.

The role of PAs in the area of language and writing, admittedly a rather extreme case, is well represented by the following sentence: ''La langue est donc un élément clé de la politique culturelle d'un pays car elle n'est pas seulement un instrument de communication . . . mais aussi un outil d'identification, un signe d'appartenance à une communauté linguistique, un élément du patrimoine national que l'État entend défendre contre les atteintes qui y sont portées'' (*language is therefore a key element of the cultural policy of a country because it is not just a communication tool . . . but also an identification means, a sign that indicates membership to a language community, an element of the national assets that the State intends to defend against the attacks that are waged against it*).

Other forms of communication, however, are or have become fairly soon after their invention of more international concern. They have invariably seen the governments as the major actors. This is the case for telegraphy, post, telephone, radio, and television.

The mail service developed quickly after the introduction of prepaid letters in the United Kingdom in 1840. A uniform rate in the domestic service for all letters of a certain weight, regardless of the distance involved, was introduced. At the international level, however, the mail service was bound by a conflicting web of postal services and regulations with up to 1200 rates. The General Postal Union (established in 1874 and renamed Universal Postal Union in 1878) defined a single postal territory where the reciprocal exchange of letter-post items was possible with a single rate for all and with the principle of freedom of transit for letter-post items.

A similar process took place for telegraphy. In less than 10 years after the first transmission, telegraphy had become available to the general public in developed countries. At the beginning telegraph lines did not cross national frontiers because each country used a different system and each had its own telegraph code to safeguard the secrecy of its military and political telegraph messages. Messages had to be transcribed, translated, and handed over at frontiers before being retransmitted over the telegraph network of the neighboring country. The first International Telegraph Convention was signed in 1865 and harmonized the different systems used. This was an important step in telecommunication, as it was clearly attractive for the general public to be able to send telegraph messages to every place where there was a telegraph network.

Following the invention of the telephone and the subsequent expansion of telephony, the Telegraph Union began, in 1885, to draw up international rules for telephony. In 1906 the first International Radiotelegraph Convention was signed. The International Telephone Consultative Committee (CCIF) set up in 1924, the International Telegraph Consultative Committee (CCIT) set up in 1925, and the International Radio Consultative Committee (CCIR) set up in 1927 were made responsible for drawing up international standards. In 1927, the union allocated frequency bands to the various radio services existing at the time (fixed, maritime and aeronautical mobile, broadcasting, amateur, and experimental). In 1934 the International Telegraph Convention of 1865 and the International Radiotelegraph Convention of 1906 were merged to become the International Telecommunication Union (ITU). In 1956, the CCIT and the CCIF were amalgamated to give rise to the International Telephone and Telegraph Consultative Committee (CCITT). Today the CCITT is called ITU-T and the CCIR is called ITU-R.

Other communication means developed without the explicit intervention of governments but were often the result of a clever invention of an individual or a company that successfully made its way into the market and became an industrial standard. This was the case for photography, cinematography, and recording. Industries in the same business found it convenient to establish industry associations, actually a continuation of a process that had started centuries before with medieval guilds. Some government then decided to create umbrella organizations—called national standards bodies—of which all separate associations were members, with the obvious exception of matters related to post, telecommunication, and broadcasting that were already firmly in the hands of governments. The first country to do so was, apparently, the United Kingdom with the establishment in 1901 of an Engineering Standards Committee that became the British Standards Institute in 1931. In addition to developing standards, whose use is often made compulsory in public procurements, these national standards bodies often take care of assessing the conformity of implementations to a standard. This aspect, obviously associated in people's minds with "quality", explains why quality is often in the titles of these bodies, as is the case for the Portuguese Standards Body IPQ (Instituto Português da Qualidade).

The need to establish international standards developed with the growth of trade. The International Electrotechnical Commission (IEC) was founded in 1906 to prepare and publish international standards for all electrical, electronic, and related technologies. The IEC is currently responsible for standards for such communication means as "receivers," audio and video recording systems, and audiovisual equipment, currently all grouped in TC 100 (Audio, Video and Multimedia Systems and Equipment). International standardization in other fields and particularly in mechanical engineering was the concern of the International Federation of the National Standardizing Associations (ISA), set up in 1926. ISA's activities ceased in 1942 but a new international organization called ISO began to operate again in 1947 with the objective "to facilitate the international coordination and unification of industrial standards." All computer-related activities are currently in the Joint ISO/IEC Technical Committee 1 (JTC 1) on Information Technology. This technical committee has achieved a very large size. About one-third of all ISO and IEC standards work is done in JTC1.

Whereas ITU and UPU are treaty organizations (i.e., they have been established by treaties signed by government representatives) and the former is an agency of the United

Nations since 1947, ISO and IEC have the status of private not-for-profit companies established according to the Swiss Civil Code.

V. THE STANDARDS BODIES AT WORK

Because "communication," as defined in this chapter, is such a wide concept and so many different constituencies with such different backgrounds have a stake in it, there is no such thing as a single way to develop standards. There are, however, some common patterns that are followed by industries of the same kind.

The first industry considered here is the telecommunication industry, meant here to include telegraphy, telephony, and their derivatives. As discussed earlier, this industry had a global approach to communication from the very beginning. Early technical differences justified by the absence of a need to send or receive telegraph messages between different countries were soon ironed out, and the same happened to telephony, which could make use of the international body set up in 1865 for telegraphy to promote international telecommunication. In the 130 plus years of its history, what is now ITU-T has gone through various technological phases. Today a huge body of "study groups" take care of standardization needs: SG 3 (Tariffs), SG 7 (Data Networks), SG 11 (Signaling), SG 13 (Network Aspects), SG 16 (Multimedia), etc.

The vast majority of the technical standards at the basis of the telecommunication system have their correspondence in an ITU-T standard. At the regional level, basically in Europe and North America, and to some extent in Japan, there has always been a strong focus on developing technical standards for matters of regional interest and preparing technical work to be fed into ITU-T. A big departure from the traditional approach of standards of worldwide applicability began in the 1960s with the digital representation of speech: 7 bits per sample advocated by the United States and Japan, 8 bits per sample advocated by Europe. This led to several different transmission hierarchies because they were based on a different building block, digitized speech. This rift was eventually mended by standards for bit rate–reduced speech, but the hundreds of billions of dollars invested by telecommunication operators in incompatible digital transmission hierarchies could not be recovered. The ATM (asynchronous transfer mode) project gave the ITU-T an opportunity to overcome the differences in digital transmission hierarchies and provide international standards for digital transmission of data. Another departure from the old philosophy was made with mobile telephony: in the United States there is not even a national mobile telephony standard, as individual operators are free to choose standards of their own liking. This contrasts with the approach adopted in Europe, where the global system for mobile (GSM) standard is so successful that it is expanding all over the world, the United States included. With universal mobile telecommunication system (UMTS) (so-called third-generation mobile) the ITU-T is retaking its original role of developer of global mobile telecommunication standards.

The ITU-R comes from a similar background but had a completely different evolution. The development of standards for sound broadcasting had to take into account the fact that with the frequencies used at that time the radio signal could potentially reach any point on the earth. Global sound broadcasting standards became imperative. This approach was continued when the use of VHF for frequency-modulated (FM) sound programs was started: FM radio is a broadcasting standard used throughout the world. The case of television was different. A first monochrome television system was deployed in the United Kingdom in the late 1930s, a different one in the United States in the 1940s, and yet a different one in Europe in the 1950s. In the 1960s the compatible addition of color information in the television system led to a proliferation of regional and national variants of television that continues until today. The ITU-R was also unable to define a single system for teletext (characters carried in unused television lines to be displayed on the television screen). Another failure has followed the attempt to define a single standard for high-definition television.

The field of consumer electronics, represented by the IEC, is characterized by an individualistic approach to standards. Companies develop new communication means based on their own ideas and try to impose their products on the market. Applied to audio-based communication means, this has led so far to a single standard generally being adopted by industry soon after the launch of a new product, possibly after a short battle between competing solutions. This was the case with the audio tape recorder, the compact cassette, and the compact disc. Other cases have been less successful: for a few years there was competition between two different ways of using compressed digital audio applications, one using a compact cassette and the other using a recordable minidisc. The result has been the demise of one and very slow progress of the other. More battles of this type loom ahead. Video-based products have been less lucky. For more than 10 years a standards battle continued between Betamax and VHS, two different types of videocassette recorder. Contrary to the often-made statement that having competition in the marketplace brings better products to consumers, some consider that the type of videocassette that eventually prevailed in the marketplace is technically inferior to the type that lost the war.

The fields of photography and cinematography (whose standardization is currently housed, at the international level, in ISO) have adopted a truly international approach. Photographic cameras are produced to make use of one out of a restricted number of film sizes. Cinematography has settled with a small number of formats each characterized by a certain level of performance.

The computer world has adopted the most individualistic approach of all industries. Computing machines developed by different manufacturers had different central processing unit (CPU) architectures, programming languages, and peripherals. Standardization took a long time to penetrate this world. The first examples were communication ports (EIA RS 232), character coding [American Standard Code for Information Interchange (ASCII), later to become ISO/IEC 646], and programming languages (e.g., FORTRAN, later to become ISO/IEC 1539). The hype of computer and telecommunication convergence of the early 1980s prompted the launching of an ambitious project to define a set of standards that would enable communication between a computer of any make with another computer of any make across any network. For obvious reasons, the project, called OSI (Open Systems Interconnection), was jointly executed with ITU-T. In retrospect, it is clear that the idea to have a standard allowing a computer of any make (and at that time there were tens and tens of computers of different makes) to connect to any kind of network, talk to a computer of any make, execute applications on the other computer, etc., no matter how fascinating it was, had very little prospect of success. And so it turned out to be, but after 15 years of efforts and thousands of person-years spent when the project was all but discontinued.

For the rest ISO/IEC JTC 1, as mentioned before, has become a huge standards body. This should be no surprise, as JTC 1 defines information technology to "include the specification, design and development of systems and tools dealing with the capture, representation, processing, security, transfer, interchange, presentation, management, organization, storage and retrieval of information.'' Just that!

While ISO and ITU were tinkering with their OSI dream, setting out first to design how the world should be and then trying to build it, in a typical top-down fashion, a group of academics (admittedly well funded by their government) were practically building the same world bottom up. Their idea was that once you had defined a protocol for transporting packets of data and, possibly, a flow-control protocol, you could develop all sorts of protocols, such as SMTP (Simple Mail Transport Protocol), FTP (File Transfer Protocol), and HTTP (HyperText Transport Protocol). This would immediately enable the provision of very appealing applications. In other words, Goliath (ISO and ITU) has been beaten by David (Internet). Formal standards bodies no longer set the pace of telecommunication standards development.

The need for other communication standards—for computers—was simply overlooked by JTC 1. The result has been the establishment of a de facto standard, owned by a single company, in one of the most crucial areas of communication: the Win32 APIs. Another case—Java, again owned by a single company—may be next in line.

VI. THE DIGITAL COMMUNICATION AGE

During its history humankind has developed manifold means of communication. The most diverse technologies were assembled at different times and places to provide more effective ways to communicate between humans, between humans and computers, and between computers, overcoming the barriers of time and space. The range of technologies include

- Sound waves produced by the human phonic organs (speech)
- Coded representations of words on physical substrates such as paper or stone (writing and printing)
- Chemical reactions triggered by light emitted by physical objects (photography)
- Propagation of electromagnetic waves on wires (telegraphy)
- Current generation when carbon to which a voltage is applied is hit by a sound wave
- Engraving with a vibrating stylus on a surface (phonograph)
- Sequences of photographs mechanically advanced and illuminated (cinematography)
- Propagation of electromagnetic waves in free space (radio broadcasting)
- Current generation by certain materials hit by light emitted by physical objects (television)
- Magnetization of a tape coated with magnetic material (audio and video recording)
- Electronic components capable of changing their internal state from on to off and vice versa (computers)
- Electronic circuits capable of converting the input value of a signal to a sequence of bits representing the signal value (digital communication)

The history of communication standards can be roughly divided into three periods. The first covers a time when all enabling technologies were diverse: mechanical, chemical, electrical, and magnetic. Because of the diversity of the underlying technologies, it was more than natural that different industries would take care of their standardization needs without much interaction among them.

In the second period, the common electromagnetic nature of the technologies provided a common theoretical unifying framework. However, even though a microphone could be used by the telephone and radio broadcasting communities or a television camera by the television broadcasting, CATV, consumer electronic (recording), or telecommunication (videoconference) communities, it happened that either the communities had different quality targets or there was an industry that had been the first developer of the technology and therefore had a recognized leading role in a particular field. In this technology phase, too, industries could accommodate their standardization needs without much interaction among them.

Digital technologies create a different challenge, because the only part that differentiates the technologies of the industries is the delivery layer. Information can be represented and processed using the same digital technologies, while applications sitting on top tend to be even less dependent on the specific environment.

In the 1980s a superficial reading of the implications of this technological convergence made IBM and AT&T think they were competitors. So AT&T tried to create a computer company inside the group and when it failed it invested billions of dollars to acquire the highly successful NCR just to transform it in no time into a money loser. The end of the story a few years later was that AT&T decided to spin off its newly acquired computer company *and* its old manufacturing arm. In the process, it also divested itself of its entire manufacturing arm. In parallel IBM developed a global network to connect its dispersed business units and started selling communication services to other companies. Now IBM has decided to shed the business because it is "noncore." To whom? Rumors say to AT&T!

The lesson, if there is a need to be reminded of it, is that technology is just one component, not necessarily the most important, of the business. That lesson notwithstanding, in the 1990s we are hearing another mermaid's song, the convergence of computers, entertainment and telecommunications. Other bloodbaths are looming.

Convergence hype apart, the fact that a single technology is shared by almost all industries in the communication business *is* relevant to the problem this chapter addresses, namely why the perceived importance of standardization is rapidly decreasing, whether there is still a need for the standardization function, and, if so, how it must be managed.

This because digital technologies bring together industries with completely different backgrounds in terms of their attitudes vis-à-vis public authorities and end users, standardization, business practices, technology progress, and handling of intellectual property rights (IPR). Let us consider the last item.

VII. INTELLECTUAL PROPERTY

The recognition of the ingenuity of an individual who invented a technology enabling a new form of communication is a potent incentive to produce innovation. Patents have existed since the 15th century, but it is the U.S. Constitution of 1787 that explicitly links private incentive to overall progress by giving the Congress the power "to promote the progress of . . . the useful arts, by securing for limited times to . . . inventors the exclusive rights to their . . . discoveries." If the early years of printing are somewhat shrouded in a cloud of uncertainty about who was the true inventor of printing and how much contributed to it, subsequent inventions such as telegraphy, photography, and telephony were duly registered at the patent office and sometimes their inventors, business associates, and heirs enjoyed considerable economic benefits.

Standardization, a process of defining a single effective way to do things out of a number of alternatives, is clearly strictly connected to the process that motivates individuals to provide better communication means today than existed yesterday or to provide communication means that did not exist before.

Gutenberg's invention, if filed today, would probably deserve several patents or at least multiple claims because of the diverse technologies that he is credited with having invented. Today's systems are several orders of magnitude more complex than printing. As an example, the number of patents needed to build a compact disc audio player is counted in the hundreds. This is why what is known as ''intellectual property'' has come to play an increasingly important role in communication.

Standards bodies such as IEC, ISO, and ITU have developed a consistent and uniform policy vis-à-vis intellectual property. In simple words, the policy tolerates the existence of necessary patents in international standards provided the owner of the corresponding rights is ready to give licenses on fair and reasonable terms and on a nondiscriminatory basis. This simple principle is finding several challenges.

A. Patents, a Tool for Business

Over the years patents have become a tool for conducting business. Companies are forced to file patents not so much because they have something valuable and they want to protect it but because patents become the merchandise to be traded at a negotiating table when new products are discussed or conflicts are resolved. On these occasions it is not so much the value of the patents that counts but the number and thickness of the piles of patent files. This is all the more strange when one considers that very often a patented innovation has a lifetime of a couple of years so that in many cases the patent is already obsolete at the time it is granted. In the words of one industry representative, the patenting folly now mainly costs money and does not do any good for the end products.

B. A Patent May Emerge Too Late

Another challenge is caused by the widely different procedures that countries have in place to deal with the processing of patent filings. One patent may stay under examination for many years (10 or even more) and stay lawfully undisclosed. At the end of this long period, when the patent is published, the rights holder can lawfully start enforcing the rights. However, because the patent may be used in a non-IEC/ISO/ITU standard or, even in that case, if the rights holder has decided not to conform to that policy, the rights holder is not bound by the fair and reasonable terms and conditions and may conceivably request any amount of money. At that time, however, the technology may have been deployed by millions and the companies involved may have unknowingly built enormous liabilities. Far from promoting progress, as stated in the U.S. Constitution of 1787, this practice is actually hampering it, because companies are alarmed by the liabilities they take on board when launching products where gray zones exist concerning patents.

C. Too Many Patents May Be Needed

A third challenge is provided by the complexity of modern communication systems, where a large number of patents may be needed. If the necessary patents are owned by a restricted number of companies, they may decide to team up and develop a product by cross-licensing the necessary patents. If the product, as in the case of the MPEG–2 standard, requires patents whose rights are owned by a large number of companies (reportedly about 40 patents are needed to implement MPEG–2) and each company applies the fair and reasonable terms clause of the IEC/ISO/ITU patent policy, the sum of 40 fair and reasonable terms may no longer be fair and reasonable. The MPEG–2 case has been resolved by establishing a "patent pool," which reportedly provides a one-stop license office for most MPEG–2 patents. The general applicability of the patent pool solution, however, is far from certain. The current patent arrangement, a reasonable one years ago when it was first adopted, is no longer able to cope with the changed conditions.

D. Different Models to License Patents

The fourth challenge is provided by the new nature of standards offered by information technology. Whereas traditional communication standards had a clear physical embodiment, with digital technologies a standard is likely to be a processing algorithm that runs on a programmable device. Actually, the standard may cease to be a patent and becomes a piece of computer code whose protection is achieved by protecting the copyright of the computer code. Alternatively, both the patent and the copyright are secured. But because digital networks have become pervasive, it is possible for a programmable device to run a multiplicity of algorithms downloaded from the network while not being, if not at certain times, one of the physical embodiments the standards were traditionally associated with. The problem is now that traditional patent licensing has been applied assuming that there *is* a single piece of hardware with which a patent is associated. Following the old pattern, a patent holder may grant fair and reasonable (in his opinion and according to his business model) terms to a licensee, but the patent holder is actually discriminating against the licensee because the former has a business model that assumes the existence of the hardware thing, whereas the latter has a completely different model that assumes only the existence of a programmable device.

E. All IPR Together

The fifth challenge is provided by yet another convergence caused by digital technologies. In the analog domain there is a clear separation between the device that makes communication possible and the message. When a rented video cassette is played back on a VHS player, what is paid is a remuneration to the holders of the copyright for the movie and a remuneration to the holders of patent rights for the video recording system made at the time the player was purchased. In the digital domain an application may be composed of some digitally encoded pieces of audio and video, some text and drawings, some computer code that manages user interaction, access to the different components of the application, etc. If the device used to run the application is of the programmable type, the intellectual property can only be associated with the bits—content and executable code—downloaded from the network.

F. Mounting Role of Content

The last challenge in this list is provided by the increasingly important role of content in the digital era. Restricted access to content is not unknown in the analog world, and it is used to offer selected content to closed groups of subscribers. Direct use of digital technologies with their high quality and ease of duplication, however, may mean the immediate loss of content value unless suitable mechanisms are in place to restrict access to those who have acquired the appropriate level of rights. Having overlooked this aspect has meant a protracted delay in the introduction of digital versatile disc (DVD), the new generation of compact disc capable of providing high-quality, MPEG–2–encoded movies.

The conclusion is the increased role of IPR in communication, its interaction with technological choices, and the advancing merge of the two components—patents and copyright—caused by digital technologies. This sees the involvement of the World Intellectual Property Organisation (WIPO), another treaty organization that is delegated to deal with IPR matters.

VIII. NOT EVERYTHING THAT SHINES IS GOLD

The challenges that have been exposed in the preceding sections do not find standardization in the best shape, as will be shown in the following.

A. Too Slow

The structure of standards bodies was defined at a time when the pace of technological evolution was slow. Standards committees had plenty of time to consider new technologies and for members to report back to their companies or governments, layers of bureaucracies had time to consider the implications of new technologies, and committees could then reconsider the issues over and over until ''consensus'' (the magic word of ISO and IEC) or unanimity (the equivalent in ITU) was achieved. In other words, standardization could afford to operate in a well-organized manner, slowly and bureaucratically. Standardization could afford to be ''nice'' to everybody.

An example of the success of the old way of developing standards is the integrated services digital network (ISDN). This was an ITU project started at the beginning of the 1970s. The project deliberately set the threshold high by targeting the transmission of two 64 kbit/sec streams when one would have amply sufficed. Although the specifications were completed in the mid-1980s, it took several years before interoperable equipment could be deployed in the network. Only now is ISDN taking off, thanks to a technology completely unforeseen at the time the project was started—the Internet. An example of failure has been the joint JTC1 and ITU-T OSI project. Too many years passed between the drawing board and the actual specification effort. By the time OSI solutions had become ready to be deployed, the market had already been invaded by the simpler Internet solution. A mixed success has been ATM standardization. The original assumption was that ATM would be used on optical fibers operating at 155 Mbit/sec, but today the optical fiber to the end user is still a promise for the future. It is only thanks to the ATM Forum specifications that ATM can be found today on twisted pair at 25 Mbit/sec. For years the CCITT discussed the 32- versus 64-byte cell length issue. Eventually, a decision for 48 bytes was made; however, in the meantime precious years had been lost and now ATM, instead of being the pervasive infrastructure of the digital network of the future, is relegated to being a basic transmission technology.

In the past a single industry, e.g., the government-protected monopoly of telecommunication, could set the pace of development of new technology. In the digital era the number of players, none of them pampered by public authorities, is large and increasing. As a consequence, standardization can no longer afford to move at a slow pace. The old approach of providing well-thought-over, comprehensive, nice-to-everybody solutions has to contend with nimbler but faster solutions coming from industry consortia or even individual companies.

B. Too Many Options

In abstract terms, everybody agrees that a standard should specify a single way of doing things. The practice is that people attending a standards committee work for a company that has a definite interest in getting one of their technologies in the standard. It is not unusual that the very people attending are absolutely determined to have their pet ideas in the standard. The rest of the committee is just unable or unwilling to oppose because of the need to be "fair" to everybody. The usual outcome of a dialectic battle lasting anywhere from 1 hour to 10 years is the compromise of the intellectually accepted principle of a single standard without changing the name. This is how "options" come in.

In the past, this did not matter too much because transformation of a standard into products or services was in many cases a process driven by infrastructure investments, in which the manufacturers had to wait for big orders from telecom operators. These eventually bore the cost of the options that, in many cases, their own people had stuffed into the standards and that they were now asking manufacturers to implement. Because of too many signaling options, it took too many years for European ISDN to achieve a decent level of interoperability between different telecommunications operators and, within the same operator, between equipment from different manufacturers. But this was the time when telecommunication operators were still the drivers of the development.

The case of the ATM is enlightening. In spite of several ITU-T recommendations having been produced in the early 1990s, industry was still not producing any equipment conforming to these recommendations. Members of the ATM Forum like to boast that their first specification was developed in just 4 months without any technical work, if not the removal of some options from existing ITU-T recommendations. Once the heavy ITU-T documents that industry, without backing of fat orders from telecom operators, had not dared to implement became slim ATM Forum specifications, ATM products became commercially available at the initiative of manufacturers, at interesting prices and in a matter of months.

C. No Change

When the technologies used by the different industries were specific to the individual industries, it made sense to have different standards bodies taking care of individual standardization needs. The few overlaps that happened from time to time were dealt with in an ad hoc fashion. This was the case with the establishment of the CMTT, a joint CCITT-CCIR committee for the long-distance transmission of audio and television signals, the meeting point of broadcasting and telecommunication, or the OSI activity, the meeting point of telecommunication and computers. With the sweeping advances in digital technologies, many of the issues that are separately considered in different committees of the different standards bodies are becoming common issues.

A typical case is that of compression of audio and video, a common technology for ITU-T, ITU-R, JTC1, and now also the World Wide Web Consortium (W3C). Instead of agreeing to develop standards once and in a single place, these standards bodies are actually running independent standards projects. This attitude not only is wasting resources but also delays acceptance of standards because it makes it more difficult to reach a critical mass that justifies investments. Further, it creates confusion because of multiple standard solutions for similar problems.

Within the same ITU it has been impossible to rationalize the activities of its ''R'' and ''T'' branches. A high-level committee appointed a few years ago to restructure the ITU came to the momentous recommendations for

1. Renaming CCITT and CCIR as ITU-T and ITU-R
2. Replacing the Roman numerals of CCITT study groups with the Arabic numerals of ITU-T study groups (those of CCIR were already Arabic)
3. Moving the responsibility for administration of the CMTT from CCIR to ITU-T while renaming it Study Group 9.

''Minor'' technical details such as ''who does what'' went untouched, so video services are still the responsibility of ITU-R SG 11 if delivered by radio, ITU-T SG 9 if delivered by cable, and ITU-T SG 16 if delivered by wires or optical fibers. For a long time mobile communication used to be in limbo because the delivery medium is radio, hence the competence of ITU-R, but the service is (largely) conversational, hence the competence of ITU-T.

D. Lagging, Not Leading

During its history the CCITT has gone through a series of reorganizations to cope with the evolution of technology. With a series of enlightened decisions the CCITT adapted itself to the gradual introduction of digital technologies first in the infrastructure and later in the end systems. For years the telecommunication industry waited for CCITT to produce their recommendations before starting any production runs.

In 1987 an enlightened decision was made by ISO and IEC to combine all computer-related activities of both bodies in a single technical committee, called ISO/IEC JTC1. Unlike the approach in the ITU, in the IEC, and also in many areas of JTC1, the usual approach has been one of endorsing de facto standards that had been successful in the marketplace.

In the past 10 years, however, standards bodies have lost most of the momentum that kept them abreast of technology innovations. Traditionally, telephony modem standards had been the purview of ITU-T, but in spite of evidence for more than 10 years that the local loop could be digitized to carry several Mbit/sec downstream, no ITU-T standards exist today for ADSL, when they are deployed by the hundreds of thousands, without any backing of ITU-T recommendations. The same is true for digitization of broadcast-related delivery media such as satellite or terrestrial media: too many ITU-R standards exist for broadcast modem. A standard exists for digitizing cable for CATV services but in the typical fashion of recommending three standards: one for Europe, one for the United States, and one for Japan. In JTC1, supposedly the home for everything software, no object-oriented technology standardization was ever attempted. In spite of its maturity, no standardization of intelligent agents was even considered. In all bodies, no effective security standards, the fundamental technology for business in the digital world, were ever produced.

E. The Flourishing of Consortia

The response of the industry to this eroding role of standards bodies has been to establish consortia dealing with specific areas of interest. In addition to the already mentioned Internet Society, whose Internet Engineering Task Force (IETF) is a large open international community of network designers, operators, vendors, and researchers concerned with the evolution of the Internet architecture and the smooth operation of the Internet, and the ATM Forum, established with the purpose of accelerating the use of ATM products and services, there are the Object Management Group (OMG), whose mission is to promote the theory and practice of object technology for the development of distributed computing systems; the Digital Audio-Visual Council (DAVIC), established with the purpose of promoting end-to-end interoperable digital audiovisual products, services, and applications; the World Wide Web Consortium (W3C), established to lead the World Wide Web to its full potential by developing common protocols that promote its evolution and ensure its interoperability; the Foundation for Intelligent Physical Agents (FIPA), established to promote the development of specifications of generic agent technologies that maximize interoperability within and across agent-based applications; and Digital Video Broadcasting (DVB), committed to designing a global family of standards for the delivery of digital television and many others.

Each of these groups, in most cases with a precise industry connotation, is busy developing its own specifications. The formal standards bodies just sit there while they see their membership and their relevance eroded by the day. Instead of asking themselves why this is happening and taking the appropriate measures, they have put in place a new mechanism whereby Publicly Available Specifications (PASs) can easily be converted into International Standards, following a simple procedure. Just a declaration of surrender!

IX. A WAY OUT

The process called standardization, the enabler of communication, is in a situation of stalemate. Unless vigorous actions are taken, the whole process is bound to collapse in the near future. In the following some actions are proposed to restore the process to function for the purpose for which it was established.

A. Break the Standardization–Regulation Ties

Since the second half of the 19th century, public authorities have seen as one of their roles the general provision of communication means to all citizens. From the 1840s public authorities, directly or through direct supervision, started providing the postal service, from the 1850s the telegraph service, and from the 1870s compulsory elementary education, through which children acquired oral and paper-based communication capabilities. In the same period the newly invented telephony, with its ability to put citizens in touch with one another, attracted the attention of public authorities, as did wireless telegraphy at the turn of the century and broadcasting in the 1920s and television in the 1930s. All bodies in charge of standardization of these communication means, at both national and international levels, see public authorities as prime actors.

Whichever were the past justifications for public authorities to play this leading role in setting communication standards and running the corresponding businesses on behalf of the general public, they no longer apply today. The postal service is being privatized in most countries, and the telegraph service has all but disappeared because telephony is ubiquitous and no longer fixed, as more and more types of mobile telephony are within everybody's reach. The number of radio and television channels in every country is counted by the tens and will soon be by the hundreds. The Internet is providing cheap access to information to a growing share of the general public of every country. Only compulsory education stubbornly stays within the purview of the state.

So why should the ITU still be a treaty organization? What is the purpose of governments still being involved in setting telecommunication and broadcasting standards? Why, if all countries are privatizing their post, telecommunication, and media companies, should government still have a say in standards at the basis of those businesses?

The ITU should be converted to the same status as IEC and ISO, i.e., a private not-for-profit company established according to Swiss Civil Code. The sooner technical standards are removed from the purview of public authorities, the sooner the essence of regulation will be clarified.

B. Standards Bodies as Companies

A state-owned company does not automatically become a swift market player simply because it has been privatized. What is important is that an entrepreneurial spirit drives its activity. For a standards body this starts with the identification of its mission, i.e., the proactive development of standards serving the needs of a defined multiplicity of industries, which I call ''shareholders.'' This requires the existence of a function that I call ''strategic planning'' with the task of identifying the needs for standards; of a function that I call ''product development,'' the actual development of standards; and of a function that I call ''customer care,'' the follow-up of the use of standards with the customers, i.e., the companies that are the target users of the standards.

A radical change of mentality is needed. Standards committees have to change their attitude of being around for the purpose of covering a certain technical area. Standards are the goods that standards committees sell their customers, and their development is to be managed pretty much with the same management tools that are used for product development. As with a company, the goods have to be of high quality, have to be according to the specification agreed upon with the customers, but, foremost, they have to be delivered by the agreed date. This leads to the first precept for standards development: *Stick to the deadline*.

The need to manage standard development as a product development also implies that there must be in place the right amount and quality of human material. Too often companies send to standards committees their newly recruited personnel, with the idea that giving them some opportunity for international exposure is good for their education, instead of sending their best people. Too often selection of leadership is based on ''balance of power'' criteria and not on management capabilities.

C. A New Standard-Making Process

The following is a list of reminders that should be strictly followed concerning the features that standards must have.

A Priori Standardization. If a standards body is to serve the needs of a community of industries, it must start the development of standards well ahead of the time the need for the standard appears. This requires a fully functioning and dedicated strategic planning function fully aware of the evolution of the technology and the state of research.

Not Systems but Tools. The industry-specific nature of many standards bodies is one of the causes of the current decadence of standardization. Standards bodies should collect different industries, each needing standards based on the same technology but possibly with different products in mind. Therefore only the components of a standard, the ''tools,'' can be the object of standardization. The following process has been found effective:

1. Select a number of target applications for which the generic technology is intended to be specified.
2. List the functionalities needed by each application.
3. Break down the functionalities into components of sufficiently reduced complexity that they can be identified in the different applications.
4. Identify the functionality components that are common across the systems of interest.
5. Specify the tools that support the identified functionality components, particularly those common to different applications.
6. Verify that the tools specified can actually be used to assemble the target systems and provide the desired functionalities.

Specify the Minimum. When standards bodies are made up of a single industry, it is very convenient to add to a standard those nice little things that bring the standard nearer to a product specification as in the case of industry standards or standards used to enforce the concept of "guaranteed quality" so dear to broadcasters and telecommunication operators because of their "public service" nature. This practice must be abandoned; only the minimum that is necessary for interoperability can be specified. The extra that is desirable for one industry may be unneeded by or alienate another.

One Functionality–One Tool. More than a rule, this is good common sense. Too many failures in standards are known to have been caused by too many options.

Relocation of Tools. When a standard is defined by a single industry, there is generally agreement about where a given functionality resides in the system. In a multi-industry environment this is usually not the case because the location of a function in the communication chain is often associated with the value added by a certain industry. The technology must be defined not only in a generic way but also in such a way that the technology can be located at different points in the system.

Verification of the Standard. It is not enough to produce a standard. Evidence must be given that the work done indeed satisfies the requirements ("product specification") originally agreed upon. This is obviously also an important promotional tool for the acceptance of the standard in the marketplace.

D. Dealing with Accelerating Technology Cycles

What is proposed in the preceding paragraphs would, in some cases, have solved the problems of standardization that started to become acute several years ago. Unfortunately, by themselves they are not sufficient to cope with the current trend of accelerating technology cycles.

On the one hand, this forces the standardization function to become even more anticipative along the lines of the "a priori standardization" principle. Standards bodies must be able to make good guesses about the next wave of technologies and appropriately invest in standardizing the relevant aspects. On the other, there is a growing inability to predict the exact evolution of a technology, so that standardization makes sense, at least in the initial phases, only if it is restricted to the "framework" or the "platform" and if it contains enough room to accommodate evolution.

The challenge then is to change the standards culture: to stress time to market, to reduce prescriptive scope, to provide frameworks that create a solution space, and to populate the framework with concrete (default) instances. Last, and possibly most important, there is a need to refine the standard in response to success or failure in the market. The concept contains contradiction: the standard, which people might expect to be prescriptive, is instead an understated framework, and the standard, which people might expect to be static, anticipates evolution.

E. Not Just Process, Real Restructuring Is Needed

Innovating the standards-making process is important but pointless if the organization is left untouched. As stated before, the only thing that digital technologies leave as specific to the individual industries is the delivery layer. The higher one goes, the less industry-specific standardization becomes.

The organization of standards bodies is currently vertical, and this should be changed to a horizontal one. There should be one body addressing the delivery layer issues, possibly structured along different delivery media, one body for the application layer, and one body for middleware.

This is no revolution. It is the shape the computer business naturally acquired when the many incompatible vertical computer systems started converging. It is also the organization the Internet world has given to itself. There is no body corresponding to the delivery layer, given that the Internet sits on top of it, but IETF takes care of middleware and W3C of the application layer.

X. CONCLUSIONS

Standards make communication possible, but standards making has not kept pace with technology evolution, and much less is it equipped to deal with the challenges lying ahead that this chapter has summarily highlighted. Radical measures are needed to preserve the standardization function, lest progress and innovation be replaced by stagnation and chaos. This chapter advocates the preservation of the major international standards bodies after a thorough restructuring from a vertical industry-oriented organization to a horizontal function-oriented organization.

ACKNOWLEDGMENTS

This chapter is the result of the experience of the author over the past 10 years of activity in standardization. In that time frame, he has benefited from the advice and collaboration of a large number of individuals in the different bodies he has operated in: MPEG, DAVIC, FIPA, and OPIMA. Their contributions are gratefully acknowledged.

Special thanks go to the following individuals, who have reviewed the chapter and provided the author with their advice: James Brailean, Pentti Haikonen (Nokia), Barry Haskell (AT&T Research), Keith Hill (MCPS Ltd.), Rob Koenen (KPN Research), Murat Kunt (EPFL), Geoffrey Morrison (BTLabs), Fernando Pereira (Instituto Superior Técnico), Peter Schirling (IBM), Ali Tabatabai (Tektronix), James VanLoo (Sun Microsystems), Liam Ward (Teltec Ireland), and David Wood (EBU).

The opinions expressed in this chapter are those of the author only and are not necessarily shared by those who have reviewed the chapter.

2

ITU-T H.323 and H.324 Standards

Kaynam Hedayat
Brix Networks, Billerica, Massachusetts

Richard Schaphorst
Delta Information Systems, Horsham, Pennsylvania

I. INTRODUCTION

The International Telephony Union, a United Nations organization, is responsible for coordination of global telecom networks and services among governments and the private sector. As part of this responsibility, the ITU provides standards for multimedia communication systems. In recent years the two most important of these standards have been H.323 and H.324.

Standard H.323 provides the technical requirements for multimedia communication systems that operate over packet-based networks where guaranteed quality of service may or may not be available. Generally, packet-based networks cannot guarantee a predictable delay for data delivery and data may be lost and/or received out of order. Examples of such packet-based networks are local area networks (LANs) in enterprises, corporate intranets, and the Internet.

Recommendation H.324 provides the technical requirements for multimedia communication systems that operate over bit rate multimedia communication, utilizing V.34 modems operating over the general switched telephone network (GSTN).

II. H.323

The popularity and ubiquity of local area networks and the Internet in the late 1980s and early 1990s prompted a number of companies to begin work on videoconferencing and telephony systems that operate over packet-based networks including corporate LANs. Traditionally, videoconferencing and telephony systems have been designed to operate over networks with predictable data delivery behavior, hence the requirement for switched circuit networks (SCNs) by videoconferencing standards such as H.320 and H.324. Generally, packet-based networks cannot guarantee a predictable delay for delivery of data, and data can be lost and/or received out of order. These networks were often deployed utilizing the Transfer Control Protocol/Internet Protocol (TCP/IP) protocol and lack of quality of service (QoS), and their unpredictable behavior was among the challenges that were faced.

Table 1 H.323 Documents

Document	Description
H.323	System architecture and procedures
H.225.0	Call signaling, media packetization and streaming
H.245	Call control
H.235	Security and encryption
H.450.1	Generic control protocol for supplementary services
H.450.2	Call transfer
H.450.3	Call forward
H.332	Larger conferences
Implementers Guide	Corrections and clarifications to the standard

Companies facing these challenges developed the appropriate solutions and championed the work on the H.323 standard within the ITU. Their goal was to introduce a standard solution to the industry in order to promote future development of the use of videoconferencing and telephony systems.

The H.323 standard was introduced to provide the technical requirements for multimedia communication systems that operate over packet-based networks where guaranteed quality of service might not be available. H.323 version 1 (V1) was finalized and approved by the ITU in 1996 and is believed to be revolutionizing the increasingly important field of videoconferencing and IP telephony by becoming the dominant standard of IP-based telephones, audioconferencing, and videoconferencing terminals. H.323 V2 was finalized and approved by the ITU in 1998, and H.323 V3 is planned for approval in the year 2000. The following sections present a general overview of the H.323 protocol and its progression from V1 to V3. The intention is to give the reader a basic understanding of the H.323 architecture and protocol. Many specific details of the protocol are not described here, and the reader is encouraged to read the H.323 standard for a thorough understanding of the protocol.

A. Documents

The H.323 standard consists of three main documents, H.323, H.225.0, and H.245. H.323 defines the system architecture, components, and procedures of the protocol. H.225.0 covers the call signaling protocol used to establish connections and the media stream packetization protocol used for transmitting and receiving media over packetized networks. H.245 covers the protocol for establishing and controlling the call.* Other related documents provide extensions to the H.323 standard. Table 1 lists currently available H.323 documents. The Implementers Guide document is of importance to all the implementers of

* Establishing the call is different from establishing the connection. The latter is analogous to ringing the telephone; the former is analogous to the start of a conversation.

H.323 systems. It contains corrections and clarifications of the standard for known problem resolutions. All of the documents can be obtained from the ITU (*www.itu.int*).

B. Architecture

The H.323 standard defines the components (endpoint, gatekeeper, gateway, multipoint controller, and multipoint processor) and protocols of a multimedia system for establishing audio, video, and data conferencing. The standard covers the communication protocols among the components, addressing location-independent connectivity, operation independent of underlying packet-based networks, network control and monitoring, and interoperability among other multimedia protocols. Figure 1 depicts the H.323 components with respect to the packet-based and SCN networks. The following sections detail the role of each component.

1. Endpoint

An endpoint is an entity that can be called, meaning that it initiates and receives H.323 calls and can accept and generate multimedia information. An endpoint may be an H.323 terminal, gateway, or multipoint control unit (combination of multipoint controller and multipoint processor). Examples of endpoints are the H.323 terminals that popular operating systems provide for Internet telephony.

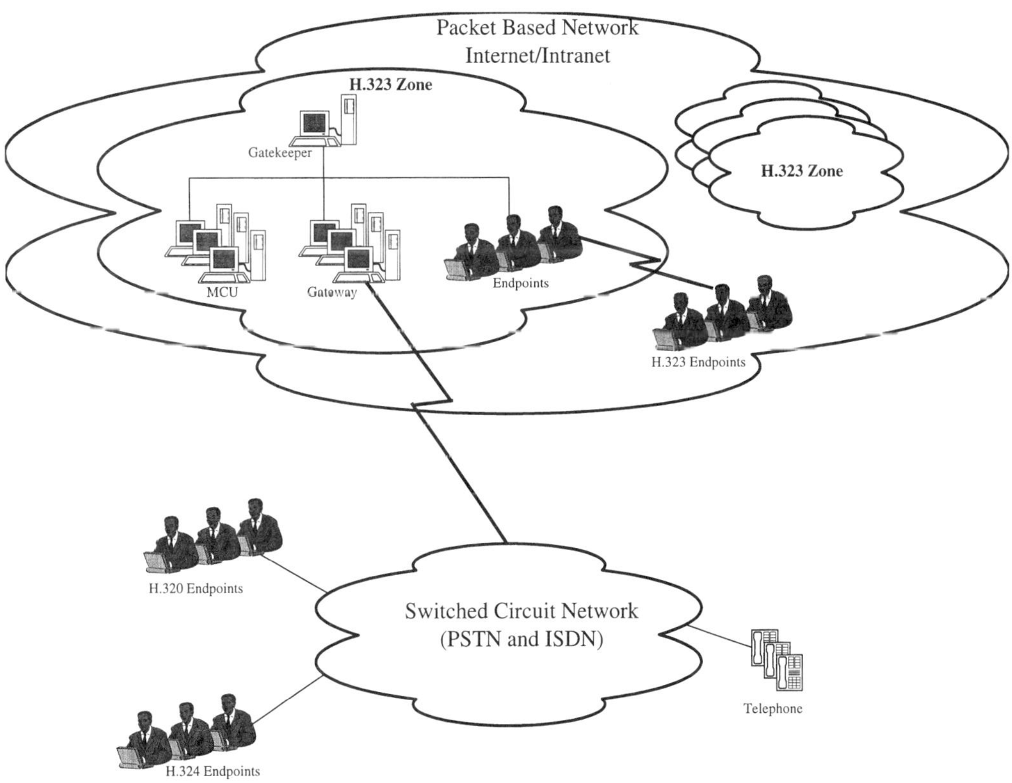

Figure 1 H.323 network.

2. Gatekeeper

An H.323 network can be a collection of endpoints within a packet-based network that can call each other directly without the intervention of other systems. It can also be a collection of H.323 endpoints managed by a server referred to as a gatekeeper. The collection of endpoints that are managed by a gatekeeper is referred to as an H.323 zone. In other words, a gatekeeper is an entity that manages the endpoints within its zone. Gatekeepers provide address translation, admission control, and bandwidth management for endpoints. Multiple gatekeepers may manage the endpoints of one H.323 network. This implies the existence of multiple H.323 zones. An H.323 zone can span multiple network segments and domains. There is no relation between an H.323 zone and network segments or domains within a packet-based network.

On a packet-based network, endpoints can address each other by using their network address (i.e., IP address). This method is not user friendly because telephone numbers, names, and e-mail addresses are the most common form of addressing. Gatekeepers allow endpoints to address one another by a telephone number, name, e-mail address, or any other convention based on numbers or text. This is achieved through the address translation process, the process by which one endpoint finds another endpoint's network address from a name or a telephone number. The address translation is achieved through the gatekeeper registration process. In this process, all endpoints within a gatekeeper's zone are required to provide their gatekeeper with identification information such as endpoint type and addressing convention. Through this registration process, the gatekeeper has knowledge of all endpoints within its zone and is able to perform the address translation by referencing its database.

Endpoints find the network address of other endpoints through the admission process. This process requires an endpoint to contact its gatekeeper for permission prior to making a call to another endpoint. The admission process gives the gatekeeper the ability to restrict access to network resources requested by endpoints within its zone. Upon receiving a request from an endpoint, the gatekeeper can grant or refuse permission based on an admission policy. The admission policy is not within the scope of the H.323 standard. An example of such a policy would be a limitation on the number of calls in an H.323 zone. If permission is granted, the gatekeeper provides the network address of the destination to the calling endpoint.

All nodes on a packet-based network share the available bandwidth. It is desirable to control the bandwidth usage of multimedia applications because of their usually high bandwidth requirement. As part of the admission process for each call, the endpoints are required to inform the gatekeeper about their maximum bandwidth. Endpoints calculate this value on the basis of what they can receive and transmit. With this information the gatekeeper can restrict the number of calls and amount of bandwidth used within its zone. Bandwidth management should not be confused with providing quality of service. The former is the ability to manage the bandwidth usage of the network. The latter is the ability to provide a guarantee concerning a certain bandwidth, delay, and other quality parameters. It should also be noted that the gatekeeper bandwidth management is not applied to the network as a whole. It is applied only to the H.323 traffic of the network within the gatekeeper's zone.

Endpoints may also operate without a gatekeeper. Consequently, gatekeepers are an optional part of an H.323 network, although their services are usually indispensable. In addition to the services mentioned, gatekeepers may offer other services such as controlling the flow of a call by becoming the central point of calls through the gatekeeper-routed call model (see Sec. II.E).

3. Gateway

Providing interoperability with other protocols is an important goal of the H.323 standard. Users of other protocols such as H.324, H.320, and public switched telephone network should be able to communicate with H.323 users. H.323 gateways provide translation between control and media formats of the two protocols they are connecting. They provide connectivity by acting as bridges to other multimedia or telephony networks. H.323 gateways act as an H.323 endpoint on the H.323 network and as a corresponding endpoint (i.e., H.324, H.320, PSTN) on the other network. A special case of a gateway is the H.323 proxy, which acts as an H.323 endpoint on both sides of its connection. H.323 proxies are used mainly in firewalls.

4. Multipoint Control Unit

The H.323 standard supports calls with three or more endpoints. The control and management of these multipoint calls are supported through the functions of the multipoint controller (MC) entity. The management includes inviting and accepting other endpoints into the conference, selecting a common mode of communication between the endpoints, and connecting multiple conferences into a single conference (conference cascading). All endpoints in a conference establish their control signaling with the MC, enabling it to control the conference. The MC does not manipulate the media.

The multipoint processor (MP) is the entity that processes the media. The MP may be centralized, with media processing for all endpoints in a conference taking place in one location, or it may be distributed, with the processing taking place separately in each endpoint. Examples of the processing of the media are mixing the audio of participants and switching their video (i.e., to the current speaker) in a conference.

The multipoint control unit (MCU) is an entity that contains both an MC and an MP. The MCU is used on the network to provide both control and media processing for a centralized conference, relieving endpoints from performing complex media manipulation. MCUs are usually high-end servers on the network.

C. Protocols

H.323 protocols fall into four categories:

- Communication between endpoints and gatekeepers
- Call signaling for connection establishment
- Call control for controlling and managing the call
- Media transmission and reception including media packetization, streaming, and monitoring

An H.323 call scenario optionally starts with the gatekeeper admission request. It is then succeeded by call signaling to establish the connection between endpoints. Next, a communication channel is established for call control. Finally the media flow is established. Each step of the call utilizes one of the protocols provided by H.323, namely registration, admissions, and status signaling (H.225.0 RAS); call signaling (H.225.0); call control (H.245); and real-time media transport and control (RTP/RTCP).

H.323 may be implemented independent of the underlying transport protocol. Two endpoints may communicate as long as they are using the same transport protocol and have network connectivity (e.g., on the internet using TCP/IP). Although H.323 is widely deployed on TCP/IP-based networks, it does not require the TCP/IP transport protocol. The only requirement of the underlying protocol is to provide packet-based unreliable transport, packet-based reliable transport, and optionally packet-based unreliable multicast transport. The TCP/IP suite of protocols closely meet all of the requirements through user datagram protocol (UDP), TCP, and IP Multicast, respectively. The only exception is that TCP, the reliable protocol running on top of IP, is a stream-oriented protocol and data are delivered in stream of bytes. A thin protocol layer referred to as TPKT provides a packet-based interface for TCP. TPKT is used only for stream-oriented protocols (e.g., SPX is a packet-oriented reliable protocol and does not require the use of TPKT). Figure 2 depicts the H.323 protocol suite utilizing the TCP/IP-based networks. As can be seen, H.323 utilizes transport protocols such as TCP/IP and operates independently of the underlying physical network (e.g., Ethernet, token ring). Note that because TCP/IP may operate over switched circuit networks such as integrated services digital network (ISDN) and plain old telephone systems using point-to-point protocol (PPP), H.323 can easily be deployed on these networks.

Multimedia applications, for proper operation, require a certain quality of service from the networks they utilize. Usually packet-based networks do not provide any QoS and packets are generally transferred with the best effort delivery policy. The exceptions are networks such as asynchronous transfer mode (ATM), where QoS is provided. Consequently, the amount of available bandwidth is not known at any moment in time, the amount of delay between transmission and reception of information is not constant, and information may be lost anywhere on the network. Furthermore, H.323 does not require any QoS from the underlying network layers. H.323 protocols are designed considering these limitations because the quality of the audio and video in a conference directly de-

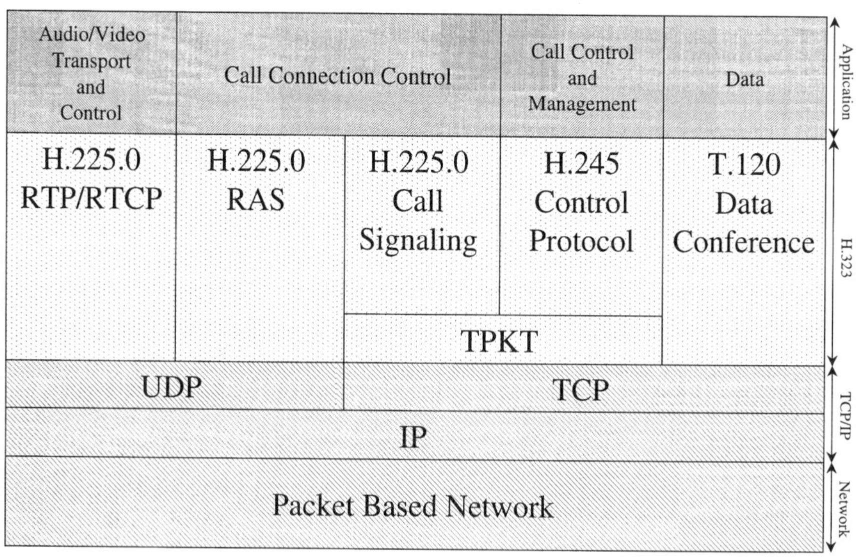

Figure 2 H.323 protocols on TCP/IP network.

pends on the QoS of the underlying network. The following four sections describe the protocols used in H.323.

1. RAS

H.225.0 RAS (registration, admissions, and status) protocol is used for communication between endpoints and gatekeepers.* The RAS protocol messages fall into the following categories: gatekeeper discovery, endpoint registration, endpoint location, admissions, bandwidth management, status inquiry, and disengage.

Endpoints can have prior knowledge of a gatekeeper through static configuration or other means. Alternatively, endpoints may discover the location of a suitable gatekeeper through the gatekeeper discovery process. Endpoints can transmit a gatekeeper discovery request message either to a group of gatekeepers using multicast transport or to a single host that might have a gatekeeper available. If multicast transport is utilized, it is possible for a number of gatekeepers to receive the message and respond. In this case it is up to the endpoint to select the appropriate gatekeeper. After the discovery process, the endpoints must register with the gatekeeper. The registration process is required by all endpoints that want to use the services of the gatekeeper. It is sometimes necessary to find the location of an endpoint without going through the admissions process. Endpoints or gatekeepers may ask another gatekeeper about the location of an endpoint based on the endpoint's name or telephone number. A gatekeeper responds to such a request if the requested endpoint has registered with it.

The admissions messages enable the gatekeeper to enforce a policy on the calls and provide address translation to the endpoints. Every endpoint is required to ask the gatekeeper for permission before making a call. During the process the endpoint informs the gatekeeper about the type of call (point-to-point vs. multipoint), bandwidth needed for the call, and the endpoint that is being called. If the gatekeeper grants permission for the call, it will provide the necessary information to the calling endpoint. If the gatekeeper denies permission, it will inform the calling endpoint with a reason for denial.

On packet-based networks the available bandwidth is shared by all users connected to the network. Consequently, endpoints on such networks can attempt to utilize a certain amount of bandwidth but are not guaranteed to succeed. H.323 endpoints can monitor the available bandwidth through various measures, such as the amount of variance of delay in receiving media and the amount of lost media. The endpoints may subsequently change the bandwidth utilized depending on the data obtained. For example, a videoconferencing application can start by utilizing 400 kbps of bandwidth and increase it if the user requires better quality or decrease it if congestion is detected on the network. The bandwidth management messages allow a gatekeeper to keep track and control the amount of H.323 bandwidth used in its zone. The gatekeeper is informed about the bandwidth of the call during the admissions process, and all endpoints are required to acquire gatekeeper's permission before increasing the bandwidth of a call at a later time. Endpoints may also inform the gatekeeper if the bandwidth of a call is decreased, enabling it to utilize the unused bandwidth for other calls. The gatekeeper may also request a change in the bandwidth of a call and the endpoints in that call must comply with the request.

Gatekeepers may inquire about the status of calls and are informed when a call is

---

* Some of the RAS messages may be exchanged between gatekeepers.

terminated. During a call the gatekeeper may require status information about the call from the endpoints through status inquiry messages. After the call is terminated, the endpoints are required to inform the gatekeeper about the termination through disengage messages.

The H.225.0 RAS protocol requires an unreliable link and, in the case of TCP/IP networks, RAS utilizes the UDP transport protocol. Gatekeepers may be large servers with a substantial number of registered endpoints, which may exhaust resources very quickly. Generally unreliable links use less resources than reliable protocols. This is one of the reasons H.225.0 RAS requires protocols such as UDP.

2. *Call Signaling*

The H.225.0 call signaling protocol is used for connection establishment and termination between two endpoints. The H.225.0 call signaling is based on the Q.931 protocol. The Q.931 messages are extended to include H.323 specific data.* To establish a call, an endpoint must first establish the H.225.0 connection. In order to do so, it must transmit a Q.931 setup message to the endpoint that it wishes to call indicating its intention. The address of the other endpoint is known either through the admission procedure with the gatekeeper or through other means (e.g., phone book lookup). The called endpoint can either accept the incoming connection by transmitting a Q.931 connect message or reject it. During the call signaling procedure either the caller or the called endpoint provides an H.245 address, which is used to establish a control protocol channel.

In addition to connection establishment and termination, the H.225.0 call signaling protocol supports status inquiry, ad hoc multipoint call expansion, and limited call forward and transfer. Status inquiry is used by endpoints to request the call status information from the corresponding endpoint. Ad hoc multipoint call expansion provides functionality to invite other nodes into a conference or request to join a conference. The limited call forward and transfer are based on call redirecting and do not include sophisticated call forward and transfer offered by telephony systems. The supplementary services (H.450 series) part of H.323 V2 provides this functionality (see Sec. II.G.2).

3. *Control Protocol*

After the call signaling procedure, the two endpoints have established a connection and are ready to start the call. Prior to establishing the call, further negotiation between the endpoints must take place to resolve the call media type as well as establish the media flow. Furthermore, the call must be managed after it is established. The H.245 call control protocol is used to manage the call and establish logical channels for transmitting and receiving media and data. The control protocol is established between two endpoints, an endpoint and an MC, or an endpoint and a gatekeeper. The protocol is used for determining the master of the call, negotiating endpoint capabilities, opening and closing logical channels for transfer of media and data, requesting specific modes of operation, controlling the flow rate of media, selecting a common mode of operation in a multipoint conference, controlling a multipoint conference, measuring the round-trip delay between two endpoints, requesting updates for video frames, looping back media, and ending a call. H.245 is used by other protocols such as H.324, and a subset of its commands are used by the H.323 standard.

---

* Refer to ITU recommendation Q.931.

The first two steps after the call signaling procedure are to determine the master of the call and to determine the capabilities of each endpoint for establishing the most suitable mode of operation and ensuring that only multimedia signals that are understood by both endpoints are used in the conference. Determining the master of the call is accomplished through the master–slave determination process. This process is used to avoid conflicts during the call control operations. Notification of capabilities is accomplished through the capability exchange procedure. Each endpoint notifies the other of what it is capable of receiving and transmitting through receive and transmit capabilities. The receive capability is to ensure that the transmitter will transmit data only within the capability of the receiver. The transmit capability gives the receiver a choice among modes of the transmitted information. An endpoint does not have to declare its transmit capability; its absence indicates a lack of choice of modes to the receiver. The declaration of capabilities is very flexible in H.245 and allows endpoints to declare dependences between them.

The start of media and data flow is accomplished by opening logical channels through the logical channel procedures. A logical channel is a multiplexed path between endpoints for receiving and transmitting media or data. Logical channels can be unidirectional or bidirectional. Unidirectional channels are used mainly for transmitting media. The endpoint that wishes to establish a unidirectional logical channel for transmit or a bidirectional channel for transmit and receive issues a request to open a logical channel. The receiving entity can either accept or reject the request. Acceptance is based on the receiving entity's capability and resources. After the logical channel is established, the endpoints may transmit and receive media or data. The endpoint that requested opening of the logical channel is responsible for closing it. The endpoint that accepted opening of the logical channel can request that the remote endpoint close the logical channel.

A receiving endpoint may desire a change in the mode of media it is receiving during a conference. An example would be a receiver of H.261 video requesting a different video resolution. A receiving endpoint may request a change in the mode for transmission of audio, video, or data with the request mode command if the transmitting terminal has declared its transmit capability. A transmitter is free to reject the request mode command as long as it is transmitting media or data within the capability of the receiver. In addition to requesting the change in the mode of transmission, a receiver is allowed to specify an upper limit for the bit rate on a single or all of the logical channels. The Flow Control command forces a transmitter to limit the bit rate of the requested logical channel(s) to the value specified.

To establish a conference, all participants must conform to a mode of communication that is acceptable to all participants in the conference. The mode of communication includes type of medium and mode of transmission. As an example, in a conference everyone might be required to multicast their video to the participants but transmit their audio to an MCU for mixing. The MC uses the communication mode messages to indicate the mode of a conference to all participants. After a conference is established, it is controlled through conference request and response messages. The messages include conference chair control, password request, and other conference-related requests.

Data may be lost during the reception of any medium. There is no dependence between transmitted data packets for audio. Data loss can be handled by inserting silence frames or by simply ignoring it. The same is not true of video. A receiver might lose synchronization with the data and require a full or partial update of a video frame. The video Fast Update command indicates to a transmitter to update part of the video data. RTCP (see Sec. II.C.4) also contains commands for video update. H.323 endpoints are required to respond to an H.245 video fast update command and may also support RTCP commands.

The H.245 round trip delay commands can be used to determine the round-trip delay on the control channel. In an H.323 conference, media are carried on a separate logical channel with characteristics separate from those of the control channel; therefore the value obtained might not be an accurate representation of what the user is perceiving. The RTCP protocol also provides round-trip delay calculations and the result is usually closer to what the user perceives. The round trip delay command, however, may be used to determine whether a corresponding endpoint is still functioning. This is used as the keep-alive message for an H.323 call.

The end of a call is signaled by the end session command. The end session command is a signal for closing all logical channels and dropping the call. After the End Session command, endpoints close their call signaling channel and inform the gatekeeper about the end of the call.

4. Media Transport and Packetization

Transmission and reception of real-time data must be achieved through the use of best effort delivery of packets. Data must be delivered as quickly as possible and packet loss must be tolerated without retransmission of data. Furthermore, network congestion must be detected and tolerated by adapting to network conditions. In addition, systems must be able to identify different data types, sequence data that may be received out of order, provide for media synchronization, and monitor the delivery of data. Real time transport protocol and real time transport control protocol (RTP/RTCP), specified in IETF RFC number 1889, defines a framework for a protocol that allows multimedia systems to transmit and receive real-time media using best effort delivery of packets. RTP/RTCP supports multicast delivery of data and may be deployed independent of the underlying network transport protocol. It is important to note that RTP/RTCP does not guarantee reliable delivery of data and does not reserve network resources, it merely enables an application to deal with unreliability of packet-based networks. The RTP protocol defines a header format for packetization and transmission of real-time data. The header conveys the information necessary for the receiver to identify the source of data, sequence and detect packet loss, identify type of data, identify the source of the data, and synchronize media. The RTCP protocol runs in parallel with RTP and provides data delivery monitoring. Delivery monitoring in effect provides knowledge about the condition of the underlying network. Through the monitoring technique, H.323 systems may adapt to the network conditions by introducing appropriate changes in the traffic of media. Adaptation to the network conditions is very important and can radically affect the user's perception of the quality of the system. Figure 3 shows the establishment of an H.323 call. Multiplexing of H.323 data over the packet-based network is done through the use of transport service access points (TSAPs) of the underlying transport. A TSAP in TCP/IP terms is a UDP or TCP port number.

5. Data Conferencing

The default protocol for data conferencing in an H.323 conference is the ITU standard T.120. The H.323 standard provides harmonization between a T.120 and an H.323 conference. The harmonization is such that a T.120 conference will become an inherent part of an H.323 conference. The T.120 conferences are established after the H.323 conference and are associated with the H.323 conference.

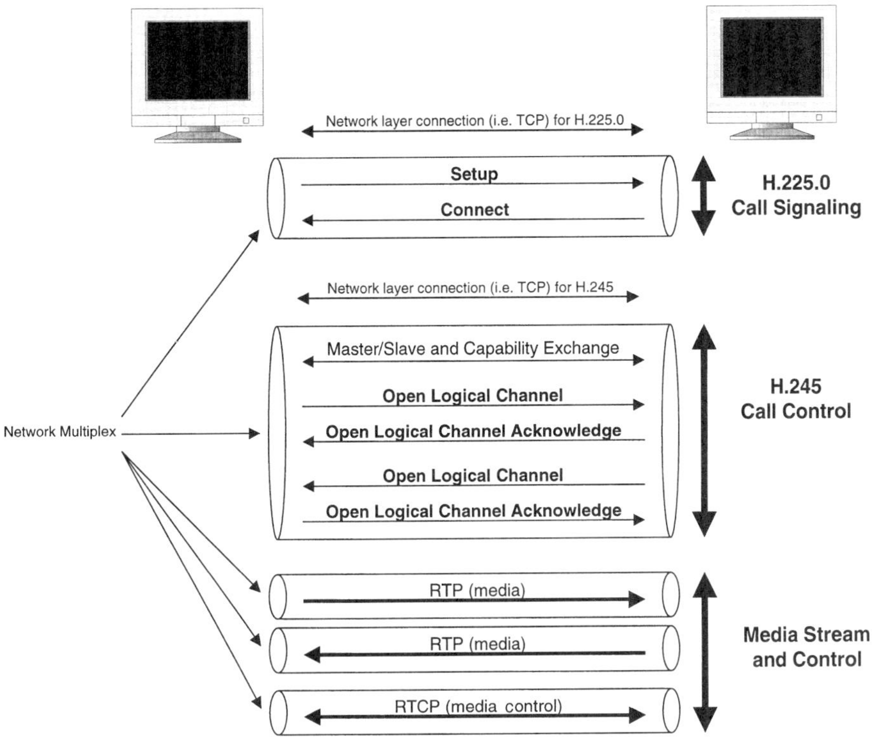

Figure 3 H.323 call setup.

To start a T.120 conference, an endpoint opens a bidirectional logical channel through the H.245 protocol. After the logical channel is established, either of the endpoints may start the T.120 session, depending on the negotiation in the open logical channel procedures. Usually the endpoint that initiated the H.323 call initiates the T.120 conference.

D. Call Types

The H.323 standard supports point-to-point and multipoint calls in which more than two endpoints are involved. The MC controls a multipoint call and consequently there is only one MC in the conference.* Multipoint calls may be centralized, in which case an MCU controls the conference including the media, or decentralized, with the media processed separately by the endpoints in the conference. In both cases the control of the conference is performed by one centralized MC. Delivery of media in a decentralized conference may be based on multicasting, implying a multicast-enabled network, or it may be based on multiunicasting, in which each endpoint transmits its media to all other endpoints in the

* An exception to this rule is when conferences are cascaded. In cascaded conferences there may be multiple MCs with one selected as master.

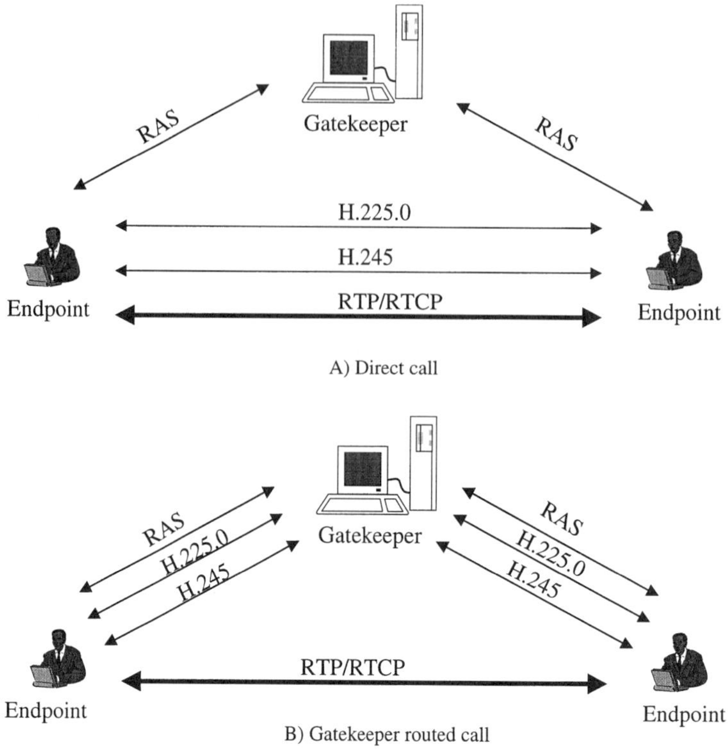

Figure 4 H.323 fast call setup.

conference separately. Figure 4 depicts different multipoint call models. In the same conference it is possible to distribute one medium using the centralized model and another medium using the decentralized model. Such conferences are referred to as hybrid.

Support of multicast in decentralized conferences is a pivotal feature of H.323. Multicast networks are becoming more popular every day because of their efficiency in using network bandwidth. The first applications to take advantage of multicast-enabled networks will be bandwidth-intensive multimedia applications; however, centralized conferences provide more control of media distribution. In addition, the resource requirements for a conference are more centralized on a single endpoint, i.e., the MCU, and not the participants. Some endpoints in the conference might not have enough resources to process multiple incoming media streams and the MCU relieves them of the task.

E. Gatekeeper-Routed and Direct Call Models

The gatekeeper, if present, has control over the routing of the control signaling (H.225.0 and H.245) between two endpoints. When an endpoint goes through the admissions process, the gatekeeper may return its own address for the destination of control signaling instead of the called endpoint's address. In this case the control signaling is routed through the gatekeeper, hence the term gatekeeper-routed call model. This call model provides control over a call and is essential in many H.323-based applications. Through this control the gatekeeper may offer services such as providing call information by keeping track of calls and media channels used in a call, providing for call rerouting in cases in which a particular user is not available (i.e., route call to operator or find the next available agent in call center applications), acquiring QoS for a call through non-H.323 protocols, or providing policies on gateway selection to load balance multiple gateways in an enterprise. Although the gatekeeper-routed model involves more delay on call setup, it has been the more popular approach with manufacturers because of its flexibility in controlling calls.

If the gatekeeper decides not to be involved in routing the control signaling, it will return the address of the true destination and will be involved only in the RAS part of the call. This is referred to as the direct call model. Figure 5 shows the two different call models. The media flow is directly between the two endpoints; however, it is possible for the gatekeeper to control the routing of the media as well by altering the H.245 messages.

F. Audio and Video Compression–Decompression

The H.323 standard ensures interoperability among all endpoints by specifying a minimum set of requirements. It mandates support for voice communication; therefore, all terminals must provide audio compression and decompression (audio codec). It supports both sample- and frame-based audio codecs. An endpoint may support multiple audio codecs, but the support for G.711 audio is mandatory: G.711 is the sample-based audio codec for digital telephony services and operates with a bit rate of 64 kbit/sec. Support for video in H.323 systems is optional, but if an endpoint declares the capability for video the system must minimally support H.261 with quarter common intermediate format (QCIF) resolution.

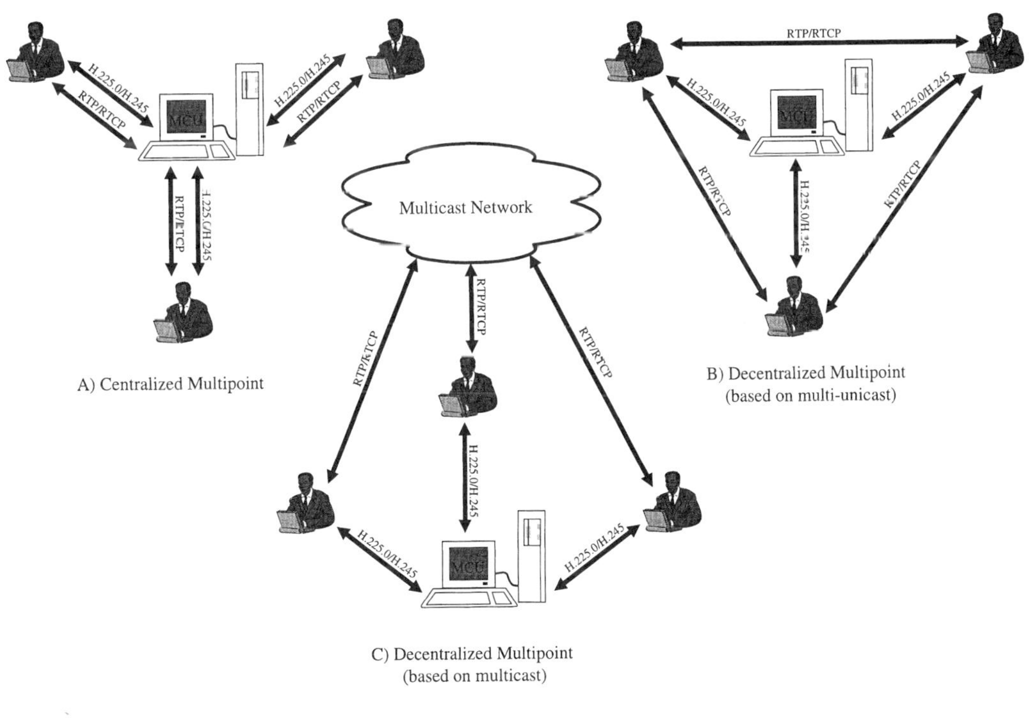

Figure 5 H.323 protocols on ATM network.

It is possible for an endpoint to receive media in one mode and transmit the same media type in another mode. Endpoints may operate in this manner because media logical channels are unidirectional and are opened independent of each other. This asymmetric operation is possible with different types of audio. For example, it is possible to transmit G.711 and receive G.722. For video it is possible to receive and transmit with different modes of the same video coding. Endpoints must be able to operate with an asymmetric bit rate, frame rate, and resolution if more than one resolution is supported.

During low-bit-rate operations over slow links, it may not be possible to use 64 kbit/sec G.711 audio. Support for low-bit-rate multimedia operation is achieved through the use of G.723.1. The G.723.1, originally developed for the H.324 standard, is a frame-based audio codec with bit rates of 5.3 and 6.4 kbit/sec. The selected bit rate is sent as part of the audio data and is not declared through the H.245 control protocol. Using G.723.1, an endpoint may change its transmit rate during operation without any additional signaling.

G. H.323 V2

The H.323 V1 contained the basic protocol for deploying audio and videoconferencing over packet-based networks but lacked many features vital to the success of its wide deployment. Among these were lack of supplementary services, security and encryption, support for QoS protocols, and support for large conferences. The H.323 V2, among many other miscellaneous enhancements, extended H.323 to support these features. The following sections explain the most significant enhancements in H.323 V2.

1. *Fast Setup*

Starting an H.323 call involves multiple stages involving multiple message exchanges. As Figure 6 shows, there are typically four message exchanges before the setup of the first media channel and transmission of media. This number does not include messages exchanged during the network connection setup for H.225.0 and H.245 (i.e., TCP connection). When dealing with congested networks, the number of exchanges of messages may have a direct effect on the amount of time it takes to bring a call up. End users are accustomed to the everyday telephone operations in which upon answering a call one can immediately start a conversation. Consequently, long call setup times can lead to user-unfriendly systems. For this reason H.323 V2 provides a call signaling procedure whereby the number of message exchanges is reduced significantly and media flow can start as fast as possible.

In H.323 the start of media must occur after the capability exchange procedure of H.245. This is required because endpoints select media types on the basis of each other's capabilities. If an endpoint, without knowledge of its counterpart's capability, provides a choice of media types for reception and transmission, it will be possible to start the media prior to the capability exchange if the receiver of the call selects media types that match its capabilities. This is the way the fast start procedure of H.323 operates. The caller gives a choice of media types, based on its capabilities, to the called endpoint for reception and transmission of media. The called endpoint selects the media types that best suit its capabilities and starts receiving and transmitting media. The called endpoint then notifies the caller about the choice that it has made so that the caller can free up resources that might be used for proposed media types that were unused. The proposal and selection of media types are accomplished during the setup and connect exchange of H.225.0. As Figure 6 shows, it is possible to start the flow of media from the called endpoint to the caller immediately after receiving the first H.225.0 message. The caller must accept reception

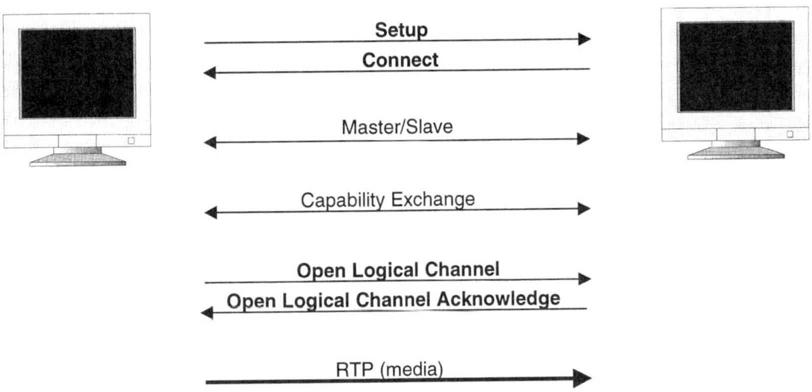

A) Media setup with normal H.323 procedure

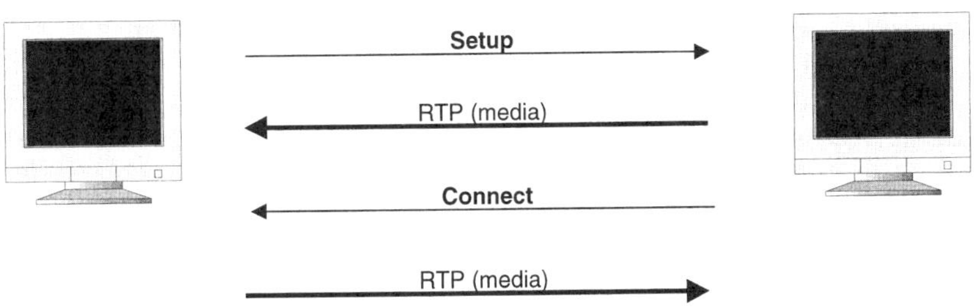

B) Media setup with fast setup procedure

Figure 6 H.323 call models.

of any one of the media types that it has proposed until the called endpoint has notified it about its selection. After the caller is notified about the called endpoint's selection, it may start the media transmission. It should be noted that there is no negotiation in the fast start procedure, and if the called endpoint cannot select one of the proposed media types the procedure fails.

2. *Supplementary Services*

The H.323 V1 supports rudimentary call forwarding, through which it is also possible to implement simple call transfer. However, the protocol for implementing a complete solution for supplementary services did not exist. The H.450.x series of protocols specify supplementary services for H.323 in order to provide private branch exchange (PBX)-like features and support interoperation with switched circuit network–based protocols. The H.450.x series of protocols assume a distributed architecture and are based on the International Standardization Organization/International Electrotechnical Commission (ISO/IEC) QSIG standards. The H.323 V2 introduced the generic control protocol for supplementary services (H.450.1), call transfer (H.450.2), and call forward (H.450.3).

3. *RSVP*

RSVP is a receiver-oriented reservation protocol from IETF for providing transport-level QoS. During the open logical channel procedures of H.245 it is possible to exchange the necessary information between the two endpoints to establish RSVP-based reservation for the media flow. The H.323 V2 supports RSVP by enabling endpoints to exchange the necessary RSVP information prior to establishing the media flow.

4. Native ATM

Transporting media over TCP/IP networks lacks one of the most important requirements of media transport, namely quality of service. New QoS methods such as RSVP are in the deployment phase but none are inherent in TCP/IP. On the other hand, ATM is a packet-based network that inherently offers QoS. Annex C of H.323 takes advantage of this ability and defines the procedures for establishing a conference using ATM adaptation layer 5 (AAL5) for transfer of media with QoS.

The H.323 standard does not require use of the same network transport for control and media. It is possible to establish the H.225.0 and H.245 on a network transport different from the one used for media. This is the property that Annex C of H.323 utilizes by using TCP/IP for control and native ATM for media. Figure 7 shows the protocol layers that are involved in a native ATM H.323 conference. It is assumed that IP connectivity is available and endpoints have a choice of using native ATM or IP. Call signaling and control protocols are operated over TCP/IP. If an endpoint has the capability to use native ATM for transmission and reception of media, it will declare it in its capabilities, giving a choice of transport to the transmitter of media. The endpoint that wishes to use native ATM for transmission of media specifies it in the H.245 Open Logical Channel message. After the request has been accepted, it can then establish the ATM virtual circuit (VC) with the other endpoint for transmission of media. Logical channels in H.323 are unidirectional, and ATM VCs are inherently bidirectional. Consequently, H.323 provides signaling to use one virtual circuit for two logical channels, one for each direction.

5. Security

Security is a major concern for many applications on packet-based networks such as the Internet. In packet-based networks the same physical medium is shared by multiple appli-

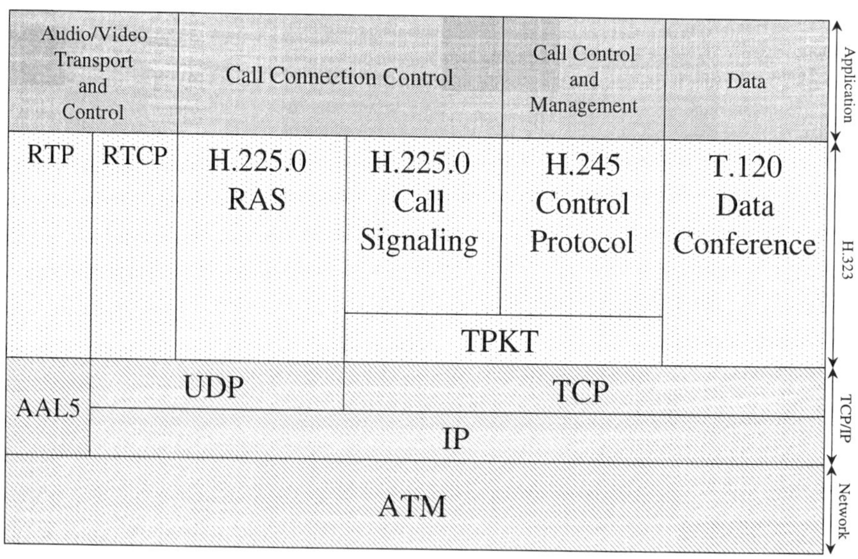

Figure 7 H.323 multipoint call models.

cations. Consequently, it is fairly easy for an entity to look at or alter the traffic other than its own. Users of H.323 may require security services such as encryption for private conversations and authentication to verify the identity of corresponding users. Recommendation H.235 provides the procedures and framework for privacy, integrity, and authentication in H.323 systems. The recommendation offers a flexible architecture that enables the H.323 systems to incorporate security. H.235 is a general recommendation for all ITU standards that utilize the H.245 protocol (e.g., H.324).

Privacy and integrity are achieved through encryption of control signaling and media. Media encryption occurs on each packet independently. The RTP information is not encrypted because intermediate nodes need access to the information. Authentication may be provided through the use of certificates or challenge–response methods such as passwords or Diffie–Hellman exchange. Furthermore, authentication and encryption may be provided at levels other than H.323, such as IP security protocol (IPSEC) in case of TCP/IP-based networks.

The H.235 recommendation does not specify or mandate the use of certain encryption or privacy methods. The methods used are based on the negotiation between systems during the capability exchange. In order to guarantee that interoperability, specific systems may define a profile based on H.235 that will be followed by manufacturers of such systems. One such example is the ongoing work in the voice over IP (VOIP) forum, which is attempting to define a security profile for VOIP systems.

6. *Loosely Coupled Conferences*

The packet-based network used by H.323 systems may be a single segment, multiple segments in an enterprise, or multiple segments on the Internet. Consequently, the conference model that H.323 offers does not put a limit on the number of participants. Coordination and management of a conference with large number of active participants are very difficult and at times impractical. This is true unless a large conference is limited to a small group of active participants and a large group of passive participants.

Recommendation H.332 provides a standard for coordinating and managing large conferences with no limit on the number of participants. H.332 divides an H.323 conference into a panel with a limited number of active participants and the rest of the conference with an unlimited number of passive participants that are receivers only but can request to join the panel at any time. Coordination involves meeting, scheduling, and announcements. Management involves control over the number of participants and the way they participate in the conference.

Large conferences are usually preannounced to the interested parties. During the preannouncement, the necessary information on how to join the conference is distributed by meeting administrators. H.332 relies on preannouncement to inform all interested parties about the conference and to distribute information regarding the conference. The conference is preannounced using mechanisms such as telephone, email, or IETF's session advertisement protocol. The announcement is encoded using IETF's session directory protocol and contains information about media type, reception criteria, and conference time and duration. In addition, the announcement may contain secure conference registration information and MC addresses for joining the panel. After the announcement, a small panel is formed using the normal H.323 procedures. These panel members can then effectively participate in a normal H.323 conference. The content of the meeting is provided to other participants by media multicasting via RTP/RTCP. Any of the passive participants, if it has received information regarding the MC of the conference, may request to join the panel at any time during the meeting. An example of an H.332 application would be a lecture given to a small number of local students (the panel) while being broadcast to other interested students across the world. This method allows active participation from both the local and the global students. The H.332 conference ends when the panel's conference is completed.

H. New Features and H.323 V3

New topics have been addressed by the ITU and new features and protocols are being added to H.323. A subset of the new features is part of H.323 V3. Some others are independently approved annexes to documents and may be used with H.323 V2. Following are brief descriptions of new topics that have been visited by the ITU since the approval of H.323 V2:

Communication Between Administrative Domains. Gatekeepers can provide address resolution and management for endpoints within their zone. Furthermore, multiple zones may be managed by one administration referred to as the administrative domain. Establishing H.323 calls between zones requires an exchange of addressing information between gatekeepers. This information exchange is usually limited and does not require a scalable protocol. As of this writing, a subset of the RAS protocol is utilized by gatekeepers. However, administrative domains may be managed independently and large deployment of H.323 networks also requires an exchange of addressing information between administrative domains. A new annex provided in H.323 defines the protocol that may be used between administrative domains regarding information exchange for address resolution.

Management Information Base for H.323. Management information is provided by defining managed H.323 objects. The definition follows the IETF simple network management protocol (SNMP) protocol.

Real-Time Fax. Real-time fax may be treated as a media type and can be carried in an H.323 session. The ITU T.38 protocol defines a facsimile protocol based on IP networks. The H.323 V2 defines procedures for transfer of T.38 data within an H.323 session.

Remote Device Control. The H.323 entities may declare devices that are remotely controllable by other endpoints in a conference. These devices range from cameras to videocassette readers (VCRs). The remote device protocol is defined by recommendation H.282 and is used in an H.323 conference. Recommendation H.283 defines the procedures for establishing the H.282 protocol between two H.323 endpoints.

UDP-Based Call Signaling. Using TCP on a congested network may lead to unpredictable behavior of applications because control over time-out and retransmission policies of TCP is usually not provided. Using TCP on servers that route call control signaling and may service thousands of calls requires large amounts of resources. On the other hand, utilizing UDP yields the control over-time out and retransmission policy to the application and requires less resources. A new annex is currently being developed to addresses carrying H.225.0 signals over UDP instead of TCP. This annex will define the retransmission and time-out policies.

New Supplementary Services. New H.450.x-based supplementary services are being introduced. These new supplementary services are call hold, call park and pickup, call waiting, and message waiting indication.

Profile for Single-Use Devices. Many devices that use H.323 have limited use for the protocol and do not need to take advantage of all that the standard offers. Devices such as telephones and faxes, referred to as single-use devices, require a well-defined and limited set of H.323 capabilities. A new annex in H.323 defines a profile with which implementation complexity for such devices is significantly reduced.

III. H.324

In September 1993 the ITU established a program to develop an international standard for a videophone terminal operating over the public switched telephone network (PSTN). A major milestone in this project was accomplished in March 1996, when the ITU approved the standard. It is anticipated that the H.324 terminal will have two principal applications, namely a conventional videophone used primarily by the consumer and a multimedia system to be integrated into a personal computer for a range of business purposes.

In addition to approving the umbrella H.324 recommendation, the ITU has completed the four major functional elements of the terminal: the G.723.1 speech coder, the H.263 video coder, the H.245 communication controller, and the H.223 multiplexer. The quality of the speech provided by the new G.723.1 audio coder, when operating at only 6.4 kbps, is very close to that found in a conventional phone call. The picture quality produced by the new H.263 video coder shows promise of significant improvement compared with many earlier systems. It has been demonstrated that these technical advances, when combined with the high transmission bit rate of the V.34 modem (33.6 kbps maximum), yield an overall audiovisual system performance that is significantly improved over that of earlier videophone terminals.

At the same meeting in Geneva, the ITU announced the acceleration of the schedule to develop a standard for a videophone terminal to operate over mobile radio networks. The new terminal, designated H.324/M, will be based on the design of the H.324 device to ease interoperation between the mobile and telephone networks.

A. H.324: Terminal for Low-Bit-Rate Multimedia Communication

Recommendation H.324 describes terminals for low-bit-rate multimedia communication, utilizing V.34 modems operating over the GSTN. The H.324 terminals may carry real-time voice, data, and video or any combination, including videotelephony.

The H.324 terminals may be integrated into personal computers or implemented in stand alone devices such as videotelephones. Support for each media type (such as voice, data, and video) is optional, but if it is supported, the ability to use a specified common mode of operation is required so that all terminals supporting that media type can interwork. Recommendation H.324 allows more than one channel of each type to be in use. Other recommendations in the H.324 series include the H.223 multiplex, H.245 control, H.263 video codec, and G.723.1 audio codec.

Recommendation H.324 makes use of the logical channel signaling procedures of recommendation H.245, in which the content of each logical channel is described when the channel is opened. Procedures are provided for expression of receiver and transmitter capabilities, so that transmissions are limited to what receivers can decode and that receivers may request a particular desired mode from transmitters. Because the procedures of H.245 are also planned for use by recommendation H.310 for ATM networks and recommendation H.323 for packetized networks, interworking with these systems should be straightforward.

The H.324 terminals may be used in multipoint configurations through MCUs and may interwork with H.320 terminals on the ISDN as well as with terminals on wireless networks. The H.324 implementations are not required to have each functional element, except for the V.34 modem, H.223 multiplex, and H.245 system control protocol, which will be supported by all H.324 terminals.

The H.324 terminals offering audio communication will support the G.723.1 audio codec, H.324 terminals offering video communication will support the H.263 and H.261 video codecs, and H.324 terminals offering real-time audiographic conferencing will support the T.120 protocol suite. In addition, other video and audio codecs and other data protocols may optionally be used via negotiation over the H.245 control channel. If a modem external to the H.324 terminal is used, terminal and modem control will be according to V.25ter.

Multimedia information streams are classified into video, audio, data, and control as follows:

Video streams are continuous traffic-carrying moving color pictures. When they are used, the bit rate available for video streams may vary according to the needs of the audio and data channels.

Audio streams occur in real time but may optionally be delayed in the receiver processing path to maintain synchronization with the video streams. To reduce the average bit rate of audio streams, voice activation may be provided.

Data streams may represent still pictures, facsimile, documents, computer files, computer application data, undefined user data, and other data streams.

Control streams pass control commands and indications between remotelike functional elements. Terminal-to-modem control is according to V.25ter for terminals using external modems connected by a separate physical interface. Terminal-to-terminal control is according to H.245.

The H.324 document refers to other ITU recommendations, as illustrated in Figure 8, that collectively define the complete terminal. Four new companion recommendations include H.263 (Video Coding for Low Bitrate Communication), G.723.1 (Speech Coder for Multimedia Telecommunications Transmitting at 5.3/6.3 Kbps), H.223 (Multiplexing Protocol for Low-Bitrate Multimedia Terminals), and H.245 (Control of Communications Between Multimedia Terminals). Recommendation H.324 specifies use of the V.34 modem, which operates up to 28.8 kbps, and the V.8 (or V.8bis) procedure to start and stop data transmission. An optional data channel is defined to provide for exchange of computer data in the workstation–PC environment. The use of the T.120 protocol is specified by H.324 as one possible means for this data exchange. Recommendation H.324 defines the seven phases of a cell: setup, speech only, modem training, initialization, message, end, and clearing.

B. G.723.1: Speech Coder for Multimedia Telecommunications Transmitting at 5.3/6.3 Kbps

All H.324 terminals offering audio communication will support both the high and low rates of the G.723.1 audio codec. The G.723.1 receivers will be capable of accepting silence frames. The choice of which rate to use is made by the transmitter and is signaled to the receiver in-band in the audio channel as part of the syntax of each audio frame. Transmitters may switch G.723.1 rates on a frame-by-frame basis, based on bit rate, audio

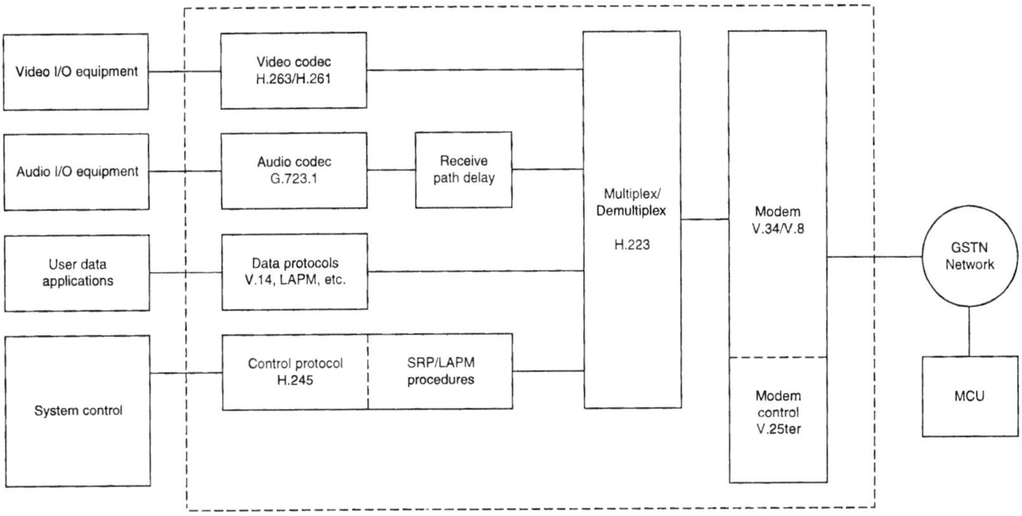

Figure 8 Block diagram for H.324 multimedia system.

quality, or other preferences. Receivers may signal, via H.245, a preference for a particular audio rate or mode.

Alternative audio codecs may also be used, via H.245 negotiation. Coders may omit sending audio signals during silent periods after sending a single frame of silence or may send silence background fill frames if such techniques are specified by the audio codec recommendation in use.

More than one audio channel may be transmitted, as negotiated via the H.245 control channel. The G.723.1 speech coder can be used for a wide range of audio signals but is optimized to code speech. The system's two mandatory bit rates are 5.3 and 6.3 kbps. The coder is based on the general structure of the multipulse–maximum likelihood quantizer (MP MLQ) speech coder. The MP-MLQ excitation will be used for the high-rate version of the coder. Algebraic codebook excitation linear prediction (ACELP) excitation is used for the low-rate version. The coder provides a quality essentially equivalent to that of a plain old telephone service (POTS) toll call. For clear speech or with background speech, the 6.3-kbps mode provides speech quality equivalent to that of the 32-kbps G.726 coder. The 5.3-kbps mode performs better than the IS54 digital cellular standard.

Performance of the coder has been demonstrated by extensive subjective testing. The speech quality in reference to 32-kbps G.726 adaptive differential pulse code modulation (ADPCM) (considered equivalent to toll quality) and 8-kbps IS54 vector sum excited linear prediction (VSELP) is given in Table 2. This table is based on a subjective test conducted for the French language. In all cases the performance of G.726 was rated better than or equal to that of IS54. All tests were conducted with 4 talkers except for the speaker variability test, for which 12 talkers were used. The symbols $<$, $=$, and $>$ are used to identify less than, equivalent to, and better than, respectively. Comparisons were made by taking into account the statistical error of the test. The background noise conditions are speech signals mixed with the specified background noise.

From these results one can conclude that within the scope of the test, both low- and high-rate coders are always equivalent to or better than IS54, except for the low-rate coder

Table 2 Results of Subjective Test for G.723.1

Test item	High rate	Low rate
Speaker variability	= G.726	= IS54
One encoding	= G.726	= IS54
Tandem	>4*G.726	>4*G.726
Level +10 dB	= G.726	= G.726
Level −10 dB	>G.726	>G.726
Frame erasures (3%)	= G.726 − 0.5	= G.726 − 0.75
Flat input (loudspeaker)	= G.726	= G.726
Flat in (loudspeaker) 2T	>4*G.726	>4*G.726
Office noise (18dB) DMOS	= IS54	= IS54
Babble noise (20 dB) DMOS	= G.726	= IS54
Music noise (20 dB) DMOS	>IS54	= IS54

with music, and that the high-rate coder is always equivalent to G.726, except for office and music background noises.

The complexity of the dual rate coder depends on the digital signal processing (DSP) chip and the implementation but is approximately 18 and 16 Mips for 6.3 and 5.3 kbps, respectively. The memory requirements for the dual rate coder are

RAM (random-access memory): 2240 16-bit words
ROM (read-only memory): 9100 16-bit words (tables), 7000 16-bit words (program)

The algorithmic delay is 30 msec frame + 7.5 msec look ahead, resulting in 37.5 msec.

The G.723.1 coder can be integrated with any voice activity detector to be used for speech interpolation or discontinuous transmission schemes. Any possible extensions would need agreements on the proper procedures for encoding low-level background noises and comfort noise generation.

C. H.263: Video Coding for Low-Bit-Rate Communication

All H.324 terminals offering video communication will support both the H.263 and H.261 video codecs, except H.320 interworking adapters, which are not terminals and do not have to support H.263. The H.261 and H.263 codecs will be used without Bose, Chaudhuri, and Hocquengham (BCH) error correction and without error correction framing. The five standardized image formats are 16CIF, 4CIF, CIF, QCIF, and SQCIF.

The CIF and QCIF formats are defined in H.261. For the H.263 algorithm, SQCIF, 4CIF, and 16CIF are defined in H.263. For the H.261 algorithm, SQCIF is any active picture size less than QCIF, filled out by a black border, and coded in the QCIF format. For all these formats, the pixel aspect ratio is the same as that of the CIF format. Table 3 shows which picture formats are required and which are optional for H.324 terminals that support video.

All video decoders will be capable of processing video bit streams of the maximum bit rate that can be received by the implementation of the H.223 multiplex (for example, maximum V.34 rate for single link and 2 × V.34 rate for double link).

Table 3 Picture Formats for Video Terminals

Picture format	Luminance pixels	Encoder		Decoder	
		H.261	H.263	H.261	H.263
SQCIF	128 × 96 for H.263[c]	Optional‡	Required[a,b]	Optional[c]	Required[a]
QCIF	176 × 144	Required	Required[a,b]	Required	Required[a]
CIF	352 × 288	Optional	Optional	Optional	Optional
4CIF	704 × 576	Not defined	Optional	Not defined	Optional
16CIF	1408 × 1152	Not defined	Optional	Not defined	Optional

[a] Optional for H.320 interworking adapters.
[b] It is mandatory to encode one of the picture formats QCIF and SQCIF; it is optional to encode both formats.
[c] H.261 SQCIF is any active size less than QCIF, filled out by a black border and coded in QCIF format.

Which picture formats, minimum number of skipped pictures, and algorithm options can be accepted by the decoder are determined during the capability exchange using H.245. After that, the encoder is free to transmit anything that is in line with the decoder's capability. Decoders that indicate capability for a particular algorithm option will also be capable of accepting video bit streams that do not make use of that option.

The H.263 coding algorithm is an extension of H.261. The H.263 algorithm describes, as H.261 does, a hybrid differential pulse-code modulation/discrete cosine transform (DPCM/DCT) video coding method. Both standards use techniques such as DCT, motion compensation, variable length coding, and scalar quantization and both use the well-known macroblock structure. Differences between H.263 and H.261 are

H.263 has an optional group-of-blocks (GOB) level.
H.263 uses different variable length coding (VLC) tables at the macroblock and block levels.
H.263 uses half-pixel (half-pel) motion compensation instead of full pel plus loop filter.
In H.263, there is no still picture mode [Joint Photographic Experts Group (JPEG) is used for still pictures].
In H.263, no error detection–correction is included such as the BCH in H.261.
H.263 uses a different form of macroblock addressing.
H.263 does not use the end-of-block marker.

It has been shown that the H.263 system typically outperforms H.261 (when adapted for the GSTN application) by 2.5 to 1. This means that when adjusted to provide equal picture quality, the H.261 bit rate is approximately 2.5 times that for the H.263 codec.

The basic H.263 standard also contained the five important optional annexes. Annexes D though H are particularly valuable for the improvement of picture quality (Annex D, Unrestricted Motion Vector; Annex E, Syntax-Based Arithmetic Coding; Annex F, Advanced Prediction; Annex G, PB-Frames; Annex H, Forward Error Correction for Coded Video Signal).

Of particular interest is the optional PB-frame mode. A PB-frame consists of two pictures being coded as one unit. The name PB comes from the name of picture types in MPEG, where there are P-pictures and B-pictures. Thus a PB-frame consists of one P-picture that is predicted from the last decoded P-picture and one B-picture that is predicted from both the last decoded P-picture and the P-picture currently being decoded. This last picture is called a B-picture because parts of it may be bidirectionally predicted from the past and future P-pictures. The prediction process is illustrated in Figure 9.

D. H.245: Control Protocol for Multimedia Communications

The control channel carries end-to-end control messages governing the operation of the H.324 system, including capabilities exchange, opening and closing of logical channels, mode preference requests, multiplex table entry transmission, flow control messages, and general commands and indications.

There will be exactly one control channel in each direction within H.324, which will use the messages and procedures of recommendation H.245. The control channel will be carried on logical channel 0. The control channel will be considered to be permanently open from the establishment of digital communication until the termination of digital communication; the normal procedures for opening and closing logical channels will not apply to the control channel.

General commands and indications will be chosen from the message set contained in H.245. In addition, other command and indication signals may be sent that have been specifically defined to be transferred in-band within video, audio, or data streams (see the appropriate recommendation to determine whether such signals have been defined).

The H.245 messages fall into four categories—request, response, command, and indication. Request messages require a specific action by the receiver, including an immediate response. Response messages respond to a corresponding request. Command messages require a specific action but do not require a response. Indication messages are informative only and do not require any action or response. The H.324 terminals will

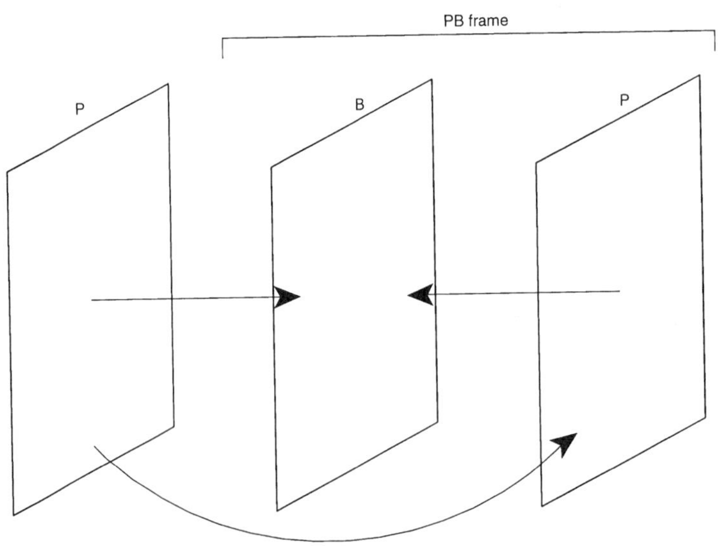

Figure 9 Prediction in PB-frames mode.

respond to all H.245 commands and requests as specified in H.245 and will transmit accurate indications reflecting the state of the terminal.

Table 4 shows how the total bit rate available from the modem might be divided into its various constituent virtual channels by the H.245 control system. The overall bit rates are those specified in the V.34 modem. Note that V.34 can operate at increments of 2.4 kbps up to 36.6 kbps. Speech is shown for two bit rates that are representative of possible speech coding rates. The video bit rate shown is what is left after deducting the speech bit rates from the overall transmission bit rate. The data would take a variable number of bits from the video, either a small amount or all of the video bits, depending on the designer's or the user's control. Provision is made for both point-to-point and multipoint operation. Recommendation H.245 creates a flexible, extensible infrastructure for a wide range of multimedia applications including storage–retrieval, messaging, and distribution services as well as the fundamental conversational use. The control structure is applicable to the situation in which only data and speech are transmitted (without motion video) as well as the case in which speech, video, and data are required.

E. H.223: Multiplexing Protocol for Low-Bit-Rate Multimedia Communication

This recommendation specifies a packet-oriented multiplexing protocol designed for the exchange of one or more information streams between higher layer entities such as data and control protocols and audio and video codes that use this recommendation.

In this recommendation, each information stream is represented by a unidirectional logical channel that is identified by a unique logical channel number (LCN). The LCN 0 is a permanent logical channel assigned to the H.245 control channel. All other logical channels are dynamically opened and closed by the transmitter using the H.245 OpenLogicalChannel and CloseLogicalChannel messages. All necessary attributes of the logical channel are specified in the OpenLogicalChannel message. For applications that require

Table 4 Example of a Bit Rate Budget for Vary Low Bit Rate Visual Telephony

Criteria	Modem bit rate (kbps)	Virtual channel (kbps) Speech	Video	Data
Overall transmission bit rate	9.6	5.3	4.3	Variable
	14.4	5.3	9.1	Variable
	[a,b]	[b]	[b]	
	28.8	6.3	22.5	Variable
	33.6	6.3	27.3	Variable
Virtual channel bit rate characteristic		Dedicated, fixed bit rate[c]	Variable bit rate	Variable bit rate
Priority		Highest priority	Lowest priority	Higher than video, lower than overhead/speech

[a] V.34 operates at increments of 2.4 kbps, that is, 16.8, 19.2, 21.6, 24.0, 26.4, 28.8, 33.6 kbps.

[b] The channel priorities will not be standardized; the priorities indicated are examples.

[c] The plan includes consideration of advanced speech codec technology such as a dual bit rate speech codec and a reduced bit rate when voiced speech is not present.

a reverse channel, a procedure for opening bidirectional logical channels is also defined in H.245. The general structure of the multiplexer is shown in Figure 10. The multiplexer consists of two distinct layers, a multiplex (MUX) layer and an adaptation layer (AL).

1. Multiplex Layer

The MUX layer is responsible for transferring information received from the AL to the far end using the services of an underlying physical layer. The MUX layer exchanges information with the AL in logical units called MUX-SDUs (service data unit), which always contain an integral number of octets that belong to a single logical channel. MUX-SDUs typically represent information blocks whose start and end mark the location of fields that need to be interpreted in the receiver.

The MUX-SDUs are transferred by the MUX layer to the far end in one or more variable-length packets called MUX-PDUs (protocol data units). The MUX-PDUs consist of the high-level data link control (HDLC) opening flag, followed by a one-octet header and by a variable number of octets in the information field that continue until the closing HDLC flag (see Fig. 11 and 12). The HDLC zero-bit insertion method is used to ensure that a flag is not simulated within the MUX-PDU.

Octets from multiple logical channels may be present in a single MUX-PDU information field. The header octet contains a 4-bit multiplexes code (MC) field that specifies, by reference to a multiplex table entry, the logical channel to which each octet in the information field belongs. Multiplex table entry 0 is permanently assigned to the control channel. Other multiplex table entries are formed by the transmitter and are signaled to the far end via the control channel prior to their use.

Multiplex table entries specify a pattern of slots each assigned to a single logical channel. Any one of 16 multiplex table entries may be used in any given MUX-PDU. This allows rapid low-overhead switching of the number of bits allocated to each logical channel from one MUX-PDU to the next. The construction of multiplex table entries and their use in MUX-PDUs are entirely under the control of the transmitter, subject to certain receiver capabilities.

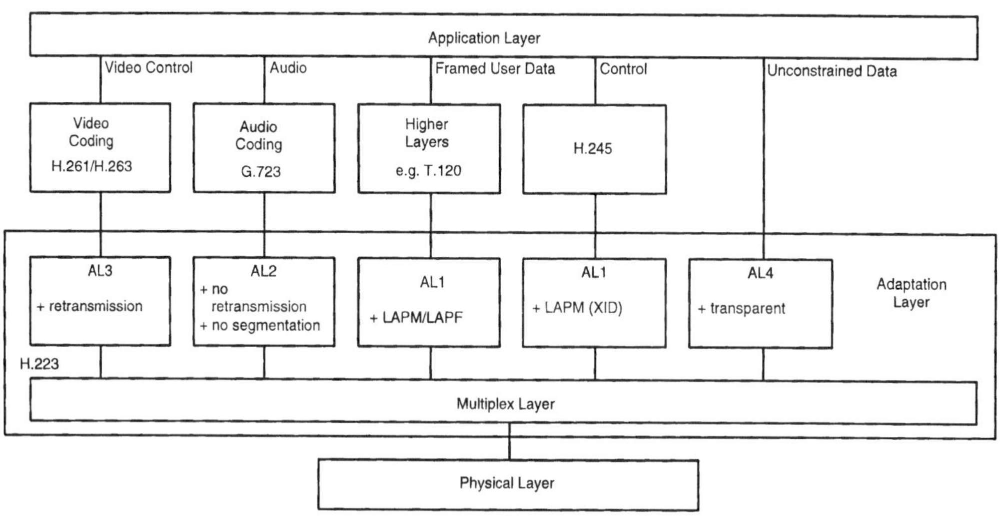

Figure 10 Protocol structure of H.223.

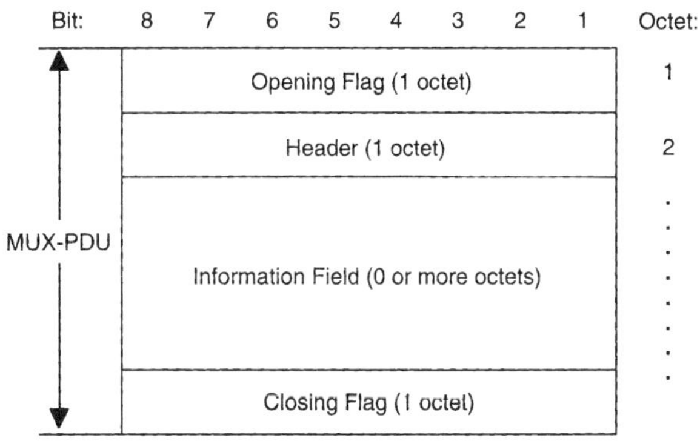

Figure 11 MUX-PDU format.

2. Adaptation Layer

The unit of information exchanged between the AL and the higher layer AL users is an AL-SDU. The method of mapping information streams from higher layers into AL-SDUs is outside the scope of this recommendation and is specified in the system recommendation that uses H.223. The AL-SDUs contain an integer number of octets. The AL adapts AL-SDUs to the MUX layer by adding, where appropriate, additional octets for purposes such as error detection, sequence numbering, and retransmission. The logical information unit exchanged between peer AL entities is called an AL-PDU. An AL-PDU carries exactly the same information as a MUX-SDU. There different types of ALs, named AL1 through AL3, are specified in this recommendation.

AL1 is designed primarily for the transfer of data or control information. Because AL1 does not provide any error control, all necessary error protection should be provided by the AL1 user. In the framed transfer mode, AL1 receives variable-length frames from its higher layer (for example, a data link layer protocol such as LAPM/V.42 or LAPF/Q.922, which provides error control) in AL-SDUs and simply passes these to the MUX layer in MUX-SDUs without any modifications. In the unframed mode, AL1 is used to transfer an unframed sequence of octets from an AL1 user. In this mode, one AL-SDU represents the entire sequence and is assumed to continue indefinitely.

AL2 is designed primarily for the transfer of digital audio. It receives frames, possibly of variable length, from its higher layer (for example, an audio encoder) in AL-SDUs and passes these to the MUX layer in MUX-SDUs, after adding one octet for an 8-bit cycle redundancy coding (CRC) and optionally adding one octet for sequence numbering.

AL3 is designed primarily for the transfer of digital video. It receives variable-length

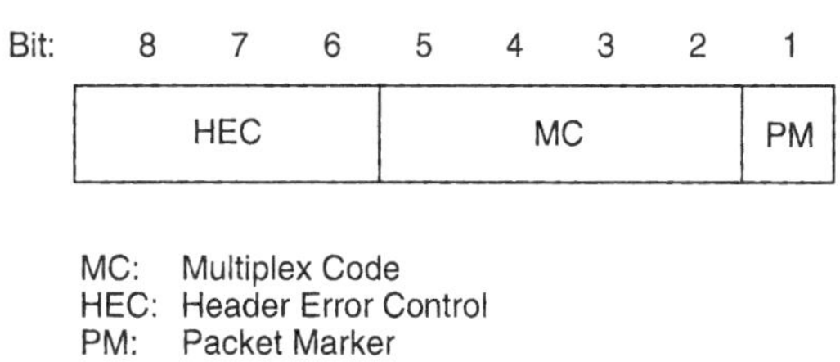

Figure 12 Header format of the MUX-PDU.

frames from its higher layer (for example, a video encoder) in AL-SDUs and passes these to the MUX layer in MUX-SDUs, after adding two octets for a 16-bit CRC and optionally adding one or two control octets. AL3 includes a retransmission protocol designed for video.

An example of how audio, video, and data fields could be multiplexed by the H.223 systems is illustrated in Figure 13.

F. Data Channel

All data channels are optional. Standardized options for data applications include the following:

T.120 series for point-to-point and multipoint audiographic teleconferencing including database access, still image transfer and annotation, application sharing, and real-time file transfer

T.84 (SPIFF) point-to-point still image transfer cutting across application borders

T.434 point-to-point telematic file transfer cutting across application borders

H.224 for real-time control of simplex applications, including H.281 far-end camera control

Network link layer, per ISO/IEC TR9577 (supports IP and PPP network layers, among others)

Unspecified user data from external data ports

These data applications may reside in an external computer or other dedicated device attached to the H.324 terminal through a V.24 or equivalent interface (implementation dependent) or may be integrated into the H.324 terminal itself. Each data application makes use of an underlying data protocol for link layer transport. For each data application supported by the H.324 terminal, this recommendation requires support for a particular underlying data protocol to ensure interworking of data applications.

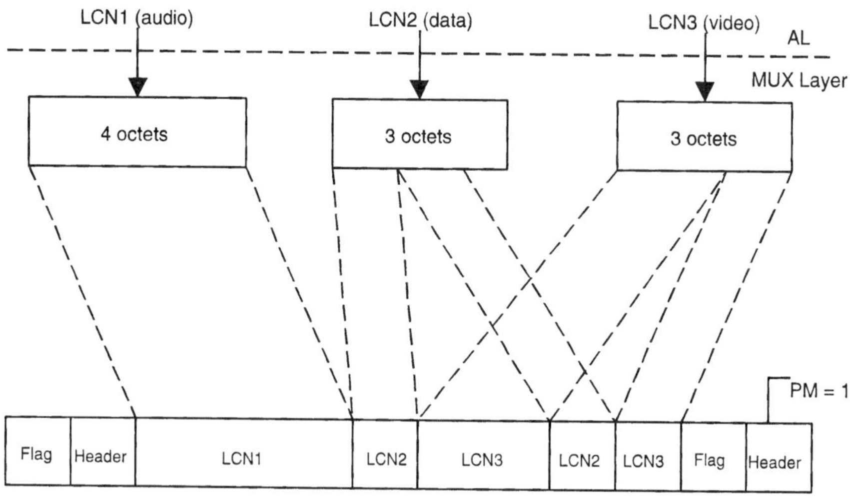

Figure 13 Information field example.

The H.245 control channel is not considered a data channel. Standardized link layer data protocols used by data applications include

Buffered V.14 mode for transfer of asynchronous characters, without error control
LAPM/V.42 for error-corrected transfer of asynchronous characters (in addition, depending on application, V.42bis data compression may be used)
HDLC frame tunneling for transfer of HDLC frames; transparent data mode for direct access by unframed or self-framed protocols

All H.324 terminals offering real-time audiographic conferencing should support the T.120 protocol suite.

G. Extension of H.324 to Mobile Radio (H.324M)

In February 1995 the ITU requested that the low bitrate coder (LBC) Experts Group begin work to adapt the H.324 series of GSTN recommendations for application to mobile networks. It is generally agreed that a very large market will develop in the near future for mobile multimedia systems. Laptop computers and handheld devices are already being configured for cellular connections. The purpose of the H.324M standard is to enable the efficient communication of voice, data, still images, and video over such mobile networks. It is anticipated that there will be some use of such a system for interactive videophone applications for use when people are traveling. However, it is expected that the primary application will be nonconversational, in which the mobile terminal would usually be receiving information from a fixed remote site. Typical recipients would be a construction site, police car, automobile, and repair site. On the other hand, it is expected that there will be a demand to send images and video from a mobile site such as an insurance adjuster, surveillance site, repair site, train, construction site, or fire scene. The advantage of noninteractive communication of this type is that the transmission delay can be relatively large without being noticed by the user.

Several of the general principles and underlying assumptions upon which the H.324M recommendations have been based are as follows.

H.324M recommendations should be based upon H.324 as much as possible.
The technical requirements and objectives for H.324M are essentially the same as for H.324.
Because the vast majority of mobile terminal calls are with terminals in fixed networks, it is very important that H.32M recommendations be developed to maximize interoperability with these fixed terminals.
It is assumed that the H.324M terminal has access to a transparent or synchronous bitstream from the mobile network.
It is proposed to provide the manufacturer of mobile multimedia terminals with a number of optional error protection tools to address a wide range of mobile networks, that is, regional and global, present and future, and cordless and cellular. Consequently, H.324M tools should be flexible, bit rate scalable, and extensible to the maximum degree possible. As with H.324, nonconversational services are an important application for H.324M.
Work toward the H.324M recommendation has been divided into the following areas of study: (1) speech error protection, (2) video error protection, (3) communi-

Table 5 Extension of H.324 to Mobile (H.324M)

	H.324 (POTS)	H.324M (mobile)	Approved
System	H.324	H.324 Annex C	1/98
Audio	G.723.1	G.723.1 Annex C	5/96
		Bit rate scalable error protection	
		Unequal error protection	
Video	H.263	H.263 Appendix II—error tracking	
		Annex K—slice structure	
		Annex N—reference picture selection	
		Annex R—independent segmented decoding	
Communication control	H.245	Mobile code points	1/98
Multiplex	H.223	H.223 annexes	1/98
		A—increase sync flag from 8 to 16 bits	
		B—A + more robust header	
		C—B + more robust payload	

cations control (adjustments to H.245), (4) multiplex or error control of the multiplexed signal, and (5) system.

Table 5 is a summary of the standardization work that has been accomplished by the ITU to extend the H.324 POTS recommendation to specify H.324M for the mobile environment.

IV. SUMMARY

This chapter has presented an overview of the two latest multimedia conferencing standards offered by the ITU. The H.323 protocol provides the technical requirements for multimedia communication systems that operate over packet-based networks where guaranteed quality of service may or may not be available. The H.323 protocol is believed to be revolutionizing the video- and audioconferencing industry; however, its success relies on the quality of packet-based networks. Unpredictable delay characteristics, long delays, and a large percentage of packet loss on a network prohibit conducting a usable H.323 conference. Packet-based networks are becoming more powerful, but until networks with predictable QoS are ubiquitous, there will be a need for video- and audioconferencing systems such as H.320 and H.324 that are based on switched circuit networks. Unlike H.323, H.324 operates over existing low-bit-rate networks without any additional requirements. The H.324 standard describes terminals for low-bit-rate multimedia communication, utilizing V.34 modems operating over the GSTN.

BIBLIOGRAPHY

1. ITU-T. Recommendation H.323: Packet-based multimedia communications systems, 1998.
2. ITU-T. Recommendation H.225.0: Call signaling protocols and media stream packetization for packet based multimedia communications systems, 1998.

3. ITU-T. Recommendation H.245: Control protocol for multimedia communication, 1998.
4. ITU-T. Recommendation H.235: Security and encryption for H-series (H.323 and other H.245-based) multimedia terminals, 1998.
5. ITU-T. Recommendation H.450.1: Generic functional protocol for the support of supplementary services in H.323, 1998.
6. ITU-T. Recommendation H.450.2: Call transfer supplementary service for H.323, 1998.
7. ITU-T. Recommendation H.450.3: Call diversion supplementary service for H.323, 1998.
8. ITU-T. Implementers guide for the ITU-T H.323, H.225.0, H.245, H.246, H.235, and H.450 series recommendations—Packet-based multimedia communication systems, 1998.
9. ITU-T: Recommendation H.324: Terminal for low bitrate multimedia communication, 1995.
10. D Lindberg, H Malvar. Multimedia teleconferencing with H.32. In: KR Rao, ed. Standards and Common Interfaces for Video Information Systems. Bellingham, WA: SPIE Optical Engineering Press, 1995, pp 206–232.
11. D Lindberg. The H.324 multimedia communication standard. IEEE Commun Mag 34(12): 46–51, 1996.
12. ITU-T. Recommendation V.34: A modem operating at data signaling rates of up to 28,8000 bit/s for use on the general switched telephone network and on leased point-to-point 2-wire telephone-type circuits, 1994.
13. ITU-T. Recommendation H.223: Multiplexing protocol for low bitrate multimedia communication, 1996.
14. ITU-T. Recommendation H.245: Control protocol for multimedia communication, 1996.
15. ITU-T. Recommendation H.263: Video coding for low bitrate communication, 1996.
16. B Girod, N Färber, E Steinbach. Performance of the H.263 video compression standard. J VLSI Signal Process. Syst Signal Image Video Tech 17:101–111, November 1997.
17. JW Park, JW Kim, SU Lee. DCT coefficient recovery-based error concealment technique and its application to MPEG-2 bit stream error. IEEE Trans Circuits Syst Video Technol 7:845–854, 1997.
18. G Wen, J Villasenor. A class of reversible variable length codes for robust image and video coding. IEEE Int Conf Image Process 2:65–68, 1997.
19. E Steinbach, N Färber, B Girod. Standard compatible extension of H.263 for robust video transmission in mobile environments. IEEE Trans Circuits Syst Video Technol 7:872–881, 1997.

3

H.263 (Including H.263+) and Other ITU-T Video Coding Standards

Tsuhan Chen
Carnegie Mellon University, Pittsburgh, Pennsylvania

Gary J. Sullivan
Picture Tel Corporation, Andover, Massachusetts

Atul Puri
AT&T Labs, Red Bank, New Jersey

I. INTRODUCTION

Standards are essential for communication. Without a common language that both the transmitter and the receiver understand, communication is impossible. In digital multimedia communication systems the language is often defined as a standardized *bitstream syntax* format for sending data.

The ITU-T* is the organization responsible for developing standards for use on the global telecommunication networks, and SG16† is its leading group for multimedia services and systems. The ITU is a United Nations organization with headquarters in Geneva, Switzerland, just a short walk from the main United Nations complex. Digital communications were part of the ITU from the very beginning, as it was originally founded for telegraph text communication and predates the 1876 invention of the telephone (Samuel Morse sent the first public telegraph message in 1844, and the ITU was founded in 1865). As telephony, wireless transmission, broadcast television, modems, digital speech coding, and digital video and multimedia communication have arrived, the ITU has added each new form of communication to its array of supported services.

* The International Telecommunications Union, Telecommunication Standardization Sector. (ITU originally meant International Telegraph Union, and from 1956 until 1993 the ITU-T was known as the CCITT—the International Telephone and Telegraph Consultative Committee.)

† Study Group 16 of the ITU-T. Until a 1997 reorganization of study groups, the group responsible for video coding was called Study Group XV (15).

The ITU-T is now one of two formal standardization organizations that develop media coding standards—the other being ISO/IEC JTC1.[‡] Along with the IETF,[§] which defines multimedia delivery for the Internet, these organizations form the core of today's international multimedia standardization activity. The ITU standards are called *recommendations* and are denoted with alphanumeric codes such as ''H.26x'' for the recent video coding standards (where ''x'' = 1, 2, or 3).

In this chapter we focus on the video coding standards of the ITU-T SG16. These standards are currently created and maintained by the ITU-T Q.15/SG16 Advanced Video Coding experts group.[¶] We will particularly focus on ITU-T recommendation H.263, the most current of these standards (including its recent second version known as H.263+). We will also discuss the earlier ITU-T video coding projects, including

Recommendation H.120, the first standard for compressed digital coding of video and still-picture graphics [1]
Recommendation H.261, the standard that forms the basis for all later standard designs [including H.263 and the Moving Picture Experts Group (MPEG) video standards] [2]
Recommendation H.262, the MPEG-2 video coding standard [3]

Recommendation H.263 represents today's state of the art for standardized video coding [4], provided some of the key features of MPEG-4 are not needed in the application (e.g., shape coding, interlaced pictures, sprites, 12-bit video, dynamic mesh coding, face animation modeling, and wavelet still-texture coding). Essentially any bit rate, picture resolution, and frame rate for progressive-scanned video content can be efficiently coded with H.263. Recommendation H.263 is structured around a ''baseline'' mode of operation, which defines the fundamental features supported by all decoders, plus a number of optional enhanced modes of operation for use in customized or higher performance applications. Because of its high performance, H.263 was chosen as the basis of the MPEG-4 video design, and its baseline mode is supported in MPEG-4 without alteration. Many of its optional features are now also found in some form in MPEG-4.

The most recent version of H.263 (the second version) is known informally as H.263+ or H.263v2. It includes about a dozen new optional enhanced modes of operation created in a design project that ended in September 1997. These enhancements include additions for added error resilience, coding efficiency, dynamic picture resolution changes, flexible custom picture formats, scalability, and backward-compatible supplemental enhancement information. (A couple more features are also being drafted for addition as future ''H.263++'' enhancements.)

Although we discuss only video coding standards in this chapter, the ITU-T SG16 is also responsible for a number of other standards for multimedia communication, including

Speech/audio coding standards, such as G.711, G.723.1, G.728, and G.729 for 3.5-kHz narrowband speech coding and G.722 for 7-kHz wideband audio coding

---

[‡] Joint Technical Committee number 1 of the International Standardization Organization and the International Electrotechnical Commission.

[§] The Internet Engineering Task Force.

[¶] Question 15 of Study Group 16 of the ITU-T covers ''Advanced Video Coding'' topics. The Rapporteur in charge of Q.15/16 is Gary J. Sullivan, the second author of this chapter.

Multimedia terminal systems, such as H.320 for integrated services digital network (ISDN) use, H.323 for use on Internet Protocol networks, and H.324 for use on the public switched telephone network (such as use with a modem over analog phone lines or use on ISDN)

Modems, such as the recent V.34 and V.90 standards

Data communication, such as T.140 for text conversation and T.120 for multimedia conferencing

This chapter is outlined as follows. In Sec. II, we explain the roles of standards for video coding and provide an overview of key standardization organizations and video coding standards. In Sec. III, we present in detail the techniques used in the historically very important video coding standard H.261. In Sec. IV, H.263, a video coding standard that has a framework similar to that of H.261 but with superior coding efficiency, is discussed. Section V covers recent activities in H.263+ that resulted in a new version of H.263 with several enhancements. We conclude the chapter with a discussion of future ITU-T video projects, conclusions, and pointers to further information in Secs. VI and VII.

II. FUNDAMENTALS OF STANDARDS FOR VIDEO CODING

A *formal* standard (sometimes also called a *voluntary* standard), such as those developed by the ITU-T and ISO/IEC JTC1, has a number of important characteristics:

A clear and complete description of the design with essentially sufficient detail for implementation is available to anyone. Often a fee is required for obtaining a copy of the standard, but the fee is intended to be low enough not to restrict access to the information.

Implementation of the design by anyone is allowed. Sometimes a payment of royalties for licenses to intellectual property is necessary, but such licenses are available to anyone under ''fair and reasonable'' terms.

The design is approved by a *consensus* agreement. This requires that nearly all of the participants in the process must essentially agree on the design.

The standardizing organization meets in a relatively open manner and includes representatives of organizations with a variety of different interests (for example, the meetings often include representatives of companies that compete strongly against each other in a market). The meetings are often held with some type of official governmental approval. Governments sometimes have rules concerning who can attend the meetings, and sometimes countries take official positions on issues in the decision-making process.

Sometimes there are designs that lack many or all of these characteristics but are still referred to as ''standards.'' These should not be confused with the formal standards just described. A de facto standard, for example, is a design that is not a formal standard but has come into widespread use without following these guidelines.

A key goal of a standard is *interoperability*, which is the ability for systems designed by different manufacturers to work seamlessly together. By providing interoperability, open standards can facilitate market growth. Companies participating in the standardization process try to find the right delicate balance between the high-functioning interoperability needed for market growth (and for volume-oriented cost savings for key components) and the competitive advantage that can be obtained by product differentiation. One way in which these competing desires are evident in video coding standardization is in the narrow scope of video coding standardization. As illustrated in Fig. 1, today's video coding standards specify only the format of the compressed data and how it is to be decoded. They specify nothing about how encoding or other video processing is performed. This limited scope of standardization arises from the desire to allow individual manufacturers to have as much freedom as possible in designing their own products while strongly preserving the fundamental requirement of interoperability. This approach provides no guarantee of the quality that a video encoder will produce but ensures that any decoder that is designed for the standard syntax can properly receive and decode the bitstream produced by any encoder. (Those not familiar with this issue often mistakenly believe that any system designed to use a given standard will provide similar quality, when in fact some systems using an older standard such as H.261 may produce better video than those using a "higher performance" standard such as H.263.)

The two other primary goals of video coding standards are maximizing *coding efficiency* (the ability to represent the video with a minimum amount of transmitted data) and minimizing complexity (the amount of processing power and implementation cost required to make a good implementation of the standard). Beyond these basic goals there are many others of varying importance for different applications, such as minimizing transmission delay in real-time use, providing rapid switching between video channels, and obtaining robust performance in the presence of packet losses and bit errors.

There are two approaches to understanding a video coding standard. The most correct approach is to focus on the bitstream syntax and to try to understand what each layer of the syntax represents and what each bit in the bitstream indicates. This approach is very important for manufacturers, who need to understand fully what is necessary for compliance with the standard and what areas of the design provide freedom for product customization. The other approach is to focus on some encoding algorithms that can be used to generate standard-compliant bitstreams and to try to understand what each component of these example algorithms does and why some encoding algorithms are therefore better than others. Although strictly speaking a standard does not specify any encoding algorithms, the latter approach is usually more approachable and understandable. Therefore, we will take this approach in this chapter and will describe certain bitstream syntax

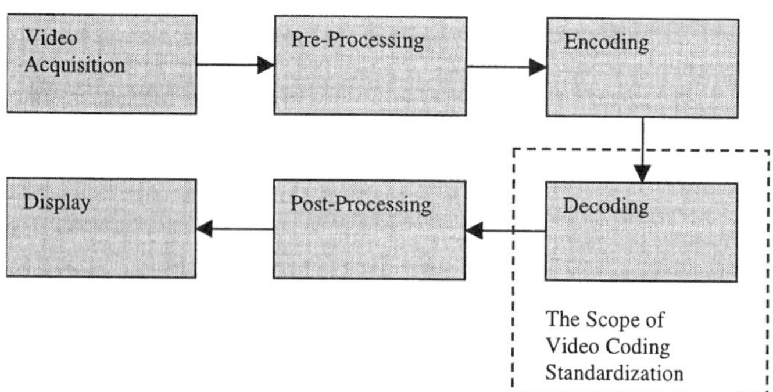

Figure 1 The limited scope of video coding standardization.

only when necessary. For those interested in a more rigorous treatment focusing on the information sent to a decoder and how its use can be optimized, the more rigorous and mathematical approach can be found in Ref. 5.

The compressed video coding standardization projects of the ITU-T and ISO/IEC JTC1 organizations are summarized in Table 1.

The first video coding standard was H.120, which is now purely of historical interest [1]. Its original form consisted of conditional replenishment coding with differential pulse-code modulation (DPCM), scalar quantization, and variable-length (Huffman) coding, and it had the ability to switch to quincunx sampling for bit rate control. In 1988, a second version added motion compensation and background prediction. Most features of its design (conditional replenishment capability, scalar quantization, variable-length coding, and motion compensation) are still found in the more modern standards.

The first widespread practical success was H.261, which has a design that forms the basis of all modern video coding standards. It was H.261 that brought video communication down to affordable telecom bit rates. We discuss H.261 in the next section.

The MPEG standards (including MPEG-1, H.262/MPEG-2, and MPEG-4) are discussed at length in other chapters and will thus not be treated in detail here. The remainder of this chapter after the discussion of H.261 focuses on H.263.

III. H.261

Standard H.261 is a video coding standard designed by the ITU for videotelephony and videoconferencing applications [2]. It is intended for operation at low bit rates (64 to 1920 kbits/sec) with low coding delay. Its design project was begun in 1984 and was originally intended to be used for audiovisual services at bit rates around $m \times 384$ kbits/sec where m is between 1 and 5. In 1988, the focus shifted and it was decided to aim at bit rates

Table 1 Video Coding Standardization Projects

Standards organization	Video coding standard	Approximate date of technical completion (may be prior to final formal approval)
ITU-T	ITU-T H.120	Version 1, 1984 Version 2 additions, 1988
ITU-T	ITU-T H.261	Version 1, late 1990 Version 2 additions, early 1993
ISO/IEC JTC1	IS 11172-2 MPEG-1 Video	Version 1, early 1993 One corrigendum, 1996
ISO/IEC JTC1 and ITU-T Jointly	IS 13818-2/ ITU-T H.262 MPEG-2 Video	Version 1, 1994 Four amendment additions and two corrigenda since version 1 New amendment additions, in progress
ITU-T	ITU-T H.263	Version 1, November 1995 Version 2 H.263+ additions, September 1997 Version 3 H.263++ additions, in progress
ISO/IEC JTC1	IS 14496-2 MPEG-4 Video	Version 1, December 1998 Version 2 additions, in progress
ITU-T	"H.26L"	Future work project in progress

Table 2 Picture Formats Supported by H.261 and H.263

	Sub-QCIF	QCIF	CIF	4CIF	16CIF	Custom picture sizes
Luminance width (pixels)	128	176	352	704	1408	≤2048
Luminance height (pixels)	96	144	288	576	1152	≤1152
Uncompressed bit rate (Mbits/sec)	4.4	9.1	37	146	584	
Remarks	H.263 only (required)	Required in all H.261 and H.263 decoders	Optional	H.261 supports for still pictures only	H.263 only	H.263 only

around $p \times 64$ kbits/sec, where p is from 1 to 30. Therefore, H.261 also has the informal name $p \times 64$ (pronounced ''*p times* 64''). Standard H.261 was originally approved in December 1990.

The coding algorithm used in H.261 is basically a hybrid of block-based translational *motion compensation* to remove temporal redundancy and *discrete cosine transform coding* to reduce spatial redundancy. It uses switching between interpicture prediction and intrapicture coding, ordered scanning of transform coefficients, scalar quantization, and variable-length (Huffman) entropy coding. Such a framework forms the basis of all video coding standards that were developed later. Therefore, H.261 has a very significant influence on many other existing and evolving video coding standards.

A. Source Picture Formats and Positions of Samples

Digital video is composed of a sequence of pictures, or frames, that occur at a certain rate. For H.261, the frame rate is specified to be 30,000/1001 (approximately 29.97) pictures per second. Each picture is composed of a number of samples. These samples are often referred to as pixels (picture elements) or simply pels. For a video coding standard, it is important to understand the picture sizes that the standard applies to and the position of samples. Standard H.261 is designed to deal primarily with two picture formats: the common intermediate format (CIF) and the quarter CIF (QCIF).* Please refer to Table 2, which summarizes a variety of picture formats. Video coded at CIF resolution using somewhere between 1 and 3 Mbits/sec is normally close to the quality of a typical videocassette recorder (significantly less than the quality of good broadcast television). This resolution limitation of H.261 was chosen because of the need for low-bit-rate operation and low complexity.

* In the still-picture graphics mode as defined in Annex D of H.261 version 2, four times the currently transmitted video format is used. For example, if the video format is CIF, the corresponding still-picture format is 4CIF. The still-picture graphics mode was adopted using an ingenious trick to make its bitstream backward compatible with prior decoders operating at lower resolution, thus avoiding the need for a capability negotiation for this feature.

It is adequate for basic videotelephony and videoconferencing, in which typical source material is composed of scenes of talking persons rather than general entertainment-quality TV programs that should provide more detail.

In H.261 and H.263, each pixel contains a luminance component, called Y, and two chrominance components, called C_B and C_R. The values of these components are defined as in Ref. 6. In particular, "black" is represented by Y = 16, "white" is represented by Y = 235, and the range of C_B and C_R is between 16 and 240, with 128 representing zero color difference (i.e., a shade of gray). A picture format, as shown in Table 2, defines the size of the image, hence the resolution of the Y component. The chrominance components, however, typically have lower resolution than luminance in order to take advantage of the fact that human eyes are less sensitive to chrominance than to luminance. In H.261 and H.263, the C_B and C_R components have half the resolution, both horizontally and vertically, of the Y component. This is commonly referred to as the 4:2:0 format (although few people know why). Each C_B or C_R sample lies in the center of four neighboring Y samples, as shown in Fig. 2. Note that block edges, to be defined in the next section, lie in between rows or columns of Y samples.

B. Blocks, Macroblocks, and Groups of Blocks

Typically, we do not code an entire picture all at once. Instead, it is divided into blocks that are processed one by one, both by the encoder and by the decoder, most often in a *raster* scan order as shown in Fig. 3. This approach is often referred to as *block-based coding*.

In H.261, a block is defined as an 8 × 8 group of samples. Because of the downsampling in the chrominance components as mentioned earlier, one block of C_B samples and one block of C_R samples correspond to four blocks of Y samples. The collection of these six blocks is called a macroblock (MB), as shown in Fig. 4, with the order of blocks as marked from 1 to 6. An MB is treated as one unit in the coding process.

A number of MBs are grouped together and called a group of blocks (GOB). For H.261, a GOB contains 33 MBs, as shown in Fig. 5. The resulting GOB structures for a picture, in the CIF and QCIF cases, are shown in Fig. 6.

C. The Compression Algorithm

Compression of video data is typically based on two principles: reduction of spatial redundancy and reduction of temporal redundancy. Standard H.261 uses a discrete cosine trans-

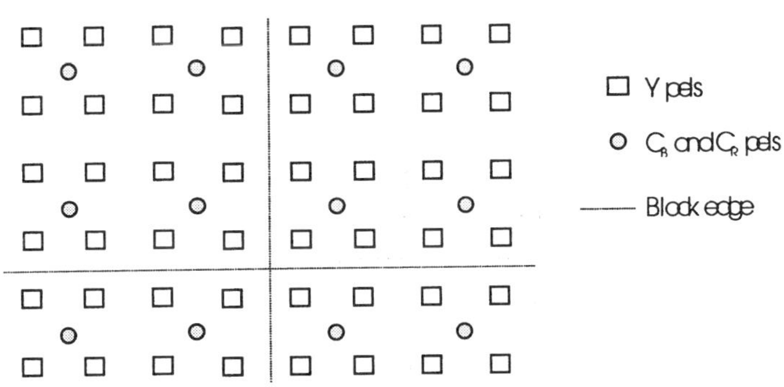

Figure 2 Positions of samples for H.261.

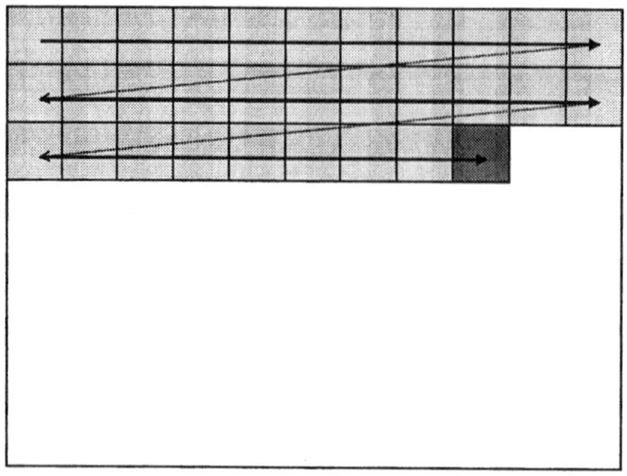

Figure 3 Illustration of block-based coding.

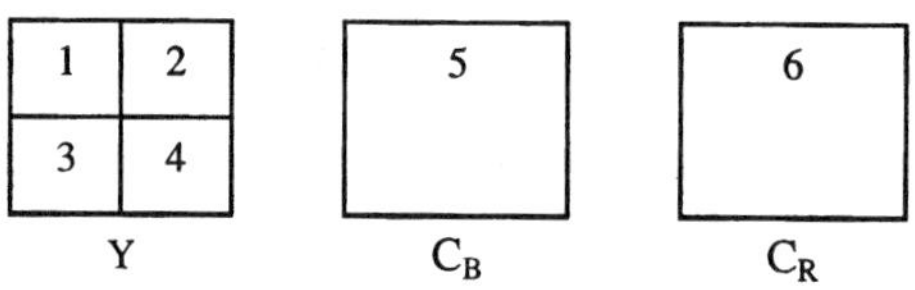

Figure 4 A macroblock (MB).

1	2	3	4	5	6	7	8	9	10	11
12	13	14	15	16	17	18	19	20	21	22
23	24	25	26	27	28	29	30	31	32	33

Figure 5 A group of blocks (GOB).

GOB 1	GOB 2
GOB 3	GOB 4
GOB 5	GOB 6
GOB 7	GOB 8
GOB 9	GOB 10
GOB 11	GOB 12

CIF

GOB 1
GOB 3
GOB 5

QCIF

Figure 6 H.261 GOB structures for CIF and QCIF.

form to remove spatial redundancy and motion compensation to remove temporal redundancy. We now discuss these techniques in detail.

1. Transform Coding

Transform coding has been widely used to remove redundancy between data samples. In transform coding, a set of data samples are first linearly transformed into a set of transform coefficients. These coefficients are then quantized and entropy coded. A proper linear transform can decorrelate the input samples and hence remove the redundancy. Another way to look at this is that a properly chosen transform can concentrate the energy of input samples into a small number of transform coefficients so that resulting coefficients are easier to encode than the original samples.

The most commonly used transform for video coding is the discrete cosine transform (DCT) [7,8]. In terms of both objective coding gain and subjective quality, the DCT performs very well for typical image data. The DCT operation can be expressed in terms of matrix multiplication by

$$\mathbf{F} = \mathbf{C}^T\mathbf{X}\mathbf{C}$$

where **X** represents the original image block and **F** represents the resulting DCT coefficients. The elements of **C**, for an 8×8 image block, are defined as

$$C_{mn} = k_n \cos\left[\frac{(2m+1)n\pi}{16}\right] \quad \text{where } k_n = \begin{cases} 1/(2\sqrt{2}) & \text{when } n = 0 \\ 1/2 & \text{otherwise} \end{cases}$$

After the transform, the DCT coefficients in **F** are quantized. Quantization implies loss of information and is the primary source of actual compression in the system. The quantization step size depends on the available bit rate and can also depend on the coding modes. Except for the intra DC coefficients that are uniformly quantized with a step size of 8, an enlarged "dead zone" is used to quantize all other coefficients in order to remove noise around zero. (DCT coefficients are often modeled as Laplacian random variables and the application of scalar quantization to such random variables is analyzed in detail in Ref. 9.) Typical input–output relations for the two cases are shown in Fig. 7.

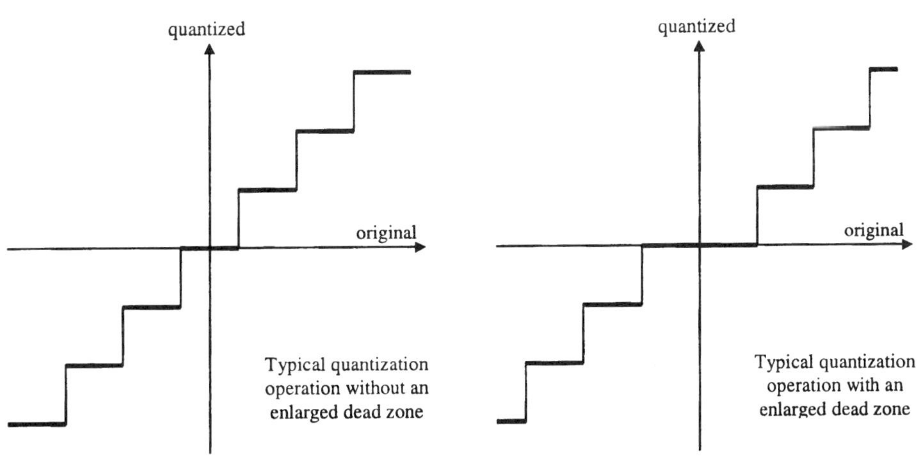

Figure 7 Quantization with and without an enlarged "dead zone."

The quantized 8 × 8 DCT coefficients are then converted into a one-dimensional (1D) array for entropy coding by an ordered scanning operation. Figure 8 shows the "zigzag" scan order used in H.261 for this conversion. Most of the energy concentrates in the low-frequency coefficients (the first few coefficients in the scan order), and the high-frequency coefficients are usually very small and often quantize to zero. Therefore, the scan order in Fig. 8 can create long runs of zero-valued coefficients, which is important for efficient entropy coding, as we discuss in the next paragraph.

The resulting 1D array is then decomposed into segments, with each segment containing one or more (or no) zeros followed by a nonzero coefficient. Let an *event* represent the pair (*run, level*), where "run" represents the number of zeros and "level" represents the magnitude of the nonzero coefficient. This coding process is sometimes called "run-length coding." Then a table is built to represent each event by a specific codeword, i.e., a sequence of bits. Events that occur more often are represented by shorter codewords, and less frequent events are represented by longer codewords. This entropy coding method is therefore called variable length coding (VLC) or Huffman coding. In H.261, this table is often referred to as a two-dimensional (2D) VLC table because of its 2D nature, i.e., each event representing a (run, level) pair. Some entries of VLC tables used in H.261 are shown in Table 3. In this table, the last bit "s" of each codeword denotes the sign of the level, "0" for positive and "1" for negative. It can be seen that more likely events, i.e., short runs and low levels, are represented with short codewords and vice versa. After the last nonzero DCT coefficient is sent, the end-of-block (EOB) symbol, represented by 10, is sent.

At the decoder, all the proceeding steps are reversed one by one. Note that all the steps can be exactly reversed except for the quantization step, which is where a loss of information arises. Because of the irreversible quantization process, H.261 video coding falls into the category of techniques known as "lossy" compression methods.

2. Motion Compensation

The transform coding described in the previous section removes spatial redundancy within each frame of video content. It is therefore referred to as *intra* coding. However, for video material, *inter* coding is also very useful. Typical video material contains a large amount of redundancy along the temporal axis. Video frames that are close in time usually have

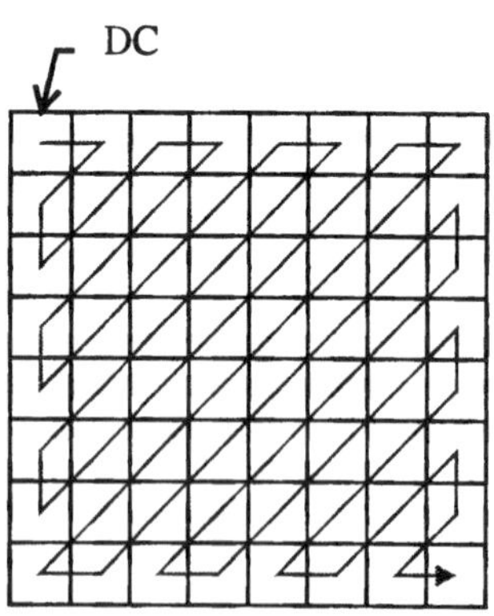

Figure 8 Scan order of the DCT coefficients.

Table 3 Part of the H.261 Transform Coefficient VLC table

Run	Level	Code
0	1	1s If first coefficient in block
0	1	11s Not first coefficient in block
0	2	0100 s
0	3	0010 1s
0	4	0000 110s
0	5	0010 0110 s
0	6	0010 0001 s
0	7	0000 0010 10s
0	8	0000 0001 1101 s
0	9	0000 0001 1000 s
0	10	0000 0001 0011 s
0	11	0000 0001 0000 s
0	12	0000 0000 1101 0s
0	13	0000 0000 1100 1s
0	14	0000 0000 1100 0s
0	15	0000 0000 1011 1s
1	1	011s
1	2	0001 10s
1	3	0010 0101 s
1	4	0000 0011 00s
1	5	0000 0001 1011 s
1	6	0000 0000 1011 0s
1	7	0000 0000 1010 1s
2	1	0101 s
2	2	0000 100s
2	3	0000 0010 11s
2	4	0000 0001 0100 s
2	5	0000 0000 1010 0s
3	1	0011 1s
3	2	0010 0100 s
3	3	0000 0001 1100 s
3	4	0000 0000 1001 1s

a large amount of similarity. Therefore, transmitting the difference between frames is more efficient than transmitting the original frames. This is similar to the concept of differential coding and predictive coding. The previous frame is used as an estimate of the current frame, and the residual, the difference between the estimate and the true value, is coded. When the estimate is good, it is more efficient to code the residual than to code the original frame.

Consider the fact that typical video material is a camera's view of moving objects. Therefore, it is possible to improve the prediction result by first estimating the motion of each region in the scene. More specifically, the encoder can estimate the motion (i.e., displacement) of each block between the previous frame and the current frame. This is often achieved by matching each block (actually, macroblock) in the current frame with

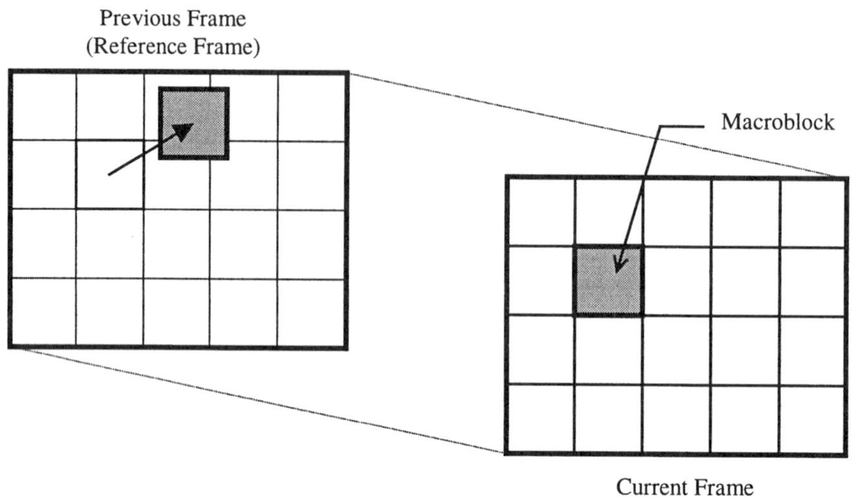

Figure 9 Motion compensation.

the previous frame to find the best matching area.* This area is then offset accordingly to form the estimate of the corresponding block in the current frame. Now, the residue has much less energy than the original signal and therefore is much easier to code to within a given average error. This process is called motion compensation (MC), or more precisely, motion-compensated prediction [10,11]. This is illustrated in Fig. 9. The residue is then coded using the same process as that of intra coding.

Pictures that are coded without any reference to previously coded pictures are called *intra* pictures, or simply I-pictures (or I-frames). Pictures that are coded using a previous picture as a reference for prediction are called *inter* or *predicted* pictures, or simply P-pictures (or P-frames). However, note that a P-picture may also contain some intra coded macroblocks. The reason is as follows. For a certain macroblock, it may be impossible to find a good enough matching area in the reference picture to be used for prediction. In this case, direct intra coding of such a macroblock is more efficient. This situation happens often when there is occlusion in the scene or when the motion is very heavy.

Motion compensation allows the remaining bits to be used for coding the DCT coefficients. However, it does imply that extra bits are required to carry information about the motion vectors. Efficient coding of motion vectors is therefore also an important part of H.261. Because motion vectors of neighboring blocks tend to be similar, differential coding of the horizontal and vertical components of motion vectors is used. That is, instead of coding motion vectors directly, the previous motion vector is used as a prediction for the current motion vector, and the difference, in both the horizontal component and the vertical component, is then coded using a VLC table, part of which is shown in Table 4. Note two things in this table. First, short codewords are used to represent small differences, because these are more likely events. Second, note that one codeword can represent up

* Note, however, that the standard does not specify how motion estimation should be done. Motion estimation can be a very computationally intensive process and is the source of much of the variation in the quality produced by different encoders.

Table 4 Part of the VLC Table for Coding Motion Vectors

MVD	Code
. . .	. . .
−7 and 25	0000 0111
−6 and 26	0000 1001
−5 and 27	0000 1011
−4 and 28	0000 111
−3 and 29	0001 1
−2 and 30	0011
−1	011
0	1
1	010
2 and −30	0010
3 and −29	0001 0
4 and −28	0000 110
5 and −27	0000 1010
6 and −26	0000 1000
7 and −25	0000 0110
. . .	. . .

to two possible values for motion vector difference. Because the allowed range of both the horizontal component and the vertical component of motion vectors is restricted to −15 to +15, only one will yield a motion vector with the allowable range.

The ±15 range for motion vector values is not adequate for high resolutions with large amounts of motion but was chosen for use in H.261 in order to minimize complexity while supporting the relatively low levels of motion expected in videoconferencing. All later standards provide some way to extend this range as either a basic or optional feature of their design. Another feature of the H.261 design chosen for minimizing complexity is the restriction that motion vectors have only integer values. Better performance can be obtained by allowing fractional-valued motion vectors and having the decoder interpolate the samples in the prior reference picture when necessary. Instead, H.261 used only integer motion vectors and had a blurring *loop filter* for motion compensation (which could be switched on or off when motion compensating a macroblock). The loop filter provided a smoother prediction when a good match could not be found in the prior picture for high frequencies in the current macroblock. The later standard designs would do away with this filter and instead adopt half-pixel motion vector capability (in which the interpolation provides the filtering effect).

3. *Picture Skipping*

One of the most effective features of H.261 is that it can easily skip entire frames of video data. Thus, when motion is too heavy to represent properly within the bit rate of the channel, H.261 encoders will simply not encode all of the pictures coming from the camera. This increases the number of bits it can spend on each picture it encodes and results in improved quality for each picture (with some loss of motion smoothness). Encoders can also skip some pictures simply to reduce the computational complexity of the encoder or to reduce the complexity needed in the decoder.

At high bit rates, picture skipping may not be necessary, so it is not supported in some standards such as MPEG-1. The inability to skip pictures makes MPEG-1 incapable of operation at bit rates much lower than its primary design region (1 to 1.5 Mbits/sec). MPEG-2 also has a limited ability to skip pictures, but picture skipping is supported seamlessly in the standards designed to support low bit rates, including H.261, H.263, and MPEG-4. Excessive picture skipping is to be avoided, as it causes annoyingly jerky motion or sometimes even a severe ''slide show'' effect that looks more like periodic still-picture coding than video. However, picture skipping is fundamental to operation at low bit rates for today's systems.

4. *Forward Error Correction*

H.261 also defines an error-correcting code that can optionally be applied to the encoded bitstream to provide error detection and correction capabilities. The code is a BCH (Bose, Chaudhuri, and Hocquengham) (511,493) code adding 20 bits of overhead (1 bit framing, 1 bit fill indication, and 18 bits of parity) to each 492 bits of data. ''Fill frames'' of 512 bits can be sent to fill the channel when sufficient video data are not generated. Its use is mandated in some environments (such as when used in the H.320 ISDN videoconferencing standard) and not supported in others (such as in the H.323 and H.324 multimedia conferencing standards).

5. *Summary*

The coding algorithm used in H.261 is summarized in block diagrams in Fig. 10 and Fig. 11. At the encoder, the input picture is compared with the previously decoded frame with motion compensation. The difference signal is DCT transformed and quantized and then entropy coded and transmitted. At the decoder, the decoded DCT coefficients are inverse DCT transformed and then added to the previously decoded picture with loop-filtered motion compensation.

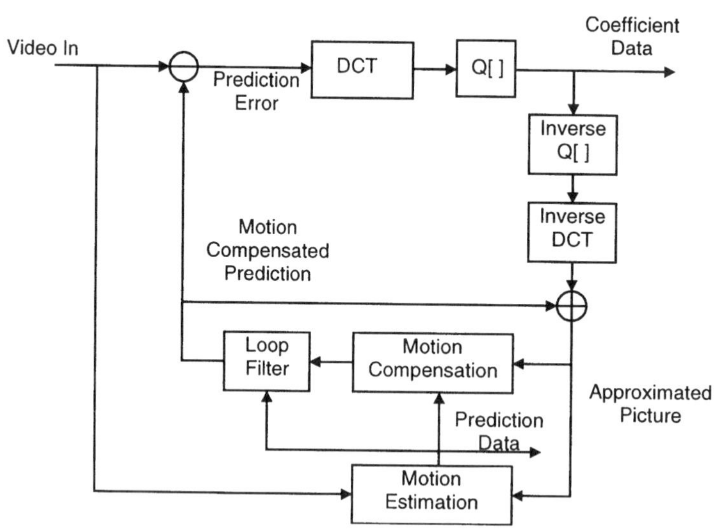

Figure 10 Block diagram of a video encoder.

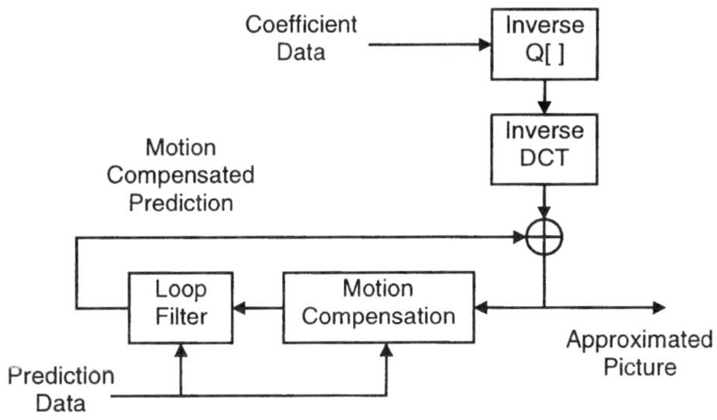

Figure 11 Block diagram of a video decoder.

D. The H.261 Reference Model

As in all recent video coding standards, H.261 specifies only the bitstream syntax and how a decoder should interpret the bitstream to decode the image. Therefore, it specifies only the design of the decoder, not how the encoding should be done. For example, an encoder can simply decide to use only zero-valued motion vectors and let the transform coding take all the burden of coding the residual. This may not be the most efficient encoding algorithm, but it does generate a standard-compliant bitstream and normally requires far fewer bits than still-picture coding of each frame.

Therefore, to illustrate the effectiveness of a video coding standard, an example encoder algorithm design is often described by the group that defines the standard. For H.261, such an example encoder is called a reference model (RM), and the latest version is RM 8 [12]. It specifies details about motion estimation, quantization, decisions for inter/intra coding and MC/no MC, loop filtering, buffering, and rate control.

IV. H.263 VERSION 1

The H.263 design project started in 1993, and the standard was approved at a meeting of ITU-T SG 15 in November 1995 (and published in March 1996) [4]. Although the original goal of this endeavor was to design a video coding standard suitable for applications with bit rates around 20 kbit/sec (the so-called very low bit rate applications), it became apparent that H.263 could provide a significant improvement over H.261 at any bit rate.

In this section, we discuss H.263. In essence, H.263 combines the features of H.261 with several new methods, including the half-pixel motion compensation first found in MPEG-1 and other techniques. H.263 can provide 50% savings or more in the bit rate needed to represent video at a given level of perceptual quality at very low bit rates (relative to H.261). In terms of signal-to-noise ratio (SNR), H.263 can provide about a 3-dB gain over H.261 at these very low rates. In fact, H.263 provides coding efficiency superior to that of H.261 at all bit rates (although not nearly as dramatic an improvement when operating above 64 kbit/sec). H.263 can also give significant bit rate savings when compared with MPEG-1 at higher rates (perhaps 30% at around 1 Mbit/sec).

H.263 is structured around a basic mode of operation informally called the ''baseline'' mode. In addition to the baseline mode, it includes a number of optional enhancement features to serve a variety of applications. The original version of H.263 had about six such optional modes, and more were added in the H.263+ project, which created H.263 version 2. The first version is discussed in this section, concentrating on a description of the baseline mode. (The second version retained all elements of the original version, adding only new optional enhancements.)

A. H.263 Version 1 Versus H.261

Because H.263 was built on top of H.261, the main structures of the two standards are essentially the same. Therefore, we will focus only on the differences between the two standards. These are the major differences between H.263 and H.261:

1. H.263 supports more picture formats and uses a different GOB structure.
2. H.263 GOB-level overhead information is not required to be sent, which can result in significant bit rate savings.
3. H.263 uses half-pixel motion compensation (first standardized in MPEG-1), rather than the combination of integer-pixel motion compensation and loop filtering as in H.261.
4. H.263 baseline uses a 3D VLC for improved efficiency in coding DCT coefficient values.
5. H.263 has more efficient coding of macroblock and block signaling overhead information such as indications of which blocks are coded and indications of changes in the quantization step size.
6. H.263 uses median prediction of motion vector values for improved coding efficiency.
7. In addition to the baseline coding mode, H.263 Version 1 provides six optional algorithmic ''mode'' features for enhanced operation in a variety of applications. Five of these six modes are not found in H.261:
   a. A mode that allows sending multiple video streams within a single video channel (the Continuous-Presence Multipoint and Video Multiplex mode defined in Annex C).
   b. A mode providing an extended range of values for motion vectors for more efficient performance with high resolutions and large amounts of motion (the Unrestricted Motion Vector mode defined in Annex D).
   c. A mode using arithmetic coding to provide greater coding efficiency (the Syntax-Based Arithmetic Coding mode defined in Annex E).
   d. A mode enabling variable block-size motion compensation and using overlapped-block motion compensation for greater coding efficiency and reduced blocking artifacts (the Advanced Prediction mode defined in Annex F).
   e. A mode that represents pairs of pictures as a single unit, for a low-overhead form of bidirectional prediction (the PB-frames mode defined in Annex G).

The optional modes of H.263 provide a means similar to the ''profile'' optional features defined for the MPEG video standards—they provide an ability to achieve enhanced performance or added special capabilities in environments that support them. H.263 also allows decoders some freedom to determine which frame rates and picture resolutions they support (Sub-QCIF and QCIF support are required in all decoders, and at least one of these two resolutions must be supported in any encoder). The limits in some decoders on picture resolution and frame rate are similar to the "levels" defined in the MPEG video standards.

In some applications such as two-way real-time communication, a decoder can send information to the encoder to indicate which optional capabilities it has. The encoder will then send only video bitstreams that it is certain can be decoded by the decoder. This process is known as a *capability exchange*. A capability exchange is often needed for other purposes as well, such as indicating whether a decoder has video capability at all or whether H.261, H.262, or H.263 video should be used. In other applications, the decoding capabilities can be prearranged (for example, by establishing requirements at the system level, as has occurred with the use of forward error correction coding for H.261 in the H.320 system environment).

B. Picture Formats, Sample Positions, and the GOB Structure

In addition to CIF and QCIF as supported by H.261, H.263 supports Sub-QCIF, 4CIF, and 16CIF (and version 2 supports custom picture formats). Resolutions of these picture formats can be found in Table 2. Chrominance subsampling and the relative positions of chrominance pels are the same as those defined in H.261. However, H.263 baseline uses a different GOB structure. The GOB structures for the standard resolutions are shown in Fig. 12. Unlike that in H.261, a GOB in H.263 is always constructed from one or more full rows of MBs.

C. Half-Pel Prediction and Motion Vector Coding

A major difference between H.261 and H.263 is the half-pel prediction in the motion compensation. This technique is also used in the MPEG standards. Whereas the motion vectors in H.261 can have only integer values, H.263 allows the precision of motion vec-

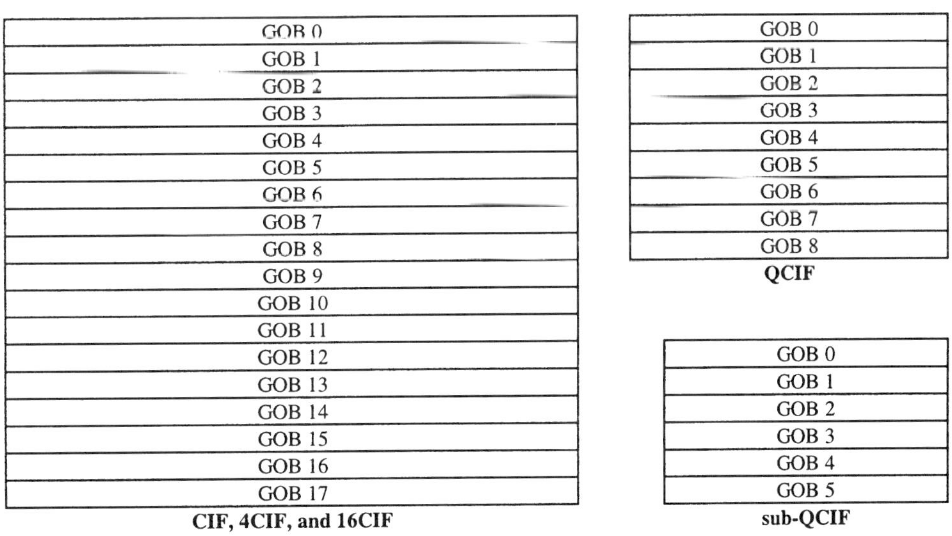

Figure 12 GOB structures for H.263.

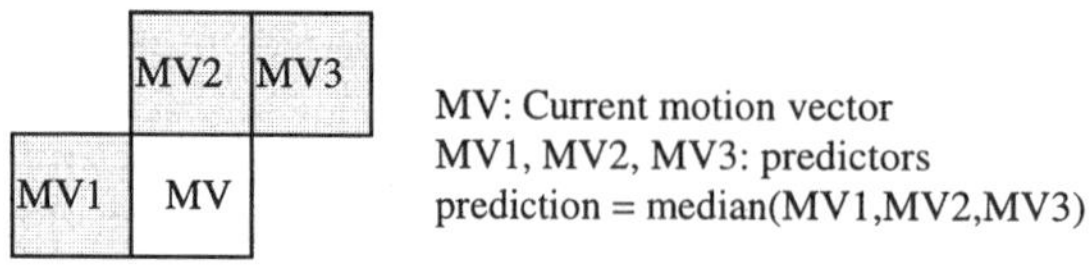

Figure 13 Prediction of motion vectors.

tors to be at multiples of half of a pixel. For example, it is possible to have a motion vector with values (4.5, −2.5). When a motion vector has noninteger values, bilinear interpolation (simple averaging) is used to find the corresponding pel values for prediction.

The coding of motion vectors in H.263 is more sophisticated than that in H.261. The motion vectors of three neighboring MBs (the left, the above, and the above-right, as shown in Fig. 13) are used as predictors. The median of the three predictors is used as the prediction for the motion vector of the current block, and the prediction error is coded and transmitted. However, around a picture boundary or at GOB boundaries that have GOB headers for resynchronization, special cases are needed. When a GOB sync code is sent and only one neighboring MB is outside the picture boundary or GOB boundary, a zero motion vector is used to replace the motion vector of that MB as the predictor. When two neighboring MBs are outside, the motion vector of the only neighboring MB that is inside is used as the prediction. These cases are shown in Fig. 14.

D. Run-Length Coding of DCT Coefficients

Recommendation H.263 improves upon the (*run, level*) coding used in H.261 by including an extra term ''last'' to indicate whether the current coefficient is the last nonzero coefficient of the block. Therefore, a 3-tuple of (*last, run, level*) represents an *event* and is mapped to a codeword in the VLC table, hence the name 3D VLC. With this scheme, the EOB (end-of-block) code used in H.261 is no longer needed. Tables 5 and 6 show some entries of the table for H.263 version 1.

E. Optional Modes of H.263 Version 1

H.263 version 1 specifies six optional modes of operation (although only four of them are commonly counted in overview descriptions such as this one). These optional features are described in the following.

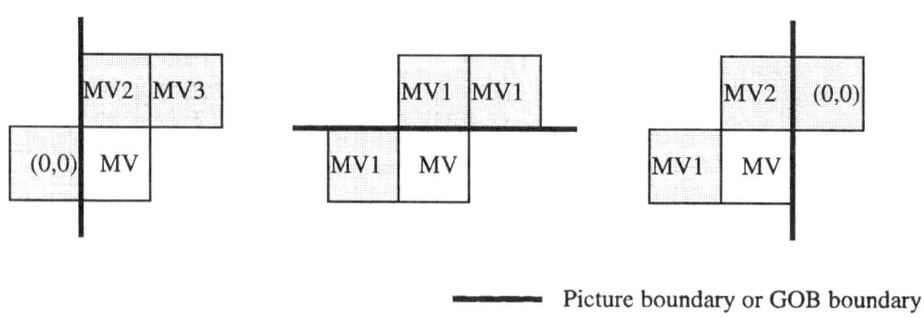

Figure 14 Motion vector prediction at picture–GOB boundaries.

Table 5 Partial VLC Table for DCT Coefficients

Last	Run	Level	Code
0	0	1	10s
0	0	2	1111s
0	0	3	0101 01s
0	0	4	0010 111s
0	0	5	0001 1111 s
0	0	6	0001 0010 1s
0	0	7	0001 0010 0s
0	0	8	0000 1000 01s
0	0	9	0000 1000 00s
0	0	10	0000 0000 111s
0	0	11	0000 0000 110s
0	0	12	0000 0100 000s
...	...	...	...

1. Continuous-Presence Multipoint Mode (CPM Mode, Annex C)

This mode allows four independent video bitstreams to be sent within a single H.263 video channel. This feature is most useful when H.263 is used in systems that require support for multiple video bitstreams but do not have external mechanisms for multiplexing different streams. Multiplexed switching between the streams is supported at the GOB level of the syntax by simply sending 2 bits in each GOB header to indicate which subbitstream it should be associated with.

2. Unrestricted Motion Vector mode (UMV mode, Annex D)

In this mode, motion vectors are allowed to point outside the picture boundary. In this case, edge pels are repeated to extend to the pels outside so that prediction can be done. Significant coding gain can be achieved with unrestricted motion vectors if there is movement around picture edges, especially for smaller picture formats such as QCIF and Sub-

Table 6 Partial VLC Table for DCT Coefficients

Last	Run	Level	Code
...	...	...	...
1	0	1	0111s
1	0	2	0000 1100 1s
1	0	3	0000 0000 101s
1	1	1	0011 11s
1	1	2	0000 0000 100s
1	2	1	0011 10s
1	3	1	0011 01s
1	4	1	0011 00s
1	5	1	0010 011s
1	6	1	0010 010s
1	7	1	0010 001s
1	8	1	0010 000s

QCIF. In addition, this mode allows a wider range of motion vectors than H.261. Large motion vectors can be very effective when the motion in the scene is heavy (e.g., motion due to camera movement), when the picture resolution is high, and when the time spacing between encoded pictures is large.

3. Syntax-Based Arithmetic Coding (SAC Mode, Annex E)

In this option, arithmetic coding [13] is used, instead of VLC tables, for entropy coding. Under the same coding conditions, using arithmetic coding will result in a bitstream different from the bitstream generated by using a VLC table, but the reconstructed frames and the SNR will be the same. Experiments show that the average bit rate saving is about 4% for inter frames and about 10% for intra blocks and frames.

4. Advanced Prediction Mode (AP Mode, Annex F)

In the advanced prediction mode, overlapped block motion compensation (OBMC) [14] is used to code the luminance component of pictures, which statistically improves prediction performance and results in a significant reduction of blocking artifacts. This mode also allows variable-block-size motion compensation (VBSMC) [15], so that the encoder can assign four independent motion vectors to each MB. That is, each block in an MB can have an independent motion vector. In general, using four motion vectors gives better prediction, because one motion vector is used to represent the movement of an 8×8 block instead of a 16×16 MB. Of course, this implies more motion vectors and hence requires more bits to code the motion vectors. Therefore, the encoder has to decide when to use four motion vectors and when to use only one. Finally, in the advanced prediction mode, motion vectors are allowed to cross picture boundaries as is the case in the Unrestricted Motion Vector mode. The Advanced Prediction mode is regarded as the most beneficial of the optional modes in the first version of H.263 (as now stated in H.263 Appendix II).

When four vectors are used, the prediction of motion vectors has to be redefined. In particular, the locations of the three "neighboring" blocks of which the motion vectors are to be used as predictors now depend on the position of the current block in the MB. These are shown in Fig. 15. It is interesting to note how these predictors are chosen. Consider the situation depicted in the upper left of Fig. 15. When the motion vector corresponds to the upper left block in an MB, note that the third predictor (MV3) is not for an area adjacent to the current block. What would happen if we were to use the motion

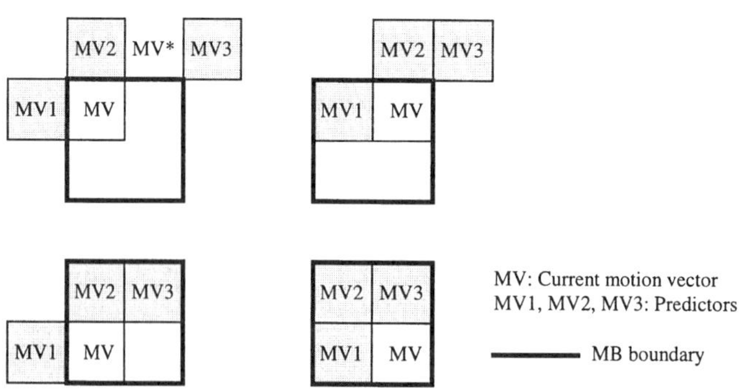

Figure 15 Redefinition of motion vector prediction.

vector of a closer block, say the one marked with MV* in Fig. 15? In that case, MV* would be very likely the same as MV2 because they belong to the same MB, and the median of the three predictors would very often be equal to MV2. Therefore, the advantage of using three predictors would be lost.

5. PB-Frames Mode (PB Mode, Annex G)

In the PB-frames mode, a PB-frame consists of two pictures coded as one unit, as shown in Fig. 16. The first picture, called the P-picture, is a picture predicted from the last decoded picture. The last decoded picture can be either an I-picture, a P-picture, or the P-picture part of a PB-frame. The second picture, called the B-picture (B for bidirectional), is a picture predicted from both the last decoded picture and the P-picture that is currently being decoded. As opposed to the B-frames used in MPEG, PB frames do not need separate bidirectional vectors. Instead, forward vectors for the P-picture are scaled and added to a small delta-vector to obtain vectors for the B-picture. This results in less bit rate overhead for the B-picture. For relatively simple sequences at low bit rates, the frame rate can be doubled with this mode with only a minimal increase in the bit rate. However, for sequences with heavy motion, PB-frames do not work as well as B-pictures. An improved version of PB frames was added (Annex M) in the second version of H.263 to improve upon the original design of this mode. Therefore, although the original (Annex G) design is effective in some scenarios, it is now mostly of historical interest only because a better mode is available. Also, note that the use of the PB-frames mode increases the end-to-end delay, so it is not suitable for two-way interactive communication at low frame rates.

6. Forward Error Correction (Annex H)

Forward error correction is also provided as an optional mode of H.263 operation. The specified method is identical to that defined for H.261 and described in Sec. III.C.4.

F. Test Model Near-Term (TMN)

As with H.261, there are documents drafted by ITU-T SG 16 that describe example encoders, i.e., the test models. For H.263, these are called TMN, where N indicates that H.263 is a near-term effort in improving H.261.

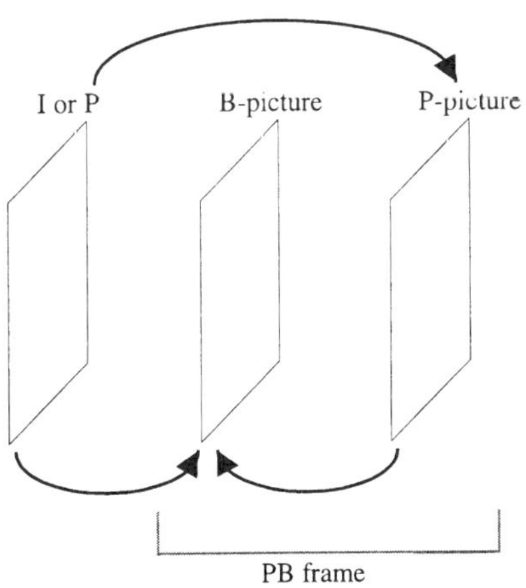

Figure 16 The PB-frames mode.

G. H.263 Version 1 Bitstream Syntax

As we mentioned earlier, an important component of a coding standard is the definition of the bitstream syntax. In fact, the bitstream syntax is all that a standard specifies. As in H.261, the bitstream of H.263 version 1 is arranged in a hierarchical structure composed of the following layers:

Picture layer
Group of blocks layer
Macroblock layer
Block layer

1. *The H.263 Version 1 Picture Layer and Start Codes*

The picture layer is the highest layer in an H.261 or H.263 bitstream. Each coded picture consists of a picture header followed by coded picture data, arranged as a group of blocks.

The first element of every picture and of the GOBs that are sent with headers is the *start code*. This begins with 16 zeros followed by a ''1.'' Start codes provide resynchronization points in the video bitstream in the event of errors, and the bitstream is designed so that 16 consecutive zeros cannot occur except at the start of a picture or GOB.

Furthermore, all H.263 picture start codes are byte aligned so that the bitstream can be manipulated as a sequence of bytes at the frame level. GOB start codes may (and should) also be byte aligned. After the picture start code, the picture header contains the information needed for decoding the picture, including the picture resolution, the picture time tag, indications of which optional modes are in use, and the quantizer step size for decoding the beginning of the picture.

2. *The H.263 Version 1 Group of Blocks Layer*

Data for the group-of-blocks (GOB) layer consists of an optional GOB header followed by data for macroblocks. For the first GOB (GOB number 0) in each picture, no GOB header is transmitted, whereas other GOBs may be sent with or without a header. (A decoder, if such a mechanism is available, can signal using an external means for the encoder always to send GOB headers.)

Each (nonempty) GOB header in H.263 version 1 contains

A GOB start code (16 zeros followed by a ''1'' bit)
A 5-bit group number (GN) indicating which GOB is being sent
A 2-bit GOB frame ID (GFID) indicating the frame to which the GOB belongs (sent in order to help detect problems caused by lost picture headers)
A 5-bit GOB quantizer step size indication (GQUANT)
If the CPM mode is in effect, a GOB subbitstream indicator (GSBI)

3. *The H.263 Version 1 Macroblock Layer*

Data for each macroblock consist of a macroblock header followed by data for blocks. The first element of the macroblock layer is a 1-bit coded macroblock indication (COD), which is present for each macroblock in a P-picture. When COD is ''1,'' no further data are sent for the macroblock. This indicates that the macroblock should be represented as an inter macroblock with a zero-valued motion vector and zero-valued transform coefficients. This is known as a ''skipped'' macroblock.

For macroblocks that are not skipped, an indication of the type of macroblock (whether it is intra or inter coded, whether the quantization step size has changed, and which blocks of the macroblock are coded) is sent next. Then, depending on the type of macroblock, zero, one, or four motion vector differences may be sent. (Intra macroblocks have no motion vector, and inter macroblocks have one motion vector in baseline mode and either one or four in advanced prediction mode.)

4. The H.263 Version 1 Block Layer

When the PB-frames mode is not in use, a macroblock consists of four luminance blocks and two chrominance blocks. If the MB is coded in intra mode, the syntax structure of each block begins with sending the DC coefficient using an 8-bit fixed-length code (to be uniformly reconstructed with a step size of 8).

In PB-frames mode, a macroblock can be thought of as being composed of 12 blocks. First, the data for six P-blocks are transmitted followed by data for the associated six B-blocks. In intra macroblocks, the DC coefficient is sent for every P-block of the macroblock. B-block data are sent in a manner similar to inter blocks.

The remaining quantized transform coefficients are then sent if indicated at the macroblock level. These are sent as a sequence of *events*. An event is a three-dimensional symbol including an indication of the run length of zero-valued coefficients, the quantized level of the next nonzero coefficient, and an indication of whether there are any more nonzero coefficients in the block. Because the (*last, run, level*) VLC symbol indicates three quantities, it is called a 3D VLC. In contrast, the events in VLC coding in H.261, MPEG-1, and MPEG-2 include only a 2D (*run, level*) combination with a separate codeword to signal end of block (EOB).

The VLC table for coding events is not sufficiently large to include all possible combinations. Instead, the rarely occurring combinations are represented by an ESCAPE code followed by fixed-length coding of the values of last, run, and level.

V. H.263 VERSION 2, OR "H.263+"

After the standardization of H.263 version 1, continuing interest in the capabilities of its basic design made it clear that further enhancements to H.263 were possible in addition to the original optional modes. The ITU-T therefore established an effort, informally known as H.263+, to meet the need for standardization of such enhancements of H.263. The result is a new second version of H.263, sometimes called H.263v2 or H.263+. The enhanced version was approved on January 27, 1998 at the January–February 1998 meeting of ITU-T SG16. The approved version contained the same technical content as the draft submitted on September 26, 1997 and retained every detail of the original H.263 version 1 design.

Like H.263 version 1 (the version approved in November 1995), H.263 version 2 is standardized for use in a wide variety of applications, including real-time telecommunication and what the ITU-T calls "nonconversational services." These enhancements provide either improved quality relative to that supported by H.263 version 1 or additional capabilities to broaden the range of applications. The enhancements in H.263 version 2 include

Features to provide a high degree of error resilience for mobile and packet network (e.g., Internet) use (along with a capability to reduce algorithmic delay)

Features for improvement of coding efficiency

Table 7 Workplan History for H.263 Version 2

April 1996	First acceptance and evaluation of proposals.
July 1996	Evaluation of proposals. Draft text created.
November 1996	Final proposals of features. Complete draft written.
February 1997	Final evaluations completed. Text written for determination.
March 1997	Determination at ITU-T SG16 meeting.
September 1997	Final draft ''white document'' submitted for decision.
January–February 1998	Decision at ITU-T SG16 meeting.

Dynamic resolution features for adapting the coded picture resolution to the scene content (along with an ability to perform global motion compensation)
Support for a very wide variety of custom video picture resolutions and frame rates
Scalability features for simultaneous multiple-bit-rate operation
Backward-compatible supplemental enhancement information for additional capabilities (chroma keying for object layering, full- and partial-picture freezes and releases, and snapshot tagging for still pictures and progressive transmission)

Because H.263v2 was a near-term solution to the standardization of enhancements to H.263, it considered only well-developed proposed enhancements that fit into the basic framework of H.263 (e.g., motion compensation and DCT-based transform coding). A history of the H.263+ milestones is outlined in Table 7.

A. Development of H.263v2

During the development of H.263v2, proposed techniques were grouped into key technical areas (KTAs). Altogether, about 22 KTAs were identified. In November 1996, after consideration of the contributions and after some consolidation of KTAs, 12 KTAs were chosen for adoption into a draft. The adopted features were described in a draft text that passed the determination process (preliminary approval) in March 1997. Combining these with the original options in H.263 resulted in a total of about 16 optional features in H.263v2, which can be used together or separately in various specified combinations. We will outline these new features in the next few sections. In addition, new test models (the latest one at press time being TMN11) have been prepared by the group for testing, simulation, and comparisons.

B. H.263+ Enhancements for Error Robustness

The following H.263v2 optional modes are especially designed to address the needs of mobile video and other unreliable transport environments such as the Internet or other packet-based networks:

1. *Slice structured mode (Annex K)*: In this mode, a ''slice'' structure replaces the GOB structure. Slices have more flexible shapes and may be allowed to appear in any order within the bitstream for a picture. Each slice may also have a specified width. The use of slices allows flexible partitioning of the picture, in contrast to the fixed partitioning and fixed transmission order required by the GOB structure. This can provide enhanced error resilience and minimize the video delay.

2. *Reference picture selection mode (Annex N)*: In this mode, the reference picture does not have to be the most recently encoded picture. Instead, any of a number of temporally previous pictures can be referenced. This mode can provide better error resilience in unreliable channels such as mobile and packet networks, because the codec can avoid using an erroneous picture for future reference.
3. *Independent segment decoding mode (Annex R)*: This mode improves error resilience by ensuring that any error in specific regions of the picture cannot propagate to other regions.
4. *Temporal, spatial, and SNR scalability (Annex O)*: See Sec. V.F.

C. H.263+ Enhancements of Coding Efficiency

Among the new optional enhancements provided in H.263v2, a large fraction are intended to improve coding efficiency, including:

1. *Advanced intra coding mode (Annex I)*: This is an optional mode for intra coding. In this mode, intra blocks are coded using a predictive method. A block is predicted from the block to the left or the block above, provided that the neighbor block is also intra coded. For isolated intra blocks for which no prediction can be found, the prediction is essentially turned off. This provides approximately a 15–20% improvement in coding intra pictures and about a 10–12% improvement in coding intra macroblocks within inter picture.
2. *Alternate inter VLC mode (Annex S)*: This mode provides the ability to apply a VLC table originally designed for intra coding to inter coding where there are often many large coefficients, simply by using a different interpretation of the level and the run. This provides up to a 10–15% improvement when coding pictures with high motion at high fidelity.
3. *Modified quantization mode (Annex T)*: This mode improves the flexibility of controlling the quantizer step size. It also reduces of the step size for chrominance quantization in order to reduce chrominance artifacts. An extension of the range of values of DCT coefficient is also provided. In addition, by prohibiting certain "unreasonable" coefficient representations, this mode increases error detection performance and reduces decoding complexity.
4. *Deblocking filter mode (Annex J)*: In this mode, an adaptive filter is applied across the 8×8 block edge boundaries of decoded I- and P-pictures to reduce blocking artifacts. The filter affects the picture that is used for the prediction of subsequent pictures and thus lies within the motion prediction loop (as does the loop filtering in H.261).
5. *Improved PB-frames mode (Annex M)*: This mode deals with the problem that the original PB-frames mode in H.263 cannot represent large amounts of motion very well. It provides a mode with more robust performance under complex motion conditions. Instead of constraining a forward motion vector and a backward motion vector to come from a single motion vector as in the first version of H.263, the improved PB-frames mode allows them to be totally independent as in the B-frames of MPEG-1 and MPEG-2.
6. *Refinements of prior features*: Some prior coding efficiency features are refined in minor ways when any other H.263+ features are in effect. These include

a. The addition of *rounding control* for eliminating a round-off drift in motion compensation
b. A simplification and extension of the unrestricted motion vector mode of Annex D
c. Adding the ability to send a change of quantizer step size along with four motion vectors in the advanced prediction mode of Annex F

D. H.263+ Dynamic Resolution Enhancements

Two modes are added in H.263v2 that provide added flexibility in adapting the coded picture resolution to changes in motion content:

1. *Reference picture resampling mode (Annex P)*: This allows a prior coded picture to be resampled, or warped, before it is used as a reference picture. The warping is defined by four motion vectors that specify the amount of offset of each of the four corners of the reference picture. This mode allows an encoder to switch smoothly between different encoded picture sizes, shapes, and resolutions. It also supports a form of global motion compensation and special-effect image warping.
2. *Reduced-resolution update mode (Annex Q)*: This mode allows the encoding of inter picture difference information at a lower spatial resolution than the reference picture. It gives the encoder the flexibility to maintain an adequate frame rate by encoding foreground information at a reduced spatial resolution while holding on to a higher resolution representation of the more stationary areas of a scene.

E. H.263+ Custom Source Formats

One simple but key feature of H.263v2 is that it extends the possible source formats specified in H.263. These extensions include

1. *Custom picture formats*: It is possible in H.263v2 to send a custom picture format, with the choices no longer limited to a few standardized resolutions. The number of lines can range from 4 to 1152, and the number of pixels per line can range from 4 to 2048 (as long as both dimensions are divisible by 4).
2. *Custom pixel aspect ratios*: This allows the use of additional pixel aspect ratios (PARs) other than the 12:11 aspect ratio used in the standardized H.261 and H.263 picture types such as CIF and QCIF. Examples of custom PARs are shown in Table 8.
3. *Custom picture clock frequencies*: This allows base picture clock rates higher (and lower) than 30 frames per second. This feature helps to support additional camera and display technologies.
4. *Additional picture sizes in CPM mode*: The original continuous-presence multipoint mode of H.263v1 appeared to support only QCIF-resolution subbitstreams. This restriction was removed in H.263v2, and new "end of subbitstream" codes were added for improved signaling in this mode. This allows the video bitstream to act as a video multiplex for up to four distinct video subbitstreams of any resolution.

Table 8 H.263+ Custom Pixel Aspect Ratios

Pixel aspect ratio	Pixel width:pixel height
Square	1:1
625-type for 4:3 pictures[a]	12:11
525-type for 4:3 pictures	10:11
625-type for 16:9 pictures	16:11
525-type for 16:9 pictures	40:33
Extended PAR	$m:n$, with m and n relatively prime and m and n between 1 and 255

[a] The value used in the standardized H.261 and H.263 picture formats (Sub-QCIF, QCIF, CIF, 4CIF, and 16CIF).

F. H.263+ Temporal, Spatial, and SNR Scalability (Annex O)

The Temporal, Spatial, and SNR Scalability mode (Annex O) supports layered-bitstream scalability in three forms, similar to MPEG-2. Bidirectionally predicted frames, such as those used in MPEG-1 and MPEG-2, are used for temporal scalability by adding enhancement frames between other coded frames. This is shown in Fig. 17. A similar syntactical structure is used to provide an enhancement layer of video data to support spatial scalability by adding enhancement information for construction of a higher resolution picture, as shown in Fig. 18. Finally, SNR scalability is provided by adding enhancement information for reconstruction of a higher fidelity picture with the same picture resolution, as in Fig. 19. Furthermore, different scalabilities can be combined together in a very flexible way.

The Reference Picture Selection mode of Annex N can also be used to provide a type of scalability operation known as Video Redundancy Coding or P-picture Temporal Scalability. This mode was described in Sec. V.B.

G. H.263+ Backward-Compatible Supplemental Enhancement Information (Annex L)

One important feature of H.263v2 is the usage of supplemental enhancement information, which may be included in the bitstream to signal enhanced display capabilities or to provide tagging information for external usage. For example, it can be used to signal a full-picture or partial-picture freeze or a freeze-release request with or without resizing. It can also be used to label a snapshot, the start and end of a video segment, and the start and end of a progressively refined picture. The supplemental information may be present in the bitstream even though the decoder may not be capable of providing the enhanced capability to use it or even to interpret it properly. In other words, unless a requirement

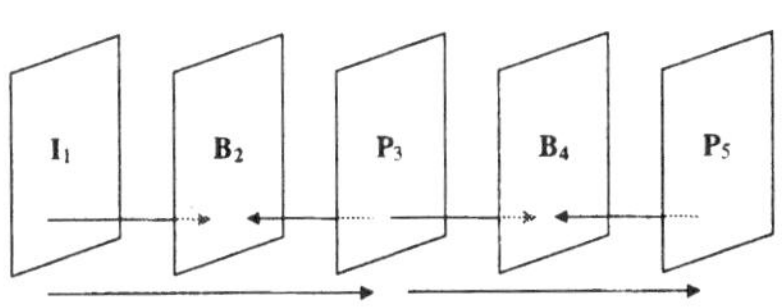

Figure 17 Temporal scalability.

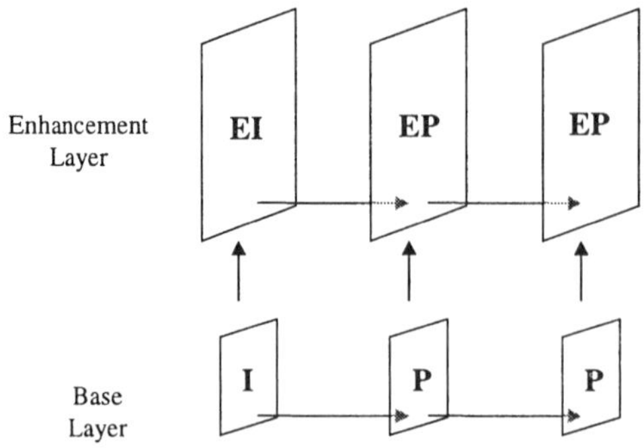

Figure 18 Spatial scalability.

to provide the requested capability has been established by external means in advance, the decoder can simply discard anything in the supplemental information. This gives the supplemental enhancement information a backward compatibility property that allows it to be used in mixed environments in which some decoders may get extra features from the same bitstream received by other decoders not having the extra capabilities.

Another use of the supplemental enhancement information is to specify chroma key for representing transparent and semitransparent pixels [16]. We will now explain this in more detail.

A Chroma Keying Information Flag (CKIF) in the supplemental information indicates that the chroma keying technique is used to represent transparent and semitransparent pixels in the decoded picture. When presented on the display, transparent pixels are not displayed. Instead, a background picture that is externally controlled is revealed. Semitransparent pixels are rendered by blending the pixel value in the current picture with the corresponding value in the background picture. Typically, an 8-bit number α is used to indicate the transparency, so $\alpha = 255$ indicates that the pixel is opaque and $\alpha = 0$ indicates that the pixel is transparent. Between 0 and 255, the displayed color is a weighted sum of the original pixel color and the background pixel color.

When CKIF is enabled, one byte is used to indicate the keying color value for each

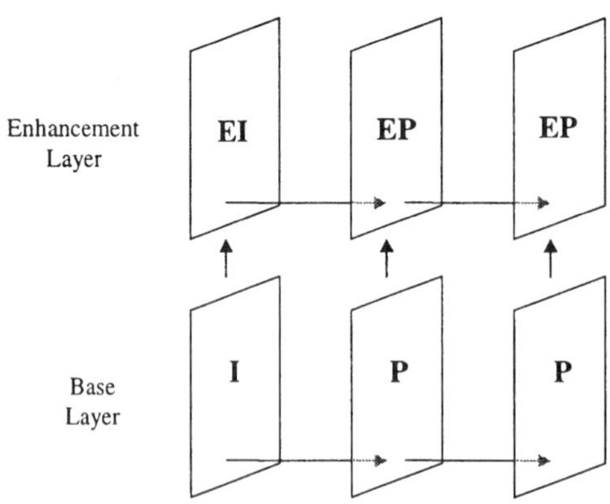

Figure 19 SNR scalability.

component (Y, C_B, or C_R) that is used for chroma keying. After the pixels are decoded, the α value is calculated as follows. First, the distance d between the pixel color and the key color value is calculated. The α value is then computed as follows:

if	$(d < T_1)$	then $\alpha = 0$
else if	$(d > T_2)$	then $\alpha = 255$
else	$\alpha = [255(d - T_1)]/(T_2 - T_1)$	

where T_1 and T_2 are the two thresholds that can be set by the encoder.

H. H.263+ Levels of Preferred Mode Support

A variety of optional modes that are all useful in some applications are provided in H.263v2, but few manufacturers would want to implement all of the options. Therefore, H.263v2 contains an informative specification of three levels of preferred mode combinations to be supported (Appendix II). Each level contains a number of options to be supported by an equipment manufacturer. This appendix is not a normative part of the standard—it is intended only to provide manufacturers with some guidelines about which modes are more likely to be widely adopted across a full spectrum of terminals and networks.

Three levels of preferred modes are described in H.263v2 Appendix II, and each level supports the optional modes specified in lower levels. In addition to the level structure is a discussion indicating that because the Advanced Prediction mode (Annex F) was the most beneficial of the original H.263v1 modes, its implementation is encouraged not only for its performance but also for its backward compatibility with H.263v1 implementations.

The first level is composed of

The Advanced Intra Coding mode (Annex I)
The Deblocking Filter mode (Annex J)
Full-frame freeze by Supplementary Enhancement Information (Annex L)
The Modified Quantization mode (Annex T)

Level 2 supports, in addition to modes supported in level 1,

The Unrestricted Motion Vector mode (Annex D)
The Slice Structured mode (Annex K)
The simplest resolution-switching form of the Reference Picture Resampling mode (Annex P)

In addition to these modes, Level 3 further supports

The Advanced Prediction mode (Annex F)
The Improved PB-frames mode (Annex M)
The Independent Segment Decoding mode (Annex R)
The Alternative inter VLC mode (Annex S)

VI. FURTHER ITU-T Q.15/SG16 WORK: H.263++ AND H.26L

When the proposals for H.263v2 were evaluated and some were adopted, it was apparent that some further new proposals might also arise that could fit into the H.263 syntactical framework but that were not ready for determination in March 1997. Therefore, Q.15/SG16 considered another round of H.263 extensions, informally called "H.263++," that would create more incremental extensions to the H.263 syntax. Two such extensions are now included in drafts adopted by the ITU-T Q.15/SG16 Advanced Video Coding experts group for near-term approval by ITU-T SG16:

Data partitioning: This is a mode of operation providing enhanced error resilience primarily for wireless communication. It uses essentially the same syntax as the Slice Structured mode of Annex K but rearranges the syntax elements so that motion vector and transform coefficient data for each slice are separated into distinct sections of the bitstream. The motion vector data can be reversibly checked and decoded, due largely to the structure of the motion vector VLC table used from Annex D. (MPEG-4 also has a data partitioning feature, but they differ in that in MPEG-4 it is the transform coefficients rather than the motion vector data that are reversibly encoded.)

Enhanced reference picture selection: This is a mode of operation that uses multiple reference pictures as found in Annex N but with more sophisticated control over the arrangement of pictures in the reference picture buffer and with an ability to switch reference pictures on a macroblock basis (rather than on a GOB basis). This feature has been demonstrated to enable an improvement in coding efficiency (typically around a 10% improvement, but more in some environments).

In addition to these drafted extensions, the H.263++ project is currently considering three other proposed key technical areas of enhancement:

Affine motion compensation, a coding efficiency enhancement technique

Precise specification of inverse DCT operation, a method of eliminating round-off error drift between an encoder and decoder

Error concealment methods, creating a precisely defined reaction in a decoder to errors in the bitstream format caused by channel corruption

The ITU-T Q.15/SG16 experts group is also working on a project called "H.26L" for defining a new video coding standard. This new standard need not retain the compatibility with the prior design that has constrained the enhancement work of H.263+ and H.263++. Proposals are now being evaluated for the H.26L project, and the Q.15/SG16 group is expected to turn its primary attention away from H.263 extensions toward this new project by about the end of 1999.

VII. CONCLUSION

By explaining the technical details of a number of important video coding standards defined by the ITU-T, we hope we have given the readers some insight to the significance of these international standards for multimedia communication. Pointers to more up-to-date information about the video coding standards described in this chapter can be found in Table 9. When this chapter was prepared, activities in H.263++ and H.26L were still in progress. It is therefore recommended that the reader check the resources in Table 9 for more recent updates on H.263++ and H.263L.

Table 9 Sources of Further Information

http://www.itu.int	International Telecommunications Union
http://standard.pictel.com	General information

REFERENCES

1. ITU-T. Recommendation H.120: Codecs for videoconferencing using primary digital group transmission, version 1, 1984; version 2, 1988.
2. ITU-T. Recommendation H.261: Video codec for audiovisual services at $p \times 64$ kbit/s, version 1, December 1990; version 2, March 1993.
3. ITU-T. Recommendation H.262/IS 13818-2: Generic coding of moving pictures and associated audio information—Part 2: Video, November 1994.
4. ITU-T. Recommendation H.263: Video coding for low bit rate communication, version 1, November 1995; version 2, January 1998.
5. GJ Sullivan, T Wiegand. Rate-distortion optimization for video compression. IEEE Signal Processing Magazine November: 74–90, 1998.
6. ITU-R. Recommendation BT.601-5: Studio encoding parameters of digital television for standard 4:3 and wide-screen 16:9 aspect ratios, October 1995.
7. N Ahmed, T Natarajan, KR Rao. Discrete cosine transform. IEEE Trans on Comput C-23: 90–93, 1974.
8. KR Rao, P Yip. Discrete Cosine Transform. New York: Academic Press, 1990.
9. GJ Sullivan. Efficient scalar quantization of exponential and Laplacian random variables. IEEE Trans Inform Theory 42:1365–1374, 1996.
10. AN Netravali, JD Robbins. Motion-compensated television coding: Part I. Bell Syst Tech J 58:631–670, 1979.
11. AN Netravali, BG Haskell. Digital Pictures. 2nd ed. New York: Plenum, 1995.
12. Description of Reference Model 8 (RM8). CCITT Study Group XV, Specialist Group on Coding for Visual Telephony. Document 525, June 1989.
13. IH Witten, RM Neal, JG Cleary. Arithmetic coding for data-compression. Commun ACM 30: 520–540, 1987.
14. MT Orchard, GJ Sullivan. Overlapped block motion compensation—an estimation-theoretic approach. IEEE Trans Image Process 3:693–699, 1994.
15. GJ Sullivan, RL Baker. Rate-distortion optimized motion compensation for video compression using fixed or variable size blocks. IEEE Global Telecom. Conference (GLOBECOM), December 1991, pp. 85–90.
16. T Chen, CT Swain, BG Haskell. Coding of sub-regions for content-based scalable video. IEEE Trans Circuits Syst Video Technol 7:256–260, 1997.

4

Overview of the MPEG Standards

Atul Puri, Robert L. Schmidt, and Barry G. Haskell
AT&T Labs, Red Bank, New Jersey

I. INTRODUCTION

In the early 1980s, in the area of multimedia, the move from the analog to the digital world began to gain momentum. With the advent of the compact disc format for digital audio, a transition from the analog world of vinyl records and cassette tapes to shiny discs with music coded in the form of binary data took place. Although the compact disc format for music did not focus much on compression, it delivered on the promise of high-fidelity music while offering robustness to errors such as scratches on the compact disc. The next step was digital video or digital audio and video together; this demanded notable advances in compression. In 1984, the International Telecommunications Union–Telecommunications [ITU-T, formerly known as The International Telephone and Telegraph Consultative Committee (CCITT)] started work on a video coding standard for visual telephony, known formally as the H.261 standard [1] and informally as the $p \times 64$ video standard, where $p = 1, \ldots 30$ (although it started out as the $n \times 384$ kbit/sec video standard where $n = 1, \ldots 5$). The H.261 standard was in a fairly mature stage in 1988 when other developments in digital video for the compact disc took place, threatening a de facto standard in 1989. Although this area was outside of the scope of ITU-T, it was within the charter of the International Standards Organization (ISO), which responded by forming a group called the Moving Picture Experts Group (MPEG) in 1988. The MPEG group was given the mandate of developing video and audio coding techniques to achieve good quality at a total bit rate of about 1.4 Mbit/sec and a system for playback of the coded video and audio, with the compact disc as the target application. The promise of MPEG was an open, timely, and interoperable practical coding standard (for video and audio) that provided the needed functionality with leading technology, low implementation cost, and the potential for performance improvements even after the standard was completed.

By September 1990, the MPEG committee advanced the standard it was originally chartered with to the fairly mature stage of the committee draft (CD), meaning that bug fixes or minor additions due to a demonstrated inadequacy to meet the objectives were the only technical changes possible. Around the same time, a set of requirements arose for a standard for digital TV demanding high-quality, efficient coding of interlaced video; thus, a new work item dubbed the second phase standard, MPEG-2, was started. The original MPEG work item, subsequently called MPEG-1 [2–4], was officially approved as the international standard (IS) in November 1992, and the second work item reached the stage of CD in November 1993. In 1993, MPEG also started work on a third item, called MPEG-4; the original agenda of MPEG-4 was very low bit rate video coding and was subsequently modified in July 1994 to coding of audiovisual objects. The MPEG-2 standard [5–7,48] was approved in November 1994, 2 years after approval of the MPEG-1 standard. More recently, some new parts such as [47] have been added to MPEG-2. The MPEG-4 standard, on the other hand, was subdivided into two parts, the basic part (version 1) and some extensions (version 2 and beyond). But even before the MPEG-4 standard could reach CD status, a new topic for the future of MPEG was chosen; the focus of the next MPEG standard was to be a multimedia content description interface, something different from traditional multimedia coding. This standard, now called MPEG-7, was started in late 1996. In the meantime, MPEG-4 version 1 (basic standard) is already way past the CD status [8–11] and is on its way to being approved as the IS, and MPEG-4 version 2 (extensions) is currently at the stage of the CD [12]. The work on MPEG-7 is now entering the development phase [13–17] and is expected to reach CD status by October 2000. Incidentally, for the purpose of simplification, we did not elaborate on the critical stage of evaluation for each standard, as well as the recently introduced final CD (FCD) stage for the MPEG-4 and the MPEG-7 standards. Also, even within each of the standards, there are many parts with their own schedules; we discuss details of each part in subsequent sections. Table 1 summarizes the various stages and schedules of the MPEG standards.

On October 1996, in recognition of MPEG's achievements in the area of standardization of video compression, MPEG was presented with the Emmy award for the MPEG-1 and MPEG-2 standards. This symbolic gesture also signified the increasing relevance of MPEG technology to the average consumer and to society at large.

Now that we have provided a historical overview of MPEG standardization, a short overview of the MPEG standardization process [18,19] and its terminology is in order before we delve into specific standards. Each MPEG standard starts by identifying its scope and issuing a call for proposals. It then enters two major stages. The first stage is a competitive stage that involves testing and evaluation of the candidate proposals to select a few top-performing proposals, components of which are then used as the starting basis for the second stage. The second stage involves collaborative development of these components via iterative refinement of the experimentation model (coding description). This is accomplished by defining a set of core experiments. Now we clarify what is actually standardized by an MPEG standard. The MPEG coding standards do not standardize the

Table 1 Development Schedule of the MPEG Standards

Standard	Started	Tests and evaluation	Committee draft (CD)/ final CD (FCD)	International standard (IS)
MPEG-1	May 1988	October 1989	September 1990	November 1992
MPEG-2	December 1990	November 1991	November 1993	November 1994
MPEG-4 Version 1	July 1993	October 1995	October 1997/March 1998	May 1999
MPEG-4 Version 2			March 1999/July 1999	February 2000
MPEG-7	November 1996	February 1999	October 2000/March 2001	September 2001

encoding methods or details of encoders. These standards only standardize the format for representing data input to the decoder and a set of rules for interpreting this data. The format for representing the data is referred to as the *syntax* and can be used to construct various kinds of valid data streams referred to as the *bitstreams.* The rules for interpreting the data (bitstreams) are called the *decoding semantics,* and an ordered set of decoding semantics is referred to as the *decoding process.* Thus, we can say that MPEG standards specify a decoding process; however, this is still different from specifying a decoder implementation. Given audio and/or video data to be compressed, an encoder must follow an ordered set of steps called the *encoding process*; this encoding process is, however, not standardized and typically varies because encoders of different complexities may be used in different applications. Also, because the encoder is not standardized, continuing improvements in quality still become possible through encoder optimizations even after the standard is complete. The only constraint is that the output of the encoding process results in a syntactically correct bitstream that can be interpreted according to the decoding semantics by a standards-compliant decoder.

While this chapter presents an overview of the high-level concepts of the various MPEG standards, several other excellent chapters in this book provide the details of the key parts of the MPEG-4 standard. Further, a chapter on MPEG-7 introduces the background and the development process of the MPEG-7 standard. Thus, after reading this chapter, the avid reader is directed to the following chapters to explore in depth the specific topic of interest.

Chapters 5 and 6 of this book for MPEG-4 Audio
Chapters 7 through 11 of this book for MPEG-4 Visual
Chapters 12 through 16 of this book for MPEG-4 Systems
Chapter 22 of this book for background on MPEG-7

The rest of the chapter is organized as follows. In Section II, we briefly discuss the MPEG-1 standard. In Section III, we present a brief overview of the MPEG-2 standard. Section IV introduces the MPEG-4 standard. In Section V, the ongoing work toward the MPEG-7 standard is presented. In Section VI, we present a brief overview of the profiling issues in the MPEG standards. Finally, in Section VII, we summarize the key points presented in this chapter.

II. MPEG-1

MPEG-1 is the standard for storage and retrieval of moving pictures and audio on a digital storage medium [19]. As mentioned earlier, the original target for the MPEG-1 standard was good-quality video and audio at about 1.4 Mbit/sec for compact disc application.

Based on the target application, a number of primary requirements were derived and are listed as follows.

- Coding of video with good quality at 1 to 1.5 Mbit/sec and audio with good quality at 128 to 256 kbit/sec
- Random access to a frame in limited time, i.e., frequent access points at every half-second
- Capability for fast forward and fast reverse, enabling seek and play forward or backward at several times the normal speed

A system for synchronized playback of, and access to, audiovisual data
The standard to be implementable in practical real-time decoders at a reasonable cost in hardware or software

Besides the preceding requirements, a number of other requirements also arose, such as support for a number of picture resolutions, robustness to errors, coding quality trade-off with coding delay (150 msec to 1 sec), and the possibility of real-time encoders at reasonable cost.

In MPEG, the work on development of MPEG-1 was organized into a number of subgroups, and within a period of about 1 year from the tests and evaluation, it reached the stable state of the CD before being approved as the final standard 2 years after the CD stage.

The MPEG-1 standard is formally referred to as ISO 11172 and consists of the following parts:

11172-1: Systems
11172-2: Video
11172-3: Audio
11172-4: Conformance
11172-5: Software

We now briefly discuss each of the three main components (Systems, Video, and Audio) [2–4] of the MPEG-1 standard.

A. MPEG-1 Systems

The Systems part of the MPEG-1 standard [2,18] specifies a systems coding layer for combining coded audio and video data. It also provides the capability for combining with it user-defined private data streams as well as streams that can presumably be defined in the future. To be more specific, the MPEG-1 Systems standard defines a packet structure for multiplexing coded audio and video data into one stream and keeping it synchronized. It thus supports multiplexing of multiple coded audio and video streams, where each stream is referred to as an elementary stream. The systems syntax includes data fields that allow synchronization of elementary streams and assist in parsing the multiplexed stream after random access, management of decoder buffers, and identification of timing of the coded program. The MPEG-1 Systems thus specifies the syntax to allow generation of systems bitstreams and semantics for decoding these bitstreams.

The Systems Time Clock (STC) is the reference time base; it operates at 90 kHz and may or may not be phase locked to individual audio or video sample clocks. It produces 33-bit time representation and is incremented at 90 kHz. In MPEG-1 Systems, the mechanism for generating the timing information from decoded data is provided by the Systems Clock Reference (SCR) fields, which indicate the time, and appear intermittently in the bitstream, spaced no more than 700 msec. The presentation playback or display synchronization information is provided by Presentation Time Stamps (PTSs), which represent the intended time of presentation of decoded video pictures or audio frames. The audio or video PTSs are samples from the common time base; the PTSs are sampled to an accuracy of 90 kHz.

To ensure guaranteed decoder buffer behavior, the MPEG Systems specifies the concepts of a Systems Target Decoder (STD) and Decoding Time Stamp (DTS). The DTS differs from the PTS only in the case of pictures that require additional reordering delay during the decoding process. These basic concepts of timing and terminology employed in MPEG-1 Systems are also common to MPEG-2 Systems.

B. MPEG-1 Video

The MPEG-1 Video standard [3,18] was originally aimed at coding video at Source Intermediate Format (SIF) resolution (352 × 240 at 30 noninterlaced frames/sec or 352 × 288 at 25 noninterlaced frames/sec) at bit rates of about 1.2 Mbit/sec. However, anticipating other applications, the MPEG-1 video syntax was made flexible to support picture sizes of up to 4096 × 4096, many frame rates (23.97, 24, 25, 29.97, 50, 59.94, 60 frames/sec), and higher bit rates. In addition, the MPEG-1 Video coding scheme was designed to support interactivity such as video fast forward, fast reverse, and random access.

The MPEG-1 Video standard specifies the video bitstream syntax and the corresponding video decoding process. The MPEG-1 Video syntax supports three types of coded frames or pictures, intra (I-) pictures, coded separately by themselves; predictive (P-) pictures, coded with respect to the immediately previous I- or P-picture; and bidirectionally predictive (B-) pictures, coded with respect to the immediately previous I- or P-picture as well as the immediately next P- or I-picture. In terms of coding order, P-pictures are causal, whereas B-pictures are noncausal and use two surrounding causally coded pictures for prediction. In terms of compression efficiency, I-pictures are the most expensive, P-pictures are less expensive than I-pictures, and B-pictures are the least expensive. However, because B-pictures are noncausal they incur additional (reordering) delay.

Figure 1 shows an example picture structure in MPEG-1 video coding that uses a pair of B-pictures between two reference (I- or P-) pictures. In MPEG-1 video coding, an input video sequence is divided into units of groups of pictures (GOPs), where each GOP typically starts with an I-picture and the rest of the GOP contains an arrangement of P-pictures and B-pictures. A GOP serves as a basic access unit, with the I-picture serving as the entry point to facilitate random access. For coding purposes, each picture is further divided into one or more slices. Slices are independently decodable entities that offer a mechanism for resynchronization and thus limit the propagation of errors. Each slice is composed of a number of macroblocks; each macroblock is basically a 16 × 16 block of luminance (or alternatively, four 8 × 8 blocks) with corresponding chrominance blocks.

MPEG-1 Video coding [3,20,21] can exploit both the spatial and the temporal redun-

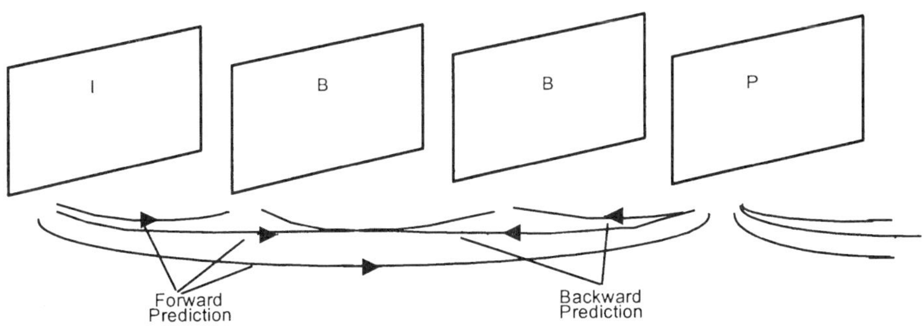

Figure 1 An example of I-, P-, and B-picture structure in MPEG-1 coding.

dancies in video scenes. Spatial redundancies are exploited by using block discrete cosine transform (DCT) coding of 8 × 8 pixel blocks, resulting in 8 × 8 blocks of DCT coefficients, which then undergo quantization, zigzag scanning, and variable length coding. A nonlinear quantization matrix can be used for weighting of DCT coefficients prior to quantization, allowing perceptually weighted quantization in which perceptually irrelevant information can be easily discarded, further increasing the coding efficiency. The zigzag scan allows scanning of DCT coefficients roughly in the order of increasing frequency to calculate runs of zero coefficients, which along with the amplitude of the next nonzero coefficient index along the run, allows efficient variable length coding. Temporal redundancies are exploited by using block motion compensation to compensate for interframe motion of objects in a scene; this results in a significant reduction of interframe prediction error.

Figure 2 shows a simplified block diagram of an MPEG-1 video decoder that receives bitstreams to be decoded from the MPEG-1 Systems demux. The MPEG-1 video decoder consists of a variable length decoder, an inter/intra DCT decoder, and a uni/bidirectional motion compensator. After demultiplexing, the MPEG-1 video bitstream is fed to a variable length decoder for decoding of motion vectors (*m*), quantization information (*q*), inter/intra decision (*i*), and the data consisting of quantized DCT coefficient indices. The inter/intra DCT decoder uses the decoded DCT coefficient indices, the quantizer information, and the inter/intra decision to dequantize the indices to yield DCT coefficient blocks and then inverse transform the blocks to recover decoded pixel blocks. If the coding mode is inter (based on the inter/intra decision), the uni/bidirectional motion compensator uses motion vectors to generate motion-compensated prediction blocks that are then added back to the corresponding decoded prediction error blocks output by the inter/intra DCT decoder to generate decoded blocks. If the coding mode is intra, no motion-compensated prediction needs to be added to the output of the inter/intra DCT decoder. The resulting decoded pictures are output on the line labeled video out.

B. MPEG-1 Audio

The MPEG-1 Audio standard [4,18] specifies the audio bitstream syntax and the corresponding audio decoding process. MPEG-1 Audio is a generic standard that does not make

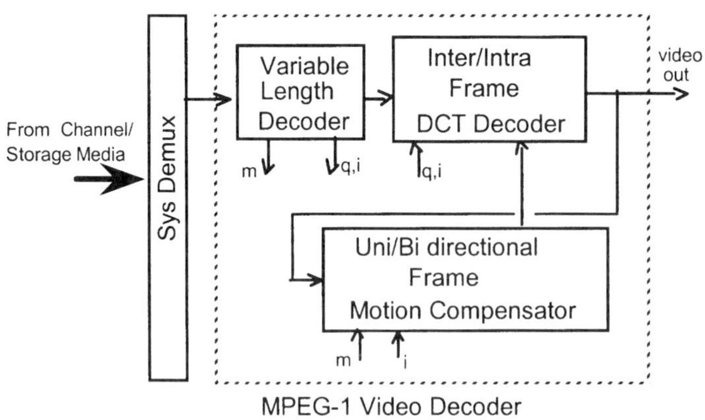

Figure 2 Systems demultiplex and the MPEG-1 video decoder.

any assumptions about the nature of the audio source, unlike some vocal-tract-model coders that work well for speech only. MPEG-1 Audio coding exploits the perceptual limitations of the human auditory system, and thus much of the compression comes from removal of perceptually irrelevant parts of the audio signal. MPEG-1 Audio coding supports a number of compression modes as well as interactivity features such as audio fast forward, fast reverse, and random access.

The MPEG-1 Audio standard consists of three layers—I, II, and III. These layers also represent increasing complexity, delay, and coding efficiency. In terms of coding methods, these layers are related, because a higher layer includes the building blocks used for a lower layer. The sampling rates supported by MPEG-1 Audio are 32, 44.2, and 48 kHz. Several fixed bit rates in the range of 32 to 224 kbit/sec per channel can be used. In addition, layer III supports variable bit rate coding. Good quality is possible with layer I above 128 kbit/sec, with layer II around 128 kbit/sec and layer III at 64 kbit/sec. MPEG-1 Audio has four modes: mono, stereo, dual mono with two separate channels, and joint stereo. The optional joint stereo mode exploits interchannel redundancies.

A polyphase filter bank is common to all layers of MPEG-1 Audio coding. This filter bank subdivides the audio signal into 32 equal-width frequency subbands. The filters are relatively simple and provide good time resolution with reasonable frequency resolution. To achieve these characteristics, some exceptions had to be made. First, the equal widths of subbands do not precisely reflect the human auditory system's frequency-dependent behavior. Second, the filter bank and its inverse are not lossless transformations. Third, adjacent filter bands have significant frequency overlap. However, these exceptions do not impose any noticeable limitations and quite good audio quality is possible. The layer I algorithm codes audio in frames of 384 samples by grouping 12 samples from each of the 32 subbands. The layer II algorithm is a straightforward enhancement of layer I; it codes audio data in larger groups (1152 samples per audio channel) and imposes some restrictions on possible bit allocations for values from the middle and higher subbands. The layer II coder gets better quality by redistributing bits to better represent quantized subband values. The layer III algorithm is more sophisticated and uses audio spectral perceptual entropy coding and optimal coding in the frequency domain. Although based on the same filter bank as used in layer I and layer II, it compensates for some deficiencies by processing the filter outputs with a modified DCT.

III. MPEG-2

MPEG-2 is the standard for digital television [19]. More specifically, the original target for the MPEG-2 standard was TV-resolution video and up to five-channel audio of very good quality at about 4 to 15 Mbit/sec for applications such as digital broadcast TV and digital versatile disk. The standard either has been deployed or is likely to be deployed for a number of other applications including digital cable or satellite TV, video on Asynchronous Transfer Mode (ATM), networks, and high-definition TV (HDTV) (at 15 to 30 Mbit/sec).

Based on the target applications, a number of primary requirements were derived and are listed as follows.

Coding of interlaced video with very good quality and resolution at 4 to 15 Mbit/sec and multichannel audio with high quality

Random access or channel switching within a limited time, allowing frequent access points every half-second
Capability for fast forward and fast reverse enabling seek and play forward or backward at several times the normal speed
Support for scalable coding to allow multiple simultaneous layers and to achieve backward compatibility with MPEG-1
A system for synchronized playback and tune-in or access of audiovisual data
At least a defined subset of the standard to be implementable in practical real-time decoders at a reasonable cost in hardware

Besides the preceding requirements, a number of other requirements also arose, such as support for a number of picture resolutions and formats (both interlaced and noninterlaced), a number of sampling structures for chrominance, robustness to errors, coding quality trade-off with coding delay, and the possibility of real-time encoders at a reasonable cost for at least a defined subset of the standard.

The MPEG-2 development work was conducted in the same subgroups in MPEG that were originally created for MPEG-1, and within a period of about 2 years from the tests and evaluation it reached the stable stage of the CD before being approved as the final standard 1 year later.

The MPEG-2 standard is formally referred to as ISO 13818 and consists of the following parts:

13818-1: Systems
13818-2: Video
13818-3: Audio
13818-4: Conformance
13818-5: Software
13818-6: Digital storage media—command and control (DSM-CC)
13818-7: Advanced audio coding (AAC) [formerly known as non–backward compatible (NBC) coding]
13818-8: 10-bit video (this work item was dropped!)
13818-9: Real-time interface
13818-10: Conformance of DSM-CC

We now briefly discuss each of the four main components (Systems, Video, Audio, and DSM-CC) [5–7,48] of the MPEG-2 standard.

A. MPEG-2 Systems

Because the MPEG-1 standard was intended for audiovisual coding for digital storage media (DSM) applications and DSMs typically have very low or negligible error rates, the MPEG-1 Systems was not designed to be highly robust to errors. Also, because the MPEG-1 Systems standard was intended for software-oriented processing, large variable length packets were preferred to minimize software overhead.

The MPEG-2 standard, on the other hand, is more generic and thus intended for a variety of audiovisual coding applications. The MPEG-2 Systems was mandated to improve error resilience and the ability to carry multiple programs simultaneously without requiring them to have a common time base. In addition, it was required that MPEG-2

Systems should support ATM networks. Furthermore, problems addressed by MPEG-1 Systems also had to be solved in a compatible manner.

The MPEG-2 Systems specification [5,18] defines two types of streams: the program stream and the transport stream. The program stream is similar to the MPEG-1 Systems stream but uses modified syntax and new functions to support advanced functionalities. Further, it provides compatibility with MPEG-1 Systems streams. The requirements of MPEG-2 program stream decoders are similar to those of MPEG-1 system stream decoders, and program stream decoders can be forward compatible with MPEG-1 system stream decoders, capable of decoding MPEG-1 system streams. Like MPEG-1 Systems decoders, program streams decoders typically employ long and variable length packets. Such packets are well suited for software-based processing in error-free environments, such as when the compressed data are stored on a disk. The packet sizes are usually in the range of 1 to 2 kbytes, chosen to match disk sector sizes (typically 2 kbytes); however, packet sizes as large as 64 kbytes are also supported. The program stream includes features not supported by MPEG-1 Systems such as hooks for scrambling of data; assignment of different priorities to packets; information to assist alignment of elementary stream packets; indication of copyright; indication of fast forward, fast reverse, and other trick modes for storage devices; an optional field for network performance testing; and optional numbering of sequence of packets.

The second type of stream supported by MPEG-2 Systems is the transport stream, which differs significantly from MPEG-1 Systems as well as the program stream. The transport stream offers robustness necessary for noisy channels as well as the ability to include multiple programs in a single stream. The transport stream uses fixed length packets of size 188 bytes, with a new header syntax. It is therefore more suited for hardware processing and for error correction schemes. Thus, the transport stream is well suited for delivering compressed video and audio over error-prone channels such as coaxial cable television networks and satellite transponders. Furthermore, multiple programs with independent time bases can be multiplexed in one transport stream. In fact, the transport stream is designed to support many functions such as asynchronous multiplexing of programs, fast access to a desired program for channel hopping, multiplexing of programs with clocks unrelated to the transport clock, and correct synchronization of elementary streams for playback; to allow control of decoder buffers during start-up and playback for constant bit rate and variable bit rate programs; to be self-describing; and to tolerate channel errors.

A basic data structure that is common to the organization of both the program stream and the transport stream data is the Packetized Elementary Stream (PES) packet. PES packets are generated by packetizing the continuous stream of compressed data generated by video and audio (i.e., elementary stream) encoders. A program stream is simply generated by stringing together PES packets with other packets containing necessary data to generate a single bitstream. A transport stream consists of packets of fixed length consisting of 4 bytes of header followed by 184 bytes of data, where data are obtained by chopping up the data in PES packets.

Demultiplexer, and a process similar to that for transport streams is followed. Also, as mentioned while briefly overviewing MPEG-1 Systems, the information about system timing is carried by a Systems Clock Reference (SCR) field in the bitstream and is used to synchronize the decoder Systems Time Clock (STC). The presentation of decoded output is controlled by Presentation Time Stamps (PTSs), which are also carried by the bitstream.

B. MPEG-2 Video

The MPEG-2 Video standard [6,18] was originally aimed at coding of video based on the ITU-R 601 4:2:0 standard (720 × 480 at 30 interlaced frames/sec or 720 × 576 at 25 interlaced frames/sec) at bit rates of about 4 to 15 Mbit/sec. However, anticipating other applications, the MPEG-2 video syntax was made flexible to support much larger picture sizes of up to 16,384 × 16,384, a number of frame rates (23.97, 24, 25, 29.97, 50, 59.94, 60 frames/sec), three chrominance formats (4:2:0, 4:2:2, and 4:4:4), and higher bit rates. In addition, the MPEG-2 Video coding scheme is designed to be syntactic superset of MPEG-1, supporting channel switching as well as other forms of interactivity such as random access, fast forward, and fast reverse.

As in the case of MPEG-1, MPEG-2 does not standardize the video encoding process or the encoder. Only the bitstream syntax and the decoding process are standardized. MPEG-2 Video coding [18,22–24] can be seen as an extension of MPEG-1 Video coding to code interlaced video efficiently. MPEG-2 Video coding is thus still based on the block motion compensated DCT coding of MPEG-1. Much as with MPEG-1 Video, coding is performed on pictures, where a picture can be a frame or field, because with interlaced video each frame consists of two fields separated in time. An input sequence is divided into groups of pictures assuming frame coding. A frame may be coded as an intra (I-) picture, a predictive (P-) picture, or a bidirectionally predictive (B-) picture. Thus, a group of pictures may contain an arrangement of I-, P-, and B-coded pictures. Each picture is further partitioned into slices, each slice into a sequence of macroblocks, and each macroblock into four luminance blocks and corresponding chrominance blocks.

In Figure 3 we illustrate both types of MPEG-2 Systems, those using program stream multiplexing and those using transport stream multiplexing. An MPEG-2 system is capable

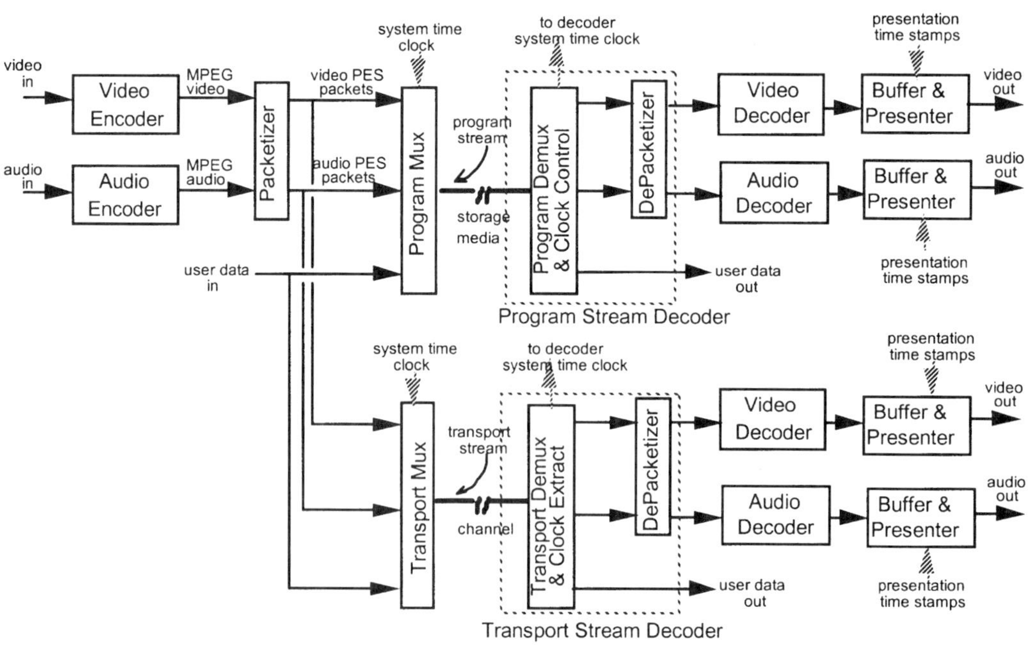

Figure 3 MPEG-2 Systems.

of combining multiple sources of user data along with MPEG encoded audio and video. The audio and video streams are packetized to form audio and video PES packets, which are sent to either a program multiplexer or a transport multiplexer, resulting in a program stream or a transport stream as the case may be. As mentioned earlier, program streams are intended for error-free environments such as DSMs whereas transport streams are intended for noisier environments such as terrestrial broadcast channels.

Transport streams are decoded by the transport demultiplexer (which includes a clock extraction mechanism), unpacketized by a depacketizer, and sent to audio and video decoders for audio and video decoding. The decoded signals are sent to respective buffer and presentation units, which output them to display device and its speaker at the appropriate time. Similarly, if Program Streams are employed, they are decoded by a Program Stream.

The MPEG-2 video encoder consists of various components such as an inter/intra frame/field DCT encoder, a frame/field motion estimator and compensator, and the variable length encoder. Earlier we mentioned that MPEG-2 Video is optimized for coding of interlaced video; this is why both the DCT coding and the motion estimation and compensation employed by the video encoder need to be frame/field adaptive. The frame/field DCT encoder exploits spatial redundancies and the frame/field motion compensator exploits temporal redundancies in the interlaced video signal. The coded video bitstream is sent to the systems multiplexer, Sys Mux, which outputs either a transport or a program stream.

Figure 4 shows a simplified block diagram of an MPEG-2 video decoder that receives bitstreams to be decoded from the MPEG-2 Systems demux. The MPEG-2 video decoder consists of a variable length decoder, an inter/intra frame/field DCT decoder, and a uni/bidirectional frame/field motion compensator. After demultiplexing, the MPEG-2 video bitstream is sent to the variable length decoder for decoding of motion vectors (*m*), quantizer information (*q*), inter/intra decision (*i*), frame/field decision (*f*), and the data consisting of quantized DCT coefficient indices. The inter/intra frame/field DCT decoder uses the decoded DCT coefficient indices, the quantizer information, the inter/intra decision, and the frame/field information to dequantize the indices to yield DCT coefficient blocks and then inverse transform the blocks to recover decoded pixel blocks (much as in the case of MPEG-1 video except for frame/field adaptation). The uni/bidirectional

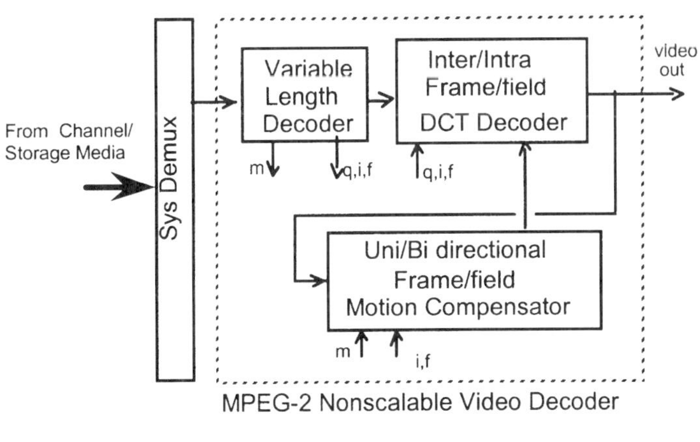

Figure 4 MPEG-2 video decoder.

frame/field motion compensator, if the coding mode is inter (based on the inter/intra decision), uses motion vectors and frame/field information to generate motion-compensated prediction of blocks (much as in the case of MPEG-1 video except for frame/field adaptation), which are then added back to the corresponding decoded prediction error blocks output by the inter/intra frame/field DCT decoder to generate decoded blocks. If the coding mode is intra, no motion-compensated prediction needs to be added to the output of the inter/intra frame/field DCT decoder. The resulting decoded pictures are output on the line labeled video out.

In terms of interoperability with MPEG-1 Video, the MPEG-2 Video standard was required to satisfy two key elements: forward compatibility and backward compatibility. Because MPEG-2 Video is a syntactic superset of MPEG-1 Video, it is able to meet the requirement of forward compatibility, meaning that an MPEG-2 video decoder ought to be able to decode MPEG-1 video bitstreams. The requirement of backward compatibility, however, means that the subsets of MPEG-2 bitstreams should be decodable by existing MPEG-1 decoders; this is achieved via scalability. Scalability is the property that allows decoders of various complexities to be able to decode video of resolution or quality commensurate with their abilities from the same bitstream. Actually, by use of B-pictures, which, being noncausal, do not feed back in to the interframe coding loop and thus can be dropped, some degree of temporal scalability is always possible. However, by nonscalable coding we mean that no special mechanism has been incorporated in the coding process to achieve scalability and only the full spatial and temporal resolution at the encoding time is expected to be decoded. A detailed discussion of scalability is beyond the scope of this chapter, so we only discuss the principle and examine a generalized scalable codec structure of Figure 5 that allows scalable coding. Further, from the various types of scalability supported by MPEG-2 Video, our generalized codec structure basically allows only spatial and temporal resolution scalabilities.

The generalized codec [18,23] supports two scalability layers, a lower layer, specifically referred to as the base layer, and a higher layer that provides enhancement of the base layer. Input video goes through a preprocessor and results in two video signals, one of which is input to the MPEG-1/MPEG-2 Nonscalable Video Encoder and the other input to the MPEG-2 Enhancement Video Encoder. Depending on the specific type of scalability, some processing of decoded video from the MPEG-1/MPEG-2 Nonscalable Video Encoder may be needed in the midprocessor before it is used for prediction in the MPEG-2 Enhancement Video Encoder. The two coded video bitstreams, one from each encoder,

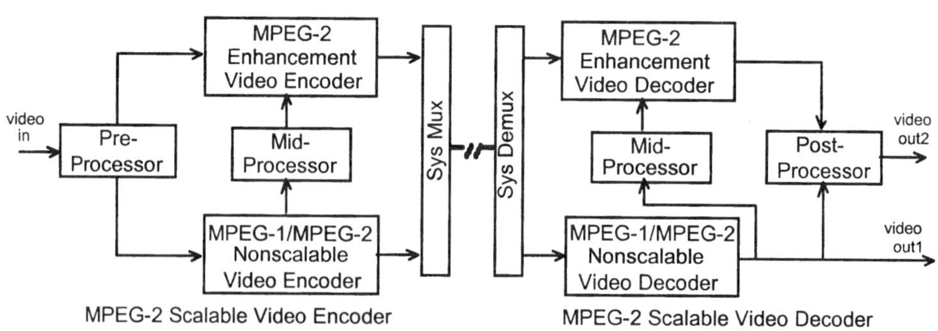

Figure 5 A generalized codec for MPEG-2 scalable video coding.

are multiplexed in the Sys Mux (along with coded audio and user data). At the decoder end, the MPEG-2 Sys Demux performs the inverse operation of unpacking from a single bitstream two substreams, one corresponding to the lower layer and the other corresponding to the higher layer. Thus, the lower layer decoding is mandatory, but the higher layer decoding is optional. For instance, if an MPEG-1/MPEG-2 Nonscalable Video Decoder is employed, a basic video signal can be decoded. If, in addition, an MPEG-2 Enhancement Video Decoder is employed, an enhanced video signal can also be decoded. Further, depending on the type of scalability, the two decoded signals may undergo further processing in a postprocessor.

Two new amendments [18] to MPEG-2 Video took place after completion of the original standard. The first amendment, motivated by the needs of professional applications, tested and verified the performance of a higher chrominance spatial (or spatiotemporal) resolution format called the 4:2:2 format. Although tools for coding this type of signal were included in the original standard, new issues such as quality after multiple generations of coding came up and needed verifying. The second amendment to MPEG-2 Video was motivated by the potential applications in video games, education, and entertainment and involved developing, testing, and verifying a solution for efficient coding of multiviewpoint signals including at least the case of stereoscopic video (two slightly different views of a scene). Not surprisingly, this involves exploiting correlations between different views of a scene, and the solution developed by MPEG-2 is a straightforward extension of the scalable video coding techniques discussed earlier.

C. MPEG-2 Audio

Digital multichannel audio systems employ a combination of p front and q back channels, for example, three front channels (left, right, center) and two back channels (surround left and surround right), to create surreal and theater-like experiences. In addition, multichannel systems can be used to provide multilingual programs, audio augmentation for the visually impaired, enhanced audio for the hearing impaired, etc. The MPEG-2 Audio standard [7,18] addresses such applications. It consists of two parts; part 3 allows coding of multichannel audio signals in a forward and backward compatible manner with MPEG-1, and part 7 does not. Here forward compatibility means that the MPEG-2 multichannel audio decoder ought to be able to decode MPEG-1 mono or stereo audio signals, and backward compatibility means that a meaningful downmix of the original five channels of MPEG-2 should be possible to deliver correct-sounding stereo when played by the MPEG-1 audio decoder. Whereas forward compatibility is not so hard to achieve, backward compatibility is a bit difficult and requires some compromise in coding efficiency. The requirement for backward compatibility was considered important at the time to allow migration to MPEG-2 multichannel from MPEG-1 stereo.

In Figure 6, a generalized codec structure illustrating MPEG-2 Multichannel Audio coding is shown. Multichannel Audio consisting of five signals, left (L), center (C), right (R), left surround (Ls), and right surround (Rs), is shown undergoing conversion by the use of a matrix operation resulting in five converted signals. Two of the signals are encoded by an MPEG-1 audio encoder to provide compatibility with the MPEG-1 standard, and the remaining three signals are encoded by an MPEG-2 audio extension encoder. The resulting bitstreams from the two encoders are multiplexed in Mux for storage or transmission.

Because it is possible to have coded MPEG-2 Audio without coded MPEG Video,

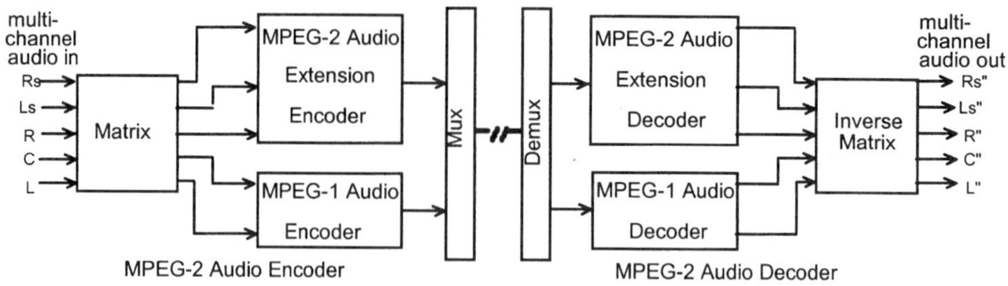

Figure 6 A generalized codec for MPEG-2 backward-compatible multichannel audio coding.

a generalized multiplexer Mux is shown. However, in an MPEG audiovisual system, the MPEG-2 Sys Mux and MPEG-2 Sys Demux are the specific mux and demux used. At the decoder, an MPEG-1 audio decoder decodes the bitstream input to it by Sys Demux and produces two decoded audio signals; the other three audio signals are decoded by an MPEG-2 audio extension decoder. The decoded audio signals are reconverted back to the original domain by using Inverse Matrix and represent approximated values indicated by L″, C″, R″, Ls″, Rs″.

Tests were conducted to compare the performance of MPEG-2 Audio coders that maintain compatibility with MPEG-1 with those that are not backward compatible. It has been found that for the same bit rate, the requirement of compatibility does impose a notable loss of quality. Hence, it has been found necessary to include a non–backward compatible (NBC) solution as an additional part (part 7) of MPEG-2, initially referred to as MPEG-2 NBC. The MPEG-2 NBC work was renamed MPEG-2 advanced audio coding (AAC), and some of the optimizations were actually performed within the context of MPEG-4. The AAC Audio supports the sampling rates, audio bandwidth, and channel configurations of backward-compatible MPEG-2 Audio but can operate at bit rates as low as 32 kbit/sec or produce very high quality at bit rates of half or less than those required by backward-compatible MPEG-2 Audio. Because the AAC effort was intended for applications that did not need compatibility with MPEG-1 stereo audio, it manages to achieve very high performance. Although a detailed discussion of the AAC technique is outside the scope of this chapter, we briefly discuss the principles involved using a simple generalized codec.

In Figure 7 we show a simplified reference model configuration of an AAC audio

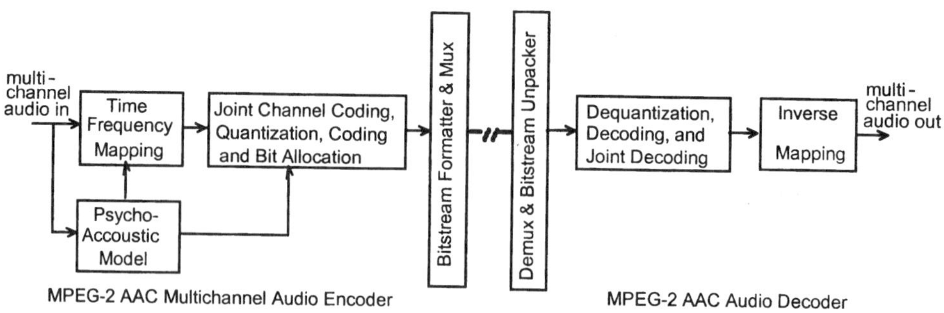

Figure 7 A generalized codec for MPEG-2 advanced audio coding (AAC) multichannel audio.

codec. Multichannel audio undergoes transformation via time-to-frequency mapping, whose output is subject to various operations such as joint channel coding, quantization, and coding and bit allocation. A psychoacoustical model is employed at the encoder and controls both the mapping and bit allocation operations. The output of the joint channel coding, quantization, and coding and bit allocation unit is input to a bitstream formatter that generates the bitstream for storage or transmission. At the decoder, an inverse operation is performed at what is called the bitstream unpacker, following which dequantization, decoding, and joint decoding occur. Finally, an inverse mapping is performed to transform the frequency domain signal to its time domain representation, resulting in reconstructed multichannel audio output.

D. MPEG-2 DSM-CC

Coded MPEG-2 bitstreams may typically be stored on a variety of digital storage media (DSM) such as CD-ROM, magnetic tape and disks, Digital Versatile Disks (DVDs), and others. This presents a problem for users when trying to access coded MPEG-2 data because each DSM may have its own control command language, forcing the user to know many such languages. Moreover, the DSM may be either local to the user or at a remote location. When it is remote, a common mechanism for accessing various digital storage media over a network is needed, otherwise the user has to be informed about type of DSM, which may not be known or possible. The MPEG-2 DSM-CC [5,18] is a set of generic control commands independent of the type of DSM that addresses these two problems. Thus, control commands are defined as a specific application protocol to allow a set of basic functions specific to MPEG bitstreams. The resulting control commands do not depend on the type of DSM or whether the DSM is local or remote, the network transmission protocol, or the operating system with which it is interfacing. The control command functions can be performed on the MPEG-1 systems bitstreams, the MPEG-2 program streams or the MPEG-2 transport streams. Examples of some functions needed are connection, playback, storage, editing, and remultiplexing. A basic set of control commands allowing these functionalities are included as an informative annex in MPEG-2 Systems, which is part 2 of MPEG-2. Advanced capabilities of DSM-CC are mandated in MPEG-2 DSM-CC, which is part 6 of MPEG-2.

The DSM control commands can generally be divided into two categories. The first category consists of a set of very basic operations such as the stream selection, play, and store commands. Stream selection enables a request for a specific bitstream and a specific operation mode on that bitstream. Play enables playback of a selected bitstream at a specific speed, direction of play (to accommodate a number of trick modes such as fast forward and fast reverse), or other features such as pause, resume, step through, or stop. Store enables the recording of a bitstream on a DSM. The second category consists of a set of more advanced operations such as multiuser mode, session reservation, server capability information, directory information, and bitstream editing. In multiuser mode, more than one user is allowed access to the same server within a session. With session reservation, a user can request a server for a session at a later time. Server capability information allows the user to be notified of the capabilities of the server, such as playback, fast forward, fast reverse, slow motion, storage, demultiplex, and remultiplex. Directory information allows the user access to information about directory structure and specific attributes of a bitstream such as type, IDs, sizes, bit rate, entry points for random access, program descriptors and others; typically, not all this information may be available through

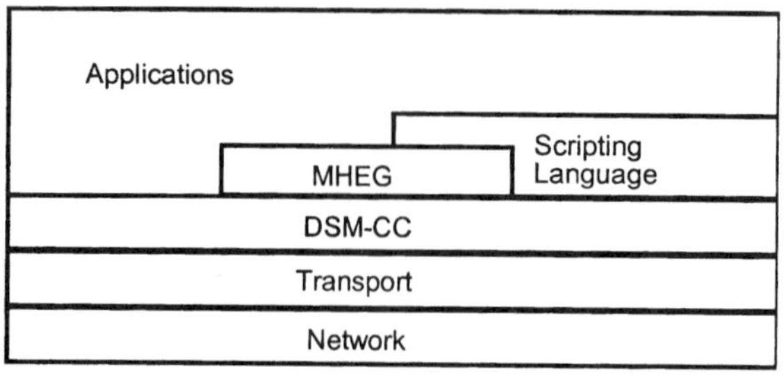

Figure 8 DSM-CC centric view of MHEG, scripting language, and network.

the application programming interface (API). Bitstream editing allows creation of new bitstreams by insertion or deletion of portions of bitstreams into others.

In Figure 8 we show a simplified relationship of DSM-CC with MHEG (Multimedia and Hypermedia Experts Group standard), scripting language, and DSM-CC. The MHEG standard is basically an interchange format for multimedia objects between applications. MHEG specifies a class set that can be used to specify objects containing monomedia information, relationships between objects, dynamic behavior between objects, and information to optimize real-time handling of objects. MHEG's classes include the content class, composite class, link class, action class, script class, descriptor class, container class, and result class. Further, the MHEG standard does not define an API for handling of objects, nor does it define methods on its classes, and although it supports scripting via the script class, it does not standardize any specific scripting language. Applications may access DSM-CC either directly or through an MHEG layer; moreover, scripting languages may be supported through an MHEG layer. The DSM-CC protocols also form a layer higher than the transport protocols layer. Examples of transport protocols are the Transfer Control Protocol (TCP), the User Datagram Protocol (UDP), the MPEG-2 program streams, and the MPEG-2 transport streams.

The DSM-CC provides access for general applications, MHEG applications, and scripting languages to primitives for establishing or deleting network connections using user–network (U-N) primitives and communication between a client and a server across a network using user–user (U-U) primitives. The U-U operations may use a Remote Procedure Call (RPC) protocol. Both the U-U and the U-N operations may employ message passing in the form of exchanges of a sequence of codes. In Figure 9 we show the scenarios

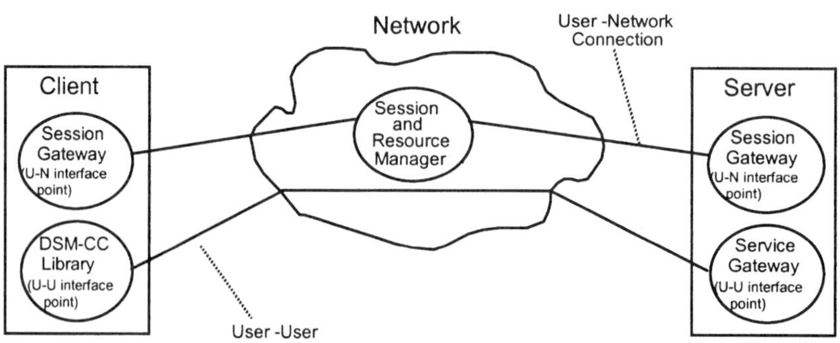

Figure 9 DSM-CC user–network and user–user interaction.

of U-N and U-U interaction. A client can connect to a server either directly via a network or through a resource manager located within the network.

A client setup is typically expected to include a session gateway, which is a user–network interface point, and a library of DSM-CC routines, which is a user–user interface point. A server setup typically consists of a session gateway, which is a user–network interface point, and a service gateway, which is a user–user interface point. Depending on the requirements of an application, both user–user connection and user–network connection can be established. Finally, DSM-CC may be carried as a stream within an MPEG-1 systems stream, an MPEG-2 transport stream, or an MPEG-2 program stream. Alternatively, DSM-CC can also be carried over other transport networks such as TCP or UDP.

IV. MPEG-4

MPEG-4 is the standard for multimedia applications [19,25,26]. As mentioned earlier, the original scope of MPEG-4 was very low bit rate video coding and was modified to generic coding of audiovisual objects for multimedia applications. To make the discussion a bit more concrete, we now provide a few examples of the application areas [18,27] that the MPEG-4 standard is aimed at.

Internet and intranet video
Wireless video
Interactive home shopping
Video e-mail and home movies
Virtual reality games, simulation, and training
Media object databases

Because these application areas have several key requirements beyond those supported by the previous standards, the MPEG-4 standard addresses the following functionalities [18,28].

Content-based interactivity allows the ability to interact with important objects in a scene. The MPEG-4 standard extends the types of interaction typically available for synthetic objects to natural objects as well as hybrid (synthetic natural) objects to enable new audiovisual applications. It also supports the spatial and temporal scalability of media objects.

Universal accessibility means the ability to access audiovisual data over a diverse range of storage and transmission media. Because of the increasing trend toward mobile communications, it is important that access be available to applications via wireless networks; thus MPEG-4 provides tools for robust coding in error-prone environments at low bit rates. MPEG-4 is also developing tools to allow fine granularity media scalability for Internet applications.

Improved compression allows an increase in efficiency of transmission or a decrease in the amount of storage required. Because of the object-oriented nature, MPEG-4 allows a very flexible adaptation of degree of compression to the channel bandwidth or storage media capacity. The MPEG-4 coding tools, although generic, are still able to provide state-of-the-art compression, because optimization of MPEG-4 coding was performed on low-resolution content at low bit rates.

As noted earlier, the MPEG-4 standard is being introduced in at least two stages. The basic standard, also referred to as version 1, has recently become an international standard in May 1999. The extension standard, also referred to as version 2, will become mature by July 1999 and is expected to become an international standard by February 2000. Version 2 technology, because it extends version 1 technology, is being introduced as an amendment to the MPEG-4 version 1 standard.

The MPEG-4 standard is formally referred to as ISO 14496 and consists of the following parts.

14496-1: Systems
14496-2: Video
14496-3: Audio
14496-4: Conformance
14496-5: Software
14496-6: Delivery Multimedia Integration Framework (DMIF)

The conceptual architecture of MPEG-4 is depicted in Figure 10. It comprises three layers: the compression layer, the sync layer, and the delivery layer. The compression layer is media aware and delivery unaware; the sync layer is media unaware and delivery unaware; the delivery layer is media unaware and delivery aware.

The compression layer performs media encoding and decoding into and from elementary streams and is specified in parts 2 and 3 of MPEG-4; the sync layer manages elementary streams and their synchronization and hierarchical relations and is specified in part 1 of MPEG-4; the delivery layer ensures transparent access to content irrespective of delivery technology and is specified in part 6 of MPEG-4. The boundary between the compression layer and the sync layer is called the elementary stream interface (ESI), and its minimum semantic is specified in part 1 of MPEG-4.

We now briefly discuss the current status of the four main components (Systems, Video, Audio, and DMIF) [8,11,12] of the MPEG-4 standard.

A. MPEG-4 DMIF

The MPEG-4 Delivery Multimedia Integration Framework (Fig. 11) [11] allows unique characteristics of each delivery technology to be utilized in a manner transparent to application developers. The DMIF specifies the semantics for the DMIF application interface

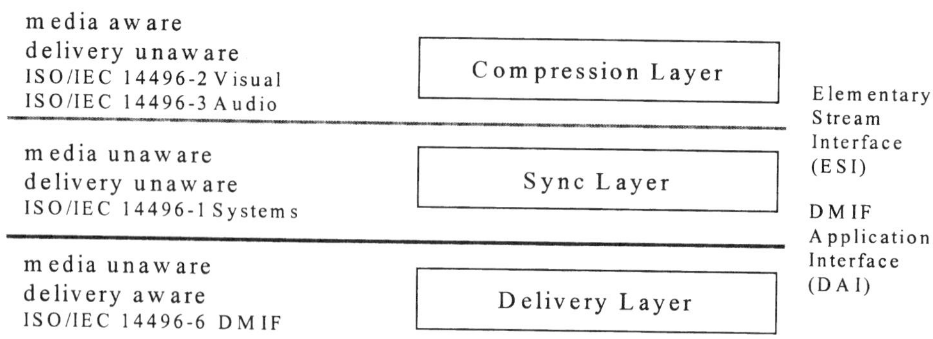

Figure 10 Various parts of MPEG-4.

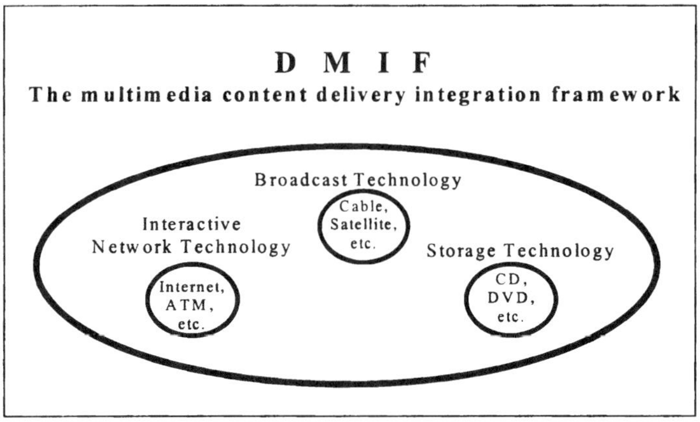

Figure 11 Integration framework for delivery technology.

(DAI) in a way that satisfies the requirements for broadcast, local storage, and remote interactive scenarios in a uniform manner. By including the ability to bundle connections into sessions, DMIF facilitates billing for multimedia services by network operators. By adopting quality of service (QoS) metrics that relate to the media and not to the transport mechanism, DMIF hides the delivery technology details from applications. These features of DMIF give multimedia application developers a sense of permanence and genericness not provided by individual delivery technology. For instance, with DMIF, application developers can invest in commercial multimedia applications with the assurance that their investment will not be made obsolete by new delivery technologies. However, to reach its goal fully, DMIF needs "real" instantiation of its DMIF application interface and well-defined, specific mappings of DMIF concepts and parameters into existing signaling technologies.

The DMIF specifies the delivery layer, which allows applications to access transparently and view multimedia streams whether the source of the streams is located on an interactive remote end system, the streams are available on broadcast media, or they are on storage media.

The MPEG-4 DMIF covers the following aspects:

DMIF communication architecture
DMIF application interface (DAI) definition
Uniform resource locator (URL) semantic to locate and make available the multimedia streams
DMIF default signaling protocol (DDSP) for remote interactive scenarios and its related variations using existing native network signaling protocols
Information flows for player access to streams on remote interactive end systems, from broadcast media or from storage media

When an application requests the activation of a service, it uses the service primitives of the DAI and creates a service session. In the case of a local storage or broadcast scenario, the DMIF instance locates the content that is part of the indicated service; in the case of interactive scenarios, the DMIF instance contacts its corresponding peer and creates a network session with it. The peer DMIF instance in turn identifies the peer application that runs the service and establishes a service session with it. Network sessions have network-wide significance; service sessions instead have local meaning. The delivery layer maintains the association between them. Each DMIF instance uses the native signaling mechanism for the respective network to create and then manage the network session (e.g., DMIF default signaling protocol integrated with ATM signaling). The application peers then use this session to create connections that are used to transport application data (e.g., MPEG-4 Systems elementary streams).

When an application needs a channel, it uses the channel primitives of the DAI, indicating the service they belong to. In the case of local storage or a broadcast scenario, the DMIF instance locates the requested content, which is scoped by the indicated service, and prepares itself to read it and pass it in a channel to the application; in the case of interactive scenarios the DMIF instance contacts its corresponding peer to get access to the content, reserves the network resources (e.g., connections) to stream the content, and prepares itself to read it and pass it in a channel to the application; in addition, the remote application locates the requested content, which is scoped by the indicated service. DMIF uses the native signaling mechanism for the respective network to reserve the network resources. The remote application then uses these resources to deliver the content.

Figure 12 provides a high-level view of a service activation and of the beginning of data exchange in the case of interactive scenarios; the high-level walk-through consists of the following steps:

Step 1: The originating application requests the activation of a service to its local DMIF instance: a communication path between the originating application and its local DMIF peer is established in the control plane 1.

Step 2: The originating DMIF peer establishes a network session with the target DMIF peer: a communication path between the originating DMIF peer and the target DMIF peer is established in the control plane 2.

Step 3: The target DMIF peer identifies the target application and forwards the service activation request: a communication path between the target DMIF peer and the target application is established in the control plane 3.

Step 4: The peer applications create channels (requests flowing through communication paths 1, 2, and 3). The resulting channels in user plane 4 will carry the actual data exchanged by the applications.

DMIF is involved in all four of these steps.

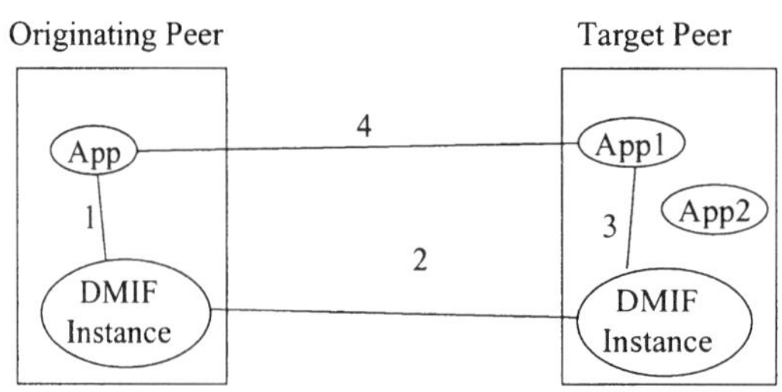

Figure 12 DMIF computational model.

B. MPEG-4 Systems

Part 1, the systems part [8] of MPEG-4, perhaps represents the most radical departure from the previous MPEG standards. The object-based nature of MPEG-4 necessitated a new approach to MPEG-4 Systems, although the traditional issues of multiplexing and synchronization were still quite important. For synchronization, the challenge for MPEG-4 Systems was to provide a mechanism to handle a large number of streams, which result from the fact that a typical MPEG-4 scene may be composed of many objects. In addition the spatiotemporal positioning of these objects forming a scene (or scene description) is a new key component. Further, MPEG-4 Systems also had to deal with issues of user interactivity with the scene. Another item, added late during MPEG-4 Systems version 1 development, addresses the problem of management and protection of intellectual property related to media content.

More precisely, the MPEG-4 Systems version 1 specification [8] covers the following aspects:

- Terminal model for time and buffer management
- Coded representation of scene description
- Coded representation of metadata—Object Descriptors (and others)
- Coded representation of AV content information—Object Content Information (OCI)
- An interface to intellectual property—Intellectual Property Management and Protection (IPMP)
- Coded representation of sync information—Sync Layer (SL)
- Multiplex of elementary streams to a single stream—FlexMux tools

Currently, work is ongoing on version 2 [12,29] of MPEG-4 Systems that extends the version 1 specification. The version 2 specification is finalizing additional capabilities such as the following.

- MPEG-4 File Format (MP4)—A file format for interchange
- MPEG-4 over Internet Protocol and MPEG-4 over MPEG-2
- Scene description—application texture, advanced audio, chromakey, message handling
- MPEG-J–Java–based flexible control of fixed MPEG-4 Systems

Although it was mentioned earlier, it is worth reiterating that MPEG-4 is an object-based standard for multimedia coding, and whereas previous standards code precomposed media (e.g., a video scene and corresponding audio), MPEG-4 codes individual video objects and audio objects in the scene and delivers in addition a coded description of the scene. At the decoding end, the scene description and individual media objects are decoded, synchronized, and composed for presentation. Before further discussing the key systems concepts, a brief overview of the architecture of MPEG-4 Systems is necessary; Fig. 13 shows the high-level architecture of an MPEG-4 terminal.

The architecture [8] shows the MPEG-4 stream delivered over the network/storage medium via the delivery layer, which includes transport multiplex, TransMux (not standardized by MPEG-4 but could be UDP, AAL 2, MPEG-2 Transport Stream, etc.), and an optional multiplex called the FlexMux. Demultiplexed streams from the FlexMux leave via the DAI interface and enter the sync layer, resulting in SL packetized elementary

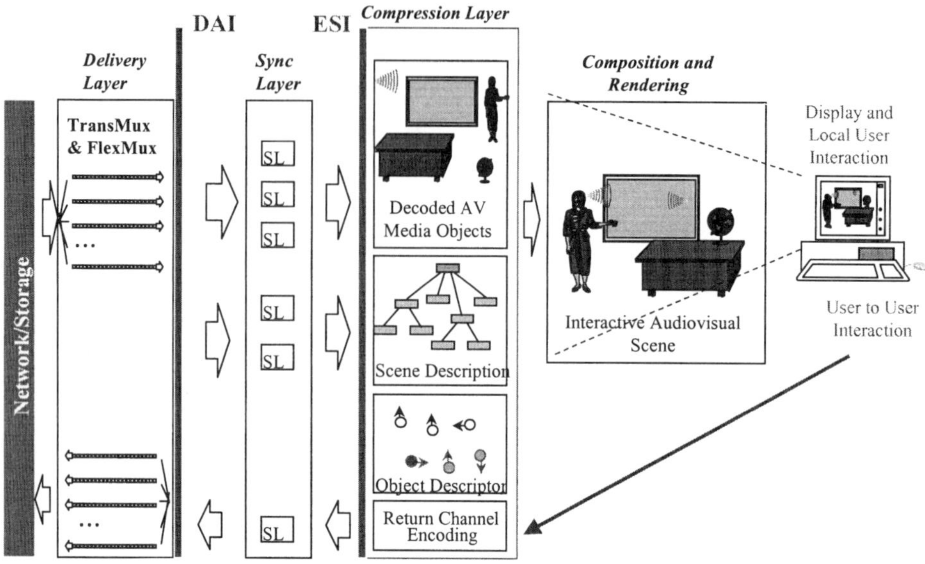

Figure 13 Architecture of an MPEG-4 terminal.

streams that are ready to be decoded. The compression layer encapsulates the functions of the media, scene description, and object descriptor decoding, yielding individual decoded objects and related descriptors. The composition and rendering process uses the scene description and decoded media to compose and render the audiovisual scene and passes it to the presenter. A user can interact with the presentation of the scene, and the actions necessary as a result (e.g., request additional media streams) are sent back to the network/ storage medium through the compression, sync, and delivery layers.

1. System Decoder Model

A key challenge in designing an audiovisual communication system is ensuring that the time is properly represented and reconstructed by the terminal. This serves two purposes: first, it ensures that ''events'' occur at designated times as indicated by the content creator, and second, the sender can properly control the behavior of the receiver. Time stamps and clock references are two key concepts that are used to control timing behavior at the decoder. Clock recovery is typically performed using clock references. The receiving system has a local system clock, which is controlled by a Phase Locked Loop (PLL), driven by the differences in received clock references and the local clock references at the time of their arrival. In addition, coded units are associated with decoding time stamps, indicating the time instance in which a unit is removed from the receiving system's decoding buffer. Assuming a finite set of buffer resources at the receiver, by proper clock recovery and time stamping of events the source can always ensure that these resources are not exhausted. Thus the combination of clock references and time stamps is sufficient for full control of the receiver. MPEG-4 defines a system decoder model (SDM) [8,12,30], a conceptual model that allows precise definition of decoding events, composition events, and times at which these events occur. This represents an idealized unit in which operations can be unambiguously controlled and characterized. The MPEG-4 system decoder model exposes resources available at the receiving terminal and defines how they can be controlled by the sender or the content creator.

The SDM is shown in Figure 14. The FlexMux buffer is a receiver buffer that can store the FlexMux streams and can be monitored by the sender to determine the FlexMux resources that are used during a session. Further, the SDM is composed of a set of decoders (for the various audio or visual object types), provided with two types of buffers: decoding and composition. The decoding buffers have the same functionality as in previous MPEG specifications and are controlled by clock references and decoding time stamps. In MPEG-2, each program had its own clock; proper synchronization was ensured by using the same clock for coding and transmitting the audio and video components. In MPEG-4, each individual object is assumed to have its own clock or object time base (OTB). Of course, several objects may share the same clock. In addition, coded units of individual objects (access units, AUs, corresponding to an instance of a video object or a set of audio samples) are associated with decoding time stamps (DTSs). Note that the decoding operation at the DTS is considered (in this ideal model) to be instantaneous. The composition buffers that are present at the decoder outputs form a second set of buffers. Their use is related to object persistence. In some situations, a content creator may want to reuse a particular object after it has been presented. By exposing a composition buffer, the content creator can control the lifetime of data in this buffer for later use. This feature may be particularly useful in low-bandwidth wireless environments. MPEG-4 defines an additional time stamp, the composition time stamp (CTS), which defines the time at which data are taken from the composition buffer for (instantaneous) composition and presentation. In order to coordinate the various objects, a single system time base is assumed to be present at the receiving system. All object time bases are subsequently mapped into the system time base so that a single notion of time exists in the terminal. For clock recovery purposes, a single stream must be designated as the master. The current specification does not indicate the stream that has this role, but a plausible candidate is the one that contains the scene description. Note also that, in contrast to MPEG-2, the resolution of both the STB and the OCRs is not mandated by the specification. In fact, the size of the OCR fields for individual access units is fully configurable.

2. *Scene Description*

Scene description [8,30,31] refers to the specification of the spatiotemporal positioning and behavior of individual objects. It allows easy creation of compelling audiovisual content. Note that the scene description is transmitted in a separate stream from the individual

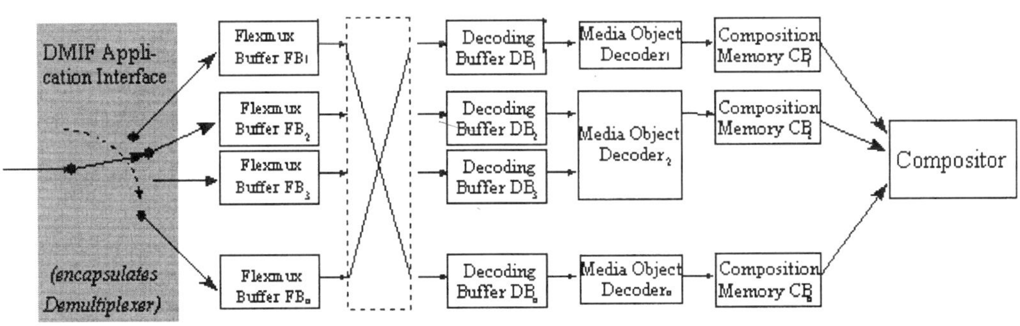

Figure 14 Systems decoder model.

media objects. This allows one to change the scene description without operating on any of the constituent objects themselves. The MPEG-4 scene description extends and parameterizes virtual reality modeling language (VRML), which is a textual language for describing the scene in three dimensions (3D). There are at least two main reasons for this. First, MPEG-4 needed the capability of scene description not only in three dimensions but also in two dimensions (2D), so VRML needed to be extended to support 2D, and second, VRML, being a textual language, was not suitable for low-overhead transmission, and thus a parametric form binarizing VRML called Binary Format for Scenes (BIFS) had to be developed.

In VRML, nodes are the elements that can be grouped to organize the scene layout by creating a scene graph. In a scene graph, the trunk is the highest hierarchical level with branches representing children grouped under it. The characteristics of the parent node are inherited by the child node. A raw classification of nodes can be made on the basis of whether they are grouping nodes or leaf nodes. VRML supports a total of 54 nodes, and according to another classification [31] they can be divided into two main categories: graphical nodes and nongraphical nodes. The graphical nodes are the nodes that are used to build the rendered scenes. The graphical nodes can be divided into three subcategories with many nodes per subcategory: grouping nodes (Shape, Anchor, Billboard, Collision, Group, Transform, Inline, LOD, Switch), geometry nodes (Box, Cone, Cylinder, ElevationGrid, Extrusion, IndexedFaceSet, IndexedLineSet, PointSet, Sphere, Text), and attribute nodes (Appearance, Color, Coordinate, FontStyle, ImageTexture, Material, MovieTexture, Normal, PixelTexture, TextureCoordinate, TextureTransform). The nongraphical nodes augment the 3D scene by providing a means of adding dynamic effects such as sound, event triggering, and animation. The nongraphical nodes can also be divided into three subcategories with many nodes per subcategory: sound (AudioClip, Sound), event triggers (CylinderSensor, PlaneSensor, ProximitySensor, SphereSensor, TimeSensor, TouchSensor, VisibilitySensor, Script), and animation (ColorInterpolator, CoordinateInterpolator, NormalInterpolator, OrientationInterpolator, PositionInterpolator, ScalarInterpolator).

Each VRML node can have a number of fields that parametrize the node. Fields in VRML form the basis of the execution model. There are four types of fields—field, eventIn field, eventOut field, and exposedField. The first field, carries data values that define characteristics of a node; the second, eventIn field, accepts incoming events that change its value to the value of the event itself (sink); the third, eventOut field, outputs its value as an event (source); and, the fourth, exposedField, allows acceptance of a new value and can send out its value as an event (source and sink). The fields that accept a single value are prefixed by SF and those that accept multiple values are prefixed by MF. All nodes contain fields of one or more of the following types: SFNode/MFNode, SFBool, SFColor/MFColor, SFFloat/MFFloat, SFImage, SFInt32/MFInt32, SFRotation/MFRotation, SFString/MFString, SFTime/MFTime, SFVec2f/MFVec2f, and SFVec3f/MFVec3f. A ROUTE provides a mechanism to link the identified source and sink fields of nodes to enable a series of events to flow. Thus, it enables events to flow between fields, enabling propagation of changes in the scene graph; a scene author can wire the (fields of) nodes together.

Binary Format for Scenes [8,31], although based on VRML, extends it in several directions. First, it provides a binary representation of VRML2.0; this representation is much more efficient for storage and communication than straightforward binary representation of VRML text as (American Standard Code for Information Interchange). Second, in recognition of the fact that BIFS needs to represent not only 3D scenes but also normal 2D (audiovisual) scenes it adds a number of 2D nodes including 2D versions of several 3D nodes. Third, it includes better support for MPEG-4–specific media such as video, facial animation, and sound by adding new nodes. Fourth, it improves the animation capabilities of VRML and further adds streaming capabilities. BIFS supports close to 100 nodes—one-half from VRML and one-half new nodes. It also specifies restrictions on semantics of several VRML nodes. Among the nodes added to VRML are shared nodes (AnimationStream, AudioDelay, AudioMix, AudioSource, AudioFX, AudioSwitch, Conditional, MediaTimeSensor, QuantizationParameter, TermCap, Valuator, BitMap), 2D nodes (Background2D, Circle, Coordinate2D, Curve2D, DiscSensor, Form, Group2D, Image2D, IndexedFaceSet2D, IndexedLineSet2D, Inline2D, Layout, LineProperties, Material2D, PlaneSensor2D, PointSet2D, Position2Dinterpolator, Proximity2DSensor, Rectangle, Sound2D, Switch2D, Transform2D), and 3D nodes (ListeningPoint, Face, FAP, Viseme, Expression, FIT, FDP). The Script node, a VRML node that adds internal programmability to the scene, has only recently been added to BIFS. However, whereas the Script node in VRML supports both Java and JavaScript as script programming languages, BIFS supports JavaScript only.

3. *Associating Scene Description with Elementary Streams*

Individual object data and scene description information are carried in separate elementary streams (ESs). As a result, BIFS media nodes need a mechanism to associate themselves with the ESs that carry their data (coded natural video object data, etc.). A direct mechanism would necessitate the inclusion of transport-related information in the scene description. As we mentioned earlier, an important requirement in MPEG-4 is transport independence [23,25]. As a result, an indirect way was adopted, using object descriptors (ODs).

Each media node is associated with an object identifier, which in turn uniquely identifies an OD. Within an OD, there is information on how many ESs are associated with this particular object (may be more than one for scalable video–audio coding or multichannel audio coding) and information describing each of those streams. The latter information includes the type of the stream, as well as how to locate it within the particular networking environment used. This approach simplifies remultiplexing (e.g., going through a wired–wireless interface), as there is only one entity that may need to be modified. The object descriptor allows unique reference to an elementary stream by an *id*; this id may be assigned by an application layer when the content is created. The transport channel in which this stream is carried may be assigned at a later time by a transport entity; it is identified by a channel association tag associated with an ES_ID (elementary stream id) by a stream map table. In interactive applications, the receiving terminal may select the desired elementary streams, send a request, and receive the stream map table in return. In broadcast and storage applications, the complete stream map table must be included in the application's signaling channel.

4. *Multiplexing*

As mentioned during the discussion of DMIF, MPEG-4 for delivery "supports" two major types of multiplex, the TransMux and the FlexMux [8,30]. The TransMux is not specified by MPEG-4 but hooks are provided to enable any of the commonly used transport (MPEG-2 transport stream, UDP, AAL 2, H.223, etc.) as needed by an application. Further, FlexMux or the flexible multiplexer, although specified by MPEG-4, is optional. This is a very simple design, intended for systems that may not provide native multiplexing services.

An example is the data channel available in GSM cellular telephones. Its use, however, is entirely optional and does not affect the operation of the rest of the system. The FlexMux provides two modes of operation, a ''simple'' and a ''muxcode'' mode. The key underlying concept in the design of the MPEG-4 multiplex is network independence. The MPEG-4 content may be delivered across a wide variety of channels, from very low bit rate wireless to high-speed ATM, and from broadcast systems to DVDs. Clearly, the broad spectrum of channels could not allow a single solution to be used. At the same time, inclusion of a large number of different tools and configuration would make implementations extremely complex and—through excessive fragmentation—make interoperability extremely hard to achieve in practice. Consequently, the assumption was made that MPEG-4 would not provide specific transport-layer features but would instead make sure that it could be easily mapped to existing such layers.

The next level of multiplexing in MPEG-4 is provided by the sync layer (SL), which is the basic conveyor of timing and framing information. It is at this level that time stamps and clock references are provided. The sync layer specifies a syntax for packetization of elementary streams into access units or parts thereof. Such a packet is called an SL packet. A sequence of such packets is called an SL-packetized stream (SPS). Access units are the only semantic entities that need to be preserved from end to end; their content is opaque. Access units are used as the basic unit for synchronization. An SL packet consists of an SL packet header and an SL packet payload. The detailed semantics of time stamps define the timing aspects of the systems decoder model. An SL packet header is configurable. An SL packet does not contain an indication of its length and thus SL packets must be framed by a low-layer protocol, e.g., the FlexMux tool. Consequently, an SL-packetized stream is not a self-contained data stream that can be stored or decoded without such framing. An SL-packetized stream does not provide the identification of the ES_ID (elementary stream ids) associated with the elementary stream in the SL packet header. As mentioned earlier, this association must be conveyed through a stream map table using the appropriate signaling means of the delivery layer. Packetization information is exchanged between an entity that generates an elementary stream and the sync layer; this relation is specified by a conceptual interface called the elementary stream interface (ESI).

C. MPEG-4 Visual

Part 2, the Visual part [9] of MPEG-4, integrates a number of visual coding techniques from two major areas [30,32–36]—natural video and synthetic visual. MPEG-4 visual addresses a number of functionalities driven by applications, such as robustness against errors in Internet and wireless applications, low-bit-rate video coding for videoconferencing, high-quality video coding for home entertainment systems, object-based and scalable object–based video for flexible multimedia, and mesh and face coding for animation and synthetic modeling. Thus, MPEG-4 visual integrates functionalities offered by MPEG-1 video, MPEG-2 video, object-based video, and synthetic visual.

More precisely, the MPEG-4 Video version 1 specification covers the following aspects:

Natural video—motion-compensated DCT coding of video objects
Synthetic video tools—mesh coding and face coding of wireframe objects
Still texture decoding—wavelet decoding of image texture objects

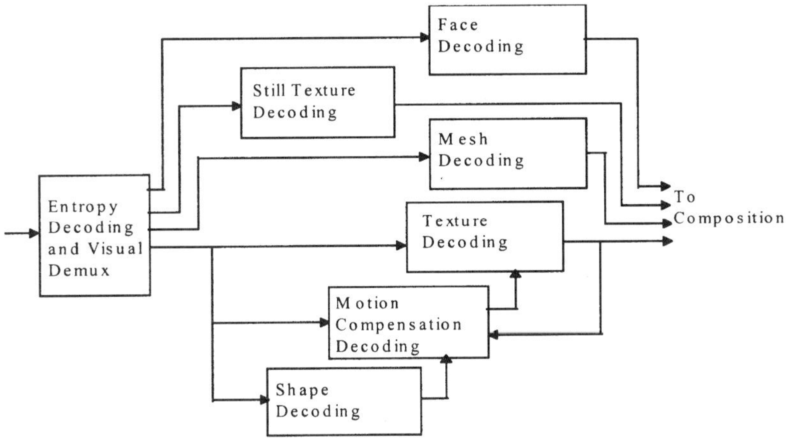

Figure 15 MPEG-4 visual decoding.

Figure 15 shows a simplified high-level view of MPEG-4 Visual decoding. The visual bitstream to be decoded is demultiplexed and variable length decoded into individual streams corresponding to objects and fed to one of the four processes—face decoding, still texture decoding, mesh decoding, or video decoding. The video decoding process further includes shape decoding, motion compensation decoding, and texture decoding. After decoding, the output of the face, still texture, mesh, and video decoding process is sent for composition.

1. *Natural Video*

In this section, we briefly discuss the coding methods and tools of MPEG-4 video; the encoding description is borrowed from Video VM8,9,12 [37–39], the decoding description follows [9]. An input video sequence contains a sequence of related snapshots or pictures, separated in time. In MPEG-4, each picture is considered as consisting of temporal instances of objects that undergo a variety of changes such as translations, rotations, scaling, and brightness and color variations. Moreover, new objects enter a scene and/or existing objects depart, leading to the presence of temporal instances of certain objects only in certain pictures. Sometimes, scene change occurs, and thus the entire scene may either be reorganized or be replaced by a new scene. Many MPEG-4 functionalities require access not only to an entire sequence of pictures but also to an entire object and, further, not only to individual pictures but also to temporal instances of these objects within a picture. A temporal instance of a video object can be thought of as a snapshot of an arbitrarily shaped object that occurs within a picture, so that like a picture, it is intended to be an access unit, and unlike a picture, it is expected to have a semantic meaning.

The concept of video objects (VOs) and their temporal instances, video object planes (VOPs) [18,37], is central to MPEG-4 video. A VOP can be fully described by texture variations (a set of luminance and chrominance values) and shape representation. In natural scenes, VOPs are obtained by semiautomatic or automatic segmentation, and the resulting shape information can be represented as a *binary shape* mask. On the other hand, for hybrid natural and synthetic scenes generated by blue screen composition, shape information is represented by an 8-bit component, referred to as *gray scale shape.*

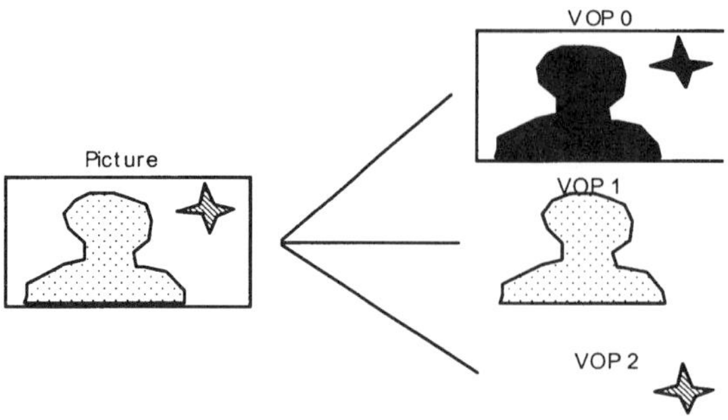

Figure 16 Semantic segmentation of picture into VOPs.

Figure 16 shows a picture decomposed into a number of separate VOPs. The scene consists of two objects (head-and-shoulders view of a human and a logo) and the background. The objects are segmented by semiautomatic or automatic means and are referred to as VOP1 and VOP2, and the background without these objects is referred to as VOP0. Each picture in the sequence is segmented into VOPs in this manner. Thus, a segmented sequence contains a set of VOP0s, a set of VOP1s, and a set of VOP2s; in other words, in our example, a segmented sequence consists of VO0, VO1, and VO2. Individual VOs are encoded separately and multiplexed to form a bitstream that users can access and manipulate (e.g., cut, paste). Together with VOs, the encoder sends information about scene composition to indicate where and when VOPs of a VO are to be displayed. This information is however, optional and may be ignored at the decoder, which may use user-specified information about composition.

Figure 17 shows an example of VOP structure in MPEG-4 video coding [18,37] that uses a pair of B-VOPs between two reference (I- or P-) VOPs. Basically, this example structure is similar to the example shown for MPEG-1/2 video coding, other than the fact that instead of pictures (or frames/fields), coding occurs on a VOP basis. In MPEG-4 Video, an input sequence can be divided into groups of VOPs (GOVs), where each GOV starts with an I-VOP and the rest of the GOV contains an arrangement of P-VOPs and B-VOPs. For coding purposes, each VOP is divided into a number of macroblocks; as in the case of MPEG-1/2, each macroblock is basically a 16 × 16 block of luminance (or alternatively, four 8 × 8 blocks) with corresponding chrominance blocks. An optional

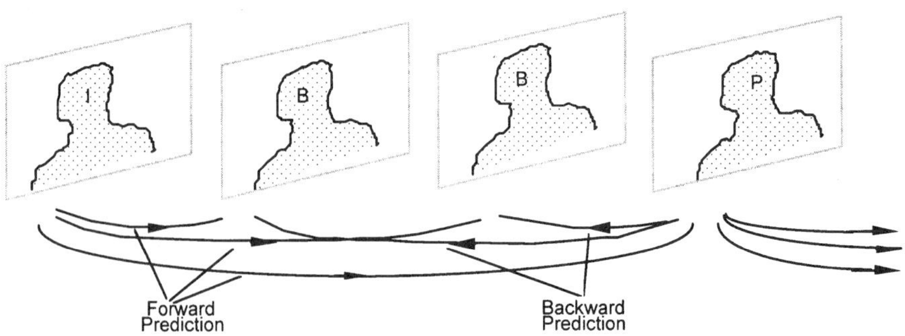

Figure 17 A VOP coding structure.

packet structure can be imposed on VOPs to provide more robustness in error-prone environments.

At a fairly high level, the coding process [9] of MPEG-4 video is quite similar to that of MPEG-1/2. In other words, MPEG-4 video coding also exploits the spatial and temporal redundancies. Spatial redundancies are exploited by block DCT coding and temporal redundancies are exploited by motion compensation. In addition, MPEG-4 video needs to code the shape of each VOP; shape coding in MPEG-4 also uses motion compensation for prediction. Incidentally, MPEG-4 video coding supports both noninterlaced video (e.g., as in MPEG-1 video coding) and interlaced video (e.g., as in MPEG-2 video coding).

Figure 18 is a simplified block diagram showing an MPEG-4 video decoder that receives bitstreams to be decoded from the MPEG-4 systems demux. The MPEG-4 video decoder consists of a variable length decoder, an inter/intra frame/field DCT decoder, a shape decoder, and a uni/bidirectional frame/field motion compensator. After demultiplexing, the MPEG-4 video bitstream is sent to the variable length decoder for decoding of motion vectors (m), quantizer information (q), inter/intra decision (i), frame/field decision (f), shape identifiers (s), and the data consisting of quantized DCT coefficient indices. The shape identifiers are decoded by the shape decoder [and may employ shape motion prediction using previous shape (ps)] to generate the current shape (cs). The inter/intra frame/field DCT decoder uses the decoded DCT coefficient indices, the current shape, the quantizer information, the inter/intra decision, and the frame/field information to dequantize the indices to yield DCT coefficient blocks inside the object and then inverse transform the blocks to recover decoded pixel blocks (much as in the case of MPEG-2 video except for shape information). The uni/bidirectional frame/field motion compensator, if the coding mode is inter (based on the inter/intra decision), uses motion vectors, current shape, and frame/field information to generate motion-compensated prediction blocks (much as in the case of MPEG-2 video except for shape information) that are then added back to the corresponding decoded prediction error blocks output by the inter/intra frame/field DCT decoder to generate decoded blocks. If the coding mode is intra, no

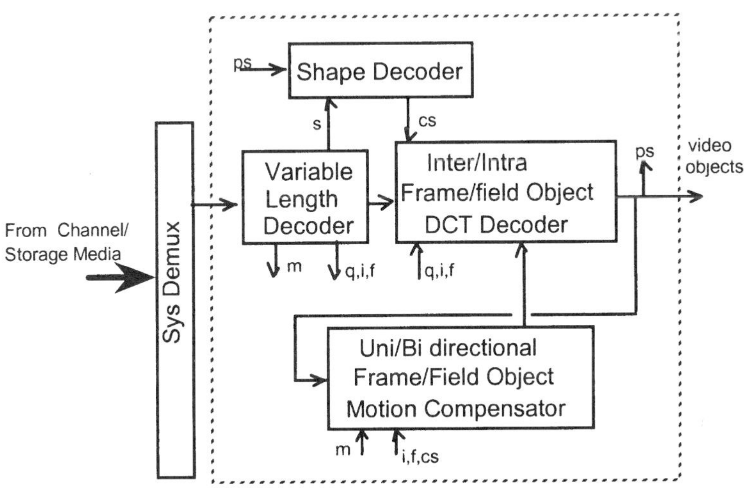

Figure 18 Systems demultiplex and the MPEG-4 video decoder.

motion-compensated prediction needs to be added to the output of the inter/intra frame/field DCT decoder. The resulting decoded VOPs are output on the line labeled video objects. Not all MPEG-4 video decoders have to be capable of decoding interlaced video objects; in fact, in a simpler scenario, not all MPEG-4 video decoders even have to be capable of decoding shape. Although we have made a gross simplification of the decoding details, conceptually, we have dealt with a complex scenario in which all major coding modes are enabled.

MPEG-4 also offers a generalized scalability framework [34,35,37,40] supporting both temporal and spatial scalabilities, the primary types of scalabilities. Scalable coding offers a means of scaling the decoder complexity if processor and/or memory resources are limited and often time varying. Further, scalability also allows graceful degradation of quality when the bandwidth resources are limited and continually changing. It even allows increased resilience to errors under noisy channel conditions. Temporally scalable encoding offers decoders a means of increasing the temporal resolution of decoded video using decoded enhancement layer VOPs in conjunction with decoded base layer VOPs. Spatial scalability encoding, on the other hand, offers decoders a means of decoding and displaying either the base layer or the enhancement layer output; typically, because the base layer uses one-quarter resolution of the enhancement layer, the enhancement layer output provides better quality, albeit requiring increased decoding complexity. The MPEG-4 generalized scalability framework employs modified B-VOPs that exist only in the enhancement layer to achieve both temporal and spatial scalability; the modified enhancement layer B-VOPs use the same syntax as normal B-VOPs but for modified semantics, which allows them to utilize a number of interlayer prediction structures needed for scalable coding.

Figure 19 shows a two-layer generalized codec structure for MPEG-4 scalability [37,38], which is very similar to the structure for MPEG-2 scalability shown in Fig. 5. The main difference is in the preprocessing stage and in the encoders and decoders allowed in the lower (base) and higher (enhancement) layers. Because MPEG-4 video supports object-based scalability, the preprocessor is modified to perform VO segmentation and generate two streams of VOPs per VO (by spatial or temporal preprocessing, depending on the scalability type). One such stream is input to the lower layer encoder, in this case, the MPEG-4 nonscalable video encoder, and the other to the higher layer encoder identified as the MPEG-4 enhancement video encoder. The role of the midprocessor is the same as in MPEG-2, either to spatially upsample the lower layer VOPs or to let them pass through, in both cases to allow prediction of the enhancement layer VOPs. The two encoded bit-

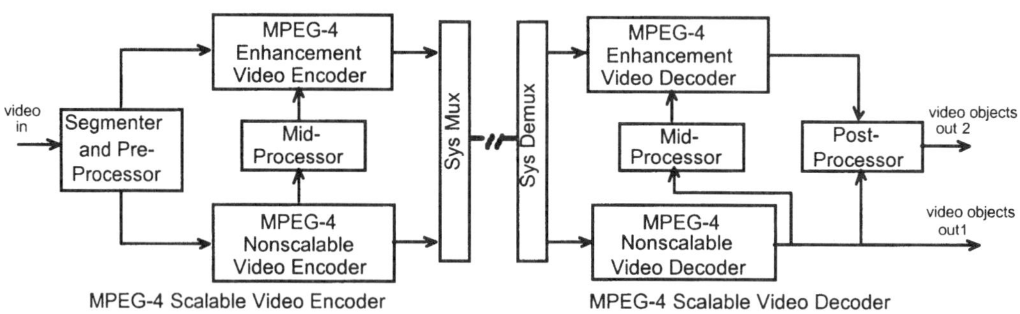

Figure 19 MPEG-4 video scalability decoder.

streams are sent to MPEG-4 Systems Mux for multiplexing. The operation of the scalability decoder is basically the inverse of that of the scalability encoder, just as in the case of MPEG-2. The decoded output of the base and enhancement layers is two streams of VOPs that are sent to the postprocessor, either to let the higher layer pass through or to be combined with the lower layer.

For simplicity, we have provided only a very high level overview of the main concepts behind MPEG-4 video.

2. Synthetic Visual

We now provide a brief overview of the tools included in the synthetic visual subpart [30,36] of MPEG-4 Visual.

Facial animation in MPEG-4 Visual is supported via the facial animation parameters (FAPs) and the facial definition parameters (FDPs), which are sets of parameters designed to allow animation of faces reproducing expressions, emotions, and speech pronunciation, as well as definition of facial shape and texture. The same set of FAPs, when applied to different facial models, results in reasonably similar expressions and speech pronunciation without the need to initialize or calibrate the model. The FDPs, on the other hand, allow the definition of a precise facial shape and texture in the setup phase. If the FDPs are used in the setup phase, it is also possible to produce the movements of particular facial features precisely. Using a phoneme-to-FAP conversion, it is possible to control facial models accepting FAPs via text-to-speech (TTS) systems; this conversion is not standardized. Because it is assumed that every decoder has a default face model with default parameters, the setup stage is necessary not to create face animation but to customize the face at the decoder. The FAP set contains two high-level parameters, *visemes* and *expressions.* A viseme is a visual correlate of a phoneme. The viseme parameter allows viseme rendering (without having to express them in terms of other parameters) and enhances the result of other parameters, ensuring the correct rendering of visemes. All the parameters involving translational movement are expressed in terms of the facial animation parameter units (FAPUs). These units are defined in order to allow interpretation of the FAPs on any facial model in a consistent way, producing reasonable results in terms of expression and speech pronunciation. The FDPs are used to customize the proprietary face model of the decoder to a particular face or to download a face model along with the information about how to animate it. The FDPs are normally transmitted once per session, followed by a stream of compressed FAPs. However, if the decoder does not receive the FDPs, the use of FAPUs ensures that it can still interpret the FAP stream. This ensures minimal operation in broadcast or teleconferencing applications. The FDP set is specified using FDP node (in MPEG-4 systems), which defines the face model to be used at the receiver.

The *mesh*-based representation of general, natural, or synthetic visual objects is useful for enabling a number of functions such as temporal rate conversion, content manipulation, animation, augmentation (overlay), and transfiguration (merging or replacing natural video with synthetic). MPEG-4 Visual includes a tool for triangular mesh–based representation of general-purpose objects. A visual object of interest, when it first appears (as a 2D VOP) in the scene, is tassellated into triangular patches, resulting in a 2D triangular mesh. The vertices of the triangular patches forming the mesh are referred to as the *node points.* The node points of the initial mesh are then tracked as the VOP moves within the scene. The 2D motion of a video object can thus be compactly represented by the motion vectors of the node points in the mesh. Motion compensation can then be achieved by texture mapping the patches from VOP to VOP according to affine transforms. Coding of video texture or still texture of object is performed by the normal texture coding tools of MPEG-4. Thus, efficient storage and transmission of the mesh representation of a moving object (dynamic mesh) require compression of its geometry and motion. The initial 2D triangular mesh is either a uniform mesh or a Delaunay mesh, and the mesh triangular topology (links between node points) is not coded; only the 2D node point coordinates are coded. A uniform mesh can be completely specified using five parameters, such as the number of nodes horizontally and the number of nodes vertically, the horizontal and the vertical dimensions of each quadrangle consisting of two triangles, and the type of splitting applied on each quadrangle to obtain triangles. For a Delaunay mesh, the node point coordinates are coded by first coding the boundary node points and then the interior node points of the mesh. By sending the total number of node points and the number of boundary node points, the decoder knows how many node points will follow and how many of those are boundary nodes; thus it is able to reconstruct the polygonal boundary and the locations of all nodes.

The *still image texture* is coded by the discrete wavelet transform (DWT); this texture is used for texture mapping or faces or objects represented by mesh. The data can represent a rectangular or an arbitrarily shaped VOP. Besides coding efficiency, an important requirement for coding texture map data is that the data should be coded in a manner facilitating continuous scalability, thus allowing many resolutions or qualities to be derived from the same coded bitstream. Although DCT-based coding is able to provide comparable coding efficiency as well as a few scalability layers, DWT-based coding offers flexibility in organization and number of scalability layers. The basic steps of a zero-tree wavelet-based coding scheme are as follows:

1. Decomposition of the texture using the discrete wavelet transform (DWT)
2. Quantization of the wavelet coefficients
3. Coding of the lowest frequency subband using a predictive scheme
4. Zero-tree scanning of the higher order subband wavelet coefficients
5. Entropy coding of the scanned quantized wavelet coefficients and the significance map

D. MPEG-4 Audio

Part 3, the Audio part [10] of MPEG-4, integrates a number of audio coding techniques. MPEG-4 Audio addresses a number of functionalities driven by applications, such as robustness against packet loss or change in transmission bit rates for Internet–phone systems, low-bit-rate coding for ''party talk,'' higher quality coding for music, improved text-to-speech (TTS) for ''storyteller,'' and object-based coding for musical orchestra synthesization. Just as video scenes are made from visual objects, audio scenes may be usefully described as the spatiotemporal combination of audio objects. An ''audio object'' is a single audio stream coded using one of the MPEG-4 Audio coding tools. Audio objects are related to each other by mixing, effects processing, switching, and delaying and may be spatialized to a particular 3D location. The effects processing is described abstractly in terms of a signal processing language (the same language used for Structured Audio), so content providers may design their own and include them in the bitstream.

More precisely, the MPEG-4 Audio version 1 specification covers the following aspects:

Low-bit-rate audio coding tools—code excited linear predictive (CELP) coding and coding based on parametric representation (PARA)
High-quality audio coding tools—time–frequency mapping techniques, AAC, and TwinVQ
Synthetic audio tools—text-to-speech (TTS) and Structured Audio

1. *Natural Audio*

Natural Audio coding [10,30] in MPEG-4 includes low-bit-rate audio coding as well as high-quality audio coding tools. Figure 20 provides a composite picture of the applications of MPEG-4 audio and speech coding, the signal bandwidth, and the type of coders used. From this figure, the following can be observed.

Sampling rates of up to 8 kHz suitable for speech coding can be handled by MPEG-4 PARA coding in the very low bit-rate range of 2 to 6 kbit/sec.
Sampling rates of 8 and 16 kHz suitable for a broader range of audio signals can be handled by MPEG-4 CELP coding in the low-bit-rate range of 6 to 24 kbit/sec.
Sampling rates starting at 8 kHz and going as high as 48 (or even 96) kHz suitable for higher quality audio can be handled by time–frequency (T/F) techniques such as optimized AAC coding in the bit rate range of 16 to 64 kbit/sec.

Figure 21 is a simplified block diagram showing integration of MPEG-4 natural audio coding tools. Only the encoding end is shown here; it consists of preprocessing, which facilitates separation of the audio signals into types of components to which a matching technique from among PARA, CELP, and T/F coding may be used. Signal analysis and control provide the bit rate assignment and quality parameters needed by the chosen coding technique.

The *PARA coder* core provides two sets of tools. The HVXC coding tools (harmonic vector excitation coding) allow coding of speech signals at 2 kbit/sec; the individual line coding tools allow coding of nonspeech signals such as music at bit rates of 4 kbit/sec

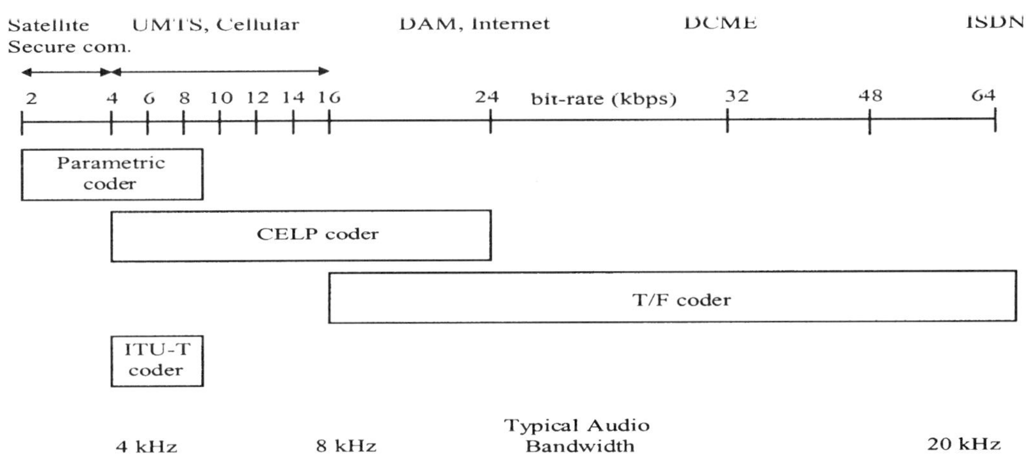

Figure 20 MPEG-4 Natural Audio coding and its applications.

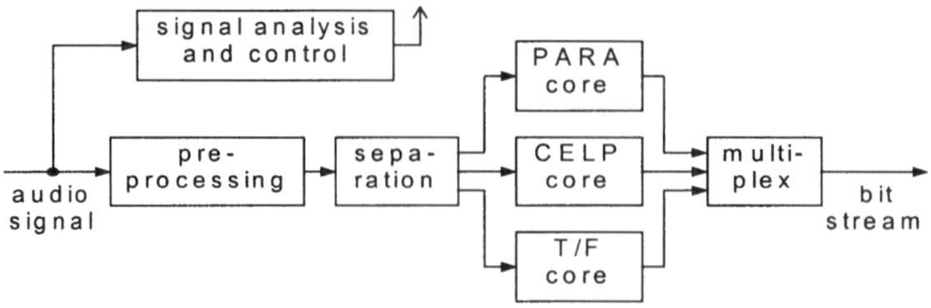

Figure 21 MPEG-4 audio encoding.

and higher. Both sets of tools allow independent change of speed and pitch during the decoding and can be combined to handle a wider range of signals and bit rates.

The *CELP coder* is designed for speech coding at two different sampling frequencies, namely 8 and 16 kHz. The speech coders using the 8-kHz sampling rate are referred to as narrowband coders and those using the 16-kHz sampling rate as wideband coders. The CELP coder includes tools offering a variety of functions including bit rate control, bit rate scalability, speed control, complexity scalability, and speech enhancement. By using the narrowband and wideband CELP coders, it is possible to span a wide range of bit rates (4 to 24 kbit/sec). Real-time bit rate control in small steps can be provided. A common structure of tools has been defined for both the narrowband and wideband coders; many tools and processes have been designed to be commonly usable for both narrowband and wideband speech coders.

The *T/F coder* provides high-end audio coding and is based on MPEG-2 AAC coding. The MPEG-2 AAC is a state-of-the-art audio compression algorithm that provides compression superior to that provided by older algorithms. AAC is a transform coder and uses a filter bank with a finer frequency resolution that enables superior signal compression. AAC also uses a number of new tools such as temporal noise shaping, backward adaptive linear prediction, joint stereo coding techniques, and Huffman coding of quantized components, each of which provides additional audio compression capability. Furthermore, AAC supports a wide range of sampling rates and bit rates, from 1 to 48 audio channels, up to 15 low-frequency enhancement channels, multilanguage capability, and up to 15 embedded data streams. The MPEG-2 AAC provides a five-channel audio coding capability while being a factor of 2 better in coding efficiency than MPEG-2 BC.

2. *Synthetic Audio*

The *TTS* conversion system synthesizes speech as its output when a text is provided as its input. In other words, when the text is provided, the TTS changes the text into a string of phonetic symbols and the corresponding basic synthetic units are retrieved from the pre-prepared database. Then the TTS concatenates the synthetic units to synthesize the output speech with rule-generated prosody. The MPEG-4 TTS not only can synthesize speech according to the input speech with a rule-generated prosody but also executes several other functions. They are as follows.

1. Speech synthesis with the original prosody from the original speech
2. Synchronized speech synthesis with facial animation (FA) tools
3. Synchronized dubbing with moving pictures not by recorded sound but by text and some lip shape information
4. Trick mode functions such as stop, resume, forward, backward without breaking

the prosody even in the applications with facial animation (FA)/motion pictures (MP)

5. Ability of users to change the replaying speed, tone, volume, speaker's sex, and age

The MPEG-4 TTS [10,30] can be used for many languages because it adopts the concept of the language code, such as the country code for an international call. At present, only 25 countries, i.e., the current ISO members, have their own code numbers, to identify that their own language has to be synthesized; the International Phonetic Alphabet (IAP) code is assigned as 0. However, 8 bits have been assigned for the language code to ensure that all countries can be assigned language code when it is requested in the future. The IPA could be used to transmit all languages. For MPEG-4 TTS, only the interface bitstream profiles are the subject of standardization. Because there are already many different types of TTS and each country has several or a few tens of different TTSs synthesizing its own language, it is impossible to standardize all the things related to TTS. However, it is believed that almost all TTSs can be modified to accept the MPEG-4 TTS interface very quickly by a TTS expert because of the rather simple structure of the MPEG-4 TTS interface bitstream profiles.

The *structured audio* coding [10,30] uses ultralow-bit-rate algorithmic sound models to code and transmit sound. MPEG-4 standardizes an algorithmic sound language and several related tools for the structured coding of audio objects. Using these tools, algorithms that represent the exact specification of a sound scene are created by the content designer, transmitted over a channel, and executed to produce sound at the terminal. Structured audio techniques in MPEG-4 allow the transmission of synthetic music and sound effects at bit rates from 0.01 to 10 kbit/sec and the concise description of parametric sound postproduction for mixing multiple streams and adding effects processing to audio scenes. MPEG-4 does not standardize a synthesis method but a signal processing language for describing synthesis methods. SAOL, pronounced "sail," stands for Structured Audio Orchestra Language and is the signal processing language enabling music synthesis and effects postproduction in MPEG-4. It falls into the music synthesis category of "Music V" languages; that is, its fundamental processing model is based on the interaction of oscillators running at various rates. However, SAOL has added many new capabilities to the Music V language model that allow more powerful and flexible synthesis description. Using this language, any current or future synthesis method may be described by a content provider and included in the bitstream. This language is entirely normative and standardized, so that every piece of synthetic music will sound exactly the same on every compliant MPEG-4 decoder, which is an improvement over the great variety of Musical Instrument Digital Interface (MIDI)-based synthesis systems.

The techniques required for automatically producing a Structured Audio bitstream from an arbitrary sound are beyond today's state of the art and are referred to as "automatic source separation" or "automatic transcription." In the meantime, content authors will use special content creation tools to create Structured Audio bitstreams directly. This is not a fundamental obstacle to the use of MPEG-4 Structured Audio, because these tools are very similar to the ones that content authors use already; all that is required is to make them capable of producing MPEG-4 output bitstreams. There is no fixed complexity that is adequate for decoding every conceivable Structured Audio bitstream. Simple synthesis methods are very low in complexity, and complex synthesis methods require more computing power and memory. As the description of the synthesis methods is under the control of the content providers, they are responsible for understanding the complexity needs of their bitstreams. Past versions of structured audio systems with similar capability have been optimized to provide multitimbral, highly polyphonic music and postproduction effects in real time on a 150-MHz Pentium computer or a simple Digital Signal Processing (DSP) chip.

V. MPEG-7

MPEG-7 is the content representation standard for multimedia information search, filtering, management, and processing [19,25,26]. The need for MPEG-7 grew because, although more and more multimedia information is available in compressed digital form, searching for multimedia information is getting increasingly difficult. A number of search engines exist on the Internet, but they do not incorporate special tools or features to search for audiovisual information, as much of the search is still aimed at textual documents. Further, each of the search engines uses proprietary, nonstandardized descriptors for search and the result of a complex search is usually unsatisfactory. A goal of MPEG-7 is to enable search for multimedia on the Internet and improve the current situation caused by proprietary solutions by standardizing an interface for description of multimedia content. MPEG-7 is intended to standardize descriptors and description schemes that may be associated with the content itself to facilitate fast and efficient search. Thus, audiovisual content with associated MPEG-7 metadata may be easily indexed and searched for. MPEG-7 aims to address not only finding content of interest in ''pull'' applications, such as that of database retrieval, but also in ''push'' applications, such as selection and filtering to extract content of interest within broadcast channels. However, MPEG-7 does not aim to standardize algorithms and techniques for extraction of features or descriptions or, for that matter, searching and filtering using these descriptions. Furthermore, it is expected that MPEG-7 will work not only with MPEG but also with non-MPEG coded content.

A number of traditional as well as upcoming application areas that employ search and retrieval in which MPEG-7 is applicable [41–43] are as follows.

Significant events—historical, political
Educational—scientific, medical, geographic
Business—real estate, financial, architectural
Entertainment and information—movie archives, news archives
Social and games—dating service, interactive games
Leisure—sport, shopping, travel
Legal—investigative, criminal, missing persons

The MPEG-7 descriptors are expected to describe various types of multimedia information. This description will be associated with the content itself, to allow fast and efficient searching for material of a user's interest. Audiovisual material that has MPEG-7 data associated with it can be indexed and searched for. This material may include still pictures, graphics, 3D models, audio, speech, video, and information about how these elements are combined in a multimedia presentation (''scenarios,'' composition information). Special cases of these general data types may include facial expressions and personal characteristics. Figure 22 shows the current understanding of the scope of MPEG-7. Although MPEG-7 does not standardize the feature extraction, the MPEG-7 description is based on the output of feature extraction, and although it does not standardize the search engine, the resulting description is consumed by the search engine.

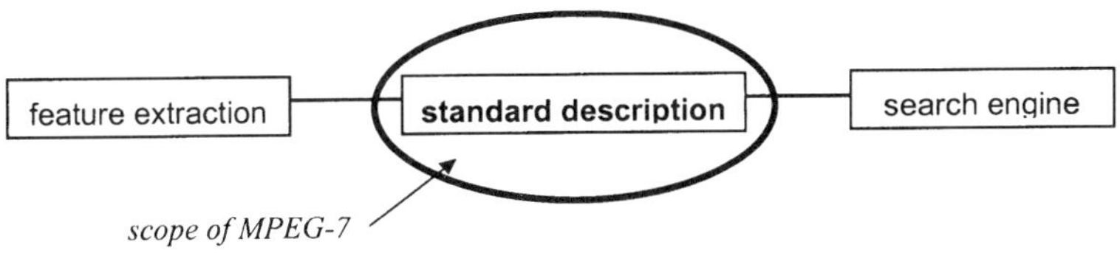

Figure 22 Scope of MPEG-7.

The words *description* and *feature* represent a rich concept that can be related to several levels of abstraction. Descriptions can vary according to the types of data, e.g., color, musical harmony, textual name, and odor. Descriptions can also vary according to the application, e.g., species, age, number of percussion instruments, information accuracy, and people with a criminal record. MPEG-7 will concentrate on standardizing a *representation* that can be used for categorization.

The detailed work plan [44] of MPEG-7 is shown in Table 2. The first working draft is expected to be complete by December 1999, and the draft international standard is expected to be complete by July 2001. The MPEG-7 standard is expected to be approved by September 2001.

We now discuss the state of MPEG-7 progress resulting from the recent evaluation [45] of the proposals according to the expected four parts [Descriptors, Description Schemes, Description Definition Language (DDL), and Systems] of the MPEG-7 standard.

A. MPEG-7 Descriptors

A descriptor (D) is a representation of a feature; i.e., the syntax and semantics of the descriptor provide a description of the feature. However, for fully representing a feature, one or more descriptors may often be needed. For example, for representing a color feature, one or more of the following descriptors may be used: the color histogram, the average of its frequency components, the motion field, and the textual description.

For descriptors, according to the outcome [45] of the evaluation process, core experiments will be needed to allow further evaluation and development of the few preselected proposals considered to be promising in the initial evaluation. Several types of descriptors, such as color, texture, motion, and shape, will be the subject of such core experiments and standardized test conditions (e.g., content, parameters, evaluation criterion) for each core experiment will be finalized. Although the core experiment framework is still in its begining phase, some progress [13] has been made regarding motion and shape descriptors.

Two core experiments on motion descriptors are being considered, the first related to the motion activity and the second related to the motion trajectory. The motion activity experiment aims to classify the intensity or pace of the action in a segment of a video

Table 2 Detailed Work Plan for MPEG-7

Call for proposals	Evaluation	Working draft	Committee draft	Draft international standard	International standard
November 1998	February 1999	December 1999	October 2000	July 2001	September 2001

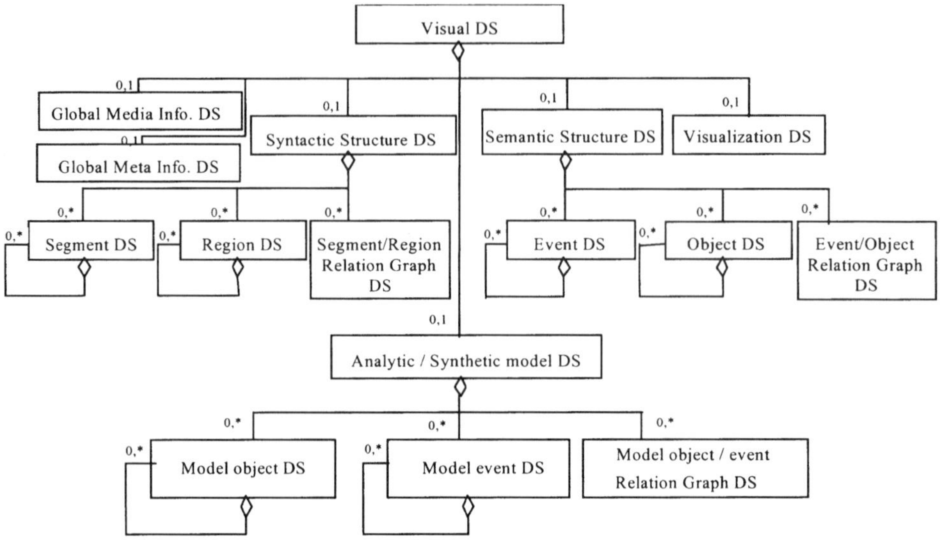

Figure 23 UML diagram of the MPEG-7 visual description scheme under development.

scene; for instance, a segment of a video scene containing a goal scored in a soccer match may be considered as highly active, whereas a segment containing the subsequent interview with the player may be considered to be of low activity. The motion trajectory experiment aims to describe efficiently the trajectory of an object during its entire life span as well as the trajectory of multiple objects in segments of a video scene.

Two core experiments on shape descriptors are also being considered, the first related to simple nonrigid shapes and the second related to complex shapes. The simple nonrigid shapes experiment expects to evaluate the performance of competing proposals based on a number of criteria such as exact matching, similarity-based retrieval, and robust retrieval of small nonrigid deformations. The complex shapes experiment expects to evaluate the performance of competing proposals based on a number of criteria such as exact matching and similarity-based retrieval.

B. MPEG-7 Description Schemes

A description scheme (DS) specifies the structure and semantics of the relationship between its components, which may be both descriptors and description schemes. Following the recommendations [45] of the MPEG-7 evaluation process, a high-level framework [14] common to all media description schemes and a specific framework on a generic visual description scheme is being designed [15]; Figure 23 shows this framework. It is composed optionally of a syntactic structure DS, a semantic structure DS, an analytic/synthetic model DS, a global media information DS, a global metainformation DS, and a visualization DS.

The syntactic structure DS describes the physical entities and their relationship in the scene and consists of zero or more occurrences of each of the segment DS, the region DS, and the DS describing the relation graph between the segments and regions. The segment DS describes the temporal relationship between segments (groups of frames) in the form of a segment tree in a scene; it consists of zero or more occurrences of the shot

DS, the media DS, and the metainformation DS. The region DS describes the spatial relationship between regions in the form of a region tree in a scene; it consists of zero or more occurrences of each of the geometry DS, the color/texture DS, the motion DS, the deformation DS, the media Information DS, and the media DS.

The semantic structure DS describes the logical entities and their relationship in the scene and consists of zero or more occurrences of an event DS, the object DS, and the DS describing the relation graph between the events and objects. The event DS contains zero or more occurrences of the events in the form of an event tree. The object DS contains zero or more occurrences of the objects in the form of an object tree.

The analytic/synthetic model DS describes cases that are neither completely syntactic nor completely semantic but rather in between. The analytic model DS specifies the conceptual correspondence such as projection or registration of the underlying model with the image or video data. The synthetic animation DS consists of the animation stream defined by the model event DS, the animation object defined by the model object DS, and the DS describing the relation graph between the animation streams and objects.

The visualization DS contains a number of view DSs to enable fast and effective browsing and visualization of the video program. The global media DS, global media information DS, and global metainformation DS correspondingly provide information about the media content, file structure, and intellectual property rights.

C. MPEG-7 DDL

The DDL is expected to be a standardized language used for defining MPEG-7 description schemes and descriptors. Many of the DDL proposals submitted for MPEG-7 evaluation were based on modifications of the extensible markup language (XML). Further, several of the proposed description schemes utilized XML for writing the descriptions. Thus, the evaluation group recommended [45] that the design of the MPEG-7 DDL be based on XML enhanced to satisfy MPEG-7 requirements. The current status of the DDL is documented in [16]; due to evolving requirements, the DDL is expected to undergo iterative refinement.

The current list of the DDL requirements is as follows.

Ability of compose a DS from multiple DSs
Platform and application independence
Unambigous grammar and the ability for easy parsing
Support for primitive data types, e.g., text, integer, real, date, time, index
Ability to describe composite data types, e.g., histograms, graphs
Ability to relate descriptions to data of multiple media types
Capability to allow partial instantiation of DS by descriptors
Capability to allow mandatory instantiation of descriptors in DS
Mechanism to identify DSs and descriptors uniquely
Support for distinct name spaces
Ability to reuse, extend, and inherit from existing DSs and descriptors
Capability to express spatial relations, temporal relations, structural relations, and conceptual relations
Ability to form links and/or references between one or several descriptions
A mechanism for intellectual property information management and protection for DSs and descriptors

D. MPEG-7 Systems

The proposals for MPEG-7 systems tools have been evaluated [45] and the work on MPEG-7 Systems is in its exploratory stages, awaiting its formal beginning. The basis of classification of MPEG-7 applications into the categories *push*, *pull*, and *hybrid* provides a clue regarding capabilities needed from MPEG-7 Systems. In push applications, besides the traditional role, MPEG-7 Systems has the main task of enabling multimedia data filtering. In pull applications, besides the traditional role, MPEG-7 Systems has the main task of enabling multimedia data browsing. A generic MPEG-7 system [46] may have to enable both multimedia data filtering and browsing. A typical model for MPEG-7 Systems is also likely to support client–server interaction, multimedia descriptions DDL parsing, multimedia data management, multimedia composition, and aspects of multimedia data presentation. This role appears to be much wider than the role of MPEG-4 Systems.

VI. PROFILING ISSUES

We now discuss the issue of profiling in the various MPEG standards. Profiling is a mechanism by which a decoder, to be compliant with a standard, has to implement only a subset and further only certain parameter combinations of the standard.

A. MPEG-1 Constraints

Although the MPEG-1 Video standard allows the use of fairly large picture sizes, high frame rates, and correspondingly high bit rates, it does not necessitate that every MPEG-1 Video decoder support these parameters. In fact, to keep decoder complexity reasonable while ensuring interoperability, an MPEG-1 Video decoder need only conform to a set of constrained parameters that specify the largest horizontal horizontal size (720 pels/line), the largest vertical size (576 lines/frame), the maximum number of macroblocks per picture (396), the maximum number of macroblocks per second (396×25), the highest picture rate (30 frames/sec), the maximum bit rate (1.86 Mbit/sec), and the largest decoder buffer size (376,832 bits).

B. MPEG-2 Profiling

The MPEG-2 Video standard extends the concept of constrained parameters by allowing a number of valid subsets of the standard organized into profiles and levels. A profile is a defined subset of the entire bitstream syntax of a standard. A level of a profile specifies the constraints on the allowable values for parameters in the bitstream.

The Main profile, as its name suggests, is unquestionably the most important profile of MPEG-2 Video. It supports nonscalable video syntax (ITU-R 4:2:0 format and I-, P-, and B-pictures). Further, it consists of four levels—low, main, high-1440, and high. Again, as the name suggests, the main level refers to TV resolution and high-1440 and high levels refer to two resolutions for HDTV. The low level refers to MPEG-1-constrained parameters.

A Simple profile is a simplified version of the Main profile that allows cheaper implementation because of lack of support for B-pictures. It does not support scalability. Further, it only supports one level, the main level.

Scalability in MPEG-2 Video is supported in the SNR, Spatial, and High profiles. The SNR profile supports SNR scalability; it includes only the low and main levels. The Spatial profile allows both SNR and Spatial scalability; it includes only the high-1440 level. Although the SNR and Spatial profiles support only the 4:2:0 picture format, the High profile supports the 4:2:2 picture format as well. The High profile also supports both SNR and Spatial scalability; it includes the main, high-1440, and high levels.

MPEG-2 Video also includes two other profiles, the 4:2:2 profile and the Multiview profile.

MPEG-2 NBC Audio consists of three profiles—the Main Profile, the Low Complexity Profile, and the Scalable Simple Profile.

C. MPEG-4 Profiling

MPEG-4 Visual profiles are defined in terms of visual object types. There are six video profiles, the Simple Profile, the Simple Scalable Profile, the Core Profile, the Main Profile, the N-Bit Profile, and the Scalable Texture Profile. There are two synthetic visual profiles, the Basic Animated Texture Profile and the Simple Facial Animation Profile. There is one hybrid profile, the Hybrid Profile, that combines video object types with synthetic visual object types.

MPEG-4 Audio consists of four profiles—the Main Profile, the Scalable Profile, the Speech Profile, and the Synthetic Profile. The Main Profile and the Scalable Profile consist of four levels. The Speech Profile consists of two levels and the Synthetic Profile consists of three levels.

MPEG-4 Systems also specifies a number of profiles. There are three types of profiles—the Object Descriptor (OD) Profile, the Scene Graph Profiles, and the Graphics Profiles. There is only one OD Profile, called the Core Profile. There are four Scene Graph Profiles—the Audio Profile, the Simple2D Profile, the Complete2D Profile, and the Complete Profile. There are three Graphics Profiles—the Simple2D Profile, the Complete2D Profile, and the Complete Profile.

VII. SUMMARY

In this chapter we have introduced the various MPEG standards. In Sec. II, we briefly discussed the MPEG 1 standard. In Sec. III, we presented a brief overview of the MPEG-2 standard. Section IV introduced the MPEG 4 standard. In Sec. V, the ongoing work toward the MPEG-7 standard was presented. In Sec. VI, we presented a brief overview of the profiling issues in the MPEG standards.

REFERENCES

1. ITU-T. ITU-T Recommendation H.261—Video codec for audiovisual services at $p \times 64$ kbit/s, December 1990.
2. MPEG{-1} Systems Group. Information Technology—Coding of Moving Pictures and Associated Audio for Digital Storage Media up to About 1.5 Mbit/s: Part 1—Systems. ISO/IEC 11172-1, International Standard, 1993.
3. MPEG{-1} Video Group. Information Technology—Coding of Moving Pictures and Associ-

ated Audio for Digital Storage Media up to About 1.5 Mbit/s: Part 2—Video. ISO/IEC 11172-2, International Standard, 1993.

4. MPEG{-1} Audio Group. Information Technology—Coding of Moving Pictures and Associated Audio for Digital Storage Media up to About 1.5 Mbit/s: Part 3—Audio. ISO/IEC 11172-3, International Standard, 1993.
5. MPEG{-2} Systems Group. Information Technology—Generic Coding of Moving Pictures and Associated Audio: Part 1—Systems. ISO/IEC 13818-1, International Standard, 1995.
6. MPEG{-2} Video Group. Information Technology—Generic Coding of Moving Pictures and Associated Audio: Part 2—Video. ISO/IEC 13818-2, International Standard, 1995.
7. MPEG{-2} Audio Group. Information Technology—Generic Coding of Moving Pictures and Associated Audio: Part 3—Audio. ISO/IEC 13818-3, International Standard, 1995.
8. MPEG{-4} Systems Group. Generic Coding of Audio-Visual Objects: Part 1—Systems. ISO/IEC JTC1/SC29/WG11 N2501, FDIS of ISO/IEC 14496-1, November 1998.
9. MPEG{-4} Video Group. Generic Coding of Audio-Visual Objects: Part 2—Visual. ISO/IEC JTC1/SC29/WG11 N2502, FDIS of ISO/IEC 14496-2, November 1998.
10. MPEG{-4} Audio Group. Generic Coding of Audio-Visual Objects: Part 3—Audio. ISO/IEC JTC1/SC29/WG11 N2503, FDIS of ISO/IEC 14496-3, November 1998.
11. MPEG{-4} DMIF Group. Generic Coding of Audio-Visual Objects: Part 6—DMIF. ISO/IEC JTC1/SC29/WG11 N2506, FDIS of ISO/IEC 14496-6, November 1998.
12. MPEG{-4} Systems Group. Text for ISO/IEC 14496-1/PDAM1. ISO/IEC JTC1/SC29/WG11 N2739, March 1999.
13. MPEG{-7} Video Group. Description of Core Experiments for MPEG-7 Motion/Shape. ISO/IEC JTC1/SC29/WG11 N2690, Seoul, March 1999.
14. MPEG{-7} Requirements Group. MPEG-7 Description Schemes Version 0.01. ISO/IEC JTC1/SC29/WG11 N2732, Seoul, March 1999.
15. MPEG{-7} Video Group. Generic Visual Description Scheme for MPEG-7. ISO/IEC JTC1/SC29/WG11 N2694, Seoul, March 1999.
16. MPEG{-7} Requirements Group. MPEG-7 DDL Development and DDL Version 0.01 Specification. ISO/IEC JTC1/SC29/WG11 N2731, Seoul, March 1999.
17. MPEG{-7} Implementation Studies Group. MPEG-7 XM Software Architecture Version 1.0. ISO/IEC JTC1/SC29/WG11 N2716, Seoul, March 1999.
18. BG Haskell, A Puri, AN Netravali. Digital Video: An Introduction to MPEG-2, New York: Chapman & Hall, 1997.
19. *http://drogo.cselt.stet.it/mpeg*, The MPEG home page, March 1999.
20. MPEG-1 Video Simulation Model Editing Committee. MPEG-1 Video Simulation Model 3. ISO/IEC JTC1/SC29/WG11 Document XXX, July 1990.
21. A Puri. Video coding using the MPEG-1 compression standard. Proceedings of International Symposium of Society for Information Display, Boston, May 1992, pp 123–126.
22. MPEG-2 Video Test Model Editing Committee. MPEG-2 Video Test Model 5. ISO/IEC JTC1/SC29/WG11 N0400, April 1993.
23. A Puri. Video coding using the MPEG-2 compression standard. Proceedings SPIE Visual Communications and Image Processing SPIE 1199:1701–1713, 1993.
24. RL Schmidt, A Puri, BG Haskell. Performance evaluation of nonscalable MPEG-2 video coding. Proceedings of SPIE Visual Communications and Image Processing, Chicago, September 1994, pp 296–310.
25. MPEG subgroups: Requirements, audio, delivery, SNHC, systems, video and test. In: R Koenen, ed. Overview of the MPEG-4 Standard. ISO/IEC JTC1/SC29/WG11 N2459, Atlantic City, October 1998.
26. BG Haskell, P Howard, YA LeCun, A Puri, J Ostermann, MR Civanlar, L. Rabiner, L Bottou, P Hafner. Image and video coding—Emerging Standards and Beyond. IEEE Trans Circuits Syst Video Technol 8:814–837, 1998.

27. MPEG{-4} Requirements Group. MPEG-4 Applications Document. ISO/IEC JTC1/SC29/WG11 N2563, Rome, December 1998.
28. MPEG{-4} Requirements Group. MPEG-4 Requirements Document Version 10. ISO/IEC JTC1/SC29/WG11 N2456, Rome, December 1998.
29. MPEG{-4} Systems Group. MPEG-4 Systems Version 2 Verification Model 6.0. ISO/IEC JTC1/SC29/WG11 N2741, March 1999.
30. A Puri, A Eleftheriadis. MPEG-4: An object-based multimedia coding standard supporting mobile applications. ACM J Mobile Networks Appl 3:5–32, 1998.
31. A Puri, RL Schmidt, BG Haskell. Scene description, composition, and playback systems for MPEG-4. Proceedings of EI/SPIE Visual Communications and Image Processing, January 1999.
32. ITU-T Experts Group on Very Low Bitrate Visual Telephony. ITU-T Recommendation H.263: Video Coding for Low Bitrate Communication, December 1995.
33. FI Parke, K Waters. Computer Facial Animation. AK Peters, 1996.
34. T Sikora, L Chiariglione. The MPEG-4 video standard and its potential for future multimedia applications. Proceedings IEEE ISCAS Conference, Hong Kong, June 1997.
35. A Puri, RL Schmidt, BG Haskell. Performance evaluation of the MPEG-4 visual coding standard. Proceedings Visual Communications and Image Processing, San Jose, January 1998.
36. J Osterman, A Puri. Natural and synthetic video in MPEG-4. Proceedings IEEE ICASSP, Seattle, April 1998.
37. MPEG-4 Video Verification Model Editing Committee. The MPEG-4 Video Verification Model 8.0 ISO/IEC JTC1/SC29/WG11 N1796, Stockholm, July 1997.
38. MPEG{-4} Video Group. MPEG-4 Video Verification Model Version 9.0. ISO/IEC JTC1/SC29/WG11 N1869, October 1997.
39. MPEG{-4} Video Group. MPEG-4 Video Verification Model Version 12.0. ISO/IEC JTC1/SC29/WG11 N2552, Rome, December 1998.
40. A Puri, RL Schmidt, BG Haskell. Improvements in DCT based video coding. Proceedings SPIE Visual Communications and Image Processing, San Jose, January 1997.
41. MPEG{-7} Requirements Group. MPEG-7 Applications Document v0.8. ISO/IEC JTC1/SC29/WG11 N2728, Seoul, March 1999.
42. MPEG{-7} Requirements Group. MPEG-7 Requirements Document v0.8. ISO/IEC JTC1/SC29/WG11 N2727, Seoul, March 1999.
43. MPEG{-7} Requirements Group. MPEG-7 Context, Objectives and Technical Roadmap. ISO/IEC JTC1/SC29/WG11 N2727, Seoul, March 1999.
44. MPEG{-7} Requirements Group. MPEG-7 Proposal Package Description (PPD). ISO/IEC JTC1/SC29/WG11 N2464, October 1998.
45. MPEG{-7} Requirements Group. Report of the Ad Hoc Group on MPEG-7 Evaluation Logistics. ISO/IEC JTC1/SC29/WG11 MPEG99/4582, Seoul, March 1999.
46. Q Huang, A Puri. Input for MPEG-7 systems work. ISO/IEC JTC1/SC29/WG11 MPEG99/4546, Seoul, March 1999.
47. MPEG{-2} Audio Group. Information Technology—Generic Coding of Moving Pictures and Associated Audio: Part 7—Advanced Audio Coding (AAC). ISO/IEC 13818-7, International Standard, 1997.
48. BG Haskell, A Puri, AN Netravali. Digital Video: An Introduction to MPEG-2. Chapman & Hall. New York, 1997.

5

Review of MPEG–4 General Audio Coding

James D. Johnston and Schuyler R. Quackenbush
AT&T Labs, Florham Park, New Jersey

Jürgen Herre and Bernhard Grill
Fraunhofer Geselshaft IIS, Erlangen, Germany

I. INTRODUCTION

Inside the MPEG–4 standard, there are various ways to encode or describe audio. They can be grouped into two basic kinds of audio services, called general audio coding and synthetic audio. General audio coding is concerned with taking pulse code–modulated (PCM) audio streams and efficiently encoding them for transmission and storage; synthetic audio involves the synthesis, creation, and parametric description of audio signals. In this chapter, we will discuss general audio coders.

A general audio coder operates in the environment of Figure 1. The encoder's input is a PCM stream. The encoder creates a bitstream that can either be decoded by itself or inserted into an MPEG–4 systems layer. The resulting bitstream can be either stored or transmitted. At the decoder end, the bitstream, after being demultiplexed from any system layer, is converted back into a PCM stream representing the audio stream at the input.

A. Kinds of Coders

Historically, most audio coders (including speech coders) have attempted to extract redundancy in order to avoid transmitting bits that do not convey information. Some examples of these coders, in quasi-historical order, are LPC (linear predictive coding) [1], DPCM (differential pulse-code modulation) [1], ADPCM (adaptive differential pulse-code modulation) [1], subband coding [2], transform coding [1], and CELP (codebook excited linear prediction) [3]. All of these coders use a model of the source in one fashion or another in order to reduce the bit rate and are called, accordingly, *source coders*. These source coders can, in general, be either lossless (i.e., they return exactly the same signal that was input to the coder) or lossy, but they are in general lossy.

1. Source Coding

Source coding is a very good way to reduce bit rate if (1) the material being coded has a good source model and (2) the source redundancy allows sufficient coding gain to pro-

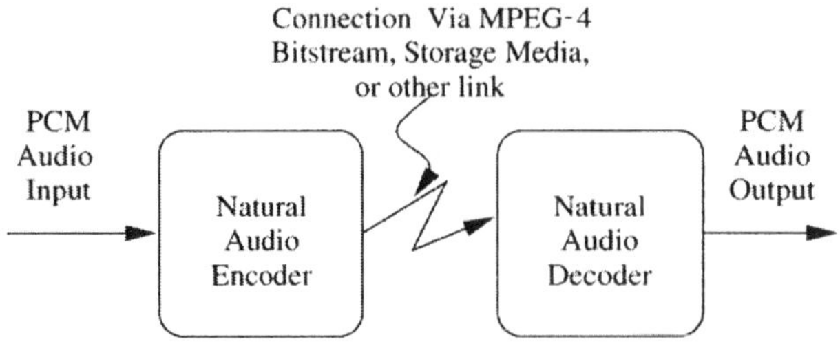

Figure 1 The environment of a general audio coder.

vide the required compression ratio. However, in the case of music, source models can be essentially arbitrary and can change abruptly (unlike speech, for which the model changes are limited by the physics of the vocal tract). In addition, the compression ratios required for efficient transmission may require more gain than a pure source coder can manage.

2. Perceptual Coding

As channel rates become lower and lower and source models no longer provide sufficient gain or become unstable, something more than source coding becomes necessary. In an audio signal, many parts of the signal are not actually audible [4–8]. They are ''masked'' by other parts of the signal or are below the absolute threshold of hearing. There is no need to send the parts of the signal that are inaudible if the signal will not be processed substantially after decoding. This is the principal means of compression in the perceptual coder. Unlike a source coder, the perceptual coder is a kind of destination coder, where the destination (i.e., human auditory system) is considered, and parts of the signal that are irrelevant are discarded. This active removal of irrelevance is the defining characteristic of the perceptual coder.

B. MPEG General Audio Coding

There are various audio coding methods in MPEG, from pure source coding techniques such as CELP to sophisticated perceptual algorithms such as advanced audio coding (AAC). The encoding techniques follow three basic block diagrams, shown in Figures 2, 3, 4, corresponding to a general LPC, a subband coder, and a perceptual coder, respectively.

In Figure 2, we show a generalized block diagram of an LPC encoder. There are many kinds of LPC encoders, hence the block diagram has a number of required and optional parts. Bear in mind that a CELP coder is a kind of LPC encoder, using a vector quantization strategy. In the diagram, the three blocks outlined by solid lines are the required blocks, i.e., a difference calculator, a quantizer, and a predictor. These three blocks constitute the basic LPC encoder. The heart of the LPC coder is the predictor, which uses the history of the encoded signal in order to build an estimate of what the next sample is. This predicted signal is subtracted from the next sample; the result, called the error signal, is then quantized; and the quantized value is stored or transmitted. The decoder is quite simple; it simply decodes the error signal, adds it to the predictor output, and puts that signal back into the predictor. If noise shaping is present, inverse noise shaping would also be applied before the signal is output from the decoder. The gain in this coder comes

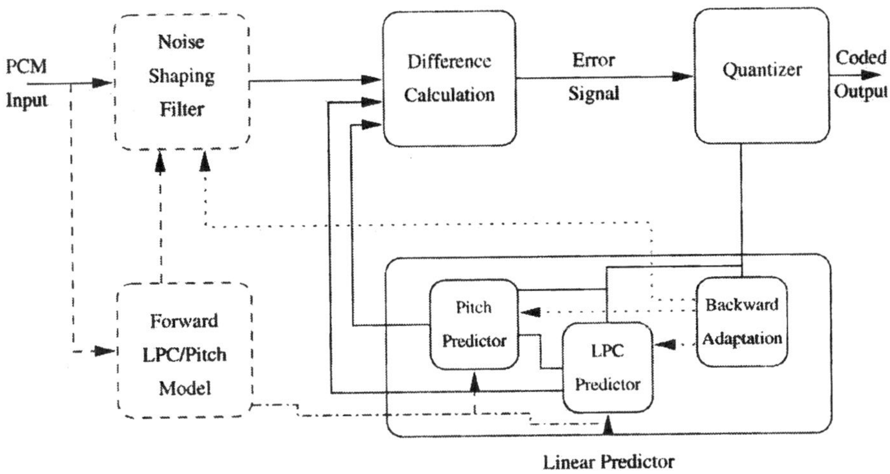

Figure 2 A general block diagram of an LPC coder.

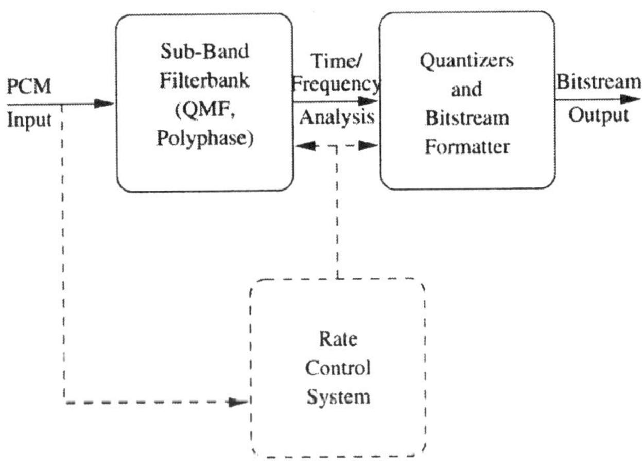

Figure 3 A general block diagram of a subband coder.

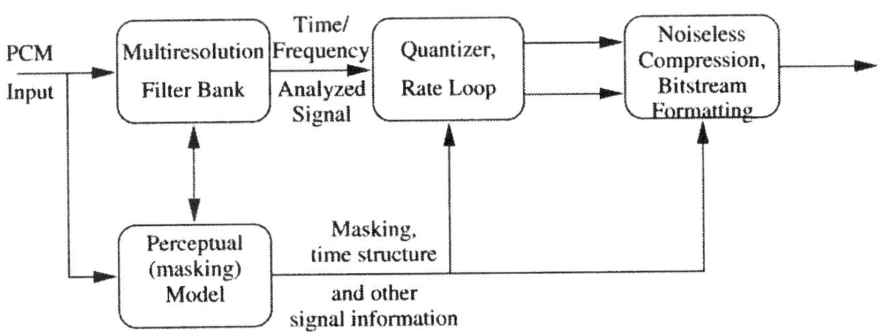

Figure 4 A general block diagram of a perceptual coder.

from the reduction in energy of the error signal, which results in a decrease in total quantization noise. Although the pitch predictor is not shown as optional, some LPC coders omit it completely or incorporate it in the LPC predictor.

The optional blocks are shown with dashed lines. These are the noise shaping filter, which can be fixed (like a fixed predictor) or adapted in a signal-dependent fashion, and the two adaptation blocks, corresponding to forward and backward adaptation. The control signals resulting from forward adaptation are shown using a dot-dashed line and those from the backward adaptation with a dotted line.

Most modern LPC coders are adaptive. As shown in the block diagram, there are two kinds of adaptation, backward and forward. The difference is that forward adaptation works on the input signal and must transmit information forward to the decoder for the decoder to recover the signal, whereas backward adaptation works on information available to the decoder and no information need be expressly transmitted. Unfortunately, the comparison is not that simple, as the response time of a backward-adaptive system must lag the signal because it can only look at the transmitted signal, whereas the forward-adaptive system can use delay to look ahead to track the signal, adapt quickly, and adapt without having to compensate for the presence of coding noise in the signal that drives the LPC adaptation. In either case, the purpose of the adaptation is to make the LPC model better fit the signal and thereby reduce the noise energy in the system.

In a subband coder, the obligatory blocks are the filter bank and a set of quantizers that operate on the outputs of the filter bank. Its block diagram is shown in Figure 3. A subband coder separates different frequencies into different bands by using a filter bank and then quantizes sets of bands differently. As the filter bank maintains the same total signal energy but divides it into multiple bands, each of which must have less than the total energy, the quantizers required in each of the frequency bands will have fewer steps and therefore will require fewer bits.

In most subband coders, the quantization is signal dependent, and a rate control system examines the signal (or analyzed signal) in order to control the quantizers. This sort of system, which was originally based on the idea of rate distortion theory [1], is also the beginning of a perceptual system, because the quantizing noise is limited to each band and no longer spreads across the entire signal bandwidth. This is modestly different from the noise shaping involved in the LPC coder, because in that coder the noise spreads across the whole band but can be shaped. In perceptual terms, the two are very similar; however, mechanisms for controlling noise in the case of subband coders are well evolved, whereas mechanisms for noise shaping in LPC, in other than speech applications, are currently rudimentary.

Finally, we come to the perceptual coder. Here, we use the term perceptual coder to refer to a coding system that calculates an express measure of a masking or just noticeable difference (JND) curve and then uses that measure to control noise injection. Figure 4 shows the high-level block diagram for a standard perceptual coder. There are four parts to the perceptual coder: a filter bank, a perceptual model, the quantizer and rate loop, and noiseless compressor–bitstream formatter.

The filter bank has the same function as the filter bank in the subband coder; i.e., it breaks the signal up into a time–frequency tiling. In the perceptual coder, the filter bank is most often switched; i.e., it has two or more different time–frequency tilings and a way to switch seamlessly between them. In MPEG-4, the filter banks are all modified discrete cosine transform (MDCT) [9] based.

The perceptual model is the heart of the perceptual coder. It takes the input signal, sometimes the filtered signal, and other information from the rate loop and the coder setup and creates either a masking threshold or a set of signal-to-noise ratios that must be met during coding. There are many versions of such a model, such as in Brandenburg and Stoll [10], as well as many variations on how to carry out the process of modeling; however, they all do something that attempts to partition frequency, and sometimes time, into something approximating the frequency, and sometimes time, resolution of the cochlea.

The rate loop takes the filter bank output and the perceptual model, arranges the quantizers so that the bit rate is met, and also attempts to satisfy the perceptual criteria. If the perceptual criteria are left unmet, the rate loop attempts to do so in a perceptually inobtrusive fashion. In most such coders, a rate loop is an iterative mechanism that attempts some kind of optimization, usually heuristic, of rate versus quality.

Finally, because the quantizers required for addressing perceptual constraints are not particularly good in the information theoretic sense, a back-end coding using entropy-coding methods is usually present in order to pack the quantizer outputs efficiently into the bitstream. This bitstream formatter may also add information for synchronization, external data, and other functions.

II. MPEG–2 AAC—ADVANCED AUDIO CODING

The ISO/IEC MPEG–2 Advanced Audio Coding (AAC) technology [11,12] delivers unsurpassed audio quality at rates at or below 64 kbps/channel. Because of its high performance and because it is the most recent of the MPEG–2 audio coding standards (effectively being developed in parallel with the MPEG–4 standard), it was incorporated directly into the MPEG–4 General Audio Standard. It has a very flexible bitstream syntax that supports multiple audio channels, subwoofer channels, embedded data channels, and multiple programs consisting of multiple audio, subwoofer, and embedded data channels. Strike AAC combines the coding efficiencies of a high-resolution filter bank, backward-adaptive prediction, joint channel coding, and Huffman coding with a flexible coding architecture to permit application-specific functionality while still delivering excellent signal compression.

AAC supports a wide range of sampling frequencies (from 8 to 96 kHz) and a wide range of bit rates. This permits it to support applications ranging from professional or home theater sound systems through Internet music broadcast systems to low (speech) rate speech and music preview systems.

A block diagram of the AAC encoder is shown in Figure 5. The blocks are as follows:

Filter bank: AAC uses a resolution-switching filter bank that can switch between a high-frequency-resolution mode of 1024 (for maximum statistical gain during intervals of signal stationarity) and a high-time-resolution mode of 128 bands (for maximum time-domain coding error control during intervals of signal nonstationarity).

TNS: The Temporal Noise Shaping (TNS) tool modifies the filter bank characteristics so that the combination of the two tools is better able to adapt to the time–frequency characteristics of the input signal [13].

Perceptual model: A model of the human auditory system that sets the quantization noise levels based on the loudness characteristics of the input signal.

Intensity and coupling, mid/side (M/S): These two blocks actually comprise three

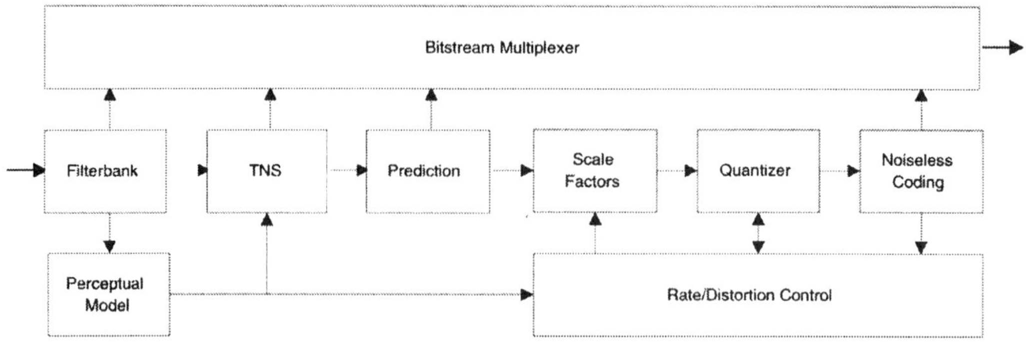

Figure 5 AAC encoder block diagram.

tools, all of which seek to protect the stereo or multichannel signal from noise imaging while achieving coding gain based on correlation between two or more channels of the input signal [14–16].

Prediction: A backward adaptive recursive prediction that removes additional redundancy from individual filter bank outputs [17].

Scale factors: Scale factors set the effective step sizes for the nonuniform quantizers.

Quantization, noiseless coding: These two tools work together. The first quantizes the spectral components and the second applies Huffman coding to vectors of quantized coefficients in order to extract additional redundancy from the nonuniform probability of the quantizer output levels. In any perceptual encoder, it is very difficult to control the noise level accurately while at the same time achieving an "optimum" quantizer. It is, however, quite efficient to allow the quantizer to operate unconstrained and then to remove the redundancy in the probability density function (PDF) of the quantizer outputs through the use of entropy coding.

Rate–distortion control: This tool adjusts the scale factors such that more (or less) noise is permitted in the quantized representation of the signal, which, in turn, requires fewer (or more) bits. Using this mechanism, the rate–distortion control tool can adjust the number of bits used to code each audio frame and hence adjust the overall bit rate of the coder.

Bitstream multiplexer: The multiplexer assembles the various tokens to form a bitstream.

This section will discuss the blocks that contribute the most to AAC performance: the filter bank, the perceptual model, and noiseless coding.

A. Analysis–Synthesis Filter Bank

The most significant aspect of the AAC filter bank is that it has high frequency resolution (1024 frequency coefficients) so that it is able to extract the maximum signal redundancy (i.e., provide maximum prediction gain) for stationary signals [1]. High frequency resolution also permits the encoder's psychoacoustic model to separate signal components that differ in frequency by more than one critical band and hence extract the maximum signal irrelevance.

The AAC analysis–synthesis filter bank has three other characteristics that are commonly employed in audio coding: critical sampling, overlap–add synthesis, and perfect reconstruction. In a critically sampled filter bank, the number of time samples input to the analysis filter per second equals the number of frequency coefficients generated per second. This minimizes the number of frequency coefficients that must be quantized and transmitted in the bitstream. Overlap-add reconstruction reduces artifacts caused by block-to-block variations in signal quantization. Perfect reconstruction implies that in the absence of frequency coefficient quantization, the synthesis filter output will be identical, within numerical error, to the analysis filter input.

When transient signals must be coded, the high-resolution or ''long-block'' filter bank is not an advantage. For this reason, the AAC filter bank can switch from high-frequency-resolution mode to high-time-resolution mode. The latter mode, or ''short-block'' mode, permits the coder to control the anticausal spread of coding noise [18]. The top panel in Figure 6 shows the window sequence for the high-frequency-resolution mode, and the bottom panel shows the window sequence for a transition from long-block to short-block and back to long-block mode.

The filter bank adopted for use in AAC is a modulated, overlapped filter bank called the modified discrete cosine transform (MDCT) [9]. The input sequence is windowed (as shown in Fig. 6) and the MDCT computed. Because it is a critically sampled filter bank, advancing the window to cover 1024 new time samples produces 1024 filter bank output samples. On synthesis, 1024 filter bank input samples produce 2048 output time samples,

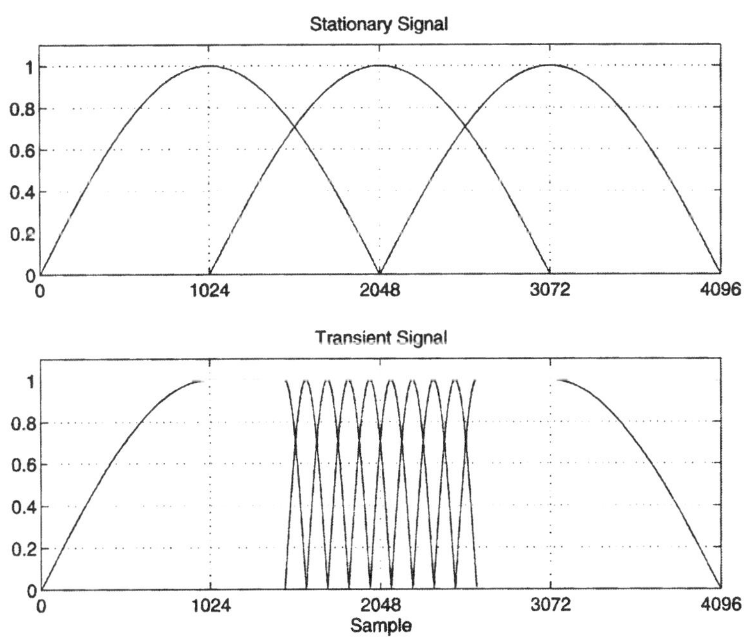

Figure 6 Window sequence during stationary and transient signal conditions.

which are then overlapped 50% with the previous filter bank result and added to form the output block.

B. Perceptual Model

The perceptual model estimates the threshold of masking, which is the level of noise that is subjectively just noticeable given the current input signal. Because models of auditory masking are primarily based on frequency domain measurements [7,19], these calculations are typically based on the short-term power spectrum of the input signal, and threshold values are adapted to the time–frequency resolution of the filter bank outputs. The threshold of masking is calculated relative to each frequency coefficient for each audio channel for each frame of input signal, so that it is signal dependent in both time and frequency. When the high-time-resolution filter bank is used, it is calculated for the spectra associated with each of the sequence of eight windows used in the time–frequency analysis. In intervals in which pre-echo distortion is likely, more than one frame of signal is considered such that the threshold in frames just prior to a nonstationary event is depressed to ensure that leakage of coding noise is minimized. Within a single frame, calculations are done with a granularity of approximately 1/3 Bark, following the critical band model in psychoacoustics. The model calculations are similar to those in psychoacoustic model II in the MPEG-1 audio standard [10].

The following steps are used to calculate the monophonic masking threshold of an input signal:

Calculate the power spectrum of the signal in 1/3 critical band partitions.
Calculate the tonelike or noiselike nature of the signal in those partitions, called the tonality measure.
Calculate the spread of masking energy, based on the tonality measure and the power spectrum.
Calculate time domain effects on the masking energy in each partition.
Relate the masking energy to the filter bank outputs.

Once the masking threshold is known, it is used to set the scale factor values in each scale factor band such that the resulting quantizer noise power in each band is below the masking threshold in that band. When coding audio channel pairs that have a stereo presentation, binaural masking level depression must be considered [14].

C. Quantization and Noiseless Coding

The spectral coefficients are coded using one quantizer per scale factor band, which is a fixed division of the spectrum. For high-resolution blocks there are 49 scale factor bands, which are approximately 1/2 Bark in width. The psychoacoustic model specifies the quantizer step size (inverse of scale factor) per scale factor band. An AAC encoder is an instantaneously variable rate coder, but if the coded audio is to be transmitted over a constant rate channel, then the rate–distortion module adjusts the step sizes and number of quantization levels so that a constant rate is achieved.

1. Quantization

AAC uses a nonlinear quantizer for spectral component x_i to produce $\hat{x}_i$

$$\hat{x}_i = \text{sign}(x_i) n \text{ int}\left(\left(\frac{|x_i|}{\sqrt[4]{2^{\text{stepsize}}}}\right)^{3/4}\right) \tag{1}$$

The main advantage of the nonlinear quantizer is that it shapes the noise as a function of the amplitude of the coefficients, such that the increase of the signal-to-noise ratio with increasing signal energy is much lower than that of a linear quantizer. The exponent stepsize is the quantized step size in a given scale factor band. The first scale factor is PCM coded, and subsequent ones are Huffman coded differential values.

2. Rate–Distortion Control

The quantized coefficients are Huffman coded. A highly flexible coding method allows several Huffman tables to be used for one spectrum. Two- and four-dimensional tables with and without sign are available. The noiseless coding process is described in detail in Quackenbush and Johnston [20]. To calculate the number of bits needed to code a spectrum of quantized data, the coding process has to be performed and the number of bits needed for the spectral data and the side information has to be accumulated.

3. Noiseless Coding

The input to the noiseless coding is the set of 1024 quantized spectral coefficients and their associated scale factors. If the high-time-resolution filter bank is selected, then the 1024 coefficients are actually a matrix of 8 by 128 coefficients representing the time–frequency evolution of the signal over the duration of the eight short-time spectra.

In AAC an extended Huffman code is used to represent n-tuples of quantized coefficients, with the Huffman code words drawn from one of 11 codebooks. The maximum absolute value of the quantized coefficients that can be represented by each Huffman codebook and the number of coefficients in each n-tuple for each codebook are shown in Table 1. There are two codebooks for each maximum absolute value, with each represent-

Table 1 Human Codebooks

Codebook index	Tuple size	Maximum absolute value	Signed values
0		0	
1	4	1	Yes
2	4	1	Yes
3	4	2	No
4	4	2	No
5	2	4	Yes
6	2	4	Yes
7	2	7	No
8	2	7	No
9	2	12	No
10	2	12	No
11	2	16 (ESC)	No

ing a distinct probability distribution function. Codebooks can represent signed or unsigned values, and for the latter the sign bit of each nonzero coefficient is appended to the codeword.

Two codebooks require special note: codebook 0 and codebook 11. Codebook 0 indicates that all coefficients within a section are zero, requiring no transmission of the spectral values and scale factors. Codebook 11 can represent quantized coefficients that have an absolute value greater than or equal to 16 by means of an escape (ESC) Human code word and an escape code that follows it in the bitstream.

III. MPEG–4 ADDITIONS TO AAC

An important aspect of the overall MPEG–4 Audio [21] functionality is covered by the so-called General Audio (GA) part, i.e., the coding of arbitrary audio signals. MPEG–4 General Audio coding is built around the coder kernel provided by MPEG–2 Advanced Audio Coding (AAC) [22], which is extended by additional coding tools and coder configurations. The perceptual noise substitution (PNS) tool and the Long-Term Prediction (LTP) tool are available to enhance the coding performance for the noiselike and very tonal signals, respectively. A special coder kernel (Twin VQ) is provided to cover extremely low bit rates. Together with a few additional tools and the MPEG–4 narrowband CELP coder, a flexible bit rate scalable coding system is defined including a variety of possible coder configurations. The following sections will describe these features in more detail.

A. Perceptual Noise Substitution

Generic audio coding predominantly employs methods for waveform coding, but other coding methods are conceivable that aim not at preserving the waveform of the input signal but at reproducing a perceptually equivalent output signal at the decoder end. In fact, relaxing the requirement for waveform preservation may enable significant savings in bit rate when parts of the signal are reconstructed from a compact parametric representation of signal features. The PNS tool [23] allows a very compact representation of noiselike signal components and in this way further increases compression efficiency for certain types of input signals.

The PNS technique is based on the observation that the perception of noiselike signals is similar regardless of the actual waveform of the stimulus provided that both the spectral envelope and the temporal fine structure of the stimuli are similar.

The PNS tool exploits this phenomenon in the context of the MPEG–4 perceptual audio coder within a coder framework based on analysis–synthesis filter banks. A similar system was proposed and investigated by Schulz [24,25]. The concept of the PNS technique can be described as follows (see Fig. 7):

In the encoder, noiselike components of the input signal are detected on a scale factor band basis.

The groups of spectral coefficients belonging to scale factor bands containing noiselike signal components are not quantized and coded as usual but omitted from the quantization–coding process.

Instead, only a noise substitution flag and the total power of the substituted spectral coefficients are transmitted for each of these bands.

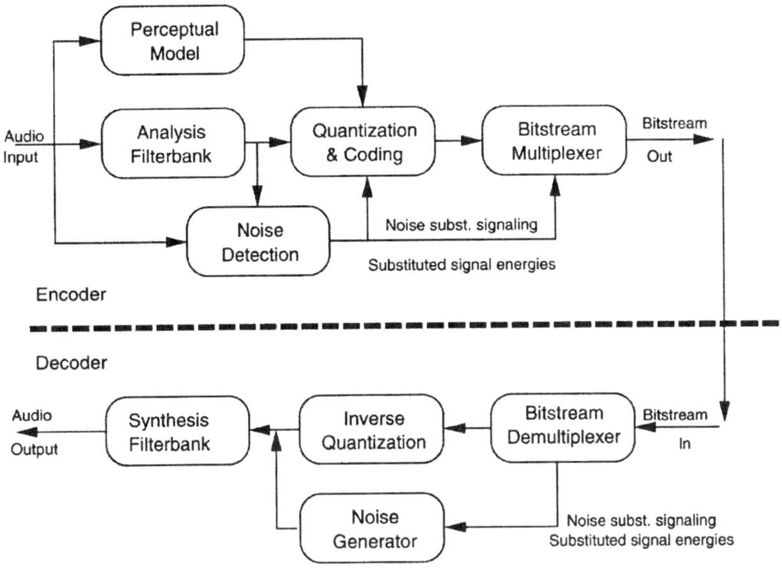

Figure 7 The principle of perceptual noise substitution.

In the decoder, pseudorandom vectors with the desired total noise power are inserted for the substituted spectral coefficients.

This approach will result in a highly compact representation of the noiselike spectral components because only the signaling and the energy information is transmitted per scale factor band rather than codebook, scale factor, and the set of quantized and coded spectral coefficients.

The PNS tool is tightly integrated into the MPEG–4 coder framework and reuses many of its basic coding mechanisms. As a result, the additional decoder complexity associated with the PNS coding tool is very low in terms of both computational and memory requirements. Furthermore, because of the means of signaling PNS, the extended bitstream syntax is downward compatible with MPEG–2 AAC syntax in the sense that each MPEG–2 AAC decoder will be able to decode the extended bitstream format as long as the PNS feature is not used.

B. Long-Term Prediction

Long-term prediction is a technique that is well known from speech coding and has been used to exploit redundancy in the speech signal that is related to the signal periodicity as manifested by the speech pitch (i.e., pitch prediction). Whereas common speech coders apply long-term prediction within the framework of a time domain coder, the MPEG–4 Audio LTP tool has been integrated into the framework of a generic perceptual audio coder; i.e., quantization and coding are performed on a spectral representation of the input signal. Figure 8 shows the combined LTP–coding system.

As shown in the figure, the LTP is used to predict the input signal based on the quantized values of the preceding frames, which were transformed back to a time domain representation by the inverse (synthesis) filter bank and the associated inverse TNS opera-

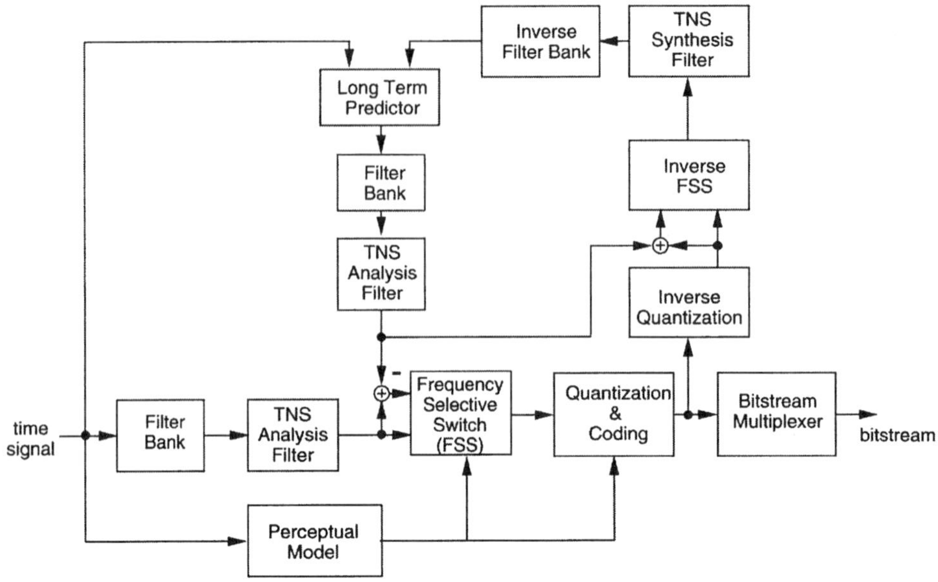

Figure 8 The LTP in the MPEG–4 GA coder.

tion. By comparing this decoded signal with the input signal, the optimal pitch lag and gain factor are determined. In the next step, both the input signal and the predicted signal are mapped to a spectral representation via an analysis filter bank and a forward TNS operation. Depending on which alternative is more favorable, coding of either the difference (residual) signal or the original signal is selected on a scale factor band basis. This is achieved by means of a so-called frequency-selective switch (FSS), which is also used in the context of the MPEG–4 GA scalable systems (see Sec. I.C.5). Because of the underlying principle, the LTP tool provides optimal coding gain for stationary harmonic signals (e.g., ''pitch pipe'') as well as some gain for nonharmonic tonal signals (e.g., polyphonic tonal instruments). Compared with the rather complex MPEG–2 AAC predictor tool, the LTP tool shows a saving of approximately one-half in both computational complexity and memory requirements.

C. Twin VQ

The Transform-Domain Weighted Interleaved Vector Quantization (Twin VQ) [26,27] is an alternative VQ-based coding kernel that is designed to provide good coding performance at extremely low bit rates (at or below 16 kbit/sec). It is used in the context of the MPEG–4 scalable GA system (see Sec. I.C.5). The coder kernel is adapted to operate within the spectral representation provided by the AAC coder filter bank [28].

The Twin VQ kernel performs a quantization of the spectral coefficients in two steps. In the first step the spectral coefficients are normalized to a specified target range and are then quantized by means of a weighted vector quantization process. The spectral normalization process includes a linear predictive coding (LPC) spectral estimation scheme, a periodic component extraction scheme, a Bark-scale spectral estimation scheme, and a power estimation scheme, which are carried out sequentially. As a result, the spectral coefficients are ''flattened''and normalized across the frequency axis. The parameters associated with the spectral normalization process are quantized and transmitted as side information. In the second step, called the weighted vector quantization process, the attened spectral coefficients are interleaved and divided into subvectors for vector quantization. For each subvector, a weighted distortion measure is applied to the conjugate structure VQ, which uses a pair of codebooks. In this way, perceptual control of the quantization distortion is achieved. The main part of the transmitted information consists of the selected codebook indices. Because of the nature of the interleaved vector quantization scheme, no adaptive bit allocation is carried out for individual quantization indices (an equal amount of bits is spent for each of the quantization indices).

The spectral normalization process includes the following steps:

LPC spectral estimation: At the first stage of the spectrum normalization process, the overall spectral envelope is estimated by means of an LPC model and used to normalize the spectral coefficients. This allows efficient coding of the envelope using line spectral pair (LSP) parameters.

Periodic component coding: If the frame is coded using one long filter bank window, ''periodic peak'' components are coded. This is done by estimating the fundamental signal period (pitch) and extracting a number of periodic peak components from the flattened spectral coefficients. The data are quantized together with the average gain of these components.

Bark-scale envelope coding: The resulting coefficients are further flattened by using a spectral envelope based on the Bark-related AAC scale factor bands. The envelope values are normalized and quantized by means of a vector quantizer with inter-frame prediction.

The weighted VQ process comprises the following steps:

Interleaving of spectral coefficients: Prior to vector quantization, the attened spectral coefficients are interleaved and divided into subvectors as shown in Figure 9. If the subvectors were constructed from spectral coefficients that were consecutive in frequency, the subvectors corresponding to the lower frequency range would require much finer quantization (more bits) than those corresponding

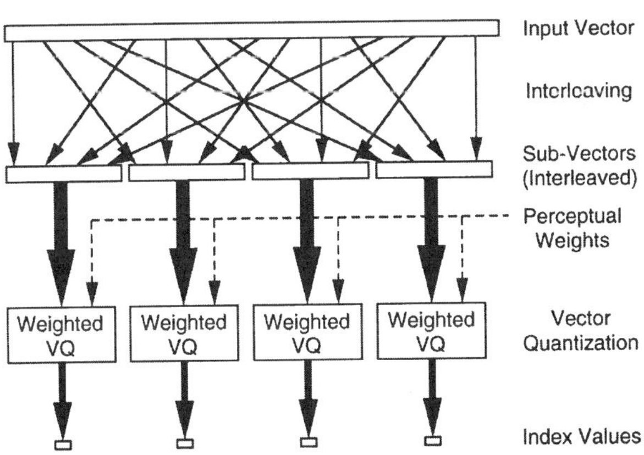

Figure 9 Twin VQ spectral coefficient interleaving.

to higher frequencies. In contrast, interleaving allows more constant bit allocation for each subvector. Perceptual shaping of the quantization noise can be achieved by applying an adaptive weighted distortion measure that is associated with the spectral envelope and the perceptual model.

Vector quantization: The vector quantization part uses a two-channel conjugate structure with two sets of codebooks. The best combination of indices is selected to minimize the distortion when two code vectors are added. This approach decreases both the amount of memory required for the codebooks and the computational demands for the codebook search.

The Twin VQ coder kernel operates at bit rates of 6 kbit/sec and above and is used mainly in the scalable configurations of the MPEG–4 GA coder.

D. MPEG-4 Scalable General Audio Coding

Today's popular schemes for perceptual coding of audio signals specify the bit rate of the compressed representation (bitstream) during the encoding phase. Contrary to this, the concept of scalable audio coding enables the transmission and decoding of the bitstream with a bit rate that can be adapted to dynamically varying requirements, such as the instantaneous transmission channel capacity. This capability offers significant advantages for transmitting content over channels with a variable channel capacity (e.g., the Internet, wireless transmission) or connections for which the available channel capacity is unknown at the time of encoding.

To achieve this, bitstreams generated by scalable coding schemes consist of several partial bitstreams that can be decoded on their own in a meaningful way. In this manner, transmission (and decoding) of a subset of the total bitstream will result in a valid, decodable signal at a lower bit rate and quality. Because bit rate scalability is considered one of the core functionalities of the MPEG–4 standard, a number of scalable coder configurations are described by the standard. In the context of MPEG–4 GA coding, the key concept of scalable coding can be described as follows (see Fig. 10):

The input signal is coded–decoded by a first coder (coder 1, the *base layer coder*) and the resulting bitstream information is transmitted as a first part of the composite scalable bitstream.

Next, the coding error signal is calculated as the difference between the encoded–

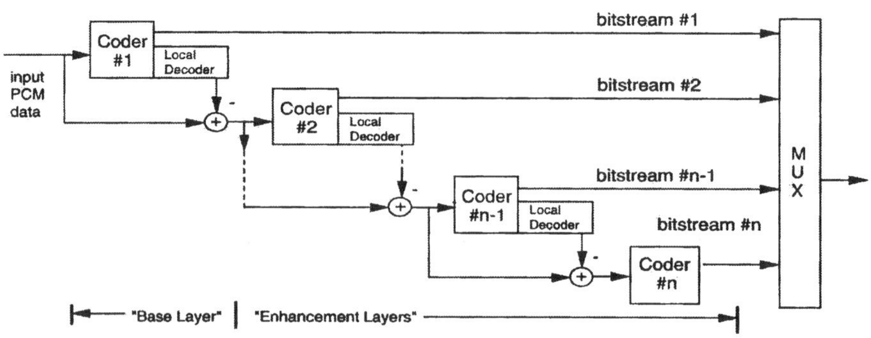

Figure 10 Basic concept of MPEG–4 scalable GA coding.

decoded signal and the original signal. This signal is used as the input signal of the next coding stage (coder 2), which contributes the next part of the composite scalable bitstream.

This process can be continued as often as desired (although, practically, no more than three or four coders are used). While the first coding stage (usually called *base layer*) transmits the most relevant components of the signal at a basic quality level, the following stages subsequently enhance the coding precision delivered by the preceding layers and are therefore called *enhancement layers*.

E. MPEG–4 Scalable Audio Coder

Figure 11 shows the structure of an MPEG–4 scalable audio coder. In this configuration, the base layer coder (called *core coder*) operates at a lower sampling frequency than the enhancement layer coder, which is based on AAC [29]:

The input signal is downsampled and encoded by the core coder. The resulting core layer bitstream is both passed on to the bitstream multiplexer and decoded by a local core decoder. The decoded output signal is upsampled to the rate of the enhancement layer encoder and passed through the MDCT analysis filter bank.

In a parallel signal path, the delay-compensated input signal is passed through the MDCT analysis filter bank.

The frequency-selective switch (FSS) permits selection between coding of spectral coefficients of the input signal and coding of spectral coefficients of the difference (residual) signal on a scale factor band basis. The assembled spectrum is passed to the AAC coding kernel for quantization and coding. This results in an enhancement layer bitstream that is multiplexed into the composite output bitstream.

Significant structural simplifications to this general scheme are achieved if both the base layer and the enhancement layer coders are filter bank–based schemes (i.e., AAC or Twin VQ). In this case, all quantization and coding are carried out on a common set of spectral coefficients and no sampling rate conversion is necessary [23]. The structure of a core based scalable decoder is shown in Figure 12: The composite bitstream is first demultiplexed into base layer and enhancement layer bitstreams.

Then the core layer bitstream is decoded, upsampled, delay compensated, and passed into the IMDCT synthesis filter bank. If only the core layer bitstream is received in a decoder, the output of the core layer decoder is presented via an optional postfilter.

If higher layer bitstreams are also available to the decoder, the spectral data are

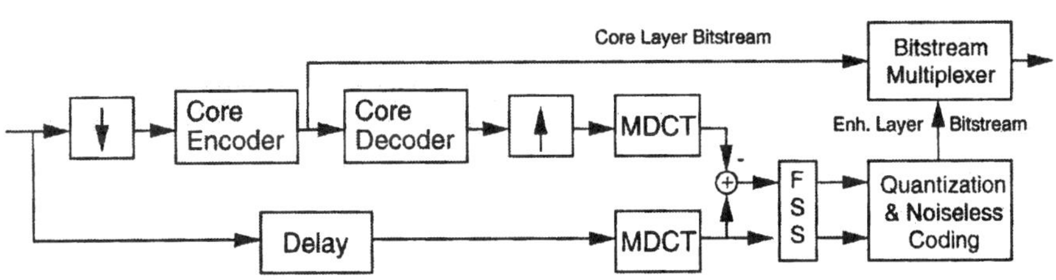

Figure 11 Structure of an MPEG–4 scalable coder.

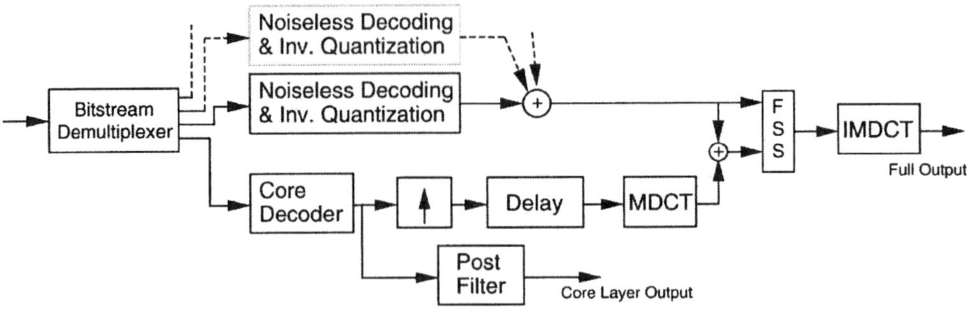

Figure 12 Structure of an MPEG–4 scalable decoder.

decoded from these layers and accumulated over all enhancement layers. The resulting spectral data are combined with the spectral data of the core layer as controlled by the the FSS and transformed back into a time domain representation.

Within the MPEG–4 GA scalable coding system, certain restrictions apply regarding the order and role of various coder types:

The MPEG–4 narrowband CELP coder 1.4.3 can be used as a core coder.

The Twin VQ coder kernel can act as a base layer coder or as an enhancement layer coder if coding of the previous layer is based on Twin VQ as well.

The AAC coder can act as both a base layer and an enhancement layer coder.

One interesting configuration is the combination of a CELP core coder and several AAC-based enhancement layers, which provides very good speech quality even at the lowest layer decoded output.

F. Mono–Stereo Scalability

Beyond the type of scalability described up to now, the MPEG–4 scalable GA coder also provides provisions for *mono–stereo scalability*. Decoding of lower layers results in a mono signal, whereas decoding of higher layers will deliver a stereo signal after decoding [30]. This useful functionality is achieved in the following way:

All mono layers operate on a mono version of the stereo input signal. The stereo enhancement layers encode the stereo signal as either an M/S (mid/side) or L/R (left/right) representation, as known from AAC joint stereo coding. When using an M/S representation, the encoded signal from the lower mono layers is available as an approximation of the mid signal.

IV. THE REST OF THE MPEG–4 NATURAL AUDIO CODER

A. Target Applications

The MPEG–4 general audio coder is the first that covers the whole range of low-bit-rate audio coding applications. MPEG–4 Audio has enough capabilities to replace virtually all of the existing audio coding standards and offers many additional functionalities not available with any other coder, such as bit rate scalability. It might be asked why such an all-in-one system is needed and whether it might be oversized for many applications; in reality, such a system is needed for upcoming communication networks. It seems very likely that telephone and computer networks will soon merge into a unified service. With the start of music and video distribution over computer networks, traditional broadcasting services will start to merge into this global communication web. In such an environment the channel to a specific end user can be anything from a low-bit-rate cellular phone connection to a gigabit computer network connected operating over fiber-optic lines. Without a universal coding system, transcoding is frequently required, e.g., when a cell phone user communicates with an end user via a future hi-fi set that normally retrieves high-quality 11-channel music material from the Internet.

B. General Characteristics

As with MPEG–1 Audio and MPEG–2 Audio, the low-bit-rate coding of audio signals is the core functionality in MPEG–4. The lower end of this bit rate range is marked by pure speech coding techniques, starting from 2 kbit/sec. The upper end reaches up to more than a 100 kbit/sec per audio channel for transparent coding of high-quality audio material, with a dynamic range, sampling frequency options, and a multichannel audio capability that exceed the standard set by today's compact disc. For example, seven-channel stereo material, at a sampling rate of 96 kHz, can be encoded with a dynamic range of more than a 150 dB if desired.

This broad range cannot be covered by a single coding scheme but requires the combination of several algorithms and tools.* These are integrated into a common framework with the possibility of a layered coding, starting with a very low bit rate speech coder and additional general audio coding layers on top of this base layer. Each layer further enhances the audio quality available with the previous layer. With such a scheme at any point in the transmission chain the audio bitstream can be adapted to the available bit rate by simply dropping enhancement layers.

In general, there are two types of coding schemes in MPEG–4 Audio. Higher bit rate applications are covered by the MPEG–4 General Audio (GA) coder 1.3, which in general is the best option for bit rates of 16 kbit/sec per channel and above for all types of audio material. If lower rates are desired, the GA coder, with a lower bit rate limit of around 6 kbit/sec, is still the best choice for general audio material. However, for speech-dominated applications the MPEG–4 speech coder is available as an alternative, offering bit rates down to 2 kbit/sec.

C. The MPEG–4 Speech Coder

1. Introduction

The speech coder in MPEG–4 Audio transfers the universal approach of the ''traditional'' MPEG audio coding algorithms to the speech coding world. Whereas other speech coder standards, e.g., G.722 [31], G.723.1 [32], or G.729 [33], are defined for one specific sam-

* A tool in MPEG–4 Audio is a special coding module that can be used as a component in different coding algorithms.

pling rate and for one to at most three different bit rates, the MPEG-4 coder supports multiple sampling rates and more than 50 different bit rate options. Furthermore, embedded coding techniques are available that allow decoding subsets of the bitstream into valid output signals. The following section gives an overview of the functionalities of the MPEG–4 speech coder.

2. Speech Coder Functionalities

a. Narrowband Speech Coding. The term narrowband speech coding usually describes the coding of speech signals with an audio bandwidth of around 3.5 kHz. The MPEG–4 coder, like other digital narrowband speech coders, supports this with a sampling rate of 8 kHz. In addition, primarily to support scalable combinations with AAC enhancement layers (see Sec. I.C.5), the MPEG–4 narrowband coder allows slightly different sampling rates close to 8 kHz, e.g., 44,100/6 = 7350 Hz. The available bit rates range from 2 kbit/sec* up to 12.2 kbit/sec. The algorithmic delay† ranges from around 40 msec for the lowest bit rates to 25 msec at around 6 kbit/sec and down to 15 msec for the highest rates. A trade-off between audio quality and delay is possible, as some bit rates are available with a different coding delay All but the lowest bit rate are realized with a coder based on CELP coding techniques. The 2 kbit/sec coder is based on the novel HVXC (see Sec. I.D.3) scheme.

b. Wideband Speech Coding. Wideband speech coding, with a sampling rate of 16 kHz, is available for bit rates from 10.9 to 21.1 kbit/sec with an algorithmic delay of 25 msec and for bit rates from 13.6 to 23.8 kbit/sec with a delay of 15 msec. The wideband coder is an upscaled narrowband CELP coder, using the same coding tools, but with different parameters.

c. Bit Rate Scalability. Bit rate scalability is a unique feature of the MPEG–4 Audio coder set, which is used to realize a coding system with embedded layers. A narrowband or a wideband coder, as just described, is used as a base layer coder. On top of that, additional speech coding layers can be added which, step by step, increase the audio quality.‡ Decoding is always possible if at least the base layer is available. Each enhancement layer for the narrowband coder uses 2 kbit/sec. The step size for the wideband coder is 4 kbit/sec. Although only one enhancement layer is possible for HVXC, up to three bit rate scalable enhancement layers (BRSELs) may be used for the CELP coder.

d. Bandwidth Scalability. Bandwidth scalability is another option to improve the audio quality by adding an additional coding layer. Only one bandwidth scalable enhancement layer (BWSEL) is possible and can be used in combination with the CELP coding tools but not with HVXC. In this configuration, a narrowband CELP coder at a sampling rate of 8 kHz first codes a downsampled version of the input signal. The enhancement layer, running with a sampling rate of 16 kHz, expands the audio bandwidth from 3.5 to 7 kHz.

3. The Technology

All variants of the MPEG–4 speech coder are based on a model using an LPC filter [34] and an excitation module [35]. There are three different configurations, which are listed

---

* A variable rate mode with average bit rates below 2 kbit/sec is also available.

† The algorithmic delay is the shortest possible delay, assuming zero processing and transmission time.

‡ For a narrowband base coder, GA enhancement layers, as well as speech layers, are possible.

Table 2 MPEG–4 Speech Coder Configurations

Configuration	Excitation type	Bit rate range (kbit/sec)	Sampling rates (kHz)	Scalability Options
HVXC	HVXC	1.4–4	8	Bit rate
CELP I	RPE	14.4–22.5	16	—
CELP II	MPE	3.85–23.8	8 and 16	Bit rate and bandwidth

in Table 2. All of these configurations share the same basic method for the coding of the LPC filter coefficients. However, they differ in the way the filter excitation signal is transmitted.

4. Coding of the LPC Filter Coefficients

In all configurations, the LPC filter coefficients are quantized and coded in the LSP domain [36,37]. The basic block of the LSP quantizer is a two-stage split-VQ design [38] for 10 LSP coefficients, shown in Figure 13. The first stage contains a VQ that codes the 10 LSP coefficients either with a single codebook (HVXC, CELP BWSEL) or with a split VQ with two codebooks for 5 LSP coefficients each. In the second stage, the quantization accuracy is improved by adding the output of a split VQ with two tables for 2 × 5 LSP coefficients. Optionally, inter-frame prediction is available in stage two, which can be switched on and off, depending on the characteristics of the input signal. The codebook in stage two contains two different sets of vectors to be used, depending on whether or not the prediction is enabled.

The LSP block is applied in six different configurations, which are shown in Table 3. Each configuration uses its own set of specifically optimized VQ tables. At a sampling rate of 8 kHz a single block is used. However, different table sets, optimized for HVXC and the narrowband CELP coder, are available. At 16 kHz, there are 20 LSP coefficients and the same scheme is applied twice using two more table sets for the independent coding of the lower and upper halves of the 20 coefficients.

Although so far this is conventional technology, a novel approach is included to support the bandwidth scalability option of the MPEG–4 CELP coder [39] (Fig. 14). The

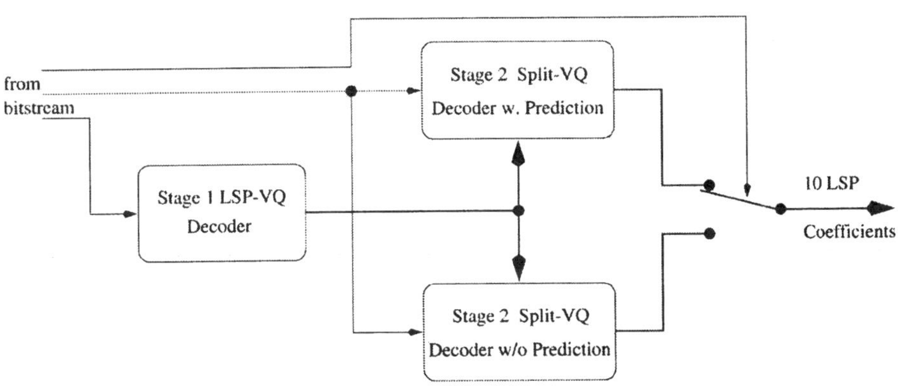

Figure 13 Two-stage split-VQ inverse LSP quantizer with optional inter-frame predictor.

Table 3 LSP-VQ Table Sizes for the Six LSP Configurations (Entries × Vector Length)

LSP configuration	Table size stage 1	Table size stage 2 (with prediction)	Table size stage 2 (without prediction)
HVXC 8 kHz	32 × 10	64 × 5 + 16 × 5	64 × 5 + 16 × 5
CELP 8 kHz	16 × 5 + 16 × 5	64 × 5 + 32 × 5	64 × 5 + 32 × 5
CELP 16 kHz (lower)	32 × 5 + 32 × 5	64 × 5 + 64 × 5	64 × 5 + 64 × 5
CELP 16 kHz (upper)	16 × 5 + 16 × 5	64 × 5 + 16 × 5	64 × 5 + 16 × 5
CELP BWSEL (lower)	16 × 10	–	16 × 5 + 64 × 5
CELP BWSEL (upper)	128 × 10	–	128 × 5 + 16 × 5

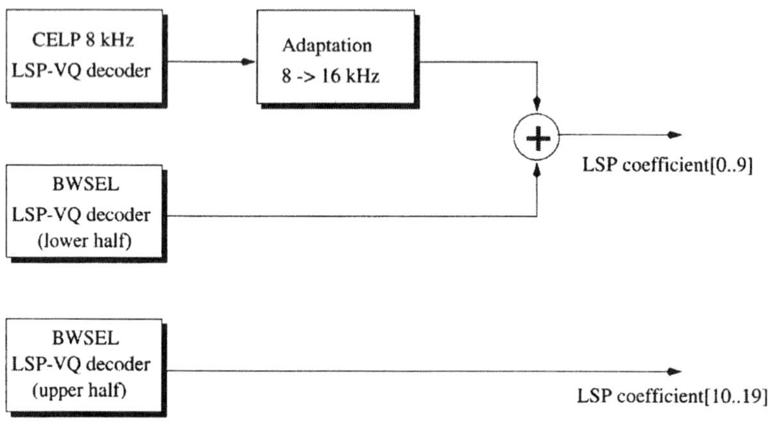

Figure 14 Bandwidth scalable LSP inverse quantization scheme.

quantized LSP coefficients of the first (narrowband) layer are reconstructed and transformed to the 16-kHz domain to form a first-layer coding of the lower half of the wideband LSP coefficients. Two additional LSP-VQ blocks provide a refinement of the quantization of the lower half of the LSP coefficients and code the upper half of the coefficients. Again, a different set of optimized VQ tables is used for this purpose.*

5. Excitation Coding

For the coding of the LPC coefficients, one common scheme is used for all configurations, bit rates, and sampling rates. However, several alternative excitation modules are required to support all functionalities:

1. *MPE*: The broadest range of bit rates and functionalities is covered by a multimode multipulse excitation (MPE) [3,39,40] tool, which supports narrow- and wideband coding, as well as bit rate and bandwidth scalability.
2. *RPE*: Because of the relatively high complexity of the MPEG–4 MPE tool if used for a wideband encoder, a regular pulse excitation (RPE) [41] module is

* For completeness: The predictor of the BWSEL specific LSP decoder blocks is applied not to the second stage only but rather to both stages together.

available as an alternative, low-complexity encoding option, resulting in slightly lower speech quality [42].

3. *HVXC*: To achieve good speech quality at very low rates, an approach quite different from MPE or RPE is required. This technique is called harmonic vector excitation coding (HVXC) [43]. It achieves excellent speech quality even at a bit rate of only 2 kbit/sec in the MPEG–4 speech coder verification tests [42] for speech signals with and without background noise. HVXC uses a completely parametric description of the coded speech signal and is therefore also called the MPEG–4 parametric speech coder. Whereas the CELP coding modes offer some limited performance (compared with the MPEG–4 GA coder) for non-speech signals, HVXC is recommended for speech only because its model is designed for speech signals.

a. Multipulse Excitation (MPE). The maximum configuration of the MPE excitation tool, comprising the base layer, all three bit rate scalable layers (BRSELs), and the bandwidth scalable BWSEL, is shown in Figure 15.

The base layer follows a layout that is often found in a CELP coder. An adaptive codebook [44] which generates one component of the excitation signal, is used to remove the redundancy of periodic signals. The nonperiodic content is represented with a multipulse signal. An algebraic codebook [45] structure is used to reduce the side information that is required to transmit the locations and amplitudes of the pulses. The base layer operates at a sampling rate of either 8 or 16 kHz. On top of this base layer, two types of

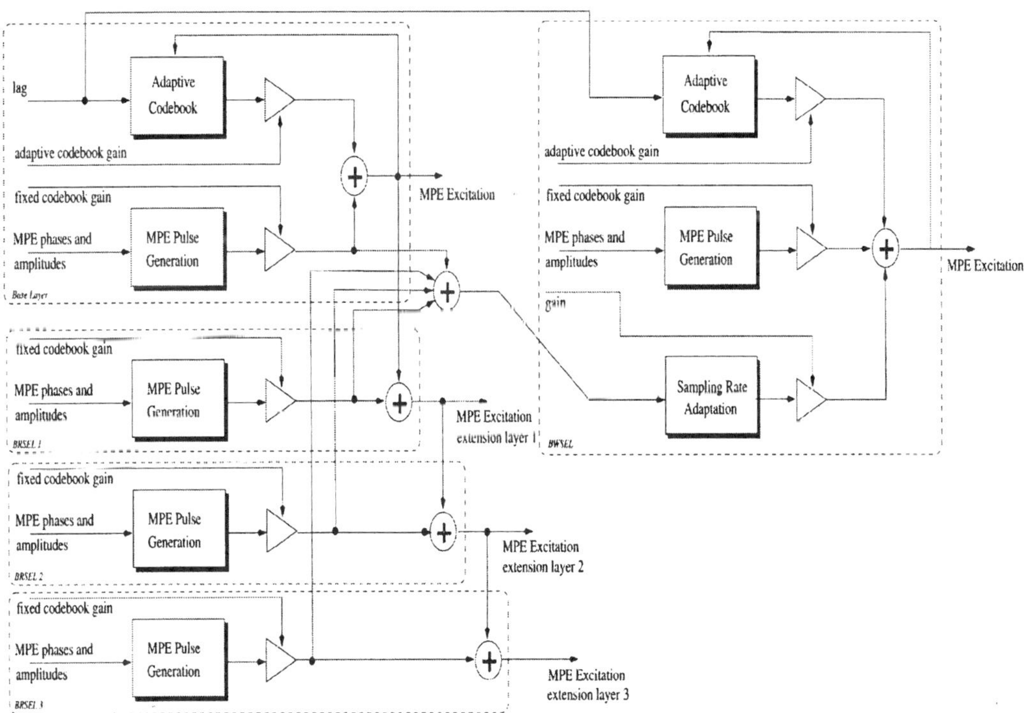

Figure 15 Block diagram of the maximum configuration of the MPE excitation tool, including bit rate (BRSEL) and bandwidth scalability (BWSEL) extension layers.

enhancement layers are possible. The BRSEL-type layers add additional excitation pulses with an independent gain factor. The combined outputs of the base layer and any number of BRSELs always share the sampling frequency of the base layer. If a BWSEL is added, with or without an arbitrary number of BRSELs, however, the excitation signal produced by the BWSEL output always represents a 16-kHz signal. In this case, the base layer and the BRSEL are restricted to a sampling rate of 8 kHz.

b. Regular Pulse Excitation (RPE). Although the MPEG–4 MPE module supports all the functionalities of the RPE tool with at least the same audio quality [42], RPE was retained in the MPEG–4 speech coder tool set to provide a low-complexity encoding option for wideband speech signals. Whereas the MPEG–4 MPE tool uses various VQ techniques to achieve optimal quality, the RPE tool directly codes the pulse amplitudes and positions. The computational complexity of a wideband encoder using RPE is estimated to be about one-half that of a wideband encoder using MPE. All MPEG–4 Audio speech decoders are required to support both RPE and MPE. The overhead for the RPE excitation decoder is minimal. Figure 16 shows the general layout of the RPE excitation tool.

c. HVXC. The HVXC decoder, shown in Figure 17, uses two independent LPC synthesis filters for voiced and unvoiced speech segments in order to avoid the voiced excitation being fed into the unvoiced synthesis filter and vice versa. Unvoiced excitation components are represented by the vectors of a stochastic codebook, as in conventional CELP coding techniques. The excitation for voiced signals is coded in the form of the spectral envelope of the excitation signal. It is generated from the envelope in the harmonic synthesizer using a fast IFFT synthesis algorithm. To achieve more natural speech quality, an additional noise component can be added to the voiced excitation signal. Another feature of the HVXC coder is the possibility of pitch and speech change, which is directly facilitated by the HVXC parameter set.

6. Postfilter

To enhance the subjective speech quality, a postfilter process is usually applied to the synthesized signal. No normative postfilter is included in the MPEG–4 specification, although examples are given in an informative annex. In the MPEG philosophy, the postfilter is the responsibility of the manufacturer as the filter is completely independent of the bitstream format.

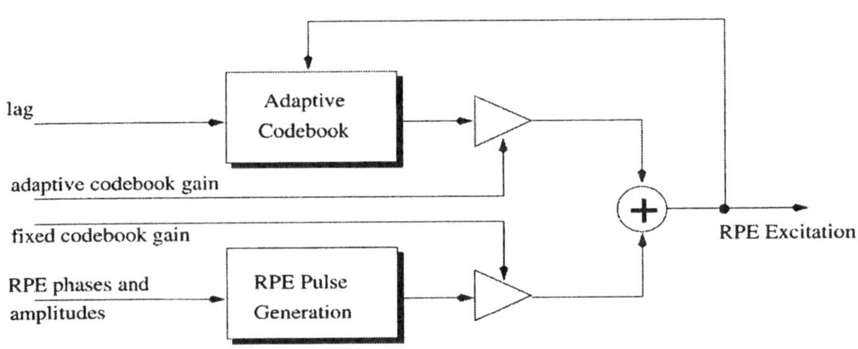

Figure 16 Block diagram of the RPE tool.

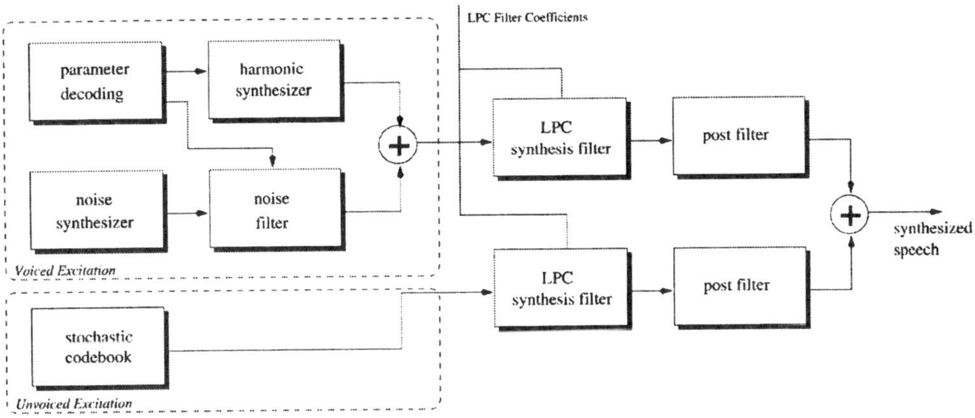

Figure 17 Block diagram of the HVXC decoder.

V. CONCLUSIONS

Based in part on MPEG–2 AAC, in part on conventional speech coding technology, and in part in new methods, the MPEG–4 General Audio coder provides a rich set of tools and features to deliver both enhanced coding performance and provisions for various types of scalability. The MPEG–4 GA coding defines the current state of the art in perceptual audio coding.

ACKNOWLEDGMENTS

The authors wish to express their thanks to Niels Rump, Lynda Feng, and Rainer Martin for their valuable assistance in proofreading and LATEX typesetting.

REFERENCES

1. N Jayant, P Noll. Digital Coding of Waveforms. Englewood Cliffs, NJ: Prentice-Hall, 1984.
2. RE Crochiere, SA Webber, JL Flanagan. Digital coding of speech in sub-bands. Bell Syst Tech J October:169–185, 1976.
3. BS Atal, JR Remde. A new model of LPC excitation for producing natural-sounding speech at low bit rates. Proceedings of the ICASSP, 1982, pp 614–617.
4. H Fletcher. In: J Allen, ed. The ASA Edition of Speech and Hearing in Communication. Woodbury, NY: Acoustical Society of America, 1995.
5. B Scharf. Critical bands. In: J Tobias, ed. Foundations of Modery Auditory Theory. New York: Academic Press, 1970, pp 159–202.
6. RP Hellman. Assymmetry of masking between noise and tone. Percept Psychophys 2:241–246, 1972.
7. Zwicker, Fastl. Psychoacoustics, Facts and Models. New York: Springer, 1990.
8. JB Allen, ST Neely. Micromechanical models of the cochlea, 1992. Phys Today, July:40–47, 1992.

9. HS Malvar. Signal Processing with Lapped Transforms. Norwood, MA: Artech House, 1992.
10. K Brandenburg, G Stoll. ISO-MPEG-1 Audio: A generic standard for coding of high quality digital audio. In: N Gilchrist, C Grewin, eds. Collected Papers on Digial Audio Bit-Rate Reduction. AES, New York, 1996, pp 31–42.
11. ISO/IEC JTC1/SC29/WG11 MPEG. International Standard IS 13818-7. Coding of moving pictures and audio. Part 7: Advanced audio coding.
12. M Bosi, K Brandenburg, S Quackenbush, L Fielder, K Akagiri, H Fuchs, M Diets, J Herre, G Davidson, Y Oikawa. ISO/IEC MPEG-2 Advanced Audio Coding. J Audio Eng Soc 45(10): 789–814, 1997.
13. J Herre, JD Johnston. Enhancing the performance of perceptual audio coders by using temporal noise shaping (TNS). 101st AES Convention, Los Angeles, November 1996.
14. BCJ Moore. An Introduction to the Psychology of Hearing. 3rd ed. New York: Academic Press, 1989.
15. JD Johnston, AJ Ferreira. Sum-difference stereo transform coding. IEEE ICASSP, 1992, pp 569–571.
16. JD Johnston, J Herre, M Davis, U Gbur. MPEG-2 NBC audio–stereo and multichannel coding methods. Presented at the 101st AES Convention, Los Angeles, November 1996.
17. H Fuchs. Improving MPEG Audio coding by backward adaptive linear stereo prediction. Presented at the 99th AES Convention, New York, October 1995, preprint 4086 (J-1).
18. J Johnston, K Brandenburg. Wideband coding perceptual considerations for speech and music. In: S Furui, MM Sondhi, eds. Advances in Speech Signal Processing. New York: Marcel Dekker, 1992.
19. MR Schroeder, B Atal, J Hall. JASA December: pp 1647–1651, 1979.
20. SR Quackenbush, JD Johnston. Noiseless coding of quantized spectral components in MPEG-2 Advanced Audio Coding. IEEE Workshop on Applications of Signal Processing to Audio and Acoustics, Mohonk Mountain House, New Paltz, NY, 1997.
21. ISO/IEC JTC1/SC29/WG11 (MPEG), International ISO/IEC 14496-3. Generic coding of audiovisual objects: Audio.
22. ISO/IEC JTC1/SC29/WG11 MPEG, International Standard ISO/IEC 13818-7. Generic coding of moving pictures and associated audio: Advanced Audio Coding.
23. J Herre, D Schulz. Extending the MPEG-4 AAC codec by perceptual noise substitution. 104th AES Convention, Amsterdam, 1998, preprint 4720.
24. D Schulz. Improving audio codecs by noise substitution. J Audio Eng Soc 44(7/8):593–598, 1996.
25. D Schulz. Kompression qualitativ hochwertiger digitaler Audio signale durch Rauschextraktion. PhD thesis, TH Darmstadt, 1997 (in German).
26. N Iwakami, T Moriya, S Miki. High-quality audio-coding at less than 64 kbit/s by using transform-domain weighted interleave vector quantization (TWINVQ). Proceedings of the ICASSP, Detroit, 1995, pp 3095–3098.
27. N Iwakami, T Moriya. Transform domain weighted interleave vector quantization (Twin VQ). 101st AES Convention, Los Angeles, 1996, preprint 4377.
28. J Herre, E Allamanche, K Brandenburg, M Dietz, B Teichmann, B Grill, A Jin, T Moriya, N Iwakami, T Norimatsu, M Tsushima, T Ishikawa. The integrated filterbank based scalable MPEG-4 Audio coder. 105th AES Convention, San Francisco, 1998, preprint 4810.
29. B Grill. A bit rate scalable perceptual coder for MPEG-4 Audio. 103rd AES Convention, New York, 1997, preprint 4620.
30. B Grill, B Teichmann. Scalable joint stereo coding. 105th AES Convention, San Francisco, 1998, preprint 4851.
31. ITU-T. Recommendation G.722:7kHz audio-coding within 64 kbit/s, 1988.
32. ITU-T. Recommendation G.723.1: Dual rate speech coder for multi-media communications transmitting at 5.3 and 6.3 kbit/s, 1996.

33. ITU-T. Recommendation G.729: Coding of speech at 8 kbit/s using conjugate-structure algebraic-code-excited linear-prediction, 1996.
34. LR Rabiner, RW Schafer. Digital Processing of Speech Signals. Englewood Cliffs, NJ: Prentice-Hall, 1978, chap 8.
35. BS Atal, MR Schroeder. Stochastic codign of speech signals at very low bit rates. In: P Dewilde, CA May, eds. Links for the Future, Science, Systems, and Services for Communications. Amsterdam: Elsevier Science, 1984, pp 1610–1613.
36. FK Soong, BH Juang. Line spectrum pair (LSP) and speech data compression. Proceedings of the ICASSP, 1984, pp 1.10.1–1.10.4.
37. WB Kleijn, KK Paliwal, eds. Speech Coding and Synthesis. New York: Elsevier Science, 1995, pp 126–132, 241–251.
38. N Tanaka, T Morii, K Yoshida, K Honma. A Multi-Mode Variable Rate Speech Coder for CDMA Cellular Systems. Proc IEEE Vehicular Technology Conf. April 1996, pp 192–202.
39. T Nomura, M Iwadare, M Serizawa, K Ozawa. A bitrate and bandwidth scalable celp coder. Proceedings of the ICASSP, 1998.
40. H Ito, M Serizawa, K Ozawa, T Nomura. An adaptive multi-rate speech codec based on MP-CELP coding algorithm for ETSI AMR standard. Proceedings of the ICASSP, 1998.
41. P Kroon, EF Deprette, RJ Sluyter. Regular-pulse excitation: A novel approach to effective and efficient multipulse coding of speech. IEEE Trans Acoust Speech Signal Process 34:1054–1063, 1986.
42. ISO/IEC JTC1/SC29/WG11 MPEG. MPEG-4 Audio verification test results: Speech codecs. Document N2424 of the October 1998 Atlantic City MPEG Meeting.
43. M Nishiguchi, J Matsumoto, S Omori, K Iijima. MPEG95/0321. Technical description of Sony IPC's proposal for MPEG-4 audio and speech coding, November 1995.
44. WB Klejn, DJ Krasinski, RH Ketchum. Improved speech quality and efficient vector quantization in SELP. Proceedings of the ICASSP 1988, pp 155–158.
45. C Laflamme, et al. 16 kbps wideband speech coding technique based on algebraic CELP. Proceedings of the ICASSP, 1991, pp 13–16.

6
Synthetic Audio and SNHC Audio in MPEG–4

Eric D. Scheirer
MIT Media Laboratory, Cambridge, Massachusetts

Youngjik Lee and Jae-Woo Yang
ETRI Switching & Transmission Technology Laboratories, Taejon, Korea

I. INTRODUCTION

This chapter describes the parts of MPEG–4 that govern the transmission of synthetic sound and the combination of synthetic and natural sound into hybrid soundtracks. Through these tools, MPEG–4 provides advanced capabilities for ultralow-bit-rate sound transmission, interactive sound scenes, and flexible, repurposable delivery of sound content.

We will discuss three MPEG–4 audio tools. The first, MPEG–4 Structured Audio, standardizes precise, efficient delivery of synthetic sounds. The second, MPEG–4 Text-to-Speech Interface (TTSI), standardizes a transmission protocol for synthesized speech, an interface to text-to-speech synthesizers, and the automatic synchronization of synthetic speech and "talking head" animated face graphics (see Chap. 11). The third, MPEG–4 AudioBIFS—part of the main Binary Format for Scenes (BIFS) framework (see Chap. 14)—standardizes terminal-side mixing and postproduction of audio sound tracks. AudioBIFS enables interactive sound tracks and three-dimensional (3D) sound presentation for virtual reality applications. In MPEG–4, the capability to mix and synchronize real sound with synthetic is termed Synthetic/Natural Hybrid Coded (SNHC) audio.

The organization of this chapter is as follows. First, we provide a general overview of the objectives for synthetic and SNHC audio in MPEG–4. This section also introduces concepts from speech and music synthesis to readers whose primary expertise may not be in the field of audio. Next, a detailed description of the synthetic audio codecs in MPEG–4 is provided. Finally, we describe AudioBIFS and its use in the creation of SNHC audio soundtracks.

II. SYNTHETIC AUDIO IN MPEG–4: CONCEPTS AND REQUIREMENTS

In this section, we introduce speech synthesis and music synthesis. Then we discuss the inclusion of these technologies in MPEG–4, focusing on the capabilities provided by synthetic audio and the types of applications that are better addressed with synthetic audio coding than with natural audio coding.

A. Relationship Between Natural and Synthetic Coding

Today's standards for natural audio coding, as discussed in Chapter 5, use perceptual models to compress natural sound. In coding synthetic sound, perceptual models are not used; rather, very specific parametric models are used to transmit sound *descriptions*. The descriptions are received at the decoding terminal and converted into sound through real-time sound *synthesis*. The parametric model for the Text-to-Speech Interface is fixed in the standard; in the Structured Audio tool set, the model itself is transmitted as part of the bitstream and interpreted by a reconfigurable decoder.

Natural audio and synthetic audio are not unrelated methods for transmitting sound. Especially as sound models in perceptual coding grow more sophisticated, the boundary between ''decompression'' and ''synthesis'' becomes somewhat blurred. Vercoe et al. [1] discuss the relationships among various methods of digital sound creation and transmission, including perceptual coding, parametric compression, and different kinds of algorithmic synthesis.

B. Concepts in Music Synthesis

Mathematical techniques that produce, or *synthesize*, sounds with desired characteristics have been a active topic of study for many years. At first, in the early 1960s, this research focused on the application of known psychoacoustic results to the creation of sound. This was regarded both as a scientific inquiry into the nature of sound and as a method for producing musical material [2]. For example, as it had been known since the 19th century that periodic sound could be described in terms of its spectrum, it was thought natural to synthesize sound by summing together time-varying sinusoids. Several researchers constructed systems that could analyze the dynamic spectral content of natural sounds, extract a parametric model, and then use this representation to drive sinusoidal or ''additive'' synthesis. This process is termed *analysis–synthesis* [3,4].

Musicians as well as scientists were interested in these techniques. As the musical aesthetic of the mid-20th century placed a premium on the exploration of novel sounds, composers became interested in using computers to generate interesting new sonic material. Composers could make use of the analysis–synthesis procedure, but rather than exactly inverting the analysis, they could modify the parameters to create special musical effects upon synthesis.

In recent years, the study of synthesis algorithms has progressed in many ways. New algorithms for efficiently generating rich sounds—for example, synthesis via frequency modulation [5]—have been discovered. Researchers have developed more accurate models of acoustic instruments and their behavior—for example, digital waveguide implementations of physical models [6]. Composers efficiently explore types of sound that are not easily realized with acoustic methods—for example, granular synthesis [7]. Many excellent books are available for further reading on synthesis methods [8].

C. Music Synthesis in Academia

There have been two major directions of synthesis development. These roughly correspond to the parallel paths of academic and industrial work in the field. The academic direction

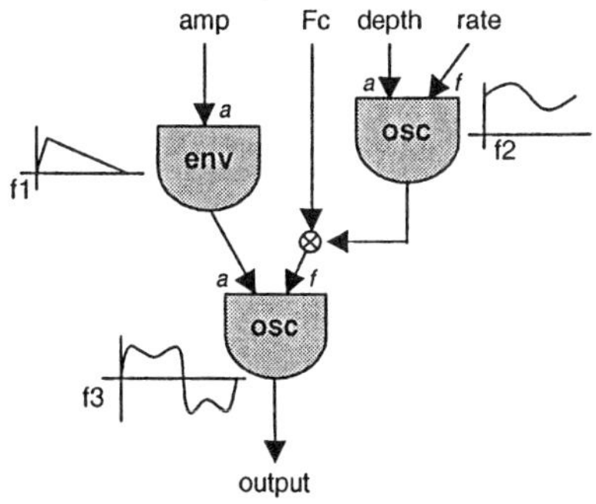


```
f1 line 128  0 0.1 0.9 0.9 0.0
f2 sine 128  1 0.1
f3 sine 2048 0 1   0.5

instr 1
  envlp   env f1 p1
  vibfreq osc f2 p3 p4
  freq    mul vibfreq p2
  sig     osc f3 envlp freq
          out sig
end
```

Figure 1 A signal-flow diagram and the associated program in a unit-generator language. This instrument maps from four parameters (labeled amp, Fc, depth, and rate in the diagram and p1, p2, p3, and p4 in the code) to a ramped sound with vibrato. Each operator in the diagram corresponds to one line of the program.

became fertile first, with Mathews' development in the early 1960s of the *unit generator* model for software-based synthesis [9,10]. In this model, sound–generation algorithms are described in a computer language specially created for the description of digital signal–processing networks. In such a language, each program corresponds to a different network of digital signal–processing operations and thus to a different method of synthesis (see Fig. 1). Computers at the time were not fast enough to execute these algorithms in real time, so composers using the technology worked in an offline manner: write the program, let the computer turn it into sound, listen to the resulting composition, and then modify the program and repeat the process.

The unit–generator music language is provably a general abstraction, in that any method of sound synthesis can be used in such a language. This model has changed little since it was developed, indicating the richness of Mathews' conception. Since the 1960s, other researchers have developed new languages in the same paradigm that are (supposed to be) easier for musicians to use or more expressively powerful [11] and that, beginning in the 1990s, could be executed in real time by the new powerful desktop computers [12].

D. Music Synthesis in Industry

The industrial development of music synthesizers began later, in the mid-1960s. Developers such as Moog, Buchla, and Oberheim built *analog synthesizers* for real-time performance. In these devices, a single method of parametric synthesis (often subtractive synthesis, in which a spectrally rich carrier signal is filtered in different ways to produce sound) was implemented in hardware, with parametric controls accessible to the musician via a front panel with many knobs. Manning [13, pp. 117–155] has presented an overview of the early development of this technology.

These devices, although simple in construction, were capable of real-time performance and proved very popular with both popular musicians and academic composers. They provided a rich palette of sonic possibilities; this was due less to the methods of synthesis they used than (as we now realize) to the subtle nonlinearities in their inexpensive components. That is, when an analog voltage-controlled oscillator is driven with a keyboard-controlled amplitude signal, the nonlinear properties of the components give the sound a particular ''feel'' or ''taste.'' It turns out to be surprisingly difficult to digitally model the behavior of these instruments accurately enough for useful musical performance. This topic is an area of active research [14–16].

With the rise of the microprocessor and the discovery of synthesis algorithms that could be implemented cheaply in digital hardware yet still produce a rich sound, hardware-based digital synthesizers began to be rapidly developed in the mid-1980s. They soon surpassed analog synthesizers in popularity. First, synthesizers based on the frequency-modulation method [5] and then ''samplers'' based on wavetable synthesis [17] became inexpensive and widely available. Later, as the personal computer revolution took hold, the hardware digital synthesizer shrank to the size of a plug-in card (or today, a single chip) and became the predominant method of sound synthesis on PCs.

It is somewhat ironic that, although the powerful technique of general-purpose software synthesis was first developed on the digital computer, by the time the digital computer became generally available, the simpler methods became the prevalent ones. General-purpose synthesis packages such as Csound [18] have been ported to PCs today, but only the academic composer uses them with any regularity. They have not made any significant impact on the broader world of popular music or audio for PC multimedia.

E. Standards for Music Synthesis

There have never been standards, de jure or de facto, governing general-purpose software synthesis. At any given time since the 1970s, multiple different languages have been available, differing only slightly in concept but still completely incompatible. For example, at the time of writing there are devotees of the Csound [18], Nyquist [19], SuperCollider [20], and CLM [21] languages. Composers who are technically capable programmers have often developed their own synthesis programs (embodying their own aesthetic notions) rather than use existing languages. Several respected music researchers [22,23] have written of the pressing need for a standard language for music synthesis.

During the mid-1980s, the Musical Instrument Digital Interface (MIDI) standard was created by a consortium of digital musical instrument manufacturers. It was quickly accepted by the rest of the industry. This protocol describes not the *creation* of sound—it does not specify how to perform synthesis—but rather the *control* of sound. That is, a synthesizer accepts MIDI instructions (through a ''MIDI port'') that tell it what note to play, how to set simple parameters such as pitch and volume, and so forth. This standard was effective in enabling compatibility among tools—keyboard controllers produced by one company could be used to control sound generators produced by another. MIDI compatibility today extends to PC hardware and software; there are many software tools available for creating musical compositions, expressing them in MIDI instructions, and communicating them to other PCs and to hardware synthesizers.

MIDI-enabled devices and a need for compabitility were among the key influences driving the use of very simplistic synthesis technology in PC audio systems. Synthesizers became *equated* with MIDI by 1990 or so; the interoperability of a new device with a musician's existing suite of devices became more important than the sound quality or flexibility it provided. Unfortunately, because MIDI has a very simple model of sound generation and control, more complex and richer models of sound were crowded out. Even when MIDI is used to control advanced general-purpose software synthesizers [24], the set of synthesis methods that can be used effectively is greatly diminished. The technical problems with MIDI have been explored at length by other authors [22,25].

Today's PC sound card is built on essentially the same model that the fixed-hardware sampler used in 1990. The musician cannot control the method of synthesis. The leading manufacturers have chosen to compete on price, rather than performance, selling the cheapest possible sound devices that provide the absolute minimum acceptable capability and sound quality. When compared with the rapid development of PC-based graphics hardware in recent years, this situation is especially alarming. The PC sound hardware has stagnated; further development seems impossible without some external motivating force.

F. Requirements and Applications for Audio Synthesis in MPEG–4

The goal in the development of MPEG–4 Structured Audio—the tool set providing audio synthesis capability in MPEG–4—was to reclaim the general-purpose software synthesis model for use by a broad spectrum of musicians and sound designers. By incorporating this technology in an international standard, the development of compatible tools and implementations is encouraged, and such capabilities will become available as a part of the everyday multimedia sound hardware.

Including high-quality audio synthesis in MPEG–4 also serves a number of important goals within the standard itself. It allows the standard to provide capabilities that would not be possible through natural sound or through simpler MIDI-driven parametric synthesis. We list some of these capabilities in the following.

The Structured Audio specification allows sound to be transmitted at very low bit rates. Many useful sound tracks and compositions can be coded in Structured Audio at bit rates from 0.1 to 1 kbps; as content developers become more practiced in low-bit-rate coding with such tools, the bit rates can undoubtedly be pushed even lower. In contrast to perceptual coding, there is no necessary trade-off in algorithmic coding between quality and bit rate. Low-bit-rate compressed streams can still decode into full-bandwidth, full-quality stereo output. Using synthetic coding, the trade-off is more accurately described as one of *flexibility* versus bit rate [26].

Interactive accompaniment, dynamic scoring, synthetic performance [27], and other new-media music applications can be made more functional and sophisticated by using synthetic music rather than natural music. In any application requiring dynamic control over the music content itself, a structured representation of music is more appropriate than a perceptually coded one.

Unlike existing music synthesis standards such as the MIDI protocol,* structured coding with downloaded synthesis algorithms allows accurate sound description and tight control over the sound produced. Allowing any method of synthesis to be used, not only those included in a low-cost MIDI device, provides composers with a broader range of options for sound creation.

There is an attractive unification in the MPEG–4 standard between the capabilities for synthesis and those used for effects processing. By carefully specifying the capabilities

* The MIDI protocol is properly only used for communicating between a control device and a synthesizer. However, the lack of efficient *audio* coding schemes has led MIDI files (computer files made up of MIDI commands) to fill a niche for Internet representation of music.

of the Structured Audio synthesis tool, the AudioBIFS tool for audio scene description (see Section V) is much simplified and the standard as a whole is cleaner.

Finally, Structured Audio is an example of a new concept in coding technology—that of the *flexible* or *downloadable* decoder. This idea, considered but abandoned for MPEG–4 video coding, is a powerful one whose implications have yet to be fully explored. The Structured Audio tool set is computationally *complete* in that it is capable of simulating a Turing machine [28] and thus of executing any computable sound algorithm. It is possible to download new audio decoders into the MPEG–4 terminal as Structured Audio bitstreams; the requirements and applications for such a capability remain topics for future research.

G. Concepts in Speech Synthesis

Text-to-speech (TTS) systems generate speech sound according to given text. This technology enables the translation of text information into speech so that it can be transferred through speech channels such as telephone lines. Today, TTS systems are used for many applications, including automatic voice–response systems (the ''telephone menu'' systems that have become popular recently), e-mail reading, and information services for the visually handicapped [29,30].

TTS systems typically consist of multiple processing modules as shown in Figure 6 (Sec. III). Such a system accepts text as input and generates a corresponding *phoneme* sequence. Phonemes are the smallest units of human language; each phoneme corresponds to one sound used in speech. A surprisingly small set of phonemes, about 120, is sufficient to describe all human languages.

The phonome sequence is used in turn to generate a basic speech sequence without *prosody*, that is, without pitch, duration, and amplitude variations. In parallel, a text-understanding module analyzes the input for phrase structure and inflections. Using the result of this processing, a prosody generation module creates the proper prosody for the text. Finally, a prosody control module changes the prosody parameters of the basic speech sequence according to the results of the text-understanding module, yielding synthesized speech.

One of the first successful TTS systems was the DecTalk English speech synthesizer developed in 1983 [31]. This system produces very intelligible speech and supports eight different speaking voices. However, developing speech synthesizers of this sort is a difficult process, because it is necessary to program the acoustic parameters for synthesis into the system. It is a painstaking process to analyze enough data to accumulate the parameters that are used for all kinds of speech.

In 1992, CNET in France developed the pitch-synchronous overlap-and-add (PSOLA) method to control the pitch and phoneme duration of synthesized speech [32]. Using this technique, it is easy to control the prosody of synthesized speech. Thus, synthesized speech using PSOLA sounds more natural; it can also use human speech as a guide to control the prosody of the synthesis, in an analysis–synthesis process that can also modify the tone and duration. However, if the tone is changed too much, the resulting speech is easily recognized as artificial.

In 1996, ATR in Japan developed the CHATR speech synthesizer [5]. This method relies on short samples of human speech without modifying any characteristics; it locates and sequences phonemes, words, or phrases from a database. A large database of human speech is necessary to develop a TTS system using this method. Automatic tools may be used to label each phoneme of the human speech to reduce the development time; typically, hidden Markov models (HMMs) are used to align the best phoneme candidates to the target speech. The synthesized speech is very intelligible and natural; however, this TTS method requires large amounts of memory and processing power.

The applications of TTS are expanding in telecommunications, personal computing, and the Internet. Current research in TTS includes voice conversion (synthesizing the sound of a particular speaker's voice), multilanguage TTS, and enhancing the naturalness of speech through more sophisticated voice models and prosody generators.

H. Applications for Speech Synthesis in MPEG–4

The synthetic speech system in MPEG–4 was designed to support interactive applications using text as the basic content type. These applications include on-demand storytelling, motion picture dubbing, and talking-head synthetic videoconferencing.

In the storytelling on demand (STOD) application, the user can select a story from a huge database stored on fixed media. The STOD system reads the story aloud, using the MPEG–4 TTSI with the MPEG–4 facial animation tool or with appropriately selected images. The user can stop and resume speaking at any moment through the user interface of the local machine (for example, mouse or keyboard). The user can also select the gender, age, and speech rate of the electronic storyteller.

In a motion picture–dubbing application, synchronization between the MPEG–4 TTSI decoder and the encoded moving picture is the essential feature. The architecture of the MPEG–4 TTS decoder provides several levels of synchronization granularity. By aligning the composition time of each sentence, coarse granularity of synchronization can easily be achieved. To get more finely tuned synchronization, information about the speaker's lip shape can be used. The finest granularity of synchronization can be achieved by using detailed prosody transmission and video-related information such as sentence duration and offset time in the sentence. With this synchronization capability, the MPEG–4 TTSI can be used for motion picture dubbing by following the lip shape and the corresponding time in the sentence.

To enable synthetic video-teleconferencing, the TTSI decoder can be used to drive the facial animation decoder in synchronization. *Bookmarks* in the TTSI bitstream control an animated face by using facial animation parameters (FAPs); in addition, the animation of the mouth can be derived directly from the speech phonemes. Other applications of the MPEG–4 TTSI include speech synthesis for avatars in virtual reality (VR) applications, voice newspapers, dubbing tools for animated pictures, and low-bit-rate Internet voice tools.

III. MPEG–4 STRUCTURED AUDIO

The tool that provides audio synthesis capability in MPEG–4 is termed the Structured Audio coder. This name originates in the Vercoe et al. [1] comparison of different methods of parametrized sound generation—it refers to the fact that this tool provides general access to any method of structuring sound. Whereas the preceding discussion of music synthesis technology mainly described tools for composers and musicians, MPEG–4 Structured Audio is, finally, a *codec* like the other audio tools in MPEG–4. That is, the standard specifies a bitstream format and a method of decoding it into sound. Although the techniques used in decoding the bitstream are those taken from the practice of general-purpose digital synthesis and the bitstream format is somewhat unusual, the overall paradigm is identical to that of the natural audio codecs in MPEG–4.

This section will describe the organization of the Structured Audio standard, focusing first on the bitstream format and then on the decoding process. There is a second, simpler, tool for using parametrized wavetable synthesis with downloaded sounds; we will discuss this tool at the end of the section and then conclude with a short discussion of encoding Structured Audio bitstreams.

A. Structured Audio Bitstream Format

The Structured Audio bitstream format makes use of the new coding paradigm known as *algorithmic structured audio*, described by Vercoe et al. [1] and Scheirer [33]. In this framework, a sound transmission is decomposed into two pieces: a set of *synthesis algorithms* that describe how to create sound and a sequence of *synthesis controls* that specify which sounds to create. The synthesis model is not fixed in the MPEG–4 terminal; rather, the standard specifies a framework for reconfigurable software synthesis. Any current or future method of digital sound synthesis can be used in this framework.

As with the other MPEG–4 media types, a Structured Audio bitstream consists of a *decoder configuration header* that tells the decoder how to begin the decoding process, and then a stream of *bitstream access units* that contain the compressed data. In Structured Audio, the decoder configuration header contains the synthesis algorithms and auxiliary data, and the bitstream access units contain the synthesis control instructions.

1. Decoder Configuration Header and SAOL

The decoder configuration header specifies the synthesis algorithms using a new unit generator language called SAOL (pronounced ''sail''), which stands for Structured Audio Orchestra Language. The syntax and semantics of SAOL are specified precisely in the standard—MPEG–4 contains the formal specification of SAOL as a language. The similarities and differences between SAOL and other popular music languages have been discussed elsewhere [34]. Space does not provide for a full tutorial on SAOL in this chapter, but we give a short example so that the reader may understand the flavor of the language.

Figure 2 shows the textual representation of a complete SAOL synthesizer or *orchestra*. This synthesizer defines one instrument (called ''beep'') for use in a Structured Audio session. Each bitstream begins with a SAOL orchestra that provides the instruments needed in that session.

The synthesizer description as shown in Figure 2 begins with a *global header* that specifies the sampling rate (in this case, 32 kHz) and control rate (in this case, 1 kHz) for this orchestra. SAOL is a two-rate signal language—every variable represents either an *audio signal* that varies at the sampling rate or a *control signal* that varies at the control rate. The sampling rate of the orchestra limits the maximum audio frequencies that may be present in the sound, and the control rate limits the speed with which parameters may vary. Higher values for these parameters lead to better sound quality but require more computation. This trade-off between quality and complexity is left to the decision of the content author and can differ from bitstream to bitstream.

After the global header comes the specification for the instrument **beep**. This instrument depends on two *parameter fields* (p-fields) named **pitch** and **amp**. The number, names, and semantics of p-fields for each instrument are not fixed in the standard; they

```
global {
  srate 32000;
  krate 1000;
}

instr beep(pitch, amp) {
  asig out;
  ksig env;
  table sound(harm,2048,1,0.5,0,0.2);

  env = kline(0,0.1,amp,dur-0.1,0);
  out = oscil(sound,pitch) * amp;
  output(out);
}
```

Figure 2 A SAOL orchestra containing one instrument that makes a ramped complex tone. Compare the syntax with the ''Music-N''–like syntax shown in Figure 1. See text for in-depth discussion of the orchestra code.

are decided by the content author. The values for the p-fields are set in the **score**, which is described in Sec. III.A.2. The instrument defines two signal variables: **out**, which is an audio signal, and **env**, which is a control signal. It also defines a stored-function table called **sound**.

Stored-function tables, also called *wavetables*, are crucial to general-purpose software synthesis. As shown in Figure 1, nearly any synthesis algorithm can be realized as the interaction of a number of oscillators creating appropriate signals; wavetables are used to store the periodic functions needed for this purpose. A stored-function table in SAOL is created by using one of several *wavetable generators* (in this case, **harm**) that allocate space and fill the table with data values. The **harm** wavetable generator creates one cycle of a periodic function by summing a set of zero-phase harmonically related sinusoids; the function placed in the table called **sound** consists of the sum of four sine waves at frequencies 1, 2, and 4 with amplitudes 1, 0.5, and 0.2, respectively. This function is sampled at 2048 points per cycle to create the wavetable.

To create sound, the **beep** instrument uses an interaction of two *unit generators*, **kline** and **oscil**. A set of about 100 unit generators is specified in the standard, and content authors can also design and deliver their own. The **kline** unit generator generates a control-rate envelope signal; in the example instrument it is assigned to the control-rate signal variable **env**. The **kline** unit generator interpolates a straight-line function between several (time, val) control points; in this example, a line segment function is specified that goes from 0 to the value of the **amp** parameter in 0.1 sec and then back down to 0 in **dur** − 0.1 sec; **dur** is a *standard name* in SAOL that always contains the duration of the note as specified in the score.

In the next line of the instrument, the **oscil** unit generator converts the wavetable **sound** into a periodic audio signal by oscillating over this table at a rate of **cps** cycles per second. Not every point in the table is used (unless the frequency is very low); rather, the **oscil** unit generator knows how to select and interpolate samples from the table in order to create one full cycle every 1/**cps** seconds. The sound that results is multiplied by the control-rate signal **env** and the overall sound amplitude **amp**. The result is assigned to the audio signal variable **out**. The last line of the instrument contains the **output** statement, which specifies that the sound output of the instrument is contained in the signal variable **out**.

When the SAOL orchestra is transmitted in the bitstream header, the plain-text format is not used. Rather, an efficient tokenized format is standardized for this purpose. The Structured Audio specification contains a description of this tokenization procedure. The decoder configuration header may also contain auxiliary data to be used in synthesis. For example, a type of synthesis popular today is ''wavetable'' or ''sampling'' synthesis, in which short clips of sound are pitch shifted and added together to create sound. The sound samples for use in this process are not included directly in the orchestra (although this is allowed if the samples are short) but placed in a different segment of the bitstream header.

Score data, which normally reside in the bitstream access units as described later, may also be included in the header. By including in the header score instructions that are known when the session starts, the synthesis process may be able to allocate resources more efficiently. Also, real-time tempo control over the music is possible only when the notes to be played are known beforehand. For applications in which it is useful to reconstruct a human-readable orchestra from the bitstream, a *symbol table* may also be included in the bitstream header. This element is not required and has no effect on the decoding process, but allows the compressed bitstream representation to be converted back into a human-readable form.

2. Bitstream Access Units and SASL

The streaming access units of the Structured Audio bitstream contain instructions that specify how the instruments that were described in the header should be used to create sound. These instructions are specified in another new language called SASL, for Structured Audio Score Language. A example set of such instructions, or *score*, is given in Figure 3.

Each line in this score corresponds to one note of synthesis. That is, for each line in the score, a different note is played using one of the synthesizers defined in the orchestra header. Each line contains, in order, a *time stamp* indicating the time at which the note should be triggered, the name of the instrument that should perform the synthesis, the duration of the note, and the parameters required for synthesis. The semantics of the parameters are not fixed in the standard but depend on the definition of the instrument. In this case, the first parameter corresponds to the **cps** field in the instrument definition in Figure

```
0.0   beep 1.0 440 0.5
1.0   beep 2.0 220 0.2
2.0   beep 1.0 264 0.5
3.0   beep 1.0 440 0.5
4.0   end
```

Figure 3 A SASL score, which uses the orchestra in Figure 2 to play four notes. In an MPEG–4 Structured Audio bitstream, each score line is compressed and transmitted as an access unit.

Figure 4 The musical notation corresponding to the SASL score in Figure 3.

2 and the second parameter in each line to the **amp** field. Thus, the score in Figure 3 includes four notes that correspond to the musical notation shown in Figure 4.

In the streaming bitstream, each line of the score is packaged as an access unit. The multiplexing of the access units with those in other streams and the actual insertion of the access units into a bitstream for transport are performed according to the MPEG–4 multiplex specification; see Chapter 13.

There are many other sophisticated instructions in the orchestra and score formats; space does not permit a full review, but more details can be found in the standard and in other references on this topic [34]. In the SAOL orchestra language, there are built-in functions corresponding to many useful types of synthesis; in the SASL score language, tables of data can be included for use in synthesis, and the synthesis process can be continuously manipulated with customizable parametric controllers. In addition, time stamps can be removed from the score lines, allowing a real-time mode of operation such as the transmission of live performances.

B. Decoding Process

The decoding process for Structured Audio bitstreams is somewhat different from the decoding process for natural audio bitstreams. The streaming data do not typically consist of ''frames'' of data that are decompressed to give buffers of audio samples; rather, they consist of parameters that are fed into a synthesizer. The synthesizer creates the audio buffers according to the specification given in the header. A schematic of the Structured Audio decoding process is given in Figure 5.

The first step in decoding the bitstream is processing and understanding the SAOL instructions in the header. This stage of the bitstream processing is similar to compiling or interpreting a high-level language. The MPEG–4 standard specifies the semantics of SAOL—the sound that a given instrument declaration is supposed to produce—exactly, but it does not specify the exact manner of implementation. Software, hardware, or dual software–hardware solutions are all possible for Structured Audio implementation; however, programmability is required, and thus fixed-hardware implementations are difficult to realize. The SAOL preprocessing stage results in a new configuration for the *reconfigurable synthesis engine*. The capabilities and proper functioning of this engine are described fully in the standard.

After the header is received and processed, synthesis from the streaming access units begins. Each access unit contains a score line that directs some aspect of the synthesis process. As each score line is received by the terminal, it is parsed and registered with the Structured Audio *scheduler* as an *event*. A time-sequenced list of events is maintained, and the scheduler triggers each at the appropriate time.

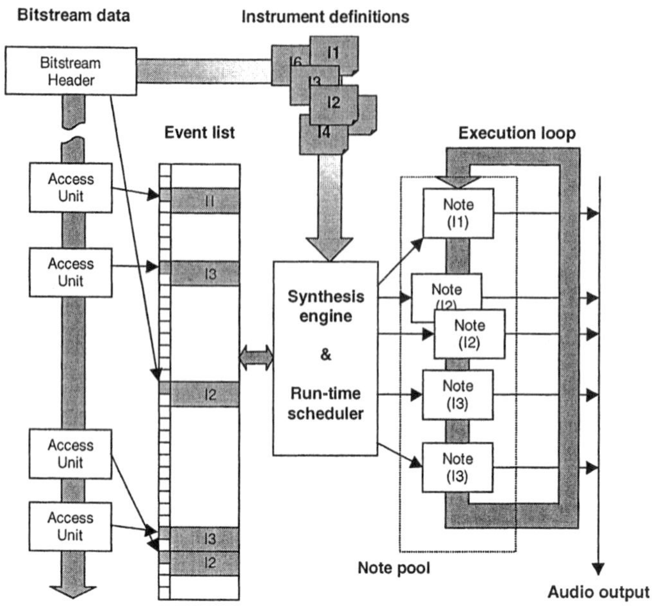

Figure 5 Overview of the MPEG–4 Structured Audio decoding process. See text for details.

When an event is triggered to turn on a note, a *note object* or *instrument instantiation* is created. A pool of active notes is always maintained; this pool contains all of the notes that are currently active (or ''on''). As the decoder executes, it examines each instrument instantiation in the pool in turn, performing the next small amount of synthesis that the SAOL code describing that instrument specifies. This processing generates one frame of data for each active note event. The frames are summed together for all notes to produce the overall decoder output.

Because SAOL is a very powerful format for the description of synthesis, it is not generally possible to characterize the specific algorithms that are executed in each note event. The content author has complete control over the methods used for creating sound and the resulting sound quality. Although the specification is flexible, it is still strictly *normative* (specified in the standard); this guarantees that a bitstream produces the same sound when played on any conforming decoder.

C. Wavetable Synthesis in MPEG–4

A simpler format for music synthesis is also provided in MPEG–4 Structured Audio for applications that require low-complexity operation and do not require sophisticated or interactive music content—karaoke systems are the primary example. A format for representing banks of wavetables, the Structured Audio Sample Bank Format of SASBF, was created in collaboration with the MIDI Manufacturer's Association for this purpose.

Using SASBF, wavetable synthesizers can be downloaded to the terminal and controlled with MIDI sequences. This type of synthesis processing is readily available today; thus, a terminal using this format may be manufactured very cheaply. Such a terminal still allows synthetic music to be synchronized and mixed with recorded vocals or other natural sounds. Scheirer and Ray [26] have presented a comparison of algorithmic (the "main profile") and wavetable synthesis in MPEG–4.

D. Encoding Structured Audio Bitstreams

As with all MPEG standards, only the bitstream format and decoding process are standardized for the Structured Audio tools. The method of encoding a legal bitstream is outside the scope of the standard. However, the natural audio coders described in Chapter 5, like those of previous MPEG Audio standards, at least have well-known starting points for automatic encoding. Many tools have been constructed that allow an existing recording (or live performance) to be turned automatically into legal bitstreams for a given perceptual coder.

This is not yet possible for Structured Audio bitstreams; the techniques required to do this fully automatically are still in a basic research stage, where they are known as *polyphonic transcription* or *automatic source separation*. Thus, for the forseeable future, human intervention is required to produce Structured Audio bitstreams. As the tools required for this are very similar to other tools used in a professional music studio today—such as sequencers and multitrack recording equipment—this is not viewed as an impediment to the utility of the standard.

IV. THE MPEG–4 TEXT-TO-SPEECH INTERFACE

Text—that is, a sequence of words written in some human language—is a widely used representation for speech data in stand-alone applications. However, it is difficult with existing technology to use text as a speech representation in multimedia bitstreams for transmission. The MPEG–4 text-to-speech interface (TTSI) is defined so that speech can be transmitted as a bitstream containing text. It also ensures interoperability among text-to-speech (TTS) synthesizers by standardizing a single bitstream format for this purpose.

Synthetic speech is becoming a rather common media type; it plays an important role in various multimedia application areas. For instance, by using TTS functionality, multimedia content with narration can easily be created without recording natural speech. Before MPEG–4, however, there was no easy way for a multimedia content provider to give instructions to an unknown TTS system. In MPEG–4, a single common interface for TTS systems is standardized; this interface allows speech information to be transmitted in the International Phonetic Alphabet (IPA) or in a textual (written) form of any language.

The MPEG–4 TTSI tool is a hybrid or multilevel scalable TTS interface that can be considered a superset of the conventional TTS framework. This extended TTSI can utilize prosodic information taken from natural speech in addition to input text and can thus generate much higher quality synthetic speech. The interface and its bitstream format are strongly scalable in terms of this added information; for example, if some parameters of prosodic information are not available, a decoder can generate the missing parameters by rule. Normative algorithms for speech synthesis and text-to-phoneme translation are not specified in MPEG–4, but to meet the goal that underlies the MPEG–4 TTSI, a decoder should fully utilize all the information provided according to the user's requirements level.

As well as an interface to text-to-speech synthesis systems, MPEG–4 specifies a joint coding method for phonemic information and facial animation (FA) parameters. Using this technique, a single bitstream may be used to control both the TTS interface and the facial animation visual object decoder. The functionality of this extended TTSI thus ranges from conventional TTS to natural speech coding and its application areas—from simple TTS to audiovisual presentation with TTS and moving picture dubbing with TTS.

The section describes the functionality of the MPEG–4 TTSI, its decoding process, and applications of the MPEG–4 TTSI.

A. MPEG–4 TTSI Functionality

The MPEG–4 TTSI has important functionalities both as an individual codec and in synchronization with the facial animation techniques described in Chapter 11. As a standalone codec, the bitstream format provides hooks to control the language being transmitted, the gender and age of the speaker, the speaking rate, and the prosody (pitch contour) of the speech. It can pause with no cost in bandwidth by transmission of a silence sentence that has only silence duration. A ''trick mode'' allows operations such as start, stop, rewind, and fast forward to be applied to the synthesized speech.

The basic TTSI format is extremely low bit rate. In the most compact method, one can send a bitstream that contains only the text to be spoken and its length. In this case, the bit rate is 200 bits per second. The synthesizer will add predefined or rule-generated prosody to the synthesized speech (in a nonnormative fashion). The synthesized speech in this case will deliver the emotional content to the listener.

On the other hand, one can send a bitstream that contains text as well as the detailed prosody of the original speech, that is, phoneme sequence, duration of each phoneme, base frequency (pitch) of each phoneme, and energy of each phoneme. The synthesized speech in this case will be very similar to the original speech because the original prosody is employed. Thus, one can send speech with subtle nuances without any loss of intonation using MPEG–4 TTSI.

One of the important features of the MPEG–4 TTSI is the ability to synchronize synthetic speech with the lip movements of a computer-generated *avatar* or talking head. In this technique, the TTS synthesizer generates phoneme sequences and their durations and communicates them to the facial animation visual object decoder so that it can control the lip movement. With this feature, one can not only hear the synthetic speech but also see the synchronized lip movement of the avatar.

The MPEG–4 TTSI has the additional capability to send facial expression bookmarks through the text. The bookmark is identified by ''⟨FAP,'' and lasts until the closing bracket ''⟩.'' In this case, the TTS synthesizer transfers the bookmark directly to the face decoder so that it can control the facial animation visual object accordingly. The FAP of the bookmark is applied to the face until another bookmark resets the FAP. Content capable of playing sentences correctly, even in trick-mode manipulations, requires that bookmarks of the text to be spoken are repeated at the beginning of each sentence. These bookmarks initialize the face to the state that is defined by the previous sentence. In such a case, some mismatch of synchronization can occur at the beginning of a sentence; however, the system recovers when the new bookmark is processed.

Through the MPEG–4 elementary stream synchronization capabilities (see Chap. 13), the MPEG–4 TTSI can perform synthetic motion picture dubbing. The MPEG–4 TTSI decoder can use the system clock to select an adequate speech location in a sentence and communicates this to the TTS synthesizer, which assigns appropriate duration for each phoneme. Using this method, synthetic speech can be synchronized with the lip shape of the moving image.

B. MPEG–4 TTSI Decoding Process

Figure 6 shows a schematic of the MPEG–4 TTSI decoder. The architecture of the decoder can be described as a collection of interfaces. The normative behavior of the MPEG–4 TTSI is described in terms of these interfaces, not the sound and/or animated faces that are produced.

In particular, the TTSI standard specifies the following:

The interface between the demux and the syntactic decoder. Upon receiving a multiplexed MPEG–4 bitstream, the demux passes coded MPEG–4 TTSI elementary streams to the syntactic decoder. Other elementary streams are passed to other decoders.

The interface between the syntactic decoder and the TTS synthesizer. Receiving a coded MPEG–4 TTSI bitstream, the syntactic decoder passes a number of different pieces of data to the TTS synthesizer. The *input type* specifies whether TTS is being used as a stand-alone function or in synchronization with facial animation or motion picture dubbing. The *control commands sequence* specifies the language, gender, age, and speech rate of the speaking voice. The *input text* specifies the character string for the text to be synthesized. Auxiliary information such as IPA phoneme symbols (which allow text in a language foreign to the decoder to be synthesized), lip shape patterns, and trick-mode commands are also passed along this interface.

The interface from the TTS synthesizer to the compositor. Using the parameters described in the previous paragraph, the synthesizer constructs a speech sound and delivers it to the audio composition system (described in Sec. V).

The interface from the compositor to the TTS synthesizer. This interface allows local control of the synthesized speech by users. Using this interface and an appropriate interactive scene, users can start, stop, rewind, and fast forward the TTS system. Controls can also allow changes in the speech rate, pitch range, gender, and age of the synthesized speech by the user.

The interface between the TTS synthesizer and the phoneme/bookmark-to-FAP converter. In the MPEG–4 framework, the TTS synthesizer and the face animation can be driven synchronously, by the same input control stream, which is the

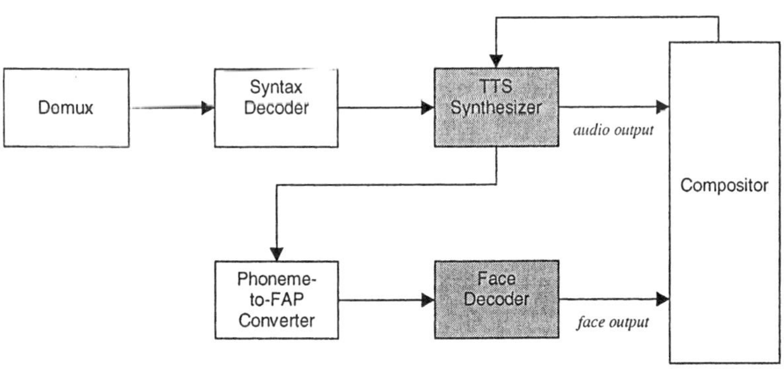

Figure 6 Overview of the MPEG–4 TTSI decoding process, showing the interaction between the syntax parser, the TTS synthesizer, and the face animation decoder. The shaded blocks are not normatively described and operate in a terminal-dependent manner.

text input to the MPEG–4 TTSI. From this input stream, the TTS synthesizer generates synthetic speech and, at the same time, phoneme symbols, phoneme durations, word boundaries, stress parameters, and bookmarks. The phonemic information is passed to the phoneme/bookmark-to-FAP converter, which generates relevant facial animation accordingly. Through this mechanism, the synthesized speech and facial animation are synchronized when they enter the scene composition framework.

V. MPEG–4 AUDIO/SYSTEMS INTERFACE AND AUDIOBIFS

This section describes the relation between the MPEG–4 audio decoders and the MPEG–4 Systems functions of elementary stream management and composition. By including sophisticated capabilities for mixing and postproducing multiple audio sources, MPEG–4 enables a great number of advanced applications such as virtual-reality sound, interactive music experiences, and adaptive soundtracks.

A detailed introduction to elementary stream management in MPEG–4 is in Chapter 13; an introduction to the MPEG–4 Binary Format for Scenes (BIFS) is in Chapter 14. The part of BIFS controlling the composition of a sound scene is called *AudioBIFS*. AudioBIFS provides a unified framework for sound scenes that use streaming audio, interactive presentation, 3D spatialization, and dynamic download of custom signal-processing effects. Scheirer et al. [35] have presented a more in-depth discussion of the AudioBIFS tools.

A. AudioBIFS Requirements

Many of the main BIFS concepts originate from the Virtual Reality Modeling Language (VRML) standard [36], but the audio tool set is built from a different philosophy. AudioBIFS contains significant advances in quality and flexibility over VRML audio. There are two main modes of operation that AudioBIFS is intended to support. We term them *virtual-reality* and *abstract-effects* compositing.

In virtual-reality compositing, the goal is to recreate a particular acoustic environment as accurately as possible. Sound should be presented spatially according to its location relative to the listener in a realistic manner; moving sounds should have a Doppler shift; distant sounds should be attenuated and low-pass filtered to simulate the absorptive properties of air; and sound sources should radiate sound unevenly, with a specific frequency-dependent directivity pattern. This type of scene composition is most suitable for ''virtual world'' applications and video games, where the application goal is to immerse the user in a synthetic environment. The VRML sound model embraces this philosophy, with fairly lenient requirements on how various sound properties must be realized in an implementation.

In abstract-effects compositing, the goal is to provide content authors with a rich suite of tools from which artistic considerations can be used to choose the right effect for a given situation. As Scheirer [33] discusses in depth, the goal of sound designers for traditional media such as films, radio, and television is not to recreate a virtual acoustic environment (although this would be well within the capability of today's film studios) but to apply a body of knowledge regarding ''what a film should sound like.'' Spatial effects are sometimes used, but often not in a physically realistic way; the same is true for the filters, reverberations, and other sound-processing techniques used to create various artistic effects that are more compelling than strict realism would be.

MPEG realized in the early development of the MPEG–4 sound compositing tool set that if the tools were to be useful to the traditional content community—always the primary audience of MPEG technology—then the abstract-effects composition model would need to be embraced in the final MPEG–4 standard. However, new content paradigms, game developers, and virtual-world designers demand high-quality sonification tools as well.

MPEG–4 AudioBIFS therefore integrates these two components into a single standard. Sound in MPEG–4 may be postprocessed with arbitrary downloaded filters, reverberators, and other digital-audio effects; it may also be spatially positioned and physically modeled according to the simulated parameters of a virtual world. These two types of postproduction may be freely combined in MPEG–4 audio scenes.

B. The MPEG–4 Audio System

A schematic diagram of the overall audio system in MPEG–4 is shown in Figure 7 and may be a useful reference during the discussion to follow.

Sound is conveyed in the MPEG–4 bitstream as several *elementary streams* that contain coded audio in the formats described earlier in this chapter and in Chapter 5. There are four elementary streams in the sound scene in Figure 7. Each of these elementary streams contains a *primitive media object*, which in the case of audio is a single-channel or multichannel sound that will be composited into the overall scene. In Figure 7, the GA-coded stream decodes into a stereo sound and the other streams into monophonic sounds. The different primitive audio objects may each make use of a different audio decoder, and decoders may be used multiple times in the same scene.

The multiple elementary streams are conveyed together in a multiplexed representation. Multiple multiplexed streams may be transmitted from multiple servers to a single MPEG–4 receiver, or *terminal*. Two multiplexed MPEG–4 bitstreams are shown in Figure 7; each originates from a different server. Encoded video content can also be multiplexed into the same MPEG–4 bitstreams. As they are received in the MPEG–4 terminal, the MPEG–4 bitstreams are demultiplexed, and each primitive media object is decoded. The resulting sounds are not played directly but rather made available for scene compositing using AudioBIFS.

C. AudioBIFS Nodes

Also transmitted in the multiplexed MPEG–4 bitstream is the BIFS scene graph itself. BIFS and AudioBIFS are simply parts of the content like the media objects themselves; there is nothing ''hardwired'' about the scene graph in MPEG–4. Content developers have wide flexibility to use BIFS in a variety of ways. In Figure 7, the BIFS part and the AudioBIFS part of the scene graph are separated for clarity, but there is no technical difference between them.

Like the rest of the BIFS capabilities (introduced in Chap. 14), AudioBIFS consists of a number of nodes that are interlinked in a scene graph. However, the concept of the AudioBIFS scene graph is somewhat different; it is termed an *audio subgraph*.

Whereas the main (visual) scene graph represents the position and orientation of

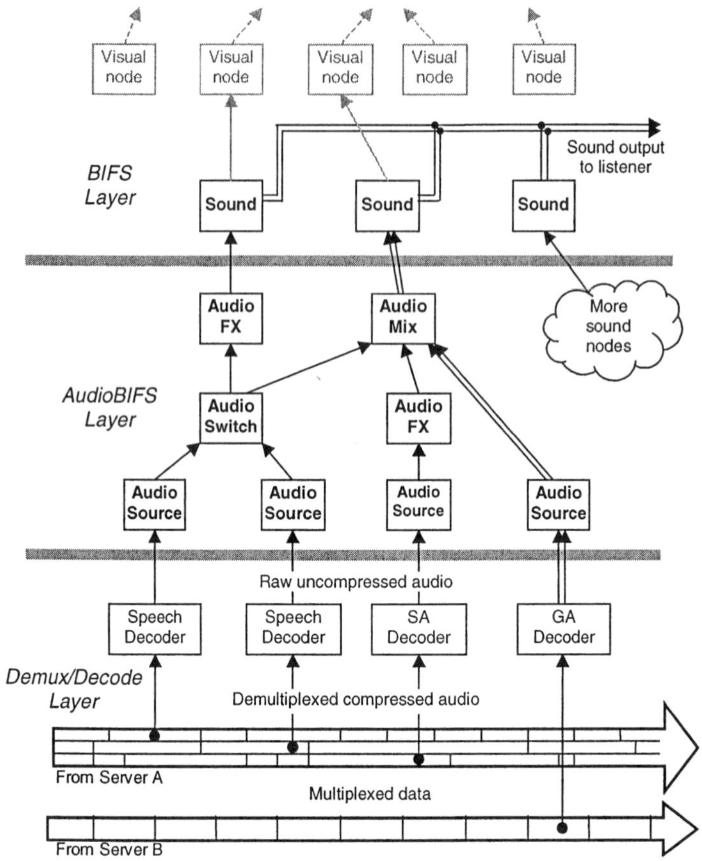

Figure 7 The MPEG–4 Audio system, showing the demux, decode, AudioBIFS, and BIFS layers. This schematic shows the interaction between the frames of audio data in the bitstream, the decoders, and the scene composition process. See text for details.

visual objects in presentation space and their properties such as color, texture, and layering, an audio subgraph represents a signal-flow graph describing digital signal-processing manipulations. Sounds flow in from MPEG–4 audio decoders at the bottom of the scene graph; each ''child'' node presents its results from processing to one or more ''parent'' nodes. Through this chain of processing, sound streams eventually arrive at the top of the audio subgraph. The ''intermediate results'' in the middle of the manipulation process are not sounds to be played to the user; only the result of the processing at the top of each audio subgraph is presented.

The **AudioSource** node is the point of connection between real-time streaming audio and the AudioBIFS scene. The **AudioSource** node attaches an audio decoder, of one of the types specified in the MPEG–4 audio standard, to the scene graph; audio flows out of this node.

The **Sound** node is used to attach sound to audiovisual scenes, either as 3D directional sound or as nonspatial ambient sound. All of the spatial and nonspatial sounds produced by **Sound** nodes in the scene are summed and presented to the listener. The semantics of the **Sound** node in MPEG–4 are similar to those of the VRML standard.

The sound attenuation region and spatial characteristics are defined in the same way as in the VRML standard to create an elliptical model of attenuation. In contrast to VRML, where the **Sound** node accepts raw sound samples directly and no intermediate processing is done, in MPEG–4 any of the AudioBIFS nodes may be attached. Thus, if an *AudioSource* node is the child node of the **Sound** node, the sound as transmitted in the bitstream is added to the sound scene; however, if a more complex audio scene graph is beneath the **Sound** node, the mixed or effects-processed sound is presented.

The **AudioMix** node allows M channels of input sound to be mixed into N channels of output sound through the use of a mixing matrix.

The **AudioSwitch** node allows N channels of output to be taken as a subset of M channels of input, where $M \leq N$. It is equivalent to, but easier to compute than, an **AudioMix** node where $M \leq N$ and all matrix values are 0 or 1. This node allows efficient selection of certain channels, perhaps on a language-dependent basis.

The **AudioDelay** node allows several channels of audio to be delayed by a specified amount of time, enabling small shifts in timing for media synchronization.

The **AudioFX** node allows the dynamic download of custom signal-processing effects to apply to several channels of input sound. Arbitrary effects-processing algorithms may be written in SAOL and transmitted as part of the scene graph. The use of SAOL to transmit audio effects means that MPEG does not have to standardize the ''best'' artificial reverberation algorithm (for example) but also that content developers do not have to rely on terminal implementers and trust in the quality of the algorithms present in an unknown terminal. Because the execution method of SAOL algorithms is precisely specified, the sound designer has control over exactly which reverberation algorithm (for example) is used in a scene. If a reverb with particular properties is desired, the content author transmits it in the bitstream; its use is then guaranteed.

The position of the **Sound** node in the overall scene and the position of the listener are also made available to the **AudioFX** node, so that effects processing may depend on the spatial locations (relative or absolute) of the listener and sources.

The **AudioBuffer** node allows a segment of audio to be excerpted from a stream and then triggered and played back interactively. Unlike the VRML node **AudioClip**, the **AudioBuffer** node does not itself contain any sound data. Instead, it records the first n seconds of sound produced by its children. It captures this sound into an internal buffer. Later, it may be triggered interactively to play that sound back.

This function is most useful for ''auditory icons'' such as feedback to button presses. It is impossible to make streaming audio provide this sort of audio feedback, because the stream is (at least from moment to moment) independent of user interaction. The backchannel capabilities of MPEG–4 are not intended to allow the rapid response required for audio feedback. There is also a special function of **AudioBuffer** that allows it to cache samples for sampling synthesis in the Structured Audio decoder. This technique allows perceptual compression to be applied to sound samples, which can greatly reduce the size of bitstreams using sampling synthesis.

The **ListeningPoint** node allows the user to set the listening point in a scene. The listening point is the position relative to which the spatial positions of sources are calculated. By default (if no **ListeningPoint** node is used), the listening point is the same as the visual viewpoint.

The **TermCap** node is not an AudioBIFS node specifically but provides capabilities that are useful in creating terminal-adaptive scenes. The **TermCap** node allows the scene graph to query the terminal on which it is running in order to discover hardware and performance properties of that terminal. For example, in the audio case, **TermCap** may be used to determine the ambient signal-to-noise ratio of the environment. The result can be used to control ''switching'' between different parts of the scene graph, so that (for example) a compressor is applied in a noisy environment such as an automobile but not in a quiet environment such as a listening room. Other audio-pertinent resources that may be queried with the **TermCap** node include the number and configuration of loudspeakers, the maximum output sampling rate of the terminal, and the level of sophistication of 3D audio functionality available.

The MPEG–4 Systems standard contains specifications for the resampling, buffering, and synchronization of sound in AudioBIFS. Although we will not discuss these aspects in detail, for each of the AudioBIFS nodes there are precise instructions in the standard for the associated resampling and buffering requirements. These aspects of MPEG–4 are normative. This makes the behavior of an MPEG–4 terminal highly predictable to content developers.

VI. SUMMARY

We have described the tools for synthetic and SNHC audio in MPEG–4. By using these tools, content developers can create high-quality, interactive content and transmit it at extremely low bit rates over digital broadcast channels or the Internet. The Structured Audio tool set provides a single standard to unify the world of algorithmic music synthesis and to drive forward the capabilities of the PC audio platform; the Text-to-Speech Interface provides a greatly needed measure of interoperability between content and text-to-speech systems.

REFERENCES

1. BL Vercoe, WG Gardner, ED Scheirer. Structured audio: The creation, transmission, and rendering of parametric sound representations. Proc IEEE 85:922–940, 1998.
2. MV Mathews. The Technology of Computer Music. Cambridge, MA: MIT Press, 1969.
3. J-C Risset, MV Mathews. Analysis of musical instrument tones. Phys Today 22(2):23–30, 1969.
4. J-C Risset, DL Wessel. Exploration of timbre by analysis and synthesis. In: D Deutsch, ed. The Psychology of Music. Orlando, FL: Academic Press, 1982, pp 25–58.
5. JM Chowning. The synthesis of complex audio spectra by means of frequency modulation. In: C Roads, J Strawn, eds. Foundations of Computer Music. Cambridge, MA: MIT Press, 1985, pp 6–29.
6. JO Smith. Acoustic modeling using digital waveguides. In: C Roads, ST Pope, A Piccialli, G de Poli, eds. Musical Signal Processing. Lisse, NL: Swets & Zeitlinger, 1997, pp 221–264.
7. S Cavaliere, A Piccialli. Granular synthesis of musical signals. In: C Roads, ST Pope, A Piccialli, G de Poli, eds. Musical Signal Processing. Lisse, NL: Swets & Zeitlinger, 1997, pp 221–264.
8. C Roads. The Computer Music Tutorial. Cambridge, MA: MIT Press, 1996.
9. MV Mathews. An acoustic compiler for music and psychological stimuli. Bell Syst Tech J 40:677–694, 1961.

10. MV Mathews. The digital computer as a musical instrument. Science 142:553–557, 1963.
11. ST Pope. Machine tongues XV: Three packages for sound synthesis. Comput Music J 17(2): 23–54.
12. BL Vercoe, DPW Ellis. Real-time CSound: Software synthesis with sensing and control. Proceedings of ICMC, San Francisco, 1990, pp 209–211.
13. P Manning. Electronic and Computer Music. Oxford: Clarendon Press, 1985.
14. T Stilson, J Smith. Alias-free digital synthesis of classic analog waveforms. Proceedings of ICMC, Hong Kong, 1996, pp 332–335.
15. T Stilson, JO Smith. Analyzing the Moog VCF with considerations for digital implementation. Proceedings of ICMC, Hong Kong, 1996, pp 398–401.
16. J Lane, D Hoory, E Martinez, P Wang. Modeling analog synthesis with DSPs. Comput Music J 21(4):23–41.
17. DC Massie. Wavetable sampling synthesis. In: M Kahrs, K Brandenburg, eds. Applications of Signal Processing to Audio and Acoustics. New York: Kluwer Academic, 1998.
18. BL Vercoe. Csound: A Manual for the Audio-Processing System. Cambridge MA: MIT Media Lab, 1996.
19. RB Dannenberg. Machine tongues XIX: Nyquist, a language for composition and sound synthesis. Comput Music J 21(3):50–60.
20. J McCartney. SuperCollider: A new real-time sound synthesis language. Proceedings of ICMC, Hong Kong, 1996, pp 257–258.
21. B Schottstaedt. Machine tongues XIV: CLM—Music V meets Common LISP. Comput Music J 18(2):20–37.
22. JO Smith. Viewpoints on the history of digital synthesis. Proceedings of ICMC, Montreal, 1991, pp 1–10.
23. D Wessel. Let's develop a common language for synth programming. Electronic Musician August:114, 1991.
24. BL Vercoe. Extended Csound. Proceedings of ICMC, Hong Kong, 1996, pp 141–142.
25. FR Moore. The dysfunctions of MIDI. Comput Music J 12(1):19–28.
26. ED Scheirer, L Ray. Algorithmic and wavetable synthesis in the MPEG-4 multimedia standard. Presented at 105th AES, San Francisco, 1998, AES reprint #4811.
27. BL Vercoe. The synthetic performer in the context of live performance. Proceedings of ICMC, Paris, 1984, pp 199–200.
28. JE Hopcroft, JD Ullman. Introduction to Automata Theory, Languages, and Computation. Reading, MA: Addison-Wesley, 1979.
29. D Johnston, C Sorin, C Gagnoulet, F Charpentier, F Canavesio, B Lochschmidt, J Alvarez, I Cortazar, D Taplas, C Crespo, J Azevedo, R Chaves. Current and experimental applications of speech technology for telecom services in Europe. Speech Commun 23:5–16, 1997.
30. M Kitai, K Hakoda, S Sagayama, T Yamada, H Tsukada, S Takahashi, Y Noda, J Takahashi, Y Yoshida, K Arai, T Imoto, T Hirokawa. ASR and TTS telecommunications applications in Japan. Speech Commun 23:17–30, 1997.
31. DH Klatt. Review of text-to-speech conversion for English. JASA 82:737–793, 1987.
32. H Valbret, E Moulines, JP Tubach. Voice transformation using PSOLA technique. Speech Commun 11:175–187, 1992.
33. ED Scheirer. Structured audio and effects processing in the MPEG-4 multimedia standard. ACM Multimedia Syst J, 7:11–22, 1999.
34. ED Scheirer, BL Vercoe. SAOL: The MPEG-4 Structured Audio Orchestra Language. Comput Music J 23(2):31–51.
35. ED Scheirer, R Väänänen, J Huopaniemi. AudioBIFS: Describing audio scenes with the MPEG-4 multimedia standard. IEEE Transactions on Multimedia 1:236–250, 1999.
36. International Organisation for Standardisation. 14472-1:1997, the Virtual Reality Modeling Language. Geneva:ISO, 1997.

7

MPEG–4 Visual Standard Overview

Caspar Horne
Mediamatics, Inc., Fremont, California

Atul Puri
AT&T Labs, Red Bank, New Jersey

Peter K. Doenges
Evans & Sutherland, Salt Lake City, Utah

I. INTRODUCTION

To understand the scope of the MPEG–4 Visual standard, a brief background of progress in video standardization is necessary. The standards relevant to our discussion are previous video standards by the International Standards Organization (ISO), which has been responsible for MPEG series (MPEG–1 and MPEG–2, and now MPEG–4) standards, and the International Telecommunications Union (ITU) which has produced H.263 series (H.263 version 1 and H.263+) standards. We now briefly discuss these standards, focusing mainly on the video part.

The ISO MPEG–1 Video standard [1] was originally designed for video on CD-ROM applications at bit rates of about 1.2 Mbit/sec and supports basic interactivity with stored video bitstream such as random access, fast forward, and fast reverse. MPEG–1 video coding uses block motion compensated discrete cosine transform (DCT) coding within a group-of-pictures (GOP) structure consisting of an arrangement of intra (I-), predictive (P-), and bidirectional (B-) pictures to deliver good coding efficiency and desired interactivity. This standard is optimized for coding of noninterlaced video only.

The second-phase ISO MPEG (MPEG–2) standard [2,3], on the other hand, is more generic. MPEG–2 is intended for coding of higher resolution video than MPEG–1 and can deliver television quality video in the range of 4 to 10 Mbit/sec and high-definition television (HDTV) quality video in the range of 15 to 30 Mbit/sec. The MPEG–2 Video standard is mainly optimized for coding of interlaced video. MPEG–2 video coding builds on the motion-compensated DCT coding framework of MPEG–1 Video and further includes adaptations for efficient coding of interlaced video. MPEG–2 Video supports interactivity functions of MPEG–1 Video as well as new functions such as scalability. Scalability is the property that enables decoding of subsets of the entire bitstream on decoders of less than full complexity to produce useful video from the same bitstream. Scalability in picture quality is supported via signal-to-noise ratio (SNR) scalability, scalability in spatial resolution by spatial scalability, and scalability in temporal resolution via temporal scalability. The MPEG–2 Video standard is both forward and backward compatible with MPEG–1; backward compatibility can be achieved using spatial scalability. Both the MPEG–1 and MPEG–2 standards specify bitstream syntax and decoding semantics only, allowing considerable innovation in optimization of encoding.

The MPEG–4 standard [4] was started in 1993 with the goal of very high compression coding at very low bit rates of 64 kbit/sec or lower. Coincidentally, the ITU-T also started two very low bit rate video coding efforts: a short-term effort to improve H.261 for coding at around 20 to 30 kbit/sec and a longer term effort intended to achieve higher compression coding at similar bit rates. The ITU-T short-term standard, called H.263 [5], was completed in 1996 and its second version, H.263 version 2 or H.263+, has been completed as well. In the meantime, the ongoing MPEG–4 effort has focused on providing a new generation of interactivity with the audiovisual content, i.e., in access to and manipulation of objects or in the coded representation of a scene. Thus, MPEG–4 Visual codes a scene as a collection of visual objects; these objects are individually coded and sent along with the description of the scene to the receiver for composition. MPEG–4 Visual [4,6,7] includes coding of natural video as well as synthetic visual (graphics, animation) objects.

MPEG–4 natural video coding been optimized [8–10] for noninterlaced video at bit rates in the range of 10 to 1500 kbit/sec and for higher resolution interlaced video formats in the range of 2 to 4 Mbit/sec. The synthetic visual coding allows two-dimensional (2D) mesh-based representation of generic objects as well as a 3D model for facial animation [6,11]; the bit rates addressed are of the order of few kbits/sec. Organizationally, within the MPEG–4 group, the video standard was developed by the MPEG–4 Video subgroup whereas synthetic visual standard was developed within the MPEG–4 Synthetic/Natural Hybrid Coding (SNHC) subgroup.

The rest of the chapter is organized as follows. In Section II, we discuss applications, requirements, and functionalities addressed by the MPEG–4 Visual standard. In Section III, we briefly discuss tools and techniques covered by the natural video coding part of MPEG–4 Visual. In Section IV, we present a brief overview of the tools and techniques in the synthetic visual part of the MPEG–4 Visual standard. Section V discusses the organization of the MPEG–4 Visual into profiles and levels. Finally, in Section VI, we summarize the key points presented in this chapter.

II. MPEG–4 VISUAL APPLICATIONS AND FUNCTIONALITIES

A. Background

The MPEG–4 Visual standard [4] specifies the bitstream syntax and the decoding process for the visual part of the MPEG–4 standard. As mentioned earlier, it addresses two major areas—coding of (natural) video and coding of synthetic visual (visual part of the SNHC work).

The envisaged applications of MPEG–4 Visual [12] include mobile video phone, information access–game terminal, video e-mail and video answering machines, Internet multimedia, video catalogs, home shopping, virtual travel, surveillance, and networked video games. We will discuss some of these applications and their requirements and functionality classes addressed by MPEG–4 Visual.

B. Video Applications and Functionalities

Digital video is replacing analog video in the consumer marketplace. A prime example is the introduction of digital television in both standard-definition and high-definition formats, which is starting to see wide employment. Another example is the digital versatile disc (DVD) standard, which is starting to replace videocassettes as the preferred medium for watching movies. The MPEG–2 video standard has been one of the key technologies that enabled the acceptance of these new media. In these existing applications, digital video will initially provide functionalities similar to those of analog video; i.e., the content is represented in digital form instead of analog form, with obvious direct benefits such as improved quality and reliability, but the content and presentation remain little changed as seen by the end user. However, once the content is in the digital domain, new functionalities can be added that will allow the end user to view, access, and manipulate the content in completely new ways. The MPEG–4 video standard provides key new technologies that will enable this. The new technologies provided by the MPEG–4 video standard are organized in a set of tools that enable applications by supporting several classes of functionalities. The most important classes of functionalities are outlined in Table 1.

The most salient of these functionalities is the capability to represent arbitrarily shaped video objects. Each object can be encoded with different parameters and at different qualities. The structure of the MPEG–4 video standard reflects the organization of video material in an object-based manner.

The shape of a video object can be represented in MPEG–4 by a binary plane or as an gray-level (alpha) plane.

The texture is coded separately from its shape. A block-based DCT transform is used to encode the texture, with additional processing that allows arbitrarily shaped object to be encoded efficiently using block transforms.

Table 1 Functionality Classes Requiring MPEG–4 Video

- **Content-based interactivity**

Coding and representing video objects rather than video frames enable content-based applications and are one of the most important new functionalities that MPEG–4 offers. Based on efficient representation of objects, object manipulation, bitstream editing, and object-based scalability allow new levels of content interactivity.

- **Compression efficiency**

Compression efficiency has been the leading principle for MPEG-1 and MPEG-2 and in itself has enabled application such as digital television (DTV) and DVD. Improved coding efficiency and coding of multiple concurrent data streams will increase acceptance of applications based on the MPEG-4 standard.

- **Universal access**

Robustness in error-prone environments allows MPEG–4 encoded content to be accessible over a wide range of media, such as mobile networks and Internet connections. In addition, temporal and spatial scalability and object scalability allow the end user to decide where to use sparse resources, which can be bandwidth but also computational resources.

Motion parameters are used to reduce temporal redundancy, and special modes are possible to allow the use of a semistatic background.
For low-bit-rate applications, frame-based coding of texture can be used, as in MPEG–1 and MPEG–2.
To increase error robustness, special provisions are taken at the bitstream level that allow fast resynchronization and error recovery.

MPEG–4 will allow the end user to view, access, and manipulate the content in completely new ways. During the development of the MPEG–4 standard much thought has been given to application areas, and several calls for proposals to industry and academia have provided a wealth of information on potential applications. Potential application areas envisioned for employment of MPEG–4 video are shown in Table 2.

C. Synthetic Visual Applications and Functionalities

In recent years, the industries of telecommunications, computer graphics, and multimedia content have seen an emerging need to deliver increasingly sophisticated high-quality mixed media in a variety of channels and storage media. Traditional coded audio–video has been spurred to address a wider range of bit rates and to evolve into streamed objects with multiple channels or layers for flexibility and expressiveness in composition of scenes. Synthetic 2D–3D graphics has moved toward increasing integration with audio–video and user interaction with synthetic worlds over networks. These different media types are not entirely convergent in technology. A coding standard that combines these media types frees the content developer to invoke the right mix of scene primitives to obtain the desired scene representations and efficiency.

MPEG–4 attracted experts from these domains to synthesize a coding and scene representation standard that can support binary interoperability in the efficient delivery of real-time 2D–3D audio and visual content. The MPEG–4 standard targets delivery of audiovisual (AV) objects and of structured synthetic 2D–3D graphics and audio over broadband networks, the Internet or Web, DVD, video telephony, etc. MPEG–4 scenes are intended for rendering at client terminals such as PCs, personal digital assistants (PDAs), advanced cell phones, and set-top boxes. The potential applications include news and sports feeds, product sales and marketing, distance learning, interactive multimedia presentations or tours, training by simulation, entertainment, teleconferencing and teaching, interpersonal communications, active operation and maintenance documentation, and more.

An initiative within MPEG–4, Synthetic/Natural Hybrid Coding or SNHC, was formed and infused within the specification parts to address concerns about coding, structuring, and animation of scene compositions that blend synthetic and natural AV objects. In part, SNHC was a response to requirements driving the visual, audio, and systems specifications to provide for

Flexible scene compositions of audio, video, and 2D–3D graphical objects
Mixing remotely streamed, downloadable, and client-local objects
Changing scene structure during animation to focus client resources on current media objects
High efficiency in transmission and storage of objects through compression
Precise synchronization of local and remote objects during their real-time animation
Scalability of content (bit rates, object resolution, level of detail, incremental rendering)

Table 2 Video Centric Application Areas of MPEG-4

• Interactive digital TV

With the phenomenal growth of the Internet and the World Wide Web, the interest in more interactivity with content provided by digital television is increasing. Additional text, still pictures, audio, or graphics that can be controlled by the end user can increase the entertainment value of certain programs or can provide valuable information that is unrelated to the current program but of interest to the viewer. Television station logos, customized advertising, and multiwindow screen formats that allow the display of sports statistics or stock quotes using data casting are prime examples of increased interactivity. Providing the capability to link and synchronize certain events with video will even improve the experience. By coding and representing not only frames of video but video objects as well, such new and exciting ways of representing content can provide completely new ways of television programming.

• Mobile multimedia

The enormous popularity of cellular phones and palm computers indicates the interest in mobile communications and computing. Using multimedia in these areas would enhance the end user's experience and improve the usability of these devices. Narrow bandwidth, limited computational capacity, and reliability of the transmission media are limitations that currently hamper widespread use of multimedia here. Providing improved error resilience, improved coding efficiency, and flexibility in assigning computational resources may bring this closer to reality.

• Virtual TV studio

Content creation is increasingly turning to virtual production techniques, extensions of the well-known technique of chroma keying. The scene and the actors are recorded separately and can be mixed with additional computer-generated special effects. By coding video objects instead of frames and allowing access to the video objects, the scenes can be rendered with higher quality and with more flexibility. Television programs consisting of composited video objects and additional graphics and audio can then be transmitted directly to the end user, with the additional advantage of allowing the user to control the programming in a more sophisticated way.

• Networked video games

The popularity of games on stand-alone game machines and on PCs clearly indicates the interest in user interaction. Most games are currently using three-dimensional graphics, both for the environment and for the objects that are controlled by the players. Addition of video objects to these games would make the games even more realistic, and, using overlay techniques, the objects could be made more lifelike. Essential is the access of individual video objects, and using-standards-based technology would make it possible to personalize games by using personal video databases linked in real time into the games.

• Streaming internet video

Streaming video over the Internet is becoming more popular, using viewing tools as software plug-ins for a Web browser. News updates and live music shows are some examples of streaming video. Here, bandwidth is limited because of the use of modems, and transmission reliability is an issue, as packet loss may occur. Increased error resilience and improved coding efficiency will improve the experience of streaming video. In addition, scalability of the bitstream, in terms of temporal and spatial resolution but also in terms of video objects, under the control of the viewer, will further enhance the experience and also the use of streaming video.

Error resilience and concealment where possible
Dynamic manipulation of content objects
Content interactivity such as conditional behavior

Synthetic 2D–3D models, their hierarchical composition into scenes, and the efficient coding of both their static and dynamic structures and attributes offer some key advantages for augmenting the MPEG–4 standard:

Inherent efficiency of highly compact, model-based, synthetic scene descriptions (rather than many snapshots of a visual or audio scene)
Trade-offs between compactness and interactivity of synthetic models rendered at the terminal versus realism and ''content values'' of audio and video
Compression of synthetic graphical and audio objects for yet further efficiency gains
High scalability through explicit coding of multiple resolutions or alternative levels of detail
Remote control of local or previously downloaded models via compressed animation streams (e.g., facial parameters, geometry, joint angles)
Media integration of text, graphics, and AV into hybrid scene compositions
Ability to add, prune, and instantiate model components within the scene moment by moment
Large ''virtual memory'' provided by networks and local high-capacity storage media to page in items and regions of interest within larger virtual worlds
Progressive or incremental rendering of partially delivered objects for quick user content inspection
Hierarchical buildup of scene detail over time or downloading of alternative levels of detail for terminal adaptation of content to rendering power
Special effects that exploit spatial and temporal coherence by manipulating natural objects with synthetics (2D mesh animation of texture)
Very low bit rate animation (e.g., client-resident or downloaded models driven by coded animations)

Specific capabilities of MPEG–4 synthetic visual [4] are outlined in later sections. These include the coded representation and animation of synthetic 2D–3D graphics (and audio), the Binary Interchange Format for Scenes [BIFS leveraging Virtual Reality Modeling Language (VRML)], face and body animation for high-efficiency communications, the integration of streaming media such as advanced text-to-speech (TTS) and face animation, animated 2D meshes, view-dependent scene processing, and the compression of 2D–3D polygonal meshes including geometry, topology, and properties with error resilience.

III. MPEG–4 VIDEO TOOLS

A. Overview

The MPEG–4 standard provides a set of technologies that can be used to offer enhanced digital audiovisual services. The standard offers an integrated solution combining video and audio with systems, graphics, network, and support for content protection. MPEG–4 video provides backward compatibility with MPEG–1 and MPEG–2 and is compatible with H.263 baseline. This means that an H.263 baseline encoded stream can be decoded by an MPEG–4 decoder.

The central concept defined by the MPEG–4 video standard is the video object (VO), which forms the foundation of the object-based representation. Such a representation is well suited for interactive applications and gives direct access to the scene contents. A video object consists of one or more layers, the video object layers (VOLs), to support scalable coding. The scalable syntax allows the reconstruction of video in a layered fashion starting from a stand-alone base layer and adding a number of enhancement layers. This allows applications, for example, to generate a single MPEG–4 video bitstream for a variety of bandwidths or computational requirements. A special case where a high degree of scalability is needed, is that in which static image data are mapped onto two- or three-dimensional objects. To address this functionality, MPEG–4 video has a special mode for encoding static textures using a wavelet transform.

An MPEG–4 video scene consists of one or more video objects that are related to each other by a scene graph. Each video object is characterized by temporal and spatial information in the form of shape, motion, and texture, combined with bitstream enhancements that provide capabilities for error resilience, scalability, descriptors, and conformance points. Instances of video objects in time are called video object planes (VOPs). For certain applications, video objects may not be desirable because of, e.g., the associated overhead. For those applications, MPEG–4 video allows coding of rectangular objects as well as arbitrarily shaped objects in a scene.

An MPEG–4 video bitstream provides a hierarchical description of a visual scene. Each level of the hierarchy can be accessed in the bitstream by special code values, the start codes. The hierarchical levels reflect the object-oriented nature of an MPEG–4 video bitstream:

Visual object sequence (VS): The complete MPEG–4 scene; the VS represents the complete visual scene.

Video object (VO): A video object corresponds to a particular object in the scene. In the simplest case this can be a rectangular frame, or it can be an arbitrarily shaped object corresponding to an object or background of the scene.

Video object layer (VOL): The VOL provides support for scalable coding. A video object can be encoded using spatial or temporal scalability, going from coarse to fine resolution. Depending on parameters such as available bandwidth, computational power, and user preferences, the desired resolution can be made available to the decoder. There are two types of video object layer: the video object layer that provides full MPEG–4 functionality and a reduced-functionality video object layer, the video object layer with short headers. The latter provides bitstream compatibility with baseline H.263.

Group of video object planes (GOV): The GOV groups together video object planes. GOVs can provide points in the bitstream where video object planes are encoded independently of each other and can thus provide random access points into the bitstream. GOVs are optional.

Video object plane (VOP): A VOP is a time sample of a video object. VOPs can be encoded independently of each other or dependent on each other by using motion compensation. The VOP carries the shape, motion, and texture information that defines the video object at a particular instant in time. For low-bit-rate applications or for applications that do not need an object-based representation, a video frame can be represented by a VOP with a rectangular shape.

A video object plane can be used in several different ways. In the most common way, the VOP contains the encoded video data of a time sample of a video object. In that case it contains motion parameters, shape information, and texture data. These are encoded using macroblocks. It can be ''empty,'' containing no video data, providing a video object that persists unchanging over time. It can also be used to encoded a sprite. A sprite is a video object that is usually larger than the displayed video and is persistent over time. It is used to represent large, more or less static areas, such as backgrounds. Sprites are encoded using macroblocks. A sprite can be modified over time by changing the brightness or by warping the video data.

A video object plane is encoded in the form of macroblocks. A macroblock contains a section of the luminance component and the spatially subsampled chrominance components. At present, there is only one chrominance form for a macroblock, the 4:2:0 format. In this format, each macroblock contains four luminance blocks and two chrominance blocks. Each block contains 8×8 pixels and is encoded using the DCT. The blocks carry the texture data of the MPEG–4 video stream. A macroblock carries the shape information, motion information, and texture information. Three types of macroblocks can be distinguished:

Binary shape: the macroblock carries exclusively binary shape information.

Combined binary shape, motion, texture: the macroblock contains the complete set of video information, shape information in binary form, motion vectors used to obtain the reference video data for motion compensation, and texture information.

Alpha shape: the macroblock contains alpha shape information. The shape information is a gray-level description of the shape and is used to obtain higher quality object descriptions. Alpha shape information is encoded in a form similar to that of the video texture luminance data.

B. Video Version 1 Tools

During the development of the MPEG–4 standard, the proposed application domain was too broad to be addressed within the original time frame of the standard. To provide industry with a standard that would provide the most important functionalities within the original time frame, MPEG–4 defined a first version of the standard that captured the most important and most mature technologies. For MPEG–4 video [8] the most important functionalities were object-based interactivity, high coding efficiency [9,13], and improved error resilience. The technologies that provide these functionalities form a large set of tools that can for conciseness be grouped together into the following classes:

Shape
Motion
Texture
Sprite
Still texture
Error resilience
Scalability
Conformance

1. Shape

Shape information is encoded separately from texture and motion parameters. The shape of an object can be presented and encoded in binary form or as gray-level (or alpha channel) information. The use of an alpha channel provides superior visual quality, but it is more expensive to encode. MPEG–4 video provides tools for binary shape coding and gray-level shape coding. Binary shape coding is performed on a 16 × 16 block basis. To encode an arbitrary shape in this fashion, a bounding rectangle is first created and is extended to multiples of 16 × 16 blocks, with extended alpha samples set to zero. The binary shape information is then encoded on a macroblock basis using context information, motion compensation, and arithmetic coding. Gray-level shape is encoded on a macroblock basis as well. Gray-level shape information has properties very similar to those of the luminance of the video channel and is thus encoded in a similar fashion using the texture tools and motion compensation tools that are used for coding of the luminance samples.

2. Motion Compensation

Motion compensation is performed on the basis of a video object. In the motion compensation process, the boundary of the reference object is extended to a bounding rectangle. The pixels that are outside the video object but inside the bounding box are computed by a padding process. The padding process basically extends the values of the arbitrarily shaped objects horizontally and vertically. This results in more efficient temporal coding. The padding process allows motion vectors in MPEG–4 video to point outside the video object. This mode of motion compensation, unrestricted motion compensation, further increases coding efficiency. Motion compensation is performed with half-pixel accuracy, with the half-pixel values computed by interpolation, and can be performed on 16 × 16, 8 × 16, or 8 × 8 blocks. For luminance, overlapped motion compensation is used, where motion vectors from two neighboring blocks are used to form three prediction blocks that are averaged together with appropriate weighting factors. Special modes can be used for interlaced and for progressive motion compensation. Motion compensation can further be performed in predictive mode, using the past reference video object, or in bidirectional mode, using both the past and the future reference video objects.

3. Texture

The texture information of a video object plane is present in the luminance, Y, and two chrominance components, (Cb, Cr) of the video signal. In the case of an I-VOP, the texture information resides directly in the luminance and chrominance components. In the case of motion-compensated VOPs, the texture information resides in the residual error remaining after motion compensation. For encoding the texture information, the standard 8 × 8 block–based DCT is used. To encode an arbitrarily shaped VOP, an 8 × 8 grid is superimposed on the VOP. Using this grid, 8 × 8 blocks that are internal to VOP are encoded without modifications. Blocks that straddle the VOP are called boundary blocks and are treated differently from internal blocks. Boundary blocks are padded to allow the use of the 8 × 8 transform. Luminance blocks are padded on a 16 × 16 basis, and chrominance blocks are padded on an 8 × 8 basis. A special coding mode is provided to deal with texture information from interlaced sources. This texture information can be more efficiently coded by separately transforming the 8 × 8 field blocks instead of the 8 × 8 frame blocks. The transformed blocks are quantized, and individual coefficient prediction can be used from neighboring blocks to reduce the entropy value of the coefficients further.

This process is followed by a scanning of the coefficients to reduce the average run length between coded coefficients. Then the coefficients are encoded by variable length encoding.

4. Sprite

A sprite is an image composed of pixels belonging to a video object that is visible throughout a video scene. For example, a sprite can be the background generated from a panning sequence. Portions of the sprite may not be visible in certain frames because of occlusion by the foreground objects or camera motion. Sprites can be used for direct reconstruction of the background or for predictive coding of the background objects. Sprites for background are commonly referred to as background mosaics in the literature. A sprite is initially encoded in intra mode, i.e., without motion compensation and the need for a reference object. This initial sprite can consequently be warped using warping parameters in the form of a set of motion vectors transmitted in the bitstream.

5. Still Texture

One of the functionalities supported by MPEG–4 is the mapping of static textures onto 2D or 3D surfaces. MPEG–4 video supports this functionality by providing a separate mode for encoding static texture information. The static texture coding technique provides a high degree of scalability, more so than the DCT-based texture coding technique. The static coding technique is based on a wavelet transform, where the AC and DC bands are coded separately. The discrete wavelet used is based on the Daubechies (9,3) tap biorthogonal wavelet transform. The transform can be applied either in floating point or in integer, as signaled in the bitstream. The wavelet coefficients are quantized and encoded using a zero-tree algorithm and arithmetic coding. A zero-tree algorithm, which uses parent–child relations in the wavelet multiscale representation, is used to encode both coefficient value and location. The algorithm exploits the principle that if a wavelet coefficient is zero at a coarse scale, it is very likely that its descendant coefficients are also zero, forming a tree of zeros. Zero trees exist at any tree node where the coefficient is zero, and all the descendants are also zero. Using this principle, wavelet coefficients in the tree are encoded by arithmetic coding, using a symbol that indicates whether a zero tree exists and the value of the coefficient.

6. Error Resilience

Another important functionality provided by MPEG–4 video is error robustness and resilience. When a residual error has been detected, the decoder has to resynchronize with the bitstream to restart decoding the bitstream. To allow this, MPEG–4 video uses a packet approach in which periodic resynchronization markers are inserted throughout the bitstream. The length of the packet is independent of the contents of the packets and depends only on the number of bits, giving a uniform distribution of resynchronization markers throughout the bitstream. In addition, header extension information can be inserted in the bitstream. This makes it possible to decode each video packet independently of previous bitstream information, allowing faster resynchronization. Error resilience is further improved by the use of data partitioning. Here the motion and macroblock header information is separated from the texture information, and a motion marker identifies the separation point. This allows better error concealment. In the case of residual errors in the texture information, the previously received motion information can be used to provide motion-compensated error concealment. To allow faster recovery after a residual error has been detected, reversible variable length codes (RVLCs) can be used. These binary codes are designed in such a way that they can be decoded in both forward and reverse directions.

A part of the bitstream that cannot be decoded in the forward direction can often be decoded in the backward direction. This makes it possible to recover a larger part of the bitstream and results in a reduced amount of lost data. The RVLCs are used only for DCT coefficient code tables.

7. Scalability

Many applications may require the video to be available simultaneously in several spatial or temporal resolutions. Temporal scalability involves partitioning of the video objects in layers, where the base layer provides the basic temporal rate and the enhancement layers, when combined with the base layer, provide higher temporal rates. Spatial scalability involves generating video layers with two spatial resolutions from a single source such that the lower layer can provide the lower spatial resolution and the higher spatial resolution can be obtained by combining the enhancement layer with the interpolated lower resolution base layer. In the case of temporal scalability, arbitrarily shaped objects are supported, whereas in the case of spatial scalability only rectangular objects are supported. In addition to spatial and temporal scalability, a third type of scalability is supported by MPEG–4 video, computational scalability. Computational scalability allows the decoding process to use limited computational resources and decode only those parts of the bitstream that are most meaningful, thus still providing acceptable video quality. Parameters can be transmitted in the bitstream that allow those trade-offs to be made.

8. Conformance

In order to build competitive products with economical implementations, the MPEG–4 video standard includes models that provide bounds on memory and computational requirements. For this purpose a video verifier is defined, consisting of three normative models, the video rate buffer verifier, the video complexity verifier, and the video reference memory verifier.

- Video rate buffer verifier (VBV): The VBV provides bounds on the memory requirements of the bitstream buffer needed by a video decoder. Conforming video bitstreams can be decoded with a predetermined buffer memory size. In the VBV model the encoder models a virtual buffer, where bits arrive at the rate produced by the encoder and bits are removed instantaneously at decoding time. A VBV-conformant bitstream is produced if this virtual buffer never underflows or overflows.
- Video complexity verifier (VCV): The VCV provides bounds on the processing speed requirements of a video decoder. Conformant bitstreams can be decoded by a video processor with predetermined processor capability in terms of the number of macroblocks per second it can process. In the VCV a virtual macroblock buffer accumulates all macroblocks encoded by the encoder and added instantaneously to the buffer. The buffer has a prespecified macroblock capacity and a minimum rate at which macroblocks are decoded. With these two parameters specified, an encoder can compute how many macroblocks it can produce at any time instant to produce a VCV-conformant bitstream.
- Video reference memory verifier (VMV): The VMV provides bounds on the macroblock (pixel) memory requirements of a video decoder. Conformant bitstreams can be decoded by a video processor with predetermined pixel memory size. The hypothetical decoder fills the VMV buffer at the same macroblock decoding rate as the VCV model. The amount of reference mem-

ory needed to decode a VOP is defined as the total number of macroblocks in a VOP. For reference VOPs (I or P) the total memory allocated to the previous reference VOP is released at presentation time plus the VCV latency, whereas for B-VOPs the total memory allocated to that B-VOP is released at presentation time plus the VCV latency.

C. Video Version 2 Tools

The second version of MPEG–4 provides additional tools [14] that enhance existing functionalities, such as improved coding efficiency, enhanced scalability, enhanced error resilience, and enhanced still texture coding, or that provide new functionalities. The list of proposed tools is given here:

- Improved coding efficiency
  - Shape-adaptive DCT
  - Boundary block merging
  - Dynamic resolution conversion
  - Quarter-pel prediction
  - Global motion compensation
- Enhanced scalability
  - Object-based spatial scalability
- Enhanced error resilience
  - Newpred
- Enhancements of still texture coding
  - Wavelet tiling
  - Error resilience for scalable still texture coding
  - Scalable shape coding for still texture coding
- New functionalities
  - Multiple auxiliary components

1. *Shape-Adaptive DCT*

The shape-adaptive DCT defines the forward and inverse transformation of nonrectangular blocks. In version 1 the 8×8 blocks that contained pixels from arbitrarily shaped objects and pixels that were outside the object, the boundary blocks, were padded before applying an 8×8 transform. The shape-adaptive DCT does not use padding but applies successive 1D DCT transforms of varying size, both horizontally and vertically. Before the horizontal transformation, the pixels are shifted vertically to be aligned to the vertical axis, and before the vertical transform the pixels are shifted horizontally to be aligned to the horizontal axis. The horizontal and vertical shifts are derived from the transmitted shape information. The shape-adaptive DCT can provide a coding efficiency gain as compared with the padding process followed by the 8×8 block–based DCT.

2. *Boundary Block Merging*

Boundary block merging is a tool to improve the coding efficiency of boundary blocks. The technique merges the texture information of two boundary blocks into a single block, which is coded and transmitted as a single texture block. At the decoder, the shape information is used to redistribute the texture information over the two boundary blocks. Two boundary blocks A and B can be merged together if there are no overlapped pixels between the shape of block A and the shape of block B when it is rotated 180°.

3. Dynamic Resolution Conversion

Dynamic resolution conversion is a tool that encodes the I- and P-VOPs in reduced spatial resolution or normal resolution adaptively. In the reduced-resolution mode, the motion-compensated interframe prediction is done by an expanded macroblock of 32 × 32 size. The texture coding first downsamples the 32 × 32 data to 16 × 16 data, followed by applying the tool set of texture coding tools that are available for normal-resolution data.

4. Quarter-Pel Prediction

Quarter-pel prediction is a tool that increases the coding efficiency of motion-compensated video objects. The quarter pel motion prediction tool enhances the resolution of the motion vectors, which results in more accurate motion compensation. The scheme uses only a small amount of syntactical and computational overhead and results in a more accurate motion description and a reduced average prediction error to be coded.

5. Global Motion Compensation

Global motion compensation is a tool to increase coding efficiency. The tool encodes global motion of a video object, usually a sprite, with a small number of motion parameters. It is based on global motion estimation and prediction, trajectory coding, and texture coding for prediction errors. It supports the following five transformation models for the warping process: stationary, translational, isotropic, affine, and perspective. The pixels in a global motion–compensated macroblock are predicted using global motion prediction. The predicted macroblock is obtained by applying warping to the reference object. Each macroblock can be predicted either from the previous video object plane by global motion compensation using warping parameters or by local motion compensation using local motion vectors as defined in MPEG–4 version 1. Coding of texture and shape is done using the tools for predictive coding as defined in MPEG–4 version 1.

6. Object-Based Spatial Scalability

The spatial scalability provided in version 1 allowed only rectangular objects. In version 2, arbitrarily shaped objects are allowed for spatial scalability. Because arbitrarily shaped objects can have varying sizes and locations, the video objects are resembled. An absolute reference coordinate frame is used to form the reference video objects. The resampling involves temporal interpolation and spatial relocation. This ensures the correct formation of the spatial prediction used to compute the spatial scalability layers. Before the objects are resampled, the objects are padded to form objects that can be used in the spatial prediction process.

7. Newpred

The newpred tool increases the error resilience for streams that can make use of a back channel. The bitstream is packetized, and a backward channel is used to indicate which packets are correctly decoded and which packets are erroneously decoded. The encoder, which receives a backward channel message, uses only the correctly decoded part for prediction in an inter-frame coding. This prevents temporal error propagation without the insertion of intra coded macroblocks and improves the picture quality in erroneous environments.

8. Wavelet Tiling

The purpose of wavelet tiling is to provide random access to part of a still picture without decoding the complete picture. The wavelet tiling tool provides this capability by dividing the image into several subimages (tiles), which are coded independently of each other, with control information. In terms of the user interaction, a decoder uses the information to find a starting point of a bitstream and then decodes only the specified subimage without decoding the whole bitstream.

9. Error Resilience for Scalable Still Texture Coding

The error resilience of the still texture coding tool can be improved by making use of resynchronization markers in the bitstreams. However, the current syntax makes it possible to emulate these markers. This tool would modify the arithmetic coder that is used to encode the coefficients to avoid emulating the resynchronization marker. In addition, the error resilience can be improved by restructuring the bitstream and using packetization of the bits. Interdependence of the packets will be removed by resetting the probability distributions of the arithmetic coders at packet boundaries.

10. Scalable Shape Coding for Scalable Still Texture Coding

The still texture coding tool defined in version 1 has no provisions for arbitrarily shaped objects. This tool extends the still texture coding tool in that respect. It defines a shape-adaptive wavelet transform to encode the texture of an arbitrarily shaped object. The shape-adaptive wavelet ensures that the number of coefficients of the transform is identical to the number of pixels that are part of the arbitrarily shaped object. To encode the shape itself it defines binary shape coding for wavelets using context-based arithmetic coding.

11. Multiple Auxiliary Components

In version 1 the shape of an object can be encoded using gray-level shape (alpha channel) coding. In version 2 the method for carrying auxiliary data in this form is generalized to include multiple auxiliary components. An auxiliary component carried in this form may represent gray-level shape (alpha channel) but may also be used to carry, e.g., disparity information for multiview video objects, depth information as acquired by a laser range finder or by disparity analysis, or infrared or other secondary texture information. Auxiliary components are carried per video object, and up to three auxiliary components per video object are supported.

IV. MPEG–4 SYNTHETIC VISUAL TOOLS

A. Overview

Synthetic visual and audio elements of MPEG–4 are distributed into the Systems, Visual, and Audio parts of the specification. The MPEG–4 SNHC objectives are met by a consistent design of AV and 2D–3D objects and the semantics of their composition and animation within the space–time continuum of a session.

The Systems specification [7,15] part provides this integration. Systems provides for demultiplexing of logical channels within an inbound physical channel, certain timing conventions for synchronizing the supported synthetic and natural media types, the structure to integrate elementary streams (downloaded or time variable) into a scene, the interfaces to user and scene events, and the control of a session.

Synthetic visual decoding tools in the Visual part of the specification provide specific elementary stream types and the decoding of the corresponding compressed models and animations. The Systems tools can then access the results of decoding by the synthetic visual tools. Systems can, for example, build up a terminal-resident model, animate specific properties of a scene composition, or synchronize synthetic visual models and animations with other stream types including AV objects in the natural video and audio tool sets.

1. Binary Interchange Format for Scenes

Systems BIFS provides for the static 2D–3D graphical object types needed for SNHC (such as a textured face model), the composition of these elements into an MPEG–4 virtual world, and a binary representation of the scene description that is very efficient (compared with the declarative textual form of VRML). The BIFS system is patterned after VRML 2.0 and provides its elementary 2D and 3D objects with parameters, scene description hierarchy including spatial localization, exposed fields for animating parameters, routes to connect MPEG–4 animation streams to the scene description, and a timing model to link object behaviors.

Relative to synthetic visual tools described later, BIFS provides some important infrastructure for animating generic parameters within the scene description and for changing the structure of the scene "on the fly." BIFS Anim features a generic, compressed type of animation stream that can be used to change data within the model of the scene. For example, this capability can be used to transmit efficiently the coded animation of the motion of objects with respect to each another. BIFS Update features a mechanism to initialize and alter the scene graph incrementally. Objects (or subgraphs) are thus activated and deactivated according to their intended persistence during a session.

2. Synthetic Visual Tools Relative to BIFS

Specific elementary stream types are tailored in MPEG–4 for synthetic visual purposes (e.g., face, 2D mesh animation). This serves to augment BIFS Anim while conforming to a consistent animation stream header and timing model including the suppression of start code emulation. In this way, synthetic visual streams can be controlled by Systems consistently, and the output of synthetic visual decoders can be routed into variable parameters within the scene description to effect the desired animation of the models.

BIFS provides some basic steps in standardizing the compression of the scene description and of generic parameter streams that animate scene variables. The synthetic visual tools under development in MPEG–4 extend this compression to specific media types required in synthetic visual scenes. Synthetic visual tools augment the BIFS compression repertoire with higher compression bitstream standards and decoder functionality that is partitioned into the Visual part of the MPEG–4 specification.

3. Summary of Synthetic Visual Tools

Face animation parameters, 2D animated meshes (version 1), body animation, and 3D model coding (version 2) provide very high compression of specific types of model data and animation streams. These include domain-specific motion variables, polygonal mesh connectivity or topology (how vertices and triangles are assembled into models), raw scene geometry (2D or 3D vertices, the dominant data consumer), and mesh radiometric properties (color values, texture map instantiation coordinates, surface normals). BIFS provides the connection between these elementary Visual stream types and the integration of the resulting decoded models and motions into a scene.

4. Content Adaptation for Synthetic Visuals

The MPEG–4 standard including its SNHC features offers new possibilities for content developers. Yet it also confers a new level of responsibility on platform vendors and can require the content designer to relinquish some control to the user. MPEG–4 scene compositions can experience more varied viewing conditions and user interaction in the terminal, compared with MPEG–1 and -2. Content developers must therefore consider in the design of content complexity not only the performance limits of conforming MPEG–4 decoders but also the level of rendering power targeted in the terminal beyond the decoding stage.

Currently, no effective approach to normative specification of the performance of scene compositions with SNHC elements in the rendering stage has been agreed upon. Consequently, the development of MPEG–4 version 2 is concerned with the quality of the delivered renderings of scenes with SNHC elements. The interoperability objective of MPEG–4 is potentially at odds with supporting a wide variety of terminal capabilities and implementations that react differently to universal content beyond the decoder.

MPEG–4 SNHC version 2 targets additional parameters in bitstreams for computational graceful degradation. CGD adds terminal-independent metrics into the stream reflecting the complexity of the content. The metrics are intended to help a terminal estimate rendering load, relative to its specific implementation of 2D–3D graphics API acceleration, to facilitate content adaptation for different terminals without a back channel, when content is designed for this purpose.

B. Synthetic Visual Version 1 tools

The version 1 synthetic visual tools were driven by the industry priorities for the integration of text, graphics, video, and audio; the maturity of the scene composition and animation technologies to meet MPEG–4 SNHC requirements; and the judged adequacy of industrial support and technology competition at points in time. A strategy was therefore adopted to build synthetic visual tools in steps toward the ultimate capabilities of MPEG–4, where a version 2 increase in functionality could build on version 1 with total backward compatibility. Hence the partitioning of capabilities:

- Version 1
  - BIFS
  - Face animation (FA)
  - 2D mesh animation
  - View-dependent scalable texture
- Version 2
  - Body animation
  - 3D model coding
  - CGD for SNHC scenes

Face animation is the capability to stream highly optimized low-bit-rate (LBR) animation parameters to activate lip movements and facial expressions of a client 2D or 3D model, typically in sync with text-to-speech (TTS) or LBR speech audio. The 2D mesh animation provides for the efficient transmission of 2D topology and motion vectors for a regular rectangular or Delaunay triangle mesh to act as a scaffolding for the manipulation and warping of texture or video (using video texture I- and P-frames or still texture). View-dependent scalable texture leverages video tools to provide incremental downloads of texture to a client as updates by a server in response to client navigation through a virtual world with a back channel.

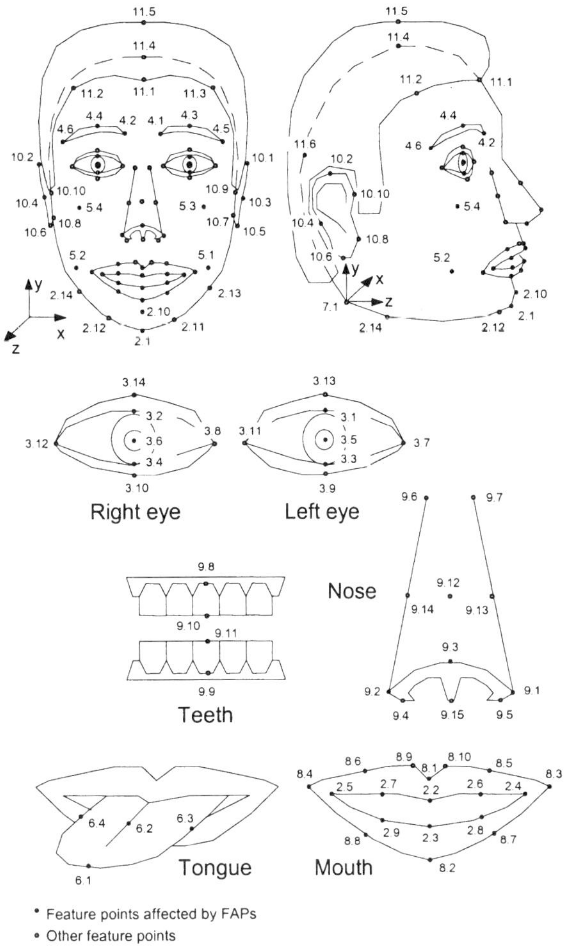

Figure 1 Face definition parameter feature set.

1. *Facial Animation*

Facial and body animations collectively provide for the remote-controlled animation of a digital mannequin. The FA coding provides for the animation of key feature control points on the face selected for their ability to support speech intelligibility and the expression of moods.

Facial animation includes a set of facial animation parameters (FAPs) that are capable of very LBR coding of the modest facial movements around the mouth, inner lip contour, eyes, jaw, etc. The FAPs are sufficient to represent the visemes of speech (temporal motion sequences of the lips corresponding to phonemes) and the expression of the face (drawing on a limited repertoire of mood states for the FAPs) (Fig. 1).

The coding methodologies for FAPs consist of two schemes motivated by different foreseen application environments:

Arithmetic coder—good efficiency, lowest lag
DCT/frame-based—best efficiency, more lag

Facial animation may be used in broadcast or more tailored point-to-point or group-to-group connections with possible interaction. Thus the system needs to accommodate the downloading of faces, client-resident models that require no setup, or calibration of a client model by sending feature control points that act initially to deform the client model into the desired shape. If FAPs are used in a broadcast application, no model customization is valid. For the other cases, facilities in BIFS are needed by FA to specify face models, to establish a mapping of FAPs into mesh vertices, and to transform FAPs for more complex purposes.

Either FAP coding scheme relies on Systems BIFS for the integration of FAP streams with other nodes in the terminal scene graph that can represent a client-resident or downloadable face model and the linkages between FAPs and the model mesh points. The mesh model of face shape and appearance needs to be specified. Face definition parameters (FDPs) are provided for this purpose. Downloaded FDPs are instantiated into a BIFS-compliant 2D or 3D model with hooks that support subsequent animation of the face mesh. Model setup supported by FDPs generally occurs once per session.

To map FAPs (which control feature points) into mesh points on a frame-to-frame basis, the face animation table (FAT) is provided to form linear combinations of FAPs for resulting perturbations of mesh points. A face interpolation transform (FIT) provides rational polynomial transforms of FAPs into modified FAPs for more complex interpretations of mesh movement in response to FAPs or to bridge in facial movements if some FAPs are missing.

FAPs are designed to represent a complete set of facial actions that can represent most of the natural facial expressions. The FAPs for controlling individual feature points involve translational or rotational movement around the resting state. The FAPs are expressed in facial animation parameter units (FAPUs). Each FAP is carefully bit limited and quantized to accomplish just the range and resolution of face movement needed. Then consistent interpretation of FAPs by any facial model can be supported.

The FAP set includes two high-level FAPs and numerous low-level FAPs:

- Viseme (visual correlate of phoneme)
  - Only static visemes included at this stage
- Expression
  - Joy, anger, fear, disgust, sadness, surprise, etc.
  - Textual description of expression parameters
  - Groups of FAPs together achieve expression
- 66 low-level FAPs
  - Drive displacement or rotation of facial feature points

Finally, the face interpolation transform (FIT) can be summarized. FIT was devised to handle more complex mappings of incoming FAPs into the actual FAP set used by the terminal. The essential functionality of FIT is the specification of a set of interpolation rules for some or all FAPs from the sender. The transmission of FIT includes the specification of the FAP interpolation graph (FIG) and a set of rational polynomial interpolation functions. The FIG provides for conditional selection of different transforms depending on what FAPs are available. This system provides a higher degree of control over the animation results and the chance to ''fill in'' FAPs from a sparse incoming set.

2. 2D Mesh Animation

The 2D mesh animation is intended for the manipulation of texture maps or video objects to create special effects (Fig. 2). These include image–video warping, dubbing, tracking, augmented reality, object transformation and editing, and spatial–temporal interactions.

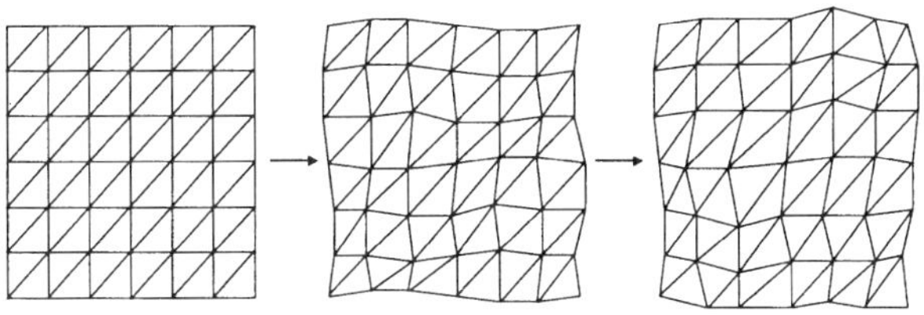

Figure 2 Warping of uniform 2D mesh.

With 2D mesh animation, spatial and temporal coherence is exploited for efficiency. The meshes are sent regularly, whereas the texture is sent only at key frames where the image content has changed enough (if at all) to merit an image update. The result is image object deformations over a sequence of frames, where update of the imagery is provided typically only when image distortion from mesh animation exceeds a threshold.

Mesh animation of image objects requires that the elementary streams of mesh setup and mesh animation be integrated with the underlying image object (video or texture). The image objects are decoded separately as often as required using the corresponding video tools. Systems BIFS provides the scene composition of mesh animation with texture maps or video objects.

The basic mesh forms supported by 2D mesh animation are uniform or regular and Delaunay (Fig. 3). Both are specifically formed from triangular meshes. These meshes are used to tessellate a 2D visual object plane with the attachment of triangular patches. Once a mesh is initialized and animated, there are no addition and deletion of nodes and no change in topology.

The regular and Delaunay meshes are extremely efficient for mesh coding and transmission, because an implicit structure can be coded by steering bits, in comparison with the vertex and triangle trees typically required to specify a general topology. Consequently, the node point order is driven by topological considerations, and the subsequent transmission of motion vectors for the nodes must follow the node order of the initial mesh transmission.

Mesh motion is coded as differential motion vectors using motion prediction related to that used in the video tools of MPEG-4 with I- and P-frame modalities. The motion vectors are, of course, two-dimensional and they are specifically scaled and quantized for manipulation of imagery in a screen or texture map space.

C. Synthetic Visual Version 2 Tools

The expected capabilities in version 2 of the synthetic visual tools are body animation and 3D model coding. Body animation was developed somewhat concurrently with face animation and has been tightly linked with FA technology (in an overall FBA activity) to leverage the work of version 1 FA. Body animation provides most of the data sets for streaming and downloading analogous to those in FA (BAPs, BDFs, and BAT). BAPs are applied to the animation of joint angles and positions for an otherwise rigid kinematic model of the human body.

3D model coding (3DMC) encompasses the compression of all the shape and appearance characteristics of 3D models as represented, for instance, in a VRML polygon model specified in an IndexFaceList (except texture maps supported by still image texture coding). Thus, 3DMC addresses the coded representation of the general topology of a triangu-

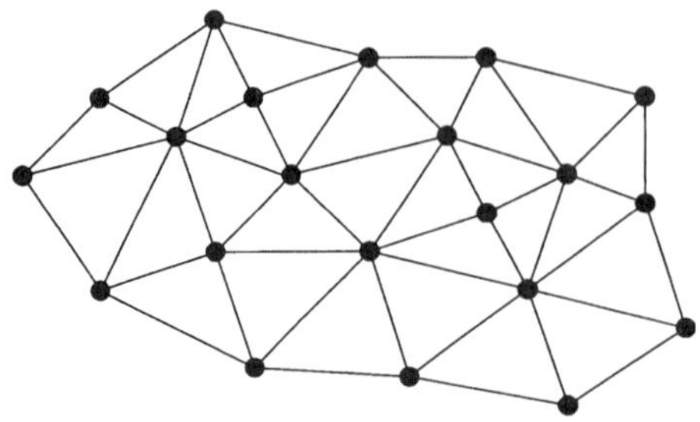

Figure 3 2D triangular mesh obtained by Delaunay triangulation.

lar mesh (2D or 3D), the geometry of vertices giving shape to the mesh, and photometric properties.

1. Body Animation

The purpose of body animation (BA) is to provide a standard way of bringing synthetic human characters or avatars to life over networks or from local streaming data sources. As with FA, BA is designed to provide for streaming of efficiently coded body animation parameters (BAPs) whose decoding can animate body models resident in the terminal. Nearly 200 BAPs are provided. If BA and FA decoders are used in harmony, a talking, gesturing human form can be the result.

In addition, considerable investigation has been made into the adequacy of the BA system to support hand signing. The rate–distortion trade-offs, quantization, and degrees of freedom (DOFs) for the joints of the hands are believed adequate to support rapid hand gestures, assuming the exact motion sequences for the component signing motions and their blending are available.

As with FA, no specific skin or clothing model for the exterior body appearance is standardized, only the kinematic structure for animation. Thus, if a polygonal model for the external appearance of the body is downloaded first, decoded BAPs will animate the discrete coordinate systems into which the body objects are partitioned in the BIFS scene graph. The result will be a moving body model whose chained rigid-body subsystems track the motions prescribed by BAPs.

To attract the greater build-out and standard animation of human forms, MPEG–4 FBA has been coordinated with the VRML Consortium Humanoid Animation Working Group with the goal that H-Anim models and MPEG–4 BA modeling and animation will conform. The same kinematic structures and naming conventions are used to ensure interoperability in model animation.

Body animation includes body definition parameters (BDPs) that provide the same ability as with faces to customize body models that are represented in the VRML-derived BIFS system including the linkage between joint coordinate systems and BAPs. The body animation table (BAT) in this case provides the ability to slave selected body surface meshes to deformation by incoming BAPs (clothing response to motion).

2. 3D Model Coding

The 3DMC addresses the compression of 2D–3D mesh structure, shape, and appearance. Mesh connectivity is coded losslessly using the so-called topological surgery method. The method makes cuts into the original mesh structure in order to form spanning trees that trace out the relationships between neighboring triangles and neighboring vertices. The larger the model, the more structural coherence can be exploited to achieve lower bit rates. For meshes of moderate size, typically 3–4 bits per triangle are needed.

Vertices are compressed using a successive quantization method, with rate–distortion trade-offs available depending on how many bits per vertex are allocated. Vertex compression uses a prediction of the location of the next vertex in the sequence based on prior vertices (assuming spatial coherence). The geometry compression method therefore relies on the vertex sequencing or vertex neighborhood indicated by the output of the topological decoding process. Different competing methods of shape prediction (polygonal, parallelogram) are being evaluated for best performance. Distortion of about 0.001 (deviation of vertices from the original mesh shape normalized to a bounding box enclosing the object) is readily achieved when 32 bits/vertex are allocated (starting with three 32-bit coordinates); 10 bits/vertex are rarely acceptable.

Coded surface properties include unit surface normals (used in shading calculations by typical 3D rendering APIs), texture $[u, v]$ coordinates (used to position and orient texture in the plane of specific polygons), and color coordinates (typically red–green–blue triples that can be referenced by one or more triangles in the mesh). Because these data all more or less represent continuously variable vector data such as the vertex $[x, y, z]$, they are subject to the use of similar coding tools that compress quantized 3D vertices. Current developments are evaluating vector symmetries, human perceptual thresholds, and such for best performance.

The 3DMC also addresses the coding of the change in detail over time for a 3D model based on the hierarchical addition and subtraction of mesh detail with respect to a base mesh. The method used is the so-called Forest Split methodology, which makes distributed cuts into a base mesh scattered over the model surface to insert mesh detail in steps. Different ratios of detail addition at each step are possible. The efficiency lost in this type of coding compared with the coding of the base mesh with topological surgery is governed by the granularity of mesh additions at each step. This coding method allows a user to adapt content to terminal rendering power or to receive model detailing in steps where rendering provides early clues about model utility.

Finally, 3DMC is looking at the prospect of progressive or incremental rendering and the segmenting of the total model coded stream (topology, geometry, properties) into more or less even-size layers or chunks that serve as error resilience partitioning. Several methods are under investigation. They all provide for start code insertion in the incremental stream at layer boundaries and suffer some efficiency loss as a consequence of making the layers separable where the spatial coherence across the whole model is interrupted.

V. PROFILES STATUS

In MPEG–4 conformance points are defined in the form of profiles and levels. Conformance points form the basis for interoperability, the main driving force behind standardization. Implementations of all manufacturers that conform to a particular conformance point are interoperable with each other. The MPEG–4 visual profiles and levels define subsets of the syntax and semantics of MPEG–4 visual, which in turn define the required decoder capabilities. An MPEG–4 visual profile is a defined subset of the MPEG–4 visual syntax and semantics, specified in the form of a set of tools, specified as object types. Object types group together MPEG–4 visual tools that provide a certain functionality. A level within a profile defines constraints on parameters in the bitstream that constrain these tools.

A. Object Types

An object type defines a subset of tools of the MPEG–4 visual standard that provides a functionality or a group of functionalities. Object types are the building blocks of MPEG–4 profiles and, as such, simplify the definition of profiles. There are six video object types (Simple, Core, Main, Simple Scalable, N-bit, and Still Scalable Texture) and three synthetic visual object types (Basic Animated 2D Mesh, Animated 2D Mesh, and Simple Face). In Table 3 the video and visual synthetic object types are described in basic terms.

B. Profiles

MPEG–4 visual profiles are defined in terms of visual object types. There are six video profiles: the Simple Profile, the Simple Scalable Profile, the Core Profile, the Main Profile, the N-Bit Profile, and the Scalable Texture Profile. There are two synthetic visual profiles: the Basic Animated Texture Profile and the Simple Facial Animation Profile. There is one hybrid profile, the Hybrid Profile, that combines video object types with synthetic visual object types. The profiles and their relation to the visual object types are described in Table 4.

C. Levels

A level within a profile defines constraints on parameters in the bitstream that relate to the tools of that profile. Currently there are 11 profile and level definitions that each constrain about 15 parameters that are defined for each level. To provide some insight into the levels, for the three most important video profiles, core, simple, and main, a subset of level constraints is given in Table 5. The macroblock memory size is the bound on the memory (in macroblock units) that can be used by the VMV algorithm. This algorithm models the pixel memory needed by the entire visual decoding process.

The profiling at level of the synthetic visual tools has endured a long process of debate about whether and how a certain quality of media experience in animating synthetic models can be ensured. For version 1 the face animation and 2D animated mesh tools exhibit Object Profiles at specified performance rates for the decoding of the FAPs and motion vectors, respectively.

In the case of FAPs, decoder performance requires that the full FAP set be decoded at specified rates, generally to ensure that resulting animations of models can achieve smooth motion and graphics update and refresh. However, after much work, no specification on model complexity or rendered pixel areas on the viewer screen has been approved, and nothing has yet been said about the quality of the models or the terminal resources that are expected. The performance of 2D meshes is centered on two alternative mesh node complexities and again is specified for certain frame rates for smooth motion.

VI. SUMMARY

In this chapter we have provided a brief overview of the applications, functionalities, tools, and profiles of the MPEG–4 Visual standard. As discussed, MPEG–4 Visual consists of

Table 3 MPEG–4 Visual Object Types

MPEG–4 visual tools	MPEG–4 visual object types								
	Simple	Core	Main	Simple scalable	*N*-bit	Scalable still texture	Basic animated	Animated 2D mesh	Simple face
Basic (I- and P-VOP, coefficient preciction, 4-MV, unrestricted MV)	●	●	●	●	●			●	
Error resilience	●	●	●	●	●			●	
Short header	●	●	●		●			●	
B-VOP		●	●	●	●			●	
P-VOP–based temporal scalability		●	●		●			●	
Binary shape		●	●		●		●	●	
Gray shape			●						
Interlace			●						
Sprite			●						
Temporal scalability (rectangular)				●					
Spatial scalability (rectangular)				●					
N-bit					●				
Scalable still texture						●	●	●	
2D dynamic uniform mesh							●	●	
2D dynamic Delaunay mesh								●	
Facial animation parameters									●

Table 4 MPEG–4 Visual Profiles

MPEG–4 visual object types	Simple	Core	Main	Simple scalable	*N*-bit	Scalable texture	Basic animated texture	Simple facial animation	Hybrid
	MPEG–4 visual profiles								
Simple	●	●	●	●	●				●
Core		●	●		●				●
Main			●						●
Simple scalable				●					
N-bit					●				
Scalable still texture			●			●	●		●
Basic animated texture							●		●
Animated 2D mesh									●
Simple face							●	●	●

not only MPEG–4 Video but also MPEG–4 Synthetic Visual. MPEG–4 Video in its version 1 includes tools to address traditional functionalities such as efficient coding and scalability for frame (rectangular video)–based coding as well as unique new functionalities such as efficient coding and scalability of arbitrary-shaped video objects and others such as error resilience. Version 2 of Video adds new tools to improve coding efficiency, scalability, and error resilience. MPEG–4 Synthetic Visual in its version 1 presents primarily brand new functionalities such as 2D mesh–based animation of generic visual objects and animation of synthetic faces. Version 2 of Synthetic Visual includes primarily tools for 3D mesh–based animation and animation of synthetic body representation. In this chapter we have only introduced these tools; in the next few chapters, many of these tools are discussed in detail.

Table 5 Subset of MPEG–4 Video Profile and Level Definitions

Profile and level		Typical scene size	Bit rate (bit/sec)	Maximum number of objects	Total macroblock memory (macroblock units)
Simple profile	L1	QCIF	64 k	4	198
	L2	CIF	128 k	4	792
	L3	CIF	384 k	4	792
Core profile	L1	QCIF	384 k	4	594
	L2	CIF	2 M	16	2,376
Main profile	L2	CIF	2 M	16	2,376
	L3	ITU-R 601	15 M	32	9,720
	L4	1920 × 1088	38.4 M	32	48,960

REFERENCES

1. MPEG–1 Video Group. Information technology—Coding of moving pictures and associated audio for digital storage media up to about 1.5 Mbit/s: Part 2—Video. ISO/IEC 11172-2, International Standard, 1993.
2. BG Haskell, A Puri, AN Netravali. Digital Video: An Introduction to MPEG–2. New York: Chapman & Hall, 1997.
3. MPEG–2 Video Group. Information technology—Generic coding of moving pictures and associated audio: Part 2—Video. ISO/IEC 13818-2, International Standard, 1995.
4. MPEG–4 Video Group. Generic coding of audio-visual objects: Part 2—Visual. ISO/IEC JTC1/SC29/WG11 N2202, FCD of ISO/IEC 14496-2, May 1998.
5. ITU-T Experts Group on Very Low Bitrate Visual Telephony. ITU-T Recommendation H.263: Video coding for low bitrate communication. December 1995.
6. J Osterman, A Puri. Natural and synthetic video in MPEG–4. Proceedings IEEE ICASSP, Seattle, April 1998.
7. A Puri, A Eleftheriadis. MPEG–4: An object-based multimedia coding standard supporting mobile applications. ACM J Mobile Networks Appl 3:5–32, 1998.
8. MPEG–4 Video Verification Model Editing Committee. The MPEG–4 video verification model 8.0. ISO/IEC JTC1/SC29/WG11 N1796, Stockholm, July 1997.
9. A Puri, RL Schmidt, BG Haskell. Improvements in DCT based video coding. Proceedings SPIE Visual Communications and Image Processing, San Jose, January 1997.
10. T Sikora. The MPEG–4 video standard verification model. Proceedings IEEE Trans. CSVT, vol 7, no 1, February 1997.
11. FI Parke, K Waters. Computer Facial Animation. AK Peters, 1996.
12. T Sikora, L Chiariglione. The MPEG–4 video standard and its potential for future multimedia applications. Proceedings IEEE ISCAS Conference, Hong Kong, June 1997.
13. A Puri, RL Schmidt, BG Haskell. Performance evaluation of the MPEG–4 visual coding standard, Proceedings Visual Communications and Image Processing, San Jose, January 1998.
14. MPEG–4 Video Group. MPEG–4 video verification model version 12.0. ISO/IEC JTC1/SC29/WG11 N2552, Rome, December 1998.
15. MPEG–4 Systems Group. Generic coding of audio-visual objects: Part 1—Systems. ISO/IEC JTC1/SC29/WG11 N2201, FCD of ISO/IEC 14496-1, May 1998.

8
MPEG-4 Natural Video Coding—Part I

Atul Puri and Robert L. Schmidt
AT&T Labs, Red Bank, New Jersey

Ajay Luthra and Xuemin Chen
General Instrument, San Diego, California

Raj Talluri
Texas Instruments, Dallas, Texas

I. INTRODUCTION

The Moving Picture Experts Group (MPEG) of the International Standardization Organization (ISO) has completed the MPEG-1 [1] and the MPEG-2 [2] standards and its third standard (which comes in two versions); MPEG-4 version 1 [3] is also complete and version 2 is currently in progress. Besides the ISO, the International Telephony Union, telecommunications section (ITU-T) has also standardized video and audio coding techniques. The ITU-T video standards are application specific (videophone and videoconferencing), whereas the MPEG standards are relatively generic. A while back, ITU-T completed its video standard, H.263 [4], a refinement of H.261, which is an earlier ITU-T standard, and it is currently finalizing H.263 version 2 [5], which extends H.263.

The MPEG-1 video standard was optimized [1,6,7] for coding of noninterlaced (progressive) video of Source Intermediate Format (SIF) resolution at about 1.5 Mbit/sec for interactive applications on digital storage media. The MPEG-2 video standard, although it is a syntactic superset of MPEG-1 and allows additional functionalities such as scalability, was mainly optimized [2,8–11] for coding of interlaced video of CCIR-601 resolution for digital broadcast applications. Compared with MPEG-1 and MPEG-2, the MPEG-4 standard brings a new paradigm as it treats a scene to be coded as consisting of individual objects; thus each object in the scene can be coded individually and the decoded objects can be composed in a scene. MPEG-4 is optimized [3,12–20] for a bit rate range of 10 kbit/sec to 3 Mbit/sec. The work done by ITU-T for H.263 version 2 [5] is of relevance for MPEG-4 because H.263 version 2 is an extension of H.263 [4,21] and H.263 was also one of the starting bases for MPEG-4. However, MPEG-4 is a more complete standard

[22] because it can address a very wide range and types of applications, has extensive systems support, and has tools for coding and integration of natural and synthetic objects.

MPEG-4 is a multimedia standard that specifies coding of audio and video objects, both natural and synthetic; a multiplexed representation of many such simultaneous objects; and the description and dynamics of the scene containing these objects. More specifically, MPEG-4 supports a number of advanced functionalities. These include (1) the ability to encode efficiently mixed media data such as video, graphics, text, images, audio, and speech (called audiovisual objects); (2) the ability to create a compelling multimedia presentation by compositing these mixed media objects by a compositing script; (3) error resilience to enable robust transmission of compressed data over noisy communication channels; (4) the ability to encode arbitrary-shaped video objects; (5) multiplexing and synchronizing the data associated with these objects so that they can be transported over network channels providing a quality of service (QoS) appropriate for the nature of the specific objects; and (6) the ability to interact with the audiovisual scene generated at the receiver end. These functionalities supported by the standard are expected to enable compelling new applications including wireless videophones, Internet multimedia, interactive television, and Digital Versatile Disk (DVD). As can be imagined, a standard that supports this diverse set of functionalities and associated applications can become fairly complex.

As mentioned earlier, the recently completed specification of MPEG-4 corresponds to version 1, and work is ongoing for version 2 of the standard, which includes additional coding tools. The video portion of the MPEG-4 standard is encapsulated in the MPEG-4 Visual part of the coding standard, which also includes coding of synthetic data such as facial animation and mesh-based coding. In this chapter, although we introduce the general concepts behind coding of arbitrary-shaped video objects, the focus of our discussion is coding of rectangular video objects (as in MPEG-1, MPEG-2, and H.263). The next chapter will focus on details of coding of arbitrary-shaped video objects and related issues.

We now discuss the organization of the rest of this chapter. In Sec. II, we introduce the basic concepts behind MPEG-4 video coding and then discuss the specific tools for increasing the coding efficiency over H.263 and MPEG-1 when coding noninterlaced video. In Sec. III, we discuss tools for increasing the coding efficiency over MPEG-2 when coding interlaced video. In Sec. IV, we introduce MPEG-4 video tools to achieve error resilience and robustness. Next, in Sec. V, we present scalability tools of MPEG-4 video. In Sec. VI, we present a summary of the tools in MPEG-4 and related standards for comparison. In Sec. VII we discuss the recommended postprocessing method (but not standardized by MPEG-4 video) to reduce blockiness and other artifacts in low-bit-rate coded video. In Sec. 8, we finally summarize the key points presented in this chapter.

II. MPEG-4 VIDEO CODING

The video portion of the MPEG-4 version 1 standard, which consists primarily of the bitstream syntax and semantics and the decoding process, has recently achieved the stable and mature stage of Final Draft International Standard (FDIS) [3]. Version 2 of the standard, which adds tools for advanced functionality, is also reasonably well understood and is currently at the stage of Committee Draft (CD). For the purpose of explaining concepts in this chapter, we borrow portions of the MPEG-4 coding description from the MPEG-4 Video VM8 [12], the MPEG-4 Video FDIS [3], and other previously published work.

A. Background

An input video sequence consists of related snapshots or pictures separated in time. Each picture consists of temporal instances of objects that undergo a variety of changes such as translations, rotations, scaling, and brightness and color variations. Moreover, new objects enter a scene and/or existing objects depart, resulting in the appearance of certain objects only in certain pictures. Sometimes, scene change occurs, and thus the entire scene may be either reorganized or replaced by a new scene. Many of the MPEG-4 functionalities require access not only to an entire sequence of pictures but also to an entire object and, further, not only to individual pictures but also to temporal instances of these objects within a picture. A temporal instance of a video object can be thought of as a snapshot of an arbitrary-shaped object that occurs within a picture, such that like a picture, it is intended to be an access unit, and, unlike a picture, it is expected to have a semantic meaning.

B. Video Object Planes and Video Objects

The concept of video objects and their temporal instances, video object planes (VOPs), is central to MPEG-4 video. A VOP can be fully described by texture variations (a set of luminance and chrominance values) and (explicit or implicit) shape representation. In natural scenes, VOPs are obtained by semiautomatic or automatic segmentation, and the resulting shape information can be represented as a *binary shape* mask. On the other hand, for hybrid (of natural and synthetic) scenes generated by blue screen composition, shape information is represented by an 8-bit component, referred to as a *gray-scale shape*. Video objects can also be subdivided into multiple representations or video object layers (VOLs), allowing scalable representations of the video object. If the entire scene is considered as one object and all VOPs are rectangular and of the same size as each picture, then a VOP is identical to a picture. In addition, an optional group of video object planes (GOV) can be added to the video coding structure to assist in random-access operations.

In Figure 1, we show the decomposition of a picture into a number of separate VOPs. The scene consists of two objects (head-and-shoulders view of a human and a logo) and the background. The objects are segmented by semiautomatic or automatic means and are referred to as VOP1 and VOP2, and the background without these objects is referred

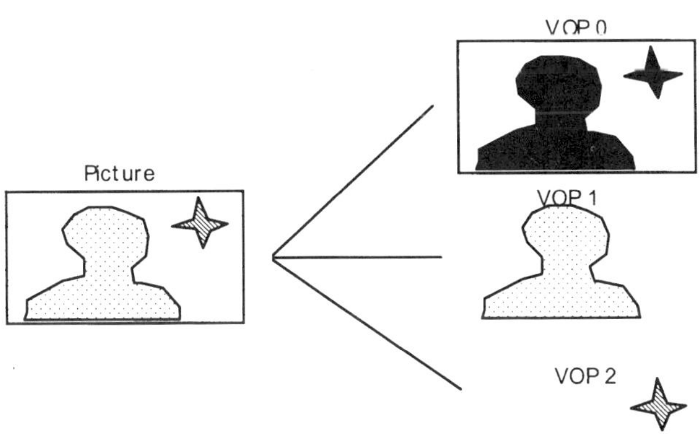

Figure 1 Semantic segmentation of a picture into VOPs.

to as VOP0. Each picture in the sequence is segmented into VOPs in this manner. Thus, a segmented sequence contains a temporal set of VOP0s, a temporal set of VOP1s, and a temporal set of VOP2s.

C. Coding Structure

The VOs are coded separately and multiplexed to form a bitstream that users can access and manipulate (cut, paste, etc.). Together with video objects, the encoder sends information about scene composition to indicate where and when VOPs of a video object are to be displayed. This information is, however, optional and may be ignored at the decoder, which may use user-specified information about composition.

In Figure 2, a high-level logical structure of a video object–based coder is shown. Its main components are a video objects segmenter/formatter, video object encoders, systems multiplexer, systems demultiplexer, video object decoders, and a video object compositor. The video object segmenter segments the input scene into video objects for encoding by the video object encoders. The coded data of various video objects are multiplexed for storage or transmission, following which the data are demultiplexed and decoded by video object decoders and offered to the compositor, which renders the decoded scene.

To consider how coding takes place in a video object encoder, consider a sequence of VOPs. MPEG-4 video extends the concept of intra (I-), predictive (P-), and bidirectionally predictive (B-) pictures of MPEG-1/2 video to VOPs, and I-VOP, P-VOP, and B-VOP, result. Figure 3 shows a coding structure that uses two consecutive B-VOPs between a pair of reference VOPs (I- or P-VOPs).

The basic MPEG-4 coding employs motion compensation and Discrete Cosine Transform (DCT)–based coding. Each VOP consists of macroblocks that can be coded as intra or as inter macroblocks. The definition of a macroblock is exactly the same as in MPEG-1 and MPEG-2. In I-VOPs, only intra macroblocks exist. In P-VOPs, intra as well as unidirectionally predicted macroblocks can occur, whereas in B-VOPs, both uni- and bidirectionally predicted macroblocks can occur.

D. INTRA Coding

MPEG-4 has made several improvements in coding of intra macroblocks (INTRA) as compared with H.263, MPEG-1/2. In particular, it supports the following:

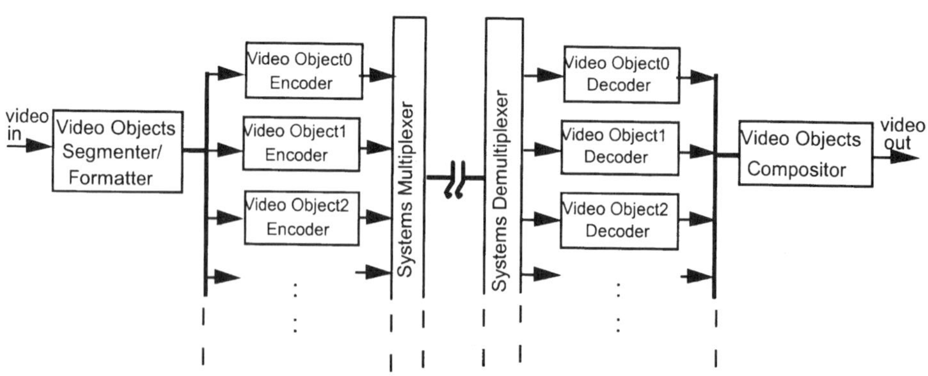

Figure 2 Logical structure of video object–based codec of MPEG-4 video.

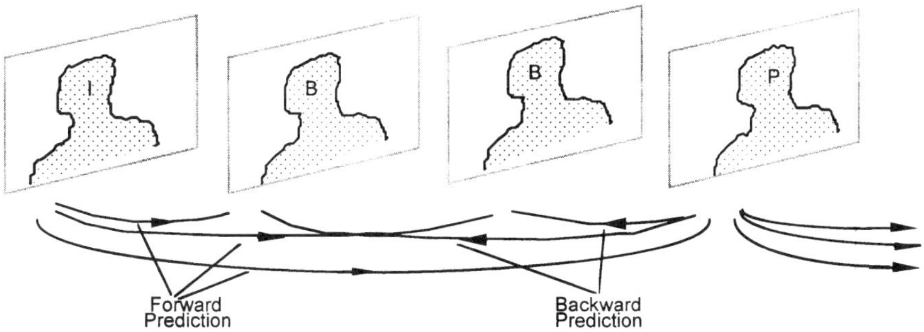

Figure 3 An example prediction structure when using I-, P-, and B-VOPs.

Prediction of the DC coefficient
Prediction of a subset of AC coefficients
Specialized coefficient scanning based on the coefficient prediction
Variable length coding (VLC) table selection
Nonlinear inverse DC quantization
Downloadable quantization matrices

We now discuss details of important combinations of these tools and specific contributions they make to overall performance improvement.

1. DC Prediction Improvement

One of the INTRA coding improvements is an advanced method for predicting the DC coefficient. For reference, H.263 does not include DC prediction, and MPEG-1 allows only simple DC prediction. The DC prediction method of MPEG-1 (or MPEG-2) is improved to allow adaptive selection of either the DC value of the immediately previous block or that of the block immediately above it (in the previous row of blocks). This adaptive selection of the DC prediction direction does not incur any overhead as the decision is based on comparison of the horizontal and vertical DC value gradients around the block whose DC value is to be coded.

Figure 4 shows four surrounding blocks of the block whose DC value is to be coded. However, only three of the previous DC values are currently being used; the fourth value is anticipated to provide a better decision in the case of higher resolution images and may

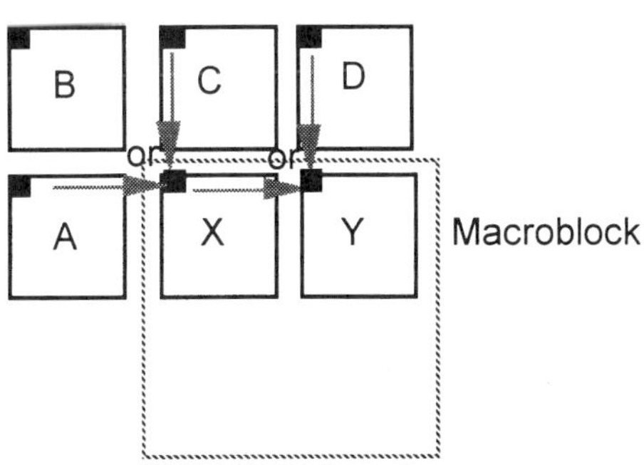

Figure 4 Previous neighboring blocks used in improved DC prediction.

be used there. Assume X, A, B, C, and D correspondingly refer to the current block, the previous block, the block above and to the left, the block immediately above, and the block above and to the right as shown.

The DC value of X is predicted by either the DC value of block A or the DC value of block C based on the comparison of horizontal and vertical gradients by use of Graham's method as follows.

The dc values ''dc'' obtained after DCT are first quantized by 8 to generate ''DC'' values; DC = dc//8. The DC prediction ($DC_{X'}$) is calculated as follows.

$$\text{If } (|DC_A - DC_B| < |DC_B - DC_C|)$$

$$DC_{X'} = DC_C$$

else

$$DC_{X'} = DC_A$$

For DC prediction, the following simple rules are used:

If any of the blocks A, B, and C are outside the VOP boundary, their DC values are assumed to take a value of 128 and are used to compute prediction values.

In the context of computing DC prediction for block X, if the absolute value of a horizontal gradient ($|DC_A - DC_B|$) is less than the absolute value of a vertical gradient ($|DC_B - DC_C|$), then the prediction is the DC value of block C; otherwise, the DC value of block A is used for prediction. This process is repeated independently for every block of a macroblock using an appropriate immediately horizontally adjacent block A and immediately vertically adjacent block C.

DC predictions are performed identically for the luminance component as well as each of the two chrominance components.

2. *AC Coefficient Prediction: Prediction of First Row or First Column*

Prediction of AC coefficients of INTRA DCT blocks is not allowed in H.263 or MPEG-1/2. With this method, either coefficients from the entire or part of the first row or coefficients from the entire or part of the first column of the previous coded block are used to predict the colocated coefficients of the current block. For best results, the number of coefficients of a row or column and the precise location of these coefficients need to be identified and adapted in coding different pictures and even within the same picture. This, however, results in either too much complexity or too much overhead. A practical solution is to use a predetermined number of coefficients for prediction; for example, we use seven ac coefficients.

On a block basis, the best direction (from among horizontal and vertical directions) for DC coefficient prediction is also used to select the direction for AC coefficient prediction. An example of the process of AC coefficient prediction employed is shown in Figure 5.

Because the improved AC prediction mainly employs prediction from either the horizontal or the vertical direction, whenever diagonal edges, coarse texture, or combinations of horizontal and vertical edges occur, the AC prediction does not work very well and needs to be disabled. Although ideally one would like to turn off AC prediction on

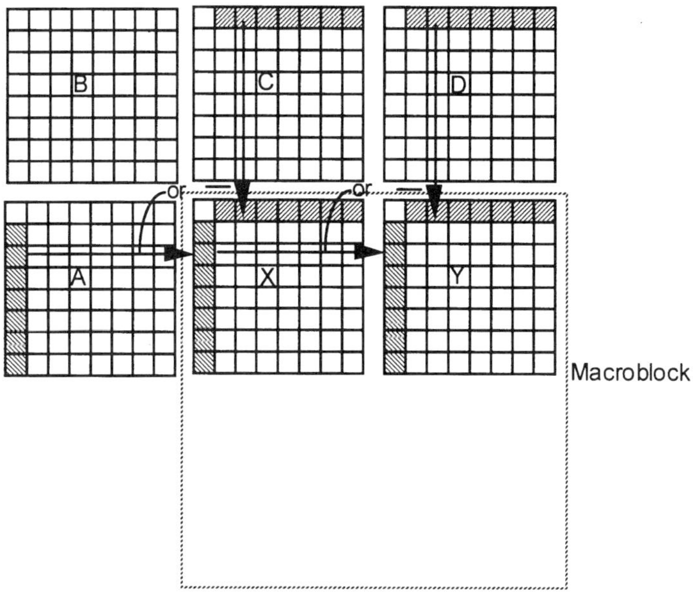

Figure 5 Previous neighboring blocks and coefficients used in improved AC prediction.

a block basis, this generates too much overhead; thus we disable AC prediction on a macroblock basis. The criterion for AC prediction enable or disable is discussed next.

In the cases in which AC coefficient prediction results in an error signal of larger magnitude as compared with the original signal, it is desirable to disable AC prediction. However, the overhead is excessive if AC prediction is switched on or off every block, so AC prediction switching is performed on a macroblock basis.

If block A was selected as the DC predictor for the block for which coefficient prediction is to be performed, we calculate a criterion, S, as follows.

$$S = \left(\sum_{i=1}^{7} |AC_{i0X}| \quad \sum_{i=1}^{7} |AC_{i0X} - AC_{i0A}| \right)$$

If block C was selected as the DC predictor for the block for which coefficient prediction is to be performed, we calculate S as follows.

$$S = \left(\sum_{j=1}^{7} |AC_{0jX}| - \sum_{j=1}^{7} |AC_{0jX} - AC_{0jC}| \right)$$

Next, for all blocks for which a common decision is to be made (in this case on a macroblock basis) a single $\sum S$ is calculated and the ACpred_flag is either set or reset to enable or disable AC prediction as follows.

If ($\sum S \geq 0$) ACpred_flag = 1

else ACpred_flag = 0

3. Scanning of DCT Coefficients

In H.263 or MPEG-1 the intra block DCT coefficients are scanned by the zigzag scan to generate run-level events that are VLC coded. The zigzag scan works well on the average and can be looked upon as a combination of three types of scans, a horizontal type of scan, a vertical type of scan, and a diagonal type of scan. Often in natural images, on a block basis, a predominant preferred direction for scanning exists depending on the orientation of significant coefficients.

We discuss the two scans (in addition to the zigzag scan) such that in coding a block (or a number of blocks, depending on the overhead incurred), a scanning direction is chosen that results in more efficient scanning of coefficients to produce (run, level) events that can be efficiently entropy coded as compared with scanning by zigzag scan alone. These two additional scans are referred to as alternate-hor and alternate-vert (used in MPEG-2 for block scanning of DCT coefficients of interlaced video) and along with the zigzag scan is shown in Figure 6.

There is, however, an important trade-off in the amount of savings that result from scan adaptation versus the overhead required in block-to-block scan selection. Furthermore, if the selection of scan is based on counting the total bits generated by each scanning method and selecting the one that produces the least bits, then complexity becomes an important issue. The key is thus that the scan selection overhead be minimized. Thus, the criterion used to decide the AC prediction direction is now also employed to indicate the scan direction. If ACpred_flag = 0, zigzag scan is selected for all blocks in a macroblock; otherwise, the DC prediction direction (hor or vert) is used to select a scan on a block basis. For instance, if the DC prediction refers to the horizontally adjacent block, alternate-vert scan is selected for the current block; otherwise (for DC prediction referring to the vertically adjacent block), alternate-hor scan is used for the current block.

4. Variable Length Coding

Neither the H.263 nor the MPEG-1 standard allows a separate variable length code table for coding DCT coefficients of intra blocks. This forces the use of the inter block DCT VLC table, which is inefficient for intra blocks. The MPEG-2 standard does allow a separate VLC table for intra blocks, but it is optimized for much higher bit rates. MPEG-4 provides an additional table optimized for coding of AC coefficients of intra blocks [4]. The MPEG-4 table is three-dimensional; that is, it maps the zero run length, the coefficient level value, and the last coefficient indication into the variable length code. For the DC coefficient, H263 uses an 8-bit Fixed Length Code (FLC), and MPEG-1 and 2 provide

0	1	2	3	10	11	12	13
4	5	8	9	17	16	15	14
6	7	19	18	26	27	28	29
20	21	24	25	30	31	32	33
22	23	34	35	42	43	44	45
36	37	40	41	46	47	48	49
38	39	50	51	56	57	58	59
52	53	54	55	60	61	62	63

a

0	4	6	20	22	36	38	52
1	5	7	21	23	37	39	53
2	8	19	24	34	40	50	54
3	9	18	25	35	41	51	55
10	17	26	30	42	46	56	60
11	16	27	31	43	47	57	61
12	15	28	32	44	48	58	62
13	14	29	33	45	49	59	63

b

0	1	5	6	14	15	27	28
2	4	7	13	16	26	29	42
3	8	12	17	25	30	41	43
9	11	18	24	31	40	44	53
10	19	23	32	39	45	52	54
20	22	33	38	46	51	55	60
21	34	37	47	50	56	59	61
35	36	48	49	57	58	62	63

c

Figure 6 (a) Alternate horizontal scan; (b) alternate vertical (MPEG-2) scan; (c) zigzag scan.

differential coding of the DC coefficient based on the previous block. MPEG-4 also uses differential coding of the DC coefficient based on the previously described improved prediction algorithm. In addition, MPEG-4 allows switching the coding of the DC coefficient from the DC table to inclusion in the AC coefficient coding for the block. This switch is based on a threshold signaled in the bitstream and is triggered by the quantization parameter for the macroblock.

5. *Coding Results for I-VOPs*

Table 1(a) and Table 1(b) present the improvements over H263 and MPEG-1 resulting from using DC and AC prediction at both Quarter Common Intermediate Format (QCIF) and CIF resolutions. The results for the improved VLC table are based on an earlier submission to MPEG, but similar results can be expected with the table used in the current standard.

As can be seen from Table 1(b), for coding of CIF resolution, intra coding tools of MPEG-4 provide around 22% improvement on the average over H.263 intra coding, irrespective of bitrates. The improvement at QCIF resolution [Table 1(a)] is a little lower but still reaches about 16% on the average over H.263 intra coding. The improvements are about half as much over MPEG-1 intra coding.

E. P-VOP Coding

As in the previous MPEG standards, inter macroblocks in P-VOPs are coded using a motion-compensated block matching technique to determine the prediction error. However, because a VOP is arbitrarily shaped and the size can change from one instance to the next, special padding techniques are defined to maintain the integrity of the motion compensation. For this process, the minimum bounding rectangle of each VOP is refer-

Table 1(a) SNR and Bit Counts of Intra Coding at QCIF Resolution

Sequence	Q p	SNR Y	Bits: H.263 intra	Bits: MPEG-1 Video intra (% reduction)	Bits: MPEG-4 Video intra (% reduction)
Akiyo	8	36.76	22,559	21,805 (−3.34)	19,604 (−13.10)
	12	34.08	16,051	15,297 (−4.69)	13,929 (−13.22)
	16	32.48	12,668	11,914 (5.95)	11,010 (−13.10)
Silent	8	34.48	27,074	26,183 (−3.29)	23,642 (−12.68)
	12	31.96	17,956	17,065 (−8.97)	15,915 (−11.36)
	16	30.42	13,756	12,865 (−6.48)	11,996 (−12.79)
Mother	8	36.23	19,050	18,137 (−4.79)	15,120 (−20.63)
	12	33.91	13,712	12,799 (−6.66)	10,583 (−22.82)
	16	32.41	11,000	10,087 (−8.30)	8,291 (−24.63)
Hall	8	35.88	29,414	28,386 (−3.49)	24,018 (−18.35)
	12	33.09	21,203	20,175 (−4.85)	17,539 (−17.28)
	16	31.13	16,710	15,682 (−6.15)	13,651 (−18.31)
Foreman	8	35.27	27,379	26,678 (−2.56)	23,891 (−12.74)
	12	32.66	19,493	18,792 (−3.60)	17,217 (−11.68)
	16	30.94	15,483	14,782 (−4.53)	13,740 (−11.26)

Table 1(b) SNR and Bit Counts of Intra Coding at CIF Resolution

Sequence	Q p	SNR Y	Bits		
			H.263 intra	MPEG-1 Video intra (% reduction)	MPEG-4 Video intra (% reduction)
Akiyo	8	38.78	61,264	55,354 (−9.65)	47,987 (−21.67)
	12	36.40	46,579	40,669 (−12.68)	36,062 (−22.58)
	16	34.69	38,955	33,045 (−15.17)	29,828 (−23.43)
Silent	8	34.49	86,219	81,539 (−5.43)	74,524 (−13.56)
	12	32.34	59,075	54,395 (−7.92)	49,218 (−16.69)
	16	30.97	45,551	40,871 (−10.27)	36,680 (−19.47)
Mother	8	38.01	54,077	47,522 (−12.12)	40,775 (−24.60)
	12	35.84	41,010	34,455 (−15.98)	29,544 (−27.96)
	16	34.47	34,127	27,572 (−19.21)	23,612 (−30.81)
Hall	8	37.30	86,025	80,369 (−6.57)	62,681 (−27.14)
	12	34.97	64,783	59,127 (−8.73)	46,897 (−27.61)
	16	33.24	53,787	48,131 (−10.52)	38,496 (−28.43)
Foreman	8	35.91	78,343	73,471 (−6.22)	67,486 (−13.86)
	12	33.76	56,339	51,467 (−8.63)	48,234 (−14.39)
	16	32.24	45,857	40,985 (−10.62)	38,819 (−15.35)

enced to an absolute frame coordinate system. All displacements are with respect to the absolute coordinate system, so no VOP alignment is necessary. The padding process for arbitrarily shaped objects is described in detail in the chapter on shape coding.

The motion vector coding in MPEG-4 is similar to that in H263; thus, either one or four motion vectors are allowed per macroblock. As in H263, the horizontal and vertical motion vector components are differentially coded, based on a prediction that is formed by the median filtering of three vector candidate predictors. These predictors (MV1, MV2, MV3) are derived from the spatial neighborhood of macroblocks or blocks previously decoded. The spatial position of candidate predictors for each block vector is depicted in Figure 7. If the macroblock is coded with only one motion vector, the top left case is used for a prediction.

Because MPEG-4 is intended to address a broader range of bit rates than H263, an f_code mechanism is used to extend the motion vector range from −2048 to +2047 in half-pel units. It also allows the motion-compensated reference to extend beyond the VOP boundary. If a pixel referenced by a motion vector falls outside the VOP area, the value of an edge pixel is used as a reference. This edge pixel is retrieved by limiting the motion vector to the last full pel position inside the decoded VOP area. Limitation of a motion vector is performed on a sample basis and separately for each component of the motion vector, as depicted in Figure 8.

The coordinates of a reference pixel in the reference VOP, (yref, xref), are related to the absolute coordinate system and are determined as follows:

$$\text{xref} = \text{MIN}(\text{MAX}(\text{xcurr} + \text{dx}, \text{vhmcsr}), \text{xdim} + \text{vhmcsr} - 1)$$

$$\text{yref} = \text{MIN}(\text{MAX}(\text{ycurr} + \text{dy}, \text{vvmcsr}), \text{ydim} + \text{vvmcsr} - 1)$$

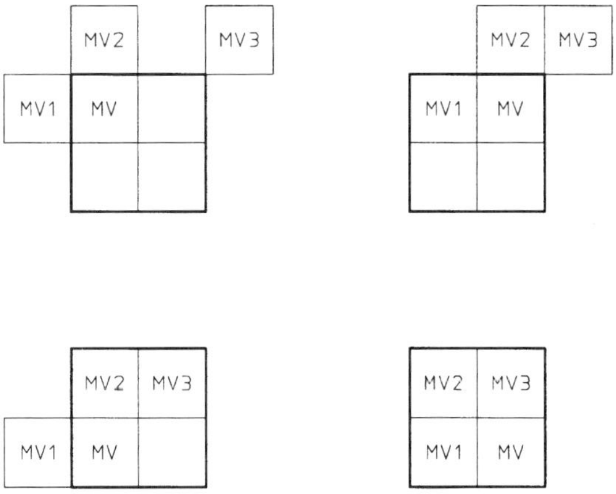

Figure 7 Definition of the candidate predictors MV1, MV2, and MV3 for each of the luminance blocks in a macroblock.

where

vhmcsr = the horizontal motion-compensated spatial reference for the VOP
vvmcsr = the vertical motion-compensated spatial reference for the VOP
(ycurr, xcurr) are the coordinates of a pixel in the current VOP
(yref, xref) are the coordinates of a pixel in the reference VOP
(dy, dx) is the motion vector
(ydim, xdim) are the dimensions of the bounding rectangle of the reference VOP

1. Overlapped Block Motion Compensation

MPEG-4 also supports an overlapped block motion compensation mode. In this mode, each pixel in a luminance block of the macroblock is computed as the weighted sum of three prediction values. In order to obtain the three prediction values, three motion vectors are used: the motion vector of the current luminance block, the motion vector from the

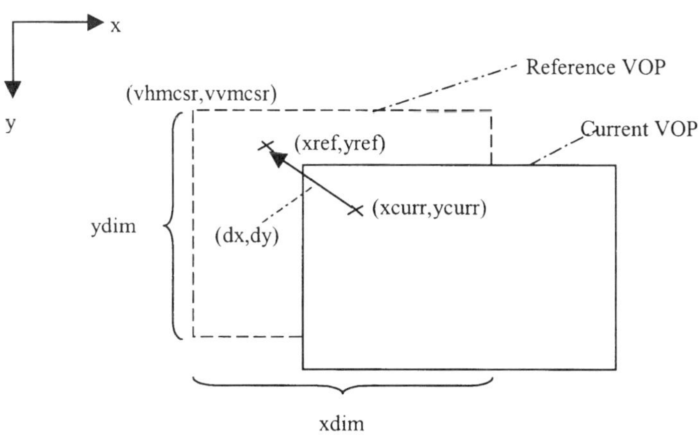

Figure 8 Unrestricted motion compensation.

nearest adjacent block above or below the current block, and the motion vector from the nearest adjacent block to the left or right of the current block. This means that for the upper half of the block the motion vector corresponding to the block above the current block is used, and for the lower half of the block the motion vector corresponding to the block below the current block is used. Similarly, for the left half of the block the motion vector corresponding to the block at the left side of the current block is used, and for the right half of the block the motion vector corresponding to the block at the right side of the current block is used. The creation of each pixel, $\bar{p}(i, j)$, in an 8×8 luminance prediction block is governed by the following equation:

$$\bar{p}(i, j) = (q(i, j) \times H_0(i, j) + r(i, j) \times H_1(i, j) + s(i, j) \times H_2(i, j) + 4)//8$$

where $q(i, j)$, $r(i, j)$, and $s(i, j)$ are the pixels from the referenced picture as defined by

$$q(i, j) = p(i + MV_x^0, j + MV_y^0)$$
$$r(i, j) = p(i + MV_x^1, j + MV_y^1)$$
$$s(i, j) = p(i + MV_x^2, j + MV_y^2)$$

Here (MV_x^0, MV_y^0) denotes the motion vector for the current block, (MV_x^1, MV_y^1) denotes the motion vector of the nearest block either above or below, and (MV_x^2, MV_y^2) denotes the motion vector of the nearest block either to the left or right of the current block as defined before.

The matrices $H_0(i, j)$, $H_1(i, j)$, and $H_2(i, j)$ are defined in Figure 9a–c, where (i, j) denotes the column and row, respectively, of the matrix.

If one of the surrounding blocks was not coded, the corresponding remote motion vector is set to zero. If one of the surrounding blocks was coded in intra mode, the corresponding remote motion vector is replaced by the motion vector for the current block. If the current block is at the border of the VOP and therefore a surrounding block is not present, the corresponding remote motion vector is replaced by the current motion vector. In addition, if the current block is at the bottom of the macroblock, the remote motion vector corresponding to an 8×8 luminance block in the macroblock below the current macroblock is replaced by the motion vector for the current block.

F. B-VOP Coding

Four modes are used for coding B-VOPs. They are the forward mode, the backward mode, the bidirectional mode, and the direct mode. Overlapped block motion compensation is

4	5	5	5	5	5	5	4
5	5	5	5	5	5	5	5
5	5	6	6	6	6	5	5
5	5	6	6	6	6	5	5
5	5	6	6	6	6	5	5
5	5	6	6	6	6	5	5
5	5	5	5	5	5	5	5
4	5	5	5	5	5	5	4

a

2	2	2	2	2	2	2	2
1	1	2	2	2	2	1	1
1	1	1	1	1	1	1	1
1	1	1	1	1	1	1	1
1	1	1	1	1	1	1	1
1	1	1	1	1	1	1	1
1	1	2	2	2	2	1	1
2	2	2	2	2	2	2	2

b

2	1	1	1	1	1	1	2
2	2	1	1	1	1	2	2
2	2	1	1	1	1	2	2
2	2	1	1	1	1	2	2
2	2	1	1	1	1	2	2
2	2	1	1	1	1	2	2
2	2	1	1	1	1	2	2
2	1	1	1	1	1	1	2

c

Figure 9 (a) H0 (current); (b) H1 (top–bottom); (c) H2 (left–right).

not used in any of the four modes. As in P-VOPs, the reference VOPs are padded to allow prediction from arbitrarily shaped references.

1. Forward Mode, Backward Mode, and Bidirectional Mode

If the forward mode is used, a single motion vector is used to retrieve a reference macroblock from the previously decoded I or P VOP. The backward mode is similar except that the motion vector is used to retrieve a reference macroblock from the future VOP, and, as the name implies, the bidirectional mode uses motion vectors for reference macroblocks in both the previous and future VOPs. These three modes are identical to those used in MPEG 1; they operate on 16×16 macroblocks only, and they do not allow motion compensation on a block basis.

2. Direct Mode

This is the only mode that allows motion vectors on 8×8 blocks. The mode uses direct bidirectional motion compensation by scaling I or P motion vectors (I-VOP assumes a zero MV) to derive the forward and backward motion vectors for the B-VOP macroblock. The direct mode utilizes the motion vectors (MVs) of the colocated macroblock in the most recently decoded I- or P-VOP. The colocated macroblock is defined as the macroblock that has the same horizontal and vertical index with the current macroblock in the B-VOP. The MVs are the block vectors of the colocated macroblock. If the colocated macroblock is transparent and thus the MVs are not available, the direct mode is still enabled by setting MVs to zero vectors.

a. Calculation of Vectors. Figure 10 shows scaling of motion vectors. The calculation of forward and backward motion vectors involves linear scaling of the colocated block in temporally next I- or P-VOP, followed by correction by a delta vector (MVDx, MVDy). The forward and the backward motion vectors are $\{(\mathrm{MVFx}[i], \mathrm{MVFy}[i]), (\mathrm{MVBx}[i], \mathrm{MVBy}[i]), i = 0, 1, 2, 3\}$ and are given in half sample units as follows.

$$
\begin{aligned}
\mathrm{MVFx}[i] &= (\mathrm{TRB} \times \mathrm{MVx}[i])/\mathrm{TRD} + \mathrm{MVDx} \\
\mathrm{MVBx}[i] &= (\mathrm{MVDx} == 0)?\ (((\mathrm{TRB} - \mathrm{TRD}) \times \mathrm{MVx}[i])/ \\
&\quad \mathrm{TRD}) : (\mathrm{MVFx}[i] - \mathrm{MVx}[i]) \\
\mathrm{MVFy}[i] &- (\mathrm{TRB} \times \mathrm{MVy}[i])/\mathrm{TRD} + \mathrm{MVDy} \\
\mathrm{MVBy}[i] &= (\mathrm{MVDy} == 0)?\ (((\mathrm{TRB} - \mathrm{TRD}) \times \mathrm{MVy}[i])/ \\
&\quad \mathrm{TRD}) : (\mathrm{MVFy}[i] - \mathrm{MVy}[i]) \qquad i = 0, 1, 2, 3
\end{aligned}
$$

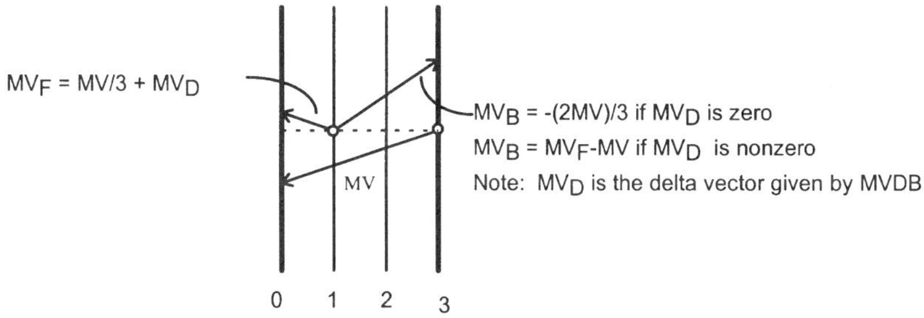

Figure 10 Direct bidirectional prediction.

where $\{(MVx[i], MVy[i]), i = 1, 2, 3\}$ are the MVs of the colocated macroblock and TRB is the difference in temporal reference of the B-VOP and the previous reference VOP. TRD is the difference in temporal reference of the temporally next reference VOP with the temporally previous reference VOP, assuming B-VOPs or skipped VOPs in between.

Generation of Prediction Blocks. Motion compensation for luminance is performed individually on 8×8 blocks to generate a macroblock. The process of generating a prediction block simply consists of using computed forward and backward motion vectors $\{(MVFx[i], MVFy[i]), (MVBx[i], MVBy[i]), i = 0, 1, 2, 3\}$ to obtain appropriate blocks from reference VOPs and averaging these blocks, in the same way as in the case of bidirectional mode except that motion compensation is performed on 8×8 blocks.

b. Motion Compensation in Skipped Macroblocks. If the colocated macroblock in the most recently decoded I-or P-VOP is skipped, the current B-macroblock is treated as the forward mode with the zero motion vector (MVFx, MVFy). If the modb equals 1, the current B-macroblock is reconstructed by using the direct mode with zero delta vector.

G. Results of Comparison

For the purposes of performance evaluation and comparisons with other standards, the subset of MPEG-4 tools listed in Table 10 is selected. The selected subset contains all (14) tools up to and including binary shape coding, as well as the tool for temporal scalability of rectangular VOPs. Further, the tools for B-VOP coding, frame/field DCT, frame/field motion, binary shape coding, and temporal scalability of rectangular VOPs are used in some experiments, whereas the remaining tools are used in all experiments.

In principle, to evaluate the statistical performance of a standard, a fixed set of test conditions and an existing standard are needed as a reference. Thus, it would appear that the comparison of statistics of the standard to be evaluated against an existing standard is sufficient. However, a major difficulty in such a comparison is that while the various standards standardize bitstream description and the decoding process, it is the differences in encoding conditions (such as motion range and search technique, quantization matrices, coding mode decisions, I/P/B structure, intra frequency, and rate control) that end up influencing the statistics. A similar choice of encoding conditions (whenever possible) for the standard being evaluated and the reference standard may be able to reduce any unnecessary bias in the statistics.

Because of the wide range of coding bit rates, resolution and formats, and other functionalities offered by MPEG-4 video, instead of evaluating its performance with respect to a single standard as reference, we perform comparisons against three video standards—H.263, MPEG-1, and MPEG-2—selecting a standard most appropriate for the conditions being tested. The purpose of the comparison is mainly to show that a single standard such as MPEG-4 video does offer the functionality and flexibility to address a number of different applications covered by the combined range of the reference standards (and beyond) while performing comparably to each of them. However, caution is needed in drawing conclusions about the superiority of one standard over another because of our limited set of comparisons and difficulty in normalizing encoding conditions.

1. Statistical Comparison with H.263

Table 2(a) presents the results of a comparison of video coding based on H.263 (TMN5) with that based on MPEG-4 video (VM8). Coding is performed using QCIF spatial resolu-

Table 2(a) Comparison of H.263 and MPEG-4 Video Coding at QCIF Resolution, 10 Hz

	H.263		MPEG-4[a]	
Sequence	Bit rate (kbit/sec)	SNR Y (dB)	Bit rate (kbit/sec)	SNR Y (dB)
News	20.16	30.19	22.20	30.69
Container	20.08	31.97	21.31	32.48
HallMonitor	20.14	33.24	22.48	34.68

[a] Rate control slightly different from that used for H.263 encoding.

tion, a temporal rate of 10 frames (or VOPs per second), and approximate bit rates of 20 kbit/sec. MPEG-4 video coding uses rectangular VOP (frame)–based coding and does not use shape coding.

Next, Table 2(b) presents the results of a comparison of video coding based on H.263 (TMN6) with that based on MPEG-4 video (VM8). Coding is performed using CIF spatial resolution and an approximate total bit rate of 112 kbit/sec. In H.263 coding, a temporal rate of 15 frames/sec is used for coding the scene. MPEG-4 video coding uses arbitrary-shaped VOPs consisting of separately coded foreground and background. The foreground object is coded at 15 VOPs/sec (at 104 kbit/sec for ''Speakers'' and for ''Akiyo'' scenes because they have stationary backgrounds and only 80 kbit/sec for the ''Fiddler'' scene, which has a moving background). The background is coded at a lower temporal rate with remaining bitrate (112 kbit/s − foreground bitrate). Further, the foreground object can be displayed by itself at lower bitrate, discarding the background object. The statistics for MPEG-4 video coding are computed on the foreground object only; for the purpose of comparison, the statistics for the (pseudo) subset of H.263 coded scene containing the foreground object are also presented.

2. *Statistical Comparison with MPEG-1 Video*

Tables 3(a) and 3(b) presents the results of a comparison of video coding based on MPEG-1 (SM3 with modified rate control) with that based on MPEG-4 video (VM8). Coding is performed using SIF spatial resolution, a temporal rate of 30 frames (or VOPs) per second, and approximate total bit rates of 1000 kbit/sec. An $M = 3$ coding structure with two B-pictures (or VOPs) as well as an intra coding distance of 15 pictures (or VOPs) is em-

Table 2(b) Comparison of H.263- and MPEG-4–Based Video Coding at CIF Resolution and 15 Hz

	H.263 complete		H.263 foreground[a]		MPEG-4 foreground[b]	
Sequence	Bit rate (kbit/sec)	SNR Y (dB)	Bit rate (kbit/sec)	SNR Y (dB)	Bit rate, (kbit/sec)	SNR Y (dB)
Speakers	112.41	35.21	—	33.34	104.11	35.11
Akiyo	112.38	40.19	—	37.50	106.53	36.89
Fiddler	112.29	34.22	—	27.34	79.65	33.57

[a] Subset (containing foreground object) of normal H.263-coded scene; however, shape not accessible.

[b] Foreground object (including shape) coded with MPEG-4; rate control slightly different from that for H.263 encoding.

Table 3(a) Comparison [18,19] of MPEG-1 and MPEG-4 Video Coding, $M = 1$, $N =$ infinity, 512 kbit/sec

	MPEG-1		MPEG-4	
Sequence	Bit rate (kbit/sec)	SNR Y (dB)	Bit rate (kbit/sec)	SNR Y (dB)
Stefan (SIF at 30)	528.1	26.07	580.1	26.61
Singer (CIF at 25)	511.9	37.95	512.0	37.72
Dancer (CIF at 25)	512.0	37.44	512.2	37.54

Table 3(b) Comparison of MPEG-1 and MPEG-4 Video Coding, $M = 3$, $N = 30$, 1024 kbit/sec

	MPEG-1		MPEG-4[a]	
Sequence (SIF at 30 Hz)	Bit rate (kbit/sec)	SNR Y (dB)	Bit rate (kbit/sec)	SNR Y (dB)
Stefan	995.9	28.42	1071.2	30.81
TableTennis	1000.3	32.83	1019.8	34.98
New York	1000.4	32.73	1053.9	35.53

[a] Rate control quite different from that used for MPEG-1 encoding.

ployed. MPEG-4 video coding uses rectangular VOP (frame)–based temporal scalability and thus supports the functionality of a lower temporal resolution layer (10 Hz) at a lower bit rate.

Besides the tools used in these comparisons, the remaining tools of MPEG-4 video allow advanced functionalities, the results of which are difficult to compare with those of traditional coding standards. For example, object-based functionalities allow possibilities of composition of a decoded scene for presentation to the user such that the composed scene may be inherently different from the original scene; in such cases the signal-to-noise (SNR) of a scene carries even less meaning, as the background used in a composed scene may be perfect but different from that in the original scene. Also, when sprite coding is used for representation (stationary or panning background), the composed scene may appear of high quality, but on a pixel-by-pixel basis there may be differences from the original sequence. Similarly, in the case of video coding under channel errors, although SNR may be useful, it may not provide sufficient information to allow selection of one error resilience technique over another. Further, the value of postprocessing or concealment is subjective and thus difficult to measure with SNR. Thus, although MPEG-4 video provides a number of advanced functionalities, their value is in many cases application specific, so it is difficult to employ traditional methods of statistical improvements in coding quality to judge their usefulness.

III. INTERLACED VIDEO CODING

Interlaced video is widely used in the television industry. It provides good picture quality under a tight bandwidth constraint. Instead of capturing and/or displaying the entire frame

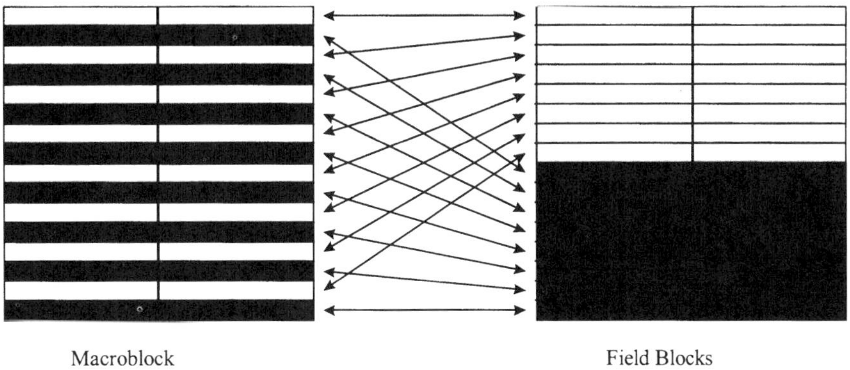

Figure 11 Top and bottom fields in a macroblock.

at one time, half of the scan lines (called a field) in a frame are captured and/or displayed at one time. Thus, two fields, called the top and bottom fields, constitute a frame. The top field consists of all even lines (counting from 0) and the bottom field is composed of all odd lines in the frame. For many applications, interlaced scan captures and/or displays fields at twice the frame rate, giving the perception of much greater temporal resolution without seriously degrading the vertical details. Also, interlaced format exploits the fact that human vision is not sensitive to high frequencies along the diagonal direction in the vertical-temporal plane.

Coding tools in MPEG-4 for interlaced video exploit the spatial and temporal redundancy by employing adaptive field/frame DCT and motion prediction, respectively. As discussed earlier, the basic coding unit in MPEG-4 is a macroblock (MB). The main effect of interlace scan for an MB is that vertical correlation in a frame is reduced when there is motion at the location of this MB of the scene because the adjacent scan lines in a frame come from different fields.

MPEG-4 provides three tools for efficiently coding MBs [20]. First, the field DCT may be applied to the MB. That is, before performing DCT the encoder may reorder the luminance lines within an MB such that the first eight lines come from the top field and the last eight lines come from the bottom field as shown in Figure 11. The purpose of this reordering is to increase the vertical correlation within the luminance 8×8 blocks and thus increase the energy packing of the 8×8 DCT in the transform domain. Second, the alternate scan may be used to replace the zigzag scan on a VOP-by-VOP basis for an interlaced VOP with lower vertical correlation (Fig. 12). Third, the key interlaced tool

0	1	2	3	10	11	12	13
4	5	8	9	17	16	15	14
6	7	19	18	26	27	28	29
20	21	24	25	30	31	32	33
22	23	34	35	42	43	44	45
36	37	40	41	46	47	48	49
38	39	50	51	56	57	58	59
52	53	54	55	60	61	62	63

a

0	4	6	20	22	36	38	52
1	5	7	21	23	37	39	53
2	8	19	24	34	40	50	54
3	9	18	25	35	41	51	55
10	17	26	30	42	46	56	60
11	16	27	31	43	47	57	61
12	15	28	32	44	48	58	62
13	14	29	33	45	49	59	63

b

0	1	5	6	14	15	27	28
2	4	7	13	16	26	29	42
3	8	12	17	25	30	41	43
9	11	18	24	31	40	44	53
10	19	23	32	39	45	52	54
20	22	33	38	46	51	55	60
21	34	37	47	50	56	59	61
35	36	48	49	57	58	62	63

c

Figure 12 (a) Alternate-horizontal scan; (b) Alternate-vertical scan; (c) zigzag scan.

for exploiting the temporal redundancy between adjacent VOPs or fields in adjacent VOPs is the macroblock-based field MC. The field MC is performed in the reordered MBs for the top-field and the bottom-field blocks (16 × 8 for luminance and 8 × 4 for chrominance). The field motion vectors for this MB are then coded and transmitted. These interlaced tools are similar to those in MPEG-2 video [2,10]. However, they provide more features and functionality in MPEG-4.

A. Adaptive Field/Frame DCT

When interlaced video is coded, superior energy compaction can sometimes be obtained by reordering the lines of the MB to form 8 × 8 luminance blocks consisting of data from one field (see Figure 11). Field DCT line order is used when, for example,

$$\sum_{i=0}^{6}\sum_{j=0}^{15}(p_{2i,j}-p_{2i+1,j})^2+(p_{2i+1,j}-p_{2i+2,j})^2>\sum_{i=0}^{6}\sum_{j=0}^{15}(p_{2i,j}-p_{2i+2,j})^2+(p_{2i+1,j}-p_{2i+3,j})^2$$

where $p_{i,j}$ is the spatial luminance data (samples or differences) just before the 8 × 8 DCT is performed. Other rules, such as the sum of absolute difference and the normalized correlation [7], can also be used for the field/frame DCT decision. The field DCT permutation is indicated by the ''dct_type'' bit having a value of 1 (a value of 0 otherwise).

When field DCT mode is used, the luminance lines (or luminance error) in the spatial domain of the MB are permuted from the frame DCT orientation to field DCT configuration as shown in Figure 11. The black regions of the diagram denote the bottom field. The resulting MBs are transformed, quantized, and VLC encoded normally. On decoding a field DCT MB, the inverse permutation is performed after all luminance blocks have been obtained from the Inverse DCT (IDCT). In the 4:2:0 format, chrominance data are not affected by this mode.

B. Alternate and Adaptive Scans of DCT Coefficients

One option available in the interlaced coding is alternate vertical scan. Scanning in MPEG-4 is a mapping process that forms a one-dimensional array from a two-dimensional array of DCT coefficients. This one-dimensional array is quantized and then coded by run length and Huffman-based variable length coding.

When the flag of alternate scan is set to zero, i.e., alternate_vertical_scan_flag = 0, MPEG-4 video uses an adaptive scanning pattern for all MBs in this VOP. The scan types (zigzag, alternate-vertical, or alternate-horizontal) are given in Figure 12. For an intra MB, if AC prediction is disabled (acpred_flag = 0), zigzag scan is selected for all blocks in the MB. Otherwise, the DC prediction direction is used to select a scan on a block basis. For instance, if the DC prediction refers to the horizontally adjacent block, alternate-vertical scan is selected for the current block; otherwise (for DC prediction referring to vertically adjacent block), alternate-horizontal scan is used for the current block. The zigzag scan is applied for all inter MBs.

The adaptive scanning pattern may not yield the most efficiency for interlaced video because the correlation in horizontal direction is usually dominant in interlaced video and the choice of adaptive scan may not follow this fact. Thus, when the flag of alternate scan is set to one for the VOP, i.e., alternate_vertical_scan_flag = 1, the alternate vertical scan is used for all intra MBs in this VOP.

C. Adaptive Field/Frame Prediction

This is an adaptive method for deciding whether a current MB of 16 × 16 pixels is divided into four blocks of 8 × 8 pixels each for MC. For coding of interlaced video, such an adaptive technique also decides whether the field MC needs to be applied to this MB. Because of its lower computational complexity as compared to other difference measures, the sum of absolute difference (SAD) is often used as a criterion for the selection of the field/frame prediction.

The decision is made by motion estimation as follows:

1. For the 16 × 16 MB, determine a motion vector (MV_x, MV_y) that results in the minimum SAD: $SAD_{16}(MV_x, MV_y)$.
2. For each 8 × 8 block in the MB, determine a motion vector (MV_{xi}, MV_{yi}) that generates the minimum SAD for this block: $SAD_8(MV_{xi}, MV_{yi})$. Thus, four SADs are obtained for this MB: $SAD_8(MV_{x1}, MV_{y1})$, $SAD_8(MV_{x2}, MV_{y2})$, $SAD_8(MV_{x3}, MV_{y3})$, and $SAD_8(MV_{x4}, MV_{y4})$.
3. For the MB, determine two best field motion vectors (MV_{x_top}, MV_{y_top}) and $(MV_{x\_bottom}, MV_{y\_bottom})$ for top and bottom fields, respectively. Obtain the minimized SADs: $SAD_{top}(MV_{x\_top}, MV_{y\_top})$, $SAD_{bottom}(MV_{x\_bottom}, MV_{y\_bottom})$ for the top and bottom fields of this MB.
4. The overall prediction mode decision is based on choosing the minimum of $SAD_{16}(MV_x, MV_y)$, $\sum_{i=1}^{4} SAD_8(MV_{xi}, MV_{yi})$, $SAD_{top}(MV_{x\_top}, MV_{y\_top}) + SAD_{bottom}(MV_{x\_bottom}, MV_{y\_bottom})$.

If the first term is the minimum, 16 × 16 prediction is used. If the second term is the smallest, 8 × 8 motion compensation is used. If the last expression is the minimum, field-based motion estimation is selected. Note that SADs are computed from forward prediction for P-VOPs and could be computed from both forward and backward prediction for B-VOPs.

D. Motion Vector Coding

For an inter MB, the motion vector must be transmitted. The motion vectors are coded differentially by using a spatial neighborhood of motion vectors already transmitted. The possible neighborhood motion vectors are candidate predictors for the differential coding.

The motion vector coding is performed separately on the horizontal and vertical components. For each component in P-VOPs, the median value of the candidate predictors for the same component is computed and then the difference between the component and median values is coded by the use of variable length codes.

For P-VOPs, the candidate predictors MV1, MV2, and MV3 are defined by Figure 13. In Figure 13, if 16 × 16 prediction is chosen, MV represents the motion vector for the current MB and is located in the first block in this MB. If field prediction is chosen, MV represents either top or bottom field motion vectors and is also located in the first block position in this MB. If 8 × 8 prediction is chosen, MV represents the motion vector for the current 8 × 8 block. The candidate predictors *MVi* (for $i = 1, 2, 3$) are generated by the following rules:

If block *i* is in a 16 × 16 prediction MB, *MVi* is the motion vector of that MB.
If block *i* is an 8 × 8 predicted block, *MVi* is the motion vector of that block.
If block *i* is in a field-predicted MB, *MVi* is derived by averaging two field motion

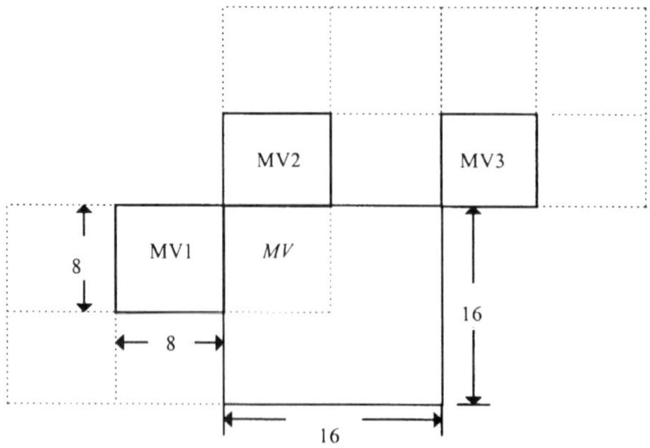

Figure 13 Definition of the candidate predictors for a block.

vectors of that field-predicted MB such that all fractional pel offsets are mapped into the half-pel displacement.

The predictors for the horizontal and vertical components are then computed from the candidates by

$$P_x = \text{Median}(MV1x, MV2x, MV3x)$$
$$P_y = \text{Median}(MV1y, MV2y, MV3y)$$

When the interlaced coding tools are used, the differential coding of both fields uses the same predictor, i.e.,

$$MVDx_{f1} = MVx_{f1} - P_x$$
$$MVDy_{f1} = MVy_{f1} - P_y$$
$$MVDx_{f2} = MVx_{f2} - P_x$$
$$MVDy_{f2} = MVy_{f2} - P_y$$

In this case, because the vertical component of a field motion vector is integer, the vertical differential motion vector encoded in the bitstream is $MVDy_{fi} = (MVy_i - \text{int}(P_y))/2$, where $\text{int}(P_y)$ means truncate P_y toward zero to the nearest integral value.

For non–direct mode B-VOPs, motion vectors are coded differentially. For the interlaced case, four more field predictors are introduced as follows: top-field forward, bottom-field forward, top-field backward, and bottom-field backward vectors. For forward and backward motion vectors the nearest left-neighboring vector of the same direction type is used as the predictor. In the case in which both the current MB and the predictor MB are field-coded MBs, the nearest left-neighboring vector of the same direction type and the same field is used as the predictor. In the case in which the current MB is located on the left edge of the VOP or no vector of the same direction type is present, the predictor is set to zero.

The possible combination of MB modes and predictors is listed in Table 4.

E. Interlaced Direct Mode Coding in B-VOPs

Interlaced direct mode is an extension of progressive direct mode and may be used whenever the MB with the same MB index as the future anchor VOP (called the colocated

Table 4 Use of Predictors

MB modes	Predictors: Frame forward	Frame backward	Top-field forward	Bottom-field forward	Top-field backward	Bottom-field backward
Frame forward	Yes	No	Yes	No	No	No
Frame backward	No	Yes	No	No	Yes	No
Frame bidirectional	Yes	Yes	Yes	No	Yes	No
Field forward	Yes	No	Yes	Yes	No	No
Field backward	No	Yes	No	No	Yes	Yes
Field bidirectional	Yes	Yes	Yes	Yes	Yes	Yes

MB) uses field motion compensation. Direct mode is an algorithm for constructing forward and backward motion vectors for a B-VOP MB from the colocated future anchor MB's motion vectors. In addition, a single delta vector (used for both fields) is added to reduce the possible prediction errors.

Interlaced direct mode forms the prediction MB separately for the top and bottom fields. The four field motion vectors are calculated from two motion vectors of the colocated MB. The top field prediction is based on the top field motion vector of the colocated MB, and the bottom field prediction is based on the bottom field motion vector of the colocated MB. The generic operation of interlaced direct mode is shown in Figure 14. The relation among the motion vectors is defined as follows.

The temporal references (TRB[i] and TRD[i]) are distances in time expressed in field periods, where i is 0 (top field of the B-VOP).

The calculation of TRD[i] and TRB[i] depends not only on the current field, reference field, and frame temporal references but also on whether the current video is top field first or bottom field first.

$$\text{TRD}[i] = 2 * (\text{T(future)//Tframe} - \text{T(past)//Tframe}) + \delta[i]$$
$$\text{TRB}[i] = 2 * (\text{T(current)//Tframe} - \text{T(past)//Tframe}) + \delta[i]$$

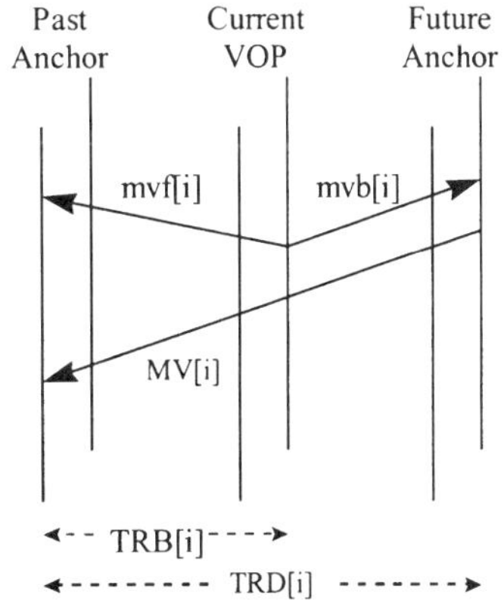

Figure 14 Interlaced direct mode.

Table 5 Selection of the Parameter δ

Future anchor VOP reference fields of the colocated macroblock		top_field_first _ 0		top_field_first _ 1	
Top field reference	Bottom field reference	Top field, δ[0]	Bottom field, δ[1]	Top field, δ[0]	Bottom field, δ[1]
0	0	0	−1	0	1
0	1	0	0	0	0
1	0	1	−1	−1	1
1	1	1	0	−1	0

where T(future), T(current), and T(past) are the cumulative VOP times calculated from modulo_time_base and vop_time_increment of the future, current, and past VOPs in display order. Tframe is the frame period determined from

$$\text{Tframe} = \text{T(first\_B\_VOP)} - \text{T(past\_anchor\_of\_first\_B\_VOP)}$$

where first_B_VOP denotes the first B-VOP following the video object layer syntax. The important thing about Tframe is that the period of time between consecutive fields that constitute an interlaced frame is assumed to be 0.5 * Tframe for purposes of scaling the motion vectors.

The parameter δ is determined from Table 5. This parameter is a function of the current field parity (top or bottom), the reference field of the co-located MB and the flag of top_field_first in the B-VOP's video object plane syntax.

Interlaced direct mode dramatically reduces the complexity of motion estimation for B-VOP and provides comparable or better coding performance [20].

F. Field-Based Padding

More efficient coding of texture at the boundary of arbitrarily shaped objects can be achieved by padding techniques. Padding generates the pixel sample outside an arbitrarily shaped object for DCT and motion prediction. In order to balance coding efficiency and complexity, padding is used instead of extrapolation. A repetitive padding process is performed, first horizontally and then vertically, by replicating and averaging the sample at the boundary of an object toward the MB boundary or averaging two boundary samples. The area remaining after repetitive padding is filled by what is called extended padding. Field padding uses the same repetitive and extended padding but the vertical padding of the luminance is conducted separately for each field. When each field is padded separately, the padded blocks retain correlation in the same field so that the high coding efficiency can be achieved.

G. Performance of Interlaced Coding Tools

Tables 6 and 7 show significant reductions in average bits per VOP when field motion compensation and field DCT are applied. The H.263 quantization method is used in these simulations.

Table 6 Field Motion Compensation (MC), $M = 1$

Sequence	Qp	Average bits per VOP M4p	M4p + field MC	% improve
FunFair	8	230,732	220,403	4.5
	12	145,613	138,912	4.6
	16	105,994	101,333	4.4
Football	8	163,467	140,826	13.6
	12	106,029	94,098	11.3
	16	79,449	72,874	8.3
Stefan	8	222,764	183,291	17.7
	12	143,813	116,210	19.2
	16	104,514	84,158	19.4

Table 7 Field DCT, $M = 3$

Sequence	Qp	Average bits per VOP M4p + field MC	M4i	% improve
FunFair	8	196,224	176,379	10.1
	12	132,447	119,081	10.1
	16	103,203	93,876	9.0
Football	8	128,355	116,800	9.0
	12	91,872	84,720	7.8
	16	75,636	70,685	6.5
Stefan	8	156,781	150,177	4.2
	12	107,934	103,196	4.4
	16	84,974	81,510	4.1

1. *Statistical Comparison with MPEG-2 Video*

Some simulation results of interlaced coding tools are listed here as references. In Table 8(a), a peak signal-to-noise ratio (SNR) comparison is provided for MPEG 2 main profile at main level (MP@ML) versus MPEG-4 progressive (M4p) and interlaced (M4i) coding. In this set of simulations, MPEG-2 test model 5 (TM5) rate control is used and the bit rate is set to 3 Mbit/sec.

Table 8(a) SNR Comparison at 3 Mbit/sec of MPEG-2 and MPEG-4, Different Quantization, Progressive and Interlaced Coding, $M = 3$ Configuration

	MPEG-2 with MPEG-2 quantization		MPEG-4 M4p with H.263 quantization		MPEG-4 M4i with H.263 quantization	
Sequence	Bit rate (Mbit/sec)	SNR Y (dB)	Bit rate (Mbit/sec)	SNR Y (dB)	Bit rate (Mbit/sec)	SNR Y (dB)
Fun Fair	3.0	28.23	3.0	29.42	3.0	29.84
Football	3.0	31.62	3.0	32.15	3.0	33.00
Stefan	3.0	28.26	3.0	28.91	3.0	29.79

Table 8(b) SNR Comparison at 3 Mbit/sec of MPEG-2 and MPEG-4, Same Quantization, All Interlaced Coding, $M = 3$ Configuration

	MPEG-2 video		MPEG-4 (M4i) video	
Sequence	Bit rate (Mbit/sec)	SNR Y (dB)	Bit rate (Mbit/sec)	SNR Y (dB)
Bus	3.00	29.99	3.00	30.29 (+0.30)
Carousel	3.00	28.10	3.01	28.33 (+0.23)

Table 8(b) presents the results of a comparison of video coding based on MPEG-2 (TM5) with that based on MPEG-4 video (VM8). Coding is performed using CCIR-601 4:2:0 spatial resolution, a temporal rate of 30 frames (or VOPs) per second, and approximate bit rates of 3 Mbit/sec. Both MPEG-2–based and MPEG-4–based video coding employ an $M = 3$ coding structure with two B-pictures (or VOPs) as well as an intra coding distance of 15 pictures (or VOPs) and interlaced coding with frame pictures and other interlaced tools.

IV. ERROR RESILIENT VIDEO CODING

A number of tools have been incorporated in the MPEG-4 video coder [22,23] to make it more error resilient. These tools provide various important properties such as resynchronization, error detection, data recovery, and error concealment. There are four new tools:

1. Video packet resynchronization
2. Data partitioning (DP)
3. Header extension code (HEC)
4. Reversible variable length codes (RVLCs)

We now describe each of these tools and their advantages.

A. Resynchronization

When the compressed video data are transmitted over noisy communication channels, errors are introduced into the bitstream. A video decoder that is decoding this corrupted bitstream will lose synchronization with the encoder (it is unable to identify the precise location in the image where the current data belong). If remedial measures are not taken, the quality of the decoded video rapidly degrades and it quickly becomes totally unusable.

One approach is for the encoder to introduce resynchronization markers in the bitstream at various locations. When the decoder detects an error, it can then hunt for this resynchronization marker and regain resynchronization. Previous video coding standards such as H.261 and H.263 [4] logically partition each of the images to be encoded into rows of macroblocks (16×16 pixel units) called groups of blocks (GOBs). A GOB corresponds to a horizontal row of macroblocks.

MPEG-4 provides a similar method of resynchronization with one important difference. The MPEG-4 encoder is not restricted to inserting the resynchronization markers only at the beginning of each row of macroblocks. The encoder has the option of dividing

Figure 15 Position of the resynchronization markers in MPEG-4 bitstream.

the image into video packets. Each video packet is made up of an integral number of consecutive macroblocks. These macroblocks can span several rows of macroblocks in the image and can even include partial rows of macroblocks. One suggested mode of operation of the MPEG-4 encoder is to insert a resynchronization marker periodically for every *K* bits. Consider the fact that when there is significant activity in one part of the image, the macroblocks corresponding to that area generate more bits than those in other parts of the image. Now, if the MPE-4 encoder inserts the resynchronization markers at uniformly spaced bit intervals (say every 512 bits), the macroblock interval between the resynchronization markers is much smaller in the high-activity areas and much larger in the low-activity areas. Thus, in the presence of a short burst of errors, the decoder can quickly localize the error to within a few macroblocks in the important high-activity areas of the image and preserve the image quality in the important areas. In the case of baseline H.263, where the resynchronization markers are restricted to be at the beginning of the GOBs, the decoder can only isolate the errors to a row of macroblocks independent of the image content. Hence, effective coverage of the resynchronization marker is reduced compared with the MPEG-4 scheme. The later version of H.263 [5,24] adopted a resynchronization scheme similar to that of MPEG-4 in an additional annex (Annex K—Slice Structure Mode).

Note that in addition to inserting the resynchronization markers at the beginning of each video packet, the encoder needs to remove all data dependences that exist between the data belonging to two different video packets. This is required because, even if one of the video packets is corrupted by errors, the others can be decoded and utilized by the decoder. In order to remove these data dependences, the encoder inserts two fields in addition to the resynchronization marker at the beginning of each video packet as shown in Figure 15. These are (1) the absolute macroblock number of the first macroblock in the video packet, *Mb. No.* (which indicates the spatial location of the macroblock in the current image), and (2) the quantization parameter, *QP*, which denotes the default quantization parameter used to quantize the DCT coefficients in the video packet. The encoder also modifies the predictive encoding method used for coding the motion vectors such that there are no predictions across the video packet boundaries.

B. Data Partitioning

After detecting an error in the bitstream and resynchronizing to the next resynchronization marker, the decoder has isolated the data in error to the macroblocks between the two resynchronization markers. Typical video decoders discard all these macroblocks as being in error and replace them with the data from the corresponding macroblocks in the previous frame to conceal the errors. One of the main reasons for this is that between two resynchronization markers, the motion and DCT data for each of the macroblocks are coded together. Hence, when the decoder detects an error, whether the error occurred in the motion part or the DCT part, all the data in the video need to be discarded. Because of the uncertainty about the exact location where the error occurred, the decoder cannot be sure that either the motion or the DCT data of any of the macroblocks in the packet are not erroneous.

Resync. Marker	Mb. No.	QP	Combined Motion and DCT Data

Figure 16 Organization of the data within the video packet.

Figure 16 shows the organization of the video data within a packet for a typical video compression scheme without data partitioning.

Within the Combined Motion and DCT Data part, each of the macroblock (MB) motion vectors and the DCT coefficients are encoded. Figure 17 shows the syntactic elements for each MB. These data are repeated for all the macroblocks in the packet. The COD is a 1-bit field used to indicate whether or not a certain macroblock is coded. The MCBPC is a variable length field and is used to indicate two things: (1) the mode of the macroblock such as INTRA, INTER, INTER4V (8×8 motion vectors), or INTRA + Q (the quantization factor is modified for this macroblock from the previous MB) and (2) which of the two chrominance blocks (8×8) of the macroblock (16×16) are coded. DQUANT is an optional 2-bit fixed length field used to indicate the incremental modification of the quantization value from the previous macroblock's quantization value or the default QP if this is the first macroblock of the video packet. CBPY is a VLC that indicates which of the four blocks of the MB are coded. Encoded MV(s) are the motion vector differences that are by a VLC. Note that the motion vectors are predictively coded with respect to the neighboring motion vectors and hence we code only the motion vector differences. DCT(s) are the 64 DCT coefficients that are actually encoded via zigzag scanning, run length encoding, and then a VLC table.

Previous researchers have applied the idea of partitioning the data into higher and lower priority data in the context of ATM or other packetized networks to achieve better error [4,10]. However, this may not be possible over channels such as existing analog phone lines or wireless networks, where it may be difficult, if not impossible, to prioritize the data being transmitted. Hence, it becomes necessary to resort to other error concealment techniques to mitigate the effects of channel errors.

In MPEG-4, the data partitioning mode partitions the data within a video packet into a motion part and a texture part separated by a unique motion boundary marker (MBM), as shown in Figure 18. This shows the bitstream organization within each of the video packets with data partitioning. Note that compared with Fig. 17, the motion and the DCT parts are now separated by an MBM. In order to minimize the differences with respect to the conventional method, we maintain all the same syntactic elements as in the conventional method and reorganize them to enable data partitioning. All the syntactic elements that have motion-related information are placed in the motion partition and all

COD1	MCBPC1	CBPY1	DQUANT1	Encoded MV(s)1	DCT Data1	.....

Figure 17 Bitstream components for each macroblock within the video packet.

Resync. Marker	Mb. No.	QP	Motion Data	MBM	DCT Data

Figure 18 Bitstream organization with data partitioning for motion and DCT data.

COD1	MCBPC1	Encoded MV(s)1	COD2	MCBPC2	Encoded MV(s)2	.....

Figure 19 Bitstream components of the motion data.

the syntactic elements related to the DCT data are placed in the DCT partition. Figure 19 shows the bitstream elements after reorganization of the motion part, and Figure 20 shows bitstream elements of the DCT part. Note that we now place COD, MCBPC and the MVs in the motion part and relegate CBPY, DQUANT, and the DCTs to the DCT part of the packet.

The MBM marks the end of the motion data and the beginning of the DCT data. It is computed from the motion VLC tables using the search program described earlier such that the word is Hamming distance 1 from any possible valid combination of the motion VLC tables. This word is uniquely decodable from the motion VLCs and gives the decoder knowledge of where to stop reading motion vectors before beginning to read texture information. The number of macroblocks (NMB) in the video packet is implicitly known after encountering the MBM. When an error is detected in the motion section, the decoder flags an error and replaces all the macroblocks in the current packet with skipped blocks until the next resynchronization marker. Resynchronization occurs at the next successfully read resynchronization marker. If any subsequent video packets are lost before resynchronization, those packets are replaced by skipped macroblocks as well. When an error is detected in the texture section (and no errors are detected in the motion section) the NMB motion vectors are used to perform motion compensation. The texture part of all the macroblocks is discarded and the decoder resynchronizes to the next resynchronization marker.

If an error is not detected in the motion or the texture sections of the bitstream but the resynchronization marker is not found at the end of decoding all the macroblocks of the current packet, an error is flagged and only the texture part of all the macroblocks in the current packet is discarded. Motion compensation is still applied for the NMB macroblocks, as we have higher confidence in the motion vectors since we got the MBM.

Hence, the two advantages of this data partitioning method are that (1) we have a more stringent check on the validity of the motion data because we need to get the MBM at the end of the decoding of motion data in order to consider the motion data to be valid and (2) in case we have an undetected error in the motion and texture but do not end on the correct position for the next resynchronization marker, we do not need to discard all the motion data—we can salvage the motion data because they are validated by the detection of MBM.

C. Reversible Variable Length Codes

As mentioned earlier, one of the problems with transmitting compressed video over error-prone channels is the use of variable length codes. During the decode process, if the decoder detects an error while decoding VLC data it loses synchronization and hence

CBPY1	DQUANT1	CBPY2	DQUANT2	...	DCT Data1	DCT Data1	...

Figure 20 Bitstream components of the DCT data.

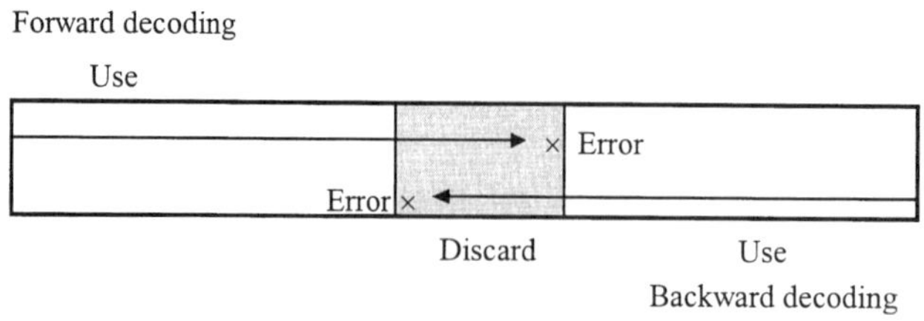

Figure 21 The data that need to be discarded can be reduced by using forward and reverse decoding property of an RVLC.

typically has to discard all the data up to the next resynchronization point. Reversible variable length codes (RVLCs) alleviate this problem and enable the decoder to better isolate the error location by enabling data recovery in the presence of errors. RVLCs are special VLCs that have the prefix property in both the forward and reverse directions. Hence, they can be uniquely decoded in both the forward and reverse directions. The advantage of these code words is that when the decoder detects an error while decoding the bitstream in the forward direction, it jumps to the next resynchronization marker and decodes the bitstream in the backward direction until it encounters an error. On the basis of the two error locations, the decoder can recover some of the data that would have otherwise been discarded. In Figure 21, only data in the shaded area are discarded; note that if RVLCs were not used, all the data between the two resynchronization markers would have to be discarded.

By proper use of training video sequences, the RVLCs can be made to match the probability characteristics of the DCT coefficients such that the RVLCs still maintain the ability to pack the bitstream compactly while retaining the error resilience properties. The MPEG-4 standard utilizes such efficient and robust RVLC tables for encoding the DCT coefficients. Note that the MPEG-4 standard does not standardize the method by which an MPEG-4 decoder has to apply the RVLCs for error resilience. Some suggested strategies are described in Refs. 12, 22, and 23. In the MPEG-4 evaluations, RVLCs have been shown to provide a significant gain in subjective video quality in the presence of channel errors by enabling more data to be recovered. Note that for RVLCs to be most effective, all the data coded using the same RVLC tables have to occur together. Hence, in MPEG-4, RVLCs are utilized in the data partitioning mode, which modifies the bitstream syntax such that all the DCT coefficients occur together and hence can be effectively coded using the RVLC tables. Currently, experiments are under way for version 2 of MPEG-4, which advocates the use of RVLCs for coding the motion vector information also.

D. Header Extension Code

Some of the most important information that the decoder needs to be able to decode the video bitstream is the header data. The header data include information about the spatial dimensions of the video data, the time stamps associated with the decoding and the presentation of the video data, and the mode in which the current video object is encoded (whether predictively coded or INTRA coded). If some of this information is corrupted because of channel errors, the decoder has no other recourse but to discard all the information belonging to the current video frame. In order to reduce the sensitivity of these data, a header extension code (HEC) is introduced into the MPEG-4 standard. In each video packet a 1-bit field called HEC bit is introduced. If this bit is set, then the important header information that describes the video frame is repeated in the video packet. By checking this header information in the video packets against the information received at the beginning of the video frame, the decoder can ascertain whether the video frame header is received correctly. If the video frame header is corrupted, the decoder can still decode the rest of the data in the video frame using the header information within the video packets.

In MPEG-4 verification tests, it was found that the use of the HEC significantly reduced the number of discarded video frames and helped achieve higher overall decoded video quality.

E. MPEG-4 Error Resilience Evaluation Criterion

All the tools considered for the MPEG-4 standard are evaluated thoroughly and independently verified by two parties before being accepted in the standard [3]. The techniques are rigorously tested over a wide variety of test sequences, bit rates, and error conditions. A set of test sequences are coded at different bit rates. The compressed bitstreams are corrupted using random bit errors, packet loss errors, and burst errors. A wide variety of random bit errors (bit error rates of 10^{-2}, 10^{-3}) and burst errors (of durations 1, 10, and 20 msec) and packet loss errors of varying lengths (96–400 bits) and error rates (10^{-2} and 3×10^{-2}) have been tested [12]. To provide statistically significant results, 50 tests are performed for each of the mentioned error conditions. In each test, errors are inserted in the bitstreams at different locations. This is achieved by changing the seed of the random number generators used to simulate the different error conditions.

For each test, the peak SNR of the video decoded from the corrupted stream and the original video is computed. Then the average of the peak SNR of the 50 runs is computed for each frame in the sequence. To evaluate the performance of these techniques, the averages of the peak SNR values generated by the error-resilient video codec with and without each error resilience tool are compared. The techniques are also compared on the basis of the number of frames discarded because of errors and the number of bits discarded. Before accepting a technique into the standard, the additional bit rate overhead that was incurred with this technique compared with the baseline is compared with the gains provided by the technique.

Techniques that consistently achieve superior performance and have been independently verified by two different parties are then accepted into the standard. Figure 22 shows the comparative performance of the various MPEG-4 error resilient tools over a simulated wireless channel. As can be seen from this figure, each of the tools, resynchronization, data partitioning, and RVLCs cumulatively add to the performance of the coder.

F. Error Resilience Summary

In this chapter we have presented an overview of the various techniques that enable wireless video transmission. International standards play a very important role in communications applications. One of the current standards that is most relevant to video applications is MPEG-4. In this section, we detailed error resilient tools that are part of MPEG-4 and enable robust image and video communication over wireless channels.

There are, however, a number of other methods that further improve the performance of a video codec that the standard does not specify. If the encoder and decoder are aware

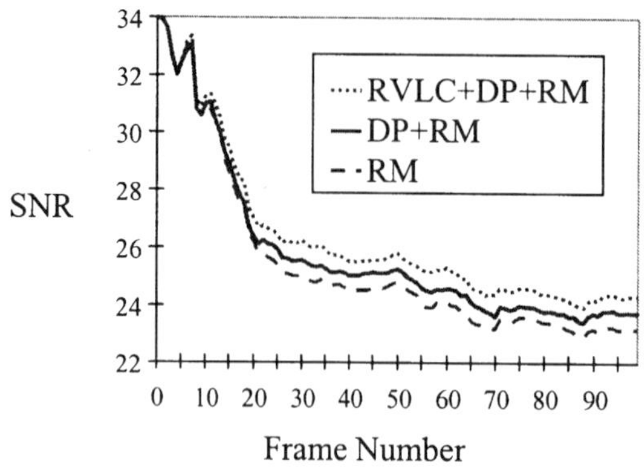

Figure 22 Performance comparison of resynchronization marker (RM), data partitioning (DP), and reversible VLC (RVLC) over a bursty channel simulated by a two-state Gilbert model. Burst durations are 1_msec and the probability of occurrence of a burst is 10^{-2}.

of the limitations imposed by the communication channel, they can further improve the video quality by using these methods. These methods include encoding techniques such as rate control to optimize the allocation of the effective channel bit rate between various parts of images and video to be transmitted and intelligent decisions on when and where to place INTRA refresh macroblocks to limit the error propagation. Decoding methods such as superior error concealment strategies that further conceal the effects of erroneous macroblocks by estimating them from correctly decoded macroblocks in the spatiotemporal neighborhood can also significantly improve the effective video quality.

This chapter has mainly focused on the error resilience aspects of the video layer. A number of error detection and correction strategies, such as forward error correction (FEC), can further improve the reliability of the transmitted video data. The FEC codes are typically provided in the systems layer and the underlying network layer. If the video transmission system has the ability to monitor the dynamic error characteristics of the communication channel, joint source–channel coding techniques can also be effectively employed. These techniques enable the wireless communication system to perform optimal trade-offs in allocating the available bits between the source coder (video) and the channel coder (FEC) to achieve superior performance.

V. SCALABLE VIDEO CODING

Scalability of video is the property that allows a video decoder to decode portions of the coded bitstreams to generate decoded video of quality commensurate with the amount of data decoded. In other words, scalability allows a simple video decoder to decode and produce basic quality video and an enhanced decoder to decode and produce enhanced quality video, all from the same coded video bitstream. This is possible because scalable video encoding ensures that input video data are coded as two or more layers, an independently coded base layer and one or more enhancement layers coded dependently, thus producing scalable video bitstreams. The first enhancement layer is coded with respect to the base layer, the second enhancement layer with respect to the first enhancement layer, and so forth.

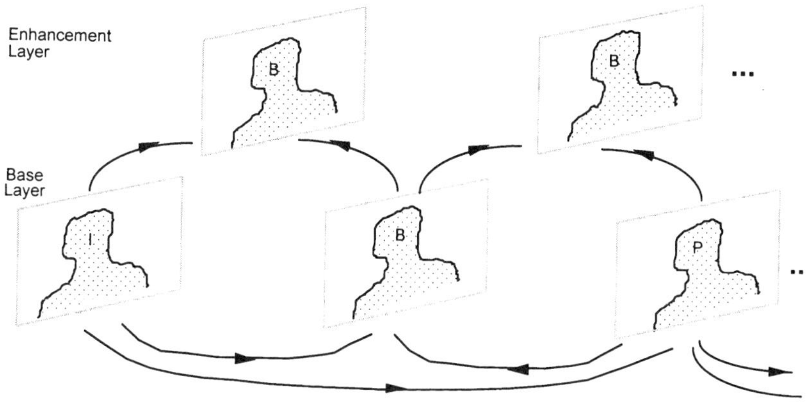

Figure 23 Temporal scalability.

Scalable coding offers a means of scaling the decoder complexity if processor and/or memory resources are limited and often time varying. Further, scalability allows graceful degradation of quality when the bandwidth resources are also limited and continually changing. It also allows increased resilience to errors under noisy channel conditions.

MPEG-4 offers a generalized scalability [12,15–17] framework supporting both temporal and spatial scalabilities, the two primary scalabilities. Temporally scalable encoding offers decoders a means of increasing the temporal resolution of decoded video using decoded enhancement layer VOPs in conjunction with decoded base layer VOPs. Spatial scalability encoding, on the other hand, offers decoders a means of decoding and displaying either the base layer or the enhancement layer output. Because the base layer typically uses one-quarter the resolution of the enhancement layer, the enhancement layer output provides the better quality, albeit requiring increased decoding complexity. The MPEG-4 generalized scalability framework employs modified B-VOPs that exist only in the enhancement layer to achieve both temporal and spatial scalability; the modified enhancement layer B-VOPs use the same syntax as normal B-VOPs but for modified semantics, which allows them to utilize a number of interlayer prediction structures needed for scalable coding

Figure 23 shows an example of the prediction structure used in temporally scalable coding. The base layer is shown to have one-half of the total temporal resolution to be coded; the remaining one-half is carried by the enhancement layer. The base layer is coded independently as in normal video coding, whereas the enhancement layer uses B-VOPs that use both an immediate temporally previous decoded base layer VOP and an immediate temporally following decoded base layer VOP for prediction.

Next, Figure 24 shows an example of the prediction structure used in spatially scalable coding. The base layer is shown to have one-quarter the resolution of the enhancement layer. The base layer is coded independently as in normal video coding, whereas the enhancement layer mainly uses B-VOPs that use both an immediate previous decoded enhancement layer VOP and a coincident decoded base layer VOP for prediction.

In reality, some flexibility is allowed in the choice of spatial and temporal resolutions for base and enhancement layers as well as the prediction structures for the enhancement layer to cope with a variety of conditions in which scalable coding may be needed. Further,

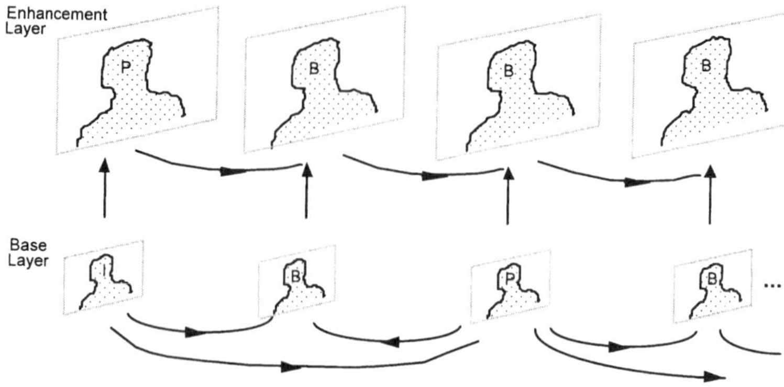

Figure 24 Spatial scalability.

both spatial and temporal scalability with rectangular VOPs and temporal scalability of arbitrarily shaped VOPs are supported. Figures 23 and 24 are applicable not only to rectangular VOP scalability but also to arbitrarily shaped VOP scalability (in this case only the shaded region depicting the head-and-shoulder view is used for predictions in scalable coding, and the rectangle represents the bounding box).

A. Coding Results of B-VOPs for Temporal Scalability

We now present results of temporal scalability experiments using several MPEG-4 standard test scenes at QCIF and CIF resolutions. For this experiment, fixed target bit rates for the combination of base and enhancement layer bit rates are chosen. All simulations use a fixed quantizer for the entire frame; however, in the case of B-VOP coding, there is an exception such that whenever ''direct'' mode is chosen, the quantizer for those macroblocks is automatically generated by scaling of the corresponding macroblock quantizer in the following P-VOP. The experiment uses I- and P-VOPs in the base layer and B-VOPs only in the enhancement layer. Two test conditions are employed: (1) QCIF sequences coded at a total of 24 kbits/sec and (2) CIF sequences coded at a total of 112 kbit/sec. In both cases, the temporal rate of the base layer is 5 frames/sec and that of the enhancement (Enh) layer is 10 frames/sec; temporal multiplexing of base and enhancement layers results in 15 frames/sec. The results of our experiments are shown in Table 9(a) for QCIF resolution and in Table 9(b) for CIF resolution.

As can be seen from Tables 9(a) and 9(b), the scheme is able to achieve a reasonable partitioning of bit rates to produce two very useful temporal scalability layers. The base layer carries one-third of the total coded frame rate at roughly half of the total assigned bit rate to both layers; the other half of the bit rate is used by the enhancement layer, which carries the remaining two-thirds of the frames.

VI. SUMMARY OF TOOLS

We have presented a summary of the MPEG-4 standard with occasional references to functionality in other standards. We now present a comparison of the tools of these various

Table 9(a) Results of Temporal Scalability Experiments with QCIF Sequences at a Total of About 24 kbit/sec

Sequence	Layer and frame rate	VOP type	Q_P	SNR Y (dB)	Avg. bits per VOP	Bit rate (kbit/sec)
Akiyo	Enh at 10 Hz	B	20	32.24	991	9.91
	Base at 5 Hz	I/P	14.14	32.33	1715	8.58
Silent	Enh at 10 Hz	B	25	28.73	1452	14.52
	Base at 5 Hz	I/P	19.02	28.93	2614	13.07
Mother and daughter	Enh at 10 Hz	B	19	32.79	1223	12.23
	Base at 5 Hz	I/P	12.08	33.05	2389	11.95
Container	Enh at 10 Hz	B	20	30.02	1138	11.38
	Base at 5 Hz	I/P	14.14	30.06	2985	14.93

Table 9(b) Results of Temporal Scalability Experiments with CIF Sequences at a Total of About 112 kbit/sec

Sequence	Layer and frame rate	VOP type	Q_P	SNR Y	Avg. bits per VOP	Bit rate (kbit/sec)
Akiyo	Enh at 10 Hz	B	27	32.85	4821	48.21
	Base at 5 Hz	I/P	22.1	32.85	5899	29.50
Mother and daughter	Enh at 10 Hz	B	27	32.91	6633	66.33
	Base at 5 Hz	I/P	22.1	32.88	8359	41.80
Silent	Enh at 10 Hz	B	29	29.07	5985	59.85
	Base at 5 Hz	I/P	24.1	29.14	9084	45.42
Container	Enh at 10 Hz	B	29	28.47	6094	60.94
	Base at 5 Hz	I/P	24.1	28.52	8325	41.63

standards and MPEG-4 [15]; Table 10 summarizes the tools in various standards being compared.

VII. POSTPROCESSING

In this section we discuss techniques by which the subjective quality of MPEG-4 video can be enhanced through nonstandard postprocessing but within the constraints of a standard MPEG-4 video bitstream.

A. Deblocking Filter

One of the problems that any DCT-based scheme has for low-bit-rate applications is blocking artifacts. Basically, when the DCT coefficient quantization step size is above a

Table 10 Comparison of Tools in Various Video Standards

Tool	H.263	MPEG-1 video	MPEG-2 video	MPEG-4 video version 1
I, P pictures (or VOPs)	Yes, pictures	Yes, pictures	Yes, pictures	Yes, VOPs
B pictures (or VOPs)	Sort of, in PB picture	Yes, pictures	Yes, pictures	Yes, VOPs
16 × 16 MC	Yes	Yes	Yes	Yes
8 × 8 and overlap block MC	Yes	No	No	Yes
Half-pel precision MC	No	Yes	Yes	Yes
8 × 8 DCT	Yes	Yes	Yes	Yes
DC prediction—intra	No	Yes	Yes	Yes, higher efficiency
AC prediction—intra	No	No	No	Yes
Nonlinear DC quant—intra	No	No	No	Yes
Quantization matrices	No	Yes	Yes	Yes
Adaptive scan and VLC	No	No	No, alternate scan	Yes
Frame/field MC—interlace	No	No	Yes	Yes
Frame/field DCT—interlace	No	No	Yes	Yes
Binary shape coding	No	No	No	Yes
Sprite coding	No	No	No	Yes
Gray-scale shape coding	No	No	No	No, but may be added
Temporal scalability of pictures	No	No	Yes	Yes, rectangular VOPs
Spatial scalability of pictures	No	No	Yes	Yes, rectangular VOPs
Temporal scalability of arbitrary VOP	No	No	No	Yes
Spatial scalability of arbitrary VOPS	No	No	No	No, but may be added
Resynch markers	Yes, GOB start code	Yes, slice start code	Yes, slice start code	Yes, flexible marker
Data partitioning	No	No	Yes, coefficients only	Yes, motion vectors and coefficients
Reversible VLCs	No	No	No	Yes
Noise reduction filter	Yes, loop filter	No	No	Yes, postfilter

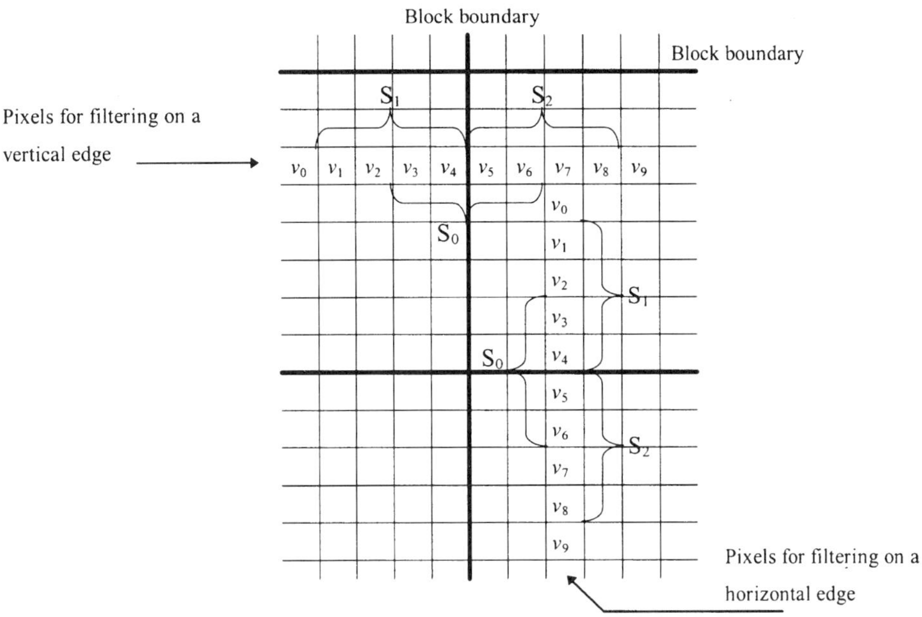

Figure 25 Boundary area around block of interest.

certain level, discontinuities in pixel values become clearly visible at the boundaries between blocks.

These blocking artifacts can be greatly reduced by the deblocking filter. The filter operations are performed along the boundary of every 8 × 8 luminance or chrominance block, first along the horizontal edges of the block and then along the vertical edges of the block. Figure 25 shows the block boundaries.

Two modes, the DC offset mode and the default mode, are used in the filter operations. The selection of these two modes is based on the following criterion.

$$A = \sum_{i=0}^{8} \phi(v_i - v_{i+1})$$

where

$$\phi(x) = \begin{cases} 1, & |x| \leq Th1 \\ 0, & \text{otherwise} \end{cases}$$

If $(A \geq Th2)$

DC offset mode is applied,

else

Default mode is applied.

The typical threshold values are $Th1 = 2$ and $Th2 = 6$. Clearly, this is a criterion for determining a smooth region with blocking artifacts due to DC offset and to assign it the DC offset mode or, otherwise, the default mode.

In the default mode, an adaptive signal-smoothing algorithm is applied by differentiating image details at the block discontinuities using the frequency information of neighbor pixel arrays, S_0, S_1, and S_2. The filtering process in the default mode is to replace the boundary pixel values v_4 and v_5 with v_4' and v_5' as follows:

$$v_4' = v_4 - d$$
$$v_5' = v_5 - d$$

and

$$d = \text{clip}(5 \times (a_{3,0}' - a_{3,0})//8, 0, (v_4 - v_5)/2) \times \delta(a_{3,0})$$

where the function clip(x, p, q) clips x to a value between p and q, and

$$\delta(x) = \begin{cases} 1, & |x| < QP \\ 0, & \text{otherwise} \end{cases}$$

where QP denotes the quantization parameter of the macroblock where pixel v_5 belongs and // denotes integer division with rounding to the nearest integer. The variable $a_{3,0}' = \text{sign}(a_{3,0}) \times \text{MIN}(|a_{3,0}|, |a_{3,1}|, |a_{3,2}|)$, where

$$\text{sign}(x) = \begin{cases} 1, & x \geq 0 \\ -1, & x < 0 \end{cases}$$

and MIN(a, b, c) selects the smallest values among a, b, and c. The frequency components $a_{3,0}$, $a_{3,1}$, and $a_{3,2}$ are computed from the following equations:

$$a_{3,0} = (2 \times v_3 - 5 \times v_4 + 5 \times v_5 - 2 \times v_6)//8$$
$$a_{3,1} = (2 \times v_1 - 5 \times v_2 + 5 \times v_3 - 2 \times v_4)//8$$
$$a_{3,2} = (2 \times v_5 - 5 \times v_6 + 5 \times v_7 - 2 \times v_8)//8$$

A very smooth region is detected for the DC offset mode. Then a stronger smoothing filter is applied in this case.

$$\text{max} = \text{MAX}(\text{v1, v2, v3, v4, v5, v6, v7, v8}),$$
$$\text{min} = \text{MIN}(\text{v1, v2, v3, v4, v5, v6, v7, v8}),$$
$$\text{if } (|\text{max} - \text{min}| < 2 \times \text{QP}) \quad \{$$

$$v_n' = \sum_{k=-4}^{4} b_k \times p_{n+k}, \quad 1 \leq n \leq 8$$

$$p_m = \begin{cases} (|v_1 - v_0| < QP)?\, v_0 : v_1, & \text{if } m < 1 \\ v_m, & \text{if } 1 \leq m \leq 8 \\ (|v_8 - v_9| < QP)?\, v_9 : v_8, & \text{if } m > 8 \end{cases}$$

$$\{b_k : -4 \leq k \leq 4\} = \{1, 1, 2, 2, 4, 2, 2, 1, 1\}//16$$

B. Deringing Filter

This filter consists of three steps, namely threshold determination, index acquisition, and adaptive smoothing. The process is as follows. First, the maximum and minimum pixel values of an 8×8 block in the decoded image are calculated. Then, for the kth block in a macroblock, the threshold, denoted by th$[k]$, and the dynamic range of pixel values, denoted by range$[k]$, are set:

$$\text{th}[k] = (\max[k] + \min[k] + 1)/2$$
$$\text{range}[k] = \max[k] - \min[k]$$

where $\max[k]$ and $\min[k]$ give the maximum and minimum pixel values of the kth block in the macroblock, respectively.

Additional processing is performed only for the luminance blocks. Let max_range be the maximum value of the dynamic range among four luminance blocks.

$$\text{max\_range} = \text{range}[k_{\max}]$$

Then apply the rearrangement as follows.

$\leftarrow$ for ($k = 1$; $k < 5$; k++) {
if (range$[k] < 32$ && max_range $> = 64$)
$\rightarrow$ th$[k]$ = th$[k_{\max}]$;
if (max_range < 16)
$\rightarrow$ th$[k] = 0$;
$\leftarrow$ }

The remaining operations are purely on an 8×8 block basis. Let rec(h, v) be the pixel value at coordinates (h, v) where $h, \text{v} = 0, 1, 2, \ldots, 7$. Then the corresponding binary index bin(h, v) can be obtained from

$$\text{bin}(h, \text{v}) = \begin{cases} 1, & \text{if rec}(h, \text{v}) \geq \text{th} \\ 0, & \text{otherwise} \end{cases}$$

The filter is applied only if all binary indices in a 3×3 window in a block are the same, i.e., all "0" indices or all "1" indices.

The recommended filter is a two-dimensional separable filter that has coefficients coef(i, j) for $i, j = -1, 0, 1$, given in Figure 26.

The coefficient at the center pixel, i.e., coef(0,0), corresponds to the pixel to be filtered. The filter output flt$'(i, j)$ is obtained from

$$\text{flt}'(h, \text{v}) = \left\{8 + \sum_{i=-1}^{1} \sum_{j=-1}^{1} \text{coef}(i, j) \times \text{rec}(h + i, \text{v} + j)\right\} // 16$$

1	2	1
2	4	2
1	2	1

Figure 26 Filter mask for adaptive smoothing.

The differential value between the reconstructed pixel and the filtered one is limited according to the quantization parameter QP. Let flt(h, ν) and flt$'(h, \nu)$ be the filtered pixel value and the pixel value before limitation, respectively.

$$\leftarrow \text{if } (\text{flt}'(h, \nu) - \text{rec}(h, \nu) > \text{max\_diff})$$
$$\text{flt}(h, \nu) = \text{rec}(h, \nu) + \text{max\_diff}$$
$$\leftarrow \text{else if } (\text{flt}'(h, \nu) - \text{rec}(h, \nu) < -\text{max\_diff})$$
$$\text{flt}(h, \nu) = \text{rec}(h, \nu) - \text{max\_diff}$$

else

$$\text{flt}(h, \nu) = \text{flt}'(h, \nu)$$

where max_diff = QP/2 for both intra and inter macroblocks.

VIII. DISCUSSION

We have presented a summary of the MPEG-4 natural video coding standard and comparisons between it and several other video coding standards in use today. Although our results demonstrate that MPEG-4 has introduced improvements in coding efficiency and statistically performs as well as or better than the current standards, its real strength lies in its versatility and the diverse applications it is able to perform. Because of its object-based nature, MPEG-4 video seems to offer increased flexibility in coding quality control, channel bandwidth adaptation, and decoder processing resource variations. The success of MPEG-4 will eventually depend on many factors such as market needs, competing standards, software versus hardware paradigms, complexity versus functionality trade-offs, timing, and profiles. Technically, MPEG-4 appears to have significant potential because of the integration of natural and synthetic worlds, computers and communication applications, and the functionalities and flexibilities it offers. Initially, perhaps only the very basic functionalities will be useful. As the demand for sophisticated multimedia grows, the advanced functionalities will become more relevant.

REFERENCES

1. MPEG-1 Video Group. Information technology—Coding of moving pictures and associated audio for digital storage media up to about 1.5 Mbit/s: Part 2—Video. ISO/IEC 11172-2, International Standard, 1993.
2. MPEG-2 Video Group. Information technology—Generic coding of moving pictures and associated audio: Part 2—Video. ISO/IEC 13818-2, International Standard, 1995.
3. MPEG-4 Video Group. Generic coding of audio-visual objects: Part 2—Visual. ISO/IEC JTC1/SC29/WG11 N1902, FDIS of ISO/IEC 14496-2, Atlantic City, November 1998.
4. ITU-T Experts Group on Very Low Bitrate Visual Telephony. ITU-T Recommendation H.263: Video coding for low bitrate communication, December 1995.
5. ITU-T Experts Group on Very Bitrate Visual Telephony. ITU-T Recommendation H.263 version 2: Video coding for low bitrate communication, January 1998.
6. MPEG-1 Video Simulation Model Editing Committee. MPEG-1 video simulation model 3. ISO/IEC JTC1/SC29/WG11 Doc. xxx, July 1990.

7. A Puri. Video coding using the MPEG-1 compression standard. Proceedings International Symposium of Society for Information Display, Boston, May 1992, pp 123–126.
8. MPEG-2 Video Test Model Editing Committee. MPEG-2 video test model 5. ISO/IEC JTC1/SC29/WG11 N0400, April 1993.
9. A Puri. Video coding using the MPEG-2 compression standard. Proceedings SPIE Visual Communication and Image Processing, SPIE vol. 1199, pp. 1701–1713, Boston, November 1993.
10. BG Haskell, A Puri, AN Netravali. Digital Video: An Introduction to MPEG-2. New York: Chapman & Hall, 1997.
11. RL Schmidt, A Puri, BG Haskell. Performance evaluation of nonscalable MPEG-2 video coding. Proceedings SPIE Visual Communications and Image Processing, Chicago, October 1994, pp 296–310.
12. MPEG-4 Video Verification Model Editing Committee. The MPEG-4 video verification model 8.0. ISO/IEC JTC1/SC29/WG11 N1796, Stockholm, July 1997.
13. MPEG-4 Video Ad Hoc Group on Core Experiments in Coding Efficiency. Description of core experiments on coding efficiency, ISO/IEC JTC1/SC29/WG11 Doc. xxxx, July–September 1996.
14. A Puri, RL Schmidt, BG Haskell. Description and results of coding efficiency experiment T9 (part 4) in MPEG-4 video, ISO/IEC JTC1/SC29/WG11 MPEG96/1320, Chicago, September 1996.
15. RL Schmidt, A Puri, BG Haskell. Results of scalability experiments, ISO/IEC JTC1/SC29/WG11 MPEG96/1084, Tampere, July 1996.
16. A Puri, RL Schmidt, BG Haskell. Improvements in DCT based video coding. Proceedings SPIE Visual Communications and Image Processing, San Jose, January 1997.
17. A Puri, RL Schmidt, BG Haskell. Performance evaluation of the MPEG-4 visual coding standard. Proceedings Visual Communications and Image Processing, San Jose, January 1998.
18. H-J Lee, T Chiang. Results for MPEG-4 video verification tests using rate control, ISO/IEC JTC1/SC29/WG11 MPEG98/4157, Atlantic City, October 1998.
19. H-J Lee, T Chiang. Results for MPEG-4 video verification tests using rate control, ISO/IEC JTC1/SC29/WG11 MPEG98/4319, Rome, December 1998.
20. B Eifrig, X Chen, A Luthra. Interlaced video coding results (core exp P-14), ISO/IEC JTC1/SC29/WG11 MPEG97/2671, Bristol, April 1997.
21. ITU-T Experts Group on Very Low Bitrate Visual Telephony. Video codec test model, TMN5, January 1995.
22. A Puri, A Eleftheriadis. MPEG-4: An object-based multimedia coding standard supporting mobile applications. ACM Mobile Networks Appl 3:5–32, 1998.
23. R Talluri. Error resilient video coding in ISO MPEG-4 standard. IEEE Commun Mag, June 1998.
24. ITU-T Recommendation H.320. Narrow-band Visual Telephone Systems and Terminal Equipment, March 1996.
25. R Talluri. MPEG-4 status and direction. Proceedings of the SPIE Critical Reviews of Standards and Common Interfaces for Video Information System, Philadelphia, October, 1997, pp 252–262.
26. F Pereira. MPEG-4: A new challenge for the representation of audio-visual information. Proceedings of the Picture Coding Symposium, PCS'96, pp 7–16. March 1996, Melbourne.
27. T Sikora. The MPEG-4 video standard verification model. IEEE Trans CSVT 7(1):1997.
28. I Moccagatta, R Talluri. MPEG-4 video verification model: Status and directions. J Imaging Technol 8:468–479, 1997.
29. T Sikora, L Chiariglione. The MPEG-4 video standard and its potential for future multimedia applications. Proceedings IEEE ISCAS Conference, Hong Kong, June 1997.
30. MPEG-4 Video Group. MPEG-4 video verification model version 9.0, ISO/IEC JTC1/SC29/WG11 N1869, Fribourg, October 1997.

31. M Ghanbari. Two-layer coding of video signals for VBR networks. IEEE J Selected Areas Commun 7:771–781, 1989.
32. M Nomura, T Fuji, N Ohta. Layered packet-loss protection for variable rate video coding using DCT. Proceedings International Workshop on Packet Video, September 1988.
33. R Talluri, et al. Error concealment by data partitioning. Signal Process Image Commun, in press.
34. MPEG-4 Video Adhoc Group on Core Experiments on Error Resilience. Description of error resilience core experiments, ISO/IEC JTC1/SC29/WG11 N1473, Maceio, Brazil, November 1996.
35. T Miki, et al. Revised error pattern generation programs for core experiments on error resilience, ISO/IEC JTC1/SC29/WG11 MPEG96/1492, Maceio, Brazil, November 1996.

9

MPEG-4 Natural Video Coding—Part II

Touradj Ebrahimi
Swiss Federal Institute of Technology (EPFL), Lausanne, Switzerland

F. Dufaux
Compaq, Cambridge, Massachusetts

Y. Nakaya
Hitachi Ltd., Tokyo, Japan

I. INTRODUCTION

Multimedia commands the growing attention of the telecommunications and consumer electronics industry. In a broad sense, *multimedia* is assumed to be a general framework for interaction with information available from different sources, including video information.

Multimedia is expected to support a large number of applications. These applications translate into specific sets of requirements that may be very different from each other. One theme common to most applications is the need for supporting interactivity with different kinds of data. Applications related to visual information can be grouped together on the basis of several features:

- Type of data (still images, stereo images, video, . . .)
- Type of source (natural images, computer-generated images, text and graphics, medical images, . . .)
- Type of communication (ranging from point-to-point to multipoint-to-multipoint)
- Type of desired functionalities (object manipulation, online editing, progressive transmission, error resilience, . . .)

Video compression standards, MPEG-1 [1] and MPEG-2 [2], although perfectly well suited to environments for which they were designed, are not necessarily flexible enough to address the requirements of multimedia applications efficiently. Hence, MPEG (Motion Picture Experts Group) committed itself to the development of the MPEG-4 standard, providing a common platform for a wide range of multimedia applications [3]. MPEG has been working on the development of the MPEG-4 standard since 1993, and finally, after about 6 years of efforts, an international standard covering the first version of MPEG-4 has been adopted [4].

MPEG-4 has been designed to support several classes of functionalities [4,5]. Chief among them are the following:

Compression efficiency: This class consists of functionalities for improved coding efficiency and coding of multiple concurrent data streams. These functionalities are required by all applications relying on efficient storage or transmission of video data. One example of such applications is video transmission over Internet Protocol (IP).

Content-based interactivity: These are functionalities to allow content-based access and manipulation of data, editing bitstreams, coding hybrid (natural and synthetic) data, and improved random access. These functionalities will target applications such as digital libraries, electronic shopping, and movie production.

Universal access: Such functionalities include robustness in error-prone environments and content-based scalability. These functionalities allow MPEG-4–encoded data to be accessible over a wide range of media, with various qualities in terms of temporal and spatial resolutions for specific objects. These different resolutions could be decoded by a range of decoders with different complexities. Applications benefiting from them are mobile communications, database browsing, and access at different content levels, scales, resolutions, and qualities.

This chapter will concentrate on the tools and functionalities in MPEG-4 natural video that go beyond a pixel-based representation of video. Other tools have been covered in the previous chapter.

A. Motivations and Background

MPEG is the working group within the International Organization for Standardization (ISO) in charge of proposing compression standards for audiovisual information. So far, MPEG has released three standards, known as MPEG-1, MPEG-2, and MPEG-4. MPEG-1 operates at bit rates up to about 1.5 Mbit/sec and targets storage on media such as CD-ROMS, as well as transmission over narrow communication channels such as the integrated services digital network (ISDN) or local area networks (LANs) and wide area networks (WANs). MPEG-2 addresses another class of coding algorithms for generic compression of high-quality video of various types and bit rates. The basic principle behind MPEG-2 algorithms is similar to that of MPEG-1, to which special features have been added to allow intrinsic coding of frames as well as fields in interlaced sequences. It also allows scalable coding of video signals by which it is possible to decode a signal with lower temporal or spatial resolutions or qualities from the same compressed bitstream. MPEG-2 mainly operates at bit rates around 1.5–35 Mbit/sec and provides higher quality video signals at the expense of more complex processing than with MPEG-1. MPEG-2 defines several profiles and levels that allow its efficient use in various applications from consumer up to professional categories. Standards such as DAVIC (Digital Audio Visual Council), DVD (digital video disk), and DVB (digital video broadcast) make use of MPEG-2 compression algorithms in their respective applications. More recently, MPEG finalized the first version of a new standard known as MPEG-4. The standard aims at providing an integrated solution for a multitude of multimedia applications, ranging from mobile videotelephony up to professional video editing, as well as Internet-like interactive communications. Because of the extensive proliferation of audiovisual information, MPEG has initiated yet another standard activity called MPEG-7, which will be used to ease the search of audiovisual content.

Since the beginning, MPEG standards have been about efficient representation of audiovisual information. Figure 1 shows how different MPEG standards may be related to each other from a data representation point of view.

The most widely used approach to representing still and moving images in the digital domain is that of pixel-based representation. This is mainly due to the fact that pixel-by-pixel acquisition and display of digital visual information are mature and relatively cheap technologies. In pixel-based representation, an image or a video is seen as a set of pixels (with associated properties such as a given color or a motion) in the same way as the physical world is made of atoms. Until recently, pixel-based image processing was the only digital representation available for processing of visual information, and therefore the majority of techniques known today rely on such a representation.

It was in the mid-1980s that for the first time, motivated by studies of the mechanism of the human visual system, other representation techniques started appearing. The main idea behind this effort was that as humans are in the majority of cases the final stage in the image processing chain, a representation similar to that of the human visual system will be more efficient in the design of image processing and coding systems. Non–pixel-based representation techniques for coding (also called second-generation coding) showed that at very high compression ratios, these techniques are superior to pixel-based representation methods. However, it is a fact that transform-based and (motion-compensated) predictive coding have shown outstanding results in compression efficiency for coding of still images and video. One reason is that digital images and video are intrinsically pixel based in all digital sensors and display devices and provide a projection of the real 4D world, as this is the only way we know today to acquire and to display them. In order to

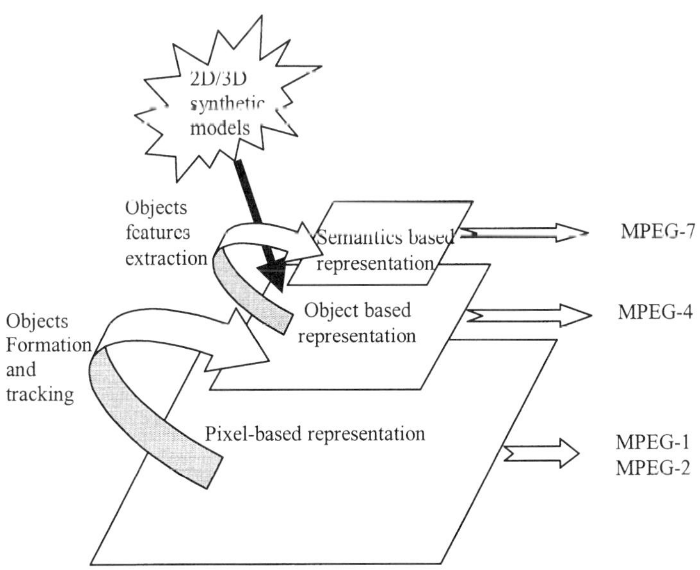

Figure 1 Relationship between MPEG standards.

use a non–pixel-based approach, pixel-based data have to be converted somehow to a non–pixel-based representation, which brings additional complexity but also other inefficiencies. Among non–pixel-based representation techniques, "object-based" visual data representation is a very important class. In object-based representation, objects replace pixels and an image or a video is seen as a set of objects that cannot be broken into smaller elements. In addition to texture (color) and motion properties, shape information is needed in order to define the object completely. The shape in this case can be seen as a force field keeping together the elements of an image or video object just as the atoms of a physical object are kept together because of an atomic force field. It is because of the force field keeping atoms of a physical object together that one can move them easily. Once you grab a corner of an object, the rest comes with it because a force field has "glued" all the atoms of the object together. The same is true in an object-based visual information representation, where the role of the force field is played by that of shape as mentioned earlier.

Thanks to this property, object-based representation brings at no cost a very important feature called interactivity. Interactivity is defined by some as the element that defines multimedia, and this is one of the main reasons why an object-based representation was adopted in the MPEG-4 standard. Clearly, because the majority of digital visual information is still in pixel-based representation, converters are needed in order to go from one representation to another. The passage from a pixel-based representation to an object-based representation can be performed using manual, semiautomatic, or automatic segmentation techniques. The inverse operation is achieved by rendering, blending, or composition. At this point, it is important to note that because the input information is pixel based, all tools used in MPEG-4 still operate in a pixel-based approach in which care has been taken to extend their operation to arbitrary-shaped objects. This is not necessarily a disadvantage, as such an approach allows easy backward–forward compatibility and easy transcoding between MPEG-4 and other standards. An exception to this statement arises when synthetic objects (2D or 3D) are added in the scene. Such objects are intrinsically non–pixel-based as they are not built from a set of pixels.

Continuing the same philosophy, one could think of yet another representation in which visual information is represented by describing its content. An example would be describing to someone a person he or she has never seen: She is tall, thin, has long black hair, blue eyes, etc. As this kind of representation would require some degree of semantic understanding, one could call it a semantics-based representation. We will not cover this representation here, as it goes beyond the scope of this chapter. It is worth mentioning that MPEG-7 could benefit from this type of representation.

II. VIDEO OBJECTS AND VIDEO OBJECT PLANES EXTRACTION AND CODING

The previous standards, MPEG-1 [1] and MPEG-2 [2], were designed mainly for the purpose of compression of audiovisual data and accomplish this task very well [3]. The MPEG-4 standard [4,5], while providing good compression performance, is being designed with other image-based applications in mind. Most of these applications expect certain basic functionalities to be supported by the underlying standard. Therefore, MPEG-4 incorporates tools, or algorithms, that enable functionalities such as scalability, error resilience, or interactivity with content in addition to compression.

MPEG-4 relies on an object-based representation of the video data in order to achieve its goals. The central concept in MPEG-4 is that of the *video object (VO)*. Each VO is characterized by intrinsic properties such as shape, texture, and motion. In MPEG-4 a scene is considered to be composed of several VOs. Such a representation of the scene is more amenable to interactivity with the scene content than pixel-based (block-based) representations [1,2].

It is important to mention that the standard will not prescribe the method for *creating* VOs. Depending on the application, VOs may be created in a variety of ways, such as by spatiotemporal segmentation of natural scenes [6–9] or with parametric descriptions used in computer graphics [10]. Indeed, for video sequences with which compression is the only goal, a set of rectangular image frames may be considered as a VO. MPEG-4 will simply provide a standard convention for describing VOs, such that all compliant decoders will be able to extract VOs of any shape from the encoded bitstream, as necessary. The decoded VOs may then be subjected to further manipulation as appropriate for the application at hand.

Figure 2 shows a general block diagram of an MPEG-4 video encoder. First, the video information is split into VOs as required by the application. The coding control unit decides, possibly based on requirements of the user or the capabilities of the decoder, which VOs are to be transmitted, the number of layers, and the level of scalability suited to the current video session. Each VO is encoded independently of the others. The multiplexer then merges the bitstreams representing the different VOs into a video bitstream.

Figure 3 shows a block diagram of an MPEG-4 decoder. The incoming bitstream is first decomposed into its individual VO bitstreams. Each VO is then decoded, and the result is composited. The composition handles the way the information is presented to the user. For a natural video, composition is simply the layering of 2D VOs in the scene.

The VO-based structure has certain specific characteristics. In order to be able to process data available in a pixel-based digital representation, the texture information for a VO (in the uncompressed form) is represented in YUV color coordinates. Up to 12 bits

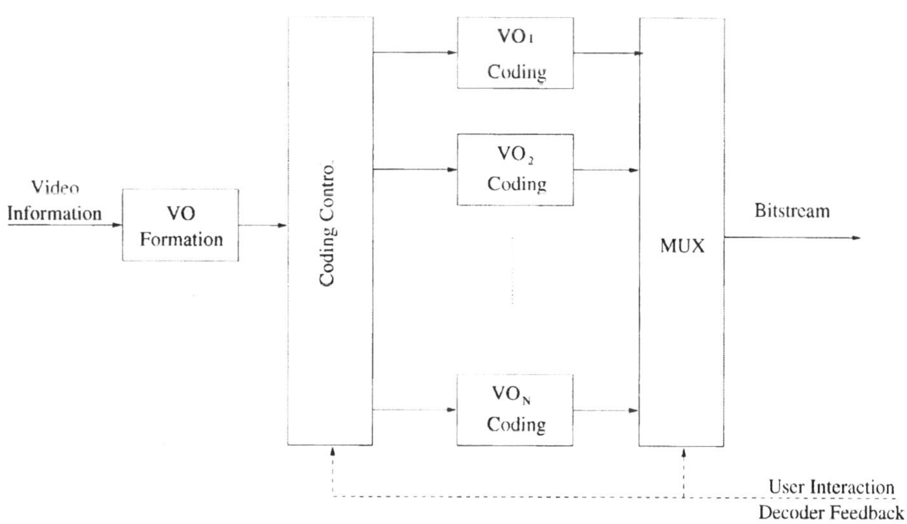

Figure 2 Structure of an MPEG-4 encoder.

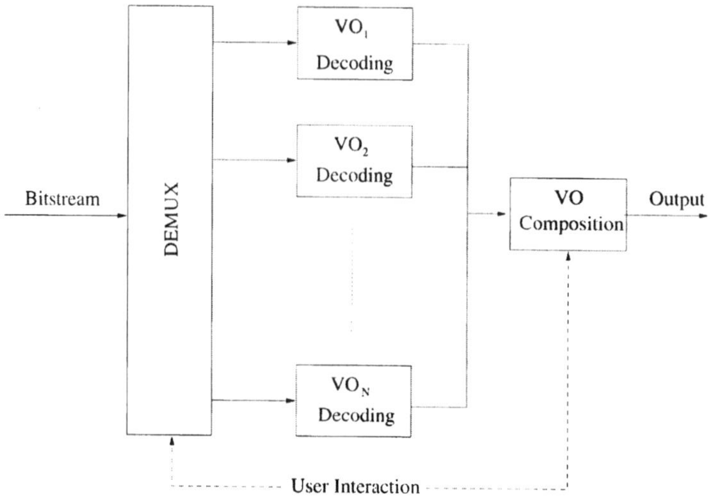

Figure 3 Structure of an MPEG-4 decoder.

may be used to represent a pixel component value. Additional information regarding the shape of the VO is also available. Both shape information and texture information are assumed to be available for specific snapshots of VOs called *video object planes (VOPs).*

Although the snapshots from conventional digital sources occur at predefined temporal intervals, the encoded VOPs of a VO need not be at the same, or even constant, temporal intervals. Also, the decoder may choose to decode a VO at a temporal rate lower than that used while encoding.

MPEG-4 video also supports sprite-based coding. The concept of sprites is based on the notion that there is more to an object (a VO, in our case) than meets the eye. A VOP may be thought of as just the portion of the sprite that is visible at a given instant of time. If we can encode the entire information about a sprite, then VOPs may be derived from this encoded representation as necessary. Sprite-based encoding is particularly well suited for representing synthetically generated scenes. We will discuss sprites in more detail later in this chapter.

A. VOP-Based Coding

For reasons of efficiency and backward compatibility, VOs are compressed by coding their corresponding VOPs in a hybrid coding scheme somewhat similar to that in previous MPEG standards. The VOP coding technique, as shown in Figure 4, is implemented in terms of *macroblocks* (blocks of 16×16 pixels). This is a design decision that leads to low-complexity algorithms and also provides a certain level of compatibility with other standards. Grouping the encoded information in small entities, here macroblocks, facilitates resynchronization in case of transmission errors. A VOP has two basic types of information associated with it: *shape* information and *texture* information. The shape information needs to be specified explicitly, because VOPs are, in general, expected to have arbitrary shapes. Thus, the VOP encoder essentially consists of two encoding schemes: one for shape and one for texture. Of course, in applications in which shape information

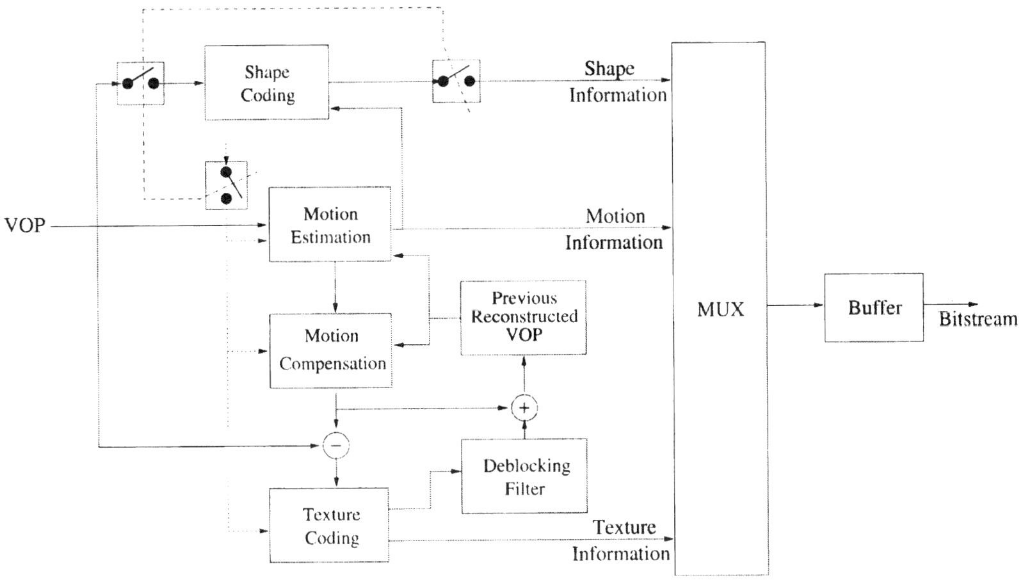

Figure 4 Block diagram of a VOP encoder.

is not explicitly required, such as when each VOP is a rectangular frame, the shape coding scheme may be disabled. The same coding scheme is used for all VOPs in a given VO.

The shape information for a VOP, also referred to as alpha-plane, is specified in two components. A simple array of binary labels, arranged in a rectangle corresponding to the bounding box of the VOP, specifies whether an input pixel belongs to the VOP. In addition, a transparency value is available for each pixel of the VOP. This set of transparency values forms what is referred to as the *gray-scale shape*. Gray-scale shape values typically range from 0 (completely transparent) to 255 (opaque).

As mentioned before, the texture information for a VOP is available in the form of a luminance (Y) and two chrominance (U,V) components. We discuss only the encoding process for the luminance (Y) component. The other two components are treated in a similar fashion. The most important tools used for encoding VOPs are discussed later.

III. SHAPE CODING

In this section we discuss the tools offered by the MPEG-4 video standard for explicit coding of shape information in arbitrarily shaped VOPs. Beside the shape information available for the VOP in question, the shape coding scheme relies on motion estimation to compress the shape information even further. A general description of the shape coding literature would be outside the scope of this chapter. Therefore, we will describe only the scheme adopted by MPEG-4 natural video standard for shape coding. Interested readers are referred to Ref. 11 for information on other shape coding techniques.

In the MPEG-4 video standard, two kinds of shape information are considered as inherent characteristics of a video object. These are referred to as binary and gray-scale shape information. By binary shape information one means label information that defines which portions (pixels) of the support of the object belong to the video object at a given time. The binary shape information is most commonly represented as a matrix of the same size as the bounding box of a VOP. Every element of the matrix can take one of two possible values, depending on whether the pixel is inside or outside the video object. Gray-scale shape is a generalization of the concept of binary shape providing the possibility to represent transparent objects.

A. Binary Shape Coding

In the past, the problem of shape representation and coding was thoroughly investigated in the fields of computer vision, image understanding, image compression, and computer graphics. However, this is the first time that a video standardization effort has adopted a shape representation and coding technique within its scope.

In its canonical form, a binary shape is represented as a matrix of binary values called a bitmap. However, for the purpose of compression, manipulation, or a more semantic description, one may choose to represent the shape in other forms such as by using geometric representations or by means of its contour.

Since its beginning, MPEG adopted a bitmap-based compression technique for the shape information. This is mainly due to the relative simplicity and higher maturity of such techniques. Experiments have shown that bitmap-based techniques offer good compression efficiency with relatively low computational complexity.

This section describes the coding methods for binary shape information. Binary shape information is encoded by a motion-compensated block-based technique allowing both lossless and lossy coding of such data. In the MPEG-4 video compression algorithm, the shape of every VOP is coded along with its other properties (texture and motion). To this end, the shape of a VOP is bounded by a rectangular window with dimensions of multiples of 16 pixels in horizontal and vertical directions. The position of the bounding rectangle is chosen such that it contains the minimum number of blocks of size 16×16 with nontransparent pixels. The samples in the bounding box and outside the VOP are set to 0 (transparent). The rectangular bounding box is then partitioned into blocks of 16×16 samples (hereafter referred to as shape blocks) and the encoding–decoding process is performed block by block (Fig. 5).

The binary matrix representing the shape of a VOP is referred to as a *binary mask*. In this mask every pixel belonging to the VOP is set to 255, and all other pixels are set to 0. It is then partitioned into *binary alpha blocks* (BABs) of size 16×16. Each BAB is encoded separately. Starting from rectangular frames, it is common to have BABs that

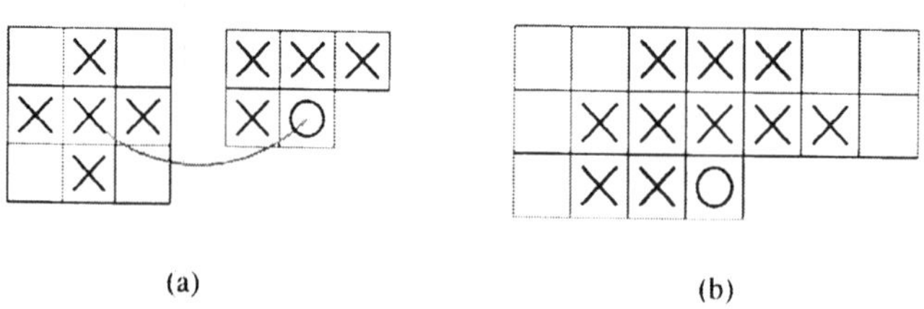

Figure 5 Context selected for InterCAE (a) and IntraCAE (b) shape coding. In each case, the pixel to be encoded is marked by a circle, and the context pixels are marked with crosses. In the InterCAE, some of the context pixels are taken from the colocated block in the previous frame.

have all pixels of the same color, either 0 (in which case the BAB is called an *All-0* block) or 255 (in which case the block is said to be an All-255 block). The shape compression algorithm provides several modes for coding a BAB. The basic tools for encoding BABs are the CAE algorithm [12] and motion compensation. InterCAE and IntraCAE are the variants of the CAE algorithm used with and without motion compensation, respectively. Each shape coding mode supported by the standard is a combination of these basic tools. Motion vectors can be computed by first predicting a value based on those of neighboring blocks that were previously encoded and then searching for a best match position (given by the minimum sum of absolute differences). The motion vectors themselves are differentially coded (the result being the motion vector difference, MVD). Every BAB can be coded in one of the following modes:

1. The block is flagged All-0. In this case no coding is necessary. Texture information is not coded for such blocks either.
2. The block is flagged All-255. Again, shape coding is not necessary for such blocks, but texture information needs to be coded (because they belong to the VOP).
3. MVD is zero but the block is not updated.
4. MVD is zero but the block is updated. IntraCAE is used for the block.
5. MVD is zero, and InterCAE is used for the block.
6. MVD is nonzero, and InterCAE is used.

The CAE algorithm is used to code pixels in BABs. The arithmetic encoder is initialized at the beginning of the process. Each pixel is encoded as follows [4,5]:

1. Compute a context number.
2. Index a probability table using this context number.
3. Use the retrieved probability to drive the arithmetic encoder for code word assignment.

B. Gray-Scale Shape Coding

The gray-scale shape information has a corresponding structure similar to that of binary shape with the difference that every pixel (element of the matrix) can take on a range of values (usually 0 to 255) representing the degree of the transparency of that pixel. The gray-scale shape corresponds to the notion of alpha plane used in computer graphics, in which 0 corresponds to a completely transparent pixel and 255 to a completely opaque pixel. Intermediate values of the pixel correspond to intermediate degrees of transparency of that pixel. By convention, binary shape information corresponds to gray-scale shape information with values of 0 and 255.

Gray-scale shape information is encoded using a block-based motion-compensated discrete cosine transform (DCT) similar to that of texture coding, allowing lossy coding only. The gray-scale shape coding also makes use of binary shape coding for coding of its support.

IV. TEXTURE CODING

In the case of I-VOPs, the term *texture* refers to the information present in the gray or chroma values of the pixels forming the VOP. In the case of predicted VOPs (B-VOPs and P-VOPs), the residual error after motion compensation is considered the texture information. The MPEG-4 video standard uses techniques very similar to those of other existing standards for coding of the VOP texture information. The block-based DCT method is adapted to the needs of an arbitrarily shaped VOP-oriented approach. The VOP texture is split into macroblocks of size 16 × 16. Of course, this implies that the blocks along the boundary of the VOP may not fall completely on the VOP; that is, some pixels in a boundary block may not belong to the VOP. Such boundary blocks are treated differently from the nonboundary blocks.

A. Coding of Internal Blocks

The blocks that lie completely within the VOP are encoded using a conventional 2D 8 × 8 block DCT. The luminance and chrominance blocks are treated separately. Thus, six blocks of DCT coefficients are generated for each macroblock.

The DCT coefficients are quantized in order to compress the information. The DC coefficient is quantized using a step size of 8. The MPEG-4 video algorithm offers two alternatives for determining the quantization step to be used for the AC coefficients. One is to follow an approach similar to that of recommendation H.263. Here, a quantization parameter determines how the coefficients will be quantized. The same value applies to all coefficients in a macroblock but may change from one macroblock to another, depending on the desired image quality or target bit rate. The other option is a quantization scheme similar to that used in MPEG-2, where the quantization step may vary depending on the position of the coefficient.

After appropriate quantization, the DCT coefficients in a block are scanned in zigzag fashion in order to create a string of coefficients from the 2D block. The string is compressed using run-length coding and entropy coding. A detailed description of these operations is given in the proceeding chapter.

B. Coding of Boundary Blocks

Macroblocks that straddle the VOP boundary are encoded using one of two techniques, *repetitive padding* followed by conventional DCT or *shape-adaptive DCT* (SA-DCT), the latter being considered only in version 2 of the standard.

Repetitive padding consists of assigning a value to the pixels of the macroblock that lie outside the VOP. The padding is applied to 8 × 8 blocks of the macroblock in question. Only the blocks straddling the VOP boundary are processed by the padding procedure. When the texture data is the residual error after motion compensation, the blocks are padded with zero values.

For intra coded blocks, the padding is performed in a two-step procedure called *low-pass extrapolation* (LPE). This procedure is as follows:

1. Compute the mean of the pixels in the block that belong to the VOP. Use this mean value as the padding value, that is,

$$f_{r,c}\big|_{(r,c)\notin \text{VOP}} = \frac{1}{N} \sum_{(x,y)\in VOP} f_{x,y} \tag{1}$$

where N is the number of pixels of the macroblock in the VOP. This is also known as mean-repetition DCT.

2. Use the average operation given in Eq. (2) for each pixel $f_{r,c}$, where r and c represent the row and column position of each pixel in the macroblock outside the VOP boundary. Start from the top left corner $f_{0,0}$ of the macroblock and proceed row by row to the bottom right pixel.

$$f_{r,c}\big|_{(r,c)\notin VOP} = \frac{f_{r,c-1} + f_{r-1,c} + f_{r,c+1} + f_{r+1,c}}{4} \tag{2}$$

The pixels considered in the right-hand side of Eq. (2) should lie within the VOP, otherwise they are not considered and the denominator is adjusted accordingly.

Once the block has been padded, it is coded in similar fashion to an internal block.

Another technique for coding macroblocks that straddle the VOP boundary is the SA-DCT technique [13]. This technique is not covered in the first version of the MPEG-4 video coding algorithm but has been planned for its version 2. In the SA-DCT–based scheme, the number of coefficients generated is proportional to the number of pixels of the block belonging to the VOP. The SA-DCT is computed as a separable 2D DCT. For example, transforming the block shown in Figure 6a is performed as follows. First, the active pixels of each column are adjusted to the top of the block (b). Then for each column, the 1D DCT is computed for only the active pixels in the column, with the DC coefficients at the top (c). This can result in a different number of coefficients for each column. The rows of coefficients generated in the column DCT are then adjusted to the left (d) before

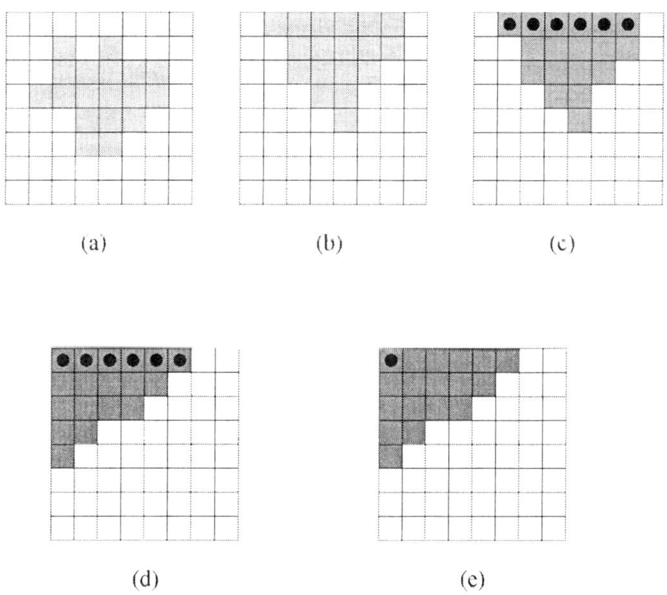

Figure 6 Steps for computing shape-adaptive DCT for an arbitrarily shaped 2D region. (a) Region to be transformed, with shaded blocks representing the active pixels; (b) top-adjusted columns; (c) column-DCT coefficients with DC coefficients marked with black spots; (d) left-adjusted rows of coefficients; (e) final 2D SA-DCT with DC coefficient marked with a black spot.

computing the row DCT. The 2D SA-DCT coefficients are laid out as shown in (e), with the DC coefficient at the top left corner. The binary mask of the shape and the DCT coefficients are both required in order to decode the block correctly. Coefficients of the SA-DCT are then quantized and entropy coded in a way similar to that explained in the previous section.

C. Static Texture Coding

MPEG-4 also supports a mode for encoding texture of static objects of arbitrary shapes or texture information to be mapped on 3D surfaces. This mode is called *static texture coding* mode and utilizes a discrete wavelet transform (DWT) to compress the texture information efficiently [14,15]. In particular, it offers a high degree of scalability both spatially and in terms of image quality.

The input texture components (luminance and chrominances) are treated separately. Each component is decomposed into *bands* by a bank of *analysis* filters. Another filter bank, the *synthesis* filters, later recombines the bands to decode and to reconstruct the texture information. The analysis and synthesis filters must satisfy certain constraints to yield perfect reconstruction. Extensive wavelet theory has been developed for designing filters that satisfy these constraints. It has been shown that the filters play an important role in the performance of the decomposition for compression purposes [16,17].

The decomposition just explained can be applied recursively on the bands obtained, yielding a decomposition tree (D-tree) of so-called *subbands*. A decomposition of depth 2 is shown in Fig. 7, where the square static texture image is decomposed into four bands and the lowest frequency band (shown on the left) is further decomposed into four subbands (1, 2, 3, 4). Therefore, subband 1 represents the lowest spectral band of the texture. At each step, the spectral domain is split into n parts, n being the number of filters in the filter bank. The number of coefficients to represent each band can therefore also be reduced by a factor of n, assuming that filters have the same bandwidth. The different bands can then be laid out as shown in Fig. 7 (right-hand side); they contain the same number of coefficients as pixels in the original texture.

It is important to note that, even though the subbands represent a portion of the signal that is well localized in the spectral domain, the subband coefficients also remain in the spatial domain. Therefore, colocated coefficients in the subbands represent the original texture at that location but at different spectral locations. The correlation between the bands, up to a scale factor, can then be exploited for compression purposes.

Shapiro [18] originally proposed a scheme for coding the coefficients by predicting

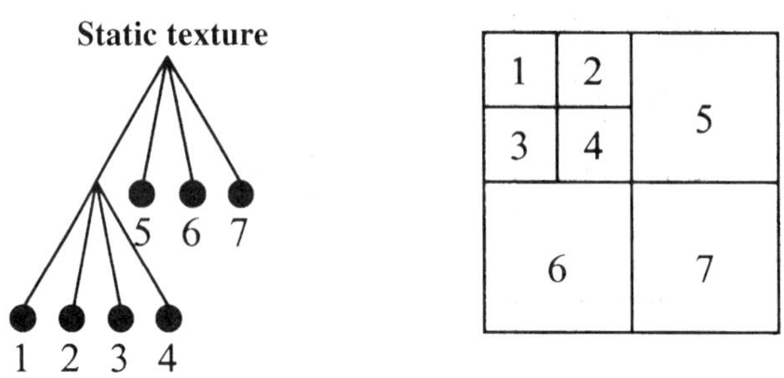

Figure 7 Illustration of a wavelet decomposition of depth 2 and the corresponding subband layout.

the position of insignificant coefficients across bands. MPEG-4 is based on a modified zero-tree wavelet coding as described in Ref. 19. For any given coefficient in the lower frequency band, a parent–child relation tree (PCR-tree) of coefficients in the subbands is built with a parent–child relationship (Fig. 8). There is one PCR-tree for each coefficient in the lowest frequency subband. Every coefficient in the decomposition can thus be located by indicating its root coefficient and the position in the PCR-tree from that root.

The problem then is to encode both the location and the value of the coefficients. This is done in two passes: the first pass locates coefficients, and the second encodes their values. As compression is lossy, the most significant coefficients should be transmitted first and the less significant transmitted later or not at all. A coefficient is considered significant if its magnitude is nonzero after quantization. The difference between the quantized and nonquantized coefficients results in so-called residual subbands. Selection and encoding of significant coefficients are achieved by iterative quantization of residual subbands. In each iteration, significant coefficients are selected, and their locations and quantized values are encoded by means of an arithmetic encoder. In subsequent iterations, the quantization is modified and the residual bands are processed in a similar manner. This results in an iterative refinement of the coefficients of the bands.

In MPEG-4, the nodes of each PCR-tree are scanned in one of two ways: depth first or band by band. At each node, the quantized coefficient QCn is considered along with the subtrees having the children nodes of QCn as root. The location of significant coefficients leads to the generation of the following symbols:

ZTR: if QCn is zero and no subtree contains any significant coefficient
VZTR: if QCn is significant but no subtree contains any significant coefficient
IZ: if QCn is zero and at least one significant coefficient can be found in the subtrees
VAL: if QCn is significant and at least one coefficient can be found in the subtrees

Once the position is encoded, the quantized coefficients are scanned in order of discovery and coded with arithmetic coding.

In addition to this, in MPEG-4, the lowest frequency is encoded predictively—differential pulse-code modulation (DPCM) with prediction based on three neighbors—and coefficients of the other bands are scanned and zero-tree encoded.

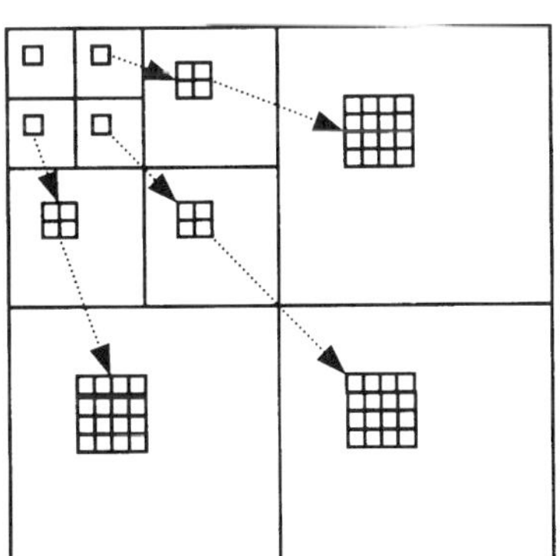

Figure 8 Parent–child relationship for the static texture compression algorithm.

V. MOTION COMPENSATED CODING

In the following sections, the motion compensation method for arbitrarily shaped VOPs is described. Because of the existence of arbitrary shapes in VOPs, the scheme shown in this section differs from the conventional block-based motion compensation for rectangular images.

A. Motion Estimation

Because the motion estimation algorithm for the encoder is not specified in the MPEG-4 standard, each encoder has the freedom to use its own algorithm. In this section, an outline of the motion estimation algorithm used for arbitrarily shaped P-VOPs in the Verification Model version 8.0 (VM8) [20] is shown as a reference algorithm. For simplification, it is assumed that the one motion vector is transmitted at most for a macroblock.

1. Motion Estimation for Texture and Gray-Scale Shape

Motion estimation for texture and gray-scale shape information is performed using the luminance values. The algorithm consists of the following three steps.

(1) *Padding of the reference VOP*: This step is applied only to arbitrarily shaped VOPs. The details of this step are given in Sec. IV.B.

(2) *Full-search polygon matching with single-pel accuracy*: In this step, the motion vector that minimizes the prediction error is searched. The search range in this step is $-2^{\text{fcode}+3} \leq MV_x, MV_y < 2^{\text{fcode}+3}$, where MV_x and MV_y denote the horizontal and vertical components of the motion vector in single pel unit and $1 \leq \text{fcode} \leq 7$. The value of fcode is defined independently for each VOP. The error measure, $SAD(MV_x, MV_y)$, is defined as

$$
\begin{aligned}
SAD(MV_x, MV_y) = \sum_{i=0}^{15} \sum_{j=0}^{15} |I(x_0 + i, y_0 + j) \\
- R(x_0 + i + MV_x, y_0 + j + MV_y)| \cdot BA(x_0 + i, y_0 + j) \\
- (NB/2 + 1) \cdot \delta(MV_x, MV_y)
\end{aligned}
\tag{3}
$$

where

(x_0, y_0) denotes the left top coordinate of the macroblock
$I(x, y)$ denotes the luminance sample value at (x, y) in the input VOP
$R(x, y)$ denotes the luminance sample value at (x, y) in the reference VOP
$BA(x, y)$ is 0 when the pixel at (x, y) is transparent and 1 when the pixel is opaque
NB denotes the number of nontransparent pixels in the macroblock
$\delta(MV_x, MV_y)$ is 1 when $(MV_x, MV_y) = (0, 0)$ and 0 otherwise
"/" denotes integer division with truncation toward zero

(3) *Polygon matching with half-pel accuracy*: Starting from the motion vector estimated in step 2, half sample search in a $\pm 0.5 \times \pm 0.5$ pel window is performed using $SAD(MV_x, MV_y)$ as the error measure. The estimated motion vector (x, y) shall stay within the range $-2^{\text{fcode}+3} \leq x, y < 2^{\text{fcode}+3}$. The interpolation scheme for obtaining the interpolated sample values is described in Sec. V.B.

2. Motion Estimation for Binary Shape

In the reference motion estimation method adopted in VM8, the motion estimation algorithm applied to the binary shape macroblocks is different from the algorithm applied to the texture macroblock. This algorithm consists of the following steps.

(1) *Calculation of the predictor*: In this step, the sum of absolute difference for the motion vector predictor is calculated. The method for obtaining the predictor is described in Sec. V.C. If the calculated sum of absolute difference for any 4×4 subblock (there are 16 subblocks in each shape macroblock) is smaller than $16 \times$ AlphaTH (AlphaTH is a parameter that specifies the target quality of the reconstructed binary shape), the predictor becomes the estimated motion vector. Otherwise, the algorithm proceeds to step 2.

(2) *Full-search block matching*: Full-search block matching with single-pel accuracy is performed within the $\pm 16 \times \pm 16$ pel window around the predictor. The sum of absolute difference without favoring the zero motion vector is used as the error measure. If multiple motion vectors minimize the sum of absolute difference, the motion vector that minimizes parameter Q is selected as the estimated motion vector. Here Q is defined as

$$Q = 2(|MVDs_x| + |MVDs_y| + 1) - \delta(MVDs_x) \quad (4)$$

where $(MVDs_x, MVDs_y)$ denotes the differential vector between the motion vector of the shape macroblock and the predictor, and the value of $\delta(MVDs_x)$ is 1 when $MVDs_x = 0$ and 0 otherwise.

If multiple motion vectors minimize the sum of absolute difference and Q, the motion vector with the smallest absolute value for the vertical component is selected from these motion vectors. If multiple motion vectors minimize the absolute value for the vertical component, the motion vector with the smallest absolute value for the horizontal component is selected.

B. Motion Compensation

The motion compensation algorithm for reconstructing the predicted VOP is normative and is strictly defined in the MPEG-4 standard. The motion compensation algorithm for texture, gray-scale shape, and binary shape information consists of the following steps:

(1) *Padding of the texture in the reference VOP*: The repetitive padding method described in Sec. IV.B is applied to the texture information of the reference VOP. This process is skipped for rectangular VOPs.

(2) *Synthesis of the predicted VOP*: The predicted VOP is synthesized using the decoded motion vectors for the texture and binary shape information. The unrestricted motion vector mode, four motion vector mode, and overlapped block motion compensation can be used for texture macroblocks in arbitrarily shaped VOPs, as well as for rectangular VOPs. For texture information, the interpolation of pixel values is performed as shown in Figure 9. The parameter rounding_control can have the value of 0 or 1 and is defined explicitly for each P-VOP. Usually, the encoder controls rounding_control so that the current P-VOP and the reference VOP have different values. By having such a control mechanism, the accumulation of round-off errors, which causes degradation of the quality of the decoded VOPs, is avoided [21].

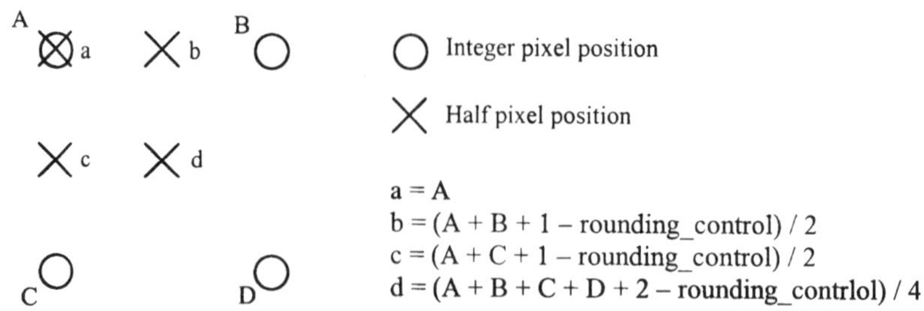

Figure 9 Pixel value interpolation.

C. Motion Coding

As in rectangular VOPs, the motion vectors are coded differentially in arbitrarily shaped VOPs. However, the method for obtaining the predictor differs between texture or gray-scale shape and binary shape. This is discussed in detail in the following.

1. Motion Coding for Texture

The difference between the motion coding methods for texture information in rectangular VOPs and arbitrarily shaped VOPs is that arbitrarily shaped VOPs may include transparent macroblocks. When the current macroblock is transparent, motion vector information is not transmitted for this macroblock. The decoder can correctly decode the coded data for this macroblock, because the shape information is transmitted before the motion information for each macroblock. When the decoded shape information indicates that the current macroblock is transparent, the decoder knows at this point that the coded data of that macroblock do not include motion vector information. The same rule applies to each 8×8 pixel block in a macroblock with four motion vectors: the motion vector information for a transparent 8×8 pixel block is not transmitted.

When the candidate predictor for motion vector prediction (see Fig. 7 of Chap. 8) is included in a transparent block (i.e., a macroblock or an 8×8 pixel block), this candidate predictor is regarded as not valid. Before applying median filtering to the three candidate predictors, the values of the candidate predictors that are not valid are defined according to the following rule:

When one and only one candidate predictor is not valid, the value of this candidate predictor is set to zero.

When two and only two candidate predictors are not valid, the values of these candidate predictors are set to value of the other valid candidate predictor.

When three candidate predictors are not valid, the values of these candidate predictors are set to zero.

2. Motion Coding for Binary Shape

Both the shape motion vectors, *MVs1*, *MVs2*, and *MVs3*, and texture motion vectors, *MV1*, *MV2*, and *MV3*, of the adjacent macroblocks shown in Fig. 10 are used as the candidate predictors for coding the motion vector of a shape macroblock. Because only single-pel accuracy motion vectors are allowed for shape macroblocks, the candidate predictors of the texture macroblocks are truncated toward 0 to an integer value. It is assumed in Figure 10 that macroblocks with one motion vector have four identical motion vectors for each of the 8×8 blocks included in it.

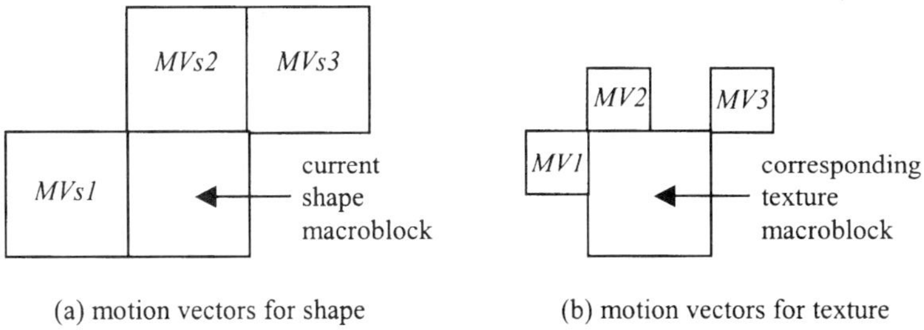

Figure 10 Candidate predictors for a shape macroblock.

The predictor is obtained by traversing *MVs1*, *MVs2*, *MVs3*, *MV1*, *MV2*, and *MV3* in this order and taking the first valid motion vector. The motion vector of a texture macroblock is not valid when the macroblock is transparent or outside the current VOP. The motion vector of a shape macroblock is valid only when this shape macroblock is coded as an inter shape macroblock.

VI. SPRITE CODING AND GLOBAL MOTION COMPENSATION

Efficient representation of temporal information is a key component in any video coding system. In video imagery, pixels are indeed most correlated in the direction of motion. The widely used model of blocks of pixels undergoing a translation is the basis of most motion-compensated techniques such as the one used in MPEG-1, -2, and -4. However, although it has achieved significant performance [22], this simple model has its limitations. In order to reach higher coding efficiency, a more sophisticated model is needed. For this purpose, a novel technique referred to as sprite coding [23] was adopted in the MPEG-4 standard. MPEG is also considering another technique referred to as global motion compensation (GMC) [24,25] for possible adoption in the MPEG-4 version 2 standard (the latter is scheduled to reach the CD stage in March 1999 [26]).

In computer graphics, a sprite is a graphic image that can be individually animated or moved around within a larger graphic image or set of images. In the context of MPEG-4, a sprite is more generally a video object modeled by a coherent motion. In the case of natural video, a sprite is typically a large composite image resulting from the blending of pixels belonging to various temporal instances of the video object. In the case of synthetic video, it is simply a graphic object. Temporal instances of the video object can subsequently be reconstructed from the sprite by warping the corresponding portion of the sprite content.

As far as video coding is concerned, a sprite captures spatiotemporal information in a very compact way. Indeed, a video sequence can be represented by a sprite and warping parameters [23]. This concept is similar to that of the mosaicking techniques proposed in Refs. 27 and 28 and to the layered representation introduced in Ref. 29. Therefore, sprite coding achieves high coding efficiency. In addition, it empowers the media consumer by enabling content-based interaction at the receiver end.

Sprite coding is most useful for encoding synthetic graphics objects (e.g., a flying logo). However, it is also well suited to encoding any video object that can be modeled by a rigid 2D motion (e.g., this assumption very often holds true for the background of a video scene). When the sprite is not directly available, such as when processing natural video, it should be constructed prior to coding. In this situation, the technique is not suitable for real-time applications because of the offline nature of the sprite generation process.

GMC [24,25] is a technique that compensates for this limitation of sprite coding. Instead of warping the sprite, GMC warps the reference VOP using the estimated warping parameters to obtain the predicted VOP used for interframe coding. GMC is most useful for encoding a video sequence with significant camera motion (e.g., pan, zoom). Because GMC does not require a priori knowledge about the scene to be encoded, it is applicable to online real-time applications.

In summary, sprite coding consists of the following steps [23]. First, a sprite is built by means of image registration and blending. The sprite and warping parameters are subsequently transmitted to the decoder side. Finally, the decoder reconstructs the video imagery by warping the sprite. On the other hand, GMC coding, which fits well in the conventional interframe coding framework, consists of interframe prediction and prediction error coding. These techniques are discussed in more detail hereafter.

A. Sprite Generation

When processing natural video, the sprite is generally not known a priori. In this case, it has to be generated offline prior to starting the encoding process. Note that the sprite generation process is not specified by the MPEG-4 standard. This section presents a technique based on the MPEG-4 Verification Model [30]. This technique is not normative.

In the following, we assume a video object that can be modeled by a rigid 2D motion, along with its corresponding segmentation mask, obtained, for instance, by means of chroma keying or automatic or supervised segmentation techniques (see Section II).

Figure 11 illustrates the process of sprite generation. It basically consists of three steps: global motion is estimated between an input image and the sprite; using this motion information the image is then aligned, i.e., warped, with the sprite; and finally the image is blended into the sprite. These steps are described more thoroughly in the following.

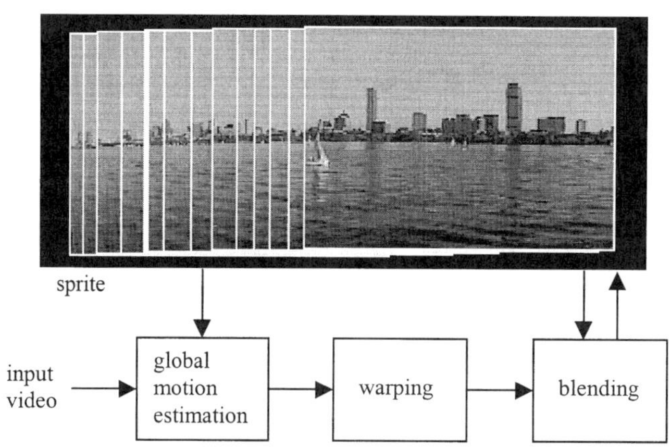

Figure 11 Sprite generation block diagram (successive images are highlighted in white for illustrative purposes).

1. Global Motion Estimation

In order to estimate the motion of the current image with respect to the sprite, global motion estimation is performed. The technique is based on hierarchical iterative gradient descent [31]. More formally, it minimizes the sum of squared differences between the sprite S and the motion-compensated image I',

$$E = \sum_{i=1}^{N} e_i^2 \quad \text{with } e_i = S(x_i, y_i) - I'(x_i', y_i') \tag{5}$$

where (x_i, y_i) denotes the spatial coordinates of the ith pixel in the sprite, (x_i', y_i') denotes the coordinates of the corresponding pixel in the image, and the summation is carried out over N pairs of pixels (x_i, y_i) and (x_i', y_i') within image boundaries. Alternatively, motion estimation may be carried out between consecutive images.

As the problem of motion estimation is underconstrained [22], an additional constrained is required. In our case, an implicit constraint is added, namely the motion over the whole image is parameterized by a perspective motion model (eight parameters) defined as follows:

$$\begin{aligned} x_i' &= (a_0 + a_1 x_i + a_2 y_i)/(a_6 x_i + a_7 y_i + 1) \\ y_i' &= (a_3 + a_4 x_i + a_5 y_i)/(a_6 x_i + a_7 y_i + 1) \end{aligned} \tag{6}$$

where $(a_0, \ldots, a_7)$ are the motion parameters. This model is suitable when the scene can be approximated by a planar surface or when the scene is static and the camera motion is a pure rotation around its optical center. Simpler models such as affine (six parameters: $a_6 = a_7 = 0$), translation–isotropic magnification–rotation (four parameters: $a_1 = a_5$, $a_2 = -a_4$, $a_6 = a_7 = 0$), and translation (two parameters: $a_1 = a_5 = 1$, $a_2 = a_4 = a_6 = a_7 = 0$) are particular cases of the perspective model and can easily be derived from it.

The motion parameters $\mathbf{a} = (a_0, \ldots, a_7)$ are computed by minimizing E in Eq. (5). Because the dependence of E on $\mathbf{a}$ is nonlinear, the following iterative gradient descent method is used:

$$\mathbf{a}^{(t+1)} = \mathbf{a}^{(t)} + \mathbf{H}^{-1}\mathbf{b} \tag{7}$$

where $\mathbf{a}^{(t)}$ and $\mathbf{a}^{(t+1)}$ denote $\mathbf{a}$ at iteration t and $t + 1$, respectively, $\mathbf{H}$ is an $n \times n$ matrix, and $\mathbf{b}$ is an n-element vector whose coefficients are given by

$$H_{kl} = \frac{1}{2} \sum_{i=1}^{N} \frac{\partial^2 e_i^2}{\partial a_k \partial a_l} \cong \sum_{i=1}^{N} \frac{\partial e_i}{\partial a_k} \frac{\partial e_i}{\partial a_l} \quad \text{and} \quad b_k = -\frac{1}{2} \sum_{i=1}^{N} \frac{\partial e_i^2}{\partial a_k} = -\sum_{i=1}^{N} e_i \frac{\partial e_i}{\partial a_k} \tag{8}$$

and n is the number of parameters of the model.

In order to improve convergence and to reduce computational complexity, a low-pass image pyramid is introduced. The gradient descent is applied at the top of the pyramid and then iterated at each level until convergence is achieved.

To ensure convergence in the presence of large displacements, an initial coarse estimate of the translation component of the displacement is computed by applying an n-step search matching algorithm at the top level of the pyramid prior to the gradient descent.

Finally, in order to limit the influence of outliers, which would otherwise bias the estimation, a robust estimator is used. For this purpose, the quadratic measure in Eq. (5) is replaced by a truncated quadratic error function defined as

$$E = \sum_{i=1}^{N} \rho(e_i) \quad \text{with} \quad \rho(e_i) = \begin{cases} e_i^2 & \text{if } |e_i| \leq T \\ 0 & \text{if } |e_i| > T \end{cases} \tag{9}$$

where T is a threshold. In other words, only the pixels for which the absolute value of the error term is below T are taken into account in the estimation process. Typically, T is computed from the data so as to exclude the samples resulting in the largest error terms $|e_i|$, for instance, eliminating the top p percent of the distribution of $|e_i|$. Note that the preceding iterative procedure remains unchanged while using this robust estimator.

2. *Warping and Blending*

Once motion is known, the image is aligned with respect to the sprite. This procedure is referred to as warping [31]. More precisely, coordinates (x_i, y_i) in the sprite are scanned, and the warped coordinates (x_i', y_i') in the image are simply computed using Eq. (6). Generally, (x_i', y_i') will not coincide with the integer-pixel grid. Therefore, $I'(x_i', y_i')$ is evaluated by interpolating surrounding pixels. Bilinear interpolation is most commonly used.

The last step to complete the sprite generation is to blend the warped image into the current sprite to produce the new sprite. A simple average can be used. Furthermore, a weighting function decreasing near the edges may be introduced to produce a more seamless sprite. However, the averaging operator may result in blurring in case of misregistration. As the sprite generation is an offline and noncausal process, memory buffer permitting, it is possible to store the whole sequence of warped images. Blending may then be performed by a weighted average, a median, or a weighted median, usually resulting in higher visual quality of the sprite when compared with the simple average.

3. *Example*

Figure 12 shows an example of sprite generated for the background of a test sequence called "Stefan" using the first 200 frames of the sequence. Note that the sequence has been previously segmented in order to exclude foreground objects. Because of camera motion, the sprite results in an extended view of the background.

Figure 12 Example of sprite.

B. Sprite Coding

This section describes how sprites are used to code video in MPEG-4. Figure 13 illustrates both the encoding and decoding processes. Using sprite coding, it is sufficient to transmit the sprite and warping parameters. The sprite is, in fact, a still image, which includes both a texture map and an alpha map. It is therefore encoded as an intraframe or I-VOP (see Chap. 8 for texture coding and Sec. III for shape coding) that can be readily decoded at the receiver end. In turns, warping parameters are coded by transmitting the trajectories of some reference points. The decoder retrieves the warping parameters from the trajectories and reconstructs the VOPs of the video imagery by warping the sprite.

As no residual signal is transmitted, high coding efficiency is achieved. Although the technique results in a wallpaper-like rendering that may not always be faithful to the original scene, it usually provides high visual quality. Finally, because local block-matching motion estimation and residual encoding–decoding are avoided, the technique has low complexity.

1. *Trajectories Coding*

Instead of directly transmitting the parameters $\mathbf{a} = (a_0, \ldots, a_7)$ of the motion model [see Eq. (6)], displacements of reference points are encoded. More precisely, for each VOP to be encoded, reference points (i_n, j_n), $n = 1, \ldots, N$, are positioned at the corners of the VOP bounding box, and the corresponding points (i'_n, j'_n) in the sprite are computed, as illustrated in Figure 14. The points (i'_n, j'_n) are quantized to half-pel precision. Finally, the displacement vectors $(u_n, v_n) = (i_n - i'_n, j_n - j'_n)$ are computed and transmitted as differential motion vectors.

MPEG-4 supports four motion models: $N = 4$ points are required for a perspective model, $N = 3$ points for affine, $N = 2$ points for translation–isotropic magnification–rotation, and $N = 1$ point for translation. At the receiver end, the decoder reconstructs the N pairs of reference points (i_n, j_n) and (i'_n, j'_n), which are then used in the warping procedure.

2. *Warping and VOP Reconstruction*

In the decoder, VOPs are reconstructed by warping the content of the sprite. Note that this warping is different from the one described in Sec. VI.A.2 in several aspects. First this warping is part of the decoding process and is thus normative. Second, the sprite is now warped toward the VOP. Furthermore, the warping is expressed in terms of the

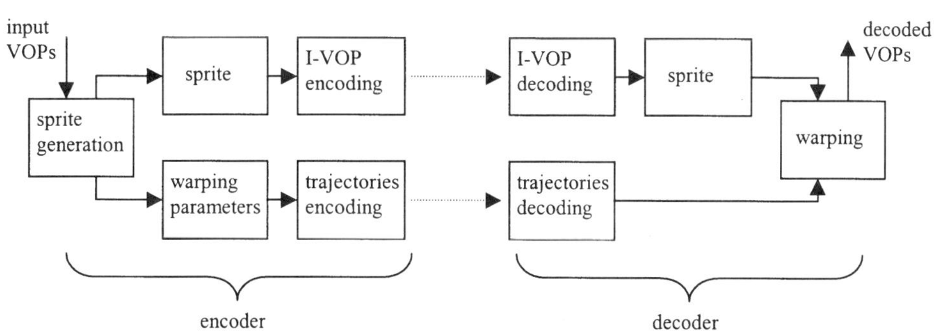

Figure 13 Sprite encoding and decoding process.

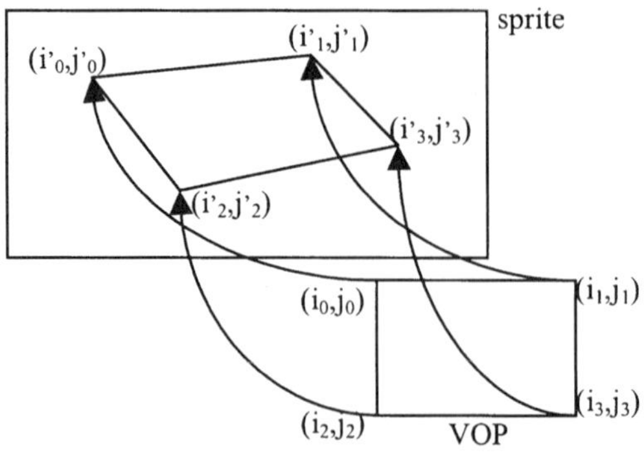

Figure 14 Reference points and trajectories coding.

N pairs of reference points (i_n, j_n) and (i'_n, j'_n) instead of the warping parameters $\mathbf{a} = (a_0, \ldots, a_7)$. Finally, in order to specify precisely the accuracy of the warping procedure, (i'_n, j'_n) are expressed as integers in $1/s$ pel accuracy, where the sprite warping accuracy s takes the values $\{2, 4, 8, 16\}$.

For instance, in the case of a perspective model, for each VOP luminance pixel (i,j), the corresponding sprite location (i', j') is determined as follows:

$$\begin{aligned} i' &= (a + bi + cj)/(gi + hj + DWH) \\ j' &= (d + ei + fj)/(gi + hj + DWH) \end{aligned} \tag{10}$$

where W and H are the width and height of the VOP, respectively, and

$$\begin{aligned} &a = Di'_0WH, \quad b = D(i'_1 - i'_0)H + gi'_1, \quad c = D(i'_2 - i'_0)W + hi'_2 \\ &d = Dj'_0WH, \quad e = D(j'_1 - j'_0)H + gj'_1, \quad f = D(j'_2 - j'_0)W + hj'_2 \\ &g = ((i'_0 - i'_1 - i'_2 + i'_3)(j'_2 - j'_3) - (i'_2 - i'_3)(j'_0 - j'_1 - j'_2 + j'_3))H \\ &h = ((i'_1 - i'_3)(j'_0 - j'_1 - j'_2 + j'_3) - (i'_0 - i'_1 - i'_2 + i'_3)(j'_1 - j'_3))W \\ &D = (i'_1 - i'_3)(j'_2 - j'_3) - (i'_2 - i'_3)(j'_1 - j'_3) \end{aligned} \tag{11}$$

The reconstructed value of the VOP pixel (i, j) is finally interpolated from the four sprite pixels surrounding the location (i', j').

3. Low-Latency Sprite

As shown in Figure 12, a sprite is typically much larger than a single frame of video. If the decoder should wait for the whole sprite to be transmitted before being able to decode and display the first VOP, high latency is introduced. In order to alleviate this limitation, MPEG-4 supports two low-latency sprite coding methods. The first method consists of transmitting the sprite piece by piece. The portion of the sprite needed to reconstruct the beginning of the video sequence is first transmitted. Remaining pieces of the sprite are subsequently sent according to the decoding requirements and available bandwidth. In the second method, the quality of the transmitted sprite is progressively upgraded. A coarsely quantized version of the sprite is first transmitted. Bandwidth permitting, finely quantized residuals are later sent to improve on the quality of the sprite. These two methods may be utilized separately or in combination and allow significant reduction of the latency intrinsic to sprite coding.

C. Global Motion Compensation

Panning and zooming are common motion patterns observed in natural video scenes. However, the existence of such motion patterns in the input scene usually causes degradation of coding efficiency to an encoder that uses conventional block matching for motion compensation. This is because a moving background leads to a large number of transmitted motion vectors, and translational motion of blocks is not the appropriate motion model for a scene with nontranslational motion (i.e., zooming, rotation, etc.). Global motion compensation (GMC) [24,25] is a motion compensation tool that solves this problem. Instead of applying warping to a sprite VOP, GMC applies warping to the reference VOP for obtaining the predicted VOP.

The main technical elements of the GMC technique are global motion estimation, warping, motion trajectory coding, LMC–GMC decision, and motion coding. Because the first three elements have already been covered, LMC–GMC decision and motion coding are discussed in the following subsections.

1. LMC/GMC Decision

In the GMC method adopted in the second version of MPEG-4 [26], each inter macroblock is allowed to select either GMC or LMC (local motion compensation, which is identical to the block-matching method used in the baseline coding algorithm of MPEG-4) as its motion compensation method. A common strategy for deciding whether to use LMC or GMC for a macroblock is to apply GMC to the moving background and LMC to the foreground. For this purpose, the following LMC–GMC decision criterion is adopted in the MPEG-4 Video VM version 12.1 [32] (for simplicity, it is assumed that one motion vector is transmitted at most for a macroblock using LMC):

$$\text{If } SAD_{GMC} - P(Qp, MV_x, MV_y) < SAD(MV_x, MV_y), \\ \text{then use GMC, otherwise use LMC} \tag{12}$$

where $SAD(MV_x, MV_y)$ is defined in Sec. V.A.1, (MV_x, MV_y) is the estimated motion vector of LMC, Qp is the quantization parameter for DCT coefficients, and SAD_{GMC} is the sum of absolute difference between the original macroblock and the macroblock predicted by GMC. Parameter $P(Qp, MV_x, MV_y)$ is defined as

$$P(Qp, MV_x, MV_y) = (1 - \delta(MV_x, MV_y))\,(N_B\,Qp)/64 \\ + 2\delta\,(MV_x, MV_y)\,(N_B/2 + 1) \tag{13}$$

where N_B, $\delta(MV_x, MV_y)$, and operator "/" are defined in Sec. V.A.1.

The purpose of having item $(1 - \delta(MV_x, MV_y))\,(N_B\,Qp)/64$ is to favor GMC, especially when the compression ratio is high (i.e., when Qp has a large value). This is because

- GMC should be favored from the viewpoint of reducing the amount of overhead information, as a macroblock that selected GMC does not need to transmit motion vector information, and
- The advantage of selecting GMC is higher at very low bit rates, as the ratio of motion information to the total coded information becomes higher as the bit rate is reduced.

2. Motion Coding

When GMC is enabled, the motion vectors of the LMC macroblocks are coded differentially as in the baseline coding algorithm of MPEG-4. Therefore, it is necessary to define the candidate predictor for GMC macroblocks in order to obtain the predictor for the motion vector of LMC macroblocks. This is achieved by calculating the averaged motion vector of the pixels in the luminance macroblock. The horizontal and vertical components of the averaged motion vector are rounded to half-pel accuracy values for use as a candidate predictor.

VII. CONCLUDING REMARKS AND PERSPECTIVES

MPEG-4 is being developed to support a wide range of multimedia applications. Past standards have concentrated mainly on deriving as compact a representation as possible of the video (and associated audio) data. In order to support the various applications envisaged, MPEG-4 enables functionalities that are required by many such applications.

The MPEG-4 video standard uses an object-based representation of the video sequence at hand. This allows easy access to and manipulation of arbitrarily shaped regions in frames of the video. The structure based on video objects (VOs) directly supports one highly desirable functionality: object scalability. Spatial scalability and temporal scalability are also supported in MPEG-4. Scalability is implemented in terms of layers of information, where the minimum needed to decode is the base layer. Any additional enhancement layer will improve the resulting image quality either in temporal or in spatial resolution. Sprite and image texture coding are two new features supported by MPEG-4.

To accommodate universal access, transmission-oriented functionalities have also been considered in MPEG-4. Functionalities for error robustness and error resilience handle transmission errors, and the rate control functionality adapts the encoder to the available channel bandwidth.

New tools are still being added to the MPEG-4 version 2 coding algorithm. Extensive tests show that this new standard achieves better or similar image qualities at all bit rates targeted, with the bonus of added functionalities.

REFERENCES

1. MPEG-1 Video Group. Information technology—Coding of moving pictures and associated audio for digital storage media up to about 1.5 Mbit/s: Part 2—Video, ISO/IEC 11172-2, International Standard, 1993.
2. MPEG-2 Video Group. Information technology—Generic coding of moving pictures and associated audio: Part 2—Video, ISO/IEC 13818-2, International Standard, 1995.
3. L Chariglione. MPEG and multimedia communications. IEEE Trans Circuits Syst Video Technol 7(1):5–18, 1997.
4. MPEG-4 Video Group. Generic coding of audio-visual objects: Part 2—Visual, ISO/IEC JTC1/SC29/WG11 N1902, FDIS of ISO/IEC 14496-2, Atlantic City, November 1998.
5. MPEG-4 Video Verification Model Editing Committee. The MPEG-4 video verification model 8.0, ISO/IEC JTC1/SC29/WG11 N1796, Stockholm, July 1997.
6. C Horne. Unsupervised image segmentation. PhD thesis, EPF-Lausanne, Lausanne, Switzerland, 1990.
7. C Gu. Multivalued morphology and segmentation-based coding. PhD thesis (1452), EPFL, Lausanne, Switzerland, 1995.
8. F Moscheni. Spatio-temporal segmentation and object tracking: An application to second generation video coding. PhD thesis (1618), EPFL, Lausanne, Switzerland, 1997.

9. R Castagno. Video segmentation based on multiple features for interactive and automatic multimedia applications. PhD thesis (1894), EPFL, Lausanne, Switzerland, 1998.
10. J Foley, A Van Dam, S Feiner, J Hughes. Computer Graphics: Principle and Practice. Reading, MA: Addison-Wesley, 1987.
11. C Le Buhan, F Bossen, S Bhattacharjee, F Jordan, T Ebrahimi. Shape representation and coding of visual objects in multimedia applications—An overview. Ann Telecommun 53(5–6):164–178, 1998.
12. F Bossen, T Ebrahimi. A simple and efficient binary shape coding technique based on bitmap representation. Proceedings of the International Conference on Acoustics, Speech, and Signal Processing (ICASSP'97), Munich, April 20–24, 1997, vol 4, pp 3129–3132.
13. T Sikora. Low complexity shape-adaptive DCT for coding of arbitrarily shaped image segments. Signal Process Image Commun no 7, November 1995.
14. CK Chui. An Introduction to Wavelets. Orlando, FL: Academic Press, 1992.
15. G Strang, T Nguyen. Wavelets and Filter Banks. Wellesley-Cambridge Press, 1996.
16. H Caglar, Y Liu, AN Akansu. Optimal PR-QMF design for subband image coding. J. Vis Commun Image Representation 4(4):242–253, 1993.
17. O Egger, W Li. Subband coding of images using asymmetrical filter banks. IEEE Trans Image Process 4:478–485, 1995.
18. JM Shapiro. Embedded image coding using zerotrees of wavelet coefficients. IEEE Trans Signal Process 41:3445–3462, 1993.
19. SA Martucci, I Sodagar, T Chiang, Y-Q Zhang. A zerotree wavelet video coder. Trans Circuits Syst Video Technol 7(1):109–118, 1997.
20. MPEG-4 Video Verification Model Editing Committee. The MPEG-4 video verification model 8.0, ISO/IEC JTC1/SC29/WG11 N1796, Stockholm, July 1997.
21. Y Nakaya, S Misaka, Y Suzuki. Avoidance of error accumulation caused by motion compensation, ISO/IEC JTC1/SC29/WG11 MPEG97/2491, Stockholm, July 1997.
22. F Dufaux, F Moscheni. Motion estimation techniques for digital TV: A review and a new contribution. Proc IEEE 83:858–879, 1995.
23. MC Lee, W Chen, CB Lin, C Gu, T Markoc, SI Zabinsky, R Szeliski. A layered video object coding system using sprite and affine motion model. IEEE Trans Circuits Syst Video Technol 7(1):130–145, 1997.
24. M Hötter. Differential estimation of the global motion parameters zoom and pan. Signal Process 16(3):249–265, 1989.
25. H Jozawa, K Kamikura, A Sagata, H Kotera, H Watanabe. Two-stage motion compensation using adaptive global MC and local affine MC. IEEE Trans Circuits Syst Video Technol 7(1): 75–85, 1997.
26. MPEG-4 Video Group. Generic coding of audio-visual objects: Part 2—Visual, ISO/IEC JTC1/SC29/WG11 N2802, FPDAM1 of ISO/IEC 14496-2, Vancouver, July 1999.
27. M Irani, S Hsu, P Anandan. Mosaic-based video compression. SPIE Proceedings Digital Video Compression: Algorithms and Technologies, vol 2419, San Jose, February 1995.
28. F Dufaux, F Moscheni. Background mosaicking for low bit rate coding. IEEE Proceedings ICIP'96, Lausanne, September 1996, pp 673–676.
29. J Wang, E Adelson. Representing moving images with layers. IEEE Trans Image Process 3: 625–638, 1994.
30. MPEG-4 Video Group. MPEG-4 video verification model version 10.0, ISO/IEC JTC1/SC29/WG11 N1992, San Jose, February 1998.
31. G Wolberg. Digital Image Warping. Los Alamitos, CA: IEEE Computer Society Press, 1990.
32. MPEG-4 Video Verification Model Editing Committee. The MPEG-4 video verification model 12.1, ISO/IEC JTC1/SC29/WG11 N2552, Rome, December 1998.

10
MPEG–4 Texture Coding

Weiping Li
Optivision, Inc., Palo Alto, California

Ya-Qin Zhang and Shipeng Li
Microsoft Research China, Beijing, China

Iraj Sodagar
Sarnoff Corporation, Princeton, New Jersey

Jie Liang
Texas Instruments, Dallas, Texas

I. OVERVIEW OF MPEG–4 TEXTURE CODING

MPEG–4 is the international standard for multimedia [1–3]. The MPEG–4 texture coding is based on wavelet coding. It includes taking a wavelet transform, forming wavelet trees, quantizing the wavelet coefficients, and entropy coding the quantized wavelet coefficients. In MPEG–4 texture coding, a biorthogonal wavelet transform is defined as the default and the syntax supports downloading a customized wavelet transform too. The wavelet coefficients are organized into wavelet trees with two possible scanning orders. Using the tree-depth scanning order, the wavelet coefficients are encoded in the order of one wavelet tree after another wavelet tree. Using the band-by-band scanning order, the wavelet coefficients are encoded in the order of one subband after another subband. In MPEG–4 texture coding, quantization of the DC band uses a uniform midstep quantizer with a dead zone equal to the quantization step size. All the higher bands are quantized by a uniform midstep quantizer with a dead zone of twice the quantization step size. The multiscale quantization scheme provides a very flexible approach to support the appropriate trade-off between layers and types of scalability, complexity, and coding efficiency for a wide range of applications. The quantized wavelet coefficients are entropy coded using adaptive arithmetic coding. There are three choices for entropy coding [4–8]. Zero-tree entropy (ZTE) coding and bilevel coding are two special cases of multiscale zero-tree entropy coding (MZTE). As discussed in the later sections on the individual steps of MPEG–4 texture coding, the different choices serve different purposes so that the user has the flexibility to choose the best combination for a given application. The cost of

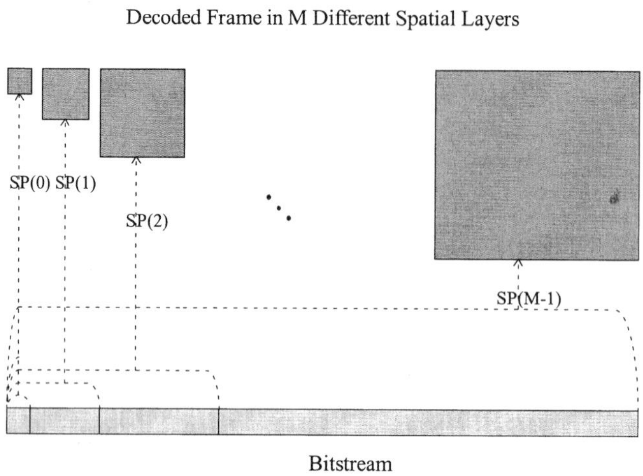

Figure 1 *M* layers of spatial scalability.

providing such a flexibility is that a compliant decoder has to implement all the combinations.

Wavelet coding in MPEG–4 texture coding provides three functionalities, namely much improved coding efficiency, scalability, and coding arbitrarily shaped visual objects. Coding efficiency is a well-understood functionality of any image coding technique. It means using the minimum number of bits to achieve a certain image quality or to achieve the best image quality for a given number of bits.

In scalable coding, a bitstream can be progressively transmitted and decoded to provide different versions of an image in terms of either spatial resolutions (spatial scalability), quality levels [quality scalability or sometimes called signal-to-noise ratio (SNR) scalability], or combinations of spatial and quality scalabilities.

Figures 1 and 2 illustrate the two scalabilities. In Figure 1, the bitstream has *M* layers

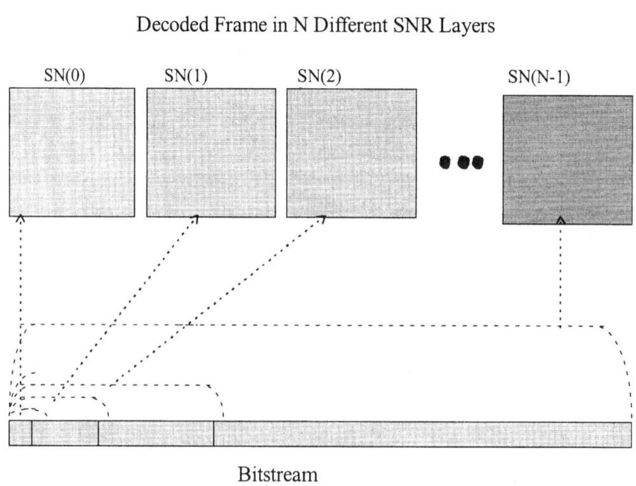

Figure 2 *N* layers of quality scalability.

of spatial scalability where the bitstream consists of *M* different segments. By decoding the first segment, the user can display a preview version of the decoded image at a lower resolution. Decoding the second segment results in a larger reconstructed image. Furthermore, by progressively decoding the additional segments, the viewer can increase the spatial resolution of the image up to the full resolution of the original image. Figure 2 shows a bitstream with *N* layers of quality scalability. In this figure, the bitstream consists of *N* different segments. Decoding the first segment provides a low-quality version of the reconstructed image. Further decoding the remaining segments results in a quality increase of the reconstructed image up to the highest quality.

Figure 3 shows a case of combined spatial–quality scalability. In this example, the bitstream consists of *M* spatial layers and each spatial layer includes *N* levels of quality scalability. In this case, both spatial resolution and quality of the reconstructed image can be improved by progressively transmitting and decoding the bitstream. The order is to improve image quality at a given spatial resolution until the best quality is achieved at that spatial resolution and then to increase the spatial resolution to a higher level and improve the quality again.

The functionality of making visual objects available in the compressed form is a very important feature in MPEG–4. It provides great flexibility for manipulating visual objects in multimedia applications and could potentially improve visual quality in very low bit rate coding. There are two parts in coding an arbitrarily shaped visual object. The first part is to code the shape of the visual object and the second part is to code the texture of the visual object (pixels inside the object region). We discuss only the texture coding part. Figure 4 shows an arbitrarily shaped visual object (a person). The black background indicates that these pixels are out of the object and not defined. The task of texture coding is to code the pixels in the visual object efficiently. There have been considerable research efforts on coding rectangular-shaped images and video, such as discrete cosine transform (DCT)–based coding and wavelet coding. A straightforward method, for example, is first to find the bounding box of the arbitrarily shaped visual object, then pad values into the

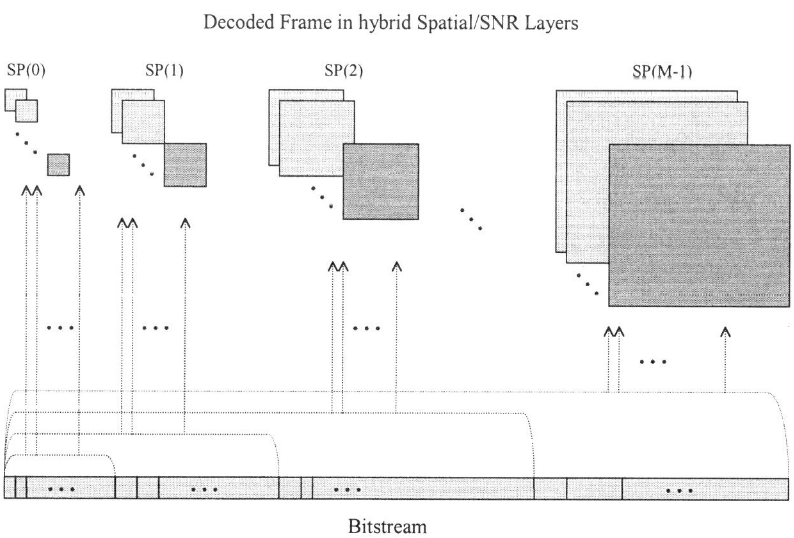

Figure 3 $N \times M$ layers of combined spatial and quality scalability.

Figure 4 An example of an arbitrarily shaped visual object.

undefined pixel positions, and code pixels inside the object together with the padded pixels in the rectangular bounding box using the conventional methods. However, this approach would not be efficient.

Therefore, during the MPEG–4 standardization process, significant efforts have been devoted to developing new techniques for efficiently coding arbitrarily shaped visual objects. Shape-adaptive DCT (SA-DCT) [9–12] is a technique for this purpose. It generates the same number of DCT coefficients as the number of pixels in the arbitrarily shaped 8×8 image block. For standard rectangular image blocks of 8×8 pixels, the SA-DCT becomes identical to the standard 8×8 DCT. Because the SA-DCT always flushes samples to an edge before performing row or column DCTs, some spatial correlation is lost. It is not efficient to apply a column DCT on the coefficients from different frequency bands after the row DCTs [13,14].

In order to provide the most efficient technique for MPEG–4 texture coding, extensive experiments were performed on the shape-adaptive wavelet coding technique proposed by Li et al. [15–18] before it was accepted. In this technique, a shape-adaptive discrete wavelet transform (SA-DWT) decomposes the pixels in the arbitrarily shaped region into the same number of wavelet coefficients while maintaining the spatial correlation, locality, and self-similarity across subbands. Because shape coding (lossy or lossless) is always performed before texture coding, the reconstructed shape is used for SA-DWT coding. A more comprehensive description of SA-DWT coding is given in Ref. 18.

II. WAVELET TRANSFORM AND SHAPE-ADAPTIVE WAVELET TRANSFORM

Wavelet transforms and their applications to image and video coding have been studied extensively [19–25]. There are many wavelet transforms that provide various features for image coding. The default wavelet filter used in MPEG–4 texture coding is the following (9,3) biorthogonal filter:

$$h[\,] = \sqrt{2}\,\{3, -6, -16, 38, 90, 38, -16, -6, 3\}/128$$
$$g[\,] = \sqrt{2}\,\{-32, 64, -32\}/128$$

where the index of $h[\,]$ is from 0 to 8 and the index of $g[\,]$ is from 0 to 2. The analysis filtering operation is described as follows:

$$L[i] = \sum_{j=-4}^{4} x[i + j]h[j + 4] \tag{1}$$

$$H[i] = \sum_{j=-1}^{1} x[i + j]g[j + 1] \tag{2}$$

where $i = 0, 1, 2, \ldots, N - 1$ and $x[\,]$ is the input sequence with an index range from -4 to $N + 3$. Because the input data sequence has an index ranging from 0 to $N - 1$, the values of $x[\,]$ for indexes less than 0 or greater than $N - 1$ are obtained by symmetric extensions of the input data. The outputs of wavelet analysis, $x_l[\,]$ and $x_h[\,]$, are generated by subsampling $L[\,]$ and $H[\,]$, respectively.

The synthesis process is the reverse of analysis. The wavelet filters $h[\,]$ and $g[\,]$ change their roles from low-pass and high-pass to high-pass and low-pass, respectively, by changing the signs of some filter taps:

$$h[\,] = \sqrt{2}\{3, 6, -16, -38, 90, -38, -16, 6, 3\}/128$$
$$g[\,] = \sqrt{2}\{32, 64, 32\}/128$$

where the index of $h[\,]$ is still from 0 to 8 and the index of $g[\,]$ is still from 0 to 2. The synthesis filtering operation is specified as follows:

$$y[n] = \sum_{i=-1}^{1} L[n + i]g[i + 1] + \sum_{i=-4}^{4} H[n + i]h[i + 4] \tag{3}$$

where $n = 0, 1, 2, \ldots, N - 1$, $L[\,]$ has an index range from -1 to N, and $H[\,]$ has an index range from -4 to $N + 3$. The values of $L[\,]$ and $H[\,]$ with indexes from 0 to $N - 1$ are obtained by upsampling $x_l[\,]$ and $x_h[\,]$, respectively. The values of $L[\,]$ and $H[\,]$ with indexes less than 0 or greater than $N - 1$ are obtained by symmetric extension of the upsampled sequences.

To describe symmetric extension, we use the terminology in MPEG–4 texture coding. Symmetric extension for obtaining the values of a sequence with indexes less than 0 is called leading boundary extension and symmetric extension for obtaining the values of a sequence with indexes greater than $N - 1$ is called trailing boundary extension. Figures 5 and 6 illustrate the different types of symmetric extension. In each figure, the arrow points to the symmetric point of the corresponding extension type. Once an extension is performed on a finite-length sequence, the sequence can be considered to have an infinite length in any mathematical equations. The samples with indexes less than 0 or greater than $N - 1$ are derived from extensions.

The wavelet coefficients from the analysis are obtained by subsampling $L[\,]$ and $H[\,]$ by a factor of 2. Subsampling can be at either even positions or odd positions. However, subsampling of low-pass coefficients and that of high-pass coefficients always have one sample shift. If the subsampling positions of low-pass coefficients are even, then the subsampling positions of high-pass coefficients should be odd, or vice versa. The subsampling process is described as follows:

$$x_l[i] = L[2i - s] \tag{4}$$
$$x_h[i] = H[2i + 1 - s] \tag{5}$$

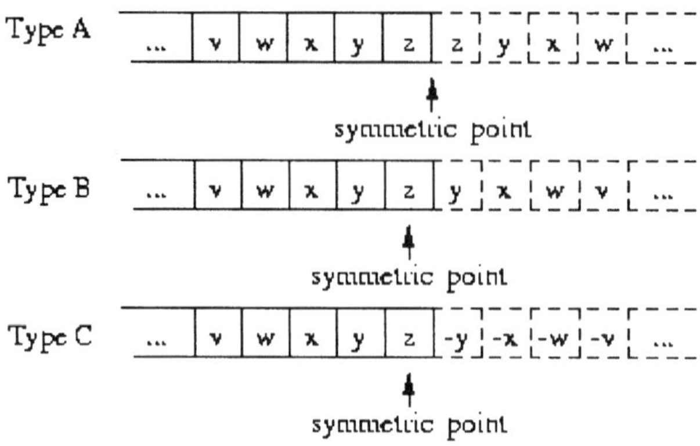

Figure 5 Symmetric extension types at the leading boundary of a signal segment.

where $x_l[i]$ and $x_h[i]$ are the low-pass and high-pass wavelet coefficients, respectively. If the low-pass subsampling positions are even, then $s = 0$. If they are odd, then $s = 1$. Note that subsampling of high-pass coefficients always has one sample advance. To perform synthesis, these coefficients are first upsampled by a factor of 2 as follows:

$$\begin{aligned} L[2i - s] &= x_l[i] \\ L[2i + 1 - s] &= 0 \end{aligned} \tag{6}$$

$$\begin{aligned} H[2i + 1 - s] &= x_h[i] \\ H[2i + s] &= 0 \end{aligned} \tag{7}$$

Again, if the low-pass subsampling positions are even, then $s = 0$. If they are odd, then $s = 1$. Assuming an N-point sequence $\{x[j], j = 0, 1, \ldots, N - 1\}$ and combining symmetric extensions, filtering, subsampling, and upsampling together, the processes of wavelet analysis and synthesis can be summarized as follows:

If N is even, the leading boundary and the trailing boundary of the sequence are extended using the type B extension shown in Figures 5 and 6, respectively. The $N/2$ low-pass wavelet coefficients $x_l[i]$, $i = s, \ldots, N/2 - 1 + s$, are generated by using Eq. (1) and Eq. (4). The $N/2$ high-pass wavelet coefficients $x_h[i]$, $i = 0, 1, \ldots, N/2 - 1$, are generated by using Eq. (2) and Eq. (5). The synthesis process begins with upsampling the low-pass and high-pass wavelet coefficients using Eq. (6) and Eq. (7), respectively. As the results, an upsampled low-pass segment $L[\,]$ and an up-sampled high-pass segment $H[\,]$ are obtained. The upsampled low-pass and high-pass segments are then extended at the leading and trailing boundaries using the type B extension shown in Figures 5 and 6 respectively. The extended low-pass and high-pass segments $L[\,]$ and $H[\,]$ are then synthesized using Eq. (3). Figure 7 illustrates this case with even subsampling for the low-pass wavelet coefficients and odd subsampling for the high-pass wavelet coefficients.

If N is odd, the leading boundary and the trailing boundary of the sequence are also extended using the type B extension shown in Figures 5 and 6, respectively. The $(N + 1)/2 - s$ low-pass wavelet coefficients $x_l[i]$, $i = s, \ldots, (N + 1)/$

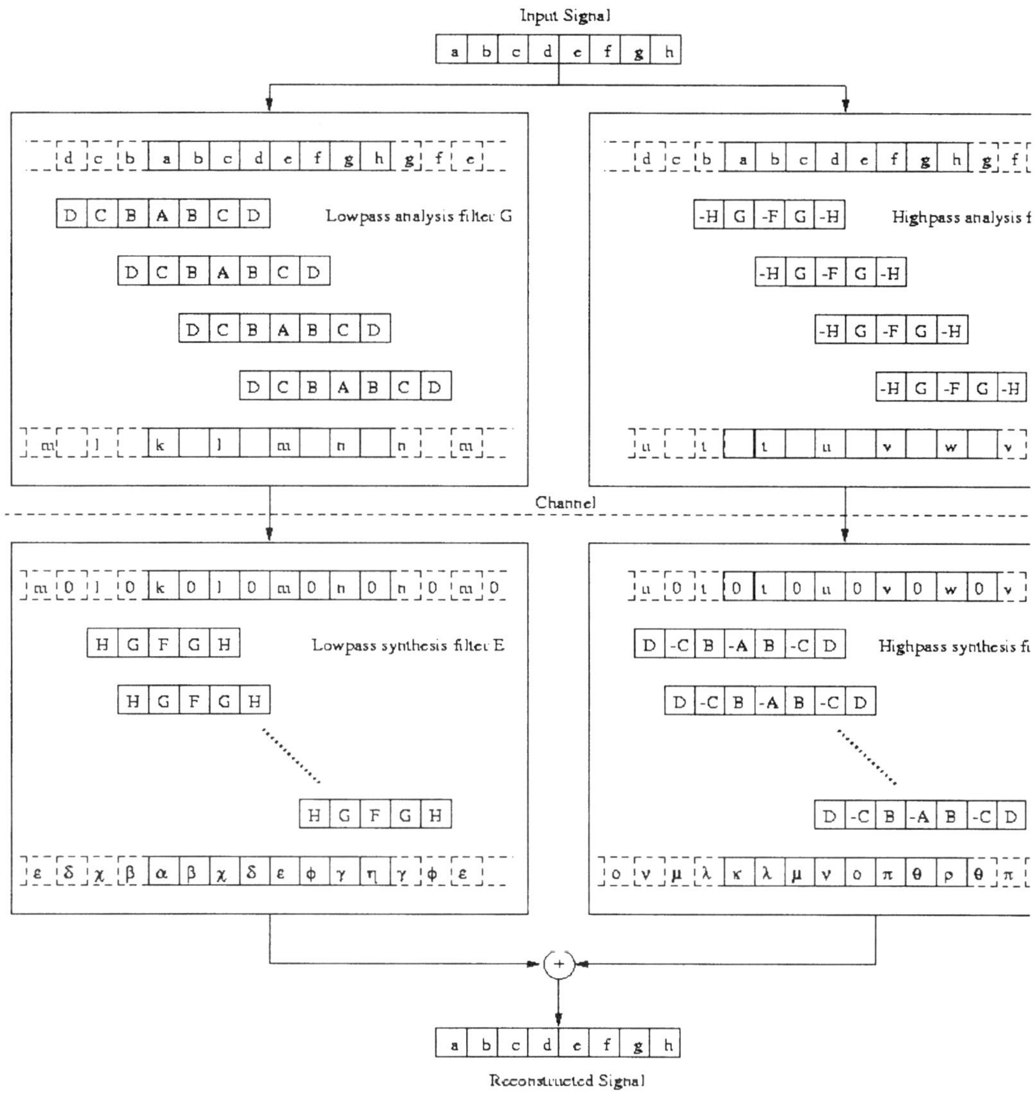

Figure 6 Symmetric extension types at the trailing boundary of a signal segment.

2 − 1, are generated by using Eq. (1) and Eq. (4). The $(N - 1)/2 + s$ high-pass wavelet coefficients $x_h[i]$, $i = 0, \ldots, (N - 1)/2 - 1 + s$, are generated by using Eq. (2) and Eq. (5). The synthesis process is the same as in the even case except that the segment length N is now an odd number. Figure 8 illustrates this case with even subsampling for the low-pass wavelet coefficients and odd subsampling for the high-pass wavelet coefficients.

In the preceding description, upsampling is performed before symmetric extension. Only type B extension is required. The MPEG–4 reference software follows this description and results in a simple implementation.

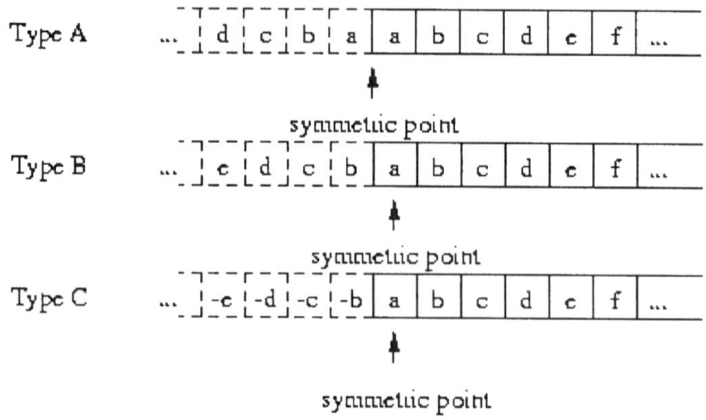

Figure 7 Analysis and synthesis of an even-length signal.

In the preceding description, we used a parameter s to distinguish two different subsampling cases. Actually, for a rectangular-shaped image, s is always 0. However, for an arbitrarily shaped visual object, a subsampling strategy is important. The SA-DWT has two components. One is a way to handle wavelet transforms for image segments of arbitrary length. The other is a subsampling method for image segments of arbitrary length at arbitrary locations. The SA-DWT allows odd-length or small-length image segments to be decomposed into the transform domain in a similar manner to the even- and long-length segments while maintaining the number of coefficients in the transform domain identical to the number of pixels in the image domain. The scale of the transform domain coefficients within each band is the same to avoid sharp changes in subbands. There are two considerations in deciding on a subsampling strategy. One is to preserve the spatial correlation and self-similarity property of the wavelet transform for the arbitrarily shaped image region. Another consideration is the effect of the subsampling strategy on the efficiency of zero-tree coding. The length-adaptive wavelet transform discussed before solves the problem of wavelet decomposition on an arbitrary-length sequence (long or short, even or odd). In the preceding discussion, we purposely leave the subsampling issue open. For each case of the arbitrary-length wavelet transform, we have two options in subsampling the low-pass and high-pass wavelet coefficients, i.e., even subsampling and odd subsampling. Different subsampling strategies have different advantages and disadvantages in terms of coding efficiency. Here, we discuss two possible subsampling strategies.

Subsampling strategy favoring zero-tree coding efficiency: Because zero-tree coding is used for entropy coding, the properties of zero-tree coding should be considered in choosing a proper subsampling strategy. One of the features of zero-tree coding is that, when all the children of a tree node are insignificant or zeros or don't-cares, the coding process does not need to continue for that tree branch from that node on. Therefore, the obvious subsampling strategy to take advantage of this property is always to allocate more valid wavelet coefficients in the lower subbands (close to roots of the wavelet tree) than in the higher subbands and have more don't-care nodes in the higher subbands.

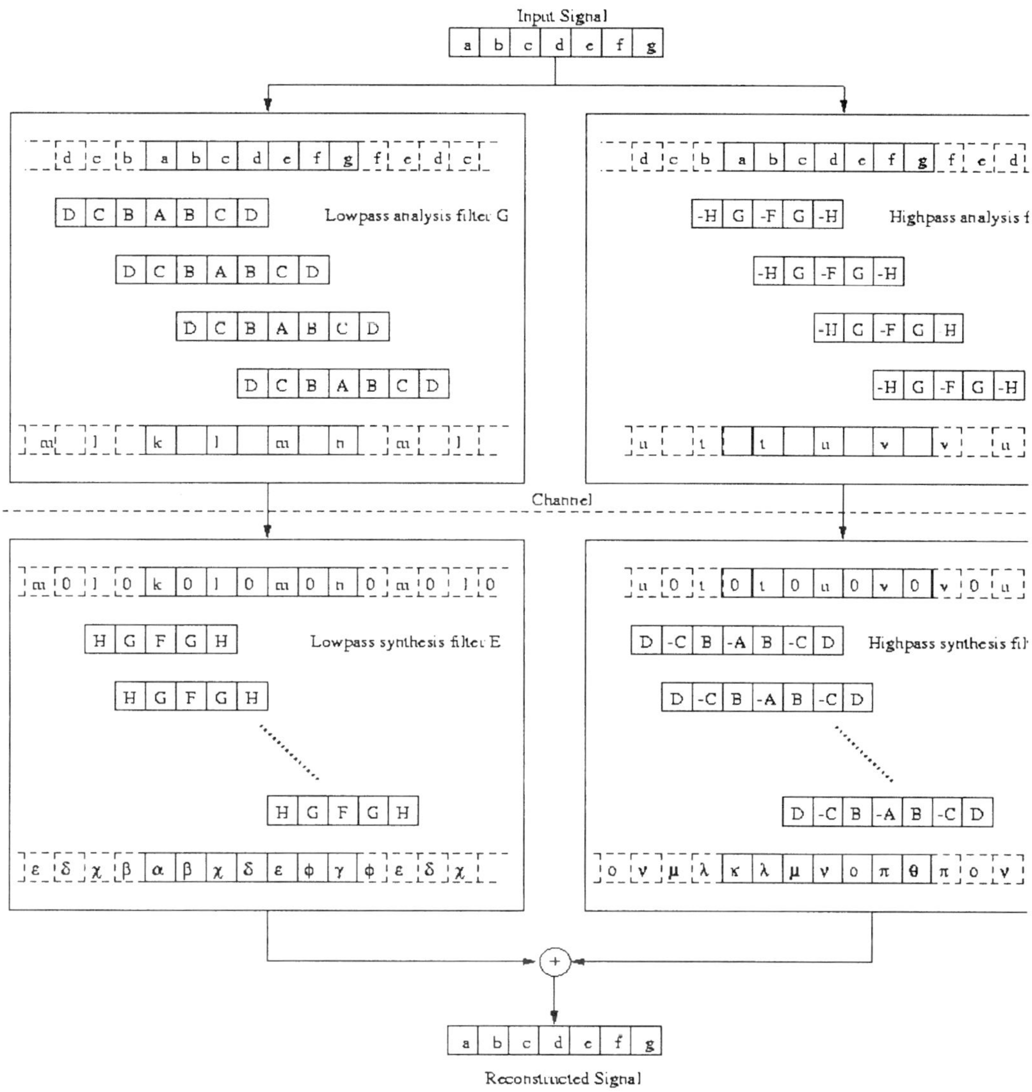

Figure 8 Analysis and synthesis of an odd-length signal.

This is achieved if the low-pass subsampling is locally fixed to even subsampling and the high-pass subsampling is locally fixed to odd subsampling. Because the signal segments in an arbitrarily shaped visual object are neither all starting from odd positions nor all starting from even positions, the phases of some of the low-pass and high-pass wavelet coefficients may be skewed by one sample when subsampling is locally fixed for all signal segments. This is not desired for wavelet decomposition in the second direction. However, because the phases of the subsampled wavelet coefficients differ by at most one sample, the spatial relations across subbands can still be preserved to a certain extent. For very low bit rate coding, zero-tree coding efficiency is more important and this subsampling strategy achieves better overall coding efficiency.

Subsampling strategy favoring signal processing gain: In contrast, another subsampling strategy does not fix even subsampling or odd subsampling locally for all the signal segments. It strictly maintains the spatial relations across the subbands by using either even subsampling or odd subsampling according to the position of a signal segment relative to the bounding box. Instead of fixing subsampling positions for each local signal segment, this strategy fixes subsampling positions of the low-pass and high-pass wavelet coefficients at global even or odd positions relative to the bounding box. Because the start position of each segment in a visual object may not be always at an even or odd position, the local subsampling in each segment has to be adjusted to achieve global even or odd subsampling. For example, we choose local even subsampling in low-pass bands and local odd subsampling in high-pass bands for all segments starting from even positions and choose local odd subsampling in low-pass bands and local even subsampling in high-pass bands for all segments starting from odd positions. This subsampling strategy preserves the spatial correlation across subbands. Therefore, it can achieve more signal processing gain. Its drawback is that it may introduce more high-pass band coefficients than low-pass band coefficients and could potentially degrade zero-tree coding efficiency.

Through extensive experiments, we decided to use the subsampling strategy favoring signal processing gain. For a rectangular region, the preceding two subsampling strategies converge to the same subsampling scheme. Another special case in SA-DWT is when $N = 1$. This isolated sample is repeatedly extended and the low-pass wavelet filter is applied to obtain a single low-pass wavelet coefficient. (Note: this is equivalent to scaling this sample by a factor K that happens to be $\sqrt{2}$ for some normalized biorthogonal wavelets.) The synthesis process simply scales this single low-pass wavelet coefficient by a factor of $1/K$ and puts it in the correct position in the original signal domain.

On the basis of the length-adaptive wavelet transform and the subsampling strategies just discussed, the two-dimensional (2D) SA-DWT for an arbitrarily shaped visual object can be described as follows:

Within the bounding box of the arbitrarily shaped object, use shape information to identify the first row of pixels belonging to the object.

Within each row, identify the first segment of consecutive pixels.

Apply the length-adaptive 1D wavelet transform to this segment with a proper subsampling strategy.

The low-pass wavelet coefficients are placed into the corresponding row in the low-pass band. The high-pass wavelet coefficients are placed into the corresponding row in the high-pass band.

Perform the preceding operations for the next segment of consecutive pixels in the row.

Perform the preceding operations for the next row of pixels.

Perform the preceding operations for each column of the low-pass and high-pass objects.

Perform the preceding operations for the low-low band object until the decomposition level is reached.

This 2D SA-DWT algorithm provides an efficient way to decompose an arbitrarily shaped object into a multiresolution object pyramid. The spatial correlation, locality, and object shape are well preserved throughout the SA-DWT. Thus, it enables multiresolution coding of arbitrarily shaped objects. This method ensures that the number of coefficients to be coded in the transform domain is exactly the same as that in the image domain. The treatment of an odd number of pixels in a segment ensures that there is not much energy in high-pass bands in a pyramid wavelet decomposition. Note that if the object is a rectangular image, the 2D SA-DWT is identical to a standard 2D wavelet transform. A more comprehensive discussion of SA-DWT, including orthogonal and even symmetric biorthogonal wavelets, is given in Ref. 18.

III. FORMATION OF WAVELET TREES

Quantization and entropy coding are applied to the wavelet coefficients resulted from the discrete wavelet transform (DWT). The DWT decomposes the input image into a set of subbands of varying resolutions. The coarsest subband is a low-pass approximation of the original image, and the other subbands are finer scale refinements. In a hierarchical subband system such as that of the wavelet transform, with the exception of the highest frequency subbands, every coefficient at a given scale can be related to a set of coefficients of similar orientation at the next finer scale. The coefficient at the coarse scale is called the parent, and all coefficients at the same spatial location and of similar orientation at the next finer scale are called children.

As an example, Figure 9 shows a wavelet tree resulting from a three-level wavelet

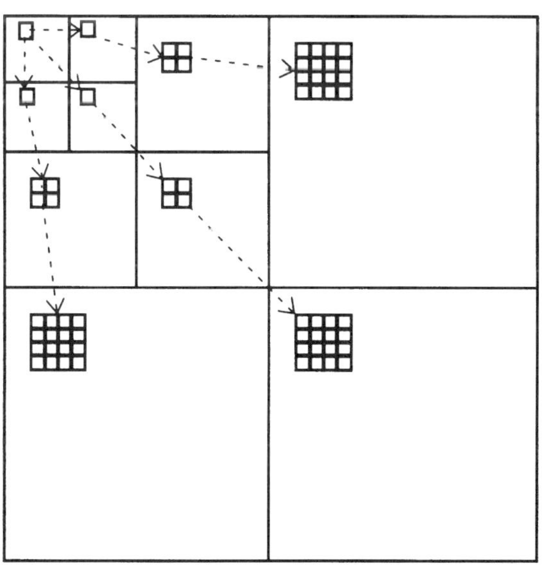

Figure 9 The parent–child relationship of wavelet coefficients.

0	1	4	67	5	6	68	69	9	10	13	14	72	73	76	77
2	3	130	193	7	8	70	71	11	12	15	16	74	75	78	79
25	88	46	109	131	132	194	195	17	18	21	22	80	81	84	85
151	214	172	235	133	134	196	197	19	20	23	24	82	83	86	87
26	27	89	90	47	48	110	111	135	136	139	140	198	199	202	203
28	29	91	92	49	50	112	113	137	138	141	142	200	201	204	205
152	153	215	216	173	174	236	237	143	144	147	148	206	207	210	211
154	155	217	218	175	176	238	239	145	146	149	150	208	209	212	213
30	31	34	35	93	94	97	98	51	52	55	56	114	115	118	119
32	33	36	37	95	96	99	100	53	54	57	58	116	117	120	121
38	39	42	43	101	102	105	106	59	60	63	64	122	123	126	127
40	41	44	45	103	104	107	108	61	62	65	66	124	125	128	129
156	157	160	161	219	220	223	224	177	178	181	182	240	241	244	245
158	159	162	163	221	222	225	226	179	180	183	184	242	243	246	247
164	165	168	169	227	228	231	232	185	186	189	190	248	249	252	253
166	167	170	171	229	230	233	234	187	188	191	192	250	251	254	255

Figure 10 The tree-depth scanning of wavelet trees.

decomposition. For the lowest frequency subband, LL3 in the example, the parent–child relationship is defined such that each parent node has three children, one in each subband at the same scale and spatial location but different orientation. For the other subbands, each parent node has four children in the next finer scale of the same orientation.

Once all the wavelet trees are formed, the next issue is a scanning order for encoding and decoding. In MPEG–4 texture coding, there are two possible scanning orders. One is called tree-depth scan and the other is called band-by-band scan. Figure 10 shows the tree-depth scanning order for a 16×16 image, with three levels of decomposition. The indices 0, 1, 2, and 3 represent the DC band coefficients that are encoded separately. The remaining coefficients are encoded in the order shown in the figure. As an example, indices 4, 5, . . . , 24 represent one tree. At first, coefficients in this tree are encoded starting from index 4 and ending at index 24. Then, the coefficients in the second tree are encoded starting from index 25 and ending at 45. The third tree is encoded starting from index 46 and ending at index 66, and so on.

Figure 11 shows that the wavelet coefficients are scanned in the subband-by-subband fashion, from the lowest to the highest frequency subbands for a 16×16 image with

0	1	4	7	16	17	28	29	64	65	68	69	112	113	116	117
2	3	10	13	18	19	30	31	66	67	70	71	114	115	118	119
5	8	6	9	40	41	52	53	72	73	76	77	120	121	124	125
11	14	12	15	42	43	54	55	74	75	78	79	122	123	126	127
20	21	32	33	24	25	36	37	160	161	164	165	208	209	212	213
22	23	34	35	26	27	38	39	162	163	166	167	210	211	214	215
44	45	56	57	48	49	60	61	168	169	172	173	216	217	220	221
46	47	58	59	50	51	62	63	170	171	174	175	218	219	222	223
80	81	84	85	128	129	132	133	96	97	100	101	144	145	148	149
82	83	86	87	130	131	134	135	98	99	102	103	146	147	150	151
88	89	92	93	136	137	140	141	104	105	108	109	152	153	156	157
90	91	94	95	138	139	142	143	106	107	110	111	154	155	158	159
176	177	180	181	224	225	228	229	192	193	196	197	240	241	244	245
178	179	182	183	226	227	230	231	194	195	198	199	242	243	246	247
184	185	188	189	232	233	236	237	200	201	204	205	248	249	252	253
186	187	190	191	234	235	238	239	202	203	206	207	250	251	254	255

Figure 11 The band-by-band scanning of wavelet trees.

three levels of decomposition. The DC band is located at the upper left corner (with indices 0, 1, 2, 3) and is encoded separately. The remaining coefficients are encoded in the order that is shown in the figure, starting from index 4 and ending at index 255.

The two different scanning orders serve different purposes. For example, the tree-depth scan requires less memory for encoding and decoding. But it requires a bitstream buffer to support progressive transmission. On the other hand, the band-by-band scan can stream out encoded wavelet coefficients without a buffer delay and requires a larger memory for encoding and decoding.

IV. QUANTIZATION

In MPEG–4 texture coding, the DC band is quantized by a uniform midstep quantizer with a dead zone equal to the quantization step size. All the higher bands are quantized by a uniform midstep quantizer with a dead zone of twice the quantization step size. The multiscale quantization scheme provides a very flexible approach to support the appropriate trade-off between layers and types of scalability, complexity, and coding efficiency for a wide range of applications. We describe some details of this scheme in this section.

The wavelet coefficients of the first spatial (and/or quality) layer are quantized with the quantizer Q0. These quantized coefficients are entropy coded and the output of the entropy coder at this level, BS0, is the first portion of the bitstream. The quantized wavelet coefficients of the first layer are also reconstructed and subtracted from the original wavelet coefficients. These residual wavelet coefficients are quantized with Q1 and entropy coded. The output of this stage, BS1, is the second portion of the output bitstream. The quantized coefficients of the second stage are also reconstructed and subtracted from the original coefficients. *N* stages of the scheme provide *N* layers of scalability. Each level represents one layer of SNR quality, spatial scalability, or combination of both.

In this quantization scheme, the wavelet coefficients are quantized by a uniform and midstep quantizer with a dead zone equal to the quantization step size as closely as possible at each scalability layer. Each quality layer and/or spatial layer has a quantization value (*Q value*) associated with it. Each spatial layer has a corresponding sequence of these *Q* values. The quantization of coefficients is performed in three steps: (1) construction of the initial quantization value sequence from input parameters, (2) revision of the quantization sequence, and (3) quantization of the coefficients.

Let n be the total number of spatial layers and $k(i)$ be the number of quality layers associated with spatial layer i. We define the total number of scalability layers associated with spatial layer i, $L(i)$, as the sum of all the quality layers from that spatial layer and all higher spatial layers:

$$L(i) = k(i) + k(i + 1) + \cdots + k(n)$$

Let $Q(m, n)$ be the Q value corresponding to spatial layer m and quality layer n. The quantization sequence (or Q sequence) associated with spatial layer i is defined as the sequence of Q values from all the quality layers from the ith spatial layer and all higher spatial layers ordered by increasing quality layer and then increasing spatial layer:

$$\begin{aligned}\mathbf{Q\_i} &= [Q\_i(0), Q\_i(1), \ldots, Q\_i(\mathrm{m})] \\ &= [Q(i, 1), Q(i, 2), \ldots, Q(i, k(i)), Q(i + 1, 1), \\ &\quad Q(i + 1, 2), \ldots, Q(i + 1, k(i + 1)), \ldots, \\ &\quad Q(n, 1), Q(n, 2), \ldots, Q(n, k(n))]\end{aligned}$$

The sequence **Q_i** represents the procedure for successive refinement of the wavelet coefficients that are first quantized in the spatial layer i. In order to make this successive refinement efficient, the sequence **Q_i** is revised before starting the quantization. Let $Q\_i(j)$ denote the jth value of the quantization sequence **Q_i**. Consider the case in which $Q\_i(j) = p\, Q\_i(j + 1)$. If p is an integer number greater than one, each quantized coefficient of layer j is efficiently refined at layer $(j = 1)$ as each quantization step size $Q\_i(j)$ is further divided into p equal partitions in layer $(j + 1)$. If p is greater than one but not an integer, the partitioning of layer $j + 1$ will not be uniform. This is due to the fact that $Q\_i(j)$ corresponds to quantization levels, which cover $Q\_i(j)$ possible coefficient values that cannot be evenly divided into $Q\_i(j + 1)$ partitions. In this case, $Q\_i(j + 1)$ is revised to be as close to an integer factor of $Q\_i(j)$ as possible. The last case is $Q\_i(j + 1) >= Q\_i(j)$. In this case, no further refinement can be obtained at the $(j + 1)$st scalability layer over the jth layer, so we simply revised $Q\_i(j + 1)$ to be $Q\_i(j)$. The revised

Table 1 Revision of the Quantization Sequence

Condition on $p = Q\_i(j)/Q\_i(j+1)$	Revision procedure
$p < 1.5$	$QR\_i(j+1) = Q\_i(j)$ (no quantization at layer $j+1$)
$p \geq 1.5$ p is integer	$QR\_i(j+1) = Q\_i(j+1)$ (no revision)
$p \geq 1.5$ p is noninteger	$q = \text{round}\,(Q\_i(j)/Q\_i(j+1))$ $QR\_i(j+1) = \text{ceil}\,(Q\_i(j)/q)$

quantization sequence is referred as **QR_i**. Table 1 summarizes the revision procedure. We then categorize the wavelet coefficients in terms of the order of spatial layers as follows:

$$S(i) = \{\text{all coefficients that } \textit{first} \text{ appear in spatial layer } i\}$$

and $$T(i) = \{\text{all coefficients that appear in spatial layer } i\}.$$

Once a coefficient appears in a spatial layer, it appears in all higher spatial layer, and we have the relationship:

$$T(1) \subset T(2) \subset \cdots \subset T(n-1) \subset T(n)$$

To quantize each coefficient in $S(i)$ we use the Q values in the revised quantization sequence, **QR_i**. These Q values are positive integers and they represent the range of values a quantization level spans at that scalability layer. For the initial quantization we simply divide the value by the Q value for the first scalability layer. This gives us our initial quantization level (note that it also gives us a double-sized dead zone). For successive scalability layers we need only send the information that represents the refinement of the quantizer. The refinement information values are called residuals and are the indexes of the new quantization level within the old level where the original coefficient values are.

We then partition the inverse range of the quantized value from the previous scalability layer in such a way that the partitions are as uniform as possible based on the previously calculated number of refinement levels, m. This partitioning always leaves a discrepancy of zero between the partition sizes if the previous Q value is evenly divisible by the current Q value (e.g., previous $Q = 25$ and current $Q = 5$). If the previous Q value is not evenly divisible by the current Q value (e.g., previous $Q = 25$ and current $Q = 10$) then we have a maximum discrepancy of 1 between partitions. The larger partitions are always the ones closer to zero. We then number the partitions. The residual index is simply the number of the partition in which the original value (which is not quantized) actually lies. We have the following two cases for this numbering:

Case I: If the previous quality level is quantized to zero (that is, the value was in the dead zone), then the residual has to be one of the $2m - 1$ values in $\{-m, \ldots, 0, \ldots, +m\}$.

Case II: If the previous quality level is quantized to a nonzero value, then (since the sign is already known at the inverse quantizer) the residual has to be one of the m values in $\{0, \ldots, m-1\}$.

The restriction of the possible values of the residuals is based solely on the relationship between successive quantization values and whether the value was quantized to zero in the last scalability pass (both of these facts are known at the decoder). This is one reason why using two probability models (one for the first case and one for the second case) increases coding efficiency.

For the inverse quantization, we map the quantization level (at the current quality layer) to the midpoint of its inverse range. Thus we get a maximum quantization error of one-half the inverse range of the quantization level we dequantize to. One can reconstruct the quantization levels given the list of Q values (associated with each quality layer), the initial quantization value, and the residuals.

V. ENTROPY CODING

Entropy coding in MPEG–4 texture coding is based on zero-tree wavelet coding, which is a proven technique for efficiently coding wavelet transform coefficients. Besides superior compression performance, the advantages of zero-tree wavelet coding include simplicity, embedded bitstream structure, scalability, and precise bit rate control. Zero-tree wavelet coding is based on three key ideas: (1) wavelet transform for decorrelation, (2) exploiting the self-similarity inherent in the wavelet transform to predict the location of significant information across scales, and (3) universal lossless data compression using adaptive arithmetic coding. In MPEG–4 texture coding, the DC band is coded separately from the AC bands. We discuss the coding method for the DC band first. Then we give a brief description of embedded zero-tree wavelet (EZW) coding [4], followed by a description of predictive EZW (PEZW) coding [7] that is used in the bilevel mode. Then, we describe the zero-tree entropy (ZTE) coding technique [5,6], which provides spatial scalability and is used in the single quantizer mode. Finally, we describe the most general technique, known as multiscale zero-tree entropy (MZTE) coding [8], which provides a flexible framework for encoding images with an arbitrary number of spatial or quality scalability levels.

The wavelet coefficients of the DC band are encoded independently of the other bands. As shown in Figure 12, the current coefficient X is adaptively predicted from three other quantized coefficients in its neighborhood, i.e, A, B, and C, and the predicted value is subtracted from the current coefficient as follows:

$$\text{if} \quad (|A - B| < |B - C|)$$
$$\hat{w} = C$$

else

$$\hat{w} = A$$
$$X = X - \hat{w}$$

If any of the neighbors, A, B, or C, is not in the image, its value is set to zero for the purpose of the prediction. For SA-DWT coding, there are some don't-care values in the DC band and they are not coded. For prediction of other wavelet coefficients in the DC band, the don't-care values are considered to be zeros.

In the bitstream, the quantization step size is first encoded; then the magnitude of the minimum value of the differential quantization indices "band_offset" and the maximum value of the differential quantization indices "band_max_value" are encoded into

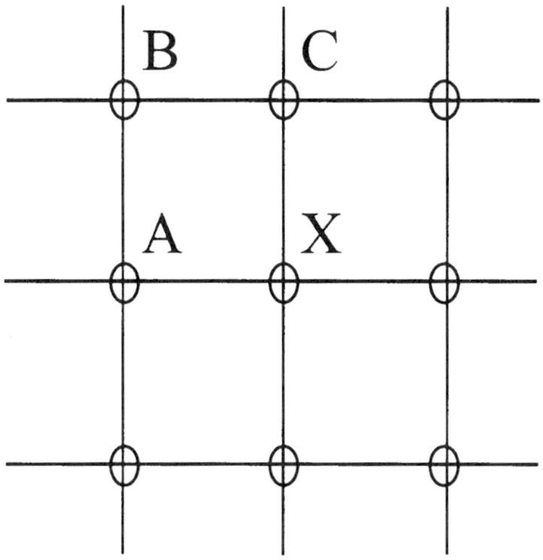

Figure 12 Adaptive predictive coding of the DC coefficients.

bitstream. The parameter ''band_offset'' is a negative integer or zero and the parameter ''band_max_value'' is a positive integer. Therefore, only the magnitudes of these parameters are coded into the bitstream. The differential quantization indices are coded using the arithmetic coder in a raster scan order, starting from the upper left index and ending at the lower right one. The model is updated with encoding a bit of the predicted quantization index to adapt to the statistics of the DC band.

EZW scans wavelet coefficients subband by subband. Parents are scanned before any of their children. Each coefficient is compared against the current threshold T. A coefficient is significant if its amplitude is greater than T. Such a coefficient is then encoded using one of the symbols *negative significant (NS)* or *positive significant (PS)*. The *zero-tree root (ZTR)* symbol is used to signify a coefficient below T with all its children also below T. The *isolated zero (IZ)* symbol signifies a coefficient below T but with at least one child not below T. For significant coefficients, EZW further encodes coefficient values using a successive approximation quantization scheme. Coding is done by bit planes and leads to the embedded nature.

Predictive EZW coding introduces several modifications that significantly improve the original EZW coder. The major improvements are as follows:

New zero-tree symbols such as VZTR, valued zero-tree root, are introduced.
Adaptive context models are used for encoding the zero-tree symbols.
The zero trees are encoded depth first and all bit planes of one zero tree are encoded before moving to the next zero tree. This significantly reduces the complexity requirement of the zero-tree coder.

The zero-tree symbols in PEZW are listed as follows:

ZTR (zero-tree root): the wavelet coefficient is zero for a given threshold and all descendants are zero.
VZTR (valued zero-tree root): the wavelet coefficient itself is nonzero but all its descendants are zero.

IZ (isolated zero): the wavelet coefficient itself is zero but not all its descendants are zero.

VAL (isolated nonzero value): the wavelet coefficient is nonzero and not all its descendants are zero.

The zero-tree symbols are encoded with context-based adaptive arithmetic coding. To reduce complexity, we do not use high-order contexts. Instead, when encoding the zero-tree symbol of a certain coefficient, we use the zero-tree symbol of the same coefficient in the *previous* bit plane as the context for the arithmetic coding. For each coefficient, the number of context models is five and only one memory access is needed to form the context. By using these simple contexts, we significantly reduced the memory requirement for storing the context models and the number of memory accesses needed to form the contexts. Our experience is that by using previous zero-tree symbols as context, we are able to capture the majority of the redundancy and not much can be gained by adding more contexts. Table 2 shows the contexts and the symbols for PEZW coding.

In addition to the zero-tree symbols, the state DZTR is used as a context. It means that the coefficient is a descendant of ZTR or VZTR in a previous bit plane. The symbol SKIP means that the encoder or decoder will skip this coefficient from now on because it is already nonzero, and no additional zero-tree symbol need to be sent. Only refinement bits need to be coded. Each scale or decomposition level and each bit plane have their context models. The arithmetic coder is initialized at the beginning of each bit plane and subband. The initialization eliminates dependences of the context models and arithmetic coders across scales and bit planes, a very important property for good error resilience performance. In addition, initialization can be done at any locations that are at the boundary of a zero tree and a resynchronization marker can be inserted, so that additional protection can be injected for a selected area of an image.

The ZTE coding is based on, but differs significantly from, EZW coding. Similar to EZW, ZTE coding exploits the self-similarity inherent in the wavelet transform of images to predict the location of information across wavelet scales. Although ZTE does not produce a fully embedded bitstream as does EZW, it gains flexibility and other advantages over EZW coding, including substantial improvement in coding efficiency, simplicity, and spatial scalability. ZTE coding is performed by assigning a zero-tree symbol to a coefficient and then coding the coefficient value with its symbol in one of the two different scanning orders described in Sec. IV. The four zero-tree symbols used in ZTE are also zero-tree root (ZTR), valued zero-tree root (VZTR), value (VAL), and isolated zero (IZ). The zero-tree symbols and quantized coefficients are then losslessly encoded

Table 2 Context Models for PEZW Coding

Context in previous bit plane	Symbols in current bit plane
ZTR	ZTR, VZTR, IZ, VAL
VZTR	SKIP
IZ	IZ, VAL
VAL	SKIP
DZTR	ZTR, VTRZ, IZ, VAL

using an adaptive arithmetic coder with a given symbol alphabet. The arithmetic encoder adaptively tracks the statistics of the zero-tree symbols and encoded values using three models: (1) *type* to encode the zero-tree symbols, (2) *magnitude* to encode the values in a bit-plane fashion, and (3) *sign* to encode the sign of the value. For each coefficient its zero-tree symbol is encoded first and, if necessary, its value is then encoded. The value is encoded in two steps. First, its absolute value is encoded in a bit-plane fashion using the appropriate probability model and then the sign is encoded using a binary probability model with "0" meaning positive and "1" meaning negative sign.

The multiscale zero-tree entropy (MZTE) coding technique is based on ZTE coding but it utilizes a new framework to improve and extend ZTE coding to a fully scalable yet very efficient coding technique. At the first scalability layer, the zero-tree symbols are generated in the same way as in ZTE coding and coded with the nonzero wavelet coefficients of that scalability layer. For the next scalability layer, the zero-tree map is updated along with the corresponding value refinements. In each scalability layer, a new zero-tree symbol is coded for a coefficient only if it was coded as ZTR or IZ in the previous scalability layer. If the coefficient was coded as VZTR or VAL in the previous layer, only its refinement value is coded in the current layer. An additional probability model, *residual*, is used for encoding the refinements of the coefficients that are coded with a VAL or VZTR symbol in any previous scalability layers. The residual model, just as the other probability models, is also initialized to the uniform probability distribution at the beginning of each scalability layer. The numbers of bins for the residual model is calculated on the basis of the ratio of the quantization step sizes of the current and previous scalability. When a residual model is used, only the magnitude of the refinement is encoded as these values are always zero or positive integers.

Furthermore, to utilize the highly correlated zero-tree symbols between scalability layers, context modeling, based on the zero-tree symbol of the coefficient in the previous scalability layer in MZTE, is used to better estimate the distribution of zero-tree symbols. In MZTE, only INIT and LEAF_INIT are used for the first scalability layer for the nonleaf subbands and the leaf subbands, respectively. Subsequent scalability layers in the MZTE use the context associated with the symbols. The different zero-tree symbol models and their possible values are summarized in Table 3. If a spatial layer is added, then the contexts of all previous leaf subband coefficients are switched into the corresponding nonleaf contexts. The coefficients in the newly added subbands use the LEAF_INIT context initially.

Table 3 Contexts and Symbols for MZTE Coding

Context for nonleaf subbands	Possible symbols
INIT	ZTR(2), IZ(0), VZTR(3), VAL(1)
ZTR	ZTR(2), IZ(0), VZTR(3), VAL(1)
ZTR DESCENDENT	ZTR(2), IZ(0), VZTR(3), VAL(1)
IZ	IZ(0), VAL(1)
Context for leaf subbands	**Possible symbols**
LEAF_INIT	ZTR(0), VZTR(1)
LEAF_ZTR	ZTR(0), VZTR(1)
LEAF_ZTR_DESCENDENT	ZTR(0), VZTR(1)

In the SA-DWT decomposition, the shape mask is decomposed into a pyramid of subbands in the same way as the SA-DWT so that we know which wavelet tree nodes have valid wavelet coefficients and which ones have don't-care values. We have to pay attention to the means of coding the multiresolution arbitrarily shaped objects with these don't-care values (corresponding to the out-of-boundary pixels or out-nodes). We discuss how to extend the conventional zero-tree coding method to the shape-adaptive case.

As discussed previously, the SA-DWT decomposes the arbitrarily shaped objects in the image domain to a hierarchical structure with a set of subbands of varying resolutions. Each subband has a corresponding shape mask associated with it to specify the locations of the valid coefficients in that subband. There are three types of nodes in a tree: zeros, nonzeros, and out-nodes (with don't-care values). The task is to extend the zero-tree coding method to the case with out-nodes. A simple way is to set those don't-care values to zeros and then apply the zero-tree coding method. However, this requires bits to code the out-nodes such as a don't-care tree (the parent and all of its children have don't-care values). This is a waste of bits because out-nodes do not need to be coded as the shape mask already indicates their status. Therefore, we should treat out-nodes differently from zeros. Although we do not want to use bits to code an out-node, we have to decide what to do with its children nodes. One way is not to code any information about the status of the children nodes of the don't-care node. That is, we always assume that it has four children to be examined further. When the decoder scans to this node, it will be informed by the shape information that this node is a don't-care node and it will continue to scan its four children nodes. In this way, all the don't-care nodes in a tree structure need not be coded. This approach performs well when there are only sparse valid nodes in a tree structure. One disadvantage of this approach is that, even if a don't-care node has four zero-tree root children, it still needs to code four zero-tree root symbols instead of one zero-tree root symbol if the don't-care value is treated as a zero. Another way is to treat an out-node selectively as a zero. This is equivalent to creating another symbol for coding some don't-care values. Through extensive experiments, we decided to use the method of not coding out-nodes. The extension of zero-tree coding to handle the SA-DWT coefficients is then given as follows.

At the root layer of the wavelet tree (the top three AC bands), the shape information is examined to determine whether a node is an out-node.

- If it is an out-node, no bits are used for this node and the four children nodes of this node are marked "to be encoded" (TBE) in encoding or "to be decoded" (TBD) in decoding.
- Otherwise, a symbol is encoded or decoded for this node using an adaptive arithmetic encoder/decoder.
  - If the symbol is either isolated zero (IZ) or value (VAL), the four children nodes of this node are marked TBE/TBD; otherwise, the symbol is either zero-tree root (ZTR) or valued zero-tree root (VZTR) and the four children nodes of this node are marked "no code" (NC).
  - If the symbol is VAL or VZTR, a nonzero wavelet coefficient is encoded or decoded for this node; otherwise, the symbol is either IZ or ZTR and the wavelet coefficient is set to zero for this node.

At any layer between the root layer and the leaf layer, the shape information is examined to determine whether a node is an out-node.

If it is an out-node, no bits are used for this node and the four children nodes of this node are marked as either TBE/TBD or NC depending on whether this node itself is marked TBE/TBD or NC, respectively. Otherwise,

If it is marked NC, no bits are used for this node and the wavelet coefficient is zero for this node and the four children nodes are marked NC.

Otherwise, a symbol is encoded or decoded for this node using an adaptive arithmetic encoder or decoder.

If the symbol is either isolated zero (IZ) or value (VAL), the four children nodes of this node are marked TBE/TBD; otherwise, the symbol is either zero-tree root (ZTR) or valued zero-tree root (VZTR) and the four nodes of this node are marked "no code" (NC).

If the symbol is VAL or VZTR, a nonzero wavelet coefficient is encoded or decoded for this node using an arithmetic encoder or decoder; otherwise, the symbol is either IZ or ZTR and the wavelet coefficient is zero for this node.

At the leaf layer, the shape information is examined to determine whether a node is an out-node.

If it is an out-node, no bits are used for this node.

Otherwise,

If it is marked NC, no bits are used for this node and the wavelet coefficient is zero for this node.

Otherwise, a wavelet coefficient is encoded or decoded for this node using an adaptive arithmetic encoder or decoder.

The same procedure is also used for coding a rectangular image with an arbitrary size if the wavelet analysis results in incomplete wavelet trees.

VI. TEST RESULTS

The MPEG–4 texture coding technique has been extensively tested and refined through the MPEG–4 core experiment process, under the leadership of Sarnoff Corporation in close collaboration with several partners, such as Sharp Corporation, Texas Instruments, Vector Vision, Lehigh University, OKI Electric Industry Co., Rockwell, and Sony Corporation. This section presents a small subset of the test results.

The images in Figures 13 and 14 were obtained by JPEG and MZTE compression schemes, respectively, at the same compression ratio of 45:1. The results show that the MZTE scheme generates much better image quality with good preservation of fine texture regions and absence of the blocking effect, compared with JPEG. The peak signal-to-noise ratio (PSNR) values for both reconstructed images are tabulated in Table 4.

Figure 15 demonstrates the spatial and quality scalabilities at different resolutions and bit rates using the MZTE compression scheme. The top two images of size of 128 × 128 are reconstructed by decoding the MZTE bitstream at bit rates of 80 and 144 kbits, respectively. The middle two reconstructed images are of size of 256 × 256 at bit rates of 192 and 320 kbits, respectively, and the final resolution of 512 × 512 at 750 kbits is shown on the bottom.

Figure 13 Result of JPEG compression.

Figure 14 Result of MZTE coding in MPEG–4 texture coding.

Table 4 Comparison of Wavelet MZTE Coding and JPEG

Compression Scheme	PSNR-Y	PSNR-U	PSNR-Y
DCT-based JPEG	28.36	34.74	34.98
Wavelet-based MZTE	30.98	41.68	40.14

Extensive experiments have also been conducted on the SA-DWT coding technique. The results have been compared with those for SA-DCT coding. The object shape is coded using the MPGE–4 shape coding tool. The test results are presented in the form of PSNR–bit rate curves, and the shape bits are excluded from the bit rate because they are independent of the texture coding scheme. Only the texture bit rates are used for comparison. The bit rate (in bits per pixel) is calculated on the basis of the number of pixels in an object with the reconstructed shape and the PSNR value is also calculated over the pixels in the reconstructed shape. Figure 16 presents the PSNR–bit rate curves. Clearly, SA-

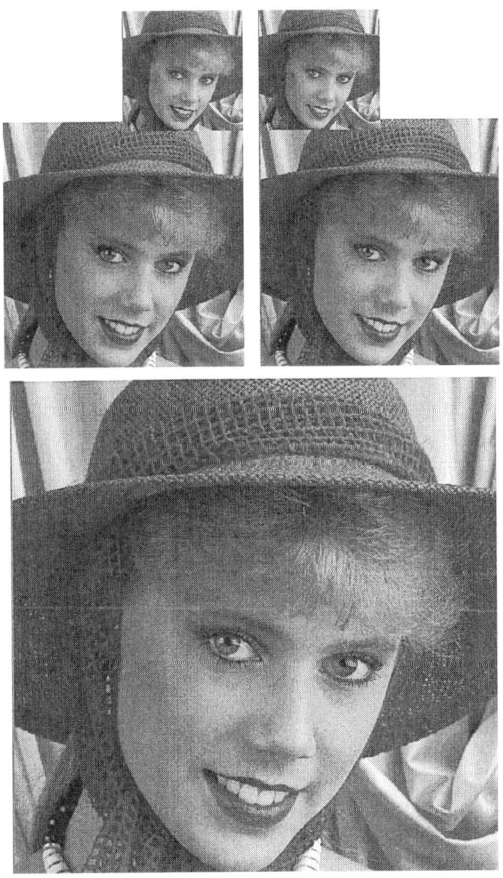

Figure 15 An example of scalability using MZTE.

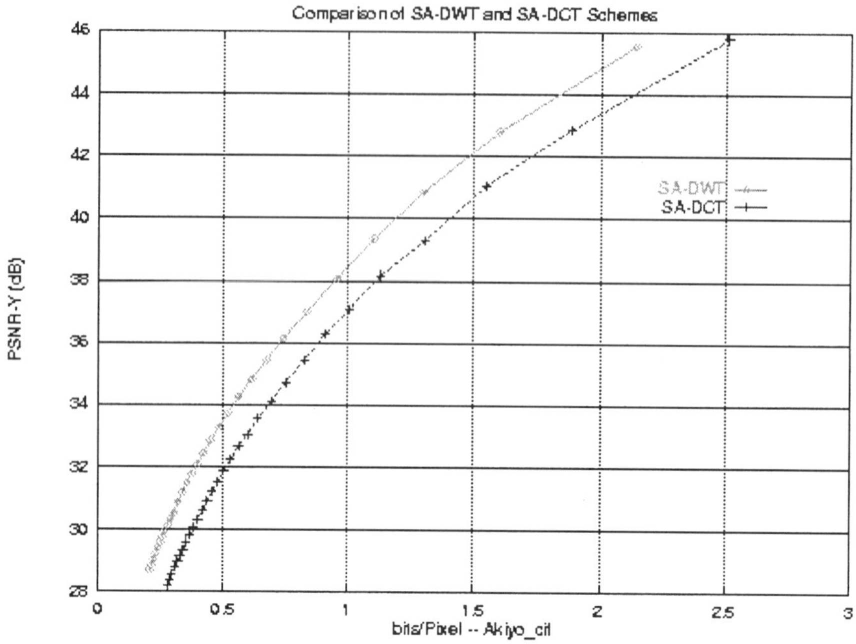

Figure 16 Comparison of SA-DWT coding with SA-DCT coding.

DWT coding achieves better coding efficiency than SA-DCT coding, which is about 1.5–2 dB lower than SA-DWT coding. Figures 17 and 18 show the reconstructed objects from SA-DCT coding and SA-DWT coding, respectively.

In summary, SA-DWT coding is the most efficient technique for coding arbitrarily shaped visual objects. Compared with SA-DCT coding, SA-DWT coding provides higher PSNR values and visibly better quality for all test sequences at all bit rate levels (in most cases, with a smaller number of total bits too).

Figure 17 Reconstructed object at 1.0042 bpp using SA-DCT coding (PSNR-Y = 37.09 dB; PSNR-U = 42.14 dB; PSNR-V = 42.36 dB).

Figure 18 Reconstructed object at 0.9538 bpp using SA-DWT coding (PSNR-Y = 38.06 dB; PSNR-U = 43.43 dB; PSNR-V = 43.25 dB).

VII. SUMMARY

In this chapter, MPEG–4 texture coding is discussed. Spatial and quality scalabilities are two important features desired in many multimedia applications. We have presented three zero-tree wavelet algorithms, which provide high coding efficiency as well as scalability of the compressed bitstreams. PEZW is an improvement on the original EZW algorithm that provides fine granularity quality scalability. Zero-tree entropy (ZTE) coding was demonstrated with high compression efficiency and spatial scalability. In the ZTE algorithm, quantization is explicit, coefficient scanning is performed in one pass, and tree symbol representation is optimized. The multiscale zero-tree entropy (MZTE) coding technique combines the advantages of EZW and ZTE and provides both high compression efficiency and fine granularity scalabilities in both spatial and quality domains. Extensive experimental results have shown that the scalability of MPEG–4 texture coding is achieved without losing coding efficiency. We also present a description of SA-DWT coding for arbitrarily shaped visual objects. The number of wavelet coefficients after the SA-DWT is identical to the number of pixels in the arbitrarily shaped visual object. The spatial correlation and wavelet transform properties, such as the locality property and self-similarity across subbands, are well preserved in the SA-DWT. For a rectangular region, the SA-DWT becomes identical to a conventional wavelet transform. The subsampling strategies for the SA-DWT coefficients are discussed. An efficient method for extending the zero-tree coding technique to coding the SA-DWT coefficients with don't-care values is presented. Extensive experiment results have shown that the shape-adaptive wavelet coding technique consistently performs better than SA-DCT coding and other wavelet-based schemes. In the JPEG-2000 November, 1997 evaluation, MPEG–4 texture coding was also rated as one of the top five schemes in terms of compression efficiency among 27 submitted proposals.

ACKNOWLEDGMENTS

The authors would like to thank Dr. Zhixiong Wu of OKI Electric Industry Co., Dr. Hongqiao Sun of Vector Vision, Inc., Dr. Hung-Ju Lee, Mr. Paul Hatrack, Dr. Bing-Bing Chai of Sarnoff Corporation, Mr. H. Katata, Dr. N. Ito, and Mr. Kusao of Sharp Corporation for their contributions to the implementation of and experiments on MPEG–4 wavelet texture coding during the MPEG–4 process.

REFERENCES

1. Y-Q Zhang, F Pereira, T Sikora, C Reader, eds. Special Issue on MPEG 4. IEEE Trans Circuits Syst Video Technol 7(1):1–4, 1997.
2. ISO/IEC JTC1/SC29/WG11 W1886. MPEG–4 requirements document v.5. Fribourg, October 1997.
3. ISO/IEC JTC1/SC29/WG11. Information technology—Generic coding of audiovisual objects, Part 2: Visual. ISO/IEC 14496-2, Final Draft International Standard, December 1998.
4. JM Shapiro. Embedded image coding using zerotrees of wavelet coefficients. IEEE Trans Signal Process 41:3445–3462, 1993.
5. SA Martucci, I Sodagar, T Chiang, Y-Q Zhang. A zerotree wavelet video coder. IEEE Trans Circuits Syst Video Technol 7(1):109–118, 1997.
6. SA Martucci, I Sodagar. Zerotree entropy coding of wavelet coefficients for very low bit rate video. IEEE International Conference on Image Processing, vol 2, September 1996.
7. J Liang. Highly scalable image coding for multimedia applications. ACM Multimedia, Seattle, October 1997.
8. I Sodagar, H Lee, P Hatrack, Y-Q Zhang. Scalable wavelet coding for synthetic natural and hybrid images. IEEE Trans Circuits Syst Video Technol, March 1999, pp 244–254.
9. T Sikora, B Makai. Low complex shape-adaptive DCT for generic and functional coding of segmented video. Proceedings of Workshop on Image Analysis and Image Coding, Berlin, November 1993.
10. T Sikora, B Makai. Shape-adaptive DCT for generic coding of video. IEEE Trans Circuits Syst Video Technol 5(1):59–62, 1995.
11. T Sikora, S Bauer, B Makai. Efficiency of shape-adaptive 2-D transforms for coding of arbitrarily shaped image segments. IEEE Trans Circuits Syst Video Technol 5(3):254–258, 1995.
12. ISO/IEC JTC1/SC29/WG11. MPEG–4 video verification model version 7.0, N1642, April 1997.
13. M Bi, SH Ong, YH Ang. Comment on 'Shape-adaptive DCT for generic coding of video.' IEEE Trans Circuits Syst Video Technol 6(6):686–688, 1996.
14. P Kauff, K Schuur. Shape-adaptive DCT with block-based DC separation and delta DC correction. IEEE Trans Circuits Syst Video Technol 8(3):237–242, 1998.
15. S Li, W Li, F Ling, H Sun, JP Wus. Shape adaptive vector wavelet coding of arbitrarily shaped texture. ISO/IEC JTC/SC29/WG11, m1027, June 1996.
16. S Li, W Li. Shape adaptive discrete wavelet transform for coding arbitrarily shaped texture. Proceedings of SPIE—VCIP'97, vol 3024, San Jose, February 12–14, 1997.
17. S Li, W Li, H Sun, Z Wu. Shape adaptive wavelet coding. Proceedings of IEEE ISCAS, vol 5, Monterey, CA, May 1998, pp 281–284.
18. S Li, W Li. Shape adaptive discrete wavelet transform for arbitrarily shaped visual object coding. IEEE Trans Circuits Syst Video Technol, Submitted.
19. M Vetterli, J Kovacevic. Wavelets and Subband Coding. Englewood Cliffs, NJ: Prentice-Hall, 1995.
20. A Arkansi, RA Haddad. Multiresolution Signal Decomposition: Transforms, Subbands, Wavelets. San Diego: Academic Press, 1996.
21. A Said, W Pearlman. A new, fast, and efficient image codec based on set partitioning in hierarchical trees. IEEE Trans Circuits Syst Video Technol 6(3):243–250, 1996.
22. D Taubman, A Zakhor. Multirate 3-D subband coding of video. IEEE Trans Image Process 3(5):572–588, 1994.

23. Z Xiong, K Ramchandran, MT Orchard. Joint optimization of scalar and tree-structured quantization of wavelet image decomposition. Proceedings of 27th Annual Asilomar Conference on Signals, Systems, and Computers, Pacific Grove, CA, November 1993, pp 891–895.
24. Y-Q Zhang, S Zafar. Motion-compensated wavelet transform coding for color compression. IEEE Trans Circuits Syst Video Technol 2(3):285–296, 1992.
25. Z Xiong, K Ramachandran, M Orchard, Y-Q Zhang. A comparison study of DCT and wavelet-based coding. ISCAS'97; IEEE Trans Circuits Syst Video Technol 1999.
26. I Witten, R Neal, J Cleary. Arithmetic coding for data compression. Commun ACM 30:520–540, 1987.
27. A Zandi, et al. CREW: Compression with reversable embedded wavelets. Data Compression Conference, March 1995.

11
MPEG-4 Synthetic Video

Peter van Beek
Sharp Laboratories of America, Camas, Washington

Eric Petajan
Lucent Technologies, Murray Hill, New Jersey

Joern Ostermann
AT&T Labs, Red Bank, New Jersey

I. INTRODUCTION

Video compression has given us the ability to transmit standard definition video using about 4 Mbit/sec. Simultaneously, the Internet coupled with public switched telephone network modems and integrated services digital network (ISDN) provide global point-to-point data transmission at a low cost but at a reliable rate of only tens of kilobits per second. Unfortunately, compression of combined audio and video for Internet applications results in low quality and limited practical application. An alternative approach to low-bit-rate visual communication is to transmit articulated object models [1,2] and animation parameters that are rendered to video in the terminal.

This chapter addresses both two-dimensional (2D) object animation and three-dimensional (3D) face animation and the coding of hybrid *synthetic–natural* objects in the context of MPEG-4. We describe the coding of two-dimensional (2D) moving-mesh models, which can be used for animation of both natural video objects and synthetic objects. This mesh-based 2D object representation allows unified coding and manipulation of and interaction with natural and synthetic visual objects. Mesh objects, as defined by MPEG-4, may for instance be applied to create entertaining or commercial Web content at low bit rates. We further describe a robust 3D face scene understanding system, face information encoder, and face animation system that are compliant with the face animation coding specification in the MPEG-4 standard. The compressed face information bit rate is only a few kilobits per second, and the face information is easily decoded from the bitstream for a variety of applications including visual communication, entertainment, and enhanced speech recognition.

A. Samples Versus Models

Objects in a scene may be rigid, jointed, or deformable. The geometry, appearance, and motion of rigid or jointed objects and light sources can be estimated and modeled from camera views. Deformable objects, such as faces, can also be modeled using a priori knowledge of human anatomy. Once the objects in a scene are modeled, the scene can be represented and reconstructed using the models and motion parameters. If the number of object parameters is much smaller than the number of pixels in the display, the scene can be represented with fewer bits than a compressed video signal. In addition, model information allows the scene to be enhanced, modified, archived, searched, or understood further with minimal subsequent processing. If the number of objects in the scene is very high (e.g., blowing leaves) or the objects are not well modeled (e.g., moving water), traditional sample-based compression may be more efficiently used at the expense of model-based functionality.

Computer-generated synthetic scenes are created using a combination of geometric models, motion parameters, light sources, and surface textures and properties. The use of model information for coding camera-generated scenes allows natural scenes to be combined with synthetic scenes using a common language. In fact, virtually all television commercials and a rapidly increasing number of film and television productions use a combination of natural and synthetic visual content. The modeling of all visual content will allow more efficient use of communication bandwidth and facilitate manipulation by the content provider, network service provider, and the consumer.

B. Overview

MPEG-4 is an emerging standard for representation and compression of natural and synthetic audiovisual content [3]. MPEG-4 will provide tools for compressing sampled and modeled content in a unified framework for data representation and network access. The methods described here are compliant with the MPEG-4 standard [4–6]. Section II describes coding and animation of 2D objects using a 2D mesh-based representation, as well as methods for model fitting. Section III describes coding and animation of 3D face objects by defining essential feature points and animation parameters that may be used with 3D wireframe models, as well as methods for face scene analysis. Section IV describes applications of the synthetic video tools provided by MPEG-4.

II. 2D OBJECT ANIMATION AND CODING IN MPEG-4

A. Introduction

Three-dimensional polygonal meshes have long been used in computer graphics for 3D shape modeling, along with texture mapping to render photorealistic synthetic images [7]. Two-dimensional mesh models were initially introduced in the image processing and video compression literature for digital image warping [8] and motion compensation [9–13]. Two-dimensional mesh-based motion modeling has been proposed as a promising alternative to block-based motion compensation, especially within smoothly moving image regions. More recently, however, the ability to use mesh models for video manipulation and special effects generation has been emphasized as an important functionality [14,15]. In this context, video motion-tracking algorithms have been proposed for forward motion modeling, instead of the more conventional frame-to-frame motion estimation methods used in video coding. In object-based mesh modeling, the shape of a video object is modeled in addition to its motion [14].

MPEG-4 mesh objects are object-based 2D mesh models and form a compact representation of 2D object geometry and deformable motion. Mesh objects may serve as a representation of both natural video objects and synthetic animated objects. Because the MPEG-4 standard does not concern itself with the origin of the coded mesh data, in principle the mesh geometry and motion could be entirely synthetic and generated by human animation artists. On the other hand, a mesh object could be used to model the motion of a mildly deformable object surface, without occlusion, as viewed in a video clip. Automatic methods for 2D model generation and object tracking from video are discussed in Section II.B. Section II.C discusses coding of mesh data and Section II.D discusses object animation with 2D mesh data in MPEG-4. Then, Section II.E shows some compression results.

B. 2D Mesh Generation and Object Tracking

In this chapter, a *2D mesh* is a planar graph that partitions an image region into triangular patches. The vertices of the triangular patches are referred to as *node points* $\mathbf{p}_n = (x_n, y_n)$, $n = 0, 1, 2, \ldots, N - 1$. Here, triangles are defined by the indices of their node points, e.g., $t_m = \langle i, j, k \rangle$, $m = 0, 1, 2, \ldots, M - 1$ and $i, j, k \in 0, 1, 2, \ldots, N - 1$. Triangular patches are deformed by the movements of the node points in time, and the texture inside each patch of a reference frame is *warped* using an affine transform, defined as a function of the node point motion vectors $\mathbf{v}_n = (u_n, v_n)$, $n = 0, 1, 2, \ldots, N - 1$. Affine transforms model translation, rotation, scaling, reflection, and shear. Their linear form implies low computational complexity. Furthermore, the use of affine transforms enables continuity of the mapping across the boundaries of adjacent triangles. This implies that a 2D video motion field may be compactly represented by the motion of the node points.

Mesh modeling consists of two stages: first, a model is generated to represent an initial video frame or video object plane (VOP); then, this mesh is tracked in the forward direction of the frame sequence or video object. In MPEG-4, a new mesh can be generated every intra frame, which subsequently keeps its topology for the following inter frames.

1. Mesh Generation

In MPEG-4, a mesh can have either *uniform* or *object-based* topology; in both cases, the topology (triangular structure) is defined implicitly and is not coded explicitly. Only the geometry (sizes, point locations, etc.) of the initial mesh is coded. A 2D uniform mesh subdivides a rectangular object plane area into a set of rectangles, where each rectangle in turn is subdivided into two triangles. Adjacent triangles share node points. The node points are spaced equidistant horizontally as well as vertically. An example of a uniform mesh is shown in Fig. 1. A uniform mesh can be used for animating entire texture frames

Figure 1 Example of a uniform mesh object overlaid with text.

or arbitrary-shaped video objects; in the latter case it is overlaid on the bounding box of the object.

A 2D object-based mesh provides a more efficient tool for animating arbitrary-shaped video objects [12,14]. The advantage of an object-based mesh is twofold; first, one can closely approximate the object boundary by the polygon formed by boundary node points; second, boundary and interior node points can be placed adaptively using a content-based procedure. The object-based mesh topology is defined by applying constrained *Delaunay triangulation* [16] to the node points, where the edges of the boundary polygon are used as constraints. Content-based procedures for adaptive node placement may start by fitting a polygon to the actual pixel-based boundary of the video object plane, e.g., by detecting high-curvature points. The selected polygon vertices become mesh boundary node points. Then, the algorithm may automatically select locations in the interior of the video object with high spatiotemporal activity, such as intensity edges or corners, to place mesh interior node points. Figure 2 shows an example of a simple object-based mesh.

Delaunay triangulation is a well-known method in the field of computational geometry and provides meshes with several nice properties [16]. For instance, the Delaunay triangulation maximizes the minimal angle between triangle edges, thus avoiding triangles that are too ''skinny.'' Restricting the mesh topology to be Delaunay also enables higher coding efficiency. Several algorithms exist to compute the Delaunay triangulation [16]; one algorithm can be defined as follows.

1. Determine any triangulation of the given node points such that all triangles are contained in the interior of the polygonal boundary. This triangulation contains $2N_i + N_b - 2$ triangles, where N_b is the number of boundary node points, N_i is the number of interior node points, and $N = N_i + N_b$.
2. Inspect each interior edge of the triangulation and, for each edge, test whether it is locally Delaunay. An interior edge is shared by two opposite triangles, e.g., $\langle a, b, c\rangle$ and $\langle a, c, d\rangle$, defined by four points p_a, p_b, p_c, and p_d. If this edge is not locally Delaunay, the two triangles sharing this edge are replaced by triangles $\langle a, b, d\rangle$ and$\langle b, c, d\rangle$.

Repeat step 2 until all interior edges of the triangulation are locally Delaunay. An interior edge, shared by two opposite triangles $\langle a, b, c\rangle$ and $\langle a, c, d\rangle$, is locally Delaunay if point p_d is outside the circumcircle of triangle $\langle a, b, c\rangle$. If point p_d is inside the circumcircle of triangle $\langle a, b, c\rangle$, then the edge is not locally Delaunay. If point p_d is exactly on

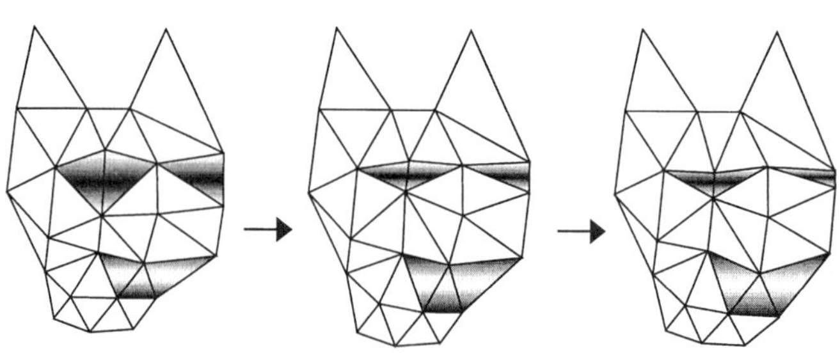

Figure 2 Example of an object-based mesh object overlaid with synthetic texture (illustrated by the shading).

the circumcircle of triangle $\langle a, b, c \rangle$, then the edge between points p_a and p_c is deemed locally Delaunay only if point p_b or point p_d is the point (among these four points) with the maximum x-coordinate, or, in case there is more than one point with the same maximum x-coordinate, the point with the maximum y-coordinate among these points. That is, the point with the maximum x-coordinate is symbolically treated as if it were located just outside the circumcircle.

2. *Object Tracking*

The motion of a 2D object can be captured by tracking the object from frame to frame using a mesh model. Mesh tracking entails estimating the motion vectors of its node points and frame-to-frame mesh propagation. Various techniques have been proposed for mesh node motion vector estimation and tracking. The simplest method is to form blocks that are centered on the node points and then use a gradient-based optical flow technique or block matching to find motion vectors at the location of the nodes. Energy minimization [17], hexagonal matching [9], and closed-form matching [11] are iterative motion vector optimization techniques that incorporate mesh connectivity and deformation constraints. Thus, a two-stage approach can be used in which an initial estimate is formed first using a general technique and is subsequently optimized using a mesh-based motion estimation technique. Hexagonal matching [9] works by iteratively perturbing single node points to a set of search locations and locally measuring the mean square error or mean absolute error of the warped intensities in the polygonal region around the node point affected by its displacement. Hierarchical mesh-based tracking methods have been proposed for uniform meshes [18] and, more recently, for object-based meshes [19], extending the two-stage approach to a multistage hierarchical framework. The latter approach is based on a hierarchical mesh representation, consisting of object-based meshes (initially Delaunay) of various levels of detail. Such hierarchical methods have been shown to improve the tracking performance by using a coarse-to-fine motion estimation procedure. That is, motion at coarser levels is used to predict the motion of a finer mesh in the hierarchy, after which the estimate is refined using generalized hexagonal matching. One of the challenges in mesh tracking, especially for high-motion video, is to ensure that the estimated node motion vectors maintain the topology of the initial mesh. If node motion vectors are estimated independently, e.g., by a gradient-based technique, then foldover of triangles may occur during mesh propagation. Thus, constraints have to be placed on the search region for estimating node motion vectors, as in Refs. 17–19, to avoid degeneracies of the topology. This search region can be defined as the intersection of a number of half-planes, where each half-plane is defined by a line through two node points connected to the moving node point [19]. Usually, motion estimation and mesh tracking involve some form of iteration to find an optimal solution while allowing large motion vectors.

C. 2D Mesh Coding

Because the initial mesh of a sequence may be adapted to image content, information about the initial mesh geometry has to be coded in addition to the motion parameters. In the case of the 2D mesh object of MPEG-4 (version 1), the initial mesh topology is restricted to limit the overhead involved. For more general 3D mesh compression schemes now proposed for standardization by MPEG-4 (version 2), see Ref. 20.

This section discusses coding of 2D mesh geometry and motion; see also Ref. 21. Mesh data are coded as a sequence of so-called mesh object planes, where each mesh

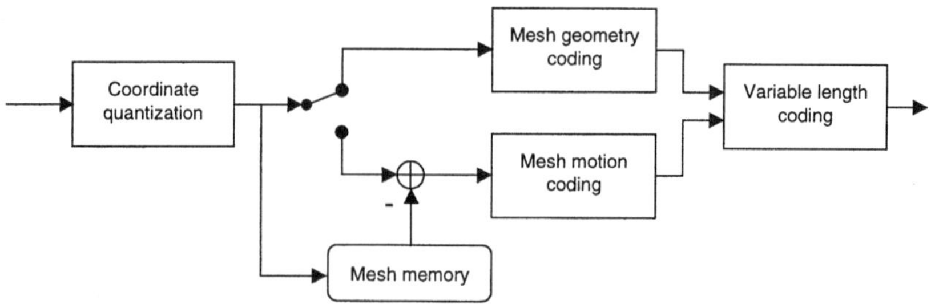

Figure 3 Overview of mesh parameter encoding.

object plane (MOP) represents the mesh at a certain time instance and an MOP can be coded in intra or predictive mode. Overviews of the encoder and decoder are shown in Figures 3 and 4. Note that in this section, a pixel-based coordinate system is assumed, where the x-axis points to the right from the origin and the y-axis points down from the origin. We assume the origin of this local coordinate system is at the top left of the pixel-accurate bounding box surrounding the initial mesh.

1. Mesh Geometry Coding: I-Planes

In the case of a uniform mesh, only the following five parameters are coded: (1) the number of node points in a row, (2) the number of node points in a column, (3) the width of a rectangle (containing two triangles) in half-pixel units, (4) the height of a rectangle (containing two triangles) in half-pixel units, and (5) the triangle orientation pattern code. The four triangle orientation patterns allowed are illustrated in Figure 5 for a uniform mesh of four by five nodes.

In the case of an object-based mesh, node points are placed nonuniformly; therefore, the node point locations $\mathbf{p}_n = (x_n, y_n)$, $n = 0, 1, 2, \ldots, N - 1$ (in half-pixel units), must be coded. To allow reconstruction of the (possibly nonconvex) mesh boundary by the decoder, the locations of boundary node points are coded first, in the order of appearance along the mesh boundary (in either clockwise or counterclockwise fashion). Then, the locations of the interior node points are coded. The x- and y-coordinates of the first boundary node point are coded by a fixed-length code. The x- and y-coordinates of all other node points are coded differentially. The coordinates of the difference vectors $\mathbf{dp}_k = \mathbf{p}_k - \mathbf{p}_{k-1}$ are coded by variable-length codes, where k indicates the order in which the node points are coded. The encoder is free to traverse the interior node points in any order that

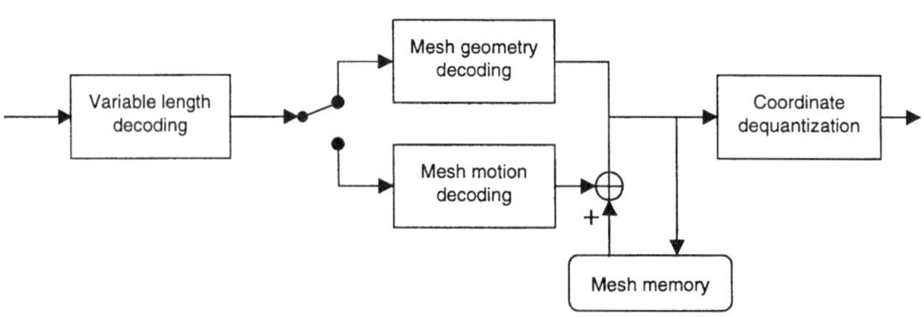

Figure 4 Overview of mesh parameter decoding.

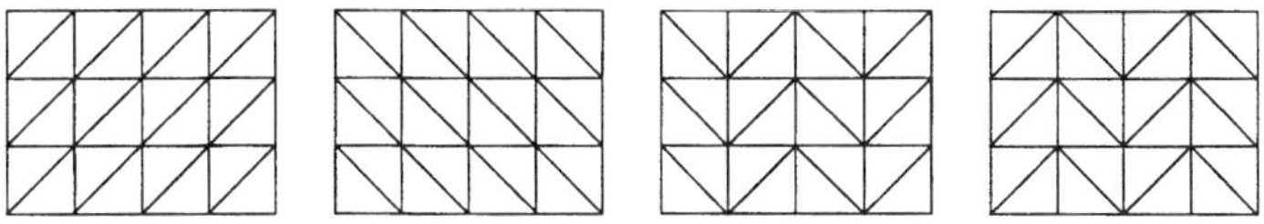

Figure 5 Uniform mesh types.

it chooses to optimize the coding efficiency. A suboptimal algorithm used here for illustration is to find the node point nearest to the last coded node point and to code its coordinates differentially; this is reiterated until all node points are coded. The traversal of the node points using this greedy strategy is illustrated in Figure 6a.

The decoder is able to reconstruct the mesh boundary after receiving the first N_b locations of the boundary nodes by connecting each pair of successive boundary nodes, as well as the first and the last, by straight-line edge segments. The N_i locations decoded next correspond to the interior mesh node points. The decoder obtains the mesh topology by applying constrained Delaunay triangulation to the set of decoded node points as defined in Sec. II.B, where the polygonal mesh boundary is used as a constraint.

2. Mesh Motion Coding: P-Planes

Each node point $\mathbf{p}_n$ of a uniform or object-based mesh has a 2D motion vector $\mathbf{v}_n = (u_n, v_n)$. A motion vector is defined from the previous mesh object plane at time t' to the current mesh object plane at time t, so that $\mathbf{v}_n = \mathbf{p}_n - \mathbf{p}'_n$. Each motion vector (except the first and second) is coded differentially and the order of coding the motion vectors is such that one can always use two previously coded motion vectors as predictors. The ordering of the node points for motion vector coding is illustrated in Fig. 6b and c.

A spanning tree of the *dual graph* of the triangular mesh can be used to traverse all mesh triangles in a *breadth-first* order as follows. Start from an initial triangle defined as the triangle that contains the edge between the top left node of the mesh and the next

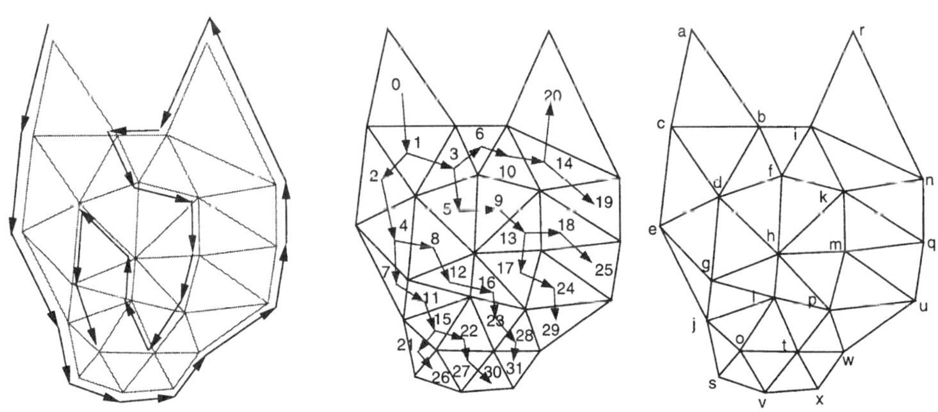

a b c

Figure 6 (a) Node point ordering for point location coding of an object-based mesh. (b and c) Mesh motion coding; (b) shows triangle ordering and triangle spanning tree; (c) shows node point ordering for motion vector coding.

clockwise node on the boundary, called the *base edge*. The top left mesh node is defined as the node n with minimum $x_n + y_n$, assuming the origin of the local coordinate system is at the top left. If there is more than one node with the same value of $x_n + y_n$, then choose the node point among these with minimum y_n. Label the initial triangle with number 0. Define the *right edge* of this triangle as the next counterclockwise edge with respect to the base edge, and define the *left edge* as the next clockwise edge with respect to the base edge. That is, for a triangle $\langle i, j, k \rangle$ with vertices ordered in clockwise order, if the edge between $\mathbf{p}_i$ and $\mathbf{p}_j$ is the base edge, then the edge between $\mathbf{p}_i$ and $\mathbf{p}_k$ is the right edge and the edge between $\mathbf{p}_j$ and $\mathbf{p}_k$ is the left edge. Now, check if there is an unlabeled triangle adjacent to the current triangle, sharing the right edge. If there is such a triangle, label it with the next available number. Then check if there is an unlabeled triangle adjacent to the current triangle, sharing the left edge. If there is such a triangle, label it with the next available number. Next, find the triangle with the lowest number label among all labeled triangles that have adjacent triangles that are not yet labeled. This triangle now becomes the *current triangle* and the process is repeated. Continue until all triangles are labeled with numbers $m = 0, 1, 2, \ldots, M - 1$.

The breadth-first ordering of the triangles as just explained also defines the ordering of node points for motion vector coding. The motion vector of the top left node is coded first (without prediction). The motion vector of the other boundary node of the initial triangle is coded second (using only the first coded motion vector as a prediction). Next, the triangles are traversed in the order determined before. Each triangle $\langle i, j, k \rangle$ always contains two node points (i and j) that form the base edge of that triangle, in addition to a third node point (k). If the motion vector of the third node point is not already coded, it can be coded using the average of the two motion vectors of the node points of the base edge as a prediction. If the motion vector of the third node point is already coded, the node point is simply ignored. Note that the breadth-first ordering of the triangles as just defined guarantees that the motion vectors of the node points on the base edge of each triangle are already coded at the time they are used for prediction, ensuring decodability. This ordering is defined by the topology and geometry of an I mesh object plane and kept constant for all following P mesh object planes; i.e., it is computed only once for every sequence of IPPPPP . . . planes.

For every node point, a 1-bit code specifies whether its motion vector is the zero vector (0, 0) or not. For every nonzero motion vector, a prediction motion vector $\mathbf{w}_k$ is computed from two predicting motion vectors $\mathbf{v}_i$ and $\mathbf{v}_j$ as follows:

$$\mathbf{w}_k = ((u_i + u_j)///2, \qquad (v_i + v_j)///2)$$

where /// indicates integer division with rounding to positive infinity and motion vectors are expressed in half-pixel units. Note that no prediction is used to code the first motion vector, $\mathbf{w}_{k0} = (0, 0)$, and only the first motion vector is used as a predictor to code the second motion vector, $\mathbf{w}_{k1} = \mathbf{v}_{k0}$. A delta motion vector is defined as $\mathbf{dv}_k = \mathbf{v}_k - \mathbf{w}_k$ and its components are coded using variable-length coding (VLC) in the same manner as block motion vectors are coded in the MPEG-4 natural video coder (using the same VLC tables). As with MPEG-4 natural video, motion vectors are restricted to lie within a certain range, which can be scaled from [−32, 31] to [−2048, 2047].

D. 2D Object Animation

Texture-mapped object animation in an MPEG-4 decoder system makes use of the object geometry and motion parameters as coded by a 2D mesh bit stream, as well as separately

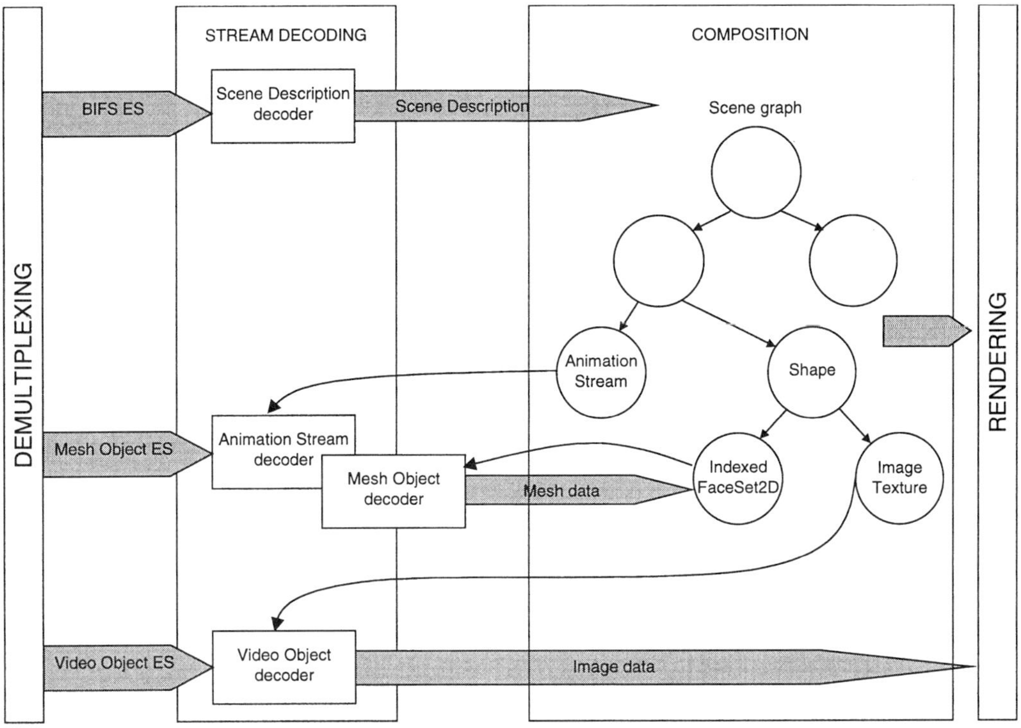

Figure 7 Overview of mesh decoding and animation in an MPEG-4 terminal. Each elementary stream (ES) is decoded separately; the decoded scene description defines a scene graph; the decoded mesh data define mesh geometry and motion; decoded video object data define the image to be texture mapped; finally, the object is rendered.

coded image texture data, as illustrated in Figure 7. Initially, a BIFS (Binary Format for Scenes) stream is decoded, containing the scene description; see Chapter 12. In the decoded scene description tree, the mesh will be specified using an IndexedFaceSet2D node and image data will be specified using an Image Texture node; see Chapter 14. The BIFS nodes are used to represent and place objects in the scene and to identify the incoming streams. Image data may be decoded by a video object decoder or scalable texture decoder; compressed binary mesh data are decoded by a mesh object decoder. The decoded mesh data are used to update the appropriate fields of the IndexedFaceSet2D node in the scene description tree. Finally, the compositor uses all the decoded data to render a texture-mapped image at regular time intervals. In MPEG-4, 2D mesh data are actually carried by a so-called BIFS-animation stream, which, in general, may contain several types of data to animate different parameters and objects in the scene (including 3D face objects). This animation stream is initially decoded by the animation stream decoder, as illustrated in Figure 7. This decoder splits the animation data for the different objects and passes the data to the appropriate decoders, in this case only the mesh object decoder. In practice, the animation stream decoder and mesh object decoder may be implemented as a single decoder. The animation stream itself is indicated in the scene description tree by an animation stream node. During the setup phase of this animation stream, a unique identifier must be passed to the terminal that determines the node to which the animation data must be streamed, in this case, the IndexedFaceSet2D node. This identifier is the ID of this node in the scene description tree, as described in Ref. 4. The appropriate fields of the IndexedFaceSet2D node are updated as described in the following.

1. The coordinates of the mesh points (vertices) are passed directly to the node, possibly under a simple coordinate transform. The coordinates of mesh points are updated every mesh object plane (MOP).
2. The coordinate indices are the indices of the mesh points forming decoded faces (triangles). The topology of a mesh object is constant starting from an intra-coded MOP, throughout a sequence of predictive-coded MOPs (until the next intra-coded MOP); therefore, the coordinate indices are updated only for intra-coded MOPs.
3. Texture coordinates for mapping textures onto the mesh geometry are defined by the decoded node point locations of an intra-coded mesh object plane and its bounding box. Let $x_{\min}$, $y_{\min}$ and $x_{\max}$, $y_{\max}$ define the bounding box of all node points of an intra-coded MOP. Then the width w and height h of the texture map must be

   $$w = \text{ceil}(x_{\max}) - \text{floor}(x_{\min}), \qquad h = \text{ceil}(y_{\max}) - \text{floor}(y_{\min})$$

   A texture coordinate pair (s_n, t_n) is computed for each node point $\mathbf{p}_n = (x_n, y_n)$ as follows:

   $$s_n = (x_n - \text{floor}(x_{\min}))/w, \qquad t_n = 1.0 - (y_n - \text{floor}(y_{\min}))/h$$

   The topology of a mesh object is constant starting from an intra-coded MOP, throughout a sequence of predictive-coded MOPs (until the next intra-coded MOP); therefore, the texture coordinates are updated only for intra-coded MOPs.
4. The texture coordinate indices (relating the texture coordinates to triangles) are identical to the coordinate indices.

An illustration of mesh-based texture mapping is given in Figure 8. This figure shows a screen snapshot of an MPEG-4 software player while playing a scene containing

Figure 8 Screen snapshot of MPEG-4 player running a scene that activates the animation stream decoder and 2D mesh decoder to render a moving and deforming image of a fish. (Player software courtesy of K. A. Oygard, Telenor, Norway and MPEG-4 player implementation group.)

Table 1 Results of Compression of 2D Mesh Sequences[a]

Sequence name	No. of nodes	No. of MOPs	MOP rate (Hz)	Geometry bits I-MOP	Motion bits per P-MOP	Overhead bits per MOP
Akiyo (object)	210	26	10	3070	547.4	42.3
Bream (object)	165	26	15	2395	1122.8	42.5
Bream (uniform)	289	26	15	41	1558.0	42.2
Flag (uniform)	289	10	12	41	1712.7	46.5

[a] The number of nodes in the mesh, the number of mesh object planes in the sequence, the frame rate, the number of bits spent on the mesh geometry in the I-MOP, the number of bits spent per P-MOP on motion vectors, and the number of bits spent per MOP on overhead.

a texture-mapped 2D mesh object. The motion of the object was obtained from a natural video clip of a fish by mesh tracking. The image of the fish was overlaid with synthetic text before animating the object.

E. 2D Mesh Coding Results

Here we report some results of mesh geometry and motion compression, using three arbitrary-shaped video object sequences (Akiyo, Bream, and Flag). The mesh data were obtained by automatic mesh design and tracking, using an object-based mesh for Akiyo and Bream and using another uniform mesh for Bream as well as for Flag. Note that these results were obtained with relatively detailed mesh models; in certain applications, synthetic meshes might have on the order of 10 node points.

The experimental results in Table 1 show that the initial object-based mesh geometry (node point coordinates) was encoded on average with approximately 14.6 bits per mesh node point for Akiyo and 14.5 bits per mesh node point for Bream. A uniform mesh geometry requires only 41 bits in total. The mesh motion vectors were encoded on average with 2.6 bits per VOP per node for Akiyo, 6.8 bits per VOP per node for the object-based Bream mesh, 5.4 bits per VOP per node for the uniform Bream mesh, and 5.9 bits per VOP node for Flag. Note that the uniform mesh of Bream and Flag contain near-zero motion vectors that fall outside the actual object—these can be zeroed out in practice.

III. 3D FACE ANIMATION AND CODING IN MPEG-4

A. Visual Speech and Gestures

Normal human communication involves both acoustic and visual signals to overcome channel noise and interference (e.g., hearing impairment, other speakers, environmental noise). The visual signals include mouth movements (lipreading); eye, eyebrow, and head gestures; and hand, arm, and body gestures. Version 1 of MPEG-4 provides an essential set of feature points and animation parameters for describing visual speech and facial gestures, which includes face definition parameters (FDPs) and face animation parameters (FAPs). Also, interfaces to a text-to-speech synthesizer are defined. Furthermore, *body* animation is intended to be standardized in version 2 of MPEG-4.

Section III.B discusses generating 3D models of human faces. Section III.C dis-

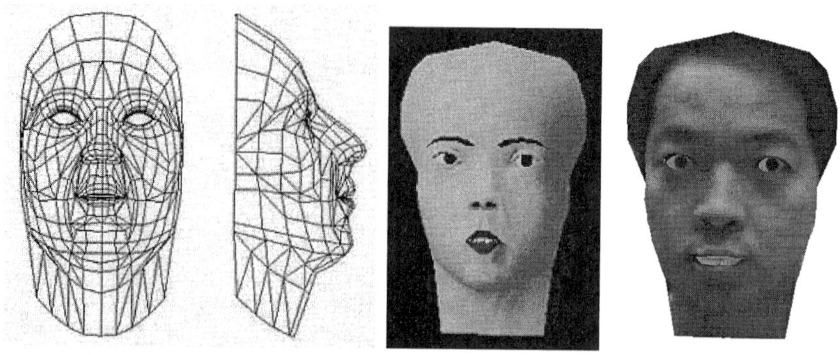

(a) (b) (c)

Figure 9 (a) Wireframe of a generic model. Fitted model with (b) smooth rendering and with (c) texture map.

cusses automatic methods for face scene analysis. Sections III.D and III.E describe the framework for coding facial animation parameters and specification of precise facial models and animation rules in MPEG-4. Section III.F describes the integration of face animation and text-to-speech synthesis. Finally, Section III.G discusses some experimental results.

B. 3D Model Generation

The major application areas of talking-head systems are in human–computer interfaces, Web-based customer service, e-commerce, and games and chat rooms where persons want to control artificial characters. For these applications, we want to be able to make use of characters that are easy to create, animate, and modify. In general, we can imagine three ways of creating face models. One would be to create a model manually using modeling software. A second approach detects facial features in the video and adapts a generic face model to the video [22]. A third approach uses active sensors and image analysis software [23,24]. Here, we give an example of fitting a generic model (Fig. 9a) to 3D range data (Fig. 10a) of a person's head to generate a face model in a neutral state [23].

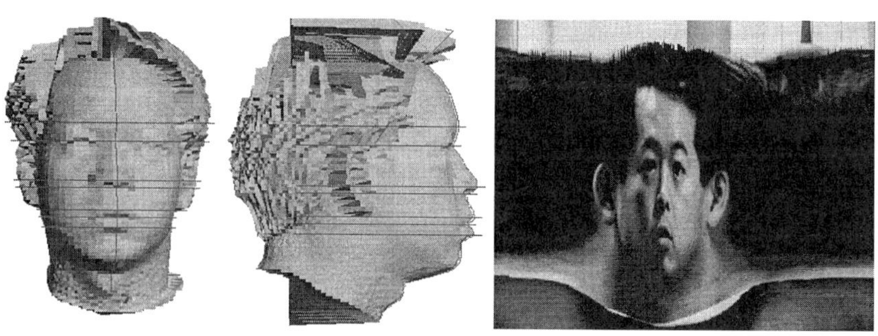

(a) (b) (c)

Figure 10 (a and b) Two views of 3D range data with profile line and feature points marked automatically on the 3D shape. (c) Texture map corresponding to the 3D range data.

Initially, the generic head model is larger in scale than the range data. Vertical scaling factors are computed from the position of feature points along the profile of the model. Each scaling factor is used to scale down one slice of the generic model to the size of the range data in the vertical direction. Then each vertex point on the model is radially projected onto the surface of the range data. Figure 9b shows the adapted model with smooth shading. Because the original model does not contain polygons to model the boundary between the forehead and the hair, it cannot be adapted precisely to the 3D range data providing significant detail in this region. This problem is overcome by texture mapping. Instead of smooth shaded polygons with prescribed colors, the color and texture of the polygon surface of the face come from a color image (Fig. 10b). The image is literally pasted onto the polygonal wireframe. Figure 9c shows the texture-mapped rendering of the head. Clearly, the resulting rendering is much more realistic.

C. Face Tracking and Feature Analysis

In this section, we describe a system for automatic face tracking, facial feature analysis, and face animation parameter estimation.

1. *Face Analysis Overview*

The nostrils are by far the easiest facial features to identify and track. They are two symmetric holes in the middle of the face that are darker than the darkest facial hair or skin and are almost never obscured by hair. Many face feature analysis applications allow camera placement suitable for viewing the nostrils (slightly below center view). No assumptions are made about the lighting conditions, skin color, eyewear, facial hair, or hairstyle. The system is robust in the presence of head roll and tilt (nodding and rotation in the image plane) and scale variations (face-to-camera distance). The system will fail during periods of excessive head rotation about the neck axis (profile view).

The primary objective of this system is to estimate accurately the inner lip contour of an arbitrary speaker under a large range of viewing conditions. An implicit system performance requirement is extremely robust tracking of the face (eyes, nostrils, and mouth). In addition, the system must reliably detect tracking failures and never falsely indicate the positions of the nostrils and mouth. Tracking failures routinely occur because of occlusion from hands, extreme head rotation, and travel outside the camera range and should be detected. Given an accurate estimate of mouth position and orientation in the image plane, mouth closure should be accurately detected.

A face recognition algorithm [25,26] periodically finds faces in static frames using morphological filtering and the relative positions of eye, nose, and mouth blobs as shown in Figure 11. This information is used to start and then verify the nostril-tracking system. A mouth window is formed using the nostril positions. A closed mouth line is combined with the resulting open mouth inner lip contours to give the final inner lip contour estimate. When the mouth is open, the mouth line lies within the inner lip contour.

2. *Static Face Image Analysis*

Prior to nostril tracking, the static face analysis subsystem uses combinations of facial features to indicate the presence of a face using only the monochrome image. This head location scheme uses a bottom-up approach, first selecting areas where facial features of head outlines might be present and then evaluating combinations of such areas to decide whether a face is present.

Figure 11 Example original and filtered static image with candidate mouth and nose lines under eye candidates on the right.

After filtering operations, an adaptive thresholding technique is used to identify the positions of individual facial features. The areas of the prominent facial features, such as eyes, mouth, eyebrows, and the lower end of the nose, are marked with blobs of connected pixels that are well separated from the rest. An example is shown in Figure 11, where mouth and nose lines are positioned below the eye candidates. Once candidate facial features are marked with connected components, combinations of these features that could represent a face have to be found. The connected components are evaluated with small classifiers that take as their inputs the sizes, ratios of distances, and orientations of the connected components.

3. *Nostril Tracking*

The scale (size) of the face is determined from the eye–eye and eye–nose–mouth distances resulting from the static face analysis subsystem. A scaled nostril-tracking window is formed approximately around the nose with a size small enough to exclude eyes and mouth yet large enough to contain the nostrils in the next frame after the fastest possible head movement. Further algorithm details can be found in Refs. 25 and 26.

Experiments with six subjects and thousands of video frames showed that the combination of static face analysis and nostril tracking was extremely robust. If the nostrils were visible, the system never failed to track the nostrils and properly place the mouth window. However, the eye location accuracy was not sufficient for gaze tracking.

4. *Mouth Detail Analysis*

Given the head position, scale, and tilt estimates from the nostril-tracking system, a window is formed around the mouth area at a fixed distance from the nostrils. After compensating the mouth image for image plane rotation and scale, a horizontal array of inner lip color thresholds are trained whenever the mouth is closed. Mouth closure is detected by the face recognition system, which looks for a single horizontal valley in the mouth region. The inner lip thresholds are taken to be 90% of the minimum color intensity in each column of pixels in the mouth window, resulting in a color threshold for each horizontal position. The inner lip contour threshold array is then applied to the mouth window. Each pixel that is below threshold is labeled as an inner mouth pixel, and isolated inner mouth pixels are removed. The teeth are then detected by forming a bounding box around the inner mouth pixels and testing all non–inner mouth pixels for teeth color given as a fixed set of RGB ranges (or YUV ranges). Then a closed contour is formed around the inner

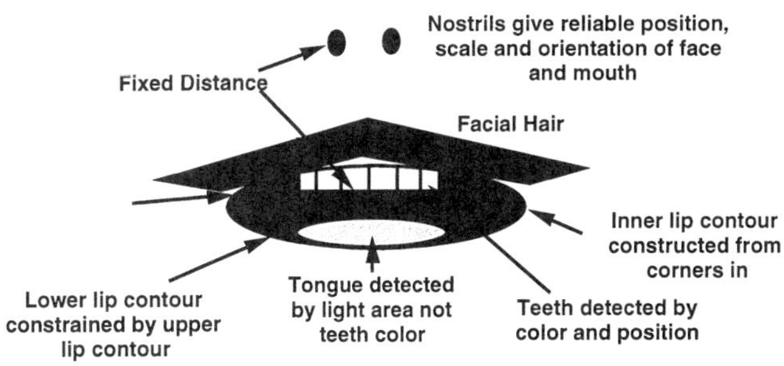

Figure 12 Illustration of detailed mouth analysis.

mouth and teeth pixels starting from the mouth corners. The upper contour is constrained to have increasing or constant height as the face centerline is approached. The lower contour is constrained to have increasing or constant distance from the upper contour. These constraints were derived from anatomical considerations. Figure 12 illustrates the mouth detail analysis process and Figure 13 shows the analysis results overlaid on a video frame.

The inner lip contour estimation algorithm is designed to position the estimated inner lip contour inside the actual inner lip contour. This is motivated by the desire for an accurate estimate of mouth closure, which is an important visual speech feature. The estimated inner lip contour is less accurate when light reflects from the teeth or tongue but is still adequate to capture essential visual speech and emotional expressions.

5. Head Pose Estimation

The precise estimation of the 3D position and orientation of the head from a single view is a difficult task that requires a priori knowledge of head size, shape, surface features, and camera parameters. A much simpler algorithm is used in this system that provides a

Figure 13 Video frame overlaid with both static and dynamic tracking face features (both shown in nostrils).

useful estimate of head pose assuming small head rotations and fixed torso position. Given these assumptions, head pitch is proportional to the vertical nose center deviation and head yaw is proportional to the horizontal nose center deviation. Head roll is assumed to be equal to the horizontal deviation of the line connecting the nostril or eye centers. Finally, all head position and orientation parameters are smoothed with a temporal low-pass filter that removes movements which are physically impossible or unlikely.

6. *Inner Lip Contour Parameter Estimation*

The face is assumed to be in the neutral position when mouth closure is detected for the first time. The position of the mouth corners, center vertical position, and vertical positions of the midpoints between the mouth center and the mouth corners are stored for comparison with subsequent frames. The inner lip facial animation parameters for a subsequent frame are computed by first normalizing the orientation and scale of the mouth image using the head roll estimate and the distance between the nostrils as a measure of scale. The horizontal and vertical deviations of the mouth corners are then measured relative to the neutral mouth corner positions. Then the vertical deviations of the upper and lower middle lip points are measured relative to the neutral face. Finally, the vertical deviations of the midpoints between the mouth center and the mouth corners are measured relative to the neutral face. The inner lip contour estimates and head orientation estimates contain noise that appears in the FAP sequences; therefore, smoothing is applied to curb the noise effects.

D. Face Animation Coding

1. *Basic Principles*

MPEG-4 is a comprehensive collection of audio and visual coding tools that are intended for a wide variety of applications including broadcast, wireless, Internet, multicast, and film or video production. In broadcast applications, the terminal is not able to request information from the encoder but must be able to decode the bitstream quickly and present the content to the user. The MPEG-4 FAPs have been designed to be independent of any particular model. Essential facial gestures and visual speech derived from a particular speaker and intended for a particular face model will produce good results on other face models unknown to the encoder. This is accomplished by defining each parameter in normalized face animation parameter units (FAPUs) as indicated in Table 2. Each of the four translational FAPUs shown in Table 2 is computed by measuring the distance between feature points on any model of a neutral face and dividing by 1024. Angular units (AUs) are 100,000ths of a radian. All MPEG-4 FAPs are defined as deviations from the neutral

Table 2 Facial Animation Parameter Units

IRISD0	Iris diameter (equal to the distance between upper and lower eyelid)	IRISD = IRISD0/1024
ES0	Eye separation	ES = ES0/1024
ENS0	Eye–nose separation	ENS = ENS0/1024
MNS0	Mouth–nose separation	MNS = MNS0/1024
MW0	Mouth width	MW = MW0/1024
AU	Angle unit	10^{-5}rad

face, which is defined as follows (note that the coordinate system is right handed and head axes are parallel to the world axes with zero angles):

> *X* direction is to the left of the face, with pitch about the *X* axis.
> *Y* direction is up, with yaw about the *Y* axis.
> Gaze is in direction of *Z* axis, with roll about the *Z* axis.
> Multiple angles for head motion are interpreted as the usual ordered Euler set.
> All face muscles are relaxed.
> Eyelids are tangent to the iris.
> The pupil is one third of IRISD0.
> Lips are in contact; the line of the lips is horizontal and at the same height as lip corners.
> The mouth is closed and the upper teeth touch the lower ones.
> The tongue is flat and horizontal with the tip of the tongue touching the boundary between upper and lower teeth (feature point 6.1 touching 9.11; see Fig. 14).

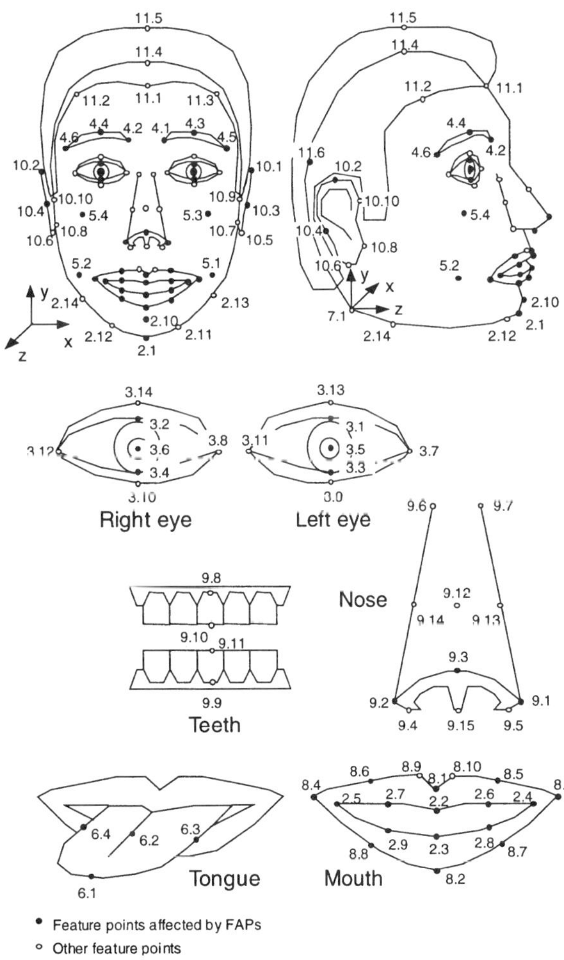

Figure 14 FDP feature point set.

2. Face Animation Parameters

Descriptions of the 68 FAPs are provided in Table 3. This table shows the FAP name, the FAP group, and the FDP feature affected (see Fig. 14), as well as the direction of positive motion. The FAP grouping is defined in Table 4. FAPs 1 and 2 are special sets of values that allow high-level visemes and facial expressions to be specified as explained in the following. The sum of two corresponding top and bottom eyelid FAPs must equal 1024 when the eyelids are closed. Inner lips are closed when the sum of two corresponding top and bottom lip FAPs equals zero. For example, (lower_t_midlip + raise_b_midlip) = 0 when the lips are closed. All directions are defined with respect to the face and not the image of the face.

3. Visemes and Expressions

A modest but relatively unambiguous and language-independent set of visemes and expressions is shown in Table 5 and Table 6. Visemes and expressions may be specified by themselves or in combination with other FAPs. Other FAPs take precedence if specified; otherwise the viseme or expression is used to inform the interpolation of missing FAPs. A special mode has also been provided that allows visemes and expressions to be associated with sets of FAP values for storage in a lookup table (LUT) in the terminal. Extremely low bit rate face animation coding is then achieved by sending only visemes or expressions that then reference other FAPs in the LUT. The inclusion of visemes in the bitstream provides a convenient means for enhancing automatic speech recognition [27–32], and coding in the network or terminal.

4. FAP Compression

The FAPs are quantized and coded by a predictive coding scheme as shown in Figure 15. For each parameter to be coded in the current frame, the decoded value of this parameter in the previous frame is used as the prediction. Then the prediction error, i.e., the difference between the current parameter and its prediction, is computed and coded by arithmetic coding. This predictive coding scheme prevents the coding error from accumulating. Each FAP has a different precision requirement. For example, the jaw movement may be represented more coarsely than the lip movement parameters without affecting the perceptual quality of the animated facial renderings. Therefore different quantization step sizes are applied to the FAPs. The bit rate is controlled by adjusting the quantization step via the use of a quantization scaling factor called FAP_QUANT, which is applied uniformly to all FAPs. The magnitude of this scaling factor ranges from 1 to 32. For noninteractive applications in which latency is not a concern, additional compression efficiency is provided using discrete cosine transform (DCT) coding of each FAP sequence in time; see Ref. 5.

5. FAP Masking

Face animation parameters can be selected for transmission using a two-level mask hierarchy. The first level contains a 2-bit code for each FAP group (see Table 4) indicating the following options:

1. No FAPs in the group are selected for transmission.
2. A group mask is sent indicating which FAPs in the group are selected for transmission. The FAPs not selected by the group mask retain their previous value if there is any previously set value (not interpolated by decoder if previously set).

Table 3 Definition of FAPs: FAP Name, Direction of Positive Motion (D), FAP Group (G) and Subgroup Number (N)[a]

No.	Name	D	G	N	No.	Name	D	G	N
1	viseme	NA	1	NA	35	raise_l_o_eyebrow	Up	4	5
2	expression	NA	1	NA	36	raise_r_o_eyebrow	Up	4	6
3	open_jaw	Down	2	1	37	squeeze_l_eyebrow	Right	4	1
4	lower_t_midlip	Down	2	2	38	squeeze_r_eyebrow	Left	4	2
5	raise_b_midlip	Up	2	3	39	puff_l_cheek	Left	5	1
6	stretch_l_cornerlip	Left	2	4	40	puff_r_cheek	Right	5	2
7	stretch_r_cornerlip	Right	2	5	41	lift_l_cheek	Up	5	1
8	lower_t_lip_lm	Down	2	6	42	lift_r_cheek	Up	5	2
9	lower_t_lip_rm	Down	2	7	43	shift_tongue_tip	Right	6	1
10	raise_b_lip_lm	Up	2	8	44	raise_tongue_tip	Up	6	1
11	raise_b_lip_rm	Up	2	9	45	thrust_tongue_tip	Forward	6	1
12	raise_l_cornerlip	Up	2	4	46	raise_tongue	Up	6	2
13	raise_r_cornerlip	Up	2	5	47	tongue_roll	Concave upward	6	3, 4
14	thrust_jaw	Forward	2	1	48	head_pitch	Down	7	NA
15	shift_jaw	Right	2	1	49	head_yaw	Left	7	NA
16	push_b_lip	Forward	2	3	50	head_roll	Right	7	NA
17	push_t_lip	Forward	2	2	51	lower_t_midlip_o	Down	8	1
18	depress_chin	Up	2	10	52	raise_b_midlip_o	Up	8	2
19	close_t_l_eyelid	Down	3	1	53	stretch_l_cornerlip_o	Left	8	3
20	close_t_r_eyelid	Down	3	2	54	stretch_r_cornerlip_o	Right	8	4
21	close_b_l_eyelid	Up	3	3	55	lower_t_lip_lm_o	Down	8	5
22	close_b_r_eyelid	Up	3	4	56	lower_t_lip_rm_o	Down	8	6
23	yaw_l_eyeball	Left	3	NA	57	raise_b_lip_lm_o	Up	8	7
24	yaw_r_eyeball	Left	3	NA	58	raise_b_lip_rm_o	Up	8	8
25	pitch_l_eyeball	Down	3	NA	59	raise_l_cornerlip_o	Up	8	3
26	pitch_r_eyeball	Down	3	NA	60	raise_r_cornerlip_o	Up	8	4
27	thrust_l_eyeball	Forward	3	NA	61	stretch_l_nose	Left	9	1
28	thrust_r_eyeball	Forward	3	NA	62	stretch_r_nose	Right	9	2
29	dilate_l_pupil	Growing	3	5	63	raise_nose	Up	9	3
30	dilate_r_pupil	Growing	3	6	64	bend_nose	Right	9	3
31	raise_l_i_eyebrow	Up	4	1	65	raise_l_ear	Up	10	1
32	raise_r_i_eyebrow	Up	4	2	66	raise_r_ear	Up	10	2
33	raise_l_m_eyebrow	Up	4	3	67	pull_l_ear	Left	10	3
34	raise_r_m_eyebrow	Up	4	4	68	pull_r_ear	Right	10	4

[a] Single letters in FAP names have the following meaning: l = left, r = right, t = top, b = bottom, i = inner, o = outer, m = middle. Note that FAPs 3, 14, 41, 42, and 47 are unidirectional; other FAPs are bidirectional.

Table 4 FAP Grouping

Group	Number of FAPs
1: Visemes and expressions	2
2: Jaw, chin, inner lowerlip, cornerlips, midlip	16
3: Eyeballs, pupils, eyelids	12
4: Eyebrow	8
5: Cheeks	4
6: Tongue	5
7: Head rotation	3
8: Outer lip positions	10
9: Nose	4
10: Ears	4

Table 5 Values for Viseme_Select

No.	Phonemes	Examples
0	NA	NA
1	p, b, m	*P*ut, *b*ed, *m*ill
2	f, v	*F*ar, *v*oice
3	T, D	*Th*ink, *th*at
4	t, d	*T*ip, *d*oll
5	k, g	*C*all, *g*as
6	tS, dZ, S	*Ch*air, *j*oin, *sh*e
7	S, z	*S*ir, *z*eal
8	N, I	*L*ot, *n*ot
9	R	*R*ed
10	A:	C*ar*
11	E	B*e*d
12	I	T*i*p
13	Q	T*o*p
14	U	B*oo*k

Table 6 Values for Expression_Select

No.	Name	Description
0	NA	NA
1	Joy	The eyebrows are relaxed. The mouth is open and the mouth corners pulled back toward the ears.
2	Sadness	The inner eyebrows are bent upward. The eyes are slightly closed. The mouth is relaxed.
3	Anger	The inner eyebrows are pulled downward and together. The eyes are wide open. The lips are pressed against each other or opened to expose the teeth.
4	Fear	The eyebrows are raised and pulled together. The inner eyebrows are bent upward. The eyes are tense and alert.
5	Disgust	The eyebrows and eyelids are relaxed. The upper lid is raised and curled, often asymmetrically.
6	Surprise	The eyebrows are raised. The upper eyelids are wide open, the lower relaxed. The jaw is opened.

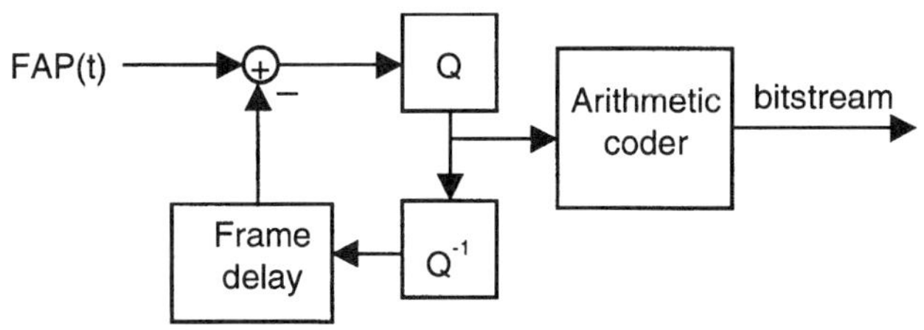

Figure 15 FAP predictive coding.

3. A group mask is sent indicating which FAPs in the group are selected for transmission. The FAPs not selected by the group mask must be interpolated by the decoder.
4. All FAPs in the group are sent.

If a given FAP is set for interpolation, the decoder is free to generate a FAP value or vertex displacement values for optimum quality. However, if one side of a left–right FAP pair is set for interpolation, the decoder must copy the value from the other side.

6. *FAP Interpolation Tables*

As mentioned in the previous section, the encoder may allow the decoder to extrapolate the values of some FAPs from the transmitted FAPs. Alternatively, the decoder can specify the interpolation rules using FAP interpolation tables (FITs). A FIT allows a smaller set of FAPs to be sent for a facial animation. This small set can then be used to determine the values of other FAPs, using a rational polynomial mapping between parameters. For example, the top inner lip FAPs can be sent and then used to determine the top outer lip FAPs. The inner lip FAPs would be mapped to the outer lip FAPs using a rational polynomial function that is specified in the FIT.

A FAP interpolation graph (FIG) is used to specify which FAPs are interpolated from other FAPs. The FIG is a graph with nodes and directed links. Each node contains a set of FAPs. Each link from a parent node to a child node indicates that the FAPs in a child node can be interpolated from those of the parent node. In a FIG, a FAP may appear in several nodes, and a node may have multiple parents. For a node that has multiple parent nodes, the parent nodes are ordered as 1st parent node, 2nd parent node, etc. During the interpolation process, if this child node needs to be interpolated, it is first interpolated from the 1st parent node if all FAPs in that parent node are available. Otherwise, it is interpolated from the 2nd parent node, and so on. An example of a FIG is shown in Figure 16. Each node has an ID. The numerical label on each incoming link indicates the order of these links. Each directed link in a FIG is a set of interpolation functions. Suppose F_1, $F_2, \ldots, F_n$ are the FAPs in a parent set and $f_1, f_2, \ldots, f_m$ are the FAPs in a child set. Then, there are m interpolation functions denoted as

$$f_1 = I_1(F_1, F_2, \ldots, F_n), f_2 = I_2(F_1, F_2, \ldots, F_n), \ldots, f_m = I_m(F_1, F_2, \ldots, F_n)$$

Each interpolation function $I_k(\,)$ is in the form of a rational polynomial

$$I(F_1, F_2, \ldots, F_n) = \sum_{i=0}^{K-1}\left(c_i \prod_{j=1}^{n} F_j^{l_{ij}}\right) \Big/ \sum_{i=0}^{P-1}\left(b_i \prod_{j=1}^{n} F_j^{m_{ij}}\right)$$

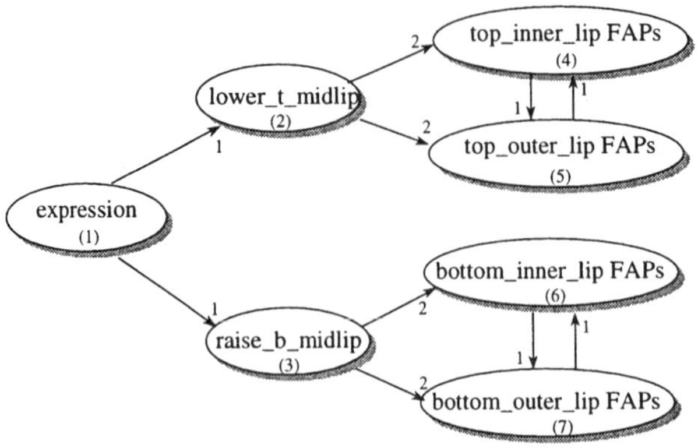

Figure 16 An FAP interpolation graph (FIG) for interpolating unspecified FAP values of the lips. If only the expression is specified, FAPs are interpolated from the expression. If all inner lip FAPs are specified, they are used to interpolate the outer lip FAPs.

where k and P are the numbers of polynomial products, c_i and b_i are the coefficients of the ith product and l_{ij} and m_{ij} are the power of F_j in the ith product. The encoder should send an interpolation function table that contains all k, P, c_i, b_i, l_{ij}, m_{ij} to the decoder for each link in the FIG.

E. Specification of Face Model and Animation Rules

In order to use face animation in the context of MPEG-4 systems [4], a BIFS scene graph has to be transmitted to the decoder. The minimum scene graph contains a *Face* node and a *FAP* node. The FAP node has children nodes *Viseme* and *Expression*, that are FAPs requiring a special syntax. The FAP decoder writes the amplitude of the FAPs into fields of the FAP, Viseme, and Expression nodes. This minimum scene graph would enable an encoder to animate the proprietary face model of the decoder. If a face is to be controlled from a TTS, an *AudioSource* node is to be attached to the face node. In order to download a face model to the decoder, the face node further requires a face definition parameter (FDP) node as one of its children. This FDP node contains the position of the feature points in the downloaded model; the scene graph of the model; and the *FaceDefTable, FaceDefMesh*, and *FaceDefTransform* nodes required to define the precise action caused by FAPs. If a profile is selected that allows the terminal to download a face, the FAP interpolation table (FIT) node may also be used as a child of the FAP node. Figure 17 shows how these nodes are related to each other.

Using the FDP node, MPEG-4 allows the encoder to specify completely the face model to animate [33]. This involves defining the static geometry of the face model in its neutral state using a scene graph, defining the surface properties, and defining the animation rules using face animation tables (FATs) that specify how this model is deformed by the facial animation parameters.

1. Neutral Face Model Using a Scene Graph

The static geometry of the head model is defined by a scene graph specified in MPEG-4 BIFS [4]. For this purpose, BIFS provides the same nodes as Virtual Reality Modeling

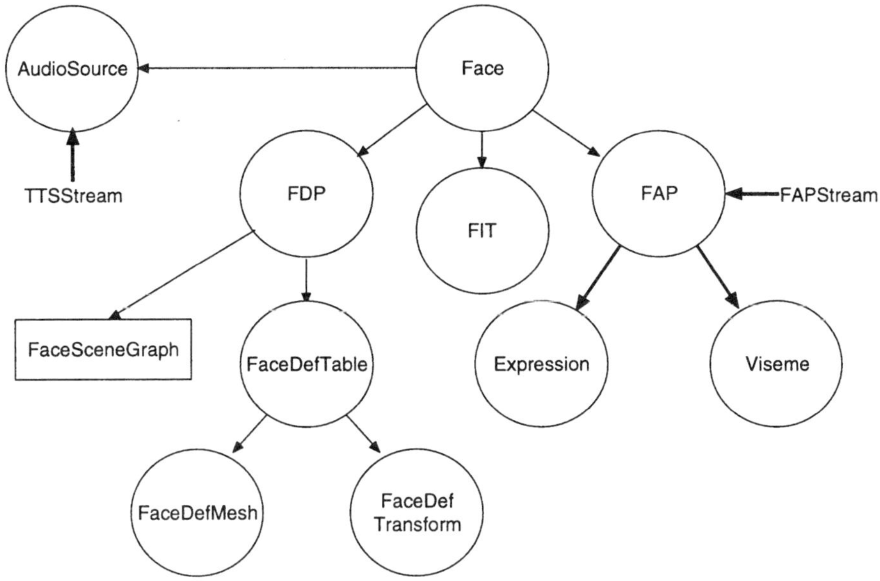

Figure 17 Nodes of a BIFS scene description tree that are used to describe and animate a face. The FaceSceneGraph contains the scene graph of the static face. The AudioSource node and the FAP node receive decoded streams.

Language (VRML) [34]. Three types of nodes are of particular interest for the definition of a static head model. A *Group* node is a container for collecting child objects: it allows building hierarchical models. For objects to move together as a group, they need to be in the same *Transform* group. The *Transform* node defines geometric 3D transformations such as scaling, rotation, and translation that are performed on its children. Nested *Transform* nodes can be used to build a transformation hierarchy. An *IndexedFaceSet* node defines the geometry (3D mesh) and surface attributes (color, texture) of a polygonal object. Texture maps can be coded with the wavelet coder of the MPEG-4 still image coder.

Figure 18 shows the simplified scene graph for a face model [35]. Nested *Transforms* are used to apply rotations about the x, y, and z axes one after another. Embedded into these global head movements are the rotations for the left and right eye. Separate IndexedFaceSets define the shape and the surface of the face, hair, tongue, teeth, left eye, and right eye, thus allowing separate texture maps. Because the face model is specified with a scene graph, this face model can easily be extended to a head and shoulder model.

2. Definition of Animation Rules Using Face Animation Tables

Face animation tables define how a model is spatially deformed as a function of the amplitude of the FAPs. This functionality is provided by three BIFS nodes: FaceDefTable, FaceDefTransform, and FaceDefMesh. Using FaceDefTransform nodes and FaceDefMesh nodes, a FaceDefTable specifies for a FAP which nodes of the scene graph are animated by it and how they are animated.

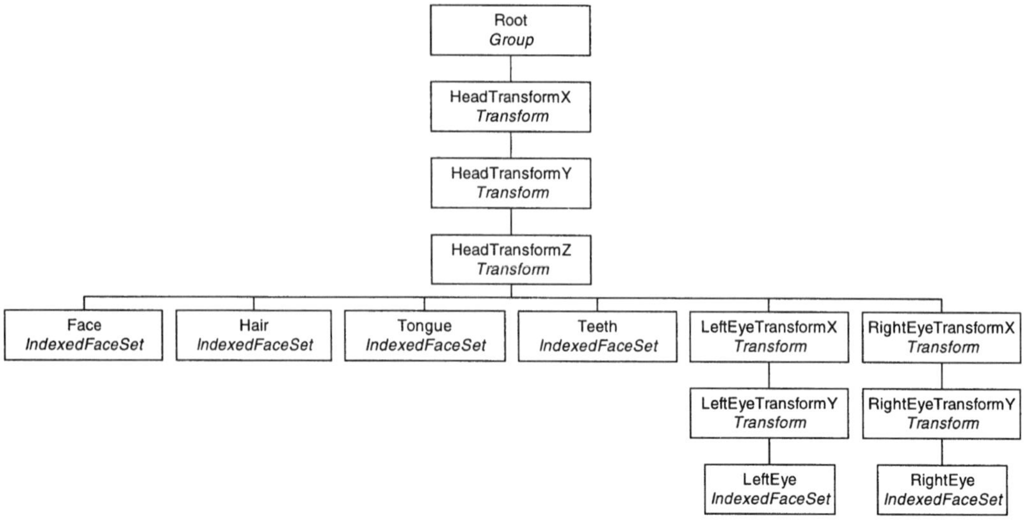

Figure 18 Simplified BIFS scene graph for a head model.

a. Animation Definition for a Transform Node. If a FAP causes a transformation such as rotation, translation, or scale, a Transform node can describe this animation. The FaceDefTable specifies a FaceDefTransform node that defines the type of transformation and a scaling factor for the chosen transformation. During animation, the received values for the FAP, the FAPU, and the scaling factor determine the actual value by which the model is transformed.

b. Animation Definition for an IndexedFaceSet Node. If a FAP causes flexible deformation of the face model, the FaceDefTable node uses a FaceDefMesh node to define the deformation of IndexedFaceSet nodes. The animation results in updating vertex positions of the affected IndexedFaceSet nodes. Moving the affected vertices as a piecewise linear function of FAP amplitude values approximates flexible deformations of an IndexedFaceSet [36,37]. The FaceDefMesh defines for each affected vertex its own piecewise linear function by specifying intervals of the FAP amplitude and 3D displacements for each interval (see Table 7 for an example). If a vertex is animated by different FAPs, the corresponding displacements are superimposed.

If $\boldsymbol{P}_m$ is the position of the mth vertex of the IndexedFaceSet in neutral state (FAP = 0) and $\boldsymbol{D}_{mk}$ is the 3D displacement that defines the piecewise linear function in the kth interval, then the following algorithm is used to determine the new position $\boldsymbol{P}'_m$ of the same vertex after animation with the given FAP value (Fig. 19):

1. Determine the interval listed in the FaceDefMesh containing the received FAP value.
2. If the received FAP is in the jth interval $[I_j, I_{j+1}]$ and $0 = I_k \leq I_j$, the new position P'_m of the mth vertex is given by

$$\boldsymbol{P}'_m = \boldsymbol{P}_m + \text{FAPU} * ((I_{k+1} - 0) * \boldsymbol{D}_{m,k} + (I_{k+2} - I_{k+1}) * \boldsymbol{D}_{m,k+1} + \cdots (I_j - I_{j-1}) * \boldsymbol{D}_{m,j-1} + (\text{FAP} - I_j) * \boldsymbol{D}_{m,j})$$

Table 7 Simplified Example of a FaceDefMesh and a FaceDefTransform Node

```
#FaceDefMesh
FAP 6 (stretch left corner lip)
    IndexedFaceSet: Face
      interval borders: −1000, 0, 500, 1000
      displacements:
        vertex 50 1 0 0, 0.9 0 0, 1.5 0 4
        vertex 51 0.8 0 0, 0.7 0 0, 2 0 0
#FaceDefTransform
FAP 23 (yaw left eye ball)
    Transform: LeftEyeX
      rotation scale factor: 0 −1 0 (axis) 1 (angle)
```

3. If FAP $> I_{max}$, then P'_m is calculated by using the equation given in 2 and setting the index $j = \max - 1$.
4. If the received FAP is in the jth interval $[I_j, I_{j+1}]$ and $I_{j+1} \leq I_k = 0$, the new position P'_m of the mth vertex is given by

$$\boldsymbol{P}'_m = \boldsymbol{P}_m + \text{FAPU} * ((I_{j+1} - \text{FAP}) * \boldsymbol{D}_{m,j} + (I_{j+2} - I_{j+1}) * \boldsymbol{D}_{m,j+1} + \cdots (I_{k-1} - I_{k-2}) * \boldsymbol{D}_{m,k-2} + (0 - I_{k-1}) * \boldsymbol{D}_{m,k-1})$$

5. If FAP $< I_1$, then P'_m is calculated by using the equation in 4 and setting the index $j = 1$.
6. If for a given FAP and ''IndexedFaceSet'' the table contains only one interval, the motion is strictly linear:

$$\boldsymbol{P}'_m = \boldsymbol{P}_m + \text{FAPU} * \text{FAP} * \boldsymbol{D}_{ml}$$

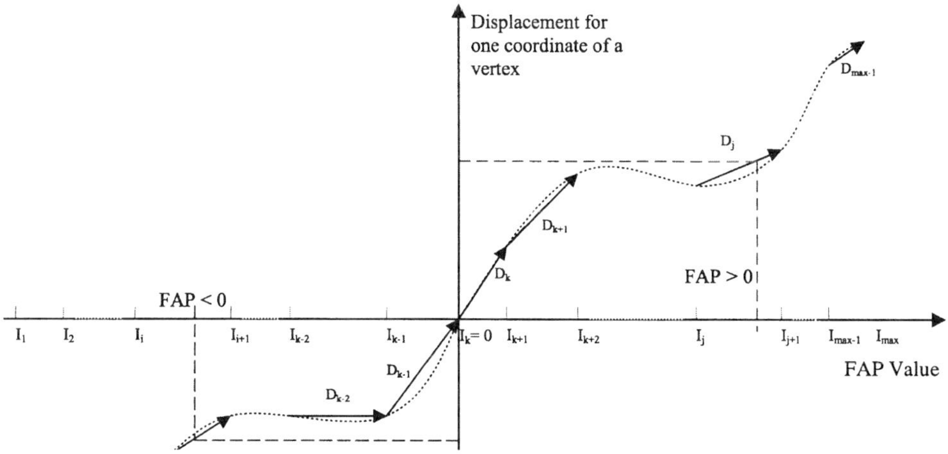

Figure 19 Piecewise linear approximation of vertex motion as a function of the FAP value.

Strictly speaking, these animation rules are not limited to faces. Using this method, MPEG-4 provides an efficient mechanism for overloading the default meaning of FAPs and animating arbitrary objects such as a torso with head, body, arms, and hands with up to 68 FAPs.

c. Example of FaceDefTable. In Table 7, two FAPs are defined by *FaceDef-Tables*: FAP 6, which stretches the left corner lip, and FAP 23, which manipulates the horizontal orientation of the left eyeball. FAP 6 deforms the IndexedFaceSet *Face*. For the piecewise linear motion function three intervals are defined: [−1000, 0], [0, 500], and [500, 1000]. Displacements are given for the vertices with indices 50 and 51. The displacements for vertex 50 are (1 0 0), (0.9 0 0), and (1.5 0 4); the displacements for vertex 51 are (0.8 0 0), (0.7 0 0), and (2 0 0). Given a FAPValue of 600, the resulting displacement for vertex 50 would be

$$\boldsymbol{P}'_{50} = \boldsymbol{P}_{50} + 500{*}(0.9\ 0\ 0)^{\mathrm{T}} + 100 * (1.5\ 0\ 4)^{\mathrm{T}} = \boldsymbol{P}_{50} + (600\ 0\ 400)^{\mathrm{T}}$$

FAP 23 updates the rotation field of the Transform node *LeftEyeX*. The rotation axis is (0, − 1, 0), and the neutral angle is 0 radian. The FAP value determines the rotation angle.

d. Face Animation Table Generation. The creation of FaceDefMesh nodes for large models can be time consuming. However, the process depicted in Figure 20 uses a FaceDefTable generator that computes these tables from a set of face models. The face model is described as a VRML file and read into the modeler. In order to design the behavior of the model for one animation parameter, the model is deformed using the tools of the modeler. The modeler may not change the topology of the model. The modeler exports the deformed model as a VRML file.

The FaceDefMesh generator compares the output of the modeler with its input, the face model in its neutral state. By comparing vertex positions of the two models, the vertices affected by the newly designed animation parameter are identified. The generator computes for each affected vertex a 3D displacement vector defining the deformation and exports this information in a FaceDefMesh table. The renderer reads the VRML file of the model and the table in order to learn the definition of the new animation parameter. Now, the renderer can use the newly defined animation as required by the animation parameters.

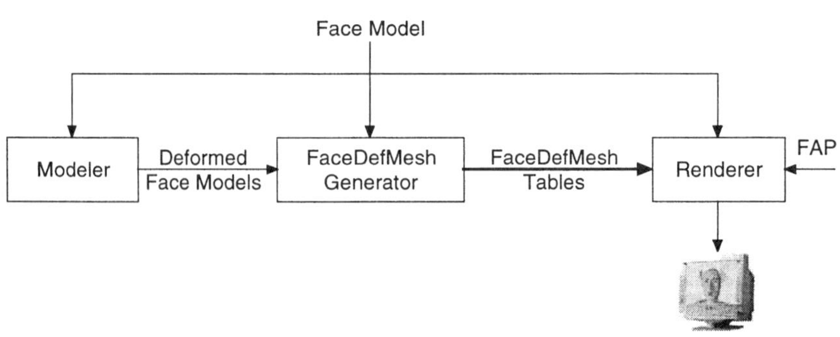

Figure 20 FaceDefMesh Interface. The modeler is used to generate VMRL files with the object in different animated positions. The generator computes one FaceDefMesh for each animation parameter.

F. Integration of Face Animation and Text-to-Speech Synthesis

MPEG-4 provides interfaces to a proprietary text-to-speech (TTS) synthesizer that allows driving a talking head from text, as illustrated in Figure 21 [23,24,38–47]. This section discusses the integration of face animation and TTS [41] allowing animation of a talking face using a TTS synthesizer. A key issue here is the synchronization of animation of facial expressions with a TTS that does not allow prediction of the time required to speak a sentence.

Given a TTS stream that contains text or prosody in binary form, the MPEG-4 terminal decodes the text and prosody information according to the interface defined for the TTS synthesizer (Fig. 21). The synthesizer creates speech samples that are handed to the compositor. The compositor presents audio and, if required, video to the user. The second output interface of the synthesizer sends the phonemes of the synthesized speech as well as start time and duration information for each phoneme to the Phoneme/Bookmark-to-FAP Converter [5,48]. The converter translates the phonemes and timing information into face animation parameters that the face renderer uses in order to animate the face model. The precise method by which the converter derives visemes from phonemes is not specified by MPEG and left to the implementation of the MPEG-4 player.

Bookmarks in the text of the TTS are used to animate facial expressions and non–speech-related parts of the face [48]. The start time of a bookmark is derived from its position in the text. When the TTS finds a bookmark in the text it sends this bookmark to the Phoneme/Bookmark-to-FAP Converter at the same time as it sends the first phoneme of the following word. The bookmark defines the *start* point and *duration* of the transition to a new FAP amplitude. The consequences are no additional delay, no look ahead in the bitstream, but no precise timing control on when the target amplitude will be reached relative to the spoken text.

An example of a TTS stream with bookmarks is shown in Figure 22 [48]. The renderer will generate the visemes associated with each word, following the timing information derived by the speech synthesizer. It will also start to deform the model to generate a joy expression with an amplitude of 100. To simulate a more natural expression, which

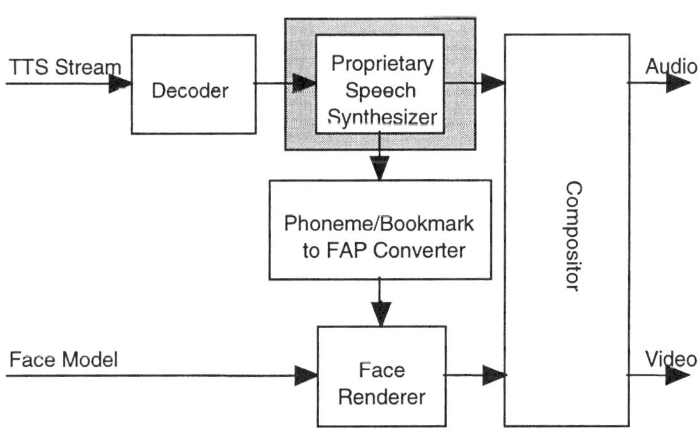

Figure 21 Block diagram showing the integration of a proprietary text-to-speech synthesizer into an MPEG-4 face animation system.

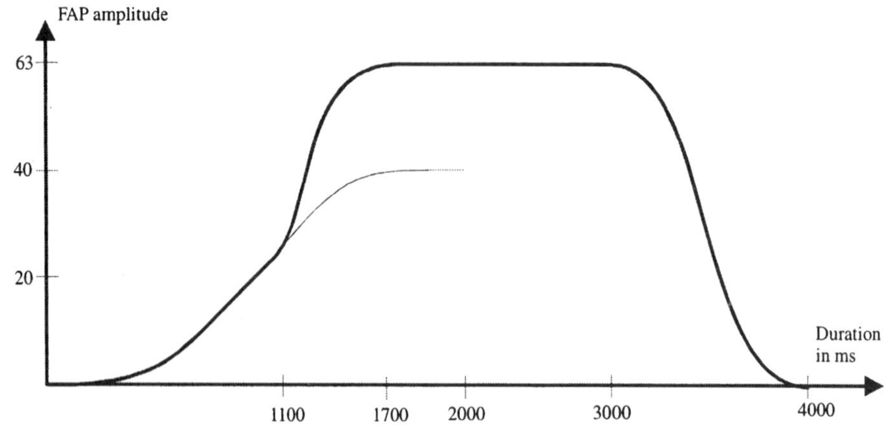

<FAP 2 1 40 1 40 2000 3>.text.<FAP 2 1 63 1 63 600 3> ....text.........<FAP 2 1 0 1 0 1000 3>...text.....

Figure 22 Example for text with bookmarks for one facial expression (joy) and the related amplitude of the animated FAP. The syntax of the bookmark is: ⟨FAP 2 (expression) 1 (joy) 40 (amplitude) 1 (joy) 40 (amplitude duration) 3 (Hermite time curve)⟩. The amplitude of joy is computed according to the bookmarks.

typically goes through three phases (onset, climax, and relax), a desired temporal behavior for an FAP can be specified in the bookmark. Three functions are defined: A linear interpolation function and a Hermite function can be specified for the transition of a FAP from its current amplitude to the target amplitude. A triangular function can be specified to increase linearly the amplitude of an FAP to the target and to decrease it back to its starting amplitude. A bookmark also specifies the desired duration to reach the FAP amplitude. If another bookmark appears before this transition interval, the renderer starts to deform the face according to the newly specified FAP information from the current position. This is illustrated in Figure 22 using Hermite functions.

G. Face Coding Results

This section discusses some FAP coding efficiency results. Figure 23 shows the bit rate as a function of the quantization parameter for the FAP sequences derived from the ‘‘Opossum2’’ sequence. An important technique in the MPEG-4 face animation coding system

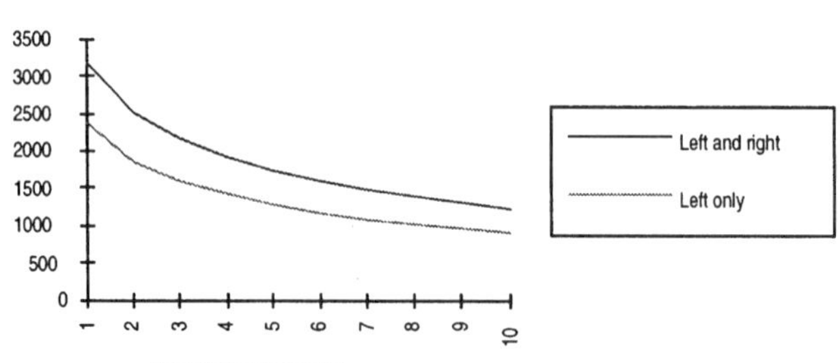

Figure 23 Coded FAP bit rates for 10 QP values with and without left–right interpolation.

is that FAPs that are not received by the terminal may be interpolated. Interpolation must be performed in the terminal if either inner or outer lip FAPs are not received and if the left or right side of a left–right FAP pair is not received. This policy allows symmetric faces to be coded at a lower bit rate by sending only one side of left–right FAP pairs. Figure 23 shows the bit rate as a function of quantization parameter for left only and left–right FAP sequences for Opossum2. Bit rate savings from left–right interpolation are about 25% (not 50%) because of midface FAPs and common coding overhead.

IV. APPLICATIONS AND SUMMARY

A. Applications of Synthetic Video Coding

MPEG-4 foresees that talking heads will play an important role in future Web applications. For example, a customized agent model can be defined for games or Web-based customer service applications. To this effect, MPEG-4 enables integration of face animation with multimedia communications and presentations and allows face animation over low-bit-rate communication channels, for point-to-point as well as multipoint connections with low delay. It has been demonstrated that face models can be successfully animated with a data rate of 300–2000 bit/sec. In many applications, the integration of face animation and text-to-speech synthesizer is of special interest. MPEG-4 defines an application program interface for a (possibly proprietary) TTS synthesizer. Using this interface, the synthesizer can be used to provide phonemes and related timing information to the face model. The phonemes are converted into corresponding mouth shapes enabling simple talking-head applications. Adding facial expressions to the talking head is achieved using bookmarks in the text. This integration allows animated talking heads driven just by one text stream.

Efficient animations of generic 2D objects are enabled in MPEG-4 by the 2D mesh coding tool. Objects to be animated may be synthetic and consist of only a few triangles; on the other hand, more complex objects may be modeled using a few hundred triangles. The corresponding data rate may vary between a few thousands and a few ten-thousands kbit per second. Simple animations can be used to enhance multimedia presentations or electronic multiplayer games. The 2D mesh tool in MPEG-4 can also be used to create special effects such as augmented reality video clips, in which natural video objects are overlaid by synthetic graphics or text, and to allow manipulation of video by the user.

MPEG-4 is well suited for applications that incorporate natural and synthetic content. As an example, the virtual travel assistant illustrated in Figure 24 could use the 2D mesh object tool for low bit-rate coding of natural scenes or synthetic backgrounds or to display animated advertising logos. A synthetic animated face could describe and point to areas of interest in the video scene. The face could be driven by a real person or by synthetic speech.

MPEG-4 (version 1) provides several profiles that combine synthetic and natural video coding tools such as scalable texture coding, mesh animation, and face animation coding. MPEG-4 also provides a profile that allows face animation parameter coding. Such profiles support the applications envisioned for MPEG-4.

B. Summary

As video analysis technology becomes more accurate and efficient, MPEG-4 is poised to enable the new forms of visual communication that incorporate both sample-based

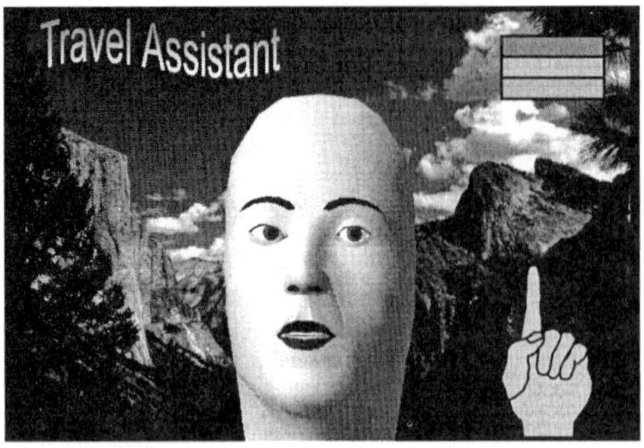

Figure 24 Application example: virtual travel assistant.

information and object-based information. This combination will simultaneously provide both higher image quality at a given bit rate and important functionalities for speech recognition, coding, indexing, storage, manipulation, and entertainment. MPEG-4 provides a comprehensive set of representation and coding tools for interactive audio, visual, and synthetic content. 2D mesh objects provide a bridge between natural and synthetic video content. Mesh objects enable animation of simple logos and other graphics or background textures and special effects generation such as moving text overlays. Face animation coding in MPEG-4 integrates video processing, coding, graphics, and animation to offer a new form of visual communication that is very low in bit rate and allows the option of visual privacy (e.g., in 3D chat rooms) when other than realistic face models are used. Version 2 of MPEG-4 will also provide body animation coding. Eventually, improvements in scene analysis and rendering will allow MPEG-4 to deliver photorealistic scenes represented as texture-mapped 2D and 3D objects at very low bit rates compared with traditional video coding techniques.

REFERENCES

1. H Musmann, M Hötter, J Ostermann. Object-oriented analysis–synthesis coding of moving images. Signal Process Image Commun 1:117–138, October 1989.
2. J Ostermann. Object-oriented analysis-synthesis coding (OOASC) based on the source model of moving flexible 3D objects. IEEE Trans Image Process 3(5):705–711, 1994.
3. L Chiariglione. MPEG and multimedia communications. IEEE Trans Circ Syst Video Tech 7:5–18, 1997.
4. ISO/IEC JTC1/SC29/WG11. Final Draft of International Standard ISO/IEC 14496-1, Coding of Audio-visual Objects: Systems, Atlantic City, Oct. 1998.
5. ISO/IEC JTC1/SC29/WG11. Final Draft of International Standard ISO/IEC 14496-2, Coding of Audio-visual Objects: Visual, Atlantic City, Oct. 1998.
6. ISO/IEC JTC1/SC29/WG11. Final Draft of International Standard ISO/IEC 14496-3. Coding of Audio-visual Objects: Audio, Atlantic City, Oct. 1998.
7. D Hearn, MP Baker. Computer Graphics, second edition, Prentice Hall, New York, 1997.
8. G Wolberg. Digital Image Warping, Computer Society Press, Los Alamitos, CA, 1990.
9. Y Nakaya, H Harashima. Motion compensation based on spatial transformations. IEEE Trans Circ Syst Video Tech 4:339–356, 1994.
10. Y Wang, O Lee, A Vetro. Use of two-dimensional deformable mesh structures for video coding, Part, I and II. IEEE Trans Circ Syst Video Tech 6:636–659, 1996.

11. Y Altunbasak, AM Tekalp. Occlusion-adaptive, content-based mesh design and forward tracking. IEEE Trans. on Image Processing, 6:1270–1280, 1997.
12. PJL van Beek, AM Tekalp. Object-based video coding using forward tracking 2-D mesh layers. Visual Communications Image Processing '97, San Jose, CA, Feb. 1997.
13. Y Wang, J Ostermann. Evaluation of mesh-based motion estimation in H.263 like coders. IEEE Trans Circuits Syst Video Technol 8:243–252, 1998.
14. AM Tekalp, P van Beek, C Toklu, B Günsel. 2D Mesh-based visual object representation for interactive synthetic/natural digital video. Proceedings of the IEEE, 86:1029–1051, 1998.
15. PE Eren, C Toklu, AM Tekalp. Special effects authoring using 2-D mesh models. IEEE Int Conference on Image Processing, Santa Barbara, CA, Oct. 1997.
16. M de Berg, M van Kreveld, M Overmars, O Schwarzkopf. Computational Geometry—Algorithms and Applications. Berlin: Springer, 1997.
17. Y Wang, O Lee. Active mesh—A feature seeking and tracking image sequence representation scheme. IEEE Trans Image Process 3:610–624, 1994.
18. C Toklu, AT Erdem, MI Sezan, AM Tekalp. Tracking motion and intensity variations using hierarchical 2-D mesh modeling. Graphical Models and Image Process 58:553–573, 1996.
19. P van Beek, AM Tekalp, N Zhuang, I Celasun, M Xia. Hierarchical 2D mesh representation, tracking and compression for object-based video. IEEE Trans Circ Syst Video Tech 9(2):353–369, March 1999.
20. G Taubin, WP Horn, J Rossignac, F Lazarus. Geometry coding and VRML. Proceedings of the IEEE, 86:1228–1243, 1998.
21. PJL van Beek, AM Tekalp, A Puri. 2-D Mesh geometry and motion compression for efficient object-based video representation. IEEE Int. Conference on Image Processing, Santa Barbara, CA, Oct. 1997.
22. M Kampmann, J Ostermann. Automatic adaptation of a face model in a layered coder with an object-based analysis–synthesis layer and a knowledge-based layer. Signal Process Image Commun 9:201–220, 1997.
23. J Ostermann, L Chen, T Huang. adaptation of a generic 3D human face model to 3D range data. VLSI Signal Process Syst Speech Image Video Tech 20:99–107, 1998.
24. M Escher, NM Thalmann. Automatic 3D cloning and real-time animation of human face, Proc Computer Animation '97, Geneva, 1997.
25. HP Graf, T Chen, E Petajan, E Cosatto. Locating faces and facial parts, Proc. Int. Workshop on Automatic Face- and Gesture-Recognition, M. Bichsel (ed.), Zürich, 1995, pp. 41–46.
26. HP Graf, E Cosatto, D Gibbon, M Kocheisen, E Petajan. Multi-Modal System for Locating Heads and Faces, Proc. 2nd Int. Conf. on Automatic Face and Gesture Recognition, IEEE Computer Soc. Press, 1996, pp. 88–93.
27. ED Petajan. Automatic Lipreading to Enhance Speech Recognition, PhD Thesis, University of Illinois at Urbana-Champaigne, 1984.
28. ED Petajan. Automatic Lipreading to Enhance Speech Recognition, Proc. Globecom Telecommunications Conference, pp. 265–272, IEEE, 1984.
29. ED Petajan. Automatic Lipreading to Enhance Speech Recognition, Proc. of the IEEE Conference on Computer Vision and Pattern Recognition, pp. 40–47, IEEE, 1985.
30. ED Petajan, NM Brooke, GJ Bischoff, DA Bodoff. An improved automatic lipreading system to enhance speech recognition, Proc Human Factors in Computing Syst pp. 19–25, ACM, 1988.
31. A Goldschen. Continuous Automatic Speech Recognition by Lipreading, PhD Thesis, George Washington University, 1993.
32. A Goldschen, O Garcia, E Petajan. Continuous optical automatic speech recognition, Proceedings of the 28th Asilomar Conference on Signals, Systems, and Computers, pp. 572–577, IEEE, 1994.
33. J Ostermann, E Haratsch. An animation definition interface: Rapid design of MPEG-4 compliant animated faces and bodies, International Workshop on synthetic—natural hybrid coding and three dimensional imaging, pp. 216–219, Rhodes, Greece, Sept. 5–9, 1997.

34. ISO/IEC 14772-1:1997, Information Technology—Computer graphics and image processing—The Virtual Reality Modeling Language—Part 1: Functional specification and UTF-8 encoding.
35. E Haratsch, J Ostermann. Parameter based animation of arbitrary 3D head models. Proceedings of Picture Coding Symposium 1997 (PCS 97), VDE Verlag, Berlin, September 1997, pp 81–84.
36. D Terzopolous, K Waters. Physically-based facial modeling, analysis and animation. J Visualization Comput Animation 1:73–80, 1990.
37. K Waters. A muscle model of animating three dimensional facial expression. Comput Graphics 22:17–24, 1987.
38. S Morishima, H Harashima. A media conversion from speech to facial image for intelligent man–machine interface. IEEE J Selected Areas Commun 9:594–600, May 1991.
39. P Kalra, A Mangili, N Magnenat-Thalmann, D Thalmann. Simulation of facial muscle actions based on rational free form deformations. Proceedings of Eurographics 92, 1992, pp 59–69.
40. K Nagao, A Takeuchi. Speech dialogue with facial displays: Multimodal human–computer conversation. Proceedings of the 32nd Annual Meeting of the Association for Computational Linguistics, 1994, pp 102–109.
41. R Sproat, J Olive. An approach to text-to-speech synthesis. In: WB Kleijn, KK Paliwal, eds. Speech Coding and Synthesis. Amsterdam: Elsevier Science, 1995.
42. FI Parke. Parameterized models for facial animation. IEEE Comput Graphics Appl 2:61–68, November 1982.
43. MM Cohen, DW Massaro. Modeling coarticulation in synthetic visual speech. In: M Thalmann, D Thalmann, eds. Computer Animation '93. Tokyo: Springer-Verlag, pp. 139–156, 1993.
44. K Waters, T Levergood. An automatic lip-synchronization algorithm for synthetic faces, Proceedings of the ACM Multimedia Conference, San Francisco, September 1994, pp 149–156.
45. C Bregler, M Covell, M Slaney. Video rewrite: Driving visual speech with audio. Proceedings of ACM SIGGRAPH 97, in Computer Graphics Proceedings, Annual Conference Series, 1997.
46. S Morishima. Modeling of facial expression and emotion for human communication system. Display 17:15–25, 1996.
47. E Cosatto, H Graf. Sample-based of photo-realistic talking heads. Computer Animation, Philadelphia, June 8–10, 1998, pp 103–110.
48. J Ostermann, M Beutnagel, A Fischer, Y Wang. Integration of talking heads and text-to-speech synthesizers for visual TTS. ICSLP 99, Australia, December 1999.

12
MPEG-4 Systems: Overview

Olivier Avaro
Deutsche Telekom–Berkom GmbH, Darmstadt, Germany

Alexandros Eleftheriadis
Columbia University, New York, New York

Carsten Herpel
Thomson Multimedia, Hannover, Germany

Ganesh Rajan
General Instrument, San Diego, California

Liam Ward
Teltec Ireland, DCU, Dublin, Ireland

I. INTRODUCTION

The concept of "Systems" in MPEG has evolved dramatically since the development of the MPEG-1 and MPEG-2 standards. In the past, Systems referred only to overall architecture, multiplexing, and synchronization. In MPEG-4, in addition to these issues, the Systems part encompasses scene description, interactivity, content description, and programmability. The combination of the exciting new ways of creating compelling interactive audiovisual content offered by MPEG-4 Systems and the efficient representation tools provided by the Visual and Audio parts promises to be the foundation of a new way of thinking about audiovisual information.

This chapter gives an overview of MPEG-4 Systems. It is structured around the objectives, history, and architecture as well as the tools and applications of MPEG-4 Systems as follows:

Objectives: This section describes the motivations and the rationale behind the development of the MPEG-4 Systems specifications. As with all MPEG activities, MPEG-4 Systems is guided by a set of requirements [1], i.e., the set of objectives that must be satisfied by the specification resulting from the work activities of the subgroup. Particular attention is given to the way the requirements of MPEG-4 Systems are derived from the principal concept behind MPEG-4, namely the coding of audiovisual objects.

History: It was not possible to match the objectives with technologies available at the time the MPEG-4 Call for Proposals was issued, because the relevant technologies were not sufficiently well developed. Hence, MPEG-4 Systems did not follow the conventional standards-building process: a competitive phase that evaluates proposals that use or extend existing technologies, followed by a collaborative phase in which the best ideas from the various proposals are integrated to match the given requirements. This section is therefore a history of ideas, describing the original MPEG-4 Systems key concepts, the definition of an appropriate architecture and set of tools, and the organization of the phased development of these tools.

Architecture: This section describes the overall structure of MPEG-4, known as the MPEG-4 Systems Architecture. A complete walkthrough of an MPEG-4 session highlights the different phases that a user will, in general, follow in consuming MPEG-4 content.

Tools: MPEG-4 is a ''toolbox'' standard, providing a number of tools, sets of which are particularly suited to certain applications. A functional description of the technology that led to the standardization of the MPEG-4 Systems tools is provided in this section. These tools are further described in the chapters that follow and are fully specified in Refs. 2 and 3.

Applications: In order to illustrate the real-world uses of the ideas described in the preceding sections, a specific section is devoted to the description of applications. For each of the application domains identified in MPEG-4, namely interactive applications, broadcast applications and conferencing applications, to name a few, the specific features of MPEG-4 Systems that enhance the application or provide new functionalities in the application context are emphasized.

Of course, MPEG-4 is not the only initiative that attempts to provide solutions in this area. Several companies, industry consortia, and even other standardization bodies have developed technologies that, to some extent, also aim to address objectives similar to those of MPEG-4 Systems. In concluding this look at MPEG-4 Systems, an overview of some of these alternative technologies is given and a comparison with the solutions provided by MPEG-4 Systems is made.

It is assumed that the reader has a general knowledge of multimedia representation and applications. However, because the architecture and tools are described at a functional level, in-depth technical knowledge is not required. Readers not familiar with the overall MPEG-4 framework may want to consult the introductory chapter of this book.

II. OBJECTIVES

A. Requirements

To understand the rationale behind this activity, a good starting point is the MPEG-4 requirements document [1]. This document gives an extensive list of the objectives that needed to be satisfied by the MPEG-4 specification. The goal of facilitating the description and coding of audiovisual objects was the primary motivation behind the development of the tools in the MPEG-4 Systems.

MPEG-4 Systems requirements may be categorized into two groups:

Traditional MPEG Systems requirements: The core requirements for the development of the systems specifications in MPEG-1 and MPEG-2 were to enable the transport of coded audio, video, and user-defined private data and incorporate timing mechanisms to facilitate synchronous decoding and presentation of these data at the client side. These requirements also constitute a part of the fundamental requirements set for MPEG-4 Systems. The evolution of the traditional MPEG Systems activities to match the objectives for MPEG-4 Systems is detailed in Sec. II.B.

Specific MPEG-4 Systems requirements: The requirements in this set, most notably the notions of audiovisual objects and scene description, represent the ideas central to MPEG-4 and are completely new in MPEG Systems. The core competencies needed to fulfill these requirements were not present at the beginning of the activity but were acquired during the standards development process. Section II.C describes these specific MPEG-4 Systems requirements.

To round out this discussion on the MPEG-4 objectives, Sec. II.D provides an answer to the question ''What is MPEG-4 Systems?'' by summarizing the objectives of the MPEG-4 Systems activity and describing the charter of the MPEG-4 Systems subgroup during its 4 years of existence.

B. Traditional MPEG Systems Requirements

The work of MPEG traditionally addressed the representation of audiovisual information. In the past, this included only natural* audio and video material. As we will indicate in subsequent sections, the types of media included within the scope of the MPEG-4 standards have been significantly extended. Regardless of the type of the medium, each one has spatial and/or temporal attributes and needs to be identified and accessed by the application consuming the content. This results in a set of general requirements for MPEG Systems:

The audiovisual information is to be delivered in a streaming manner, suitable for live broadcast of such content. In other words, the audiovisual data are to be transmitted piece by piece, in order to match the delivery of the content to clients with limited network and terminal capabilities. This is in stark contrast to some of the existing scenarios, the World Wide Web, for example, wherein the audiovisual information is completely downloaded onto the client terminal and then played back. It was thought that such scenarios would necessitate too much storage on the client terminals for applications envisaged by MPEG.

Typically, the different components of an audiovisual presentation are closely related in time. For most applications, audio samples with associated video frames have to be presented together to the user at precise instants in time. The MPEG representation needs to allow a precise definition of the notion of time so that data received in a streaming manner can be processed and presented at the right instants in time and be temporally synchronized with each other.

Finally, the complete management of streams of audiovisual information implies the need for certain mechanisms to allow an application to consume the content. These include mechanisms for unambiguous location of the content, iden-

* ''Natural,'' in this context, is generally understood to mean representations of the real world that are captured using cameras, microphones, and so on, as opposed to synthetically generated material.

tification of the content type, description of the dependencies between content elements, and access to the intellectual property information associated with the data.

In the MPEG-1 and MPEG-2 standards, these requirements led to the definition of the following tools:

1. *Systems Target Decoder (STD)*: The Systems Target Decoder is an abstract model of an MPEG decoding terminal that describes the idealized decoder architecture and defines the behavior of its architectural elements. The STD provides a precise definition of time and recovery of timing information from information encoded within the streams themselves, as well as mechanisms for synchronizing streams with each other. It also allows the management of the decoder's buffers.
2. *Packetization of streams*: This set of tools defines the organization of the various audiovisual data into streams. First, definition of the structure of individual streams containing data of a single type is provided (e.g., a video stream), followed by the multiplexing of different individual streams for transport over a network or storage in disk files. At each level, additional information is included to allow the complete management of the streams (synchronization, intellectual property rights, etc.).

All these requirements are still relevant for MPEG-4. However, the existing tools needed to be extended and adapted for the MPEG-4 context. In some cases, these requirements led to the creation of new tools, more specifically:

1. *Systems Decoder Model (SDM)*: The MPEG-4 streams can be different from the ones in the traditional MPEG-1 and MPEG-2 models. For example, MPEG-4 streams may have a bursty data delivery schedule. They may be downloaded and cached before their actual presentation to the user. Moreover, to implement the MPEG-4 principle of ''create once, access everywhere,'' the transport of content does not need to be (indeed, should not be) integrated into the model. These new aspects, therefore, led to a modification of the MPEG-1 and MPEG-2 models resulting in the MPEG-4 Systems Decoder Model.
2. *Synchronization*: The MPEG-4 principle of create one, access everywhere is easier to achieve when all of the content-related information forms part of the encoded representation of the multimedia content. This also includes synchronization information. The range of bit rates addressed by MPEG-4 is broader than in MPEG-1 and MPEG-2 and can be from a few kbit/sec up to several Mbit/sec. These observations led to the definition of a flexible tool, called the *sync layer*, to encode the synchronization information.
3. *Packetization of streams*: On the delivery side, most of the existing networks provide ways for packetization and transport of streams. Therefore, beyond defining the modes for the transport of MPEG-4 content on the existing infrastructures, MPEG-4 Systems did not see a need to develop any new tools for this purpose. However, due to the possibly unpredictable temporal behavior of MPEG-4 data streams as well as the possibly large number of such streams in MPEG-4 applications, MPEG-4 Systems developed an efficient and simple multiplexing tool to enhance the transport of MPEG-4 data: the *FlexMux* tool. Finally, MPEG-4 Systems completes the toolbox for transport and storage of MPEG-4 content by providing a file format [3].

C. MPEG-4 Specific Systems Requirements

The foundation of MPEG-4 is the coding *of audiovisual objects*. As per MPEG-4 terminology, an audiovisual object is the representation of a natural or synthetic object that has an audio and/or visual manifestation. Examples of audiovisual objects include a video sequence (perhaps with shape information), an audio track, an animated 3D face, speech synthesized from text, and a background consisting of a still image.

The advantages of coding audiovisual objects can be summarized as follows:

It facilitates interaction with the content. At the client side, users can be given the possibility to access, manipulate, or activate specific parts of the content.

It improves reusability and coding of the content. At the content creation side, authors can easily organize and manipulate individual components and reuse existing material. Moreover, each type of content can be coded using the most effective algorithms. Artifacts related to joint coding of heterogeneous objects (e.g., graphics overlaid on natural video) disappear.

It allows content-based scalability. At various stages in the authoring-delivery-consumption process, content can be ignored or adapted to match bandwidth, complexity, or price requirements.

In order to be able to use these audiovisual objects in a presentation, additional information needs to be transmitted to the client terminals. The individual audiovisual objects are only a part of the presentation structure that an author wants delivered to the consumers. Indeed, for the presentation at the client terminals, the coding of audiovisual objects needs to be augmented by the following:

1. The coding of information that describes the spatiotemporal relationships between the various audiovisual objects present in the presentation content. In MPEG-4 terminology, this information is referred to as the *Scene Description* information.
2. The coding of information that describes how time-dependent objects in the scene description are linked to the streamed resources actually transporting the time-dependent information.

These considerations imply additional requirements for the overall architectural design, which are summarized here:

Object description: In addition to the identification of the location of the streams, other information may need to be attached to streamed resources. This may include the identification of streams in alternative formats or a scalable stream hierarchy that may be attached to the object or description of the coding format of the object.

Content authoring: The object description information is conceptually different from the scene description information. It will therefore have different life cycles. For example, at some instant in time, object description information may change (such as the intellectual property rights of a stream or the availability of new streams) while the scene description information remains the same. Similarly, the structure of the scene may change (e.g., changing the positions of the objects) while the streaming resources remain the same.

Content consumption: The consumer of the content may wish to obtain information about the content (e.g., the intellectual property attached to it or the maximum bit rate needed to access it) before actually requesting it. The consumer then only needs to receive information about object description, not about the scene description.

Besides the coding of audiovisual objects organized spatiotemporally, according to a scene description, one of the key concepts of MPEG-4 is the idea of interactivity, meaning that the content reacts upon the action of a user. This general idea is expressed in three specific requirements:

1. The user should be able to manipulate the scene description as well as the properties of the audiovisual objects that the author wants to expose for interaction.
2. It should be possible to attach behavior to audiovisual objects. User actions or other events, such as time, trigger these behaviors.
3. Finally, in case a return channel from the client to the server is available, the user should be able to send back information to the server that will act upon it and eventually send updates or modification of the content.

Sections IV and V explore how these requirements have been addressed in the MPEG-4 Systems specifications.

D. What Is MPEG-4 Systems?

The main concepts that were described in this section are depicted in Figure 1. The mission, therefore, of the MPEG-4 Systems activity may be summarized by the following sentence: ''Develop a coded, streamable representation for audiovisual objects and their associated time-variant data along with a description of how they are combined.''

More precisely, in this sentence:

''Coded representation'' should be seen in contrast to ''textual representation.'' Indeed, all the information that MPEG-4 Systems contains (scene description, object description, synchronization information) is binary encoded for bandwidth efficiency.

''Streamable'' should *not* be seen in contrast to ''stored,'' because storage and transport are dealt with in a similar and consistent way in the MPEG-4 framework. It should rather be seen in contrast to ''downloaded.'' Indeed, MPEG-4 is built on the concept of streams that have a temporal extension and not on the concept of files of finite size.

''Elementary audiovisual sources along with a description of how they are combined'' should be seen in contrast to ''individual audio or visual streams.'' MPEG-4 Systems does not deal with the encoding of audio or visual information but only with the information related to the combinations of streams: combination of audiovisual objects to create an interactive audiovisual scene, synchronization of streams, multiplexing of streams for storage or transport. The term ''description'' here reinforces the fact that the combination involves explicit content authoring.

III. HISTORY

A. Introduction

This section describes the MPEG-4 Systems activity from a historical perspective. This description deals with a history of ideas rather than a history of the people who were the

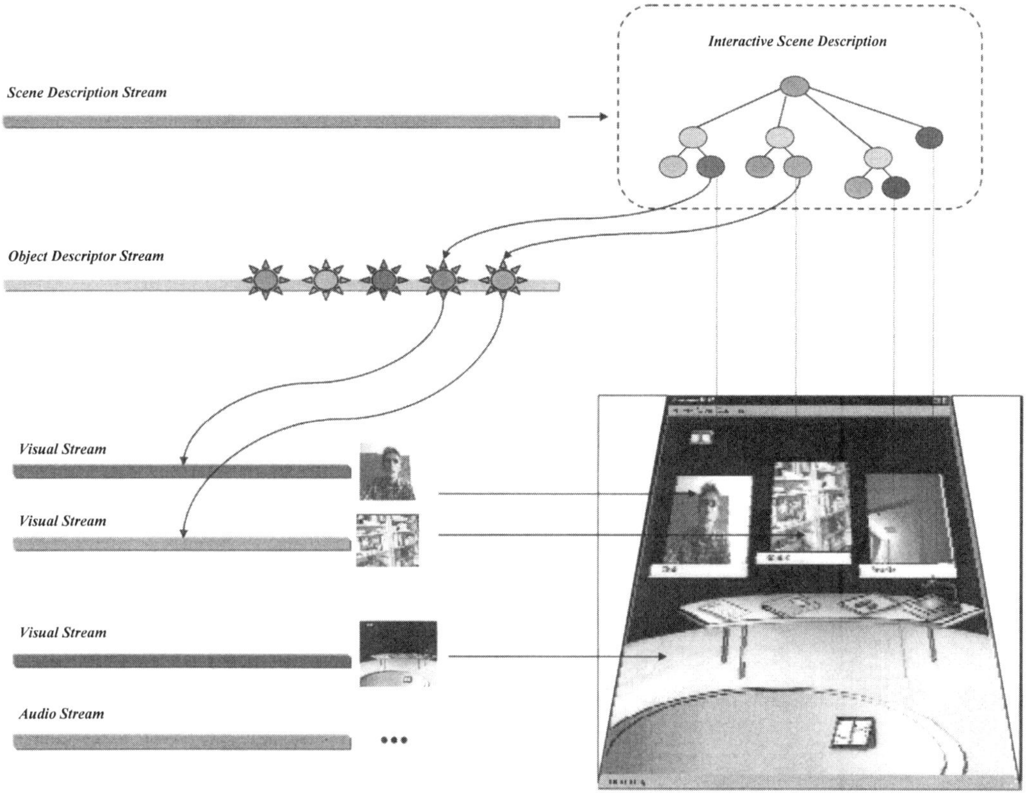

Figure 1 MPEG-4 systems principles.

promoters of these ideas. Such an exercise always presents difficulties and dangers: the different perceptions of individuals involved, putting more emphasis on some points and less on others, or the possible lack of complete information. The views presented here represent the perspective of a few of the people involved in the process, and this should be borne in mind by the reader.

The mapping between tools and objectives, as documented in Sec. II, was possible to describe only a posteriori, when the standardization process was about to be completed. More precisely, mapping the MPEG-4 Systems objectives with technologies available at the time the MPEG-4 Call for Proposal was issued was not possible because no such technology existed in a sufficiently mature state. Indeed, there has been, during the development process of MPEG-4, a continuous refinement of the objectives as the technologies that allowed their realization were identified.

This refinement happened in two ways:

1. Leading concepts, derived from the general MPEG-4 idea of ''coding of audio-visual objects,'' implied consideration of specific technologies.
2. Available or reachable technologies within the MPEG-4 time frame implied clarification, new organization, or reformulation of the objectives.

Thus, the development of MPEG-4 Systems did not completely follow the ''ideal'' process: a collaborative development that integrated available technologies to meet established requirements.

This section is structured as follows. Section III.B identifies the leading ideas and their evolution. Section III.C describes the transformation of the ideas into tools and their actual realization with specific technologies. Finally, Sec. III.D describes how the development of tools was organized in phases.

B. The Leading Ideas

1. Object-Oriented Design

The MPEG-4 project started in September 1993, at the same time that object-oriented design and programming were becoming a commonly used design methodology. Still, most of the MPEG community, with background in digital signal processing, was using the C programming language and the core of the standardization activity was the definition of ''on-the-wire'' formats. This included

1. The definition of a *syntax*, the grammar that a stream of bits should follow in order to be a ''valid'' stream. A pseudo-C language was used to describe this grammar. But the grammar of this language was not precisely defined.
2. Definition of *semantics*, the process that should be followed to reconstruct coded information from the syntax elements. Textual descriptions or pseudo-C code was used to detail the semantics of the syntax elements.

In 1994, MPEG had not yet cogently formulated its ideas on the coding of audiovisual objects. One of the more innovative ideas being discussed at that time was that of defining a future-proof ''extensible standard,'' with built-in capabilities to describe more than just audiovisual content. The vision involved being able to describe new audiovisual representations or new algorithms as they were developed. Because MPEG was mainly oriented toward the definition of syntax, it was commonly thought that a flexible syntax would be sufficient to allow such descriptions. Inspired by the emergence of object-oriented programming, this idea led to the definition of a first tool named OODL (Object-Oriented Descriptive Language), which is described in Ref. 4. The tool was to be used to describe the syntax of bitstreams in an object-oriented fashion.

It was discovered rather quickly that such a description did not provide much new functionality. At best, it provided a description of the syntax of the transmitted data. Also, the C++ approach used in OODL was too broad to define easily a workable grammar for syntax description. Indeed, the syntactic approach was a myopic view of what eventually would become the MPEG-4 paradigm of coding of audiovisual objects.

Still, the development of a syntactic description tool was pursued. A new language, called SDL (Syntactic Description Language) and first proposed in November 1995 [5], fixed the deficiencies of the C++ approach. This language had a clean and well-defined grammar and could considerably simplify the parsing stage of multimedia data as well as the documentation of standards [6]. SDL is now used in the MPEG-4 specifications for the syntactic description of MPEG-4 streams. It is sufficiently C++-like that it can be easily understood and written by those with a knowledge of C or C++ but has in-built support for common operations and notations used in the definition of syntax. The SDL is part of the MPEG-4 Systems tools and is described in Section V.D.

During 1995, much effort took place within MPEG-4 Systems* to clarify the MPEG-4 requirements and the architecture, in general, and MPEG-4 Systems in particular, specifically in terms of requirements for the description of audiovisual content. These requirements then led to a new idea: programmability. Still, from that time until the end of the project, object-oriented design was firmly established in MPEG-4 Systems and gave birth to rich concepts such as audiovisual objects.

2. Programmability

The functional limitations of the OODL led to further investigation of the requirements and technology that could solve the issues raised by the objective of audiovisual content description. These investigations converged on three levels of description capabilities, or flexibility (''Flex'') levels, and the associated underlying architecture. These levels of description capabilities, finally described in July 1996 [7], are summarized here.

1. *Flex_0*: The description of the coded representation of an audio or visual entity is equivalent to identifying one from a multitude of predefined standardized formats for each content type.
2. *Flex_1*: The description of the coded representation of the audio or visual entity is transmitted to the receiver using a standardized parametric language, detailing the relevant combinations of standardized components named tools.
3. *Flex_2*: The description of the coded representation of the audio or visual entity is transmitted in the form of executable code. New algorithms would be downloadable to improve the decoding of the existing object types or to introduce new object types.

Flex_0 did not seem at first to solve any MPEG-4 specific problem or to give any compelling functionality: the limited number of standardized algorithms or relevant ongoing standardization activities did not justify a specific Flex_0 activity. Still, a few years later, with the rich proliferation of MPEG-4 objects, the organization and description of the MPEG-4 algorithms became relevant, and the Object Descriptor Protocol can be seen as an incarnation of the Flex_0 capability.

Flex_1, applied to the description of audio or visual coding algorithms, quickly appeared to be impractical. First, the parametric description needed to be designed from scratch, and several attempts to specify such languages, although laudable, failed. In fact, it became clear that there was no practical way to implement a Flex_1 approach without, essentially, covering the Flex_2 capabilities as well. Second, it appeared that the expressiveness of a parametric language would have been too poor to describe efficiently new interesting coding algorithms. Finally, the interest in describing coding algorithms within the bitstream declined, as explained later on in this chapter.

Flex_2 appeared to be the more reasonable and powerful choice. Indeed, at that time, the first downloadable executable codes or scripts were becoming popular. Early MPEG-4 Systems documents in November 1995 [8] mention ScriptX, MCODE, Java, etc.

* At that time, the MPEG-4 Systems activity took place in an ad hoc group of MPEG named MSDL. The acronym meant first MPEG-4 Syntax Description Language. As the focus of the group evolved, the acronym reflected less and less the reality of the activity. Therefore, the ad hoc group became in July 1996 an MPEG-4 subgroup of its own, MPEG-4 Systems, and MSDL was used to design unambiguously and appropriately the MPEG-4 Syntactic Description Language.

The path for developing the specification was reasonably clear: architecture of a terminal, definition of APIs and run-time engine, and definition of a format for programmatic code transmission.

The development of a Java-based Flex_2 architecture depicted in Figure 2 [9,10] considerably sped up the convergence of the MPEG-4 Systems activity. It led to a better definition of the various components that needed an MPEG-4 multimedia terminal, namely (de) multiplexing, synchronization, decoding, and rendering, and provided the basis of the current MPEG-4 architecture. Indeed, one of the most important concepts of MPEG-4 crystallized from these developments: the concept of scene description.

After the Flex_2 architectural issues were settled, several critical developments followed:

1. The Audio and Video subgroups were actively defining tools according to the now commonly accepted MPEG-4 idea of coding of audiovisual objects; identifying the various elementary objects to represent, and associating with each type of such objects the best coding algorithms. In such cases, downloadable decoders would provide only marginal performance gain in comparison with these standardized algorithms to the detriment of interoperability. Therefore, in November 1996, the Flex_2 activity shifted focus to deal only with scene description, bringing it to the forefront of the Systems group attention.

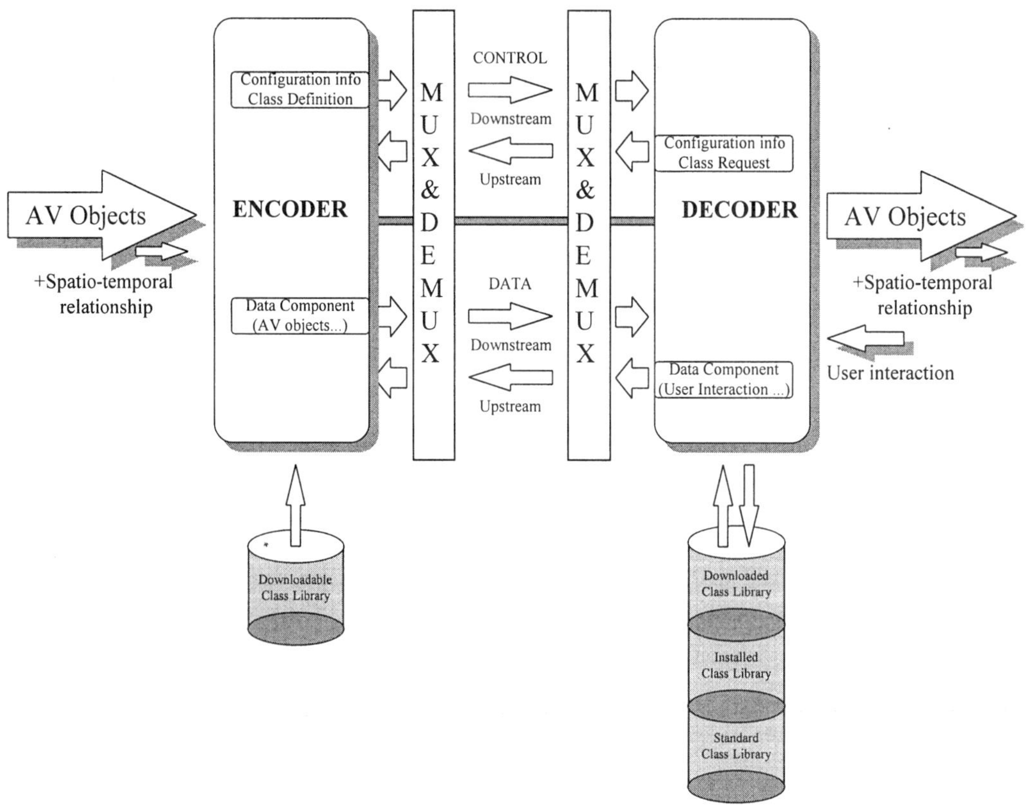

Figure 2 Architecture of a Flex_2 terminal.

2. The capabilities of various programmatic languages, especially Java, were being seriously questioned. One of the more critical aspects was the lack of predictable performance when running the interpreted Java code on virtual machines that were hosted on diverse hardware platforms. The first virtual machines were seriously handicapped by unpredictable garbage collection, loose notion of time, and inefficient graphical capabilities. Moreover, there were no guarantees that these features would be improved in the future. MPEG's experts decided it was not acceptable to take such risks with what was considered to be a core technology of MPEG-4.
3. The activities of the SNHC (Synthetic and Natural Hybrid Coding) group were gaining momentum, exposing MPEG to new ideas and bringing new expertise from the computer graphics area into the MPEG community. The most important concept was the one of parametric and explicit scene description. MPEG's SNHC group was developing specifications for the integration of text and graphical content, working on technology referred to as MITG, Media Integration of Text and Graphics. Meanwhile, the Virtual Reality Modeling Language (VRML) Consortium had developed scene description technology for 3D graphical worlds [11]. Both of these initiatives were quite different from the Flex_2 approach, in which the scene description was to be implicitly contained in a procedural language. The VRML specification had been examined as early as November 1995 [8] but was proposed as a possible technology for MPEG-4 Systems only at the end of 1996 [12].

These facts merely served to emphasize the concept of scene description and, ironically, to herald the demise of the Flex_2 programmable approach that gave birth to it, along with many key concepts of MPEG-4.

C. Transformation of the Ideas into Tools

1. The VRML Legacy

The final decision, based on the facts already described, was made in February 1997, in favor of a VRML-based parametric approach for scene description. The MPEG community was going back to a known and secure ground: the definition of syntax and semantics for the representation of content, in this case, the scene description information.

The VRML structure provided the scene description and client-side interaction but severely lacked support for streaming, timing and synchronization, 2D, and integration of 2D and 3D. This was a vast set of requirements that had to be covered, in which, luckily, MPEG had expertise from its core industries.

The MPEG-4 Systems group subsequently followed a linear process that led to the specification of *BIFS*, the Binary Format for Scenes, which is described further in Section V.B as well as in Chapter 14 of this book.

2. Programmable Extensions

After the decision to develop a parametric format for the representation of the scene description information, it was still clear from the VRML experience that some kind of programmability would be needed, in particular to

1. Extend the possible audiovisual object behavior, e.g., to include state information or to perform calculations in between input and output events, and

2. Provide interfaces for manipulation of the BIFS scene structure and dynamics via downloaded programmatic code, allowing in particular adaptability of the content to networks and terminal characteristics.

The elements needed in order to develop an architecture supporting such programmable extensions existed within the MPEG-4 Systems subgroup, having been acquired during the Flex_2 developments and the collaboration with VRML. They were identified as follows: (1) definition of a run-time environment, (2) definition of specific APIs, and (3) definition of a format for the downloadable code. This led to the development of the MPEG-J* tool [13,14], described in Sec. V.C and in Chapter 15 of this book.

3. Management of Elementary Streams

Parallel to the development of the ideas of object orientation and programmability, the MPEG-4 Systems subgroup also developed the needed mechanisms for the complete management of streams carrying MPEG-4 content.

Indeed, the core competences required for development of a framework for elementary streams management were already present, and MPEG's experts had already successfully delivered the specification of MPEG-1 and MPEG-2 Systems. Although it was tempting for MPEG-4 to adopt the approach followed by these two standards, this would have merely duplicated existing work. Still, the evolution of the elementary stream management activity is somewhat more straightforward than that of the scene description. The primary ideas underpinning the work were the following:

1. *A transport agnostic format*: MPEG-4 Systems chose not to define a new format for the transport of the data but rather to use existing ones. The MPEG-4 Systems architecture defined a clean separation between the transport and the content [15]. Still, the nature of the MPEG-4 streams and the possibly large numbers of objects led to the creation of a lightweight multiplexing tool to be used by any transport mechanism: the FlexMux tool, described in Section V.E.
2. *Synchronization information within the content*: Some network architectures provide means for tight synchronization of media and others do not. Following a requirement that MPEG-4 be transport agnostic, the MPEG-4 architecture defined its own layer for the carriage of synchronization information. The progressive separation of transport-related information and content-related information led to the design of the sync layer,† described in Section V.D.
3. *Object descriptor protocol*: As soon as the scene description and the elementary stream framework matured, the need to link the scene description with time-dependent information became crucial‡. This led to the concept of audiovisual objects being associated with elementary streams using object descriptors. From April 1997, where the object descriptor structure first appeared [16], to the final standard in December 1998, the tool has been continuously enriched (e.g., with

---

* The MPEG-J activity was formerly known as AAVS: Adaptive Audio-Visual Session.

† The sync layer was previously named the adaptation layer. The "sync layer" terminology appeared in March 1998, but the functionality of the layer was identified and specified in November 1996.

‡ Indeed, this need was already identified and solved in the Flex_2 architecture with specific syntactic elements.

intellectual property management and protection information) and the architecture of the protocol consolidated. Finally, it became a key and independent piece of the MPEG-4 architecture, providing a future-proof framework for the description and identification of content streams.

Still, in order to be able to deliver MPEG-4 content in existing environments, the transport of MPEG-4 content on existing communication infrastructures had to be defined. Three key environments have been addressed by MPEG-4 starting from April 1997: (1) MPEG-4 content over MPEG-2 Systems, (2) MPEG-4 content over Internet Protocol (IP) networks, and (3) storage of MPEG-4 content in files for further playback, editing, or streaming.

D. The Development of the Tools in Phases

Since the beginning of the MPEG-4 Systems activity, validation of the architecture and specifications via software implementations has been a constant preoccupation. The first attempts at a Flex_2 architecture implementation provided important information on the possible advantages and limitation [17] of an approach based on the Java language alone. The report from the implementation activity also generated continuous feedback on the specifications (e.g., as in Ref. 18). The implementation activity is described further in Chapter 16.

Later, the implementation activity allowed the maturity of the various MPEG-4 Systems tools to be verified. Indeed, because the maturity of the technology of the various tools was not the same, the MPEG-4 project, in general, and MPEG-4 Systems, in particular, were organized in two phases: version 1 and version 2.

Figure 3 summarizes the tools developed during the time frame of the MPEG-4

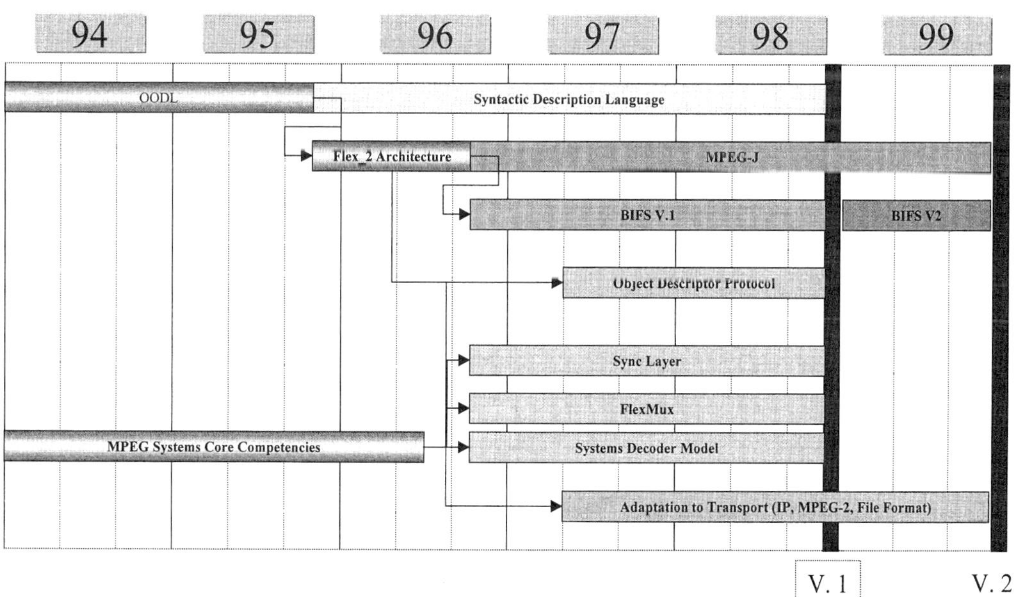

Figure 3 MPEG-4 Systems tools.

Systems activity. The tools provided in version 1 already contain the majority of the functionality of MPEG-4 Systems and allow the development of compelling multimedia applications. Version 2 adds to the version 1 toolbox new tools and new functionalities. Version 2 tools are not intended to replace any of the version 1 tools. On the contrary, version 2 is a completely backward-compatible extension of version 1.

The development of the MPEG-4 Systems activity may appear complex, and indeed, to a certain extent, it was. The following reasons may be at the roots of this complexity but also of its quality:

1. *Meaningful technology*: It took time to identify a meaningful level of standardization to solve the problems stated at a general level in the MPEG-4 requirements document. It also took time for the right people to get involved in the process and to develop this technology.
2. *Time-to-market technology*: During the development process, some technical options had to be eliminated in favor of delivering technology with a shorter time to market.
3. *Environment-friendly technology*: The MPEG-4 Systems philosophy has been to respect the existing infrastructure and tools, to extend the existing concepts in order to match the specific needs of MPEG-4, and to allow the application of MPEG-4 to new environments.

IV. ARCHITECTURE

The overall architecture of an MPEG-4 terminal is depicted in Figure 4. Starting at the bottom of the figure, we first encounter the particular storage or transmission medium. This refers to the lower layers of the delivery infrastructure (network layer and below, as well as storage). The transport of the MPEG-4 data can occur on a variety of delivery systems. This includes MPEG-2 transport streams, UDP over IP, AAL2 on ATM, an MPEG-4 (MP4) file, and the DAB multiplexer.

Most of the currently available transport layer systems provide native means for multiplexing information. There are, however, a few instances where this is not the case (e.g., GSM data channels). In addition, the existing multiplexing mechanisms may not fit MPEG-4 needs in terms of low delay, or they may incur substantial overhead in handling the expected large number of streams associated with an MPEG-4 session. As a result, the FlexMux tool can optionally be used on top of the existing transport delivery layer.

Regardless of the transport layer used and the use (or not) of the FlexMux option, the delivery layer provides to the MPEG-4 terminal a number of elementary streams. Note that not all of the streams have to be downstream (server to client); in other words, it is possible to define elementary streams for the purpose of conveying data back from the terminal to the transmitter or server. In version 1 of MPEG-4, there is no normative support for the structure or semantics of this upstream data. Version 2 standardizes both the mechanisms with which the transmission of such data is triggered at the terminal and its formats as it is transmitted back to the sender.

In order to isolate the design of MPEG-4 from the specifics of the various delivery systems, the concept of the DMIF Application Interface (DAI) (Chapter 13) was defined. This interface defines the process of exchanging information between the terminal and the delivery layer in a conceptual way, using a number of primitives. It should be pointed

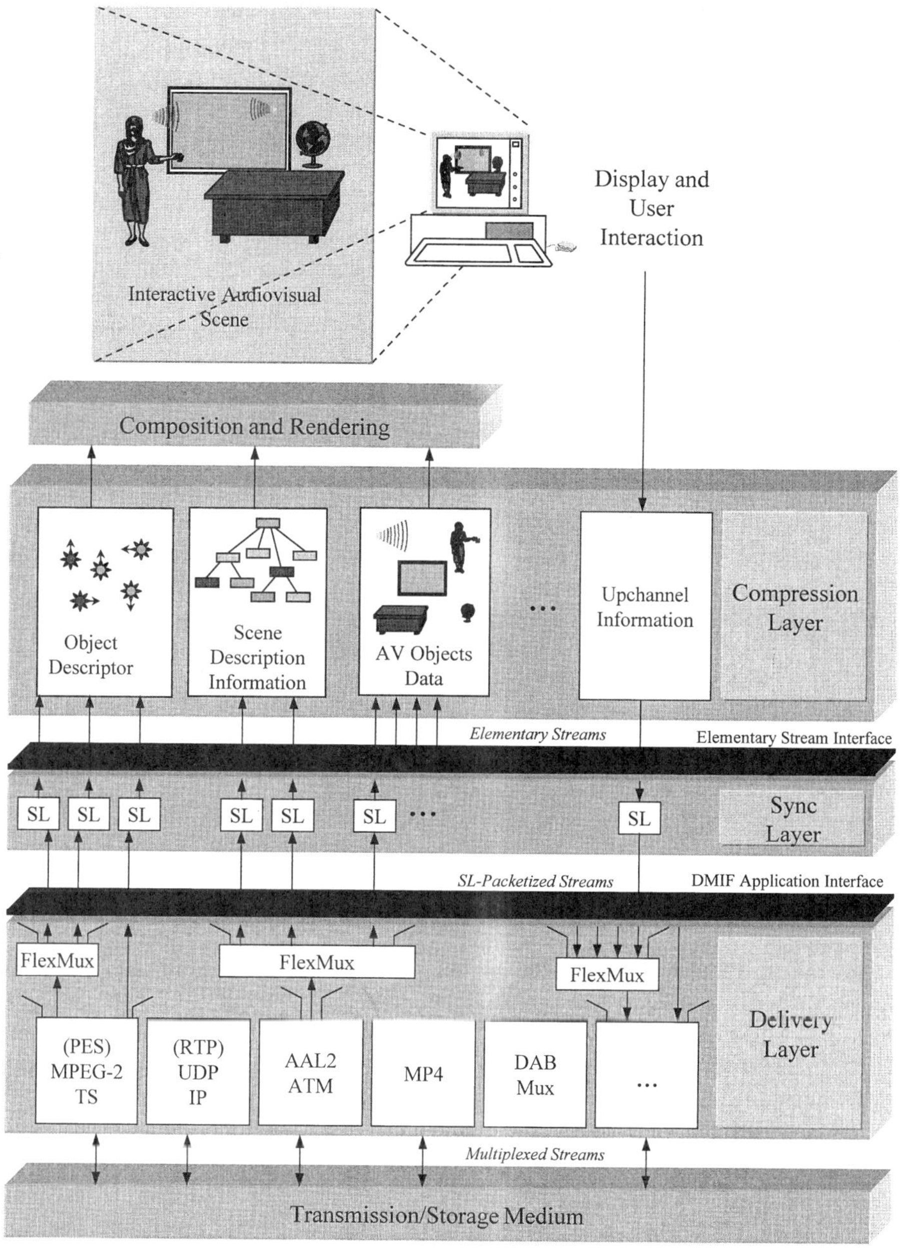

Figure 4 MPEG-4 Systems architecture.

out that this interface is nonnormative; actual MPEG-4 terminal implementations do not need to expose such an interface.

The DAI defines procedures for initializing an MPEG-4 session and obtaining access to the various elementary streams that are contained in it. These streams can contain a number of different types of information: audiovisual object data, scene description information, control information in the form of object descriptors, as well as meta-information that describes the content or associates intellectual property rights to it.

Regardless of the type of data conveyed in each elementary stream, it is important to provide a common mechanism for conveying timing and framing information. The sync layer (SL) is defined for this purpose. It is a flexible and configurable packetization facility that allows the inclusion of timing, fragmentation, and continuity information on associated data packets. Such information is attached to data units that constitute complete presentation units, e.g., an entire video frame or an audio frame. These are called *access units*. An important feature of the SL is that it does not contain frame demarcation information; in other words, the SL header contains no packet length indication. This is because it is assumed that the delivery layer that processes SL packets will make such information available. Its exclusion from the SL thus eliminates duplication.

The SL is the sole mechanism of implementing timing and synchronization mechanisms in MPEG-4. The fact that it is highly configurable allows the use of several different models. At one end of the spectrum, traditional clock recovery methods using clock references and time stamps can be used. It is also possible to use a rate-based approach (access unit rate rather than explicit time stamps). At the other end of the spectrum, it is possible to operate without any clock information—data is processed as soon as it arrives. This would be suitable, for example, for a slide show presentation. The primary mode of operation, and the one supported by the currently defined conformance points of the specification, involves the full complement of clock recovery and time stamps. By defining a system decoder model, this makes it possible to synchronize the receiver's clock to the sender's as well as manage the buffer resources at the receiver.

From the SL information we can recover a time base as well as a set of elementary streams. These streams are sent to their respective decoders that process the data and produce composition units (e.g., a decoded video frame). For the receiver to know what type of information is contained in each stream, control information in the form of object descriptors is used. These descriptors associate sets of elementary streams to one audio or visual object, define a scene description stream, or even point to an object descriptor stream. These descriptors, in other words, are the means by which a terminal can identify the content being delivered to it. Unless a stream is described in at least one object descriptor, it is impossible for the terminal to make use of it.

At least one of the streams must be the scene description information associated with the content. The scene description information defines the spatial and temporal position of the various objects, their dynamic behavior, as well as any interactivity features made available to the user. As mentioned before, the audiovisual object data is actually carried in its own elementary streams. The scene description contains *pointers* to object descriptors when it refers to a particular audiovisual object. We should stress that it is possible that an object (in particular, a synthetic object) may be fully described by the scene description. As a result, it may not be possible to associate an audiovisual object uniquely with just one syntactic component of MPEG-4 Systems. As detailed in Sec. V.B, the scene description is tree structured and is heavily based on VRML structure.

A key feature of the scene description is that, because it is carried in its own elementary stream(s), it can contain full timing information. This implies that the scene can be dynamically updated over time, a feature that provides considerable power for content creators. In fact, the scene description tools provided by MPEG-4 also provide a special lightweight mechanism for modifying parts of the scene description in order to effect animation (BIFS Anim). This is accomplished by coding, in a separate stream, only the parameters that need to be updated.

The system's *compositor* uses the scene description information, together with decoded audiovisual object data, in order to render the final scene that is presented to the user. It is important to note that the MPEG-4 Systems architecture does not define how information is to be rendered. In other words, the standard does not detail mechanisms through which the values of the pixels to be displayed or audio samples to be played back can be uniquely determined. This is an unfortunate side effect of providing synthetic content representation tools. Indeed, in this case it is not possible to define rendering without venturing into issues of terminal implementation. Although this makes compliance testing much more difficult (requiring subjective evaluation), it allows the inclusion of a very rich set of synthetic content representation tools.

The scene description tools provide mechanisms for capturing user or system events. In particular, it allows the association of events with user operations on desired objects that can, in turn, modify the behavior of the stream. Event processing is the core mechanism by which application functionality and differentiation can be provided. In order to provide flexibility in this respect, MPEG-4 allows the use of ECMAScript (also known as JavaScript) scripts within the scene description. Use of scripting tools is essential in order to access state information and implement sophisticated interactive applications. Version 2 of MPEG-4 also defines a set of Java language application programming interfaces (APIs; MPEG-J) through which access to an underlying MPEG-4 engine can be provided to Java applets (called MPEG-lets). This complementary set of tools can form the basis for very sophisticated applications, opening up completely new ways for audiovisual content creators to augment the use of their content. The inclusion of a normative specification for back-channel information in version 2 closes the loop between the terminal and its server, vastly expanding the types of applications that can be implemented on an MPEG-4 infrastructure.

It is important to point out that, in addition to the new functionalities that MPEG-4 makes available to content consumers, it provides tremendous advantages to content creators as well. The use of an object-based structure, with composition performed at the receiver, considerably simplifies the content creation process. Starting from a set of coded audiovisual objects, it is easy to define a scene description that combines these objects in a meaningful presentation. A similar approach is essentially used in HTML and Web browsers, allowing even nonexpert users to create their own content easily. The fact that the content's structure survives the process of coding and distribution also allows its reuse. For example, content filtering and/or searching applications can easily be implemented using ancillary information carried in object descriptors (or its own elementary streams, as described in Sec. V.A). Also, users themselves can easily extract individual objects, assuming that the intellectual property information allows them to do so.

In the following section the different components of this architecture are described in more detail.

V. TOOLS

A. Stream Management: The Object Description Framework

The Object Description Framework provides the ''glue'' between the scene description and the streaming resources—the elementary streams—of an MPEG-4 presentation, as indicated in Fig. 1. Unique identifiers are used in the scene description to point to the *object descriptor*, the core element of the object description framework. The object descriptor is

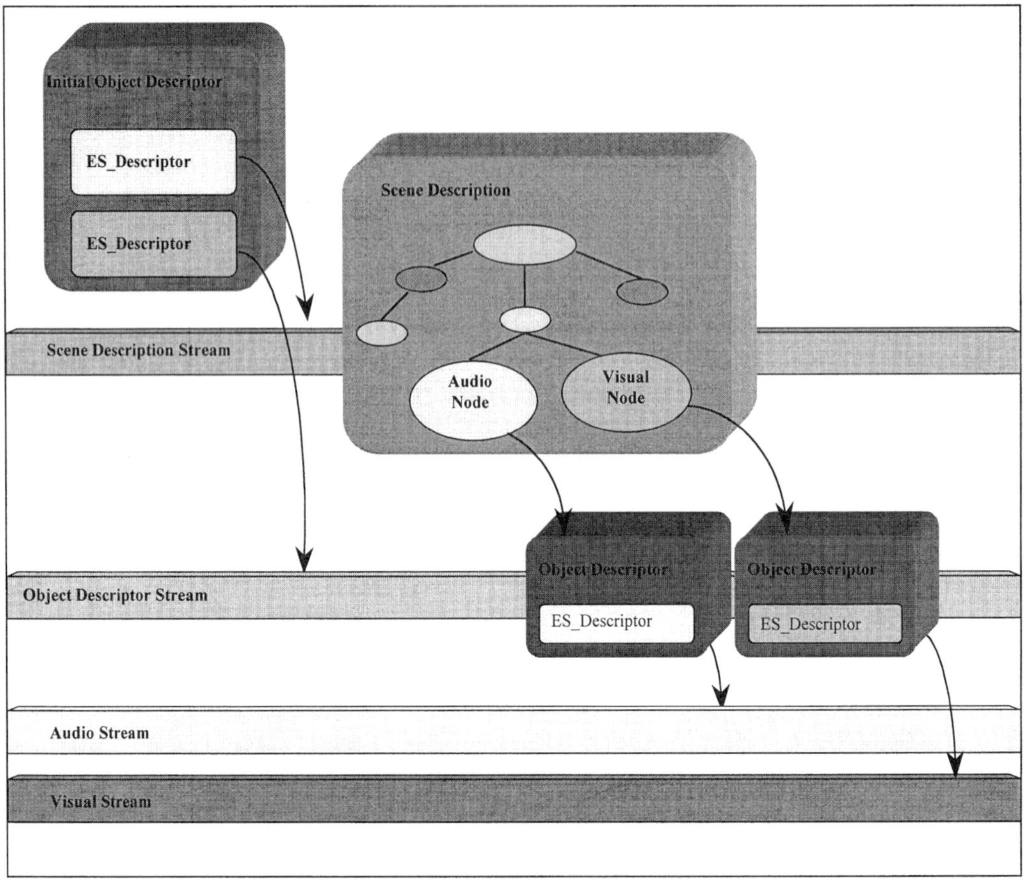

Figure 5 The initial object descriptor and the linking of elementary streams to the scene description.

a container structure that encapsulates all of the setup and association information for a set of elementary streams. A set of subdescriptors, contained in the object descriptor, describe the individual elementary streams, including the configuration information for the stream decoder as well as the flexible sync layer syntax for this stream. Each object descriptor, in turn, groups a set of streams that are seen as a single entity from the perspective of the scene description.

Object descriptors are transported in dedicated elementary streams, called object descriptor streams, that make it possible to associate timing information with a set of object descriptors. With the appropriate wrapper structures, called *OD commands*, around each object descriptor, it is possible to update and remove each object descriptor in a dynamic and timely manner. The existence or the absence of descriptors determines the availability (or lack thereof) of the associated elementary streams to the MPEG-4 terminal.

The *initial object descriptor*, a derivative of the object descriptor, is a key element necessary for accessing the MPEG-4 content. It conveys content complexity information in addition to the regular elements of an object descriptor. As depicted in Fig. 5, the initial object descriptor usually contains at least two elementary stream descriptors. One of the descriptors must point to a scene description stream and the others may point to an object descriptor stream. This object descriptor stream transports the object descriptors for the elementary streams that are referred to by some of the components in the scene description. Initial object descriptors may themselves be transported in object descriptor streams because they allow content to be hierarchically nested, but they may as well be conveyed by other means, serving as starting pointers to MPEG-4 content.

In addition to providing essential information about the relation between the scene description and the elementary streams, the object description framework provides mechanisms to describe hierarchical relations between streams, reflecting scalable encoding of the content and means of indicating multiple alternative representations of content. Furthermore, descriptors for textual information about content items, called *object content information* (OCI), and descriptors for intellectual property rights management and protection (IPMP) have been defined. The latter allow conditional access or other content control mechanisms to be associated with a particular content item. These mechanisms may be different on a stream-by-stream basis and possibly even a multiplicity of such mechanisms could coexist.

A single MPEG-4 presentation, or program, may consist of a large number of elementary streams with a multiplicity of data types. The object description framework has been separated from the scene description to account for this fact and the related consequence that service providers may possibly wish to relocate streams in a simple way. Such relocation may require changes in the object descriptors; however, this will not affect the scene description. Therefore object descriptors improve content manageability.

A detailed discussion of the object description framework is presented in Chapter 13. The reader is referred to the MPEG-4 Systems specification [2] for the syntax and semantics of the various components of this framework and their usage within the context of MPEG-4.

B. Presentation Engine: BIFS

MPEG-4 specifies a Binary Format for Scenes (BIFS) that is used to describe scene composition information: the spatial and temporal locations of objects in scenes, along with their attributes and behaviors. Elements of the scene and the relationships between them form the scene graph that must be coded for transmission. The fundamental scene graph elements are the ''nodes'' that describe audiovisual primitives and their attributes, along with the structure of the scene graph itself. BIFS draws heavily on this and other concepts employed by VRML* [11].

Designed as a file format for describing 3D models and scenes (''worlds'' in VRML terminology), VRML lacks some important features that are required for the types of multimedia applications targeted by MPEG-4. In particular, the support for conventional video and audio is basic, and the timing model is loosely specified, implying that synchronization in a scene consisting of multiple media types cannot be guaranteed. Furthermore, VRML worlds are often very large. Animations lasting around 30 sec typically consume several megabytes of disk space. The strength of VRML is its scene graph description

* VRML was proposed and developed by the VRML Consortium and their VRML 2.0 specification became an ISO/IEC International Standard in 1998.

Consider a simple coding scheme with the following tags:

`<begin>`	- beginning of record
`<end>`	- end of record
`<break>`	- end of element in record
`<string>`	- text string follows
`<number>`	- number follows

We wish to use this scheme code a record consisting of first name, last name and phone number, for example:

First name:	Jim
Last name:	Brown
Phone:	777 1234

With no knowledge context we would need to code this as:

```
<start><string>Jim<break><string>Brown<break><number>7771234<break><end>
```

If the context is known, i.e. we know that the structure of the record is "string, string, number" we do not have to spend bits specifying the type of each element:

```
<start>Jim<break>Brown<break>7771234<end>
```

Figure 6 Simple example of the use of context in efficient coding.

capabilities, and this strength has been the basis upon which MPEG-4 scene description has been built.

BIFS includes support for almost all of the nodes in the VRML specifications. In fact, BIFS is essentially a superset of VRML, although there are some exceptions. BIFS does not support the PROTO and EXTERNPROTO nodes in version 1 of the MPEG-4 specifications, nor does it support the use of Java language in the *Script* nodes (BIFS supports only ECMAScript). BIFS does, however, expand significantly on VRML's capabilities in ways that allow a much broader range of applications to be supported. Note that a fundamental difference between the two is that BIFS is a *binary* format, whereas VRML is a *textual* format. So, although it is possible to design scenes that are compatible with both BIFS and VRML, transcoding of the representation formats is required.

Here, we highlight the functionalities that BIFS adds to the basic VRML set. Readers unfamiliar with VRML might find it useful first to acquire some background knowledge from Ref. 11.

1. *Compressed binary format*: BIFS describes an efficient binary representation of the scene graph information. The coding may be either lossless or lossy. The coding efficiency derives from a number of classical compression techniques plus some novel ones. The knowledge of context is exploited heavily in BIFS. This technique is based on the fact that when some scene graph data has been previously received, it is possible to anticipate the type and format of data to be received subsequently. This technique is illustrated in Figure 6. Quantization of numerical values is supported, as well as the compression of 2D meshes as specified in MPEG-4 Visual [19].
2. *Streaming*: BIFS is designed so that the scene may be transmitted as an initial scene followed by time stamped modifications to the scene. For dynamic scenes that change over time, this leads to a huge improvement in memory usage and

reduced latency compared with equivalent VRML scenes. The *BIFS Command* protocol allows replacement of the entire scenes; addition, deletion and replacement of nodes and behavioral elements in the scene graph; and modification of scene properties.

3. *Animation*: A second streaming protocol, *BIFS Anim*, is designed to provide a low-overhead mechanism for the continuous animation of changes in numerical values of the components in the scene. These streamed animations provide an alternative to the interpolator nodes supported in both BIFS and VRML. The main difference is that interpolator nodes typically contain very large amounts of data that must be loaded in the scene and stored in memory. By streaming these animations, the amount of data that must be held in memory is reduced significantly. Second, by removing this data from the scene graph that must be initially loaded, the amount of data that must be processed (and therefore the time taken) in order to begin presenting the scene is also reduced.
4. *2D primitives*: BIFS includes native support for 2D scenes. This facilitates content creators who wish to produce low-complexity scenes, including the traditional television and multimedia industries. Many applications cannot bear the cost of requiring decoders to have full 3D rendering and navigation. This is particularly true where hardware decoders must be of low cost, as for television set-top boxes. However, rather than simply partitioning the multimedia world into 2D and 3D, MPEG-4 BIFS allows the combination of 2D and 3D elements in a single scene.
5. *Enhanced audio*: VRML provides simple audio support. This support has been enhanced within BIFS, which provides the notion of an audio scene graph where the audio sources, including streaming ones, can be mixed and nodes interface to the various MPEG-4 audio objects [20]. Audio content can even be processed and transformed with special procedural code to produce various sounds effects as described in Chapter 6 of this book.
6. *Facial animation*: BIFS provides support at the scene level for the MPEG-4 Facial Animation decoder [19]. A special set of BIFS nodes expose the properties of the animated face at the scene level, making it a full participant of the scene that can be integrated with all BIFS functionalities.

Version 2 provides additional BIFS functionalities. These functionalities extend BIFS and do not replace existing BIFS functionalities with other tools, following the MPEG rule "one functionality, one tool." Therefore, version 2 decoders supporting version 1 tools are able to decode version 1 content constructed with these tools.

The following BIFS tools are supported in version 2:

All the VRML constructs that were not supported in version 1. The major ones are PROTO and EXTERNPROTO, allowing more efficient compression of the scene description, more flexible content authoring, as well as a robust mechanism to define BIFS extension.

Specification of BIFS interfaces for new version 2 audio and visual media such as synthetic body description and animation or 3D model coding.

Specification of Advanced Audio BIFS for more natural sound source and sound environment modeling. These new audio functionalities include modeling of air absorption, modeling of the audio response of the visual scene, and more natural modeling of distance-dependent attenuation and sound source directivity.

Specification of BIFS nodes to define parametric interfaces between the scene description and elements outside the scene for server-based interactivity and for sensing time from other elementary streams.

C. Application Engine: MPEG-J

The BIFS Scene Description framework, described in the previous section, offers a parametric methodology for scene representation in addition to efficiently coding it for transmission over the wire. Version 2 of the MPEG-4 standard also offers a programmatic environment in addition to this parametric capability. This programmatic environment seeks to extend content creators' ability to incorporate complex controls and data processing mechanisms in their programs along with the parametric scene representations and elementary media data. At the presentation end, the MPEG-J environment intends to enhance the end user's ability to interact with the content.

Figure 7 depicts the architecture for the enhanced MPEG-4 terminal. The lower part of the diagram shows the parametric MPEG-4 system, also known as the presentation engine, and includes the DMIF Application Interface (DAI), the elementary stream decoders, as well as the scene compositor. The MPEG-J system, also known as the applications engine, is in the upper part of this diagram and enhances the operation of the presentation engine. During the course of the development of the requirements and specifications for a programmable MPEG-4 architecture, it was decided that Java was the language of choice. Thus, the applications engine includes a run-time environment, sets of Java-based APIs, and a Java virtual machine (the Java execution engine), as indicated in Figure 7.

The Java-based MPEG application, denoted in the figure as Java *MPEG-let*, is received by the client MPEG-4 terminal, in Java bytecode form, over a separate elementary stream. This stream is transported using the same underlying delivery infrastructure as all

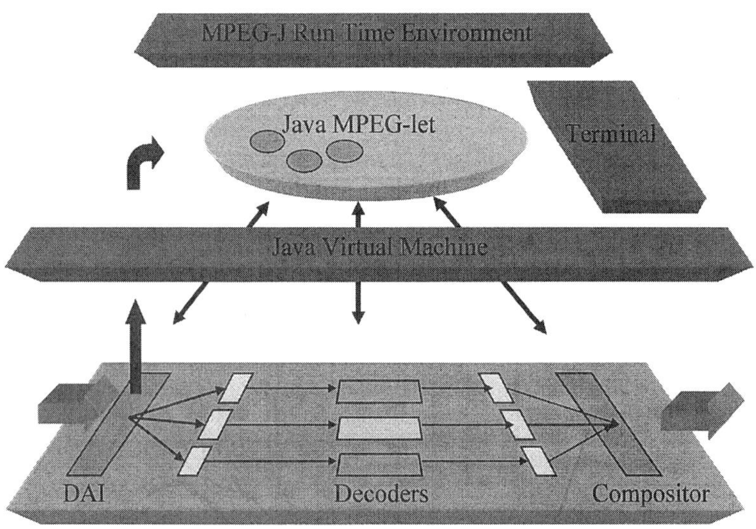

Figure 7 Architecture of an MPEG-J terminal.

other MPEG-4 data. The run-time environment in the receiving terminal provides the MPEG-let with access to the Java packages and APIs, as well as to the various components of the presentation engine. In addition to the APIs and packages provided as a part of one of the Java language platforms, new APIs and packages have been defined within the framework of MPEG-J development.

The collection of scene graph APIs provides access to the MPEG-4 scene graph that may be received in a parametric form. These APIs are intended to facilitate scene graph inspection and management and scene graph modification (node updates or deletions, field value modifications, route updates or deletions, etc.). These APIs are provided in addition to the parametric methods of performing similar actions on the scene graphs. In fact, they may also be used to implement the parametric scene graph commands received via one of the elementary streams.

The resource manager APIs are to be used by the MPEG-J applications for monitoring and allocating the terminal's resources. This, for example, includes processing of events arising because of changes in the availability of the terminal's resources (e.g., media decoders and renderers).

The terminal capability APIs may be used by the MPEG-J applications to interrogate and obtain the terminal's configuration and capabilities. An application aware of the terminal's environment may use this information to adapt its execution accordingly. For example, an application may tailor its visual and audio decoding and presentation formats to be commensurate with the capabilities of the presentation devices available at the client terminal. Some of the capabilities may be dynamic in nature, e.g., the available memory resources. By dynamically monitoring the availability of such resources, an intelligent application can modify its behavior and manage its execution to be resource efficient.

High-level applications may also want some level of control of the elementary stream decoders. The media APIs define interfaces that allow monitoring and modification of the media decoding parameters. Some other examples of these are interfaces to start, suspend, and stop the media decoding processes. However, the media APIs are not intended to support user-defined media decoders that may be downloaded and executed on the client terminals.

The MPEG-J architecture also includes network APIs, i.e., means of interacting with the underlying network protocols. These APIs are in support of the DMIF specifications, which are detailed in Ref. 21. Some of the functionalities supported include network queries (information on DMIF usage) and channel controls (enable or disable one or more elementary stream channels).

D. Timing and Synchronization: The Systems Decoder Model and the Sync Layer

The MPEG-4 Systems Decoder Model (SDM) is conceived as an adaptation of its MPEG-2 predecessor. The System Target Decoder in MPEG-2 is a model that precisely describes the temporal and buffer constraints under which a set of elementary streams may be packetized and multiplexed. Because of the generic approach taken toward stream delivery—which includes stream multiplexing—MPEG-4 chose not to define multiplexing constraints in the SDM. Instead, the SDM assumes the concurrent delivery of an arbitrary number of—already demultiplexed—elementary streams to the decoding buffers of their respective decoders. A constant end-to-end delay is assumed between the encoder output and the input to the decoding buffer on the receiver side. This leaves the task of handling the delivery jitter (including multiplexing) to the delivery layer.

Timing of streams is expressed in terms of decoding and composition time of individual access units within the stream. Access units are the smallest sets of data to which individual presentation time stamps can be assigned (e.g., a video frame). The decoding time stamp indicates the point in time at which an access unit is removed from the decoding buffer, instantaneously decoded, and moved to the composition memory. The composition time stamp allows the separation of decoding and composition times, to be used, for example, in the case of bidirectional prediction in visual streams. This idealized model allows the encoding side to monitor the space available in the decoding side's buffers, thus helping it, for example, to schedule ahead-of-time delivery of data. Of course, a resource management for the memory for decoded data would be desirable as well. However, it was acknowledged that this issue is strongly linked with memory use for the composition process itself, which is considered outside the scope of MPEG-4. Therefore, management of composition buffers is not part of the model.

Time stamps are readings of an object time base (OTB) that is valid for an individual stream or a set of elementary streams. At least all the streams belonging to one audiovisual object have to follow the same OTB. Because the OTB in general is not a universal clock, object clock reference time stamps can be conveyed periodically with an elementary stream to make it known to the receiver. This is, in fact, done on the wrapper layer around elementary streams, called the sync layer (SL). The sync layer provides the syntactic elements to encode the partitioning of elementary streams into access units and to attach both decoding and composition time stamps as well as object clock references to a stream. The resulting stream is called an SL-packetized stream. This syntax provides a uniform shell around elementary streams, providing the information that needs to be shared between the compression layer and the delivery layer in order to guarantee timely delivery of each access unit of an elementary stream.

Unlike the packetized elementary stream (PES) of its predecessor (MPEG-2), the sync layer of MPEG-4 does not constitute a self-contained stream but rather a packet-based interface to the delivery layer. This takes into account the properties of typical delivery layers such as IP, H.223, or MPEG-2 itself, into which SL-packetized streams are supposed to be mapped. There is no need to encode either unique start codes or the length of an SL packet within the packet, because synchronization and length encoding are already provided by the mentioned delivery layer protocols.

Furthermore, MPEG-4 has to operate at both very low and rather high bit rates. This has led to a flexible design of the sync layer elements, making it possible to encode time stamps of configurable size and resolution, as required in a specific content or application scenario. The flexibility is made possible by means of a descriptor that is conveyed as part of the elementary stream descriptor that summarizes the properties of each (SL-packetized) elementary stream.

E. The Transport of MPEG-4 Content

Delivery of MPEG-4 content is a task that is supposed to be dealt with outside the MPEG-4 Systems specification. All access to delivery layer functionality is conceptually obtained only through a semantic interface called the DMIF Application Interface (DAI). It is specified in Part 6 of MPEG-4, Delivery Multimedia Integration Framework (DMIF) [21]. In practical terms, this means that the specification of control and data mapping to underlying transport protocols or storage architectures is to be done jointly with the respective organization that manages the specification of the particular delivery layer. For example, for the case of MPEG-4 transport over IP, development work is done jointly with the Internet Engineering Task Force (IETF).

An analysis of existing delivery layer properties showed that there might be a need for an additional layer of multiplexing in order to map the occasionally bursty and low-bit-rate MPEG-4 streams to a delivery layer protocol that exhibits fixed packet size or too much packet overhead. Furthermore, the provision of a large number of delivery channels may have a substantial burden in terms of management and cost. Therefore, a very simple multiplex packet syntax has been defined, called the FlexMux. It makes it possible to multiplex a number of SL-packetized streams into a self-contained FlexMux stream with rather low overhead. It is proposed as an option to designers of the delivery layer mappings but not as a conformance point for MPEG-4.

Initially, FlexMux streams have also served as a trivial file format that was used in the MPEG-4 reference software implementation project (referred to as IM1 for historical reasons). However, it has been noted that a file format should also provide a fair amount of meta-information about the content in order to allow indexing, fast search, and random access into such a file. Furthermore, it should be easy to repurpose the file content for delivery over various delivery layers. Especially because of the latter requirement, a file format is a key element for the exchange of MPEG-4 content. MPEG therefore provides in version 2 its own file format based on QuickTime, providing rich meta-information in a very flexible way.

As mentioned before, specification of MPEG-4 delivery on IP is a joint issue for MPEG and the IETF. The control and data encapsulation portions of this task are treated in different IETF working groups. A great deal of progress has already been made on data encapsulation in RTP, the real-time transport protocol, which is the focal point for delivery of streaming multimedial content on the Internet. A second version of an Internet draft specifying the mapping of SL packets to RTP packets has been published. Control of MPEG-4 content access will probably use the session description protocol (SDP) [22]; however, some extensions are needed to cater for the dynamic character of MPEG-4 content.

The second widespread delivery infrastructure is based, of course, on MPEG-2 Systems [23], used in both digital broadcasting and storage-based playback (digital video disk, DVD). In that case, it is MPEG itself that has to amend its MPEG-2 standard to cater for the needs of MPEG-4. This process has already led to a so-called proposed draft amendment for MPEG-2 Systems. It defines a set of descriptors that provide for the signaling of the presence of MPEG-4 content, both in the form of individual MPEG-4 elementary streams and as complete MPEG-4 presentations. The MPEG-4 SL-packetized streams themselves are encapsulated in the packetized elementary stream (PES) syntax, which is the common syntax that allows exchange between digital broadcast and storage applications. Not only may MPEG-2 content and MPEG-4 content coexist within an MPEG-2 multiplex but also it is possible to reference the MPEG-2 content in the MPEG-4 scene description. This allows backward-compatible enhancement of current digital broadcast services with MPEG-4 content.

F. MPEG-4 Systems Syntactic Description Language

SDL originated from the need to formalize bitstream representation. In the past, MPEG as well as all other standardization bodies used ad hoc techniques combining pseudo-C, tabular, or textural information in order to describe the syntax of their data. This made the task of understanding the specification and transforming it into running code unnecessarily difficult. In other domains, software tools are successfully used to model data as well as generate automatically code that can operate on them (e.g., Sun ONC RPC or OMG CORBA).

SDL (or Flavor, Formal Language for Audio-Visual Object Representation, as it is referred to by its originators [6]) was designed to be an intuitive and natural extension of the typing system of C++ and Java. Whenever differences appear between C++ and Java, a common denominator approach has been followed (e.g., only single inheritance is supported).

SDL has been explicitly designed to follow (as much as possible) a declarative approach to bitstream syntax specification. Developers specify how the data is laid out on the bitstream and do not detail a step-by-step procedure that parses it. As a result, SDL has no methods or functions.

SDL is based on the notion of parsable variables; it is the proper definition of these variables that defines the bitstream syntax. Parsable variables include a parse length specification immediately after their type declaration (e.g., int(3) a;). This provides in a single place the name of the variable, its type, as well as the bitstream representation information. This allows, for example, a variable to be used as the length of another variable (e.g., int(a) b;). In contrast to regular variables, such variables may be declared more than once. SDL also includes all the usual flow control constructs (if-else, switch, for, do, and while); however, these appear in the data declaration part of a class. Other basic features include bitstring literals (e.g., 0b0010.010), arrays with dynamic sizes and partial arrays (declaration of *part* or slice of an array), as well as support for variable-length coding representations.

Classes are the fundamental type structure of SDL. The parsable objects defined in a class will be present in the bitstream in the order in which they are declared. A class is considered parsable if it contains at least one variable that is parsable. SDL classes are able to access external information by using *parameter types*. These act in the same way as formal arguments in function or method declarations. SDL supports bitstream-level polymorphism for parsable classes by introducing the concept of an *object identifier* or *tag*, which is a unique parsable variable associated with each class that allows its resolution. SDL also supports the notion of abstract polymorphic classes. Here the base class is assigned a tag value of zero and cannot appear in the bitstream. This allows the entire tag space to be used for derived classes. This type of polymorphism is used for the syntax of the object descriptor framework and allows for a flexible and extensible bitstream.

For a complete description of SDL, see Refs. 2 and 6 or visit the Flavor web site [24]. A translator is available for converting SDL source to regular C++ or Java code [25]. It reads a set of SDL source files and generates a set of C++ or Java files that contain declarations of all SDL classes. The translator also generates for each class a put( ) and a get( ) method. The get( ) method is responsible for reading a bitstream and loading the class variables with their appropriate values, and the put( ) method does the reverse. For parsing operations, the only task required by the programmer is to declare an object of the class type at hand, and then call its get( ) method with an appropriate bitstream. Although the same is also true for the put( ) operation, the application developer must also load all class member variables with their appropriate values before the call is made. The translator can also automatically generate tracing code, something very useful for debugging (especially when bitstream exchange is performed for compliance verification).

We should point out that SDL is not a normative part of the MPEG-4 Systems specification, in the sense that it does not have to be part of a compliant implementation. It is used only as a documentation tool that can provide unambiguous specification of media representation schemes and also convert them to running code with ease and efficiency. These two features can significantly reduce the time it takes to convert a specification into a running application.

VI. APPLICATIONS

A. Introduction

The previous MPEG standards, MPEG-1 and MPEG-2, have found widespread use in the digital multimedia industries. Streaming video over the Internet as well as video CDs conform to the MPEG-1 standards. Digital broadcast TV as well as DVDs were driven by the emergence of MPEG-2. It is expected that the technologies developed in MPEG-4 will revolutionize the digital multimedia industry.

At the time the MPEG-4 version 1 standard was released in early 1999, several real-time implementations were already demonstrating prototypes of applications based on the standard. This section illustrates the use of MPEG-4 features in several application scenarios that have been demonstrated by these prototypes.

Three domains of applications are described. Section VI.B describes the scenario of an application using MPEG-4 Systems in a Web-like environment. Section VI.C presents a virtual conferencing space with GroupWare. Finally, Sec. VI.D covers interactive broadcast applications. It is to be noted that the list of applications that may use MPEG-4 technologies is not limited to the ones described here. These examples are merely intended to illustrate possible uses of MPEG-4 Systems technologies in the multimedia industries.

For each of the application scenarios, we discuss how MPEG-4 Systems enhances the application or provides new functionality in the given application context.

B. Streaming Content over the Net

In this application scenario, the client terminal is a multimedia terminal connected to the Internet. An example of such a terminal is a personal computer (PC) with multimedia features. The MPEG-4–based application may be received over the network from a remote server. The application may also be a local application resident on a CD-ROM or on a DVD-ROM.

The users of this application first enter a specific address, perhaps a universal resource locator (URL), and are connected to the server. The URL may point to either a remote site or to a file on the local system. As a result of establishing a connection to the server, a ‘‘page’’ appears on the client terminal's screen, as indicated in Figure 8, screen shot 1. This page looks very much like an HTML page with Java applet–based animations. In fact, the page is not created as a result of receiving an HTML file but is actually an MPEG-4, application. It is retrieved faster than users are accustomed to from the usual Web because of the compressed format of the data. Moreover, the animations of the content on the screen (scrolling effects, interactive pop-up menus, etc.) are not realized by downloaded code or by embedded Java applets but are generated by processing the lightweight BIFS parametric animation instructions that are efficiently streamed from the

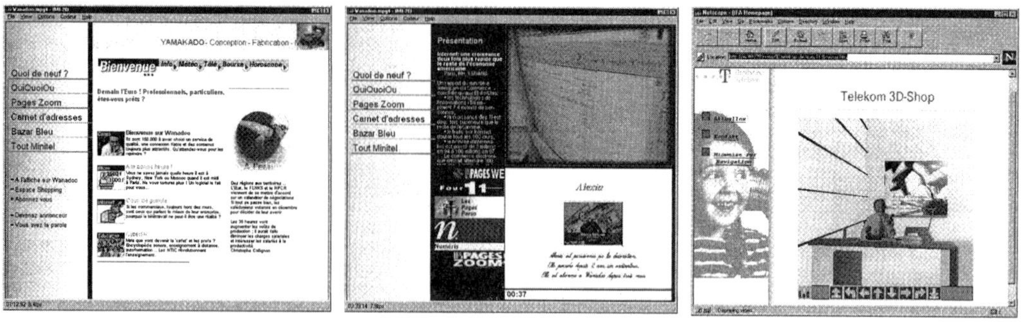

Figure 8 Snapshots of MPEG-4 streaming content over the Net application.

server. These features enable users to begin interacting with the content sooner than they would with conventional HTML-based applications.

Following one of the hyperlinks, the user navigates to a page indicated in Figure 8, screen shot 2. This new page contains streamed audio and video, tightly synchronized together. Moreover, scrolling text and slides also appear with tight synchronization with the audiovisual presentation.

The MPEG-4 browser may well be configured as a ''plug-in'' for a standard web browser. Still, when accessing an MPEG-4 site, the user can safely receive the content without the browser crashing because of either memory or processing power limitations. The user may also be confident that he has the complete set of required MPEG-4 tools to decode the content thanks to the definition of well-defined conformance points, namely *MPEG-4 profiles and levels.*

Finally, after having collected information about a product of interest, the user may decide to receive more information about it via a direct communication with a vendor. He may then, as depicted in Fig. 8, screen shot 3, enter the vendor's 3D virtual shop, navigate in it, further examine products modeled in 3D, and finally, by triggering a button, start a real-time communication with the vendor. MPEG-4 Systems provides the tools for such integration of content with the notions of mixed 2D–3D scenes. Real-time presentations of streamed content, such as a 2D segmented video from the vendor, can easily be included in the scene through updates of the scene description and the object descriptors.

C. Multimedia Conferencing

In this application, the terminal may be a multimedia PC (equipped with camera, microphone, and speakers), or it can be part of a high-end videoconferencing system. In fact, it can very well be the case that these two kinds of terminals are communicating in the same virtual environment. Indeed, the notion of shared communication space is the main idea of this application, and the use of MPEG-4 Systems to represent the shared data is the technical choice.

The user connects to the conference site in the same way as to a normal Web site, i.e., through a specific address or URL. He then receives the MPEG-4 data representing the shared space. Other participants in this multimedia conference session may already be connected to each other, and therefore the shared space contains streamed data representing them. Still, the application supports more than simply observing the shared space:

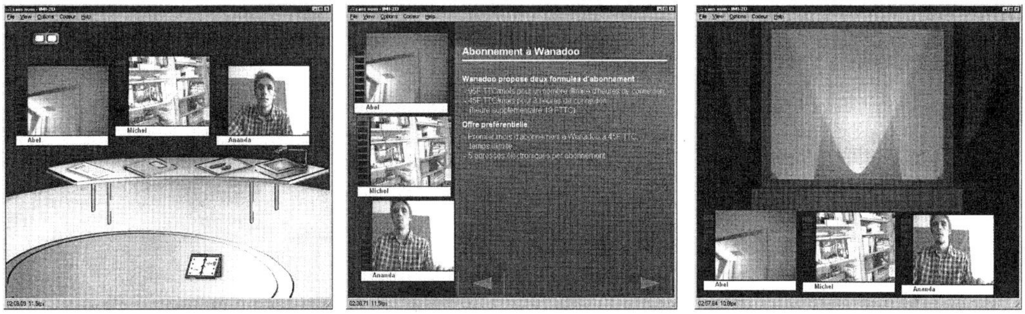

Figure 9 Snapshots of an MPEG-4 multimedia conferencing application.

the user may send his own representation into the shared space using, for example, audiovisual data streams captured from a camera and microphone.

The dynamic nature of the scene representation allows the adaptation of the virtual world to the different tasks that need to be performed during the conference. The simple audiovisual conference, as shown in Figure 9, screen shot 1, can be enhanced with data conferencing. Figure 9, screen shot 2, shows a presentation of slides along with the audiovisual data. Figure 9, screen shot 3, finally demonstrates integration of a broadcast film in a virtual theater, wherein the participants can comment on the events while the film is running.

This application highlights two key features of MPEG-4 Systems:

1. *Dynamic scene representation*: the representation of the shared space can be dynamically updated with new content (streams, text pages, other conference rooms, etc.) in a manner conformant with MPEG-4 specifications.
2. *Scalability*: in such an application, the scalable representations provided by object descriptors may be very useful. For example, a participant connected to a mobile network may not have enough bandwidth to receive the complete virtual scene shared by colleagues connected to a local high-speed network. The representation of the virtual scene can scale itself down for the mobile participant to be an audio-only conference. The other participants in the conference, on the high-speed network, may see him as a 3D avatar (a synthetic face) animated by his voice inputs.

D. Interactive Broadcast

In this scenario, the MPEG-4 receiver may be the traditional home set-top box, a part of a high-end home theater, and is connected to a high-bandwidth broadcast network at its input end. The receiver could also be a conventional multimedia terminal connected to the broadcast network. With the advent of digital broadcasts, the broadcast networks are not limited to the conventional satellite or cable networks that were the only available options until recently. The Internet can now be considered to be an instance of a broadcast network too. The key concept of MPEG-4, ''create once, access everywhere,'' and the tools that support it allow content creators and service providers to make their content available across the entire range of available delivery systems.

In order to receive the broadcast content, the user needs to be connected to the

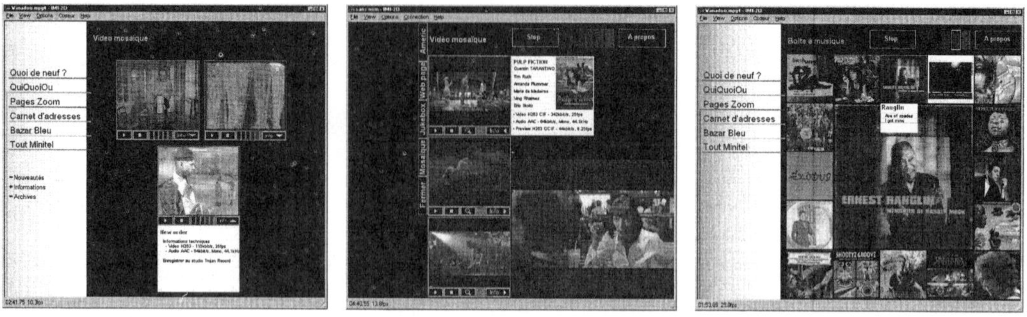

Figure 10 Snapshots of an MPEG-4 interactive broadcast application.

broadcast server. Without delving into the details of how this is done, we assume that the user has managed to tune to the broadcast channel or program of his choice. Chapter 13 goes into some detail on how this may be accomplished. The client terminal acquires the necessary scene description information for this channel. The means for acquisition of this scene description is detailed in Chapter 13. The necessary audio and visual streams, constituents of the program on the tuned channel, are acquired and the user is now able to watch and possibly interact with the program.

Figure 10 shows simple features such as interactive enhancement of the broadcast streams. They can easily be extended to support many applications, such as interactive home shopping, enriched documentary programming, advanced electronic services and program guides, interactive advertisements, interactive entertainment such as sports programs or quiz shows, viewing of Web-like content, and demographically focused programming.

A key feature here is the compressed nature of the scene description information. The gain factor is indeed directly linked to the amount of content that you can deliver in such broadcast carousel. Another important feature is the built-in support of intellectual property management and protection in MPEG-4 Systems, allowing the intellectual property rights on the content to be respected. Finally, the definition of the transport of MPEG-4 content on top of MPEG-2 Systems will allow a smooth and backward-compatible evolution from existing MPEG-2 broadcast applications to fully interactive television.

VII. CONCLUDING REMARKS

A. MPEG-4 Systems and Competing Technologies

This section completes the description of MPEG-4 Systems by trying to make a fair comparison between the tools provided by MPEG-4 Systems and the ones that can be found—or will be found in the near future—in the marketplace. Indeed, in parallel with the development of the MPEG-4 specifications, other standardization bodies and industry consortia have developed tools and applications that address some of the MPEG-4 Systems objectives. We emphasize that our focus is on standard and publicly available specifications rather than on individual products. Exceptions to this rule are specifications that were designed by individual corporations but were made publicly available for the purposes of wider adoption.

Technical issues aside, the mere fact of being proprietary is a significant disadvantage in the content industry when open alternatives exist. With the separation of content production, delivery, and consumption stages in the multimedia pipeline, the MPEG-4 standard will enable different companies to develop separately authoring tools, servers, or players, thus opening up the market to independent product offerings. This competition is then very likely to allow fast proliferation of content and tools that will interoperate.

1. Transport

As stated in Section V.E, MPEG-4 Systems does not specify or standardize a transport protocol. In fact, it is designed to be transport agnostic. However, in order to be able to utilize the existing transport infrastructures (e.g., MPEG-2 transport or IP networks), MPEG-4 defines an abstraction of the delivery layer with specific mappings of MPEG-4 content on existing transport mechanisms [21]. However, there are two exceptions in terms of the abstraction of the delivery layer: the MPEG-4 File Format and the FlexMux tool.

There are several available file formats for storing, streaming, and authoring multimedia content. Among the ones most used at present are Microsoft's ASF, Apple's QuickTime, and RealNetworks' file format (RMFF). The ASF and QuickTime formats were proposed to MPEG-4 in response to a call for proposals on file format technology. QuickTime has been selected as the basis for the MPEG-4 file format (referred to as MP4) [3]. The RMFF format has several similarities to QuickTime (in terms of object tagging using four-character strings and the way indexing information is provided).

The MPEG-4 proposal for lightweight multiplexing (FlexMux) addresses some MPEG-4 specific needs as described in Section V.E. The same kinds of requirements have been raised within the IETF with regard to delivery of multimedia content over IP networks. With the increasing number of streams resident in a single multimedia program, with possibly low network bandwidths and unpredictable temporal network behavior, the overhead incurred by the use of RTP streams and their management in the receiving terminals is becoming considerable. IETF is therefore currently investigating a generic multiplexing solution that has requirements similar to those of the MPEG-4 FlexMux. We expect that, with the close collaboration between IETF and MPEG-4, a consistent solution will be developed for the transport of MPEG-4 content over IP networks.

2. Streaming Framework

With the specifications of the object description framework and the sync layer, MPEG-4 Systems provides a consistent framework for the efficient description of content and the means for its synchronized presentation at client terminals. At this juncture, this framework, with its flexibility and dynamics, does not have any equivalents in the standards arena. A parallel could be drawn with the combination of RTP and SDP; however, such a solution is Internet specific and cannot be applied directly on other systems (e.g., digital cable or DVDs).

3. Multimedia Content Representation

The competition between MPEG-4 Systems and the ''outside world'' tools may appear tougher at first glance when considering the format of the multimedia content representation. This competition can appear at three levels that are successively analyzed in this section: architecture, functionality, and popularity.

A number of serious competitors to MPEG-4 BIFS base their syntax architecture on an XML [26] syntax, whereas MPEG-4 bases its syntax architecture on VRML using a binary, SDL-described form. The main difference between the two is that XML is a general-purpose text-based description language for tagged data and VRML with SDL provides a binary format for a scene description language.

An advantage of an XML-based approach is ease of authoring; documents can easily be generated using a text editor. However, for delivery over a finite bandwidth medium, a compressed representation of multimedia information is without a doubt the best approach from the bandwidth efficiency point of view. This is the problem addressed and solved by MPEG-4. A textual representation of the content may still be useful at the authoring stage. Such textual representations for MPEG-4 content, based on XML, were under consideration at the time this book was published. Another approach would be to define new XML semantics that will leverage the functionality developed by MPEG-4 and then define a new binary mapping, possibly different from BIFS.

Indeed, the competition is mainly at the level of the semantics, i.e., the functionality provided by the representation. At the time the MPEG-4 standard was published, several specifications were providing semantics with an XML-compliant syntax to solve multimedia representation in specific domains. For example, the W3C HTML-NG was redesigning HTML to be XML compliant [27], the W3C SMIL working group has produced a specification for 2D multimedia scene descriptions [28], the ATSC/DASE BHTML specifications were working at providing broadcast extensions to HTML-NG, the W3D X3D requirements were investigating the use of XML for 3D scene description, whereas W3C SVG was standardizing scalable vector graphics also in an XML-compliant way [29].

MPEG-4 is built on a true 3D scene description, including the event model, as provided by VRML. None of the XML-based specifications currently available reaches the sophistication of MPEG-4 in terms of composition capabilities and interactivity features. Furthermore, incorporation of the temporal component in terms of streamed scene descriptions is a nontrivial matter. MPEG-4 has successfully addressed this issue, as well as the overall timing and synchronization issues, whereas alternative approaches are lacking in this respect.

It may very well happen that all these specifications, at some stage, completely leverage the MPEG-4 functionality. Still, in January 1999, there was no evidence that this will happen. Indeed:

1. Alternative frameworks are at the stage of research and specification development, while MPEG-4 is at the stage of standard verification and deployment. The eventual alternative frameworks may appear in the market too late for significant adoption.
2. The alternative frameworks may not bring any significant functional added value compared with the existing MPEG-4 standards. Indeed, there is no evidence that the alternative frameworks will be able to leverage efficiently all of the advantages of MPEG-4 specifics, including compression, streaming, and synchronization.
3. MPEG-4 is the fruit of a 5-year collaborative effort on an international scale aimed at the definition of an integrated framework. The fragmented nature of the development of XML-based specifications by different bodies and industries certainly hinders integrated solutions. This may also cause a distorted vision of the integrated, targeted system as well as duplication of functionality. There is no evidence that real integration can be achieved by the alternative frameworks.

Finally, the industry is already providing compelling solutions for the representation of multimedia content. Indeed, functionalities such as those offered by MPEG-4 have already appeared in products such as games and CD-ROM authoring applications, long before the standard was even finalized. There is always a time lag between the functionality of the more advanced products in a domain and relevant open technology standards. However, as soon as a standard appears that solves the needs of the industries involved, a range of diverse interoperable product offerings and their market deployment is made possible. Each player may then invest in its core business, seeking to gain competitive advantage in areas not covered by the standard. It is therefore very unlikely that a proprietary format for multimedia representation can compete successfully, in the long run, with an open standard such as MPEG-4 when universal access is required at both national and international levels.

B. The Key Features of MPEG-4 Systems

In summary, the key features of MPEG-4 Systems can be stated as follows:

1. MPEG-4 Systems first provides a consistent and complete architecture for the coded representation of the desired combination of streamed elementary audiovisual information. This framework covers a broad range of applications, functionality, and bit rates. However, not all of the features need to be implemented in a single device. Through profile and level definitions, MPEG-4 Systems establishes a framework that allows consistent progression from simple applications (e.g., an audio broadcast application with graphics) to more complex ones (e.g., a virtual reality home theater).
2. MPEG-4 Systems augments this architecture with a set of tools for the representation of the multimedia content, namely a framework for object description (the OD framework), a binary language for the representation of multimedia interactive 2D and 3D scene description (BIFS), a framework for monitoring and synchronizing elementary data stream (the SDM and the SyncLayer), and programmable extensions to access and monitor MPEG-4 content (MPEG-J).
3. Finally, MPEG-4 Systems completes this set of tools by defining an efficient mapping of the MPEG-4 content on existing delivery infrastructures. This mapping is supported by the following additional tools: an efficient and simple multiplexing tool to optimize the carriage of MPEG 4 data (FlexMux), extensions allowing the carriage of MPEG-4 content on MPEG-2 and IP systems, and a flexible file format for authoring, streaming, and exchanging MPEG-4 data.

The MPEG-4 Systems tools can be used separately in some applications. But MPEG-4 Systems also guarantees that they will work together in an integrated way, as well as with the other tools specified within the MPEG-4 standards.

C. Future Work

In order to begin this discussion, we pose the question: Are there any application requirements that are satisfied either partially or not at all by the current MPEG-4 Systems specification? From a functional perspective, it appears that most of the Systems requirements, as contained in the MPEG-4 Requirements documents, are met by the specifications in versions 1 and 2 of the standard. There will probably be addenda and corrigenda to augment or correct the standard, but at this time it does not appear that there is a need for new critical elements.

However, there are two areas of work that may need to be addressed in the near future by MPEG-4 Systems:

Better integration with trends of the World Wide Web and, more notably, with the XML-based content representation schemes. As described in Section VII.A.3, it may be useful to provide an XML-compliant representation of BIFS for usage at the authoring stage. It should be noted that this will not be a new format for the transport of the multimedia content. The format for the delivery will remain BIFS. It will be a textual format for MPEG-4 content authoring that will be well integrated with the format that the authors are used to operating with.

A second possible area of work is related to mixed communication, broadcast and interactive applications. These applications range from point-to-point multimedia communication and multimedia conferencing up to shared distributed virtual communities. Because MPEG-4 Systems has been mostly dedicated to broadcast applications, this field of activity may not be completely covered, in terms of content as well as transport.

Lastly, the new MPEG work item, *MPEG-7* [30], may need to leverage the core competences from MPEG Systems, in terms of architecture design, management of elementary streams, and multimedia scene description.

ACKNOWLEDGMENTS

The MPEG-4 Systems specification reflects the results of teamwork within a worldwide project in which many people invested enormous time and energy. The authors would like to thank them all and hope that the experience and results achieved at least matched the level of their expectations.

Among the numerous contributors to the MPEG-4 Systems projects, three individuals stand out for their critical contributions in the development of MPEG-4: Cliff Reader is the person who originally articulated the vision of MPEG-4 in an eloquent way and led it through its early formative stages; Phil Chou was the first to propose a consistent and complete architecture that could support such a vision; and Zvi Lifshitz, by leading the IM1 software implementation project, was the one who made the vision manifest itself in the form of a real-time player.

Finally, the authors would like to thank Ananda Allys (France Telecom—CNET) and Andreas Graffunder (Deutsche Telekom—Berkom) for providing some of the pictures used in this chapter, the European project MoMuSys for its support of Olivier Avaro and Liam Ward, the National Science Foundation and the industrial sponsors of Columbia's ADVENT Project for their support of Alexandros Eleftheriadis, the German project MINT for its support of Carsten Herpel, and the General Instrument Corporation for their support of Ganesh Rajan.

REFERENCES

1. ISO/IEC JTC1/SC29/WG11 N2562. MPEG-4 requirements document, December 1998.
2. ISO/IEC 14496-1. Coding of audio-visual objects: Systems, final draft international standard, ISO/IEC JTC1/SC29/WG11 N2501, October 1998.

3. ISO/IEC JTC1/SC29/WG11 N2611. MPEG-4 systems version 2 WD 5.0, December 1998.
4. ISO/IEC JTC1/SC29/WG11 MPEG 94/301. Description of the syntactic descriptive language for MPEG-4, Syntax Ad-Hoc Group.
5. A Eleftheriadis. ISO/IEC JTC1/SC29/WG11 MPEG95/M0546. A syntactic description language for MPEG-4, Dallas, November 1995.
6. A Eleftheriadis. Flavor: A language for media representation. Proceedings, ACM Multimedia '97 Conference, Seattle, November 1997, pp 1–9.
7. ISO/IEC JTC1/SC29/WG11 N1295. MPEG-4 requirements document version 1, July 1996.
8. ISO/IEC JTC1/SC29/WG11 N1111. MSDL specification, version 0.1, November 1995.
9. M Agrawala, A Beers, N Chaddha, PA Chou, RM Gray, R Joshi, M Vishwanath. ISO/IEC JTC1/SC29/WG11 MPEG95/M0544, A video synthesizer tool and related objects, November 1995.
10. P Chou. ISO/IEC JTC1/SC29/WG11 MPEG95/M0745, MSDL architecture, Firenze, March 1996.
11. ISO/IEC 14772-1. The Virtual Reality Modeling Language, 1997, http://www.vrml.org/Specifications/VRML97.
12. P Chou, J Signès. ISO/IEC JTC1/SC29/WG11 MPEG95/M1441, Report of the Ad Hoc Group on MPEG-4 Architecture Evolution, Maceio, December 1996.
13. R Agarwal, S Battista, P Shah. ISO/IEC JTC1/SC29/WG11 MPEG97/M2146, Report of the AHG on systems VM, April 1997.
14. J Courtney, P Shah, J Webb, G Fernando, V Swaminathan. ISO/IEC JTC1/SC29/WG11 MPEG97/M2566, Adaptive audio-visual session format, July 1997.
15. F Seytter. ISO/IEC JTC1/SC29/WG11 MPEG97/M1376, 2-layer multiplex for MPEG-4, November 1996.
16. ISO/IEC JTC1/SC29/WG11 N1693. Text of systems VM version 4.0, April 1997.
17. S Battista, F Casalino, M Quaglia. ISO/IEC JTC1/SC29/WG11 MPEG97/M1994, Implementation scenarios for an MPEG-4 player, April 1997.
18. Z Lifshitz, K Oygard. ISO/IEC JTC1/SC29/WG11 MPEG97/M3112, Systems software implementation AHG report, January 1998.
19. ISO/IEC 14496-2. Coding of audio-visual objects: Visual, final draft international standard, ISO/IEC JTC1/SC29/WG11 N2502, October 1998.
20. ISO/IEC 14496-3. Coding of audio-visual objects: Audio, final draft international standard, ISO/IEC JTC1/SC29/WG11 N2503, October 1998.
21. ISO/IEC 14496-6. Coding of audio-visual objects: Delivery multimedia integration framework, final draft international standard, ISO/IEC JTC1/SC29/WG11 N2506, October 1998.
22. M Handley, V Jacobson. IETF RFC 2327, SDP: Session description protocol, April 1998.
23. ISO/IEC 13818-1. Generic coding of moving pictures and associated audio information—Part 1: Systems.
24. Flavor Web Site, http://www.ee.columbia.edu/flavor.
25. ISO/IEC 14496-5. Coding of audio-visual objects: Reference software, final draft international standard, ISO/IEC JTC1/ SC29/WG11 N2505, October 1998.
26. T Bray, J Paoli, CM Sperberg-McQueen, eds. Extensible Markup Language (XML) 1.0. February 10, 1998. http://www.w3.org/TR/REC-xml.
27. XHTML 1.0: The Extensible HyperText Markup Language. A reformulation of HTML 4.0 in XML 1.0. W3C Working Draft, February 24, 1999. http://www.w3.org/TR/WD-html-in-xml/.
28. Synchronized Multimedia Integration Language. (SMIL) 1.0 Specification W3C Recommendation, June 15, 1998. http://www.w3.org/TR/REC-smil/.
29. Scalable Vector Graphics (SVG) Specification. W3C Working Draft, February 11, 1999. http://www.w3.org/TR/WD-SVG/.
30. MPEG Requirements Group. MPEG-7 Requirements Document. ISO/IEC JTC1/SC29/WG11 N2461, October 1998.

13

MPEG-4 Systems: Elementary Stream Management and Delivery

Carsten Herpel
Thomson Multimedia, Hannover, Germany

Alexandros Eleftheriadis
Columbia University, New York, New York

Guido Franceschini
CSELT, Torino, Italy

I. INTRODUCTION

MPEG-4 Systems [1] defines the overall architecture of MPEG-4 and provides the means that allow the combined use of the elements defined in the Audio [2] and Visual [3] parts of the MPEG-4 specification. In the past, the Systems layer in MPEG specifications was concerned exclusively with system architecture, synchronization, and multiplexing. In MPEG-4, its scope is significantly expanded. Traditional issues such as synchronization and multiplexing are attacked from completely different viewpoints because of the object-based nature of the design and the need to cater to a broad application space. New concepts are introduced, such as the binary scene description Binary Format for Scenes (BIFS), which also defines the interactive or dynamic aspects of the content [4], and object descriptors that characterize the content of streams and identify their location. These, among other features, result in a very flexible representation framework that can be easily applied to a large array of applications, from Internet based to broadcast TV. The objective of this chapter is to discuss in particular the MPEG-4 Systems tools for the management of elementary streams, as well as the strongly related issue of their delivery across communication systems of different types. These tools, together with the scene description, cover the entire arsenal of MPEG-4 Systems tools.

MPEG-4 is the first standard that addresses content as a set of audiovisual objects. Objects are presented, manipulated, and transported as individual entities. This is achieved by a set of tools distributed over all the parts of the standard. First, the media compression schemes are object oriented so that they can represent elemental audiovisual entities (e.g., arbitrarily shaped visual objects). The scene description expresses how individual audiovisual objects are to be composed together for presentation on the user's screen and speakers. Finally, some information about the corresponding elementary streams that convey the data is needed. The term elementary stream management is used to refer to the entire set of functionalities that are needed to describe, express relations between, and effect synchronization among such data streams.

Before going to any level of detail in describing MPEG-4, it is important to clarify some terms that are used throughout the specification and this chapter. The term *audiovisual object* refers to an audio or visual entity that participates as an individual element in a scene. Objects can be either natural or synthetic. Synthetic objects are generated by the graphics and synthesized sound operations available in the scene description. The data for natural objects are conveyed separately, in different ''channels.'' Synthetic objects may be animated using the BIFS-Anim tool. In that case, the corresponding coded information will be conveyed in its own channel. Some objects may, however, be fully described within the scene description stream itself. As a result, the term audiovisual object, although conceptually clear, cannot be uniquely associated with just one feature or syntactic element of MPEG-4.

As a general paradigm in MPEG-4, all information is conveyed in a streaming manner. The term *elementary stream* refers to data that fully or partially contain the encoded representation of a single audio or visual object, scene description, or control information. The stream may contain part of the data when multiple streams are used for the representation of a single audiovisual object as, e.g., in multichannel audio or scalable video coding. It is also possible to have multiple streams for the delivery of the scene description information as well. In other words, elementary streams are the conceptual delivery pipes of MPEG-4; they are mapped to actual delivery channels using mechanisms that are described in detail later on.

Elementary streams or groups thereof are identified and characterized by *object descriptors*. We should point out that the term ''object'' here refers not just to audiovisual objects but to any type of MPEG-4 information. This includes the scene description, audiovisual objects data, and object descriptor streams themselves. Although this can be a source of confusion, the term is used here for historical reasons. Object descriptors, then, allow us to specify that a particular set of elementary streams jointly contain, for example, the compressed representation of an audio signal. The information contained will include the format of the data (e.g., MPEG-1 layer III audio) as well as an indication of the resources necessary for decoding (e.g., profile on level indication). Most important, it will contain the information necessary to identify directly or indirectly the location of the data, in terms of either an actual transport channel or a universal resource locator (URL). An indirect approach is necessary so as not to use network- and/or protocol-specific addressing.

There is a clear separation between scene description and stream description. In particular, the scene description contains no information about the streams that are needed to reconstruct a particular audiovisual object, whereas the stream description contains no information related to how an object is to be used within a scene. This separation facilitates editing and general manipulation of MPEG-4 content. The link between the two is the object descriptor; the scene description contains pointers to object descriptors, which in turn provide the necessary information for assembling the elementary stream data required to decode and reconstruct the object at hand.

Even though there is a strong relation between scene and stream descriptions, they are conceptually quite different. The scene description is consciously authored by the content creator; the stream description mostly follows from general content creator preferences, default settings of an editing tool, or even service provider constraints and rules. Even though this suggests that stream description is mundane, it is an essential component of the overall MPEG-4 architecture. It is also a convenient place to add helpful metainformation to streams. Finally, as we will see, even the stream description itself is conveyed in the form of elementary streams.

The actual packaging of the elementary streams as defined by MPEG does not depend on a specific delivery technology. MPEG-4 defines a *sync layer* that just packetizes elementary streams in terms of *access units* (e.g., a frame of video or audio data) and adds a header with timing and other information. This is done in a uniform manner for all different stream types in order to ease identification and processing of these fundamental entities in each stream. All further mappings of the streams to delivery protocols are handled by the delivery layer and, hence, are to be defined outside the core MPEG-4 standard. For example, transport of MPEG-4 content over the real-time transport protocol (RTP) is to be defined under the auspices of the International Engineering Task Force (IETF).

The organization of this chapter is as follows. First, we describe the object descriptor framework that allows the identification and characterization of a related set of elementary streams. It can be considered as a guide through the potentially large set of streams that may belong to a sophisticated MPEG-4 presentation. This guide must be consulted before it is made possible to present any of the content conveyed within the streams. Then, the issue of synchronization is discussed. Because many streams are quasi-concurrently generated and consumed, time stamping is employed here—not surprisingly—to align them on the temporal axis. In this context, the system decoder model is introduced; it describes in an abstract way how the notion of time is defined in MPEG-4 and how some of the buffer memory in the system is modeled and managed. MPEG-4 maintains the traditional broadcast model of timing with clock recovery (as in MPEG-2). However, in certain network environments (e.g., the Internet) the time base could be expressed on the basis of a global clock made available to multiple transmitters via means outside MPEG (e.g., the Network Time Protocol—NTP), thus allowing synchronization of multiple sources.

Next, the concept of a separate and abstract delivery layer that incorporates the actual transport or storage of the elementary stream data is introduced. In order to cater to the needs of different delivery systems (e.g., stored file, broadcasting, and the Internet), MPEG-4 does not specify a transport layer. This is in direct contrast to MPEG-2, which defined the transport stream as the underlying transport multiplexer. The concepts of the MPEG-4 delivery layer are mostly documented in the Delivery Multimedia Integration Framework (DMIF) [5] specified in Part 6 of MPEG-4 but also involve some ongoing work in and outside the MPEG community.

Equipped with these building blocks, we finally present some walkthroughs to illustrate the process of accessing MPEG-4 content consisting of many different elementary streams, possibly even originating from different locations.

II. OBJECT DESCRIPTION FRAMEWORK

A. Basic Elements

The object description framework consists of a few major building blocks, called descriptors, that are hierarchically structured. The highest level descriptor is the object descriptor (OD). It serves as a shell that aggregates a number of other descriptors. Most prominent among these are the elementary stream descriptors (ES descriptors), which describe individual streams. Additional auxiliary information can be attached to an OD either to describe the content conveyed by these streams in a textual form (object content information, OCI) or to do intellectual property management and protection (IPMP). Finally, rules are needed for how these descriptors may be used and, more important, how the descriptors are actually communicated via so-called OD commands. We first introduce the semantics of the descriptors themselves and later on discuss the implications that arise from the fact that they are conveyed in a streaming fashion.

B. Object Descriptors

An object descriptor is merely a shell that groups a number of descriptive components that are linked to an MPEG-4 scene through a single node of the scene description. The major components are depicted in Figure 1. The linking to the scene is achieved in a two-stage process. First, there is a numeric identifier, called the object descriptor ID (OD_ID), that labels the object descriptor and is referenced by the scene description. This links the object descriptor to the scene. The second stage involves the actual binding of elementary streams identified and described by the elementary stream descriptors included in this object descriptor, using another identifier, called ES_ID, which is part of the elementary stream descriptor.

In the simplest case, an OD contains just one ES descriptor that identifies, for example, the audio stream that belongs to the AudioSource node by which this OD is referenced, as illustrated in Figure 2a. The same object descriptor may as well be referenced from two distinct scene description nodes (Fig. 2b). On the other hand, within a single OD it is also possible to have two or more ES descriptors, for example, one identifying a low-bit-rate audio stream and another one identifying a higher bit rate stream with the same content (Fig. 3a). In that case the terminal (or rather the user) has a choice between two audio qualities. Specifically, for audio it is also possible to have multiple audio streams with different languages that can be selected according to user preferences (Fig. 3b). In

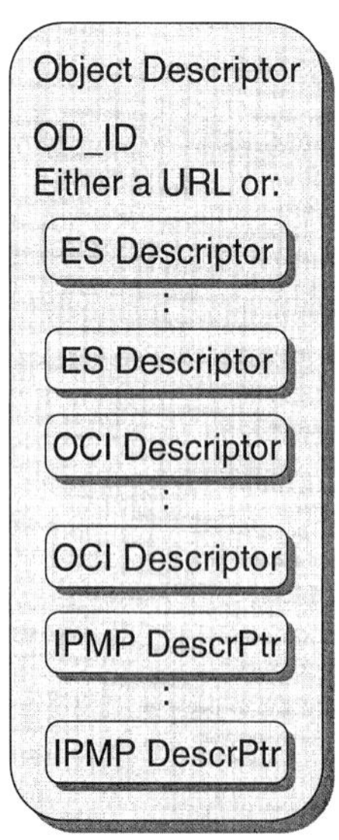

Figure 1 Main components of the object descriptor.

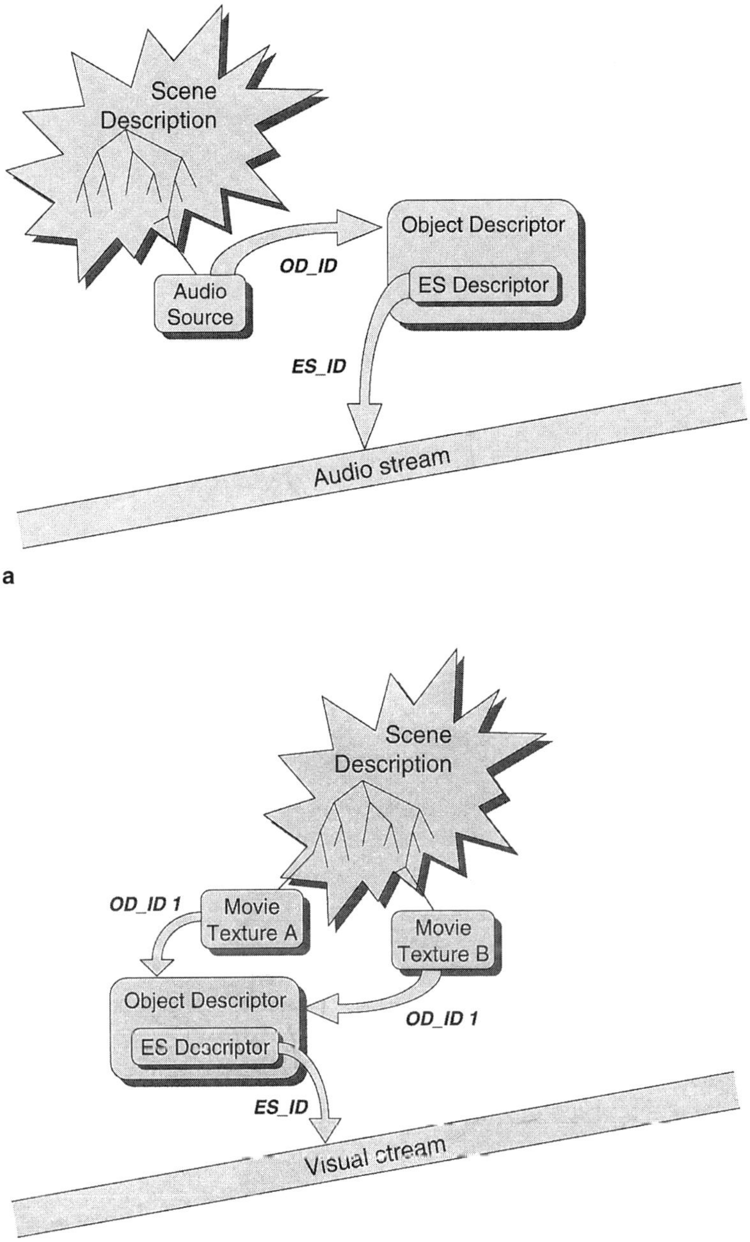

Figure 2 ES reference through object descriptor (a) from one scene description node and (b) from two scene description nodes.

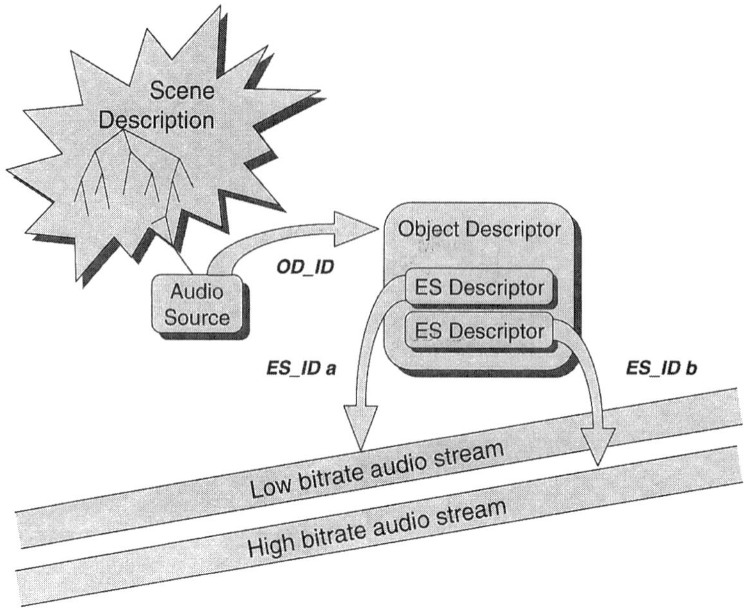

a

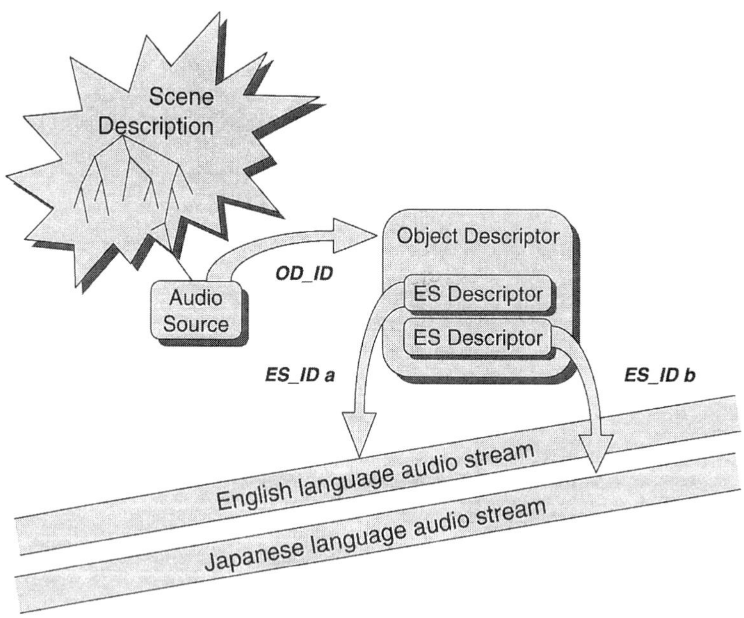

b

Figure 3 Reference to multiple ESs through one object descriptor (a); audio application: multiple languages (b).

general, all kinds of different resolution or different bit rate streams with the same audio or visual content that a content provider may want to offer could be advertised in this object descriptor.

Even though grouping of streams within one object descriptor is allowed, there are restrictions in that streams that are not supposed to be referenced all from the same scene description node must be referenced through distinct object descriptors. As an example, an AudioSource node and a MovieTexture node that (obviously) refer to different elementary streams have to utilize two distinct ODs as shown in Figure 4.

Finally it is possible to describe within one object descriptor a set of streams that corresponds to a scalable, or hierarchical, encoding of the data that represents an audiovisual object. In that case it is necessary to signal not only the properties of the individual elementary streams but also their interdependences. These may be trivial as indicated in Figure 5a, where each stream just depends on the previous one, or more complex as shown in Figure 5b, where the same base layer stream is referenced by two other streams, providing, for example, quality improvements in the temporal domain (temporal scalability) and in the spatial domain (spatial scalability). In this example the order in which spatial and temporal enhancements have to be applied would be arbitrary. The precise instructions on how the decoder has to handle each individual stream in such a case is incorporated in a DecoderSpecificInfo subdescriptor that is contained in each ES descriptor.

Hierarchical dependences are allowed to exist only between the ES descriptors that are included in a single object descriptor. If the same elementary stream is to be referenced from two different scene description nodes within different contexts, say as a single quality stream and as the base layer of an audiovisual object encoded in a scalable fashion, then there have to be two different object descriptors; however, both of them will include the same ES descriptor pointing to the base layer stream (Fig. 6).

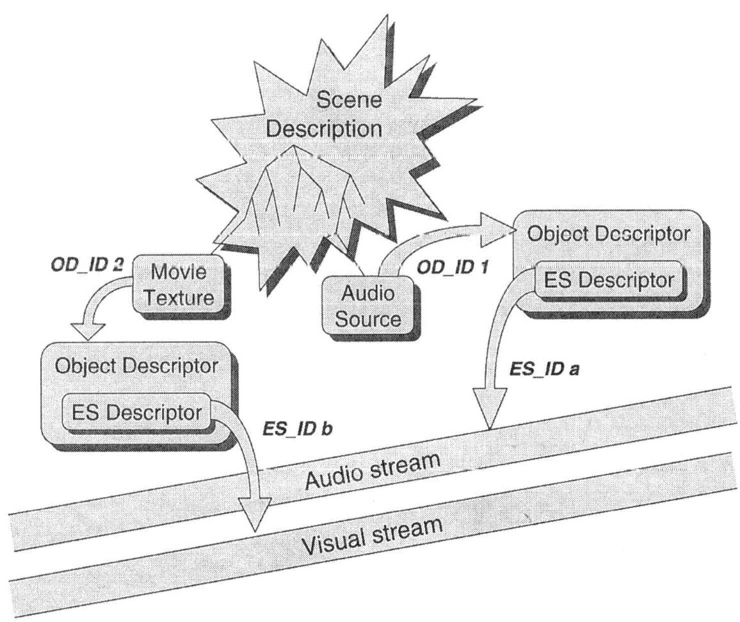

Figure 4 Different scene description node types need different object descriptors.

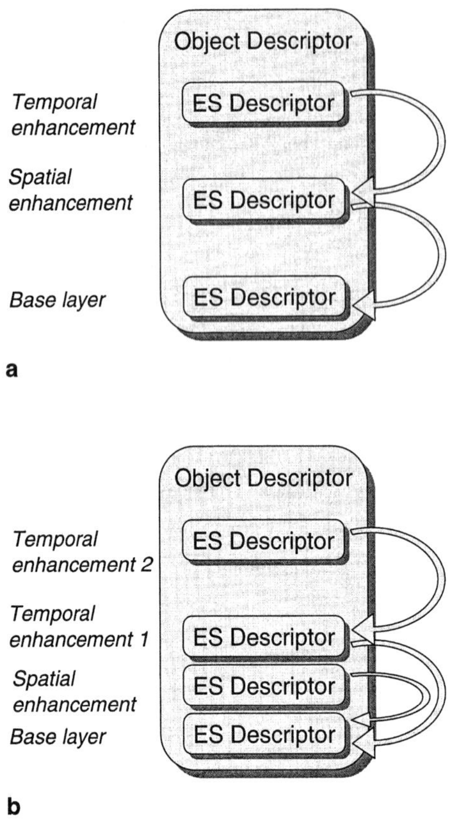

Figure 5 Simple (a) and more complex (b) ES dependence indication.

Additional descriptors in the OD may carry further information that refers to the whole object. Such information includes object content information (Fig. 7) and intellectual property management and protection information as detailed further in the following.

Instead of providing information about the media object in place, the object descriptor may contain only a URL that points to another object descriptor at a remote location. In that case an object descriptor has to be retrieved from that location according to the rules implied by the actual URL syntax used. The content of that object descriptor will be used instead of any locally provided information. Only the object descriptor ID remains the same as before, because it is the element that is referenced by the scene description, which need not be aware of the fact that this object descriptor is retrieved from a remote location.

C. The Initial Object Descriptor

In order to "bootstrap" an MPEG-4 presentation, it is necessary to identify the elementary streams that contain the scene description and the associated object descriptors. A special OD called the initial OD (IOD) is defined in order to convey that information. In addition to the elements of the regular object descriptor, the IOD contains information about the complexity of the overall presentation, expressed in profile and level indications. The IOD

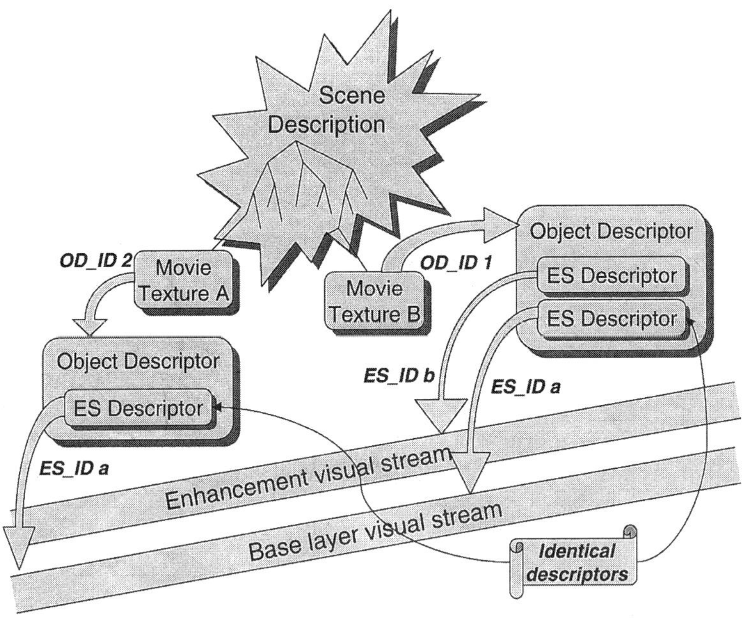

Figure 6 Referencing the same ES in different contexts.

is usually communicated out of band, e.g., during the session initialization phase before any elementary stream channels are set up. However, because content may be hierarchically nested, using the Inline node of the scene description, IODs may also occur in place of regular object descriptors in that special case, making it possible to insert complexity information not only for the overall presentation but also for subscenes.

D. Elementary Stream Descriptor

Each elementary stream descriptor contains all descriptive information available for a single elementary stream, as illustrated in Figure 8. It identifies this stream with a numeric ES_ID and an optional URL. The ES_IDs allow the location of this stream by number. This is possible only within a specific name scope. The URLs are more flexible and allow referring to streams by a unique name. In contrast to the case of an object descriptor, if a URL is present in an ES descriptor, does means not that the descriptive information about this stream is retrieved from a remote location but rather that the stream itself can be located using this URL. All the descriptive information is still present in the local ES descriptor. Again, the reason for the presence of a URL is to facilitate the description of distributed content.

The indication of stream dependences mentioned before is signaled in this descriptor, as well as the stream priority, which is a relative indication that may be used to group streams according to their priority on the delivery layer.

The DecoderConfig descriptor is a mandatory subdescriptor of each ES descriptor and contains all information that is needed to select and initialize a media decoder for

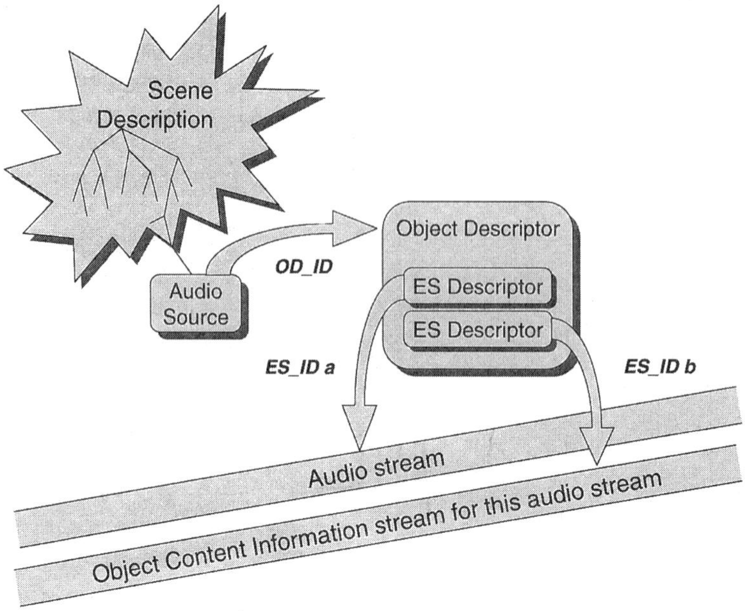

a

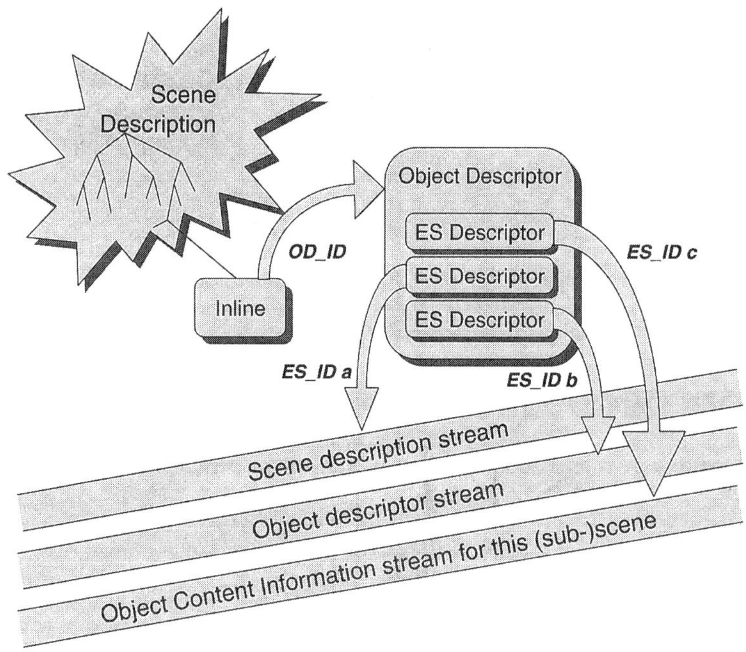

b

Figure 7 Attaching an OCI stream to a media stream (a) or a whole subscene (b).

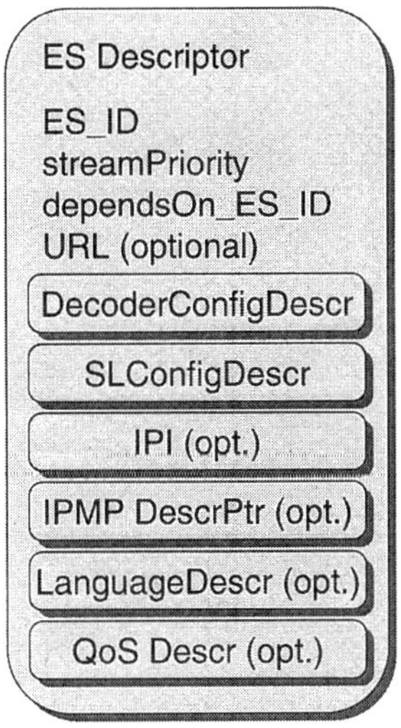

Figure 8 Main components of the ES descriptor.

this elementary stream. Because an elementary stream descriptor may also describe control streams with object-related information flowing upstream, in fact, it might also be the configuration of the associated control command *encoder* that is described here. Note, however, that actual back-channel signaling protocols are not included in version 1 of MPEG-4.

The DecoderConfig descriptor includes an indication of an object type and a stream type as well as information about average and maximum bit rate and the size required for the decoding buffer in the receiving terminal.

The stream type indication in the DecoderConfig descriptor just identifies the most basic aspect of the stream, i.e., whether it is a visual stream, audio stream, scene description stream, object descriptor stream, clock reference stream, or object content information stream. More detailed information about the stream is provided by the object type indication, which defines the permissible tools and parameters of the compression scheme for this object. In the language of the visual part of MPEG-4, it should be called "object layer type" rather than "object type." MPEG-4 defines a number of object types for both visual and audio objects.

Another important element of the DecoderConfig descriptor is the embedded decoder-specific configuration information that is conceptually passed on to the media decoder selected and initialized on the basis of stream type and object type indication. This information corresponds to what classically is referred to as "high-level headers" in previous audiovisual compression standards (e.g., "sequence headers" in MPEG-2 Video). This also explains why the content of this descriptor for visual and audio media types is actually specified in Parts 2 and 3 of MPEG-4, respectively.

Like the previous descriptor, the mandatory SLConfig descriptor carries configuration information, this time for the sync layer (SL) configuration of this elementary stream. Since the features of the SL are discussed in Sec. III of this chapter we defer a detailed discussion of the elements of the SLConfig descriptor to that section as well.

E. Auxiliary Descriptors and Streams

Apart from grouping media streams, an object descriptor makes it possible to point to auxiliary streams associated to this set of media streams. There are three types of such streams. First, semantic information about the content of the audiovisual object that is fed by these streams may be conveyed by means of *object content information*. Then, *intellectual property management and protection* information may be attached. Finally, *clock reference streams* may be used to convey time base information (see Sec. III). Another type of auxiliary information is provided by the *quality of service (QoS) descriptors* that may be attached to ES descriptors. And, of course, the very design of the descriptors as self-describing entities constitutes a generic *extension mechanism* that may be used by the International Standardization Organization (ISO) or specific applications to attach arbitrary descriptive information to an audiovisual object.

1. Object Content Information

Object content information basically consists of a set of *OCI descriptors* to communicate a number of features of the audiovisual object that is constructed by means of the elementary streams grouped in a given object descriptor. There are descriptors with keywords, possibly to be used with search engines, a textual description of the content, language and content rating information, as well as creation dates and names for both the content item and the authors of this object content information.

These descriptors may be included directly in object descriptors to indicate static OCI properties. Because all elementary streams that are collected within one object descriptor are supposed to refer to the same content item, i.e., the same audiovisual object, there is no OCI for specific elementary streams, with the exception that elementary streams may have a language descriptor in order to be able to associate a set of different language versions of an audio object with one object descriptor and, hence, one scene description node.

In case the OCI changes over the lifetime of the media streams associated with this object descriptor, it is possible to attach an *OCI stream* to the object descriptor. The OCI stream conveys a set of *OCI events* that are qualified by their start time and duration. This means that an MPEG-4 presentation may be further decomposed semantically on the basis of the actual content presented.

The OCI streams as well as OCI descriptors may be attached to either audio, visual, or scene description streams. This influences their scope. When attached to object descriptors with audio or video streams, OCI obviously just describes those streams. When attached to a scene description stream, OCI describes everything that is, in turn, described by the scene description. In this case, OCI streams (rather than single descriptors) may be most useful, because a scene description typically conveys a sequence of semantically meaningful audiovisual events.

Finally, it may be noted that OCI is a first step toward MPEG-7 in the sense that the expectation for MPEG-7 is that it will provide much more extensive semantic information about media content in a standardized format. It is thus likely that OCI will be only a temporary solution to be superseded or extended by the results of this new standardization effort.

2. Intellectual Property Management and Protection

The IPMP framework consists of a fully standardized *intellectual property identification (IPI) descriptor* as well as *IPMP descriptors* and *IPMP streams*, which are standardized shells with nonnormative (i.e., not standardized) content.

The IPI descriptor occurs as an optional part of an ES descriptor and is a vehicle to convey standardized identifiers for content, such as ISBN (International Standard Book Number), ISMN (International Standard Music Number), or DOI (Digital Object Identifier) if so desired by the content author or distributor. If multiple audiovisual objects within one MPEG-4 session are identified by the same IPI information, the IPI descriptor may just consist of a pointer to another elementary stream (i.e., its ES_ID) that carries the IPI information.

The core IPMP framework consists of the IPMP descriptors and IPMP streams. It has been engineered in a way that should allow the coexistence of multiple IPMP systems that govern the conditional access to specific content streams or entire presentations. This has been done in acknowledgment of the fact that service providers are not willing to adopt a single IPMP system. Of course, the chosen design will work perfectly as well if the whole world agrees on one type of IPMP system, but it will not break if there is no convergence.

IPMP descriptors convey proprietary information that may help to decrypt (content) elementary streams or that contain authorization or entitlement information evaluated by an—equally proprietary—IPMP subsystem in the receiving terminal. IPMP descriptors have IDs that may be assigned to individual vendors of IPMP systems, so that descriptors for different systems may be attached to content without conflict.

IPMP descriptors are not included directly in object descriptors or ES descriptors but are conveyed independently. Instead, IPMP descriptor pointers that point to IPMP descriptors with a specific ID are placed inside either object descriptors or ES descriptors. If placed in an object descriptor, it means that the IPMP descriptor is relevant for all streams described by this OD. If placed in an ES descriptor, it will be valid only for this stream.

IPMP streams have been designed as a complementary method to convey—possibly quite similar—information in a streaming fashion. This may be confusing at first glance, because IPMP descriptors can also be updated over time. IPMP descriptors, however, are conveyed inside an object descriptor stream, while IPMP streams, obviously, are separate. This has the advantage that IPMP information can more easily be kept separate from the original MPEG-4 information. Furthermore, the object descriptor parser, which, at least conceptually, is not interested in IPMP information, is not bothered by data that it just has to pass through. An example of a terminal with an IPMP system and all the possible control points is depicted in Figure 9. The meaning of ''control point'' is IPMP system specific and might translate, for example, to decrypting or enabling of data flows or to reports about the data flow.

3. Quality of Service Descriptor

The *quality of service (QoS) descriptor* is an additional descriptor that may occur in ES descriptors. It aims to qualify the requirements that a specific elementary stream has on the QoS of the transport channel for this stream. The most obvious QoS parameter is the stream priority and a possibility to signal predefined QoS scenarios, most notably ''guaranteed'' and ''best effort'' channels. Apart from that, unfortunately it remains difficult to agree on a universal set of QoS parameters that is valid for a variety of transport networks. Therefore a generic set of QoS_Qualifiers has been adopted that needs to be assessed and possibly extended in the scope of specific applications.

QoS descriptors at the ES level have an obvious use in interactive scenarios, where the receiving terminal may select individual elementary streams on the basis of their traffic (signaled in DecoderConfigDescriptor) and their QoS requirements as well as the associ-

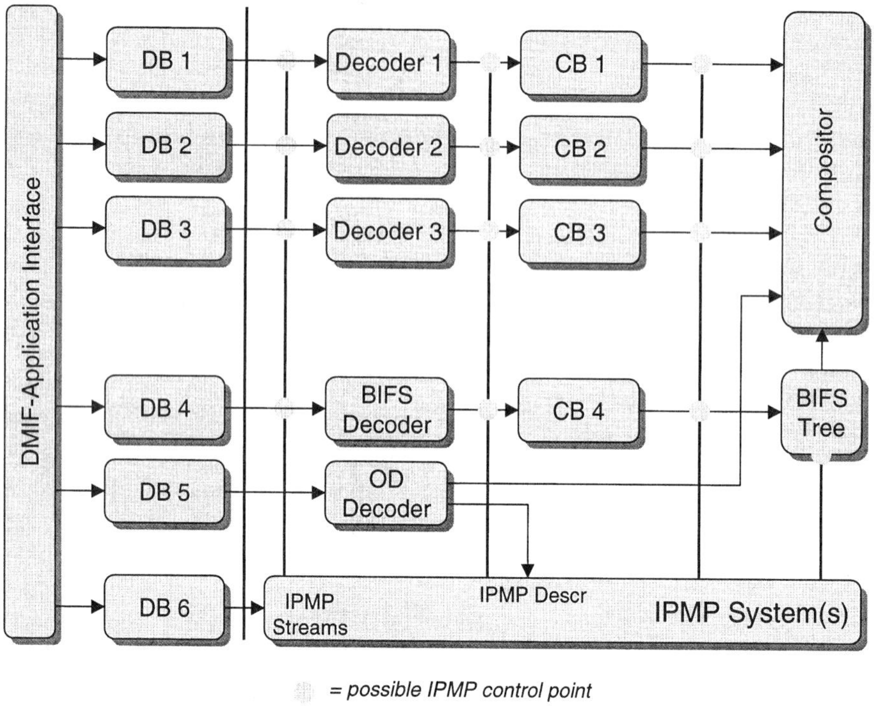

Figure 9 IPMP system in an MPEG-4 terminal.

ated communication cost. However, QoS descriptors are allowed as well in multicast and local retrieval scenarios. They may also be of interest in heterogeneous networks, if a network bridge needs to be made aware of the requirements of individual elementary streams that will, in turn, allow more intelligent processing by this bridge. However, transmission of QoS information for delivery channels that carry a set of multiplexed elementary streams may prove to be more important in practice.

F. Streaming Object Descriptors

We now explore the process by which object descriptors actually reach their destination. It has already been mentioned that they are streamed, similarly to an audio, visual, or scene description stream. In order to provide flexibility, ODs are not just put in an elementary stream one after the other; instead, a lightweight *object descriptor protocol* has been defined to encapsulate object descriptors in *OD commands*. These commands allow update or removal of a set of object descriptors or individual elementary stream descriptors at a specific point in time. In the same way, IPMP descriptors can also be updated or removed.

The timing aspect is very important, because time stamps on object descriptors can be used to indicate at which point in time the terminal is expected to be ready to receive data for a specific, newly set up elementary stream. The time stamp associated with such a command is placed on the sync layer (see Section III), as with any other elementary stream.

Updates of object descriptors usually mean that additional elementary streams show up in the updated OD. How to use this feature is largely left to users. For example, a service provider could use this feature to communicate to the receiver that a higher bit rate version of the same content is now available, maybe because server load or network load has been reduced enough to make this possible. There is no differentiation, however, between an OD update and sending a completely new OD. So, whenever in the middle of a presentation new audiovisual objects enter a scene, the associated elementary streams may be made known to the receiving terminal by a corresponding OD update command with the new object descriptors.

Because object descriptors and scene description constitute absolutely vital information for an MPEG-4 player, it is recommended that reliable transport channels be used for both, at least in unicast applications. In multicast applications this may not always be possible. This data, however, can be periodically repeated or conveyed out of band to enable random access to the multicast MPEG-4 session. In that respect, object descriptor information is comparable to program-specific information (PSI) or service information (SI) in MPEG-2 applications.

G. Linking BIFS with OD Streams and Scoping

As discussed earlier, object descriptors and the scene description are conceptually different and therefore separated; nevertheless, a strong link exists between the two. This is also evident from the rule that an object descriptor that describes one or more scene description streams must also describe the related object descriptor streams. So, an object descriptor is the glue between both. In fact, it will be an *initial* object descriptor at the initial access point to content. Object descriptors attached to Inline nodes that also contain pointers to both object descriptor streams and scene description streams can either be initial object descriptors or ordinary object descriptors (see Fig. 10).

As long as MPEG-4 content consists of just a single object descriptor stream, an associated scene description stream, and a number of audiovisual streams as needed by

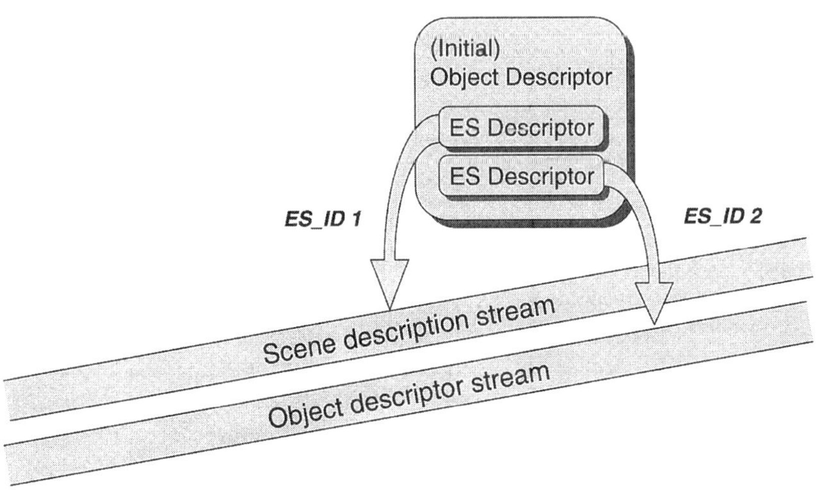

Figure 10 (Initial) object descriptor as reference to scene description and OD streams.

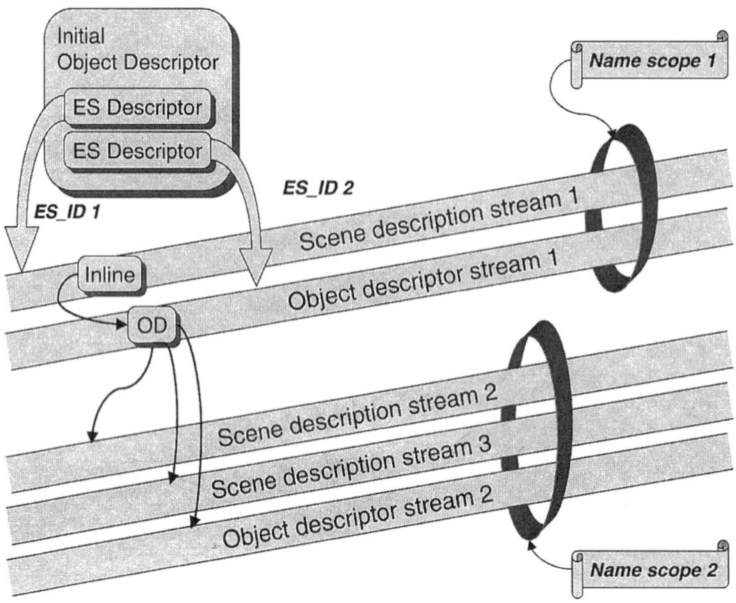

Figure 11 Scene description and OD streams for a hierarchical scene (using Inline node).

the scene, it is quite easy to know the scope of OD_IDs, ES_IDs, and other identifiers used by both scene description and object descriptor streams. They all live in a single name space or scope.

Now, what if there are multiple object descriptor streams? This may occur in two ways. First, the use of Inline nodes in the scene description forces additional scene description streams into existence. Also, there is usually no scene description stream without an object descriptor stream, unless the additional scene does not refer to any media streams. Second, as discussed earlier, there may be multiple scene description and object descriptor streams that are associated within one object descriptor. In that case, there is still only one Inline node but multiple scene description streams are linked to it. The use of this scenario will be explored further later on.

For these two scenarios there is a simple name scoping rule that is inspired by the semantics of the Inline node: all scene description and object descriptor streams that are associated with a single object descriptor constitute a single name scope for the identifiers used by them, as they will all be attached to the scene through a single Inline node. Object descriptor streams and scene description streams that are announced through different object descriptors have different name scopes, as illustrated in Figure 11. Note again that instead of talking about an Inline node, we could as well refer to the top (or only) scene description. In that case the object descriptor in question would be the initial object descriptor.

H. Grouping Streams

Equipped with these precise scoping rules for descriptor identifiers, the interesting question is how to use this facility to improve the structure of content. In particular, we want to be able to place a large number of elementary streams into groups. Even though initial MPEG-4 applications may require only a small number of streams, the design allows efficient management of content consisting of a large number of streams. As hardware and software capabilities improve, it is only natural that the sophistication of content will also increase.

A number of application scenarios are conceivable; among them are the delivery of differentiated quality content to different user groups, delivery of different portions of the content to different user groups, and reception of content originating from different sources. Grouping is quite relevant for multicast applications, where it should be easy to remove parts of the content somewhere "on the way," and in point-to-point applications it should also be possible to negotiate the desired subset of elementary streams between client and server. Let us briefly examine how these scenarios can be handled with the object description framework.

1. Grouping Streams for Content Partitioning

If a content provider needs to make a preview version of his or her content, so that limited information is available independently of the full version, this is possible without sending any part of the content twice. In other words, a receiver of the full-fledged version processes all incoming streams, whereas a receiver of the preview needs just a few of them.

First, the scene description has, of course, to be split appropriately. Parts of the scene that should go both in preview and full version, e.g., consisting of some audio and still images, have to form one stream. The scene description for the enhancements, e.g., additional video and a high-detail mesh graphic, forms the second stream. Both scene description streams are assumed to live in the same name space. Therefore, full flexibility exists to cross-reference between both parts of the scene description. The ability to cross-reference is important if the scene has interactive elements, i.e., may be manipulated by the user.

Now, two object descriptor streams can be set up, one conveying only ODs for the media streams required by the preview content (audio and still images), the other one conveying ODs for the additional media streams referenced by the complete content (video and mesh graphics). This scenario is illustrated in Figure 12, where media streams 1 and 2 correspond to the basic content and media streams 3, 4, and 5 convey the additional information.

2. Grouping Streams for Location

A special kind of content partitioning occurs in applications that present content originating from various locations at one receiving terminal. The most obvious application here is a videoconference. In that case, Inline nodes would usually be used in the scene description because the individual subscenes are essentially independent and do not need to share name spaces for their OD_ID and ES_ID identifiers. Obviously, there will be at least one object descriptor stream and one scene description stream plus the needed media streams originating from each content source, as shown in Figure 13 (with some more abstraction for better readability). Apart from that, this scenario is quite similar to the previous one where the different portions of content come from the same source.

3. Grouping Streams for Content Quality Scaling

We have seen that an individual object descriptor may contain pointers to multiple elementary streams that form a scalable encoding of visual or audio data. Actually, even the graphics objects included as part of a scene description stream can be scalably encoded

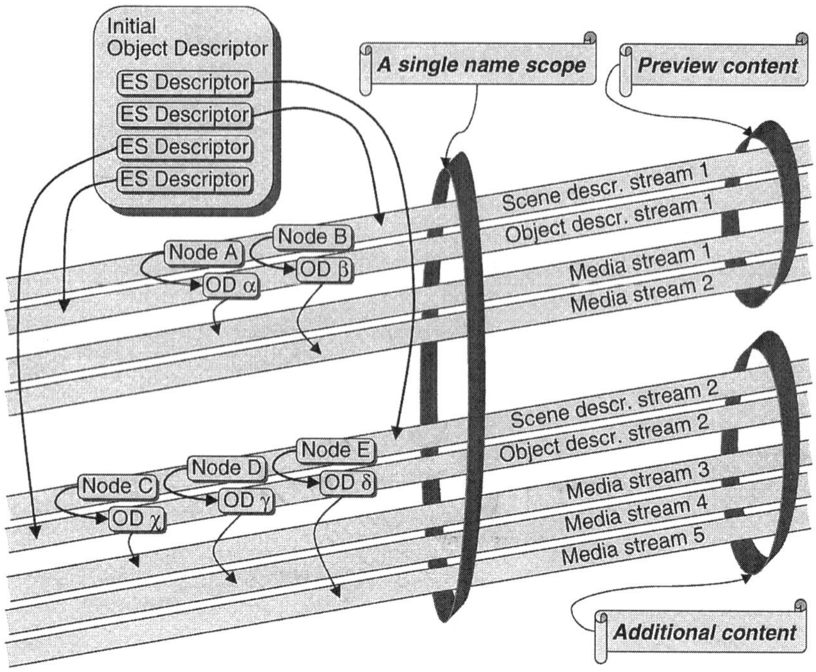

Figure 12 Object scalability with a single name scope.

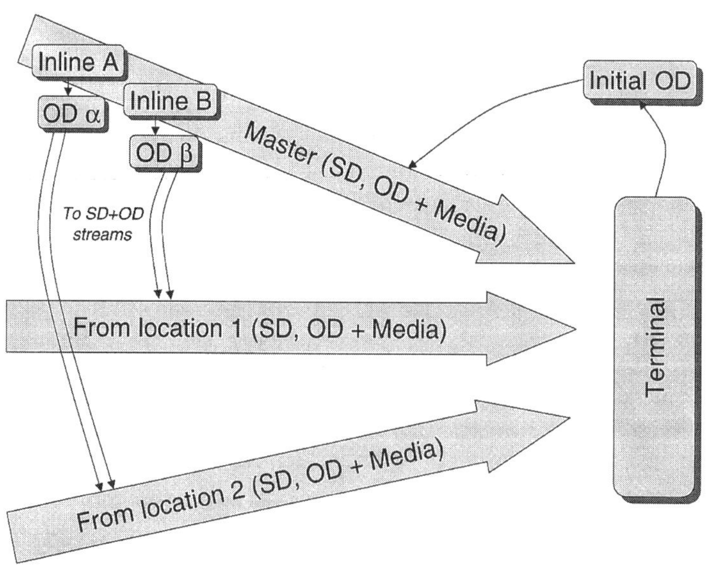

Figure 13 Content from different locations.

to some extent, providing graphics with different levels of detail and hence generating two or more scene description streams for the same content.

Now assume that a low-quality and, hence, low-bit-rate version of the content will be made available to everyone and the higher bit rate version will be provided only on request (e.g., if there is enough bandwidth available or if the customer is willing to pay for it). In this case it is possible to form a group of streams that contains the low-quality version of the content and another group that contains the high-quality upgrade. It is also possible to have two object descriptor streams, splitting each object descriptor related to an individual audiovisual object appropriately, so that even this overhead information related to the high-quality upgrade version will not be sent to receivers that receive only the low-quality version.

In the example in Figure 14, media streams 1 and 2 carry the basic, low-quality information of the content. Object descriptors in OD stream 1 just have the ES descriptors that point to those streams. The object descriptors with the same OD_IDs in OD stream 2, on the other hand, have only the ES descriptors that point to the enhancement streams conveyed in media streams 3, 4, and 5. A terminal interested in the basic quality just has to receive and process the first stream group, but for the higher quality the enhancement streams have to be processed as well. Those two stream groups can now be processed differently on the delivery layer, i.e., by a server, as desired. One option here is to transmit the basic quality in a more reliable way than the enhancement in order to improve the quality in case of transmission errors.

Note that for clarity only one scene description stream is shown in the figure. As mentioned, it is also possible to have a base and an enhancement layer for the scene description.

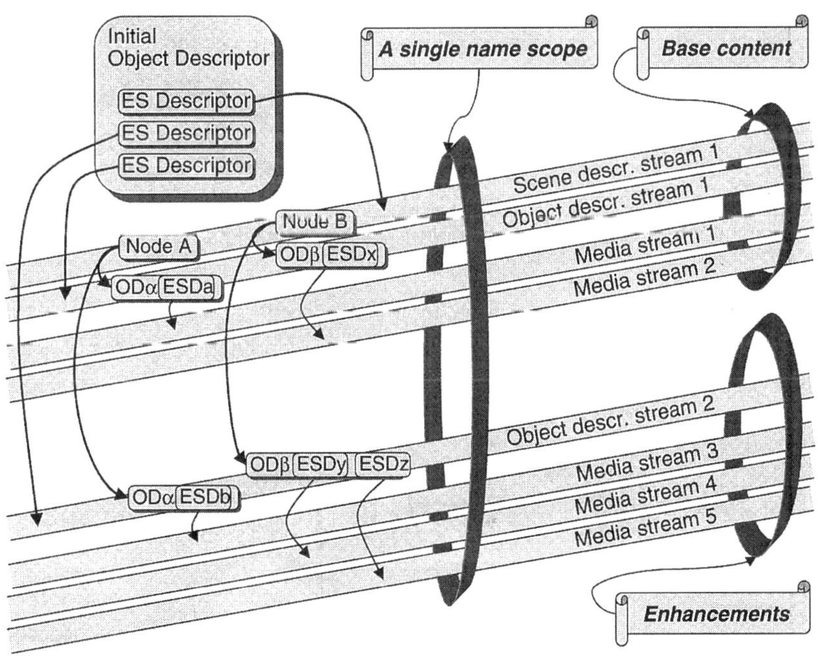

Figure 14 Quality scalability with a single name scope.

4. *Handling of Grouping Information on the Delivery Layer*

The grouping relations expressed through object descriptors have to be evaluated by a server in order to multiplex the streams together appropriately. This process is out of the scope of MPEG but may well be facilitated by the use of the MPEG-4 file format (which is currently being finalized) as well as by extending it in server-specific ways.

It may be desirable that MPEG-4 agnostic service agents in a network are enabled to remove portions of the content easily, e.g., if there is not enough bandwidth available or if the user does not wish to pay for them. Therefore, the multiplexed bitstreams need to be flagged appropriately. Such procedures depend on the actual delivery layer as well as the actual application and are therefore not in the scope of the core MPEG-4 standard.

I. Content Complexity and Selecting from a Variety of Incoming Streams

It has already been mentioned that initial object descriptors convey some indication of profiles and levels of the content referenced by them. These scalar indications allow an immediate decision by the receiving terminal about whether it is able to decode and present the content being received. Because of the potential complexity in terms of the number of scenes, it has also been made possible to indicate such complexity only for the ''current subscene,'' i.e., excluding parts of the scene that are included through Inline nodes.

In the absence of profile and level indications or at the discretion of the receiving terminal, it is also possible to evaluate the decoder configuration, stream priorities, bit rate requirements, and the dependences signaled in the ES descriptors. This allows the receiving terminal in an interactive application to request the delivery of a meaningful subset of streams for each media object, so that the computational resources of the terminal are not exceeded. Apart from these resource-driven considerations, of course, the terminal or the user needs to evaluate the scene description in order to decide which subset of the media objects advertised in an object descriptor stream is relevant.

J. Multicasting MPEG-4 Content: Object Descriptor Usage

Multicast or broadcast applications are characterized by the fact that clients may randomly tune into a running presentation. To accommodate this, usually all the crucial information for the configuration of elementary stream decoders is repeated periodically. Such a classical (e.g., MPEG-2–like) scenario is depicted in Figure 15a and shows that each stream follows its own schedule for inserting some configuration information.

In MPEG-4 the object descriptors convey the basic information that is required to gain access and be able to process the media streams that form part of an MPEG-4 presentation. The configuration information is part of the DecoderConfig descriptor and, therefore, is automatically repeated if the ODs themselves are repeated. This repetition is necessary in both broadcast and multicast applications as shown in Figure 15b.

Even in a multicast scenario, it is recommended that the object descriptors be conveyed on a rather reliable channel, which has the positive side effect that crucial decoder configuration data might be received more reliably than the media data.

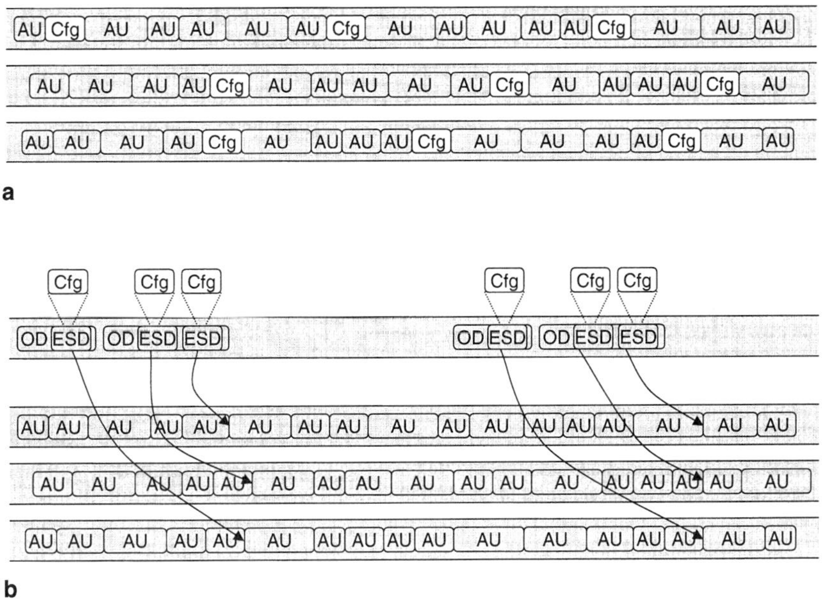

Figure 15 Periodic repetition of configuration information for broadcast: MPEG-2 approach (a) and using object descriptors (b).

K. Considerations for Distributed Content Handling

Similarly to MPEG-2, MPEG-4's architecture is based on an abstract, idealized receiver model. This allows precise resolution of issues that have to do with performance, timing and delay, and so on. In practice, implementors make sure that their designs approximate this idealized design or, more appropriately, that their implementations compensate for the imperfections of the environment on which they operate.

In contrast to MPEG-2, however, MPEG-4 allows the construction of content of which different parts may originate from different locations. Furthermore, MPEG-4 makes no assumptions about the type of underlying communications infrastructure (IP, asynchronous transfer mode [ATM], broadcast, etc.). It is then impossible to ensure that use of MPEG-4's distributed content capabilities will result in seamless content presentation in all circumstances.

We expect that content creators, jointly with both content and communication services providers, will create content cognizant of the environment on which it will operate, leveraging the available resources in order to provide a user experience of maximum quality. This is similar to the practice of tuning of hypertext-markup language (HTML) pages so that they provide visually rich content as well as reasonable download times over telephone lines. At any rate, the expectation is that the evolution of the Internet and other suitable transport networks will gradually solve these issues.

L. Authoring Stream and Scene Descriptions

Scene descriptions are intended for manipulation only by content creators and associated software tools. As a result, the encoding of scene descriptions is quite sophisticated and is based on extensive node coding tables. In contrast, stream descriptions are not only created by content creation tools but also can be modified by service provider software such as servers or gateways. A simple example in which this is necessary is when a server performs real-time reorganization or relocation of a set of streams, in which case the ES_IDs may have to be rewritten. Another example is ''cookie'' generation. To facilitate this type of processing, all descriptors follow simple byte-oriented and byte-aligned encoding, so that the computational overhead is minimized. We should point out that the multiple levels of indirection that are found in the various parts of MPEG-4 Systems are instrumental in facilitating these types of operations. For example, it is possible to remultiplex streams *without* modifying any bit of the content by reorganizing the mapping of ES_IDs to actual transport layer channels (stream map table).

III. SYNCHRONIZATION OF ELEMENTARY STREAMS

The synchronization of elementary streams is accomplished by the well-known concepts of time stamps and clock references, as used, for example, in MPEG-2 Systems [6]. These concepts are summarized in the following section, which discusses the system decoder model (SDM). This model is used to define the buffer and timing behavior of an idealized MPEG-4 terminal. Subsequently, the mechanism for packetizing elementary streams and conveying timing information, i.e., the sync layer, is introduced.

A. The Theory: System Decoder Model

An SDM is used to specify the behavior of a receiving MPEG-4 terminal in terms of a timing model and a buffer model. In MPEG-4, unlike MPEG-1 or MPEG-2, the SDM receives individual elementary streams from the delivery layer through the DMIF-Application Interface (DAI; see Sec. IV.A.4). Because there is no mandatory single protocol stack below the DAI that accomplishes the multiplexed delivery of elementary streams, it is not easily possible to extend the model to cover multiplexing in a meaningful way. MPEG-4 just puts requirements on the end-to-end delivery of data through the DAI, most prominently; a constant end-to-end delay. It is considered a task for the delivery layer below the DAI to guarantee this constant delay by suitable means. As a side effect, MPEG-4 specifies neither the additional end-to-end delay induced by multiplexing nor the associated delivery jitter. Such extensions of the SDM may, however, be developed in the context of specific MPEG-4 application scenarios.

The system decoder model as outlined in Figure 16 consists of the DAI, a number of decoding buffers, decoders, composition memories, and the compositor. The core entity for the purpose of the SDM is the *access unit* (AU). Each elementary stream is partitioned in a sequence of such AUs. The semantic meaning of an AU is determined by individual media encoders and is not relevant for the SDM or for the Systems perspective as such, with one important exception: The AU is the smallest entity with which timing information can be associated. The syntax of the sync layer (SL), described in the second part of this section, makes it possible to encode both AU boundaries and the timing information associated with AUs.

The DAI supplies access units or parts thereof to the *decoding buffer* that stores the access units until the decoding time. At that point in time the SDM assumes instantaneous decoding of the access unit, removal of the access unit data from the decoding buffer, and appearance of the decoded representation of the access unit in the associated composition

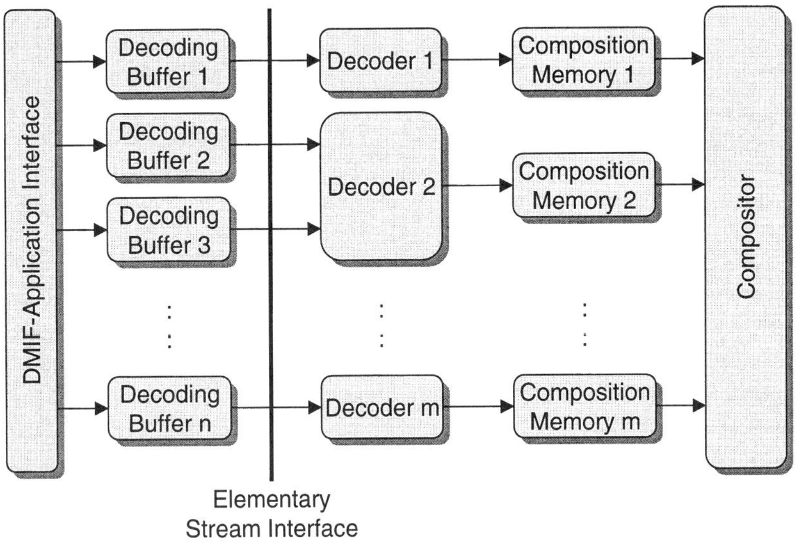

Figure 16 System Decoder Model.

memory. With this model it is possible for the *encoding* terminal to know how much decoding buffer resources are available in the receiving terminal for a specific stream at any point in time.

The content of decoding buffers is consumed by *decoders*. In case of hierarchically encoded audiovisual objects, a decoder may be connected to multiple decoding buffers, as indicated with decoder 2 in Figure 16. A decoder outputs the decoded data to one *composition memory*. The decoded data is grouped in *composition units*. The relation between access units and composition units need not be one to one but is assumed to be known for each specific decoder type. Each composition unit of decoded data is available for composition starting at an indicated composition time, known either implicitly or through an explicit composition time stamp, and ending at the composition time of the subsequent composition unit. The amount of composition memory required for specific audiovisual objects is not modeled by the SDM. A complete model with practical use would actually need to consider both the memory needed at the output of individual decoders and the memory needed as consequence of the transforms to be applied to this output according to the scene description.

The timing model postulates *object time bases* (OTBs) to which the timing of elementary streams will adhere. Such a time base is actually established at the sending side. Because in the general case it cannot be assumed that sender and receiver are locked to the same clock, samples of the time base can be communicated to the receiver by means of *object clock reference* (OCR) time stamps. The receiver can then estimate and, therefore, reconstruct the speed of the encoder clock by observing the arrival times of such OCRs.

In general, audiovisual objects whose elementary streams have different OTBs may be combined into a single presentation. A typical case would be a multipoint videoconference where it is not required that all sources run on the same time base. It is therefore not easily possible to synchronize contributions from different locations with each other in a strict sense, as there is no common, absolute, clock. However, of course, it is still necessary to synchronize the audio and video from one location to achieve lip-synch. Therefore, the terminal has to recover the time bases from all contributing sources in order to present associated audio and video composition units at the right point in time.

Such points in time are signaled by means of *decoding time stamps* and *composition time stamps*. They convey the intended decoding time* of an access unit and the intended composition time of a composition unit, respectively. These times are expressed in terms of the OTB applicable for the given elementary stream. Often, decoding and composition times will be the same. Different decoding and composition times for an access unit and the resulting composition unit(s) may be used with bidirectional predictive coding in video when the transmission order and composition order of video object planes are reversed. Use of these two time stamps to offset a known noninstantaneous decoding duration is conceivable; however, this is problematic, because two different receiver implementations most likely would not have the same decoding duration.

Only a minimum of assumptions is made about the compositor for the purposes of defining the SDM. This includes the assumption that the compositor instantaneously samples the content of each composition memory of the MPEG-4 terminal. If the composition frame rate is high, visual composition units can actually be accessed more than one time until the composition time of the subsequent composition unit has been reached.

B. The Bits and Pieces: The Sync Layer

The compression layer cares only for individual media streams and defines a binary representation specific to such media types. The delivery layer cares only to convey data packets from sender to receiver. The sync layer supplies the information that is needed by both the compression and delivery layers and therefore constitutes a kind of glue between them. The sync layer provides a flexible syntax that encodes all relevant properties of access units, as they are defined by the compression layer, and allows the mapping of complete or partial access units into a delivery layer protocol.

The atomic entity of the sync layer is an *SL packet*. Such SL packets are exchanged with the delivery layer. The sequence of SL packets terminating in one decoding buffer of a decoder is termed an *SL-packetized stream*. This naming is somewhat awkward according to the common understanding of the notion stream, because actually it is not always possible and not intended to concatenate SL packets back to back and then obtain a parsable stream. In order to avoid double encoding of the packet length information, the task of framing SL packets has instead been left to the delivery layer.

The SL packets serve a double purpose. First, they allow fragmentation of access units in a content-agnostic way during adaptation to a delivery layer. Second, this fragmentation may as well be guided by the encoder. In that case SL packets are a convenient

---

* The model does not explicitly state what to do when an access unit has not arrived completely in the decoding buffer at its decoding time, because in that case the model is violated and there is no normative behavior. However, there are typically two ways to deal with such a situation: Either the AU is considered to be lost, i.e., in error, and the terminal continues from that stage, or the terminal "freezes" the internal clock (and, hence, possibly the presentation) in order to wait a little longer, assuming that it has enough memory to buffer all the other incoming data during that time. More flexible ways to indicate tolerances on decoding times during content authoring are discussed for version 2 of MPEG-4.

way to store the resulting SL-packetized stream including these fragmentation hints. This second purpose is motivated by real-world scenarios in which it is often beneficial if an encoder knows about some characteristics of the delivery layer, most prominently the size of the maximum transfer unit (MTU), i.e., the largest packet size that will be conveyed without fragmentation. If the compression layer is able to provide self-contained packets smaller than the MTU size, according to the application layer framing (ALF) principle [7], this usually leads to improved error resilience. In case of MPEG-4 video, such packets correspond to *video_packets*, and access units, of course, correspond to complete *video_object_planes*.

The sync layer syntax is flexible in that it can be configured individually for each elementary stream by means of the SLConfig descriptor, a mandatory component of the ES descriptor. This descriptor allows the selection of the length of many syntactic fields according to the requirements of this individual stream. For example, a very low bit rate audio stream may require time stamps that consume few bits, whereas a higher bit rate video stream may need very precise decoding time stamps. Under some conditions, time stamps can even be completely omitted if the access unit duration and the decoding time of the first access unit are known.

Each SL packet may contain a *sequence number* to discover lost SL packets, a clock reference time stamp, and various flags to indicate the presence of padding, start and end of access units, and the presence of a random access entry point. The SL packets that contain the beginning of an access unit may have additional information about this AU, namely decoding and composition time stamp as well as the length of the AU or the instantaneous bit rate of this stream.

In many applications multiple media encoders will actually use the same object time base, so that clock reference time stamps are needed in only one of the generated SL-packetized streams. Therefore the SLConfig descriptor may simply provide a reference to another ES that conveys the clock reference for the current stream. It is even possible to create an elementary stream for the sole purpose of conveying clock references (with no media payload) in order to separate the two functionalities completely.

IV. DELIVERY OF ELEMENTARY STREAMS

A. The Delivery Multimedia Integration Framework (DMIF)

1. Stream Delivery Before MPEG-4

MPEG-1 and MPEG-2 considered the transport of encoded content as part of their specifications. MPEG-1 defined how to store data on a file, because the intended application space involved primarily playback from CD-ROM devices. MPEG-2 expanded the scope to include, in addition to local retrieval, broadcast over cable, over-the-air, or satellite links. In all the broadcast scenarios, no protocol stack was available. As a result, two different flavors of MPEG-2 Systems where specified: program stream (PS) for the local retrieval and transport stream (TS) for the broadcast scenario.

MPEG-2 TS was designed and optimized for ''raw'' transport media having no particular support for the carriage of structured data (e.g., satellite links or coaxial cables). The MPEG-2 TS thus includes the essential data-link layer features that were not offered by the target delivery technologies; today it can be considered as the basic transport protocol to carry digital signals, mostly television, over these infrastructures. However, it is difficult and/or inefficient to transport MPEG-2 content over networks other than those mentioned. The carriage of MPEG-2 TS over IP (or ATM) networks requires either removal of the data-link layer replicated functionality (difficult) or acceptance of a duplication of them (inefficient). Examples of the mentioned features are sync bytes and the use of a fixed packet length (for TS).

2. The MPEG-4 Approach to Stream Delivery

In contrast to its predecessors, MPEG-4 has been designed to adapt to multiple operating scenarios (local and remote retrieval, broadcast or multicast) and delivery technologies and protocols. The design choice was to abstract the functionality that the delivery layer has to provide in order to focus the core MPEG-4 Systems activity on the common features. A separate part of MPEG-4, called DMIF (Delivery Multimedia Integration Framework), deals with the details that are specific to different delivery technologies or protocols. The demarcation line that separates the scope of Systems and DMIF is called DAI (DMIF-Application Interface) and represents a conceptual interface. The separation of common and delivery specific tasks has allowed improvement of the comprehension and classification of the aspects related to the multimedia delivery and elaboration of a clean, layered model. The MPEG-4 reference software [8], described in Chapter 16, makes use of this layered model and demonstrates some of the advantages of this approach.

3. Architecture

The DMIF communication architecture as depicted in Figure 17 is such that applications do not have to be concerned with the underlying communications methods. The implementation of DMIF takes care of the operating scenario as well as of the delivery technology details and presents a common interface to the application, the DAI.

The DMIF model states that the originating application—in the terminology of the MPEG-4 Systems specification this corresponds to the receiving terminal—always accesses the multimedia content through the DMIF-Application Interface. The application requests are processed by a DMIF filter, which determines the actual DMIF instance that should serve the request based on the URL that the application supplies. The application

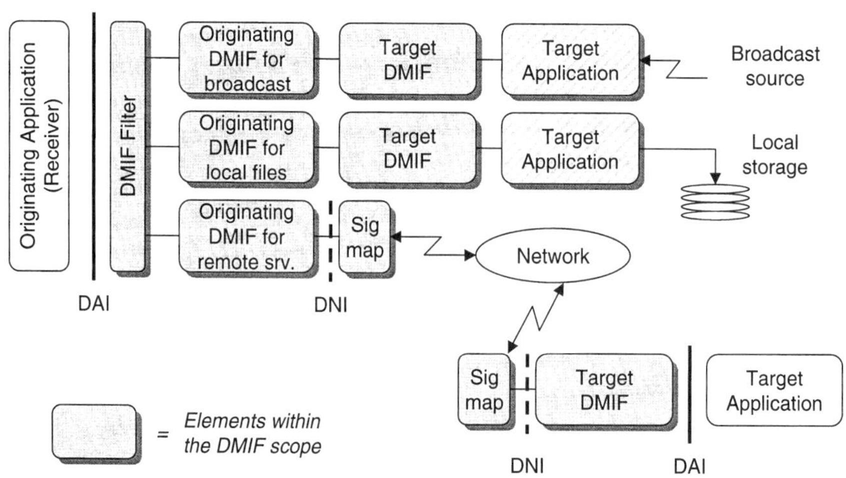

Figure 17 DMIF system model.

will not have any direct knowledge of the DMIF instance actually activated; this is completely transparent. An application can also concurrently use more than one DMIF instance (e.g., for concurrent use of IP and native ATM).

A DMIF instance provides the application with a uniform view of a service. The provision of a service is the result of a cascade of modules that are labeled in Figure 17 as Originating DMIF, Target DMIF, and Target Application. Depending on the operational scenario, the distribution of these modules and their interfaces vary considerably. For broadcast and local retrieval scenarios, all such modules are hosted by the terminal and can therefore be implemented as a single monolith. For remote retrieval scenarios, these three modules are instead located on different machines and a communication protocol is needed to make them interoperate. The details of this communication protocol may depend on the underlying network capabilities, because the connection management differs significantly among IP, ATM, and mobile networks.

The originating DMIF module is meant to work in cooperation with the target DMIF module to provide a session-level service. The distinction between originating and target DMIF modules in the local retrieval and broadcast scenario is a bit artificial but is kept for consistency with the remote retrieval scenario. The distinction of a target application module, which is obvious for the remote retrieval scenario, is essential as well in the other two cases: with this modeling it is indeed possible to maintain that DMIF is application agnostic, as all the MPEG-4 Systems and, hence, MPEG-4 content-related parameters are delivered transparently from originating to target application. Thus, the DMIF model can be applied consistently to all scenarios.

In other words, the target application is always considered as external to DMIF, even if an implementation of a DMIF instance for local retrieval or broadcast would probably not distinguish between DMIF module and target application module.

An implementation should conveniently make use of the DMIF architecture in case the support of multiple delivery technologies is a goal. If, however, it is targeted to a single delivery technology, the adoption and implementation of the DMIF architecture may not be beneficial. In these cases the same external behavior can be obtained with a simpler software architecture, where, e.g., there is no notion of a DMIF filter and of DAI primitives.

4. The DMIF-Application Interface

The *DMIF-Application Interface (DAI)* is a semantic API that formalizes the demarcation line between the content-related elements and tools of the MPEG-4 specification that are documented in Part 1 of the standard and the delivery-related elements documented in Part 6 (DMIF) of MPEG-4.

The DAI has been defined in such a way as to abstract not just the delivery technology details but also the operating scenario; thus, at this API local and remote retrievals are no different from multicast or broadcast. An MPEG-4 browser using this API would therefore be able to access and present multimedia content uniformly and independently of the operational scenario.

In contrast to the rest of the MPEG-4 specification, the DMIF part addresses both receiver and sender parts. The advantages of building a generic MPEG-4 browser based on the DAI are quite clear, but the adoption of the DAI at a server is less obvious. Actually, DMIF includes this side for two reasons:

Uniformity (especially when describing the behavior through complete walk-throughs)

Completeness, when considering that applications other than simple MPEG-4 browsers might be using this model (e.g., conferencing applications that need both the receiver and sender roles). In any case, it should be stressed that the adoption of the DAI is an implementation matter that does not affect interoperability between equipment.

The DAI does not impose any programming language or syntax (e.g., the exact format for specifying a particular parameter or the definition of reserved values). Moreover the DAI provides only the minimal semantics for defining the behavior of DMIF.

A real implementation of the DAI needs more than what is specified in DMIF; for example, it would need methods to initialize, reset, reconfigure, destroy, query the status, register services, and so on. These features depend heavily on the programming language used as well as the specifics of the implementation. For example, synchronous versus asynchronous implementation, need for call-back functions, and event handling versus polling all require different implementation architectures. Because such details have no impact on the DMIF model, they have been kept out of the scope of DMIF.

The DMIF-Application Interface comprises the following classes of primitives:

Service primitives, which deal with the control plane and allow the management of service sessions (ServiceAttach and ServiceDetach)

Channel primitives, which deal with the control plane and allow the management of channels (ChannelAdd and ChannelDelete)

Data primitives, which deal with the user plane and serve the purpose of transferring data through channels (Data and UserData, for ''real'' data and ''control'' data)

Through this set of primitives, it is possible to access content seamlessly in various, even coexisting, operation scenarios.

5. Relation Between DMIF and the MPEG-4 Systems Specification

DMIF has had some impact on the rest of the MPEG-4 specification, which is minimal in one sense but on the other hand extremely important to guarantee future evolution. The DAI, as seen from an MPEG-4 Systems perspective, actually implies the definition of the exact behavior of an MPEG-4 browser. By means of this interface, walkthroughs explaining the access to MPEG-4 content in Systems need not define the detailed, delivery-specific actions and can be maintained unique and general. Also, in DMIF, all walkthroughs start at the DAI, so that all delivery technologies are analyzed consistently.

Even though it is optional in implementations, the DAI thus serves the purpose of making the specification clearer and of making sure that different delivery technologies can be used without affecting the Systems features. It can also guide developers, because the definition of this conceptual interface was not trivial anyway, particularly in understanding the scope of the various identifiers, and so as to avoid the temptation of concentrating on, for example, an MPEG-2 delivery mechanism and derive rules that would have been valid only in that environment.

As far as the ''bits on the wire'' are concerned, DMIF has allowed the formalization of how cross-references between elementary streams should be resolved. In order to manage cross-references uniformly, it has been decided to make use of a common syntax: the URL. Thus, cross-references in object descriptors, which actually point to additional content components, have to be expressed in the form of URLs. Such URLs are part of the information delivered, so DMIF has a real impact here. Moreover, such URLs have a precise meaning: they point not to individual content but to a ''service'' that hides the content. Put differently, a service is represented in MPEG-4 Systems either by an initial object descriptor or by an object descriptor that contains a URL.

As a conclusion, DMIF only minimally affects MPEG-4 Systems, but by making sure that all the elements are used coherently, it is extremely helpful in guiding the development of future specifications such as MPEG-4 over IP or over MPEG-2.

B. Delivery Scenarios for MPEG-4 Content

The following elements are needed to describe an MPEG-4 scene fully and locate its ingredients, independently of the delivery scenario:

- An initial object descriptor that serves as the starting point for content access
- Scene description streams, which describe how a scene is composed and refer to its ingredients through pointers to object descriptors
- Object descriptor streams, which describe each individual elementary stream in a scene, including its unique identification
- Delivery specific ''stream map tables'' that map the identifiers (ES_ID or URL) to actual physical locations of streams

Now, what are the differences between the delivery scenarios? Although the precise mappings for most of these scenarios still await standardization inside and outside MPEG, several general observations can be made.

1. *Broadcast Scenario*

In a broadcast or multicast scenario the service is represented by a bundle of streams that are delivered over a set of channels. These streams are sent independently of whether or not a specific receiver needs them. All information needed for content access must either be repeated regularly in band or be made available on request by separate means. This includes all the elements mentioned earlier.

The DMIF instance has to read and continuously update, if necessary, the stream map table in order to be able to satisfy application requests to the service, for example, to make available some of the transmitted streams. For any given broadcast DMIF instance, specific syntax for DMIF URLs is needed that makes it possible to identify and access the initial object descriptor.

2. *Local Retrieval Scenario*

In a local retrieval scenario, the service is represented by a set of files that cooperatively provide the required content. The files are accessible by the application as needed. The information needed for content access has to be part of the file format and, indeed, the MPEG-4 file format that is currently being developed will specify a mechanism to describe the stream map table and to convey the initial object descriptor.

The DMIF instance has to read the stream map table in order to be able to satisfy application requests to the service, for example, to make available some of the streams stored in one of the files. For any given local retrieval DMIF instance, a specific syntax for DMIF URLs is needed that makes it possible to identify and access the initial object descriptor.

3. Remote Retrieval Scenario

In a remote retrieval scenario, the service is defined by the server. When the application requests a particular service, identified through a specific DMIF URL, the originating DMIF instance is supposed to parse the initial part of the URL and determine the server; the target DMIF instance in turn will identify the application executive running the service requested, and a logical connection between originating and target applications is established. The server will provide the initial OD for that service that will be returned to the requesting application but no stream map table. The stream map table will instead be generated incrementally on demand, as a result of the application requesting a stream: this is due to the fact that the stream map table provides a map between elementary streams and physical resources and that in this operational scenario such resources are made available only on demand.

4. Heterogeneous Scenarios

Heterogeneous scenarios are already partly supported by the current DMIF specification. Heterogeneity is supported at a terminal, meaning that several different DMIF instances for different delivery technologies and operational scenarios may coexist and even run concurrently. This enables distributed content with significant flexibility.

Heterogeneity is not yet supported within a connection itself. In other words, it is not currently possible to deliver MPEG-4 content over a sequence of different networks, e.g., an MPEG-2 TS broadcast link followed by an IP multicast link.

C. The Data Plane: Delivery of the Elementary Stream Data

In general, MPEG decided not to venture too much into the domain of multiplexing, which is the data plane part of the delivery layer functionality. MPEG will just specify the adaptations necessary to define the encapsulation of MPEG-4 SL-packetized elementary streams in delivery layer protocols. However, in some environments the use of a special MPEG-4 multiplexing tool called FlexMux (described in the next section) may be of value to reduce multiplex overhead or delay. With the exception of the FlexMux streams resulting from application of this tool, one SL-packetized elementary stream is mapped to one transport channel.

This section discusses approaches to the adaptation of MPEG-4 streams to such transport channels, termed payload format specifications, using the examples of the Internet and MPEG-2 transport environments. The last example discusses storage formats that are seen as conceptually very similar to transmission environments. All such adaptations eventually require standardization within the community that oversees the development or maintenance of a transport specification, e.g., the IETF in the case of the Internet and MPEG itself in the case of MPEG-2 transport.

Any delivery layer protocol into which SL-packetized streams are mapped has to provide framing of SL-packets; i.e., it has to encode the packet size implicitly or explicitly. Furthermore, in order to respond to quality of service requests by streams, a variety of different error protection mechanisms may be desirable in lossy delivery environments.

1. The FlexMux Tool

The FlexMux is a multiplexer with a simple packet syntax created for low-bit-rate, low-delay streams, such as object descriptor, scene description, animation, or speech streams.

It is regarded not as a true transport level protocol but rather as a tool that should be used if the cost in terms of management load, overhead, or delay to set up and use transport channels for each individual elementary stream would be too high. This may be the case if a presentation consists of dozens of audiovisual objects with a similar amount of corresponding elementary streams.

The two variations of FlexMux packet structure are depicted in Figure 18. In *simple mode* the header consists of an *index* that corresponds to the FlexMux channel, or stream number, and the packet length in bytes, both 8 bit, limiting the number of streams and payload size to 256 each. The *MuxCode mode* is active for index values greater than or equal to 240. It further reduces the overhead by conveying a priori how the payload of a single FlexMux packet is shared between multiple streams. This mode has an initial cost: each of the 16 possible templates, called MuxCode table entries, needs to be conveyed before it can be used. MuxCode table entries should be conveyed by the same protocol that sets up this transport channel. If needed, MuxCode table entries may as well be changed dynamically. In order to maintain the correct state and to discover transmission errors, a version number is used that must match between the current packet and the MuxCode table entry. The sequence of FlexMux packets that flows through one transport channel is termed the FlexMux stream.

2. *Transport on the Internet*

There are several options for transport of real-time data on the Internet. The real-time transport protocol [9] is evolving as one of the most used protocols. Issues of organizing real-time content into sessions and their control are addressed by accompanying protocols [10,11]. Adaptation of MPEG-4 to RTP is currently being discussed by MPEG and the IETF [12,13].

Other options such as transport of MPEG-4 data on the hypertext transport protocol (HTTP) are not considered here, as HTTP is deemed to be a download protocol rather than a real-time transport protocol. Of course, a self-contained MPEG-4 file could be conveyed via HTTP. Similarly, direct transport of SL-packetized streams on user datagram protocol (UDP) is not currently considered for standardization, even though this would be a perfectly viable scenario that has already been successfully demonstrated. However, the expectation is that RTP will be the focal point for the integration of MPEG-4 and non–MPEG-4 media types.

Both RTP and the MPEG-4 SL syntax provide ways to label a payload with a time stamp and enable loss detection through sequence numbers. Whereas RTP has been designed for computational efficiency by respecting byte, word, or even 32-bit boundaries

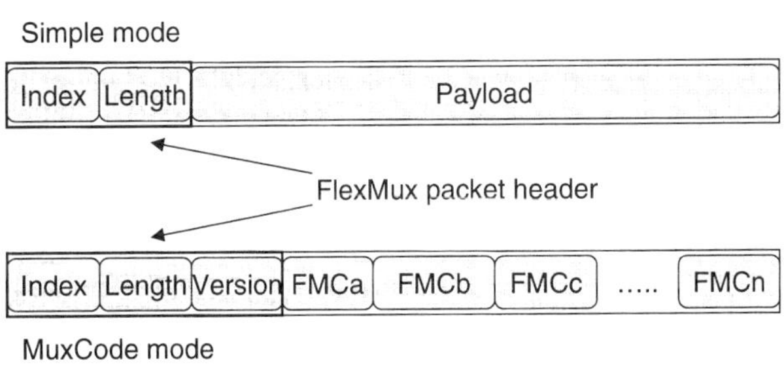

Figure 18 FlexMux packet definition.

for its syntactic elements, the SL syntax is focused more on flexibility and coding efficiency. Both expect the payload to be a semantically meaningful application data unit to facilitate stream-specific loss recovery according to the ALF (Application Level Framing) principle [7]. Therefore, wherever possible, one elementary stream should be mapped into one RTP stream. In this case one SL packet will be conveyed per RTP packet. The redundancy between the RTP time stamp and sequence number fields and the SL packet header elements decodingTimeStamp and sequenceNumber can be reduced by removing them from the SL packet header. This, however, requires processing of the SL packets during the RTP encapsulation process. Because an MPEG-4 presentation may well consist of dozens of media streams, an additional mapping of multiple elementary streams into one RTP stream is proposed, using FlexMux.* Both the overhead for individual RTP streams and even more the management load induced by the parallel RTCP (RTP Control Protocol, also defined in Ref. 9) streams and by the increased number of lower level (IP) protocol packets suggest that the ALF principle should optionally be ignored in this case by bundling multiple MPEG-4 elementary streams into one RTP session. This is achieved by mapping a complete FlexMux stream in one RTP session. The encapsulation process becomes simpler in this case, as only the fixed FlexMux packet header needs to be parsed to ensure that an integer number of FlexMux packets are put in each RTP packet. This, of course, comes at the cost that time stamps and sequence numbers can no longer be shared between SL and RTP layers. Another disadvantage that needs to be considered is the fact that loss of one RTP packet might result in lost data from multiple elementary streams.

3. *Transport in MPEG-2 Transport and Program Streams*

MPEG-2 transport streams (TSs) and program streams (PSs) provide the transport encapsulation for real-time digital broadcast data and digital video disk (DVD), respectively. The specification also includes all the necessary descriptors (program specific information, PSI) to identify the services communicated within a TS or PS. This section briefly discusses several options for encapsulating MPEG-4 streams in MPEG-2; the description of and access to MPEG-4–based services using PSI are addressed in Sections IV.B.1 and IV.D.3, respectively.

In the case of MPEG-2 TSs a single SL-packetized stream may be mapped to a sequence of transport packets labeled by a specific packet identifier (PID). The TS packets have a fixed size. Because SL packets have a variable size, some means of adaptation is needed. The MPEG-2 PES (packetized elementary stream) syntax could be used; however, it is not very efficient because each SL packet would have at least a 6-byte overhead. Instead, each SL packet may be prepended by a 1-byte length indication. A pointer mechanism may be used to indicate the start of the first SL packet within a transport packet, in the same way as it is done for PSI sections.

This method may be inefficient if an SL-packetized stream has a rather low bit rate. In order to minimize the transmission delay, the sender may need to send SL packets before an amount of 183 bytes has been collected, which is the payload length of a transport packet minus 1 byte for the mentioned pointer. This results in padding being applied, i.e., wasted bandwidth.

* IETF is in the process of defining a generic RTP multiplex. Depending on the properties of this multiplex specification, it may turn out that it could be used instead of FlexMux.

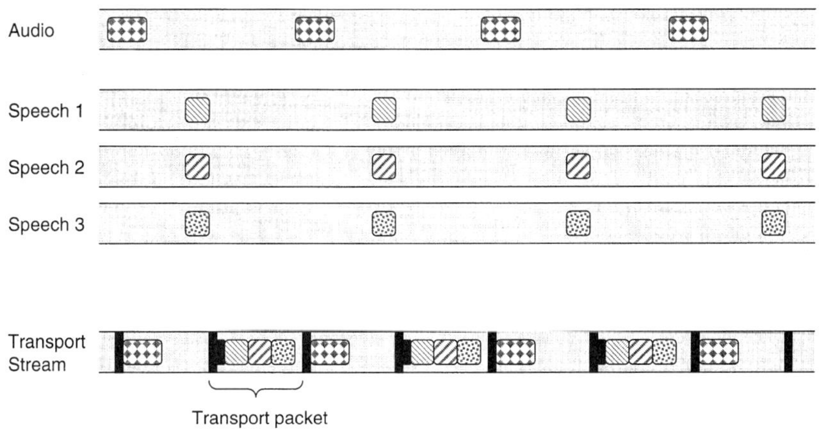

Figure 19 Mapping MPEG-4 streams in an MPEG-2 transport stream.

Figure 19 shows a visualization of an audio and a number of speech objects and how these objects produce SL packets of data over time. These SL packets are multiplexed in a sequence of transport packets shown at the bottom. The data packets of the single audio stream on top occur at regular intervals but are much less than 183 bytes in length. In order to keep the delay low, each audio SL packet is immediately put in a transport packet with a given PID and padding is applied.

To address this inadequacy, a FlexMux stream can be mapped to a PID in a similar way. FlexMux is particularly useful if the data production for the streams multiplexed into the FlexMux stream is well correlated in time, as indicated for the three speech streams in the middle of Figure 19. The bandwidth waste can be reduced enormously using the FlexMux MuxCode mode.

First, one SL packet of each stream is concatenated in one FlexMux packet; i.e., only one FlexMux packet header is needed for all three SL packets. Then each FlexMux packet is put in one transport packet (with a different PID than the audio stream, of course), leading to much less padding compared with the transport packets of the single audio stream. Again, the pointer mechanism mentioned before could be used to indicate the start of the first FlexMux packet within the transport packet, if it is desired to have them unaligned with TS packet boundaries.

Despite the lower bandwidth efficiency, as already mentioned, current standardization efforts favor the use of the MPEG-2 PES as an intermediate layer between the SL-packetized stream and the transport stream. The main reason for this approach is that in this case the same encapsulation rules may be applied for both transport stream and program stream. Even if PES is used, either a single SL-packetized stream may be encapsulated, mapping one SL packet to one PES packet, or a FlexMux stream may be encapsulated, mapping an integer number of SL packets into one PES packet.

4. Storage Format for MPEG-4 Presentations

The storage of MPEG-4 presentations seems to be an issue that is not directly related to the previously described mappings to transport formats. However, conceptually it is not much different. MPEG-4 presentations may consist of a large number of elementary streams and, as before, it has to be decided whether each of the streams is to be stored in a separate file or whether streams share one file, potentially through the use of FlexMux. It is not intended to use the file format directly for transmission of MPEG-4 streams. A translation to a transmission format that is appropriate for the target transport network should be done.

The issue of storage of MPEG-4 presentations has been deemed very important and a special call for proposals was issued. A number of responses were received [14] that may be broadly categorized into two quite different basic approaches to the issue. The first approach proposed a simple storage format that consisted of FlexMux streams, preceded by a map to associate the ES_IDs of the elementary streams to the numbers of the FlexMux channels carrying them. At the beginning of the file a single object descriptor identifies either the object descriptor stream(s) and scene description stream(s) within the file or, if only one elementary stream is stored in the file, it identifies this stream. This approach deliberately ignores the fact that such a storage format is not particularly optimized for application in a media server or editing environment. Such a storage format is thought of as a pure interchange format that may be converted to more specialized storage formats in specific applications.

Proposals of the second type attempted to remedy this situation by providing more side information to the file, for example, for indexing, in order to obtain a more versatile file format. Apple's QuickTime file format has been chosen as a starting point of a number of proposals in this category. It is currently being adapted to the needs of MPEG-4 and will be included in the second version of the MPEG-4 standard that is expected to be approved at the end of 1999.

D. The Control Plane: DMIF Signaling and Its Mapping to Concrete Delivery Layers

1. *Signaling in Generic Interactive Networks: The DMIF Signaling Protocol*

The DMIF Signaling Protocol is a generic session level protocol meant to provide a number of features that are not readily available with current techniques and that could be more and more relevant in the future. Such features are QoS and resource management and operation over a variety of networks, even heterogeneous networks.

The DMIF Signaling Protocol is somehow analogous to the file transfer protocol (FTP): a session is first opened, and then streaming content is requested on that session, as opposed to files in the case of FTP. In other words, with FTP a list of files is requested and immediately downloaded within the session, whereas with the DMIF Signaling Protocol a list of streams is requested and pointers to the channels carrying them are returned.

This protocol makes use of resource descriptors. Resource descriptors are a generic, easily extensible mechanism that makes any kind of resource equally manageable by the protocol. Thus, IP port numbers and addresses are not different from ATM virtual channel (VC), virtual path (VP), or MPEG-2 TS resources. It is also envisioned that resource descriptors may be used in the future to require transcoders or mixers, although there is currently no fully specified walkthrough for such a scenario.

The DMIF Signaling Protocol mostly derives from the DSM-CC User-Network Protocol, a protocol specified in Part 6 of the MPEG-2 specification [15] and meant to provide a signaling mechanism for heterogeneous networks. It differs from DSM-CC U-N in that it does not discriminate between client and server roles, thus allowing for not just one-way communication, and because it focuses on homogeneous rather than heterogeneous networks, at least in MPEG-4 version 1.

It is probable that in the short term the DMIF Signaling Protocol will be rarely used, as it does not bring practical advantages over existing tools such as real-time streaming protocol (RTSP) as far as unidirectional flow of streaming content over the Internet is concerned. When, however, QoS control will be feasible, especially across different network types, the advantages of the DMIF Signaling Protocol will become evident—even more so if charging policies for the use of wide area networks (WANs) become more sophisticated then just flat rate and adequate QoS support is offered, e.g., for premium services.

2. Signaling in an Internet Context

Object descriptors are signaled, according to MPEG-4 Systems, through the Object Descriptor Protocol and carried in the object descriptor stream. In theory, however, other tools could deliver the object descriptor information to an MPEG-4 receiver.

As far as we can foresee right now, given the fast evolution of the Internet protocols, signaling for MPEG-4 content in the Internet may be based on SDP, the Session Description Protocol [11]. SDP is itself comparable to some extent to the object descriptors in MPEG-4. Therefore, it is worth analyzing briefly whether SDP can completely take the role of the object descriptors in an Internet scenario.

SDP makes it possible to advertise media streams that belong to a multimedia session, including some level of descriptive information about the stream and an indication of the transport channel (UDP port) through which this media stream is conveyed. However, there is no explicit advertisement of hierarchical or alternative representations of a single audiovisual object and, of course, there is no predefined way to associate specific streams with such audiovisual objects. There is also no means of associating intellectual property rights information or object content information with individual audiovisual objects, nor can the number of streams in the session be dynamically modified.

Of course, it is conceivable to extend SDP in this direction, but assuming that the Internet will not be the only medium on which MPEG-4 content will be released, it seems preferable to keep the MPEG-defined object descriptors to describe the elements of an MPEG-4 session and to use SDP at the delivery layer simply to point to this session as a whole. In that case SDP can be used to convey the initial object descriptor as well as the stream map table (using a syntax that remains to be defined). Using SDP as a descriptor for an MPEG-4 presentation allows making use of other infrastructure that is currently being defined for content access. The Session Initiation Protocol [16] and Session Announcement Protocol [17] that are meant to announce multimedia events both use SDP descriptors. Similarly, RTSP [10] has a DESCRIBE command that conveys an SDP descriptor. Finally, SDP descriptors may also be conveyed via HTTP or e-mail, using multipurpose internet mail extensions (MIME) types.

3. Signaling in an MPEG-2 Context

An MPEG-4 application within a digital TV broadcast scenario of course does not use client–server interaction, as may be suggested by both the earlier DAI description and the abstract content access procedure described in Sec. V. Still, this procedure can conceptually be followed by de facto reading the initial object descriptor for an MPEG-4 session from a specific MPEG-4 descriptor in a PMT (program map table) entry. Additional descriptors, corresponding to the stream map table, are needed to establish the correspondence between the ES_IDs that identify elementary streams within the session and the transport channel consisting of either a PID or a PID plus a FlexMux channel number. This approach is currently proposed for standardization as an extension of MPEG-2 Systems.

V. MPEG-4 CONTENT ACCESS PROCEDURE WALKTHROUGH

To conclude this chapter, we will now outline the MPEG-4 content access procedure, using the building blocks previously described. This access may occur in different ways, depending on the application context. The details differentiating the different access procedures are confined to the delivery layer, below the DAI, and are shown indented in the following. The first level of the walkthrough focuses on the MPEG-4 Systems–related behavior.

It is assumed that the MPEG-4 receiver application is already active (it is referenced as the originating application). How it has been activated is outside the scope of the MPEG-4 specifications. It is also assumed that the URL corresponding to a service has been made available to it, by means that are also outside the scope of the MPEG-4 specifications (e.g., a link on the Web).

Given these preconditions, the following steps will take place (bearing in mind that the usage of the DAI is instrumental to the description of the walkthrough but is not mandatory in a compliant implementation):

1. The originating application requests the service by passing a URL through the DA_ServiceAttach ( ) primitive at the DAI; as a result a service session is established and the initial object descriptor is returned. In the delivery layer the following steps take place:

1.1. The URL is examined by the DMIF filter, which determines the originating DMIF Instance in charge of providing the service. (URL schemes identifying the access protocol for MPEG-4 content are not yet standardized. They might be of the form ''dudp://myhost.com/myservice'' or ''dmcast://yourhost.com/yourservice'' for DMIF signaling protocol over UDP or a to-be-defined protocol for IP multicast, respectively.)

1.2. The originating DMIF instance contacts the target DMIF instance (trivially, in the case of local retrieval and broadcast scenarios; using the DMIF signaling protocol—or an equivalent mechanism—in the case of remote interactive scenarios).

1.3. The target DMIF Instance in turn identifies and contacts the target application; it also establishes a service session with it.

1.3.1. In the case of local retrieval and broadcast scenarios, the stream map table is loaded by the target application module as conveyed by the particular delivery technology.

1.3.2. In the case of remote interactive scenarios, a network session between originating and target DMIF is established, with network-wide significance, and locally mapped by each DMIF peer to the locally meaningful service session.

1.4. The target application finally locates the service and, if existing, returns a positive answer that is forwarded back to the originating application; the answer includes the initial object descriptor for the requested service.

2. The ES descriptors in this initial OD identify the primary OD stream(s) and scene description stream(s) for this service.
3. The originating application selects the ES_IDs (or URLs, if any) of the OD and scene description streams that it requires.
4. The originating application requests delivery of the streams with these ES_IDs (or URLs) through the DA_ChannelAdd( ) primitive at the DAI; handles to the channels carrying the respective streams are returned. In the delivery layer the following steps take place:

4.1. The DA_ChannelAdd( ) primitive is processed by the originating DMIF Instance responsible for that service. The elementary stream identifier is opaque to the originating and target DMIF instances and is transparently delivered to the target application.

4.2. The target application, by means that depend on the specific delivery technology (in case of local retrieval and broadcast scenarios) or on the server implementation (in case of remote interactive scenarios), locates the desired streams.

4.2.1. In the case of local retrieval and broadcast scenarios, the streams are located by looking at the stream map table loaded by the target application module at the time the service was attached.

4.2.2. In the case of remote interactive scenarios the peer DMIF instances set up the needed network connections for the requested channels using the appropriate protocol (such as the DMIF signaling protocol over UDP). This actually corresponds to creating entries in the stream map table. The QoS parameters that have been exposed at the DAI in association with each requested stream may influence the setup of network connections, depending on the criteria that the sender side adopts to aggregate multiple elementary streams into, e.g., a single socket (by means of the MPEG-4 FlexMux).

5. The originating application requests to play the data, using the DA_UserCommand() primitive at the DAI. The command is transparently delivered by the DMIF instances to the target application, which is supposed to act accordingly.
6. The target application starts delivering the stream (which in a broadcast scenario may actually mean simply "starts reading data from the network"), and the originating application accesses data through the DA_Data() primitive at the DAI.
7. The originating application parses arriving BIFS and OD data.
8. The originating application selects ES_IDs of the media streams that it requires.
9. The originating application requests delivery of the streams with these ES_IDs (continue at step 4).

As can be seen, the MPEG-4 content acquisition requires multiple stages between originating and target applications. In some scenarios this might lead to the definition of "well-known" ES_IDs, so that some streams can be acquired without having to go through the entire procedure. Furthermore, the process deliberately ignores questions of delays induced by establishing access to various, potentially distributed resources. This has to be taken care of by making all relevant object descriptors available sufficiently ahead of the time at which the connection to the ESs is actually needed.

VI. CONCLUDING REMARKS

The framework of the elementary stream management in the MPEG-4 standard has been described. Its basic building blocks are the object description framework and the means of synchronizing elementary streams. Object descriptors are the means of linking audiovisual objects in the scene description to the elementary streams that carry their encoded information. These descriptors contain a set of various other descriptors in order to describe fully the properties of the streams and their associations. Use of time-stamped object descriptor commands allows the creation of very dynamic content in which objects—and thus streams—may be added or removed on the fly.

Timing and synchronization have been considerably extended compared with MPEG-2. In addition to the traditional combination of clock references and time stamps that allow clock recovery and audiovideo synchronization, MPEG-4 allows several other modes: rate based, global clock, or no timing information at all.

Most interesting is the approach to the delivery layer, where MPEG-4 makes minimal assumptions about the underlying communication infrastructure. In this way, it provides a design that can be easily adapted for use within a broad range of existing networks. Particular emphasis is placed on Internet-based delivery and MPEG-2 TS delivery, as they are expected to be the dominant means of bringing MPEG-4 content to users. The use of the DMIF concept and the conceptual service interface provided by the DAI enable proper definition of the operation of an MPEG-4 terminal in various access scenaria (file based, broadcast, or client–server).

As a result of this flexibility, there are several details in the periphery of MPEG-4 content delivery that still remain undefined. This is because they can be addressed only within the standardization bodies responsible for the transport layers concerned. There is already considerable progress in these areas, however, and we expect that complete solutions will be defined soon after the official release of the MPEG-4 specification.

A second version of the standard is already anticipated for the end of 1999 featuring, among other tools, the storage format as well as possibly a set of Java-based APIs that will render MPEG-4 content and players extremely flexible in a broader application context.

ACKNOWLEDGMENTS

This chapter reflects the work of many people around the world who dedicated their creative abilities to the success of MPEG-4 Systems and Delivery. We would like to thank all of them for their active involvement and dedication to the goals of MPEG-4.

REFERENCES

1. Coding of audio-visual objects: Systems, ISO/IEC 14496-1 Final Draft International Standard, ISO/IEC JTC1/SC29/WG11 N2501, November 1998.
2. Coding of audio-visual objects: Audio, ISO/IEC 14496-3 Final Draft International Standard, ISO/IEC JTC1/SC29/WG11 N2503, November 1998.
3. Coding of audio-visual objects: Visual, ISO/IEC 14496-2 Final Draft International Standard, ISO/IEC JTC1/SC29/WG11 N2502, November 1998.

4. The Virtual Reality Modeling Language—International Standard ISO/IEC 14772-1, 1997 (http://www.vrml.org/Specifications/VRML97).
5. Coding of audio-visual objects: Delivery multimedia integration framework, ISO/IEC 14496-6 Final Draft International Standard, ISO/IEC JTC1/SC29/WG11 N2506, November 1998.
6. Generic coding of moving pictures and associated audio: Systems, ISO/IEC 13818-1, 1994.
7. DD Clark, DL Tennenhouse. Architectural considerations for a new generation of protocols. Proceedings, ACM SIGCOMM, Philadelphia, September 1990, pp 200–208.
8. Coding of Audio-Visual Objects: Reference Software, ISO/IEC 14496-5 Final Draft International Standard, ISO/IEC JTC1/SC29/WG11 N2505, November 1998.
9. H Schulzrinne, S Casner, R Frederick, V Jacobson. RTP: A transport protocol for real–time applications, RFC 1889, 1996.
10. H Schulzrinne, A Rao, R Lanphier. Real time streaming protocol (RTSP), RFC 2326, 1998.
11. V Jacobson, M Handley. SDP: Session description protocol, RFC 2327, 1998.
12. H Schulzrinne, D Hoffman, M Speer, R Civanlar, A Basso, V Balabanian, C Herpel. RTP payload format for MPEG-4 elementary streams. Internet-Draft draft-ietf-avt-rtp-mpeg4-00.txt (work in progress).
13. C Herpel, ed. Architectural considerations for carriage of MPEG-4 over IP network, ISO/IEC JTC1/SC29/WG11 N2615, December 1998.
14. A Basso, H Kalva, A Puri, A Eleftheriadis, RL Schmidt. The MPEG-4 file format: An advanced, multifunctional standard for new generation multimedia content. IEEE Trans Circuits Syst Video Technol Special Issue on Synthetic Natural Hybrid Coding, 1999 (in press).
15. ISO/IEC 13818-6, Generic coding of moving pictures and associated audio: DSM-CC, ISO/IEC 13818-6, 1996.
16. E Schooler, H Schulzrinne, M Handley. SIP: Session initiation protocol. Internet-Draft draft-ietf-mmusic-sip-11.txt (work in progress).
17. M Handley, V Jacobson. SAP: Session announcement protocol. Internet-Draft draft-ietf-mmusic-sap-00.txt (work in progress).

14

MPEG-4: Scene Representation and Interactivity

Julien Signès
France Telecom Inc., Brisbane, California

Yuval Fisher
Institute for Nonlinear Science, University of California at San Diego, La Jolla, California

Alexandros Eleftheriadis
Columbia University, New York, New York

I. INTRODUCTION

MPEG-4 is a digital bitstream protocol that allows the encoding of audio, video, and natural or synthetic objects. This means that it is a specification that allows various kinds of multimedia content to be represented, stored, and transmitted digitally. The Binary Format for Scenes (BIFS) is the MPEG-4 component that is used to compose, mix, and interact MPEG-4 objects [1]. MPEG-4's BIFS is not deep, but it is unintuitive. Thus, this chapter serves as a road map for understanding its structure, usage, capabilities, and limitations.

Although an MPEG-4 scene has, for the most part, a structure inherited from the Virtual Reality Modeling Language (VRML 2.0 [2]), its explicit bitstream representation is completely different. Moreover, MPEG-4 adds several distinguishing mechanisms to VRML: data streaming, scene updates, and compression. MPEG-4 uses a client–server model. An MPEG-4 client (or player, or browser) contacts an MPEG-4 server, asks for content, receives the content, and renders the content. The ''content'' can consist of video data, audio data, still images, synthetic 2D or 3D data, or all of these. The way all this different data is to be combined at the receiver for display on the user's screen or playback in the user's speakers is determined by the scene description. The content and the scene description are typically streamed, which means that the client gets little bits of data and renders them as needed. For example, a video can be played as it arrives from the server. This is in notable contrast to VRML, where a scene is first completely transferred from the server to the client and only then rendered, leading to long latency. MPEG-4's streaming capability, thus, allows faster rendering of content.

Once a scene is in place, the server can further modify it by using MPEG-4's scene update mechanism, called BIFS-Command. This powerful mechanism can be used to do many things; for example, an avatar can be manipulated by the server within a 3D scene, a video channel can be switched from the server side, and news headlines can be updated on a virtual reality marquee. This mechanism can also be used to transmit large scenes progressively, thus reducing bandwidth requirements.

Finally, by combining the updating and streaming mechanisms, BIFS allows scene components to be animated. BIFS animation (or BIFS-Anim) consists of an arithmetic coder that sends a stream of differential steps to almost any scene component. This functionality is not different from what can be achieved with scene updates, but it has better coding efficiency when the equivalent BIFS commands would be numerous.

This chapter is organized as follows: the first part discusses MPEG-4's scene description capabilities. It discusses the design goals of MPEG-4 and how BIFS fits into the rest of the MPEG-4 set of tools. It details the supported features that an author can utilize when authoring an MPEG-4 scene. The second part of the chapter discusses in detail how BIFS scenes are compressed and transmitted. It discusses the compression techniques used and provides a few sample results. Finally, we conclude with a discussion on the use of the BIFS framework within the context of interactive audiovisual applications. Some familiarity with the overall MPEG-4 architecture is assumed; the reader is referred to the MPEG-4 Systems overview chapter in this book.

II. BIFS AND AUTHORING

A. Design Goals

MPEG-4 provides high-compression encoding for many different types of multimedia "objects": audio and speech streams, video (frame-based as well as arbitrarily shaped), 2D and 3D meshes, still textures, synthetic audio, etc. (see Refs. 3 and 4 for the detailed coding algorithms used for individual objects). These objects are encoded and transmitted to the receiver in separate streams. The function of BIFS is to support the mixing and composition of MPEG-4 objects for presentation to the user. It also enables interaction with them and the creation, animation, and modification of objects in a streaming environment. BIFS is thus a central piece of the MPEG-4 object-based architecture.

During the BIFS design process, the MPEG Systems group considered several alternatives. The World Wide Web uses static text presentation based on the hypertext markup language (HTML). Soon after its birth, a group of 3D enthusiasts came up with the Virtual Reality Markup Language, later changed to the Virtual Reality Modeling Language with, thankfully, the same acronym VRML. VRML 1.0 allowed 3D scenes to be constructed easily, but it was static and text based. Static means that the scenes did not contain moving components or any significant user interaction; text based means that the scenes had a textual representation that was easy for humans to author or modify but was not very compact. The situation changed a bit with VRML 2.0 (henceforth known as just VRML), which introduced much more user interaction and dynamic behavior into scenes. VRML 2.0 was still text based, and this gave MPEG-4 enthusiasts a brief but important calling.

MPEG-4 considered VRML 2.0 as a basis for the BIFS format because it provided a good foundation for representation of mixed media scenes with absolute positioning, versus a page-based layout as in HTML. It also provided the basis for interaction and behaviors. However, it was not designed initially for inclusion in a streaming or broadcast environment, was very inefficiently represented, and provided no 2D composition framework. All these are scene description features that are now provided by BIFS. Although MPEG-4 uses the scene structure of VRML and its 3D nodes, MPEG-4 also provides many new features. VRML features used in MPEG-4 will be only briefly summarized; the reader should refer to the VRML 97 standard for detailed specifications [2]. Our focus is on providing more details on the new MPEG-4 specific features.

Compression and streaming are two very important goals of the BIFS tool. Scene representation data are usually smaller than textures and video or audio streams, but this is not true in many cases; for example, when the content becomes complex, scene description can represent a significant amount of data to download (1 MB and over) that are worth compressing. This is also true when considering any of the following situations: low-bit-rate connections (1 MB of data takes more than 5 min to download on a 28.8 k bit/sec modem), broadcast applications in which the initial scene data typically have to be repeated every half second (100 kB of scene occupies a bandwidth of 1.6 Mbit/sec in that case, 16 Mbit/sec for a 1-MB scene), or communication applications in which scene data have to be transmitted by ''live'' sources. In all of these situations, BIFS provides a very attractive solution. In this part, we are concerned with the set of features available to content creators. A detailed discussion of the compression algorithms used for BIFS is provided in the second part, including experimental results.

B. BIFS and the MPEG-4 Architecture

An MPEG-4 system can be conceptually decomposed into several components, as shown in Figure 1. Starting at the bottom of the figure, we can identify the following layers:

Transport layer: The transport layer is media unaware and delivery aware. MPEG-4 does not define any specific transport layer. Rather, MPEG-4 media can be transported on existing transport layers such as Real Time Transport Protocol (RTP), TCP, MPEG-2 transport streams, H.323, or asynchronous transfer mode (ATM).

Sync or elementary stream layer: This component of the system is in charge of the synchronization and buffering of compressed media. The sync layer deals with elementary streams. Elementary streams are a key notion in MPEG-4. A complete MPEG-4 presentation transpors each medium or object in a different elementary stream. Elementary streams are composed of *access units* (e.g., a frame of video), packetized into sync layer (SL) packets. Some media may be transported in several elementary streams, for instance, if scalability is involved. This layer is media unaware and delivery unaware and talks to the transport layer through the DMIF application interface (DAI) [5]. The DAI is only a nonnormative conceptual abstraction of the transport layer and, in addition to the usual session setup and stream control functions, enables setting the quality-of-service requirements for each stream. The DAI is network independent.

Media or compression layer: This is the component of the system performing the decoding of the media: audio, video, etc. Media are extracted from the sync layer through the elementary stream interface. Note that BIFS itself also has to go through the compression layer and then be decoded. Another special MPEG-4 medium type is object descriptors (ODs). An OD is a structure similar to a universal resource location (URL), containing pointers to elementary streams. Typically, however, these pointers are not to remote hosts but to elementary streams that have already been received by the client. ODs also

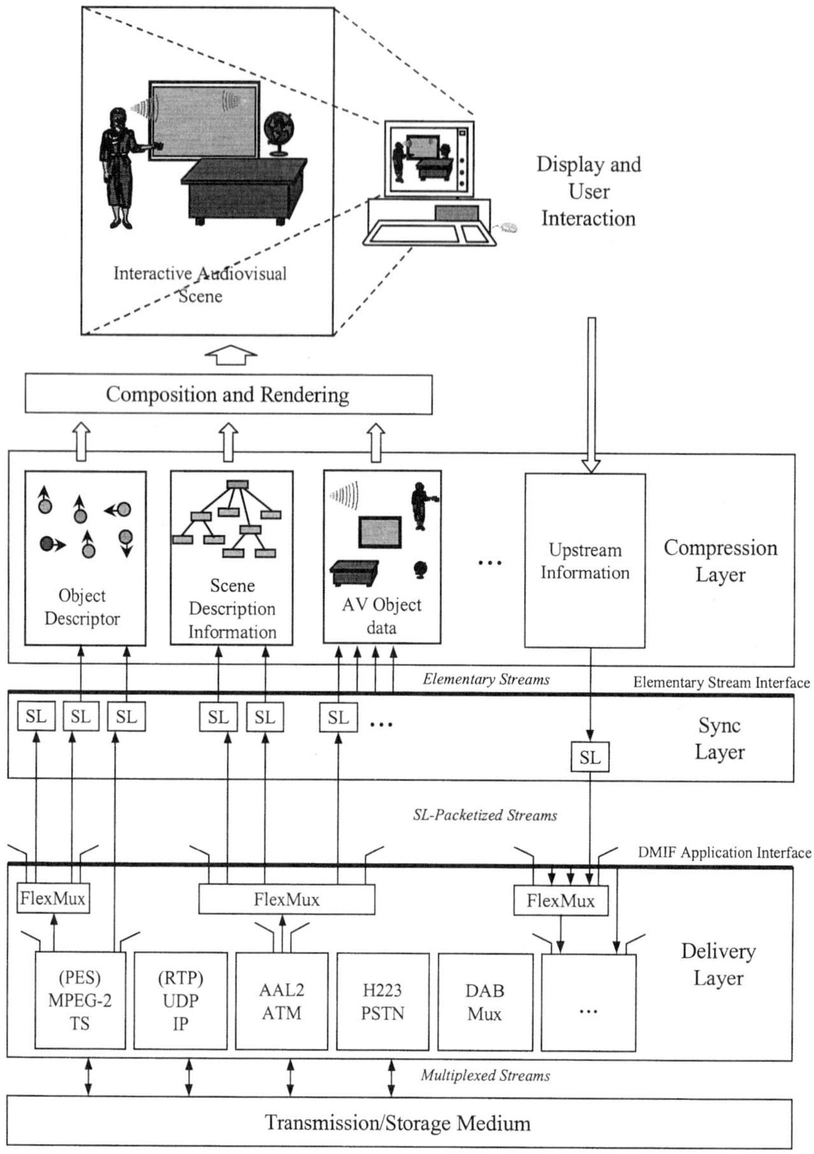

Figure 1 The architecture of a typical MPEG-4 terminal.

contain additional information such as quality-of-service parameters. This layer is media aware but delivery unaware.

1. Scene Description Streams

The BIFS scene description in MPEG-4 is transported as a special object of type "sceneDescription". The (Figure 2) object descriptor for a scene description carries, as does any object descriptor, the specific information for the BIFS decoder. This "specificInfo"

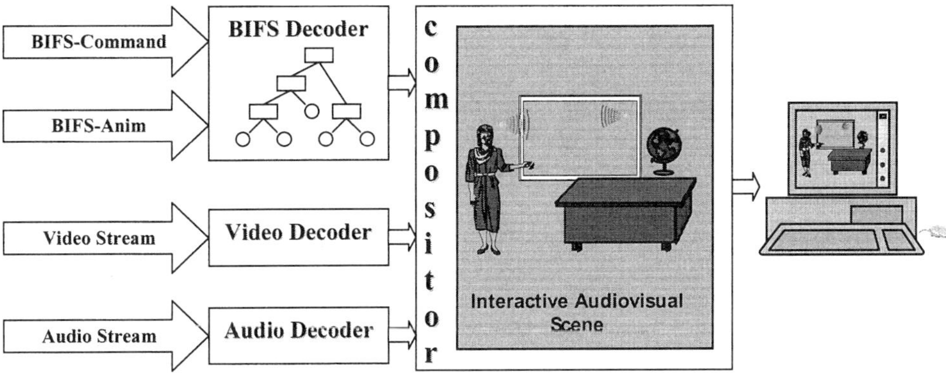

Figure 2 The two different scene description stream subtypes (Command and Anim).

descriptor describes the parameters useful for the configuration and initialization of the BIFS decoder. The configuration in particular specifies one of the two subtypes for 'sceneDescription':

BIFS-Command: This is the stream responsible for progressive loading and modification of the scene.

BIFS-Anim: This stream type is responsible for animation of the scene content. Details are provided in the following section.

2. BIFS and the Toolkit Approach of MPEG-4

MPEG-4 is a large and complex standard because it addresses a huge problem space: the creation and efficient delivery of compelling interactive audiovisual content across a variety of networks. MPEG-4, however, defines profiles in order to best match specific application areas with subsets of its specifications. The same approach has been used very effectively with MPEG-2. Not all applications need the entire set of MPEG-4 tools, and it would be a burden to force terminals to implement fully features that they do not require. MPEG-4 thus defines a limited number of subsets of visual, audio, and system tools, to which specific applications can conform. Within each profile, levels may be defined to match the varying complexity of different terminals.

Within the context of this toolkit approach, BIFS can also be seen as a separate tool. Whereas full MPEG-4 implementations would get the maximum benefit from the standard in terms of features and capabilities, it is certainly possible for a given application to comply to certain BIFS profiles and not use other parts of the MPEG-4 specification.

C. BIFS Scene Description Features

1. Scene Components: Nodes, Fields, and Events

MPEG-4 uses the VRML node, fields, and events types to represent the scene elements. A node typically represents an object in the scene. For instance, a Transform node represents the spatial transformation, applied to all its children nodes. MPEG-4 and VRML scenes are both composed of a collection of nodes arranged in a hierarchical tree. MPEG-4 has roughly 100 nodes that fall into seven categories. Each node consists of a list of fields that define the particular behavior of the node. For example, a Sphere node has a radius field that specifies the size of the sphere, whereas the Transform node has fields such as *translation* or *rotation* to represent the transformation to apply to geometry. The node fields are labeled as being of type *field*, *eventIn*, *eventOut*, or *exposedField*. The *field* label is used for values that are set only when instantiating the node. Fields that can receive incoming events have the *eventIn* label, whereas fields that emit events are labeled as *eventOut*. Finally, some fields may set values but also receive or emit events, in which case they are labeled as *exposedField*. An additional important feature of the node and field structure is that node fields receive default values. When instantiated, the default values are assumed if not specified explicitly in the bitstream. The following example is the semantic declaration of the Viewpoint node.

Viewpoint {

eventIn	SFBool	**set_bind**	
exposedField	SFFloat	**fieldOfView**	0.785398
exposedField	SFBool	**jump**	TRUE
exposedField	SFRotation	**orientation**	0, 0, 1, 0
exposedField	SFVec3f	**position**	0, 0, 10
field	SFString	**description**	""
eventOut	SFTime	**bindTime**	
eventOut	SFBool	**isBound**	

}

There are 11 basic field data types for representing the basic data types: Boolean, integer, floating point, color, two- and 3-dimensional vectors, etc. The node fields sometimes represent one value, as in the radius of the Sphere node, or many values, as in a list of vertices that define a polygon. Thus, each basic data type can be represented as a single-value type, denoted by type names that begin with an "S," or as a multiple-value type, denoted by type names that begin with an "M." For example, the floating point radius of the Sphere node has type SFFloat, whereas the collection of points within three dimensions in a Coordinate node that can be used to define a polygon has the type MFVec3f. Table 1 shows the list of MPEG-4 types.

2. Scene Structure

The scene is constructed as a directed acyclic graph of nodes. The scene graph contains structuring elements, also called *grouping nodes*, that construct the scene structure. The children of these grouping nodes represent the multimedia objects in the scene and are called *children nodes*. Among children nodes, a specific subtype plays an important role: the *bindable children nodes*. They represent the node types for which only one instance of the node type can be active at a time in the scene. A typical example of that is the viewpoint for a 3D scene. A 3D scene may contain multiple viewpoints or "cameras," but only one can be active at a time. *Interpolator nodes* are another subtype of children nodes that represent interpolation data to perform key frame animation. The last of the important nodes that compose the scene are the *Sensor nodes*, which sense the user and environment changes for authoring interactive scenes. Details of these types of scene features are described further in the following sections.

Table 1 MPEG-4 Field Types

Basic data type	Example	Single field type	Multiple field type
Floating point value	1.0	SFFloat	MFFloat
Integer value	1	SFInt32	MFInt32
Time value (64-bit float)	0.1	SFTime	MFTime
Boolean value	TRUE	SFBool	*None*
Color (RGB triplet)	100 (represents red)	SFColor	MFColor
3-vector	0.510	SFVec3f	MFVec3f
2-vector	0.25	SFVec2f	MFVec2f
Rotation (direction + angle)	0013.1415	SFRotation	MFRotation
Image	See VRML spec	SFImage	*None*
String values	"Hello World"	SFString	MFString
Nodes	Sphere {. . .}	SFNode	MFNode

3. *Composition*

a. The 3D Scene Graph. MPEG-4 uses the VRML scene graph structure with nodes and fields. 3D objects can be hierarchically transformed using a full 4 × 4 matrix transform that propagates in a graph-based representation of spatial object positions. The *Transform* node is the basic construction node for the 3D scene graph. The basic unidirectional tree structure can become a more complex graph when nodes are being reused throughout the scene. MPEG-4 adopts the VRML DEF mechanism to tag objects that can be referenced to. The USE mechanism makes it possible to point to DEFed nodes in the scene graph. Because BIFS uses a binary encoding, the DEF names are not text strings but integers, called *node identifiers or IDs*.

Once the scene graph is instantiated, composition is the operation of putting all scene objects into the same spatiotemporal space for presentation to the end user. For this transformation to occur, 3D scenes use a perspective projection, controlled by the active Viewpoint node in the scene.

b. The 2D Scene Graph. Although 2D composition can theoretically be considered as a special case of 3D, MPEG-4 offers native "true" 2D interfaces for the following reasons:

- Authoring 2D content is best done by 2D primitives rather than using restricted 3D primitives.
- A great deal of MPEG-4 content will use only 2D and does not require the more complex 3D interfaces.
- 2D composition requires a different kind of projection, which can be interpreted as a 3D orthogonal projection. However, this is best interpreted as a simple 2D scene composition with depth.
- Using true 2D primitives enables a more compressed representation.

The 2D scene graph is based on two basic operations:

- Spatial transformation, which contains 2D translation, rotation, and scaling. These transformations are controlled by the *Transform 2D* node.

```
Transform2D {
    eventIn         MFNode     addChildren
    eventIn         MFNode     removeChildren
    exposedField    SFVec2f    center             0, 0
    exposedField    MFNode     children           []
    exposedField    SFFloat    rotationAngle      0.0
    exposedField    SFVec2f    scale              1, 1
    exposedField    SFFloat    scaleOrientation   0.0
    exposedField    SFVec2f    translation        0, 0
}
```

Ordering of children nodes to control their apparent depth. The order of drawing the objects is controlled by the *OrderedGroup* node. The basic principle is to specify the order in which children nodes should be drawn. To this effect, each child receives a floating point specifying the drawing order.

```
OrderedGroup {
    eventIn         MFNode     addChildren
    eventIn         MFNode     removeChildren
    exposedField    MFNode     children           []
    exposedField    MFFloat    order              []
}
```

For producing page-like documents, MPEG-4 also proposes a way to lay out 2D elements on a page with various constraints. The Form node enables control of the alignment of 2D graphics primitives by assigning alignment constraints to objects or groups of objects. Finally, the Layout node enables scrolling of its children in various modes.

The 2D scenes use a default mode for mapping the composed scene to the rendering area: the origin of the screen is in the middle, the coordinate of the upper right corner is $(1.0, AR^{-1})$, and while the coordinate of the lower left corner is $(-1.0, -AR^{-1})$. AR is the aspect ratio, the ratio between the real width and height of the screen. When using square pixels, this is equivalent to the ratio of the numbers of pixels in width and height. In this way, a square will be drawn as a square and a circle as a circle. Figure 3 illustrates the 2D scene coordinate system.

c. Mixing 2D and 3D Composition. One of the major contributions of MPEG-4 in terms of composition is the capability to mix 2D and 3D graphics primitives seamlessly. For all 3D formats, the notion of 2D is unknown. 2D can be interfaced to the 3D

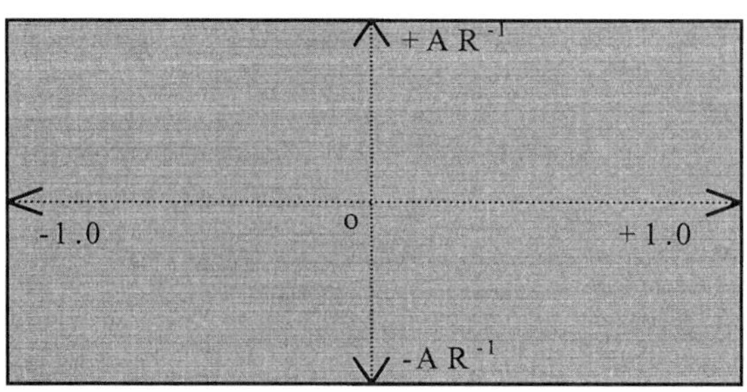

Figure 3 The 2D coordinate system (AR, aspect ratio).

scene either by restricting the 3D primitives to 2D (very inefficient in terms of rendering, compression, and authoring) or by interfacing to the external 3D presentation engine from the 2D engine (also inefficient and complex, often because of major incompatibilities in the event models). MPEG-4 has solved this issue by providing three ways to integrate 2D and 3D graphics presentations:

(1). The direct inclusion of 2D primitives inside the 3D space. This enables drawing 2D primitives directly inside a local xOy plane of a 3D scene graph. For instance, the following example shows a rectangle displayed in the xOy plane of the 3D Transform T:

```
DEF T Transform {
    translation 10.0 0.0 0.0
    rotation 1 1 0 0.57
    children
        Shape { geometry Rectangle {... . }}
}
```

(2). A simple way to compose transparent regions of the screen containing a 2D or 3D scene. A layer is a rectangle in which a scene is clipped. (This notion generalizes the concept of HTML frames.) These layers can be overlapped in the 2D rendering space of the screen, using the Transform2D node. The design of the Layer2D and Layer3D nodes enables control of the stacks of bindable children nodes for each layer. In this way, it is possible, for instance, to show the same scene from multiple viewpoints. Another advantage of the Layer node is that it enables sharing behaviors between 2D and 3D scenes. Figure 4 shows a typical structure of an MPEG-4 scene graph using layers. Figure 5 shows some examples of how to use layers to design a ''heads-up display'' or to overlap a 3D logo on a video. In order to be able to show the same scene with different children

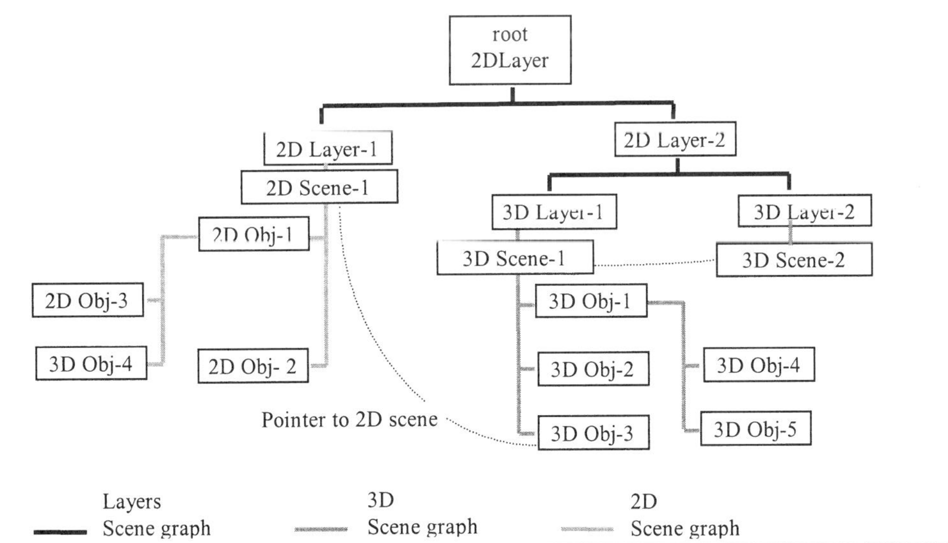

Figure 4 A typical MPEG-4 scene graph structure.

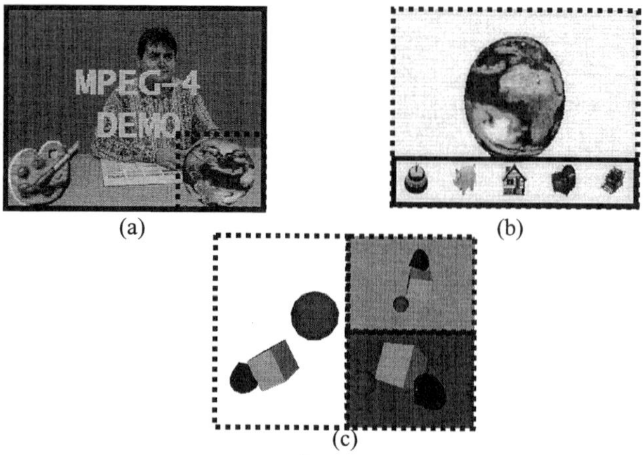

Figure 5 Layer examples (2D layers are delimited by plain lines, 3D layers by dotted lines): a 3D logo on top of a video (a), 2D interfaces to a 3D world (b), and a scene viewed from three different viewpoints (c) in the same BIFS scene.

nodes bound, the layer actually contains the stack of bindable children nodes. To each bindable child node type there corresponds a field of the layer that contains, at any instant, the active bindable child node. In this way, the same scene under two different layers may use the same bindable child node. The following illustrates the Layer3D node semantics and shows an example of sharing the same scene with multiple backgrounds using layers.

Layer3D {

eventIn	MFNode	**addChildren**	
eventIn	MFNode	**removeChildren**	
exposedField	MFNode	**children**	NULL
exposedField	SFVec2f	**size**	−1, −1
exposedField	SFNode	**background**	NULL
exposedField	SFNode	**fog**	NULL
exposedField	SFNode	**navigationInfo**	NULL
exposedField	SFNode	**viewpoint**	NULL

}

Example. In the following example, the same scene is used in two different Layer2D nodes. However, one scene is initially viewed with background b1, the other with background b2. When the user clicks on the button 1 object, all layers are set with background b3.

```
OrderedGroup{
   children [
      Transform2D { # A set of transforms to translate and scale the layer
         ...
         children [
            Layer2D {
               background DEF b1 Background2D {…}
```

```
            # It is possible to define the bindable children node directly in
            # the corresponding field
            children [
               DEF MYSCENE Transform2D {
                  children [
                     DEF b3 Background2D {…} # a shared background
                     DEF TS TouchSensor{}
                     DEF button1 Shape{..}    # button 1
                             # The objects of my scene
                  ]
               }
            ]
         }
      ]
   }
   Transform2D {
            # Another set of transforms to translate and
            # scale the layer
      children [
         Layer2D {
            children [
            DEF b2 Background2D{…} # It is possible to define the bindable
                                   # children node in the children field.
                                   # b2 is initially bound since it is the
                                   # first background 2D in the children
                                   # field OF the parent Layer2d
            Transform2D USE MYSCENE
            ]
         }
      ]
   }
]
}
```

(3). A third option for mixing 2D and 3D scenes is to use the *Composite Texture 2D* and *Composite Texture3D* nodes to texture a rendered scene on any geometry. Composite Texture nodes are designed in the same way as the layers for holding the stacks of bindable children nodes. Figure 6 shows some examples of using Composite Texture nodes.

d. Sound Composition. The VRML sound model is based on the concept of linking sound rendering to the geometry and the user point of view. MPEG-4 enhanced this model by attaching physical properties to material in the scene and by defining some environmental sound rendering parameters.

MPEG-4 also proposes another sound composition concept, which is not based on physical modeling. MPEG-4 specifies a real notion of an audio scene graph for performing audio composition at the terminal. This composition applies high-quality signal processing transformations to the input audio streams to produce the output audio stream. A simple feature is the mixing of sound sources. More advanced features include the possibility to produce special effects on existing ''natural'' audio streams.

Figure 6 Composite Texture2D example. The 2D scene is projected on the 3D cube.

The audio scene graph consists of the audio sources that come from the output of the audio decoders. When a visual and audio scene is presented, audio nodes are placed in the spatial 2D or 3D scene graph by using the *Sound2D* or *Sound* node. The following nodes are available for sound composition:

AudioBuffer	The audio buffer stores the output of a rendered audio scene. All children of the audio buffer are rendered into the buffer of the AudioBuffer.
AudioDelay	The AudioDelay node allows sounds to be started and stopped under temporal control. The start time and stop time of the child sounds are delayed or advanced accordingly.
AudioMix	The AudioMix node mixes any number of input sources and outputs any number of mixed sources. A matrix controls the mixing parameters.
AudioFX	The AudioFX node is used to allow arbitrary signal-processing functions defined using structured audio tools to be included and applied to its children.
AudioSwitch	The AudioSwitch node is used to select a subset of audio channels from the child nodes specified.

4. *Synthetic and Natural Objects*

a. 2D Graphics. In order to integrate 2D graphics primitives seamlessly within a 3D environment, the design of the 2D primitives has strictly followed the design of their 3D counterparts. For instance, 2D graphics primitives include Rectangle and Circle, which correspond to Box and Sphere in 3D.

Circle	Draws a circle with a radius at the center of the local coordinate system.
Curve2D	The Curve2D uses Bezier interpolation to draw a 2D curve.
IndexedFaceSet2d	Used to render 2D mesh primitives. Designed in a similar fashion to its 3D counterpart, IndexedFaceSet.
IndexedLineSet2D	Used to render 2D poly-lines. Designed in a similar fashion to its 3D counterpart, IndexedFaceSet.
PointSet2D	A set of points in the 2D space. Equivalent to PointSet in 3D.
Rectangle	A 2D rectangle, equivalent to the 3D box.
Text	For 2D, the VRML Text node is used.

b. 3D Graphics. MPEG-4 uses VRML's graphics primitives, such as meshes, spheres, or cones. MPEG-4 also uses the texturing and lighting model of VRML 2.0, as well as VRML's text primitive.

c. Video and Textures. To present video and textures with the pixel-by-pixel accuracy that results from the output of a video or texture decoder, MPEG-4 provides a new geometry node, the BitMap node. The BitMap node is a rectangle that always faces the viewer. The rectangle is not affected by rotations or scaling but only by translations. The rectangle size is a scaled version of the original texture pixel size. By default, the scale (−1.0, −1.0) means that that the rectangle is presented in the original pixel size of the texture.

MPEG-4 also uses the texturing mechanisms for still or moving pictures with the ImageTexture, PixelTexture, and MovieTexture nodes. The following example shows a semitransparent video presented to the user at its original pixel size.

Example. To specific semitransparent video:

```
Shape {
   appearance Appearance {
      texture MovieTexture {  // Visual object
         ...
      }
      material Material2D {
         transparency  0.5   // semi-transparent
      }
   }
   geometry Bitmap {}
}
```

d. Sound. MPEG-4 provides the ability to play streaming or nonstreaming sounds sources. With the Sound2D and Sound nodes, it is possible to place a sound source in a 2D or 3D scene. It is also possible to place a ''microphone'' in the 3D scene with the Listening Point node. The following table summarizes the audio capabilities.

AudioClip	The VRML AudioClip can be used to play small, nonstreaming sound effects.
AudioSource	Represents the source of the audio, which can be streaming or nonstreaming.
Listening Point	Represents a ''microphone'' placed in a 3D scene to render the sound sources heard at this particular point.
Sound	The Sound node attaches a sound to the 3D scene graph. The sound is transformed with the scene graph at the location where it is inserted.
Sound2D	The Sound2D node is used to attach a sound to a 2D scene graph.

5. Local Interaction and Behaviors

The event model of BIFS uses the VRML concept of ROUTEs to propagate events between scene elements. ROUTEs are connections that assign the value of one field to another field. As is the case with nodes, ROUTEs can be assigned a ''name'' in order to be able to identify specific ROUTEs for modification or deletion. Because the encoding is binary, naming is based on the use of integer identifiers called ROUTE IDs.

ROUTEs combined with interpolators can cause animation in a scene. For example, the value of an interpolator is ROUTEd to the rotation field in a Transform node, causing the nodes in the Transform node's children field to be rotated as the values in the corresponding field in the interpolator node change with time.

Many interpolator nodes are inherited directly from VRML: ColorInterpolator, CoordinateInterpolator, NormalInterpolator, OrientationInterpolator, PositionInterpolator, ScalarInterpolator. For 2D composition, specific interpolators have been added to BIFS: CoordinateInterpolator2D, PlaneSensor2D, PositionInterpolator2D, ProximitySensor2D.

User inputs can be detected with sensors. Sensors generate events when the user interacts with an object or a group of objects. Sensors that take into account user inputs are Anchor, CylinderSensor, PlaneSensor, Planesensor2D, SphereSensor, TouchSensor, and DiscSensor. Other sensors react to changes in the environment: Collision, ProximitySensor, Visibility-Sensor. Finally, TimeSensor reacts to changes of time and drives animations.

For performing interaction and key frame animation, it is necessary for the BIFS scene to contain a TimeSensor and interpolation data. Interpolation data can be quite voluminous, and this is why MPEG-4 provides streamed animation, as detailed in Section II.C.7.

6. Streaming Scene Description Updates: BIFS-Command

MPEG-2 was primarily a standard addressing broadcast video applications. Although MPEG-4 also makes it possible to deliver broadcast applications, MPEG-4 addresses other domains such as consultation and communication applications.

To fit this requirement, a very important concept developed within MPEG-4 BIFS is that the application itself can be seen as a temporal stream. This means that the presentation or the scene itself has a temporal dimension. On the Web, the model used for multimedia presentations is that a scene description (for instance, an HTML page or a VRML scene) is downloaded once and then played locally. In the MPEG-4 model, a BIFS presentation, which describes the scene itself, is delivered over time. The basic model is that an initial scene is loaded and can then receive further updates. In fact, the initial scene loading itself is considered an update. The concept of a scene in MPEG-4, therefore, encapsulates the elementary stream(s) that conveys it over time.

The mechanism by which BIFS information is provided to the receiver over time is the BIFS-Command protocol (also known as BIFS-Update), and the elementary stream that carries it is thus called the BIFS-Command stream. BIFS-Command conveys commands for the replacement of a scene, addition or deletion of nodes, modification of fields, etc. For example, a ''ReplaceScene'' command becomes the entry point for a BIFS stream, exactly in the same way as an intra frame serves as a random access point for video. A BIFS-Command stream can be read from the Web as any other scene, potentially containing only one ReplaceScene command, but it can also broadcast as a ''push'' stream or even exchanged in a communications or collaborative application.

BIFS commands come in four main functionalities: scene replacement, node/field/route insertion, node/value/route deletion, and node/field/value/route replacement. More details about the encoding of these commands are provided in Sec. III.D. From a functional perspective, the commands enable the following operations.

Replacing the entire current scene with a new one. When a BIFS Replace Scene command is received, the whole context is reset, and a new scene graph corresponding to the new BIFS scene is constructed.

Insertion command: This command comes in three subtypes:

Node insertion: Nodes can be inserted into the children field of grouping nodes. This command allows the insertion of a node at the beginning, end, or indexed position in the list of children nodes of an already existing node.

Indexed field insertion: With this update command, a generic field is inserted into a specified position in a multiple-value field.

ROUTE insertion: This can be used to enable user interaction or other dynamic functionality in the scene by linking event source and sink fields in the scene.

Deletion command: This command also comes in three subtypes, analogous to those of the insertion command:

Node deletion: Simply deletes the identified node. Note that it is possible to delete a node that does not have an ID (i.e., it has not been DEFed) by using the IndexedValue deletion command (see below).

IndexedValue Deletion: This command allows the deletion of a specified entry in a multiple-value field.

ROUTE deletion: Simply deletes a ROUTE.

Replacement command: This command also comes in the standard three flavors, plus one more for replacing single-value field values:

Node replacement: Replaces an existing node in the scene with a new one.

Field replacement: Changes the value of a field within a node. This command can be used, for example, to change a color, a position, or the vertices of a mesh or to switch an object on and off.

IndexedValue replacement: This command is very similar to the field replacement command, except that the field referred to is a multiple-valued one, and hence the command also includes indexing information in order to identify the particular value that should be replaced.

ROUTE replacement: Replaces an existing ROUTE with a new one.

7. Streaming Animations: BIFS-Anim

In Section II.C.5 we described how it is possible to trigger a key frame animation by downloading interpolator data in the scene along with an appropriate TimeSensor. This can be used, for example, to modify the fields of a Transform2D node in order to move its children around. An alternative method is to use BIFS-Command to update the fields of a Transform node. These methods are well suited for small animations. However, when the source of animation is a live one or when better compression is sought, MPEG-4 provides an alternative way to stream animation to the scene with the BIFS-Anim tool. BIFS-Anim enables optimal compression of the animation of all parameters of a scene: 2D and 3D positions, rotations, normals, colors, scalar values, etc.

The BIFS-Anim framework works as follows:

1. The BIFS scene is loaded with objects that have been DEFed, i.e., assigned a unique node ID.
2. An Animation mask is loaded, containing the list of nodes and fields to be animated. When multiple fields are animated, it is possible to select the field to be animated.
3. The animation stream itself is streamed, containing animation frames in time-stamped access units.

8. Facial Animation

The MPEG-4 Visual specification (Part 2 of MPEG-4) defines the syntax for the representation of facial and body animation. At the Systems level, some nodes are provided to generate face and body geometries in the scene graph. Face and body animation streams are carried in the same framework as BIFS-Anim. This enables animating of faces along with all other parameters of the scene.

BIFS provides (in version 1) the following nodes for managing the face animation:

Face	The Face node is used to define and animate a face in the scene.
FaceDefMesh	The FaceDefMeshNode allows the deformation of an IndexedFaceSet as a function of the amplitude of a facial animation parameter (FAP) as specified in the related FaceDefTable node.
FaceDefTables	The FaceDefTables node defines the behavior of a facial animation parameter on a downloaded face model.
FaceDefTransform	The FaceDefTransform node defines which field (rotation, scale, or translation) of a Transform node (faceSceneGraphNode) of FaceSceneGraph (defined in an FDP node) is updated by a facial animation parameter and how the field is updated.
FAP	This node defines the current look of the face by means of expressions and facial animation parameters.
FDP	The FDP node defines the face model to be used at the terminal.
FIT	The FIT node allows a smaller set of FAPs to be sent during a facial animation. This small set can then be used to determine the values of other FAPs, using a rational polynomial mapping between parameters.

9. System-Related Features

a. Programmability. MPEG-4 Systems supports scripting in the same way VRML does. However, whereas VRML allows both inlined JavaScript (recently standardized as ECMAScript) and URLs pointing to externally available Java class files, MPEG-4 currently allows only JavaScript to be used. Use of a scripting capability is essential in order to implement complex application functionalities that involve complex state information.

MPEG (as of this writing) is currently defining a Java layer, called MPEG-J, in order to provide additional programmatic control of the terminal to application developers. The model enhances the external application interface (EAI) used in VRML and is expected to be included in version 2 of MPEG-4 in January 2000. In addition to controlling the scene, an MPEG-J program can access and assess system resources and act on the network and decoder interfaces.

b. Scalability and Scene Control. One of the challenging issues when dealing with complex and rich multimedia content is to ensure that the content will be rendered according to the author's intentions regardless of the client's platform. The diversity of client platforms makes it difficult to know how the content will behave. Rather than providing normative composition, MPEG-4 provides hints for the terminal to present the "best" content according to the terminal's capabilities.

In particular, MPEG-4 provides two ways to control the scene degradation in order to adapt it according to specific environment parameters. The TermCap node contains a list of system capabilities that can be used to drive a degradation of the scene through a script or Java code. The TermCap capabilities are the following:

Frame rate
Color depth
Screen size
Graphics hardware
Audio output format
Maximum audio sampling rate
Spatial audio capability
CPU load
Memory load

Each capability can have five levels. At any time during the scene graph traversal, the current capacity for a given category can be estimated and the scene graph can be modified adequately according to the status of the system. A typical example of this is to use different scene versions according to the frame rate available in the system.

Another hint for the content creator to perform controlled scene degradation to fit best the terminal capabilities is to use object priority to degrade the objects in the scene adequately. In MPEG-4, it is possible to assign to each object (sound, video BIFS scene) a global priority. The terminal can use this information to adapt the content intelligently following the content creator's hints.

c. The Timing and Execution Model of BIFS. As with any other content type in MPEG-4, BIFS obtains its ''clock'' through the elementary stream by which it is delivered. The BIFS stream can use object clock references to recover its clock and composition time stamps to determine when a particular BIFS-Command (or BIFS-Anim) instruction is to be executed. Detailed information regarding the timing and buffering model of MPEG-4 can be found in the chapter in this book on elementary stream management and delivery.

The SFTime values contained in node fields are expressed in units of the stream clock. When a node becomes valid at a Composition Time Stamp (CTS) as a result of a BIFS-Command, this is the origin of SFTime for the nodes contained in this BIFS command, expressed in seconds. For instance, if a MovieTexture node becomes active at 10.0s (CTS = 10.0 secs) in the presentation, its startTime field will specify the number of seconds since 10.0s. In this way, it is possible to synchronize scene-related events with the streams that carry the presentation.

Based on these strong timing constraints, the execution model of the BIFS compositor needs to take into account all changes to the scene at the correct CTS and at any given instant of rendering. In the case of MPEG-4, the compositor needs to take into account all sources of events into the scene:

BIFSCommand streams
BIFS-Anim streams
ROUTEs
MPEG-J events

d. Name Space Rules. As mentioned earlier, nodes and ROUTEs in BIFS can receive unique identifiers or IDs. In MPEG-4 terminals, the following scoping rules apply for the ID name spaces:

A BIFS-Command stream shares the name space across the whole stream.

An inlined BIFS-Command stream (i.e., a BIFS-Command stream acccessed through an inline node) opens a new name space.

Multiple BIFS-Command streams pointed at through the same Object Descriptor structure share the same name space.

The last rule allows building multiple streams of content sharing the same name space. The object descriptor hence gathers streams that share the same name space. By grouping streams under the same Object Descriptor, the content creators can signal their intention to allow the pieces of content to share the name space. Obviously, the implication is that the nodes receive a unique identifier across all BIFS-Command elementary streams sharing the same object descriptor. This functionality is very useful for building large content packages.

10. Summary of MPEG-4 Version 1 Features

The scene features are presented in terms of nodes grouped as semantic categories as shown in Table 2.

11. Future Features

In this section, we list some of the features that will be offered by MPEG-4 version 2 (see Ref. 7).

a. Server Interaction. One of the key features of version 2 MPEG-4 BIFS is enabling the transmission of data from the terminal presentation back to the server. This may be useful in particular for many applications such as stream control and initiating transactions. To provide this functionality, MPEG is considering adding a ServerCommand node that will enable interactively triggering messages back to the server. The contents of the command will be left completely application dependent and be carried back to the server using a normative syntax. The command sent back to the server will also include the node ID of the ServerCommand node that triggered the command.

b. Enhanced Sound. In version 1, MPEG-4 provides the ability to enhance considerably the basic VRML 2.0 sound model. In version 2, MPEG-4 will go further and provide new sound features:

1. Enhanced physical sound-rendering model so that more natural sound source and sound environment modeling is made possible in BIFS. The functionality of the new sound source includes modeling of air absorption and more natural distance-dependent attenuation, as well as sound source directivity modeling. In version 2, BIFS also takes into account the response of the environment so that the sound can be rendered to correspond to the visual parts of the scene.
2. An enhanced sound-rendering model, based on geometry-independent perceptual parameters. These perceptual parameters provide high-level interfaces to control the ''aural aspect'' of the sound rendering. Parameters such as the source presence, heaviness, envelope, brilliance, and warmth can be controlled.

c. Scene Extensions: Prototypes. The BIFS version 2 specification provides ways to encode PROTO and EXTERNPROTO. These scene constructs enable the definition of new interfaces to collections of BIFS code. For example, a button PROTO can be constructed that accepts a string label as input. The button is rendered regularly but with the parameter value substituted at the appropriate portion of the scene.

Table 2 Nodes Grouped as Semantic Categories

Type of nodes	Function	BIFS nodes
Grouping nodes	Grouping nodes have a field that contains a list of children nodes. Each grouping node defines a corrdinated space for its children. This coordinates space is relative to the coordinate space of the nodes of which the group node is a child. Such a node is called a parent node. This means that transformations accumulate down the scene graph hierarchy. Grouping nodes are ordered in four subcategories: nodes usable for both 2D and 3D grouping, 2D specific nodes, 3D specific nodes, and Audio specific nodes.	Group Inline OrderedGroup Switch Form Layer2D Layout Transform2D Anchor Billboard Collision Layer3D LOD Transform AudioBuffer AudioDelay AudioFX AudioMix AudioSwitch
Interpolator nodes	Interpolator nodes perform linear interpolation for key frame animation. They receive as an input a key and output a value interpolated according to the key value and the reference points value. Interpolator nodes are classified in three categories: nodes usable for both 2D and 3D interpolation, 2D specific nodes, and 3D specific nodes.	ColorInterpolator ScalarInterpolator PositionInterpolator2D CoordinateInterpolator2D CoordinateInterpolator NormalInterpolator OrientationInterpolator PositionInterpolator
Sensor nodes	Sensor nodes detect events in their environment and fire events. For instance, a TouchSensor detects a click of a mouse and a ProximitySensor detects that the user entered a region of the space. Sensor nodes are classified in three categories: nodes usable for both 2D and 3D sensores, 2D specific nodes, and 3D specific nodes. Interpolator, Sensor, and ROUTE statements enable the design of interactive scenes.	Anchor TimeSensor TouchSensor DiscSensor PlaneSensor2D ProximitySensor2D Collision CylinderSensor PlaneSensor ProximitySensor SphereSensor VisibilitySensor

Table 2 (Continued)

Type of nodes	Function	BIFS nodes
3D geometry nodes	Represent a geometry object. Geometry nodes are classified in two categories: nodes usable for 2D specific scenes and 3D specific scenes. Note that all 2D geometry can also be used in 3D scenes.	BitMap Circle Curve2D IndexedFaceSet2D IndexedLineSet2D PointSet2D Rectangle Text
		Box Cone Cylinder ElevationGrid Extrusion IndexedFaceSet IndexedLineSet PointSet Sphere
Bindable children nodes	These nodes represent features of the scene for which exactly one instance of a node can be active at any instant. For example, in a 3D scene, exactly one Viewpoint node is always active. For each node type, a stack of nodes is stored. The active node is put on the top of the stack. Activating a particular node can be triggered through events. The 2D specific nodes are listed followed by 3D specific nodes.	Background2D Background, Fog ListeningPoint NavigationInfo, Viewpoint
Children nodes	These are direct children of a grouping node. They can represent geometry (Shape), sound nodes, lighting parameters, interpolators, sensors, or grouping nodes. This category contains All grouping nodes All sensor nodes All interpolator nodes All bindable children nodes	AnimationStream Conditional Face QuantizationParameter Script Shape TermCap Valuator WorldInfo
		Sound2D
	Children nodes are classified in three categories: nodes usable for both 2D and 3D children, 2D specific nodes, and 3D specific nodes.	DirectionalLight PointLight Sound SpotLight

Table 2 (Continued)

Type of nodes	Function	BIFS nodes
Media-related nodes	These nodes enable inclusion of media in the scene: audio, video, animation, or update of scenes.	Anchor AnimationStream AudioClip AudioSource Background Background2D ImageTexture Inline MoveTexture
FBA nodes	FBA nodes are nodes related to face and body animation. They contain on child node (Face), and the rest are attributes for the Face node.	Face FaceDefMesh FaceDefTables FaceDefTransform FDP FIT Viseme
Miscellaneous attributes	Attributes are features of the children nodes that are represented by specific nodes, except FBA, media, or geometry specific attributes. Attributes nodes are classified in three categories: nodes usable for both 2D and 3D attributes, 2D specific nodes, and 3D specific nodes	Appearance Color FontStyle PixelTexture Coordinate2D Material2D Coordinate Material Normal TextureCoordinate TextureTransform
Top nodes	Top nodes are the nodes that can be put at the top of an MPEF-4-scene.	Group Layer2D Layer3D OrderedGroup

In addition, by assigning an interface coding Table to the PROTO or EXTERNPROTO nodes, BIFS version 2 allows the addition of parameters for controlling these new interfaces with BIFS-Anim and BIFS-Command. In this way, it is possible to control the new interface from a stream. It is also possible to assign quantization parameters to the prototypes fields in order to allow efficient compression.

d. Body Animation. Whereas version 1 allows only the animation of facial and body models, version 2 will add the capability to define the model itself. The body will be defined in the scene by compliant H-Anim definitions and will use a specific, optimized encoding algorithm. Furthermore, as with the facial animation, the animation stream will be integrated in the BIFS-Anim framework.

e. MPEG-J. MPEG-J is a set of application programming interfaces (APIs) that allow Java code to communicate with an MPEG-4 player engine. By combining MPEG-4 media with safe executable code, content creators may embed complex control and data processing mechanisms with their media data to manage the operation of the audiovisual session intelligently.

MPEG-J defines the following interfaces:

Scene graph API (similar to the VRML External Authoring Interface [EAI]) that enables scene control
Network API to control the up- or downloading of content
Decoder API to control media decoders
Devices API to control various input devices
System capability APIs in order to be able to assess the status of the system in terms of memory, CPU load, etc.

Figure 7 depicts the architecture of an MPEG-4 engine with MPEG-J overlaid. Note that the Java code is not involved in the data pipeline (from the demultiplexer encapsulated by the DMIF application interface to the decoding buffers, decoders, composition buffers, and the compositor itself). This ensures that high levels of performance can be maintained at these critical processing elements.

D. MPEG-4 BIFS Profiles and Levels

In order to properly partition the different functionalities provided by MPEG-4 Systems to match specific application requirements, the specification defines a number of profiles and levels. These are subsets of the specification to which applications can conform to without compromising interoperability while at the same time allowing an implementation complexity commensurate with the targeted application environment.

BIFS contains nodes that pertain to four different functionalities: scene graph construction, video object–related nodes, audio object–related nodes, and graphics-related nodes. For example, a Layer2D node is a scene graph node, whereas AudioClip is obviously audio related. The presence of media-related nodes (including graphics) depends on the profile selected media profile. Separate profiles are defined for the scene graph itself.

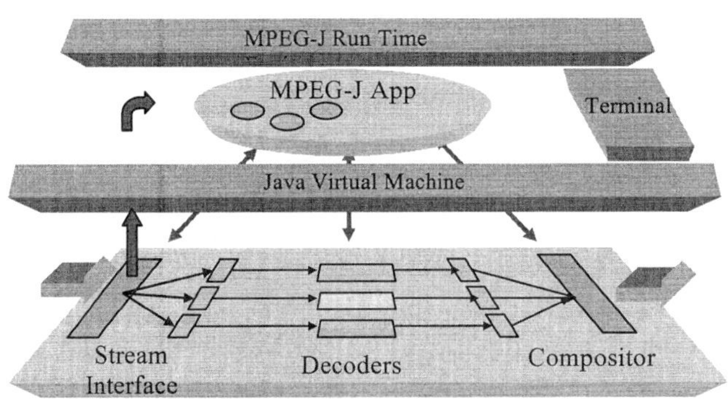

Figure 7 The MPEG-J framework.

With respect to audio objects, the AudioClip and AudioSource nodes are included in all profiles. For visual objects, ImageTexture, Background2D, Background, and MovieTexture are included in almost all visual profiles (with the exception of the Scaleable Texture and Simple Face profiles).

For graphics, three different profiles are defined: Simple2D, Complete2D, and Complete. Simple2D involves only Appearance, Bitmap, and Shape. Complete2D involves all the remaining nodes that are valid in a 2D-only environment, and Complete involves all of the graphics-related nodes.

For the scene graph, four different profiles are defined: Audio, Simple2D, Complete2D, and Complete. These parallel the graphics profiles, with the exception of Audio, which was defined to cater to the needs of audio-only MPEG-4 applications.

In addition to profiles, MPEG-4 defines a set of levels, which constrain the range values of certain parameters to ensure realistic performance expectations from implementations. The interested reader is referred to the MPEG-4 Systems specification for more information on profile and level definitions.

III. BIFS COMPRESSION AND STREAMING

A. The BIFS Format Design Goals

The binary format for MPEG-4 scene description attempts to balance several considerations. Compression is, of course, the ultimate goal, but compression conflicts with extensibility, ease of parsing, and a simple specification. The BIFS protocol is a compromise between these goals.

For compression, BIFS uses a compact representation for the scene components. For example, when a scene parameter is specified, the minimal number of bits needed to distinguish that parameter from others is used. This scheme is also used for specifying the scene contents. For example, 18 different components are used to specify the geometry of a scene: spheres, cones, polygons, etc. These are indexed and specified using 5 bits. Although this is not as efficient as possible, it makes parsing and specification simpler.

The actual parameter values associated with the scene parameters have a different quantization scheme that is actually part of the scene and thus under the control of the scene author. By default, scene parameter values are not quantized at all; they are stored in their native format (e.g., 32 bits for floats and integers, 1 bit for Booleans). The scene parameters are classified into different categories, and the values in each category can be linearly quantized using quantization parameters (maximum and minimum values, along with the number of quantization bits) that are specified locally in the scene. The categories consist of parameters that should have similar values, e.g., scaling values and 3D coordinate values. This scheme balances the utility of having local quantization control with the cost of specifying the quantization parameters.

BIFS uses a different quantization scheme in the case of animation. Animation allows scene parameter values to be modified as a function of time. The scene author specifies which node parameters should be animated by associating these parameters with a stream of input values. The stream consists of a sequence of initial values (intra frames) and successive difference values (P-frames) that are arithmetically encoded. This scheme allows highly efficient encoding of animation values, which can typically compose a large part of a scene. The following sections discuss these notions in more detail.

B. BIFS Representation of VRML

The MPEG-4 scene structure is heavily dependent on VRML. Where VRML is a bit short on functionality, for example, two-mensional content, MPEG-4 extends it in a VRMLesque way. But whereas VRML is concerned only with scene description, MPEG-4 is also concerned with compact representation.

1. *Nodes and Fields*

BIFS and VRML scenes are both composed of a collection of nodes arranged in a hierarchical tree. Each node consists of a list of fields that define the particular behavior of the node. For example, a Sphere node has a radius field that specifies the size of the sphere. There are roughly 20 basic field types for representing the basic data types: Boolean, integer, floating point, two- and three-dimensional vectors, time, normal vectors, rotations, colors, URLs, strings, images, and other more arcane data types such as scripts. Many nodes have fields that hold other nodes—this is what gives the scene its tree structure. Whereas VRML has only two node types, SFNode and MFNode, MPEG-4 has a rigidly typed collection of nodes that specifies exactly which nodes can be contained within which other nodes. A node may have more than one type, however, when it can be contained as a child node in different contexts.

For example, the Shape node is used to include a geometric shape in the scene. It has a geometry field that holds any node with type SFGeometryNode, of which there are 18, and an appearance field that holds nodes of type SFAppearanceNode, of which there is only one. An example is shown in Figure 8. In this example, the geometry field holds a Cube node that is of size 1 in each direction; the appearance field of Shape node holds an Appearance node, which in turn holds a Material node that defines the color associated with the Cube (in this case red).

The example in Figure 8 gives a textual description of a portion of a scene that is based on VRML. MPEG-4, however, specifies only a binary description of a scene. The translation from a textual to a binary description is almost direct. As an example, consider the simple scene portion in Figure 9 and the following explanation of its BIFS representation.

The (simplified) BIFS representation of the scene in Figure 9 would consist of the following:

1. A header that contains some global information about the encoding.
2. A binary value representing the Transform node.
3. A bit specifying that the fields of the Transform node will be specified by their index rather than in an exhaustive list.

```
Shape {
      geometry Cube { size 1 1 1}
      appearance Appearance {
            material Material {
                  diffuseColor 1 0 0
            }
      }
}
```

Figure 8 A Shape node with two fields, appearance and geometry, containing other nodes.

4. The index for the ''translation'' field.
5. A binary encoding of the SFVec3f value 1 0 0. (Because no quantization is defined here, this encoding consists of three 32-bit values. During decoding, the decoder knows the type of the field it is reading and thus knows how many bits to read and how to interpret them.)
6. The index of the ''children'' field of the Transform node.
7. The binary representation of the Shape node, which is:
    7.1. A binary value for the Shape node.
    7.2. A bit specifying that all of the fields of the Shape node and their values will be listed sequentially rather than by index–value pairs.
    7.3. A binary representation for the Cube node, which is:
        7.3.1. A binary value for the Cube node.
        7.3.2. A bit specifying that the fields of the Cube will be specified by index.
        7.3.3. The index of the ''size'' field.
        7.3.4. A binary encoding of the SFVec3f value 111.
        7.3.5. A bit specifying that no more fields for the Cube node will be sent.
    7.4. A binary value for the Appearance node, followed by its encoding, omitted here.
8. A bit terminating the list of fields for the Transform node.

In BIFS, the binary code for a node is an index into a list of nodes of that type. That is, it is an index into the list of all node types that can appear within the currently encoded field. Because a node may have more than one type, its binary code is dependent on where in the tree it appears. This is discussed in more detail in Sec. III.C.1.

Both VRML and BIFS have a mechanism for reusing nodes. For example, once a wheel is defined as a collection of geometric nodes collected inside a Group node, it is possible to reuse the wheel elsewhere in the scene rather than defining it explicitly wherever it is to appear. In VRML, this is done using the DEF and USE keywords. A node is given a DEF name, and it is inserted into the scene graph wherever the same name appears after a USE statement. In BIFS, a bit is used to determine whether the node has an ID, and when this bit is set, it is followed by a certain number of bits that hold an

```
Transform {
  Translation 1 0 0
  Children [
        Shape {
              geometry Cube { size 1 1 1}
              appearance Appearance {
                    material Material {
                          diffuseColor 1 0 0
                    }
              }
        }
  ]
}
```

Figure 9 A simple VRML scene consisting of a red cube centered at (1,0,0).

integer ID value. The number of bits used in the scene to specify IDs is specified in a header that is sent at the beginning of the scene (in fact, contained in the decoder configuration information within the object descriptor through which the BIFS stream was accessed).

2. ROUTEs and Dynamical Behavior

BIFS reproduces the same ROUTE mechanism that VRML has, extending it with the addition of ROUTE IDs. In BIFS, an extra bit can be set in the binary representation of a ROUTE that specifies when the ROUTE has an ID that, when the bit is true, follows immediately after. As with node IDs, the number of bits used to specify ROUTE IDs is passed in a header that is sent at the beginning of the scene.

In order to allow some dynamical behavior in the scene, VRML introduces ROUTEs. These are connections that assign the value of one field to another field. ROUTEs combined with interpolators (nodes that can generate a sequence of values) can cause motion in a scene. For example, the value of an interpolator is ROUTEd to the rotation field in a Transform node, causing the nodes in the Transform node's children field to be rotated in real time. In order to distinguish between input and output fields, each field can have one of four modes: Field, ExposedField, EventIn, EventOut. These modes correspond, respectively, to private fields that cannot be ROUTEd at all, fields that can be both input and output in a ROUTE, fields that can only accept input from a ROUTE, and fields that can only serve as output for a ROUTE. The VRML ROUTE syntax has the form

ROUTE ⟨OutNodeID⟩.outFieldName TO ⟨InNodeID⟩.inFieldName

This syntax means that values from the field with name "outFieldName" in the node with DEF ID ⟨OutNodeID⟩ will be mapped to the field with name "inFieldName" in the node with DEF ID ⟨InNodeID⟩. In MPEG-4, the same data are sent to specify a ROUTE, except that the IDs are integer values and the fields are also indexed numerically rather than referenced by name.

Figure 10 shows a VRML scene in which a Text node is moved across the screen repeatedly. This might be part of a scene in which news headlines are scrolled across the bottom of a larger scene. Although the ROUTE mechanism has no problem in causing the text to scroll nicely across the screen, the scene is otherwise static. If the news is old, it is not possible to change it from the server side. However, the BIFS update mechanism allows precisely this. With BIFS update, it is possible to send commands to insert, delete, or replace nodes or field values in the scene. The specific syntax of BIFS update is discussed in Sec. III.D, but we can think of it textually, as shown in Figure 11. After this command is sent to the MPEG-4 client, the scene would look almost identical, except with the new recent news headline replacing the previous field in the Text node. (In fact, it is not necessary to send a whole new text node. The same effect can be achieved by replacing only the string field of the text node.)

3. Animation Using Interpolators and BIFS Animation

VRML scenes can display a wide range of complicated dynamical behavior. One way this is achieved is by the use of interpolator nodes that convert an increasing sequence of time events into a sequence of interpolated positions. Figure 12 shows a scene in which a ball bounces along steps repeatedly. The ball position is encoded by using the position

```
Group {
    DEF position Transform {
          translation 1 0 0
          children [
             Shape {
               geometry DEF MyHeadline Text {
                      string "news headline here"
                 }
          ]
    }
}
DEF PosInterp PositionInterpolator {
    Key [0 1]
    KeyValue [1 0 0, -1 0 0] # move from x=1 to x=-1
}
DEF MyTime TimeSensor {
    loop TRUE
    stopTime -1
}
ROUTE MyTime.fraction_changed TO PosInterp.key
ROUTE PosInterp.keyValue TO position.translation
```

Figure 10 A VRML scene portion in which a Text node is moved across the screen using a ROUTE. A TimeSensor node generates successive values that are ROUTEd to the key field of a PositionInterpolator, which converts these input values to interpolated SFVec3f values in the key Value field. This field is then ROUTEd to the translation field of a Transform node, causing the children in the Transform node to be repositioned.

interpolator shown in Figure 13. The position interpolator interpolates each ROUTEd time value into a position using the values in its key and key Value fields. When the motion is complicated, these fields must contain many floating point numbers and thus require a lot of memory.

This is one way of introducing animation into the scene, but it suffers from several drawbacks. First, the animation is fixed—it is part of the scene and cannot be modified. Second, the whole scene has to be downloaded, including the lengthy interpolator fields, before it can be rendered. MPEG-4 combines streaming and updates to create an animation functionality that overcomes these drawbacks.

```
Update {
      time 10
      action "Replace"
      id "MyHeadline"
      data Text  {
            string "More recent news headline here"
            }
}
```

Figure 11 A textual representation of a BIFS update that replaces the node with ID ''headline'' with a new text node.

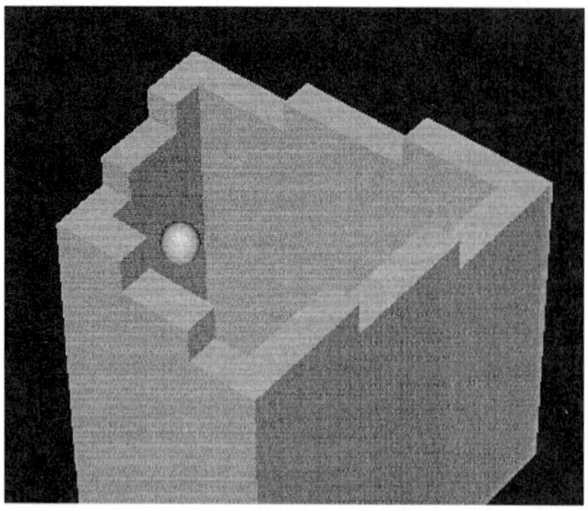

Figure 12 An animated scene in which a ball bounces on steps. The animation is generated using interpolators that modify the position of the ball in time. (Scene composed by Diganta Saha.)

```
DEF Ball-Tform-Move Transform {
  children [
.
.
.
        DEF Ball-Move-Timer TimeSensor {loop TRUE . . . . . . },
        Shape {
            appearance Appearance {
               material Material {
                   specularColor     .2 .2 .2
                   diffuseColor      1 1 0
               }
            }
            geometry Sphere{}
        }
    ]
}
.
.
.

DEF Ball-Mover PositionInterpolator {
  key [ 0, .03125, .0625, .09375, . . . . . ]
  keyValue [ 0 0 0, 0 0 2, 0 -2 4, 0 -2 6, 0 -4 8, . . . . . ]


.
.
.
ROUTE Ball-Move-Timer.fraction_changed TO Ball-Mover.set_fraction
ROUTE Ball-Mover.value_changed TO Ball-Tform-Move.set_translation
```

Figure 13 A portion of the scene shown in Fig. 14 containing the PositionInterpolator node that causes the ball to move around the stairs. By ROUTing the Ball-Mover output values to the translation of the Ball-Tform-Move node, the ball is animated in the scene.

In BIFS animation, a field of a node is selected for animation. A separate stream is opened and field values are sent to the field, causing it to animate the scene. The field values in the stream can consist of initial values (intra frames), used to set or reset the field values, and difference values (predictive frames), used to modify successively the previous field value. The intra and predictive frames are further arithmetically coded to give a highly compressed representation of the animated values.

It is possible to convert automatically VRML scenes that use interpolators to animated MPEG-4 scenes that use the BIFS animation mechanism, and this leads to a considerable reduction in scene size.

C. BIFS Compression

BIFS provides an efficient compression algorithm for MPEG-4 scenes. Because MPEG-4 defines a superset of VRML nodes and structures, BIFS yields a compressed representation of standard VRML scenes as well. Although the protocol is efficient, it represents a trade-off between bitstream efficiency on the one hand and parsing complexity, (relatively) simple specification, and extensibility on the other. In fact, a compressed scene can sometimes be further compressed using existing data compression tools. This is true because some data, e.g., strings, contain redundancy that is not eliminated using BIFS coding (BIFS encodes strings as a collection of characters). The following sections discuss the components of scene coding in MPEG-4.

1. Context Dependence

BIFS takes advantage of context dependence in order to compress the scene efficiently at the node and the field level. Whereas VRML treats all nodes as being of type SFNode, BIFS makes use of the fact that only certain nodes can appear as children of other nodes. This allows the binary specification of the children nodes to be more efficient.

BIFS introduces the concept of a node data type (NDT). There are some 30 different NDTs, and each node belongs to one or more of them. The NDT table consists of a list of nodes for each NDT. In each node data type, the node receives a fixed-binary-length local 1D value that corresponds to its position in the NDT table. For example, the Shape node can appear in both a 3D context and a 2D context. It thus has both the SF3DNode and SF2DNode node data type (in fact, as with all nodes, it also has the global SFWorld-Node NDT). In a 3D context, the Shape node can appear in the children field of a Transform node—this field has the type MF3DNode and it accepts nodes of type SF3DNode. In a 2D context, the Shape node can appear in the children field of a Transform2D node—this field has type MF2DNode. The binary ID code for a Shape node thus depends on the context in which it appears. When it is an SF3DNode, it has a 6-bit binary ID value of 100100 (decimal value 36), because it occupies position number 36 in the list of 48 total SF3DNodes. In the 2D case, its ID is represented by 5 bits with value 10111. There are only 30 different SF2DNodes, requiring only 5 bits to specify them all.

This node representation is almost optimal in an information theoretic sense. It does not make sense to apply entropy coding techniques based on the probabilistic distribution of nodes in scenes—such data do not currently exist. Moreover, as the ID value 0 is reserved as an escape value for future extensions and a fractional bit arises from the fact that the number node in each category is not a power of 2, the coding is slightly suboptimal in any case.

At the field level, BIFS introduces types that do not exist in VRML:

The SFURL type is used for representing URLs. VRML uses a string to encode URLs, but BIFS requires a finer resolution and allows URLs to be encoded either as strings or as references to stream IDs (object descriptors).

The SFScript type is used to encode scripts. VRML uses URLs to hold scripts, which are typically either externally accessible Java class files or in-lined JavaScript (or, more precisely, ECMAScript). In BIFS, only JavaScript is supported and it is represented using a binary format. The SFScript type allows either an MPEG-4 URL to be encoded (which allows full compatibility with VRML) or a binary script representation.

The SFBuffer type is used to encode buffers in the MPEG-4 Conditional node. This special field holds a binary representation of a BIFS-Update command that can be triggered using the ROUTE mechanism.

2. The Node Coding Tables

Each node is associated with a node coding table (NCT). Each node's NCT holds information related to how the fields of the node are coded. The NCT specifies the type of values that each field can hold. For example, the NCT for the Transform2D node specifies that its children field has type MF2DNode and thus can hold a list of SF2DNode nodes. The NCT also specifies the other field types that can occur, for example, SFFloat and SFInt32.

The NCT also specifies an index for each field according to one of four usage categories:

DEF—used for defining field values when a node is transmitted. This corresponds to the Field and ExposedField modes, as these are the only modes that have values that can be specified.

IN—used for data that can be modified using BIFS updates or ROUTEs. This corresponds to the EventIn and ExposedField modes.

OUT—used for the EventOut and ExposedField modes, that is, fields that can be used as input values for ROUTEs.

DYN—a subset of the IN category, used for BIFS-Animation, discussed in Sec. III.E.

With the creation of these four categories, a BIFS scene always references fields with the minimal number of bits needed. For example, the IndexedFaceSet node is used to hold a collection of polygons that form a 3D object. It has 13 fields that can be defined when it is created, and thus each field requires 4 bits when it is indexed. However, this node has only four fields that can output values to a ROUTE, so that only 2 bits are needed to specify which field is being used when this node is routed from. The IndexedFaceSet node has 8 fields that can accept ROUTEd values or be updated using BIFS-Command, and thus these protocols need only 3 bits to specify the modified field.

The NCT also defines a quantization type for each field that holds numerical values. This is the topic of the next section.

3. Quantization of Node Fields

Quantizing BIFS scenes is achieved using the QuantizationParameter node. This node affects the quantization of field values for all nodes that appear hierarchically after or beneath it (i.e., its siblings and children). Using a node to convey the quantization parameters makes use of the already existing BIFS framework to encode the parameters efficiently and takes advantage of specific reuse mechanisms (VRML DEF/USE) in order to reuse locally existing parameters. This mechanism also allows content creators to fine-tune the compression of their content when it has special redundant characteristics.

BIFS quantization is complex because no clear statistics can be derived from the source data. Unlike the case of video and audio streams, there is not a clear redundancy of information in the signal. The problem is to find the best trade-off between the declaration cost of the local quantization parameters using the QuantizationParameter node and the gain brought by this quantization. To overcome this difficulty, every numerical field in each node is classified into one of 14 quantization categories (see Table 3). The idea is that each category, for example, ''Position 3D,'' will contain data in a similar range of values. A value range and number of bits for a linear quantizer for each quantization type are specified in the QuantizationParameter node. When a field is to be encoded, its quantization type is looked up in the NCT for the field's node. The values of the field are then coded using the number of bits and value range specified for that quantization type in the QuantizationParameter node.

The QuantizationParameter node has several other features. It has a useEfficientFloat field that holds a Boolean value. When this value is true and when floating point values are not quantized in one of the quantization categories described in the table, they will be quantized using MPEG-4–specified ''efficientFloat'' coding that has less resolution than IEEE 755 and requires fewer bits per number, especially for 0 values. The details of this coding are beyond the scope of this chapter, but the idea is that a small number of bits is used to specify how many bits are used for the exponent and mantissa of the floating point value. Values with small exponents and easily specifiable mantissas can then be stored using a smaller number of bits overall. However, the coding is limited to a 15-bit mantissa and a 7-bit exponent.

The QuantizationParameter node has another Boolean field, isLocal, that specifies

Table 3 BIFS Quantization Categories

Category	Usage
None	No quantization
Position3D	Used for 3D positions of objects
Position2D	Used for 2D positions of objects
Color	Used for colors and color intensities
TextureCoordinate	Used for texture coordinates
Angle	Used for angles
Scale	Used for scales in transformations
Interpolator Keys	Used for interpolator keys, MFFloat values
Normals	Used for normal vectors
Rotations	Used for SFRotations
ObjectSize 3D	Used for 3D object sizes
ObjectSize 2D	Used for 2D object sizes
Linear Quantization	NCT holds max, min, and number of bits
Coord Quantization	Used for lists of coordinates of points, color, and texture coordinates that are indexed, e.g., in IndexedFaceSet

that the quantization parameters specified in the node should apply only to the next node in the scene tree. This design allows the declaration of quantization parameters to be factored and applies it to the maximum amount of data.

D. BIFS-Command

The BIFS-Command protocol (also known as BIFS-Update) allows the server to modify the current scene. It is a powerful tool that can be used to manipulate the scene remotely, modify portions of it, or download scene components progressively in order to reduce bandwidth requirements. BIFS commands come in four main functionalities, summarized in Figure 14 and detailed below in the following. There are four top-level functionalities, specified using 2 bits:

Replace the whole scene with this new scene.
Insertion command.
Deletion command.
Replacement command.

The insertion, deletion, and replacement commands can insert, delete, and replace nodes, indexed values of MF fields, and ROUTEs. These three options are encoded using 2 bits. The replacement command can also replace a field value, which makes that command completely efficient in its use of the 2 bits.

Figure 15 shows a textual representation of a BIFS update command that can be applied to the scene in Fig. 13. After this command is sent to the MPEG-4 client, the scene would look almost identical, except with the new recent news headline replacing the previous field in the Text node.

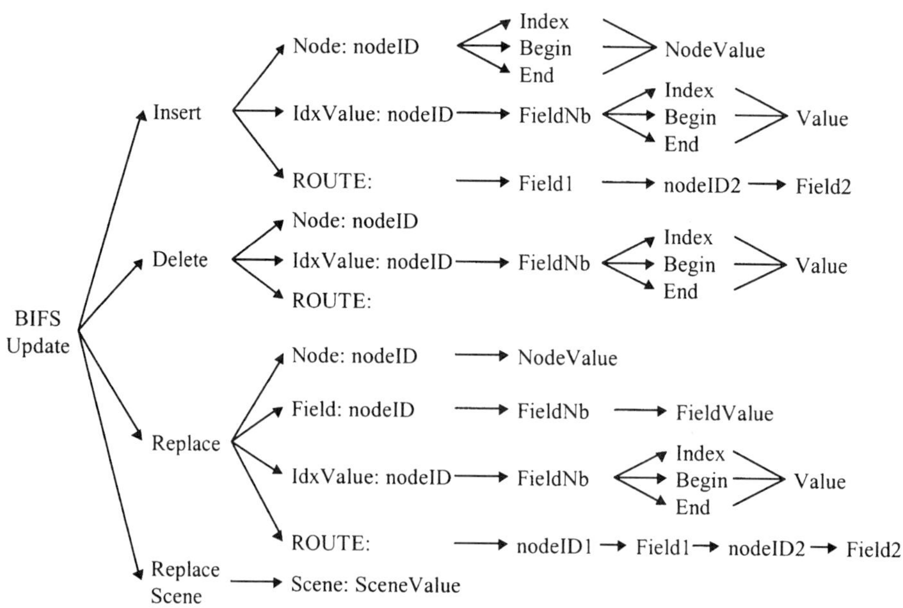

Figure 14 BIFS-Command types.

```
Update {
       time 10
       action "Replace"
       what "Node"
       id "MyHeadline"
       data Text  {
              string "More recent news headline here"
              }
}
```

Figure 15 Textual representation of a scene update.

The (simplified) binary representation of this update would consist of the following:

1. Two bits (with binary representation 10) indicating that the update is a replacement command.
2. Two bits (with binary representation 00) specifying that the update is replacing a Node (as opposed to an MFField element or a ROUTE).
3. A number of bits specifying the node ID of the node that is to be replaced. The number of bits is specified in a header that is sent at the top of the scene. In the example, the node ID is the string ''MyHeadline,'' and this must be mapped uniquely to an integer in the BIFS representation.
4. A binary representation of the Text node that contains the new node. The node type is specified by sending an index into the NDT table. In this case, the SFWorldNode NDT that contains all of the MPEG-4 nodes is used. We discuss this further in the following.

The effect achieved in the example in Fig. 15 can also be produced by just replacing the value of the string field of the Text node. In this case the binary representation of the replacement command would consist of the following:

1. Two bits (with binary representation 10) indicating that the update is a replacement command.
2. Two bits (with binary representation 01) specifying that the update is replacing a field.
3. A number of bits specifying the node ID of the node that is to be replaced.
4. Two bits (with binary representation 00) specifying that the string field is to be replaced. There are four replaceable (IN) fields in the Text node, and string is the first. Note that in order to know this the decoder must first read the node ID, look up the node type of the node with this ID, and use the node coding tables to determine how many IN fields this node has.
5. A binary representation of string field. Because the decoder knows which field is to be replaced, it can look in the node coding table to determine the field type and hence know what type of data to expect.

The BIFS commands are coded as a tree as illustrated in Figure 14. ''FieldNb'' indicates the number of bits to be used to indicate the position of the value within the field.

E. BIFS-Animation

The BIFS-Animation (or BIFS-Anim) protocol uses an arithmetic coder applied to a sequence of difference values that are computed from a sequence of animated parameter values. The binary animation data consist of an animation mask that contains a collection of elementary masks, one for each node that is animated. This is followed by a stream of animation frames that contain the animation data. Only updatable nodes (i.e., nodes with IDs) can be animated, because the elementary masks use these IDs to specify which nodes are animated. Each elementary mask also specifies which of the animated nodes' fields arc animated and initial quantization parameters for the animation. Finally, when the animated fields are multiple-value fields, the elementary mask specifies the indices of the elements that are to be animated. Only fields that have a DYN index defined in the node coding tables can be animated. These dynamic fields may be of the following types:

SFInt32/MFInt32
SFFloat/MFFloat
SFRotation/MFRotation
SFColor/MFColor
SFVec2f/MFVec2f
SFVec3f/MFVec3f

The animation frames specify the values of the animated fields. An animation frame can contain new values for a subset of the animated fields at a specified time. That is, all the fields or a selection of fields can be modified at each specified time. The field values can be sent in Intra (the absolute value is sent) and Predictive modes (the difference between the current and previous values is sent).

In intra mode, the field values are quantized using the quantization parameters defined for the animated field in the associated elementary mask. In predictive mode, the difference from the previous sample is computed and then entropy coded. Figure 16 summarizes the decoding steps.

The quantization is similar to that used in the BIFS-Scene compression. In BIFS-Animation, as for BIFS scene compression, the notion of quantization category is used. The following animation categories are possible:

Position 3D
Positions 2D
Reserved
Color
Reserved

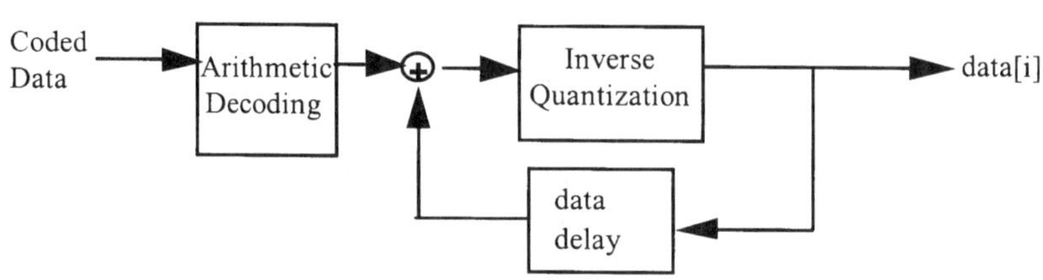

Figure 16 The four steps of the BIFS-Anim predictive decoding scheme: arithmetic decoding, inverse quantization, delay, and compensation.

Angle
Float
BoundFloat
Normals
Rotation
Size 3D
Size 2D
Integer

However, the animation categories correspond more to data types than to a semantic grouping, because quantization parameters are not shared but declared individually for each field to be animated. Each category has a specific syntax for declaring its quantization parameters: min and max values, number of bits, in intra and predictive modes. Computing the difference of quantized values is direct except for normalized vectors such as rotations and normals. Rotations are coded using quaternions, and a specific prediction mechanism enables coding of rotations and normals in a predictive mode, just as with the other data types. The entropy coding uses an adaptive arithmetic encoder, which avoids the issue of unknown data statistics.

F. BIFS Usage

The BIFS tools available in MPEG-4 are a mixed blessing. On the one hand, they allow scenes to be represented, modified, and animated efficiently. On the other hand, the tools are complicated and a repertoire of tricks is only now in the process of being developed. The following sections list a few tricks and suggestions for using these tools.

1. BIFS Scene Usage

The main complication in using BIFS well is the use of the QuantizationParameter node. This node can reduce the size of a scene significantly, even when used naively. This is because the native encoding of floating point and integer values typically has considerably more resolution than is needed for graphic scenes.

The most direct and naive way to use a QuantizationParameter node is to compute the maximum and minimum values for each of the quantization types in the scene, decide on an acceptable error, and insert one QuantizationParameter node at the top of the scene graph with the corresponding quantization parameter values. Almost all scenes should at least use this technique, as the cost of the QuantizationParameter node is minimal. (The exceptions are very small scenes in which the cost of specifying the QuantizationParameter node is greater than the bits saved in the scene.)

A more complicated technique for using the QuantizationParameter node involves grouping scene components that contain similar ranges of values and inserting a QuantizationParameter node at the top of each such grouping. For example, a scene that consists of two houses separated by some distance could benefit from a QuantizationParameter node at the top of the grouping nodes that hold each house. This technique essentially breaks the scene into smaller scenes and applies the naive quantization approach to each scene. The process of breaking up a scene and determining where to place QuantizationParameter nodes is something of an art, although it is possible to automate this function.

Finally, when possible, the QuantizationParameter node can be DEFed and USEd, so that the cost of specifying the quantization parameters is very low. In the preceding example of a scene containing two houses, if each house has about the same geometry, the first can be quantized normally and the second can be quantized using only a reference to the QuantizationParameter node used on the first house.

2. BIFS Command Usage

The capability of BIFS-Command to modify a scene can be used for real-time update of content, for example, news headlines. It can also be used to modify the user's experience by interactively changing the user's position or viewpoint or by sending status or chat messages between different users on a server (or even directly between users).

Figure 17 shows an example of using BIFS-update to load a scene progressively. In this technique, a large scene is broken into parts. The parts that are viewable from the initial viewpoint are sent as part of the initial scene tree, but the other parts are sent using BIFS-update only later: when the bandwidth allows, when they are required in the scene, etc. This mechanism can significantly reduce latency during the initial scene loading.

3. BIFS-Animation Usage

BIFS-Animation is a fairly direct method. Although it can be used on scenes to animate objects in complicated ways, it can also be generated from existing VRML content by automatic conversion of Interpolator nodes into the animation protocol. In this way, BIFS-Animation can compress scenes more efficiently than using Interpolator nodes combined with QuantizationParameter nodes. Figure 18 illustrates the use of the BIFS-Animation stream.

G. Sample Compression Results

In this section we give some results of compressing VRML files (in ASCII text representation) into the MPEG-4 BIFS representation.

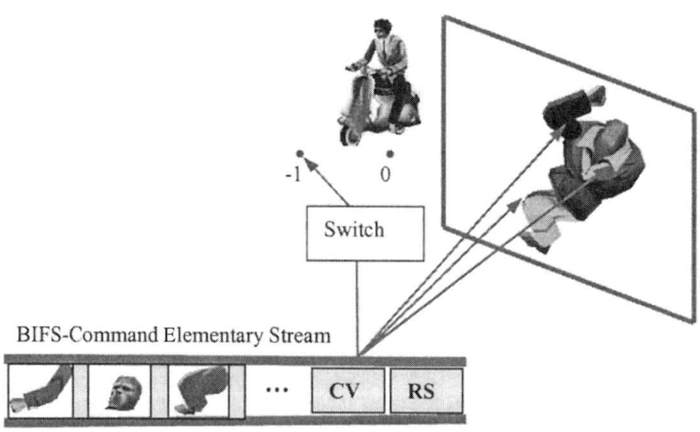

Figure 17 The BIFS-Command protocol. Building the scene graph progressively with Insert Object BIFS commands. The motorcyclist is added after the initial load and is placed off screen for future use by changing the value of the switch around it (CV). The replace scene (RS) command replaces the whole scene graph with the initial new scene.

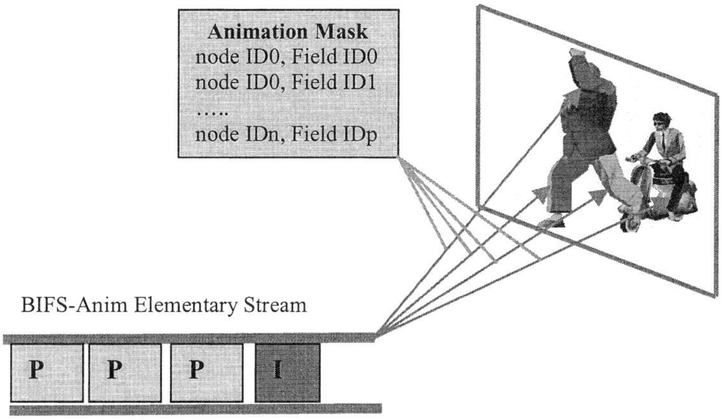

Figure 18 The BIFS-Anim protocol: animating the position and the viewpoint. As with video, intra (I) frames are used for random access (tune-in) and/or error recovery. Predictive (P) frames are differentially encoded for improved compression.

The sample files, listed in Table 4, were chosen to have a range of characteristics:

The size of the files ranges from 4 kb to 2.3 Mb.
The content involves 2D and 3D scenes.
The scenes can be static (only geometry) or dynamic and interactive.

For some of the files containing dynamic content, the Scene (i.e., the VRML tree containing the geometry) was separated from the animation (i.e., the interpolators) in order to take advantage of the BIFS-Anim protocol. Note that Movie1 and Floops are both 3D cartoons containing animation, but in the case of Floops, the animation information

Table 4 Characteristics of Sample VRML Files Used for Comparing BIFS Representation

File	Description	VRML file (kB)	VRML scene (kB)	VRML animation (kB)	Duration (sec)
Intro	3D cartoon	1077	347	730	42.5
Prolog	3D cartoon	1293	621	672	69.3
Movie 1	3D cartoon	2370	968	1402	113
Finale	3D cartoon	538	412	126	13
Movie2	3D cartoon	228	117	111	18
Skeleton	3D animated character	72	34	38	5.3
Floops	3D cartoon	1324	1324		
Fishswim	3D animated character	37	37		
Store	3D static scene	155	155		
Tree	3D-simple geometry (no IFS)	80	80		
Channel10	2D interactive content	100	100		
Meteo1	2D interactive content	4	4		
Flight	2D interactive content	62	62		

Table 5 Sample Compression Results from BIFS-Scene Compression

File	VRML scene file size (kB)	BIFS-Scene: no QP File size (kB)	Ratio	BIFS-Scene: 1 QP File size (kB)	Ratio	BIFS-Scene: Many QPs File size (kB)	Ratio
Intro	347	101	3.4	48	7.3	45	7.7
Prolog	621	176	3.5	88	7.0	81	7.7
Movie1	968	222	4.4	110	8.8	103	9.4
Finale	412	122	3.4	62	6.6	58	7.1
Movie2	117	30	3.9	15.6	7.5	13.2	8.9
Skeleton	34	8.7	3.9	4.5	7.6	2.9	11.7
Floops	1324	284	4.7	182	7.3	95	13.9
Fishswim	37	17	2.2	6.7	5.5	4.4	8.4
Store	155	43	3.6	26	6.0	14	11.1
Tree	80	4.9	16.3	4.9	16.3	4.9	16.3
Channel10	100	3.2	31.3	2.9	34.5	2.9	34.5
Meteo1	4.0	0.325	11.6	0.285	14.0	0.285	14.0
Flight	62	5.5	11.3	4.7	13.2	4.7	13.2

(contained in the interpolators) was not extracted, and the whole scene was simply encoded using plain BIFS.

1. Scene Compression Schemes

The scenes were compressed using BIFS and BIFS-Command, with results shown in Table 5 and Table 6, using quantization parameters that resulted in no apparent error. Table 5 shows only BIFS-Command data, while Table 6 shows both BIFS-Command and BIFS-Anim data. Four different compression schemes were used:

No QP: No QuantizationParameter node is used in the scene. However, the floats in the scene are coded using the ''efficient float'' format, using roughly 16 bits each.

Table 6 Combined BIFS and BIFS-Animation Compression Results

File	VRML file file size (kB)	BIFS File size (kB)	Ratio	Winzip File size (kB)	Ratio	BIFS vs. Winzip gain (%)
Intro	1077	75	14.4	179	6.0	+140
Prolog	1293	135	9.6	243	5.3	+80
Movie1	2370	187	12.7	422	5.6	+130
Finale	538	63	8.5	92	5.8	+50
Movie2	228	19.5	11.7	46	5.0	+140
Skeleton	72	5.4	13.3	13	5.5	+140
Floops	1324	95	13.9	124	10.7	+30
Fishswim	37	4.4	8.4	6.0	6.2	+40
Store	155	14	11.1	23	6.7	+60
Tree	80	4.9	16.3	6.2	12.9	+30
Channel10	100	2.9	34.5	4.9	20.4	+70
Meteo1	4.0	0.285	14.0	0.890	4.5	+210
Flight	62	4.6	13.5	3.0	20.7	−30

1 QP: Only one QuantizationParameter Node is inserted at the top of the scene tree. Therefore all numerical fields are linearly quantized.

Many QPs: Several QuantizationParameter nodes are placed in the tree at selected places and with parameters set to take advantage of the local error control.

Winzip: The file is ''WinZipped''; an ASCII compression format is used to compress the VRML file.

2. Animation Compression Scheme

The scenes containing animation were compressed using the BIFS-Animation protocol, using between 10 and 15 frames per second, producing good visual results. These results are shown in Table 7.

3. Comments

The first compression scheme used for the Scene (''noQP'') with ''efficient floats'' shows the efficiency of the MPEG-4 binary format in coding the scene tree. Quantized floats in this case are typically represented using 15 bits. The average compression ratio is 3 to 4. The scene graph compression efficiency is well illustrated with content such as Tree (3D simple geometry), Channel10, and Meteo1 (mostly 2D Text and simple geometry). The compression ratio reaches 15 or more. Other content with more geometry is better compressed when using quantization.

The second compression scheme used (''1 QP'') shows the effect of quantization on the geometric data in the scene. Quantized values in this case are represented by roughly 8 to 10 bits. The compression rate then reaches 6 to 8.

The third compression scheme (''Many QPs'') shows the effect of locally optimized quantization of the geometric data. Quantized values in this case use about 5 to 10 bits, and the compression rates rise to the range of 8 to 12.

For the VRML content in which the interpolators that generated the animation were extracted and converted to BIFS-Animation, the compression rates were roughly 15–25. When BIFS-Animation is combined with the BIFS-Scene compression, the best results are achieved. These are (typically) considerably better than the Winzip results.

These compression results are not the last word for MPEG-4 compression. MPEG-4 contains tools that can be used on typical VRML-like content to increase considerably the compression of the scene representation. These are face-and-body animation, which represent face and/or body models that can be animated very efficiently. MPEG-4 also has 3D mesh coding that allows 3D objects to be efficiently represented using lossy com-

Table 7 Results of BIFS-Animation Compression

File	Duration	VRML animation file size (kB)	BIFS-Anim File size (kB)	BIFS-Anim Ratio	Bit rate (kb/sec)
Intro	42.5	730	30	24.3	5.6
Prolog	69.3	672	54	12.4	6.2
Movie1	113.0	1402	84	16.7	5.9
Finale	13.0	126	5.3	23.8	3.3
Movie2	18.0	111	6.3	17.6	2.8
Skeleton	5.3	38	2.5	15.2	3.8

pression schemes. Finally, version 2 MPEG-4 BIFS compression will provide improved compression for multiple fields, using an algorithm similar to BIFS-Anim with predictive encoding, as well as PROTO, which can also improve compression in some specific contexts by reusing parts of the scene multiple times.

IV. CONCLUDING REMARKS

BIFS provides a powerful framework for creating interactive audiovisual content. It was not designed to cater to just a single application (as, e.g., MPEG-2 does) but rather provides a very rich environment on which content creators can unleash their creativity. Of course, a wide range of applications was considered in its design to ensure that the needs of users and content developers would be satisfied [6]. This most likely is its biggest strength.

The following is a brief, and hardly exhaustive, list of the types of applications that could be delivered on MPEG-4 enabled terminals:

1. ''Broadcast'' applications. These are possible even without server interactivity (that is, a back channel from the terminal to the server is not needed). This also implies playback from a mass storage device, such as a DVD.
   a. Electronic program guides
   b. Interactive advertising
   c. Interactive access to value-added ancillary content (e.g., immediate access to score information for sports programs)
   d. Interactive participation in quiz shows
   e. Computer-based training
   f. Corporate presentations
   g. Games
2. Interactive applications. These may involve an asymmetric back channel. All of the preceding applications can be enhanced via server interactivity. The list can be expanded to include, for example,
   a. E-commerce
   b. Online training
   c. Interactive on multiplayer games
   d. Polling
3. Communication applications. These involve a symmetric back channel and include, for example,
   a. Person-to-person communication
   b. Multiperson collaboration environments

MPEG-4 fills a very important gap in today's technology landscape, especially from a standardization point of view. Although a number of individual solutions that partially address MPEG-4's domain already exist (proprietary as well as standards based), none can provide all of its features in a single environment. Considering the ever-increasing penetration of interactive content in our daily lives, as well as the continuous increase in access bandwidth and availability of high-quality content, MPEG-4 can provide a much-needed catalyst that will help propel the interactive content industry (across Internet, DVD, and interactive TV environments) forward.

ACKNOWLEDGMENTS

The authors would like to thank Shout Interactive Inc. (www.shoutinteractive.com) for providing the visual material that appears in Figure 17 and Figure 18. Yuval Fisher would like to thank Rockwell Corporation for their support.

REFERENCES

1. ISO/IEC 14496-1. Coding of audio-visual objects: Systems, final draft international standard, ISO/IEC JTC1/SC29/WG11 N2501, October 1998.
2. ISO/IEC 14772-1. The Virtual Reality Modeling Language, 1997, http://www.vrml.org/Specifications/VRML97.
3. ISO/IEC 14496-2. Coding of audio-visual objects: Visual, final draft international standard, ISO/IEC JTC1/SC29/WG11 N2502, October 1998.
4. ISO/IEC 14496-3. Coding of audio-visual objects: Audio, final draft international standard, ISO/IEC JTC1/SC29/WG11 N2503, October 1998.
5. ISO/IEC 14496-6. Coding of audio-visual objects: Delivery multimedia integration framework, final draft international standard, ISO/IEC JTC1/SC29/WG11 N2506, October 1998.
6. ISO/IEC JTC1/SC29/WG11 N2562. MPEG-4 Requirements Document, December 1998.
7. ISO/IEC JTC1/SC29/WG11 N2611. MPEG-4 Systems version 2 WD 5.0, December 1998.

15

Java in MPEG-4 (MPEG-J)

Gerard Fernando and Viswanathan Swaminathan
Sun Microsystems, Menlo Park, California

Atul Puri and Robert L. Schmidt
AT&T Labs, Red Bank, New Jersey

Gianluca De Petris
CSELT, Torino, Italy

Jean Gelissen
Nederlandse Phillips Bedrijven, Eindhoven, The Netherlands

I. INTRODUCTION

MPEG-J is a collection of Java application programming interfaces (APIs) with which applications can be developed to interact with the platform and the content. In the context of MPEG-J, the platform is a device such as a set-top box or a PC with Java packages conforming to a well-defined Java platform. The Java-based application consists of Java byte code, which may be available from a local source, such as a hard disk, or it may be loaded from a remote site over a network. The MPEG-J Java byte code will be available as a separate elementary stream. The term ''presentation engine'' has been used to refer to the MPEG-4 System. What MPEG J brings is programmatic control through the ''application engine,'' which enhances the presentation engine by providing added interactive capability.

In this chapter we shall deal with the basic architectural assumptions involved in the development of MPEG-J. In addition, we shall give some details of the MPEG-J specific Java APIs that have been developed. We shall also provide details of a few sample applications. The reader wishing to obtain more information on MPEG-J is referred to the MPEG-4 Systems version 2 standard [1].

A. Standards and Systems Utilizing Java

There are groups within Digital Video Broadcasting (DVB), ATSC, and DAVIC that are developing Java APIs. These activities are mainly directed toward the entertainment and TV market and are highly focused on achieving that goal. The MPEG-J effort differs from these efforts in that it is not focused on one specific market. In addition to including applications for the entertainment–TV market, it is intended for applications ranging from 3D multiuser gaming to electronic information guides for business and home use.

We now discuss the organization of the rest of this chapter. In Sec. 2, we present a brief overview of the basics of MPEG-4 Systems version 1. In Sec. III, we briefly discuss the high-level architecture of the MPEG-J system. In Sec. IV, we describe the Java APIs that make the MPEG-J system programmable. Next, in Sec. V, we discuss the application areas of MPEG-J. The summary in Sec. 6 covers the key points presented in this chapter.

II. MPEG-4 SYSTEMS

MPEG-4 Systems provides representation for synthetic and natural audio–video information. The inclusion of synthetic audio–video content in MPEG-4 Systems is a departure from the model of MPEG-1 Systems and MPEG-2 Systems, in which only natural audio–video content representation was addressed. Furthermore, these earlier standards were concerned only with representation of temporal attributes. There was no need for representing spatial attributes in a scene. The audio–video model was very simple—a given elementary stream covered the whole scene. The introduction of audio–video objects in MPEG-4 meant that spatial attributes in the scene had to be correctly represented. The MPEG-4 audio–video content has temporal and spatial attributes that need to be correctly represented at the point of content generation (also known as encoding) and also correctly presented at the player or decoder. Representation of spatial information is carried out with a parametric approach to scene description. This utilizes Virtual Reality Modeling Language (VRML). However, this has been extended to provide features missing from VRML. The key extensions to VRML are for streaming, timing, and integration of 2D and 3D objects. These extensions are all included in the Binary Format for Scenes (BIFS) specification.

The correct representation of temporal attributes is essentially not different from the methods used in MPEG-1 Systems and MPEG-2 Systems to achieve the same ends. For these earlier standards the temporal attributes were utilized for two purposes: audio–video synchronization (lip synchronization) and the provision of system clock information to the decoder to help buffer management. Because of the possibility of significantly more diverse types of elementary streams being included in MPEG-4 Systems, the representation of temporal attributes has become more complex. But, as mentioned earlier, the fundamental method for representation of temporal attributes is essentially the same as for MPEG-1 Systems and MPEG-2 Systems.

There have been two significant additions to MPEG-4 Systems:

IPMP (Intellectual Property management and protection): The experts who developed the IPMP framework within MPEG-4 Systems correctly noted the complexity of MPEG-4 Systems and the diversity of potential applications. The IPMP methods required are as diverse as these applications. The IPMP framework has been designed to provide maximum flexibility so that the application builder can develop the most appropriate domain-specific IPMP solution.

The second addition to MPEG-4 Systems is not strictly an addition; instead, it is a fundamental shift in design assumptions. In the MPEG-1 Systems and MPEG-2 Systems standards the specifications extended monolithically from the packetization layer all the way to the transport layer. For example, the

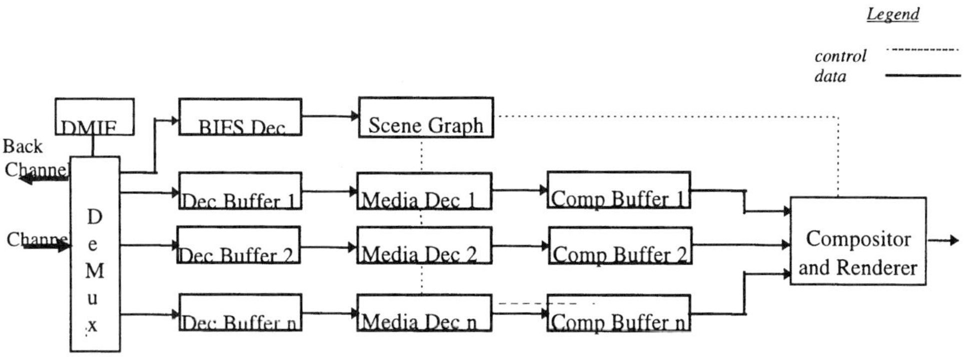

Figure 1 MPEG-4 Systems version 1.

MPEG-2 Systems transport stream specification defined the packetization of elementary streams (i.e., PES layer) as well as the transport layer. With MPEG-4 Systems this restriction has been relaxed. The transport layer is not defined normatively, as it is correctly perceived that this is very application specific. It is left to other standards-setting bodies to define the transport layer for their respective application areas. One such body is the IETF (Internet Engineering Task Force), which will be defining standards for the transport of MPEG-4 streams over the Internet.

These key features of MPEG-4 Systems have been finalized in version 1. The main effort since then has been in adding key functionalities. These are

MPEG-J: Java APIs with which applications can be developed to interact with the platform and the content
MP4: MPEG-4 file format for which Quicktime has been the basis.
Advanced BIFS: Extensions to MPEG-4 Systems version 1 BIFS

The MPEG-4 Systems version 1 player is shown in Figure 1. It is also referred to as the presentation engine. The main categories of components on the main data path are the demultiplexer, media decoders, and the compositor and renderer. Between these three sets of components there are decoder buffers and composition buffers, respectively. The MPEG-4 Systems decoder model has been developed to provide guidelines for platform developers. The BIFS data are extracted from the demultiplexer and used to construct the scene graph.

III. MPEG-J ARCHITECTURE

The MPEG-J system allows a combination of MPEG-4 media and safe executable code so that the content creators can embed complex control mechanisms within their media data to manage the operation of the audiovisual session intelligently. Figure 2 shows the architecture of an MPEG-J system including the MPEG-4 version 1 system (lower half of the drawing). More precisely, the MPEG-J system, also called the application engine, specifies interfaces for control of the MPEG-4 version 1 system referred to as the presentation engine.

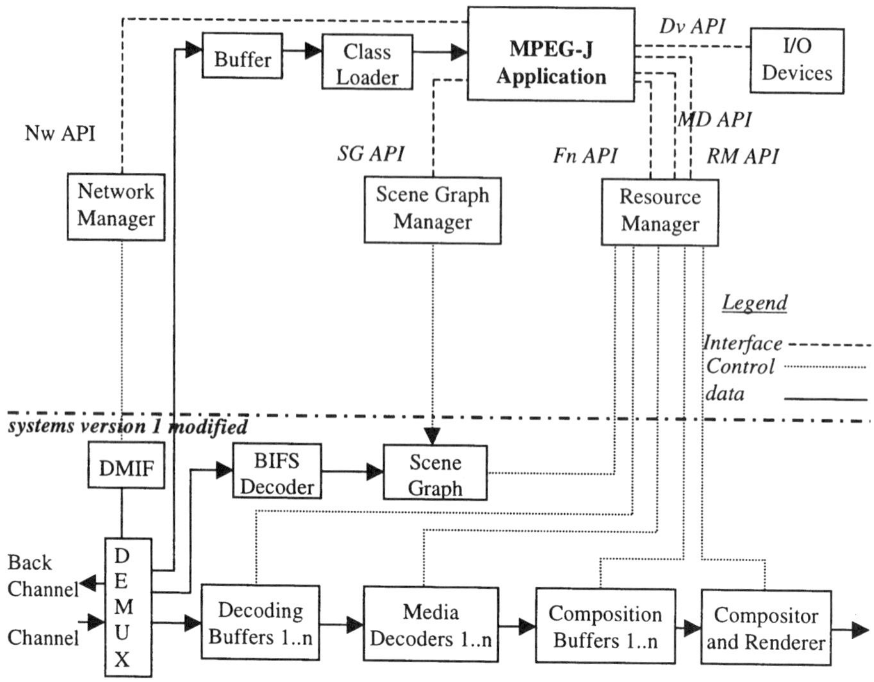

Figure 2 Architecture of the MPEG-J system.

The MPEG-J system consists of the class loader (and buffer), the I/O devices, the network manager, the scene graph manager, and the Resource Manager. The network manager interfaces the MPEG-J application with the Delivery Multimedia Integration Framework (DMIF); the scene graph manager interfaces the MPEG-J application with the scene graph; and the resource manager interfaces the MPEG-J application with the scene graph, the decoding buffers, the media decoders, the composition buffers, and the compositor and renderer. The I/O devices allow users to interact directly with and control the MPEG-J application.

The resource manager manages system resources allowing the regulation of overall performance; this includes controlling the media decoders, functionalities, and resources. The scene graph manager allows direct control of the BIFS scene graph (which specifies scene description for composition), including turning off portions of the scene graph to better match the available processing–memory to the needed resources. The network manager provides a means for the MPEG-J application to send requests or receive data through the DMIF.

A. Java Streaming

Application programs based on the MPEG-J APIs are made available to the MPEG-4 player in the form of an MPEG-4 elementary stream. Each time an MPEGlet (a remote Java MPEG-J application implementing the MPEGlet interface) is received, execution starts as a new thread. Timely delivery of Java byte code may be required for MPEGlets. The Java streaming method in MPEG-J provides the necessary functionality for this. The

Java stream header, defined in the MPEG-4 Systems version 2 standard, is attached to each class file and object data at the point of content generation and transmission. Once these data are packetized, then any time-critical transport mechanism, such as Real Time Transport Protocol or MPEG-2 transport stream, may be used to transport the data.

Java streaming in MPEG-J provides the following features:

Compression: Java byte code can be optionally compressed for bandwidth efficiency using the zip compression mechanism.

Class dependence: For a given class there may be other classes on which it depends, and these other classes need to be loaded first. Similarly, before an object can be instantiated, the class of which it is an instance needs to be loaded first. The Java Archive (JAR) mechanism in Java does provide this functionality. However, there can be problems with this solution, particularly if the Java data are transported over lossy channels. In such a situation, the loss of a single packet can make the whole JAR package unusable. The Java streaming technology in MPEG-J provides an alternative. A class dependence mechanism is provided in the Java stream header. All of the dependent classes for a given class are listed in the header. From a delivery and implementation perspective, these dependent classes need to be loaded before the class can be loaded.

IV. MPEG-J API

An MPEG application (local or remote) will use well-defined sets of interfaces (MPEG-J APIs) to interact with the underlying MPEG-4 player. These Java APIs are implemented on all compliant MPEG-J players. The details of the required implementation of these APIs can be found in the MPEG-4 Systems version 1 standard (ISO/IEC 14496-1). It is expected that the MPEG-J application (local or remote) can call methods defined in the MPEG-J APIs and those from a minimal set of Java packages. The MPEG-J APIs are functionally divided into the following categories.

A. Terminal Capabilities

The resource manager APIs help to manage resources on an MPEG-J terminal during an MPEG session. These APIs can be used to adapt an MPEG session to a given terminal. The main components of the resource manager APIs are the resource manager interface and the event model defined by the resource manager.

1. *Functionality Supported*

Access to the capability manager
Access to the decoder with any node
Providing a list of available decoders of a specific type
Access to the priority associated with a decoder
Event model with downloadable event handlers
Events mechanism for stream and decoder problems
Events mechanism for renderer problems

2. Key Features

The resource manager interface provides access to the capability manager (discussed in detail in the next section), renderer, decoders, and their priorities. This interface also defines methods for gaining access to a decoder given a node in the scene graph and change it to another available decoder. The media decoder and the decoder functionality APIs rely on the resource manager to obtain the instance of a decoder.

The resource manager APIs also define decoder and renderer events. The corresponding interfaces for event generators and listeners (event handlers) are also defined. The resource manager implicitly defines an event model for graceful degradation. For each decoder the resource manager would have an instantiation of a class that implements MPDecoder or a subinterface. These decoder instances generate the events corresponding to different situations. These events can handle events if necessary in addition to the default event handlers in the application. The MPEG-J application can receive the event handlers as byte code in the bitstream.

B. Network (Nw) API

The terminal capabilities API provides a standardized way for an MPEGlet to query the MPEG-4 terminal for its capabilities. It enables an MPEGlet to become aware of the capabilities of the terminal it is running on. This facilitates an MPEGlet or an application to adapt its own execution and the execution of various components, as they may be configured and running in the MPEG-4 terminal. The terminal capabilities APIs are responsible for providing access to the static and dynamic capabilities and the profile and level information.

1. Functionality Supported

Access to information about the profiles and levels
Access to static capabilities
Access to dynamic capabilities

2. Key Features

The need to allow the MPEGlet to discover at run time the MPEG profile and level led to the definition of a set of APIs for this specific purpose. These APIs are part of the so-called static capabilities, because these capabilities cannot change during the lifetime of the MPEGlet and are specific to a particular kind of terminal.

The profile–level capabilities are not the only ones that can be retrieved. Availability of information about the status of the terminal (CPU load, memory, etc.) can be used to design complex applets, which can modify the scene according to the available resources. For example, the content creator may decide to ask the user which parts of a scene he or she wants to switch off, in order to satisfy the real-time constraint.

The information about the terminal mentioned above is called dynamic because it may change during the lifetime of the MPEGlet.

The separation between static and dynamic terminal capabilities has been reflected in the API. Static capabilities are defined in the StaticCapability interface, and the DynamicCapability interface defines the dynamic capabilities. The TerminalObserver interface allows the MPEGlet that implements this interface to be notified when terminal capabilities change. Information about the terminal profile and level is retrieved from the TerminalProfileManager.

C. Resource Manager (RM) API

Although the API contained in the standard Java package java.net allows general-purpose control of the network features of the terminal, an MPEGlet can interact with a higher abstraction of the network, which is compliant with what the DMIF part of the standard has already defined. In this way the MPEGlet is independent of the kind of connection that the decoder is using to retrieve the MPEG-4 stream. Compliance with the DAI (DMIF application interface) does not imply definition of the whole functionality provided by this specification; definition of the whole functionality would have threatened consistency of the MPEGlet with the underlying player.

1. Functionality Supported

For the just stated reason, the functionality provided by the current API is limited to the following groups:

Network query. This feature allows the retrieval of statistical information about the DMIF resources used by the MPEG-4 player, such as the current session and its channels, and their QoS (quality of service).

Channel control. Through the channel control, an MPEGlet can disable or enable existing channels through which elementary streams are conveyed. This feature is needed to write MPEGlets that implement graceful degradation under limited or time-varying resources.

2. Key Features

The class ChannelController allows minimal control over channel behavior through a couple of primitives, which allow enabling and disabling a specified channel belonging to a specified session.

The class DMIFMonitor allows an application to retrieve specific information from MPEG-4 sessions and channels. It also facilitates monitoring of the network resources. The classes ChannelDescriptor and SessionDescriptor collect data specific to channels and sessions, respectively.

D. Media Decoder (MD) API

The decoder APIs facilitate basic control (e.g., start, stop, pause, resume) of all the installed decoders in an MPEG session. A list of decoders that are currently installed and those available can be obtained through the resource manager. The resource manager also provides the instance of the decoder associated with a node in the scene graph.

1. Functionality Supported

- Starting, stopping, pausing, and resuming a decoder
- Decoder attaching to and detaching from elementary streams
- Getting the type and other attributes of a decoder

2. Key Features

The MPDecoder is an interface that abstracts the most generic decoder. The MPDecoder APIs facilitates attaching decoders to and detaching them from elementary streams. It also provides access to the type, vendor, and an instance number of the decoder. A number of interfaces that abstract decoders such as scalable video decoder (SclVideoDecoder) and structured audio decoder (SclAudioDecoder) that are subinterfaces of MPDecoder are also defined. These subinterfaces provide access and control to specialized features of that particular decoder.

E. Scene Graph (SG) API

The SceneGraph APIs provide a mechanism by which MPEG-J applications access and manipulate the scene used for composition by the BIFS player. It is a low-level interface, allowing the MPEG-J application to monitor events in the scene and modify the scene tree in a programmatic way.

1. *Functionality Supported*

Access a DEFed node
Manipulate a DEFed node
Remove a DEFed node
Change a DEFed node
Does not allow node creation
Monitor events in a scene

It should be noted that the current definition of MPEG-J does not allow creation of nodes in the scene. Furthermore, only nodes that have been instanced with DEF are accessible to the MPEG-J application.

2. *Key Features*

Because Java is a complete programming language, it would be possible to allow an MPEG-J application substantial control over the composition and presentation of an MPEG-4 session. However, it was decided that for security reasons, access to the BIFS scene graph would be limited to the mechanisms provided by the BIFS node tables. Consequently, an MPEG-J application cannot create any new nodes for the scene, and it cannot access node fields that are not defined in the BIFS scene.

a. Events Events in the BIFS scene graph are identified by the two interface classes, EventIn and EventOut. These classes reflect exactly the input and output IDs defined in the normative node coding tables found in Annex H of the systems portion of the MPEG-4 specification ISO/IEC 14496-1. For instance, the ''scale'' exposed field of the Transform BIFS node has an inID = 5 and an outID = 3, based on Annex H of ISO/IEC 14496-1. There is also an EventOutListner class, which allows monitoring of events.

EventIn The EventIn interface class contains an interface class definition for each node type defined in MPEG-4 Systems. These definitions enumerate all of the exposedField and eventIn field types in the node in the order in which they are defined in the ISO/IEC specification 14496-1.

EventOut Likewise, the EventOut interface class contains an interface class definition for each node type defined in MPEG-4 Systems. These definitions enumerate all of the exposedField and eventOut field types in the node in the order in which they are defined in the ISO/IEC specification 14496-1.

EventOutListener The scene graph API also provides an EventOutListner interface, which can be used by the scene graph manager to identify a field value change when an eventOut is triggered.

b. Field Values. The scene graph APIs provide an interface for tagging objects that can return a field value. Like VRML, MPEG supports two general field types. SFField is used for single-value fields and MFField is used for multiple-value fields. The supported SFField types are extended directly from the FieldValue interface; the Multiple field types are extended through the MFFieldValue interface.

c. Scene Management. The following interfaces are used to facilitate programmatic control over the MPEG-4 terminal's native scene.

SceneManager The SceneManager interface is the interface that allows access to the native scene. It contains methods for adding and removing a SceneListner. In order to access the BIFS scene graph, the SceneManager requires an instance of the scene, which it obtains through notification on a SceneListner instance. This method is the only normative way for an MPEG-J application to obtain a scene instance.

SceneListner The SceneListner contains a notify method that can be called by the SceneManager when the BIFS scene has changed. The notify method contains arguments to indicate the nature of the change and an updated Scene instance. Currently, three states can be passed through the scene listener. They indicate that the scene is ready, it has been replaced, or it has been removed.

Scene The Scene interface acts as a proxy for the BIFS scene. It contains a getNode method, which returns a node proxy for the desired node in the scene. If the requested node does not exist, it throws a BadParameterException, and if the scene is no longer valid, it throws an InvalidSceneException.

Node The Node interface acts as a proxy for a BIFS node in the scene graph. As previously mentioned, only nodes that have been instanced by a DEF identifier are available to the MPEG-J application. Three methods are available in the Node proxy for monitoring output events. The getEventOut method reads the current value of an eventOut or exposedField of this node. There are also methods for adding and removing an EventOutListner. All three of these methods throw a BadParameterException if they fail.

The fourth method contained in the Node interface is the sendEventIn method. This is the only method available to the application for modifying the BIFS scene. It updates the value of the eventIn or exposedField of the node. It is a synchronous call that will not return until the field is updated in the scene.

F. Functionality

The Functionality API provides mechanisms for controlling the media decoding process. A number of useful predefined functionalities become possible under user control.

1. *Functionality Supported*

Because the functionality APIs are dependent on other parts of the MPEG-J application engine (media decoder, resource manager, scene graph API), they will most likely change as these APIs mature. Following is a list of currently envisioned functionalities:

ProgressiveImage
HotObject
DirecDecoding
Transparency
TrickDecoding

2. Key Features

Interfaces are defined one each for all the preceding functionalities. The details of each of these interfaces are presented here:

a. ProgressiveImage. ProgressiveImage is an interface that triggers the action of progressive refinement of quality of the image being decoded under user control. This interface extends SclTextureDecoder, which is used to decode image texture streams.

b. HotObject. HotObject is an interface that triggers the action of enhancement of a visual object provided that the object is a hot object. Thus, hot objects have some enhancement streams associated with them that are triggered when needed. This interface extends SclVideoDecoder, which is used to decode base and enhancement streams.

c. DirecDecoding. DirecDecoding is an interface that allows creation of visual objects that are directionally sensitive (in the form of prequantized directions for now). This interface is most easily explained by assuming a bitstream composed of a number of static visual Ops (video object planes) coded as an AV object such that, depending on the user interaction, Ops corresponding to one or more viewpoints are decoded as needed.

d. Transparency. This interface allows formation of visual objects with transparency information. It also supports selective decoding of objects under user control. The user may not need to decode all objects or portions of an object because of bandwidth or computing resources. If certain objects or portions of an object are hidden, they may not need to be decoded. The portions of objects not to be shown can be marked with the transparency color, which can be extracted to form transparency information. This process can also be used to highlight or mark objects.

e. TrickDecoding. This interface allows selective decoding of audiovisual objects for playback in trick mode (fast-forward, fast-reverse) under user control.

V. SAMPLE APPLICATION AREAS

A. Electronic Information Guide

This application area includes informational guides such as electronic program guide, electronic display guide (e.g., user-controlled flexible Bloomberg-like TV), and electronic navigational guide (e.g., maps, routes, hotel or car rental locations).

It is expected that clients with a wide range of capabilities will use channels with a wide range of bandwidth capabilities.

Adaptation to a wide range of clients and networks, adaptation to a wide range of varying client capabilities, and displaying a subset of the original scene graph according to the user's preference and geographic location are some of the requirements of such electronic information guide applications.

In these types of applications, the client displays a portion of the scene graph from a single coded representation received in the bitstream. This subset for, e.g., only the maps corresponding to the user's geographic location or interest needs to be shown in an electronic navigational guide application. The quality of the picture shown, size, etc. can be adapted according to the current client and channel capabilities.

In the context of MPEG-J, this application area relies on the use of user interaction, scene control, and resource management capabilities.

B. Enriched DTV

The MPEG-J APIs enable the development of enriched Digital Television (DTV) content. This is a significant departure from today's television viewing experience. With MPEG-J the viewer can interact with the platform and the content. In this application scenario the platform is a set-top box with Java packages conforming to a well-defined Java platform. The audio–video content will be delivered via an MPEG-2 transport stream. The MPEG-2 transport stream can also be used as the transport layer for carrying MPEG-J along with any other MPEG-4 data. The MPEG-J applications will have access to the DTV functionality, and because MPEG-J is an open interface, it is an attractive option to content developers and service providers.

Keeping the interfaces open while prescribing distribution and presentation formats facilitates embedding a large variety of applications in the broadcast stream. Applications such as electronic program guide, interactive travel magazine, and real estate presentation can be developed with MPEG-J so that they can be executed on DTV set-top boxes.

C. Multipurpose Multimedia Services

This application area addresses services such as multimedia yellow pages that employ audiovisual databases. These multipurpose multimedia services, with their associated databases, are expected to allow ease of content repurposing such that a wide variety of connection bandwidths, channel conditions, and clients and a dynamically changing environment on a client can be handled. Multipurpose databases may contain a variety of types of media objects such as MPEG-4 coded speech, music, synthetic audio, image objects, video objects, and synthetic visual objects.

Multipurpose multimedia services have a number of specific requirements as follows:

Ease of adaptation to client capabilities so that from a single coded representation of a scene, a client can recover and display content consistent with its capabilities.

Ability to facilitate graceful degradation to channel conditions and bandwidth so that the user can recover the high-priority information and the lower priority information may be discarded.

Ease of adaptation to time-varying client-side resources such as when a user may be running multiple applications concurrently.

Functionalities that allow a user to interact efficiently with multimedia data to see various behaviors.

Here is a list of specific example scenarios in which some of these requirements apply:

In *wireless* applications, a device of limited power and decoding complexity may have to extract useful subsets of a scene from a single coded representation on the server. The bandwidth available for content access may be low, and moreover it may fluctuate in a harsher operating environment.

In *Internet* applications, client-side PC or notebooks may have a wide range of processing and memory capabilities. In addition, network connections with a wide range of speeds may be available to access the database. The service (and database design) should enable adaptation of performance to client-side capabilities.

In *consumer* applications, a consumer may have to repurpose the multimedia content to include portions of it in a multimedia e-mail to a friend, forward it to a real estate agent, etc. without reencoding it.

Overall, in terms of MPEG-J, this application area relies on the use of the resource management, media decoding, extended functionalities, user interaction, and scene control capabilities.

VI. SUMMARY

In this chapter we have introduced MPEG-J, a programmatic system that uses Java to control an MPEG-4 Systems version 1–based player. We first presented a brief overview of the basics of MPEG-4 Systems version 1. We then briefly discussed the high-level architecture of the MPEG-J system. Next, we described the Java application program interfaces (APIs) that make the MPEG-J system programmable. Finally, in Section V, we discussed the application areas of MPEG-J and the issues of profiling.

REFERENCES

1. MPEG-4 Systems. Coding of audio-visual objects: Systems version 2, ISO/IEC 14496-1 PDAM-1.
2. MPEG-4 Systems. Coding of audio-visual objects: System version 1, ISO/EC JTC1/SC29/WG11 N2201, final draft international standard 14496-1 version 1, October 1998.
3. A Puri, A Eleftheriadis. MPEG-4: An object-based multimedia coding standard supporting mobile applications. ACM J Mobile Networks Appl 3:5–32, 1998.
4. G Fernando, A Puri, V Swaminathan, RL Schmidt, P Shah, K Deutsch. Architecture and API for MPEG-J, ISO/IEC JTC1/SC29/WG11 MPEG98/3545 rev2.0, July 1998.
5. A Puri, RL Schmidt, and BG Haskell. Scene description, composition, and playback systems for MPEG-4. Proceedings Visual Communications and Image Processing, San Jose, January, 1999.
6. MPEG-4 Systems. MPEG-4 Systems. version 2 verification model 6.0, ISO/IEC JTC1/SC29/WG11 N2741, March 1999.

16
MPEG-4 Players Implementation

Zvi Lifshitz
Triton R&D Ltd., Jerusalem, Israel

Gianluca Di Cagno and Guido Franceschini
CSELT, Torino, Italy

Stefano Battista
bSoft, Macerata, Italy

I. INTRODUCTION

A. History

The MPEG-4 player project was launched in March 1997, one-half year before the standard specifications arrived at committee draft status. The project had several objectives: to verify the applicability of the specifications, to demonstrate the capabilities of the standard, and to serve as a reference for developers. The project focused on the Systems part of the standard, as the component that turns a collection of media decoders into a comprehensive *player*.

For this purpose the MPEG committee established two ad hoc groups, Systems Software Implementation 1 and 2, known as IM1 and IM2. Each of the groups attracted a handful of volunteers, representing diverse organizations from around the globe, who committed to contribute to the group mission. IM1 proposed a C++-based implementation, whereas IM2 would use Java. Of these two groups, only IM1 persisted (the other would revive a year later when MPEG-J was introduced), and it is still active at the time these lines are written. This chapter describes the IM1 player.

At the start of the IM1 project, the task was not easy. First, the time constraints were tight. The group was challenged to present a working prototype, along with proper demonstration content, within 3 months—at the following MPEG meeting. Second, the standard specifications that were supposed to be implemented were a year and a half ahead of their final status, which meant the group had to design an implementation flexible enough to allow smooth adoption of future modifications. Third, the IM1 project was presumed to be a platform on which implementations of other parts of the standard, such as media decoders and Delivery Multimedia Integration Framework [DMIF (transport)] stacks, would be easily integrated. Last, but not least, this effort would be conducted by a team of geographically dispersed engineers, usually a recipe for a perpetuum mobile convergence process.

In order to cope successfully with the challenges, the IM1 team applied highly modular and open architecture, published detailed and vigilant application programming interface (API) documents, and managed tight control on schedule. The result of its work is presented in this chapter.

B. The Players

The IM1 project produced several MPEG-4 players. One is a two-dimensional player, which may handle the 2D profile of the standard. The other is a three-dimensional player, which handles the complete specifications of both 2D and 3D nodes but requires 3D graphic hardware. Lately there was a contribution of another 3D player. All the players are based on a single framework code called the Core, which is essentially a platform-independent set of modules. The players implement platform-dependent Compositors, built over the Core.

This chapter describes the architecture of the various players. Section II explains the overall structure of the players. It also describes the process of creation of content for these players and the tools that were used. Section III depicts the Core components. Section IV and V describe the 2D and the 3D Compositors. Section VI introduces the DMIF part of the project and its role in client–server scenarios. A summary and reference list conclude this chapter.

II. THE IM1 SOFTWARE PLATFORM

A. Architecture Overview

The IM1 software platform comprises several modules, each developed and tested independently. The main modules are as following:

The *Core*. As its name suggests, this module sits at the core of the system. It includes submodules for controlling the execution of the player (the Executive), sync layer (SL) processing, buffer and time management, Binary Format for Scenes (BIFS) and object descriptor (OD) decoding, and the creation and manipulation (as result of BIFS updates and routes) of the scene tree.

The *Compositor*. This is the ''consumer'' of the scene tree. It continuously traverses the tree, fetches composition units from composition buffers, and composes and renders the scene.

The *Decoders*. Straightforwardly, these modules read access units from decoding buffers, decompress them, and store the decoded data in composition buffers.

The *DMIF* stack. Provides transport and demultiplex services to the players. In other words, the DMIF stack feeds the Core with SL packets.

IPMP systems. Optional modules that provide Intellectual Property Management and Protection services. IPMP systems may require predecoding (e.g., decryption) or postdecoding (e.g., watermarking) processing.

The *frame application*. This is the facade through which the user interacts with the players. It provides the frame for the scene view and handles user interaction operations such as mouse commands.

Figure 1 illustrates the modular design of the player software.

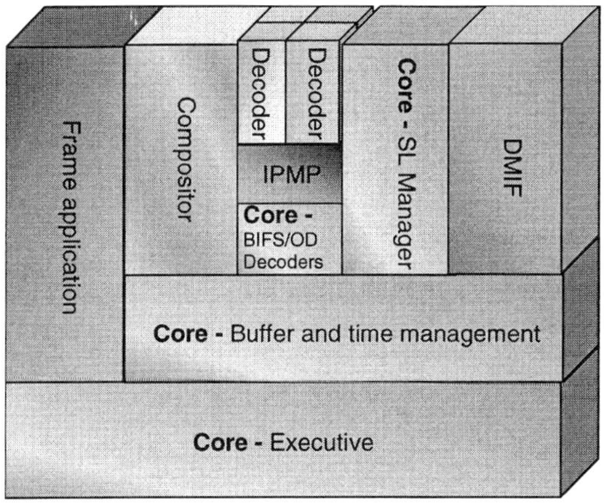

Figure 1 The MPEG-4 player software modules.

B. The APIs

The glue between the player components is a set of APIs. There are source-level and binary-level APIs. The source-level APIs serve for assembling a complete MPEG-4 Systems engine from several parts. This construction requires recompilation of source files. During the development process, the IM1 team used these APIs internally to distribute the work between individual engineers. There are also binary-level APIs, which enable the development of ''drop-and-play'' plug-in dynamic-link libraries (DLLs). Those APIs allow complete independent development of non-Systems modules, such as decoders, DMIF stacks, and IPMP systems. Here is a complete list of the APIs:

Source-level APIs:

The *Executive API*. Allows the frame application to control the execution of the player and to forward events such as user commands. It also provides a callback mechanism to pass events from the Core to the frame application.

The *Presenter API*. The Core uses this API to control the execution of the Compositor.

The *Scene Graph API*. Provides the glue between the BIFS decoder and the Compositor. The BIFS decoder uses it to create and manipulate the scene graph, while the API provides access methods to the Compositor.

The *Buffer and Time Management API*. Enables the interaction between various modules and the buffer management module for storing data units in and retrieving them from the buffers. The buffer management module includes a mechanism to control the timing of the store and fetch operations. This mechanism is the basis for media synchronization.

Binary-level APIs:

The *Decoder API*. Enables independent development of decoders DLLs.

The *DMIF API*. Provides the means for development of DMIF stacks.

The *IPMP API*. Allows the construction and integration of Intellectual Property Management and Protection systems.

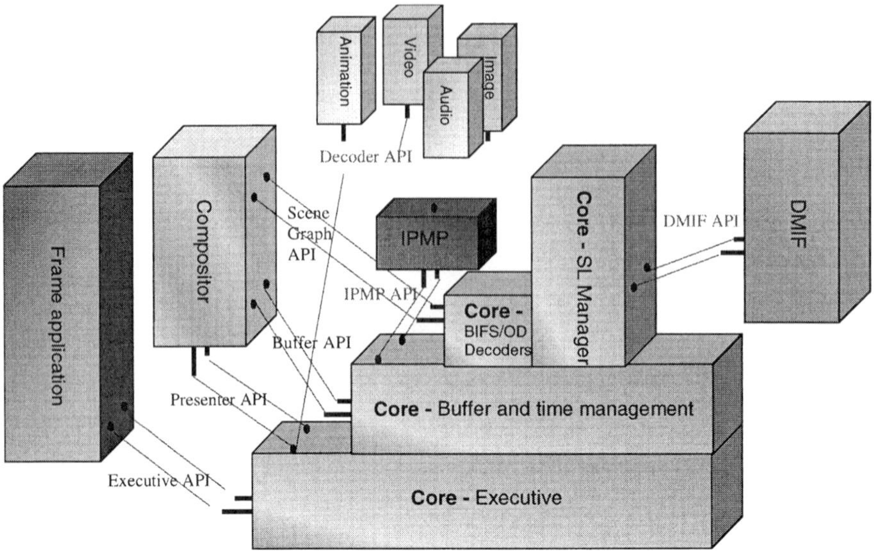

Figure 2 The MPEG-4 player APIs.

Figure 2 illustrates the role of the various APIs in the modular architecture of the player software.

C. MPEG-4 Coding Process and Tools

The discussion so far was all about the player, but players need content to play, and the IM1 group could wait for nobody to create this content. Video and audio streams could be generated by encoding tools made available by other MPEG-4 groups. Because the focus of the group was on the Systems part of the standard, it also allowed itself to use non-MPEG audiovisual streams. But there was a need to create scene and object descriptions in MPEG-4 format and to multiplex all into files that would be locally accessible or transmitted over networks. For that purpose the BifsEnc (BIFS and OD encoder) and Mux tools were created.

BifsEnc is an application that takes a Virtual Reality Modeling Language (VRML)–like text scene description and compiles it into a BIFS description. For handling BIFS update commands, a few non-VRML directives were invented, such as *AT time INSERT node INTO field* or *REPLACE field BY value* (see a reference to the complete list at the end of this chapter). In addition, the input text file may include OD commands, such as *UPDATE OD ObjectDescriptor*. The text description of the OD elements borrows the VRML syntax. The descriptors and fields look like VRML nodes. *BifsEnc* knows how to distinguish these entities, encode them properly, and store them as a separate OD stream.

The nice thing about this program is that it did not require much work. This is because of the tricky use of the BIFS–OD classes and macros, as explained in the next section, that inspired a great deal of reuse of the player's Core code.

The other tool, the *Mux*, is used to wrap access units by SL data and pack media and Systems (BIFS–OD) streams into one file. The file format follows the *TRIF* specifications, i.e., the blueprint for trivial MPEG-4 file format described in the Systems specifica-

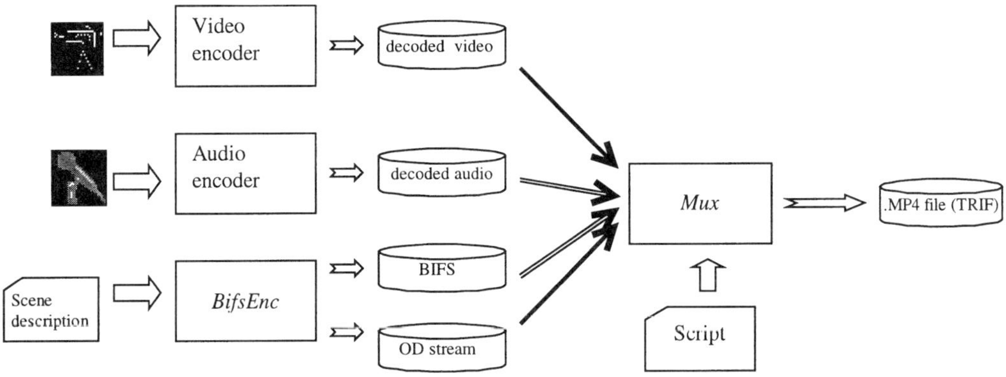

Figure 3 Content creation process.

tions. The structure is simple: a header made of an initial object descriptor, followed by a stream map table, which maps each stream ID to a channel number, followed by the data in form of SL packets, which are interleaved according to the FlexMux specifications.

The Mux input is a script file containing a set of object descriptors. The first descriptor is usually the initial object descriptor. The other descriptors describe the elementary streams. The syntax of the descriptors is identical to that used by BifsEnc (therefore allowing, again, code reuse). The text format of these descriptors allows a few extra fields, required by the Mux application, such as the location of the elementary streams and the modules that are used to parse them.

As just mentioned, elementary streams need to be parsed in order to determine their partition into access units and to extract the timing information associated with the units. The syntax of each stream depends on the encoding tool that created them. Therefore, the Mux tool needs to include a parsing module for each stream syntax. Adding parsing modules in easy with the *MUX API*. The object descriptors in the script file ought to contain fields that assign the proper parsing module for each stream. Again, the full Mux and MUX API specifications are referenced at the end of this chapter.

Figure 3 illustrates the IM1 content creation process.

III. THE CORE

A. The Core Components

As illustrated in Fig. 1, the Core module is constructed of several submodules.

The *Executive*. This component controls the execution of the player, instantiates other modules, and handles the data flow between the modules.

The *Sync Layer Manager*. Receives (through DMIF) data packets of elementary streams, decodes SL information, and dispatches access units to decoding buffers.

The *Buffer Manager*. Controls the flow of data through buffers. The buffers used by the system are time aware; i.e., units may be fetched from buffers at specific times, depending on the time stamps attached to the units. Thus, this module manages both memory and synchronization.

The *OD Decoder*. Parses the OD streams and passes the extracted data to the Executive for execution.

The *BIFS Decoder*. Parses scene description streams, creates the scene tree, and updates it according to BIFS-update and BIFS-animation commands included in these streams.

The *Scene Manager* (not visible in Fig. 1). This module provides the infrastructure for the scene-graph operation. It handles most of its generic actions, such as processing of routes, firing of events when eventOut fields are modified, notification to the node when any of its fields are changed, and execution of Conditional and Inline nodes. It also implements basic functionalities of a few classes that are used as base classes for specific nodes—GroupNode for all grouping nodes, StreamConsumer for nodes that consume elementary streams, and Root for the subset of grouping nodes that are allowed as scene roots.

B. The IM1 Classes

As a C++ project, the IM1 software is organized by classes. Figure 4 shows the main classes used by the IM1 players, the flow of data through the classes, and the correlation between the classes and the player modules.

Figure 4 is a rudimentary illustration of the most basic classes. A detailed description of all the classes used in the project is outside the scope of this book (see references at

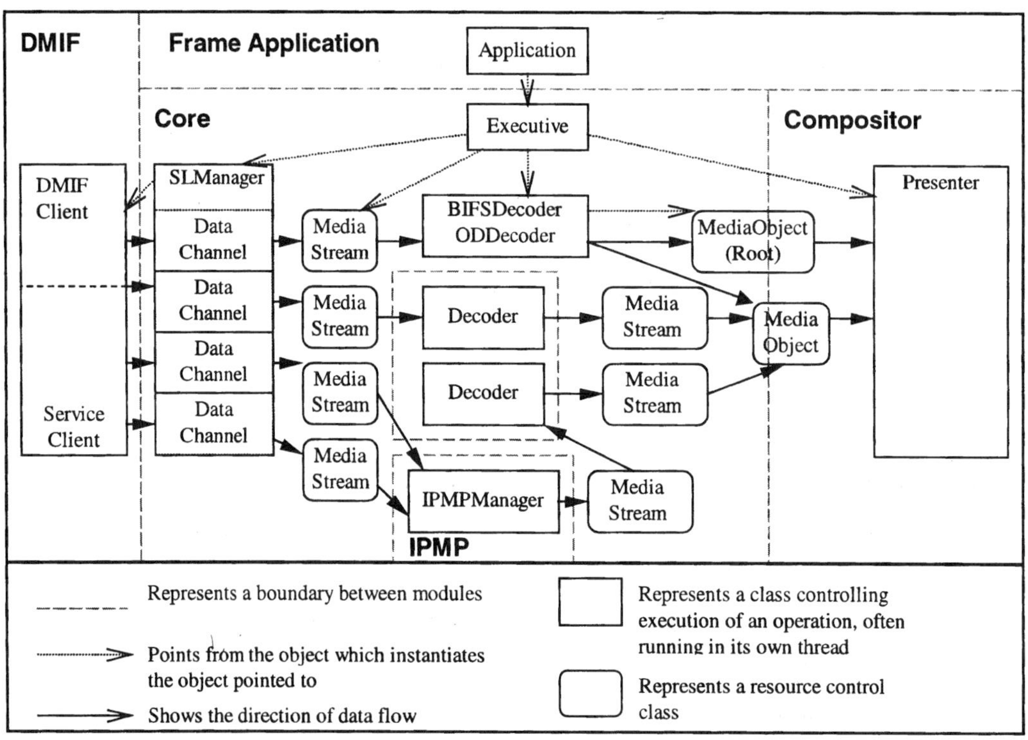

Figure 4 The IMI classes.

the end of this chapter). However, understanding the functionality of the basic classes is instrumental for this chapter, and a general description is provided hereinafter. We will describe classes used by all the modules, not just the Core, as it is impossible to understand the Core without knowing how it interacts with the classes implemented by other modules.

1. Frame Application Class

Application. An abstract base class extended by specific implementations of frame application. It provides the means for callback notification from the Core-Executive to the frame application, though the following virtual methods:

OnConnect	Notifies when the main service is connected and an initial object descriptor is received.
OnFish	Called when the primary service has been disconnected.
OnError	Reports encountered errors.

2. Core-Executive Class

Executive. The primary class of the Executive module, which controls the execution of the player. It includes the following methods:

Start	Called by the frame application to start a session on a given address.
Stop	Stops a session.
OnUserEvent	The frame application passes user commands such as mouse clicks to this method.
OnConnect	The sync layer manager calls this method and passes it an initial OD when a new service is established.
OnFinish	Called by the SL manager when a service is disconnected.

3. Core–Sync Layer Manager Classes

SLManager. An object of this class is instantiated for each active service. It executes service operations on behalf of the Executive and handles control callbacks received through the DAI (DMIF Application Interface). Some notable methods are the following:

Connect	To attach a service in a given address.
Close	To detach a service.
AttachStream	Called by Executive to configure a stream whose ObjectDescriptor is known.
RequestChanels	Requests DMIF to open all the streams that have been configured before.
OnServiceAttach	A callback method used by DMIF to notify completion of a service attach request.
OnServiceDetach	DMIF notifies a service detach through this callback method.
OnChannelAdd	Called by DMIF when a channel is added. In reaction, the Core instantiates a *DataChannel* object and returns its address to DMIF. DMIF subsequently uses this object to notify the Core of incoming streaming data on that channel.

DataChannel. An object of this class is instantiated by *SL Manager* for each played stream. It is where the sync layer processing, i.e., the parsing of SL headers, is performed. Its main methods are the following:

Config	Configure the SL header information using the *SL Config* data from the *ES_Descriptor*.
OnPacketReceived	DMIF calls this callback method to deliver a received SL packet. The method parses the SL headers, constructs the access units, and stores them in the decoding buffer, along with their timing information.

4. Core–Buffer Manager Classes

MediaStream. This class controls the buffers used by the player. It is used for both decoding buffers and composition buffers and, if an IPMP system is used, also the buffers required for IPMP processing. All data received by the player is delivered from one module to another, i.e., from the sync layer manager to (a predecoding *IPMP* system to) the decoder to (a postdecoding IPMP system) to the Compositor, over buffers controlled by this class. Therefore the player data flow can be viewed as a dynamic graph of processing units connected with MediaStreams.

MediaStreams have clock objects attached to them. With these and with the time stamp information stored in the buffers along the data units, it is possible to time the fetch operation on the buffer, as explained subsequently. Synchronization is achieved when several buffers share a single clock object.

These are the primary methods of MediaStream:

Allocate	Allocates a block of memory in the buffer for storing a unit. To save memory copy instructions, a unit can be stored incrementally into the buffer, and the available block may be reallocated when necessary.
Reallocate	Reallocates a large block for the unit.
Dispatch	Notifies that the complete unit was stored and attaches time information to it.
FetchNext	Fetches the next unit from the buffer. This method is executed immediately, regardless of the time information attached to the unit.
Fetch	Performs a time-aware fetch. According to its arguments, the function may block the calling process or not. When a blocking fetch is required, the calling process is blocked until the time is ripe for the next unit to be fetched. A nonblocking fetch will return the next unit immediately, provided its time has arrived. Otherwise, it will still return immediately, but with a null pointer, without fetching any unit.
Release	Releases the unit previously fetched, thus acknowledging that the unit is no longer required, and its memory can be taken when needed for a new unit.

ClockReference. An instance of this class is attached to each MediaStream and is used for performing time-aware fetches as just explained. Streams that need to be synchronized flow over MediaStreams that share a single ClockReference.

Initially, *ClockReferences* run on the machine time, but they can be adjusted when the sync layer receives OCRs (object clock references). Several methods exist for manipulating ClockReferences:

SetTime	Sets the clock with an OCR.
SetGranularity	Set the effective granularity of the clock.

Adjust	Does a posteriori clock adjustment. For instance, if the first OCR with the value 0 is parsed by the SL at time *x* and that OCR was immediately followed by an access unit with CTS = 0, then because decoding takes some processing time, the player will not be able to play the unit on time and synchronization will not be achieved. In order to fix that, the Compositor would take the time when the unit was actually rendered. If that time is $x+d$, it would adjust the clock by *d*. From then on all streams timed by this clock object would know that adjustment of *d* is required for rendering.

5. Core–BIFS Decoder Classes

SysCoder–BIFSDecoder. *BIFSDecoder* is the class in charge of BIFS parsing. It extends a more general *SysCoder* class, which implements the following encoding–decoding routines:

ParseStream	Starts a thread to parse an entire stream.
ParseInt	Reads an integer number of a given number of bits.
ParseFloat	Reads a 32-bit float number.
ParseDouble	Reads a 64-bit float number.
EncodeInt	Stores an integer number in a given number of bits.
EncodeFloat	Stores a 32-bit float number.
EncodeDouble	Stores a 64-bit float number.

BIFSDecoder extends *SysCoder* by adding the following method:

ParseUpdateFrame	Called from *ParseStream* to parse each access unit.

MediaObject. *MediaObject* is the base class for all BIFS nodes. Although it is basically part of the Scene Manager module, we mention it here because each node is responsible for parsing itself. Therefore most of the BIFS decoder functionality is actually implemented by this class. This class will be discussed later in more detail.

6. Core–OD Decoder Classes

SysCoder–ODCoder–ODDecoder. *ODDecoder* is the class in charge of OD stream decoding. It is derived from *ODCoder*, which in turn extends the SysCoder class mentioned earlier. ODCoder adds two methods to those implemented by SysCoder:

ReadDescLength	Parses the size field of an expandable SDL class.
StoreDescLength	Stores the size field of an expandable SDL class.

The ODDecoder adds the following method:

ParseODFrame	Called from *ParseStream* to parse each OD access unit.

BaseDescriptor. *BaseDescriptor* is the base class for all the OD protocol descriptors. Each derived class contains a set of *SDLField*-derived member fields. As explained below subsequently, *SDLField* is a base class for SDL class fields that can parse or encode themselves. Each BaseDescriptor-derived class contains a table of the offsets (i.e., the value of the *offsetof* macro) and the names of its fields as a static member variable. With this table, it is possible to have one generic parsing–encoding function for many of the descriptors. The most important class method is

DescEncDec	The generic method for encoding, decoding, and size calculation of a descriptor. The method is applied to all descriptors that have a ''simple'' encoding scheme, i.e., when all the fields of the descriptor are encoded in a row, without any conditional or loop statements. When the encoding is not simple, this method is overloaded by derived classes.

SDLField. *SDLField* is the base class for all the fields of an OD class, namely *SDLInteger*, *SDLIntegerArray*, and the templates *SDLClass<DescriptorName>* and *SDLClassArray<DescriptorName>*. All these classes overload the *EncDec* function, which is used to decode, encode, or calculate the size of an encoded field. Thanks to the structure of SDLField and this function, a nice trick of having the same code doing both encoding and decoding could be implemented.

7. Core–Scene Manager Classes

The sheet here is too short for describing all the Scene Manager classes and their full functionality. We will give only a glimpse of the two most essential classes—*MediaObject* and *NodeField*. The reader is invited to read the complete IM1 reference if she needs to do some implementation work of her own.

MediaObject. *MediaObject* is the base class for all BIFS nodes. A great deal of the player functionality is achieved by MediaObjects operating upon themselves. A MediaObject can parse itself (and encode itself in the BifsEnc tool), update itself (when a BIFS-Update command is received), animate itself (when a BIFS-Anim command is received), render itself (and its children), and a lot more.

MediaObjects use a field-table technique, similar to what you encountered before with BaseDescriptor. Each derived class contains a set of NodeField-derived member fields. *NodeField* is a base class for node fields that can parse, encode, and perform other nice things on themselves. Each MediaObject-derived class contains a table of the offsets of the fields (i.e., the value of the *offsetof* macro), their names, and some other field information, as a static member variable. With this table, it is possible to have generic parsing, encoding, update, and animation functions for all the nodes.

Some of MediaObject notable methods are the following:

Parse	The node parses itself.
Render	The node renders itself. The function is overloaded for each node type.
animate	The node animates itself.
InsertValue	Inserts a value into one of its vector fields.
DeleteValue	Deletes a value from one of its vector vields.
ReplaceValue	Replaces the value of one of its fields.
AddRoute	Add a route originating from one of its fields.
ReplaceNode	Replaces itself by another node.
Delete	Deletes itself.

NodeField. *NodeField* is the base class for all the fields of a BIFS node. A set of template classes extends this class, such as *SF<class>*, *SFNode<NodeClass>*, *SFNodeType<type>*, *MF<class>*, *MFNode<NodeClass>*, and *MFNodeType<type>*. All these classes overload the *Parse* method (and a few others), so each field can be parsed appropriately. These are some other methods:

Animate	The field changes its value according to some animation parameters.
Dequantize	The field calculates its value according to quantization information.
operator=	The assignment operator is overloaded, so whenever a field is changed, all routes originating from it are activated automatically and events defined for it are fired.

8. Compositor Class

PresenterBase–Presenter. *PresenterBase* is the base class for all Compositors. Each player implementation implements the derived *Presenter* class in its own way. This requires, among other things, the implementation of the following virtual methods:

Init	Compositor initialization.
Render	Scene rendering.
OnUserEvent	Passed a user command received by the Executive.
Terminate	Terminates the compositor.

9. Decoders Class

Decoder. A base class for the implementation of media decoders. Media decoder would implement the following methods:

Setup	Given decoder-specific information, the method perform a setup procedure.
SetInputStream	Sets the input buffer to the given MediaStream object.
SetOutputStream	Sets the output buffer to the given MediaStream object.
Start	Starts (on a dedicated thread) to fetch access units from the input buffer, decode them, and dispatch them to the output buffer.
Stop	Stops decoding.

A decoding development kit is available that provides a skeleton implementation of decoders in a class, derived from *Decoder*, called *DecoderImp*. DecoderImp implements the preceding methods. The developer of a specific decoder needs to extend it by, as a minimum, implementing the Setup method and one additional one—Decode. The *Decode* method accepts a pointer to a compressed access unit, decodes it, and stores the decoded composition unit in a given output block.

10. IPMP Class

IPMPManager. A base class for the implementation of IPMP systems. Derived classes would implement the following methods:

Init	System initialization.
SetInputStream	Sets the input buffer for the protected media stream to the given MediaStream object.
SetIPMPStream	Sets the input buffer for the IPMP stream to the given MediaStream object.
SetOutputStream	Sets the output buffer to the given MediaStream object.
Start	Starts (on a dedicated thread) to fetch media units from the input buffer, synchronize them with IPMP messages from the IPMP buffer, process them, and dispatch them to the output buffer.
Stop	Stops decoding.

An IPMP development kit is available that provides a skeleton implementation of IPMP systems in a class, derived from *IPMPManager*, called IPMPManagerImp. *IPMPManagerImp* implements the preceding methods. The developer of a specific system needs to extend it by, as a minimum, implementing the DecodeIPMP method. The *DecodeIPMP* method accepts a pointer to a protected media unit, a matching IPMP unit, decodes the media unit, and stores the decoded data unit in a given output block.

11. DMIF Classes

A general discussion of DMIF implementation in IMI is given out in Section VI. Here we describe the classes used by the Core to interface with the DMIF module. This is in fact an implementation of the DAI.

DMIFClient. *DMIFClient* is an interface object through which the Core accesses the DMIF module. A DMIF module needs to implement this interface by overloading the following pure methods:

ServiceAttach	Receives a URL identifying a local or remote service and establishes connection with this service.
ServiceDetach	Terminates a connection to a service.
ChannelsAdd	Request the service to open the channels according to a list of user-data items.
ChannelsDelete	Asks the service to stop sending data on the given channels and close them.

DMIFClientSink. All callback notifications from the DMIF module to the Core are made through the *DMIFClientSink* interface. The Core implements this interface by overloading the *OnServiceAttach*, *OnServiceDetach*, and *OnChannelAdd* pure methods. The Core's SLManager class, described earlier, is, among other things, an implementation of DMIFClientSink, and those methods are described there.

Channel. An object of a class implementing the *Channel* interface is instantiated by the Core for every open channel. The DMIF module uses this object to pass incoming data packets to the application by calling the object's *OnPacketReceived* method. The *DataChannel* class, described in Section III.B.3, is the Core's implementation of Channel.

IV. THE 2D PLAYER

A. Introduction

The Im 2D player project started on March 1997, when the Iml ad hoc group got its first mandate. It consists of a full software implementation of a real-time 2D browser conforming to the MPEG-4 Systems standard, designed to be portable on different platforms, even though in the code-writing phase the chosen platform was the Intel Pentium–Windows NT. The code is written in C++ and the compiler used is the Microsoft Visual C++ that has evolved since the start of the project from the 4.2 to the 6.0 version. The rendering libraries used are the Microsoft Direct X, in particular Direct Draw for 2D bitmaps rendering, and Direct Sound to render the audio nodes. It is built on top of the Iml core software framework that provides MPEG-4 stream demultiplexing, BIFS decoding, scene graph construction, decoders development kit, DMIF development kit, and media buffers management. Implementing an MPEG-4 browser using the Im1 core framework means building an application module (see Fig. 1) that provides the user graphic interface, then implementing a private scene graph attached to the one provided by the core (not necessary but useful as I will explain later), implementing the nodes functionality, and handling overall scene redrawing and audio capabilities. In the following, I will explain the details of the main application module, the scene graph construction and traversal, the presenter main loop, and the visual and audio renderer features. At the time of this writing, work is in progress to align the Im1 2D player to the MPEG-4 Systems Final Draft International Standard (FDIS) spec. This description covers the CD version of the player with some of the changes that already have been made to align to the latest specification.

B. Implementation

1. Application Module

The application user interface makes use of the Microsoft Foundation Class Library (MFC) framework. It is very simple because the creation of a complete graphical user interface falls outside the scope of the project. It consists of a window with a few menu options to start a session by reading a file or start a network session and an option to change the frame rate (from 25 to 5 fps), which affects the number of times the presenter composes the scene. The link between the MFC framework and the Im1 framework is made through a class, CMPEG4PlayerApp, which derives from both the MFC CWinApp and the Core Application class. When the application is launched, the MFC OnInitialUpdate() function creates the Executive object and stores the pointer in the CMPEG4PlayerApp class. When a session is then started, the string indicating the service is passed to the Executive::Start function and the browser then waits for the acknowledgment (OnConnect() function). If the session is correctly started and the BIFS and OD decoders decode, the scene Presenter thread starts composing the scene and rendering, calling the Visual Renderer function. The visual render is initialized with the handle of the application window. User events are passed from the Windows queue of messages to the Executive, which in turns propagates this information to the Presenter that stores the events in a list then processed by the function ProcessEvents.

2. Scene Graph Creation and Traversal

The scene tree, as constructed by the BifsDecoder, consists of a tree of objects that derive from the core MediaObject class. The Render() method of these objects is empty. Attached to every object in this tree is a proxy object. It is these proxy objects that contain the necessary data and methods to perform the traversals of the scene, which are required to render the scene. Every MediaObject BIFS node implemented in the 2D player is derived using the macro OVERLOAD_NODE included in the Im1 framework, which enables a derived node to be instantianted instead of the MediaObject associated with a particular node code. This derived node overloads the method CreateRenderingObject to create a proxy object for this node.

This object contains part of implementation details of the associated BIFS node, in particular all the platform-independent features of this node. The scene is also traversed through these proxy objects. This is extremely useful because it introduces a separation between the BIFS nodes and their implementation details. Because the BIFS nodes files are generated automatically from a template as soon as the standard evolves, the pros of this approach are evident.

All the proxy objects derive from a class, Traversable, that defines the interface the Presenter will use for the scene traversal.

These are the most important methods:

virtual void PreRender (Effects const& effects)

This function will recursively traverse the scene tree, propagating to the leaves the information gathered through the effects struct. For example, a drawable node such as a rectangle will receive its position from the ancestor transforms or the switch on or off state from its switch node parent. All the nodes but the sensors have to implement this function because the default implementation does nothing. The grouping nodes will have to call the function for all children, and the leaves nodes will have to create a rendering context structure containing all the information the visual and audio renderer will need to render the node. This includes the node bounding box, very useful for correctly handling drawable object overlaps, user interaction, and rendering process optimization. During this process every drawable node in this way registers with the Visual Renderer, which will then draw the objects when the prerendering step is finished.

virtual BOOL IsSensor() {return FALSE;}

This is the default implementation; the sensor nodes will return TRUE. This is useful in the prerendering step. The group children are first visited to test whether a sensor is present and then visited through the prerender function containing in its argument the pointer of the sensor found, which will be added to the rendering context of each drawable. In this way, when the user clicks on a object, the object is found using its bounding box contained in the rendering context, and if the sensor node pointer points to a sensor that event is then propagated.

3. *Presenter*

The Presenter thread loops continuously until a open session is closed. It processes user events and calls the Visual Renderer to draw the scene. The Im1 framework supplies the interface of a Presenter class in the class PresenterBase, and the 2D Player overloads this class implementing in particular the Render() method.

This method performs the following steps a number of times in a second, depending on the frame rate chosen by the user. If the frame rate is set to N, the presenter presents the scene every $1000/N$ msec. These are the steps performed.

Process the events received from the Executive. These are typically events caused by user interaction; processing these events may cause the fields of some nodes to be changed as a result of routes being triggered.

Update the TimeSensor nodes in the scene. This action can cause some change in the scene graph.

Lock the scene graph. At this point the scene graph is locked to prevent other threads from changing the scene while the composition is going on.

PreRender the scene, calling the PreRender () method of the root proxy object. Build a list of rendering contexts containing pointers to the graph drawable nodes and composition information as bounding boxes, pointers to TouchSensors, and switch on and off information.

Draw the scene. Call the Visual Renderer to draw all the objects that need to be drawn.

Unlock the scene graph.

The composition at this point of time has finished. Calculate the difference between the time of the next composition and the time elapsed, and sleep if the difference is positive.

Restart.

4. Visual Renderer

The visual renderer performs the rendering of all the graphic objects of a scene. It consists of an abstract class that defines the interface and a derived class that implements the rendering functions using the Microsoft Direct Draw technology. The thread performing the rendering is the presenter thread. The rendering process consists of two steps. In the first phase, called the prerendering phase, the scene graph is traversed to collect all the drawable nodes of the scene in a list owned by the visual renderer that contains all the information needed to perform the drawing of each object. The second step is the real drawing phase; the drawable list is traversed to find the objects that need to be redrawn totally or partially.

This is done in the DrawScene () function that contains the logic of the scene drawing. The purpose of this function is to draw the scene in a optimized way, i.e., redrawing only the part of the screen that has changed ''in respect to the scene before.'' The visual render has a list of drawableContext classes that contains at every screen drawing the list of all the drawables present in the scene graph. During the prerendering step a drawableContext class is created for every drawable node and is then added to the drawableContext list of the visual renderer. During the drawing phase, this list is examined to build the part of the screen that needs to be redrawn and the rect that needs to be cleaned. In order to do so, the process starts from an empty ''needsDrawingRect'' and a empty ''transparentRect.'' A drawable bounding box is added to the needsDrawingRect if its geometry or appearance or position has changed. If this node is transparent (i.e., it does not completely fill its bounding box), its bounding box is also added to the transparentRect.

Then the transparent area is cleaned and the objects are drawn with the needsDrawingRect as input to the draw functions if they need to be redrawn, intersect with the transparent area, or intersect with the area redrawn so far. It is important to note that this process assumes the objects are ordered in a back-to-front order in the visual renderer list. When the objects are traversed in the prerendering phase, they are put in the drawable list in the order in which they are traversed (top left approach). Because the OrderedGroup node has a field that specifies the drawing order of its children, its children are traversed and inserted in the rendering list in the order specified. In this way, it can be assumed that at the end of the prerendering step the objects in the rendering list are in a back-to-front order.

The rectangle to be redrawn is first drawn on a secondary surface synchronously; then the secondary surface is copied on the first in an a-sync way and the DrawScene () function returns.

The browser graphics capabilities include YUV, RGB images display as well as YUVA, RGBA, and basic graphics primitives.

5. Audio Renderer

The audio renderer has been designed to be platform independent. It is defined in terms of an abstract base class (AudioRenderer) that defines the interfaces and classes that derive from this to obtain actual platform-specific implementations. Microsoft Direct Sound Technology has been used in the implementation of a Windows NT audio renderer.

Table 1 Supported Nodes in the 2D Player

Node	Fields, transformations, and appearance supported	Status
AnimationStream	All	Test.
Appearance	Fields are specified for each *geometry* node	Exer.
AudioSource	F: startTime (only initial value), stopTime, url	Exer.
Background2D	F: url	Exer.
Bitmap	F: scale	Exer
Circle	F: radius T: <translation, scaling> A: <diffuseColor, transparency (0.0/1.0), filled, shadow, width>	Exer.
Conditional	F: active, reverseActivate, buffer, isActive	Exer.
ColorInterpolator	F: set_fraction, key, keyValue,	Test.
Curve2D	F: points T: <translation, scaling, rotation> A: <diffuseColor, transparency (0.0/1.0), shadow, width>	Test.
FontStyle	Fields depend on the implementation of the *Text* and *StreamingText* nodes	Exer.
Group	F: children	Exer.
ImageTexture	F: url T: <translation>	Exer.
Inline2D	F: url	Test.
IndexLineSet2D	F: coord T: <translation, scaling, rotation> A: <diffuseColor, transparency (0.0/1.0), shadow, width>	Test.
Layout		
Material2D	Fields are specified for each *geometry* node	Exer.
MovieTexture	All	Test.
OrderedGroup	All	Exer.
PointSet2D	F: Color, coord T: <translation, scaling, rotation> A: <diffuseColor, transparency, shadow>	Test.
PositionInterpolator2D	F: Set_fraction, key, keyValue,	Test.
PlaneSensor2D	F: Enabled, trackPoint_changed	Exer.
Rectangle	F: Size T: <translation, scaling, rotation> A: <diffuseColor, transparency (when filled, not rotated and no shadow, the full range of values are supported, otherwise only 0.0/1.0), filled, shadow, width>	Test.
ScalarInterpolator	F: set_fraction, key, keyValue,	Test.
Shape	F: appearance, geometry	Exer.
Sound2D	F: source	Exer.
Switch	F: choice, whichChoice	Test.
Text	F: string, Length, FontStyle T: <translation> A: <diffuseColor, transparency (0.0/1.0)> FontStyle: <family (only windows font names), size, style,>	Test.
TimeSensor	F: cycleInterval, enabled, loop, startTime, stopTime, cycleTime, fraction_changed, isActive, time	Test.
TouchSensor	F: enabled, hitPoint_changed, hitTexCoord_changed, isActive, isOver, touchTime	Exer.
Transform2D	F: Children The supported transformation fields depend on the implementation of the *drawable* nodes	Exer.
Valuator	All fields implemented	Test.
WorldInfo	F: Info, title	Impl.

The audio renderer runs in its own thread. It is launched from the presenter thread as soon as the scene begins. It performs the rendering of mono and stereo audio samples as the audio decoders attached to the platform produce them. Multiple audio sources are supported, and each source can be a different format (sample rate, bits per sample, and number of channels).

Basically, for each source it gets data from the audio composition buffer and puts it in a Direct Sound secondary buffer set to the output frequency of the decoded data. This is done continuously, filling the circular buffer every fixed number of milliseconds so that the buffer is never empty. All the secondary buffers are then mixed in the primary buffer automatically by the Direct Sound engine. Before starting to play a sound, every secondary buffer is full to three-quarters of its capacity to avoid "playing silence."

The first time the preRender() function of an audio node is called, the audio renderer function AddSourceNode is called so that the node is registered with the audio renderer. This function creates a "streaming object" (class StreamInfo), i.e., a class that contains a Direct Sound secondary buffer and all the functions to access the buffer. This StreamInfo class is then added to the list of secondary buffers the audio renderer has. Every audio node exports a GetData() function to get audio samples from the associate decoder composition buffer. The audio renderer calls the GetData() function of all the registered nodes, asking for a given number of bytes (multiple of the number of bytes required for a single sample), and writes data in the associated secondary buffer using the WriteData() function provided by the StreamInfo interface. Finally, the DataUsed() function of the audio node is called to release the fetched data. The number of bytes returned can be less than that asked for. If fewer bytes were returned to the audio renderer than expected, the audio renderer will ask for more. If zero bytes are returned, the audio renderer will assume that no data can be given and will insert silence.

Figure 5 The 2D player—a screen shot.

6. Nodes Supported

Table 1 lists the nodes supported at the time of writing. For each node the fields implemented are reported, together with the status of the implementation. (Test. means tested, Exer. means exercised, and Impl. means implemented.)

7. Sample

Figure 5 shows a snapshot of a simple MPEG-4 application. An MPEG-4 Video shaped girl is dancing before a waterfall background. This is a very simple MPEG-4 scene composed of two bitmap nodes (one for the background and one for the girl). The user can interact with the scene via a PlaneSensor2D associated with the dancing girl Shape; by clicking on the dancing girl the user can drag her into another position.

V. THE 3D PLAYER

A. Introduction

The structure of the software for the 3D player consists of two separate hierarchies of classes for the "core nodes" and for the "rendering nodes." The structure is coherent with the design choice of isolating the functionality related to the management of the scene description (building and accessing the nodes of the scene graph with their fields) and the functionality related to displaying video and graphics and playing sounds.

Most of the software for the rendering part of the 3D player has been developed by Karl Anders Øygard for Telecom Norway Research (Telenor).

The software has been designed and implemented to work on different platforms: personal computers with Windows or LINUX and workstations with UNIX. This goal has been achieved adopting OpenGL as the library for rendering video together with 2D and 3D graphics objects. Because implementations of OpenGL are available for Windows, LINUX, and UNIX, the same rendering code can easily be reused on all platforms. Concerning audio, the implementation for the PC platform relies on DirectSound, which is specific to Windows and hence would require rewriting the audio rendering code to port on a different sound library.

The two class hierarchies for the nodes are perfectly symmetrical, with grouping of the VRML and MPEG-4 nodes according to their common functionality. The top of the hierarchy is the MediaObject class (with its corresponding MediaObjectProxy class for rendering), from which all other VRML and MPEG-4 nodes derive. Other intermediate base classes have been added, clustering together nodes with common utilities such as Lights, Textures, Sensors, and Grouping Nodes. The following scheme depicts the common class hierarchy for "core" and "rendering" nodes; each class such as Box has a corresponding rendering class BoxProxy.

```
MediaObject
  Bindable
    Viewpoint
    WorldInfo
    ....
  DragSensor
    CylinderSensor
    PlaneSensor
    ...
```

```
GeometryNode
  Box
  Sphere
  ...
InterpolatorNode
  CoordinateInterpolator
  PositionInterpolator
  ...
Light
  DirectionalLight
  PointLight
  ...
TextureNode
  ImageTexture
  MovieTexture
  ...
GroupNode
  Collision
  Group
  ...
```

B. Composition and Rendering

Two phases are required for displaying video and audio objects: composition (traversing the scene graph to compute the transformations and other attributes of the scene nodes) and rendering (calling the graphics and audio libraries to translate the nodes into rendered objects). In the 3D player for Windows, the overall application is based on the standard structure of an MFC user interface: from a single user window a file containing an MPEG-4 multiplexed stream can be opened and played. The classes are named after the acronym for man–machine interface (MMI):

CMmiApp
CMmiView
CMmiDoc

The other classes, specific to the 3D player, involved in the rendering process are those common to the platform-independent part of the application and the derived classes that customize the rendering to the specific libraries (OpenGL and DirectSound):

Executive
Presenter (OpenGLPresenter)
VisualRenderer (OpenGLVisualRenderer)
AudioRenderer (DirectSoundAudioRenderer).

The mechanism controlling composition and rendering to update the playback of the scene is based on a Windows timer. The timer is initialized in the CMmiView: :OnInitialUpdate method to fire an event every 40 msec. The WM_TIMER event is handled by the method CMmiView: :OnTimer, which removes the event from the event queue and calls the method CMmiView: :PaintScene. This method checks the clock time actually elapsed from the beginning to the end of the rendering cycle and calls in turn the Executive: :Paint method. The method calls OpenGLPresenter: :Paint, implemented in the class that specializes the generic presenter class.

The following excerpts in pseudocode describe this mechanism.

```
void CMmiView::PaintScene()
{
  DWORD timeBefore = GetTickCount();
  if (m_nTimer != NULL)
    Executive::Paint(pvInfo);
  DWORD timeAfter = GetTickCount();
}

void Executive::Paint(void * pvInfo)
{
  OpenGLPresenter::Paint(pvInfo);
}

void OpenGLPresenter::Paint(void * pvInfo)
{
  OpenGLVisualRenderer::Paint((RECT *) pvInfo);
  m_dwLoop++;
}

void OpenGLVisualRenderer::Paint(RECT * poRect)
{
  Executive::LockScene();
  MediaObject * poScene = Presenter::GetScene();
  if (poScene != NULL)
  {
    DWORD dwTime = Presenter::GetTime();
    TimeSensorProxy::UpdateTimeSensors();
    if (poActiveViewpoint != NULL)
    if (poActiveBackground != NULL)
      BackgroundProxy::RenderBackground(this, dwTime);
    if (poActiveFog != NULL)
      FogProxy::Enable(this, dwTime);
    Matrix oTransform;
    Viewpoint::ViewpointToWorldMatrix(dwTime, oTransform);
    ::glLoadMatrixf(oTransform);
    MediaObjectProxy::RenderLights(this, dwTime);
    CullFlag oCullFlag = PARTIALLY_INSIDE;
    poScene->CullOrRender(oTransform, oCullFlag, this, dwTime);
  }
  Executive::UnlockScene();
  VisualRenderer::Paint(poRect);
}
```

The rendering of all nodes in the scene graph is completed by recursively calling the CullOrRender method in the MediaObjectProxy objects. If the scene object is completely or partially inside the viewing volume (frustum), the MedaObjectProxy: :Render method is called; otherwise the method returns without any rendering.

```
BOOL MediaObjectProxy::CullOrRender(
    const Matrix & roTransform,
    const CullFlag & roCullFlag,
    OpenGLVisualRenderer * poRenderer,
    DWORD dwTime)
{
```

```
  if (roCullFlag == COMPLETELY_INSIDE)
    return Render(roTransform, roCullFlag, poRenderer, dwTime);
  else
  {
    CullFlag oCullFlag = InsideFrustum(roTransform, poRenderer);
    if (oCullFlag != COMPLETELY_OUTSIDE)
      return Render(roTransform, roCullFlag, poRenderer, dwTime);
  }
  return TURE;
}
```

The three phases of construction, rendering, and destruction of the objects involved in the 3D player operation are depicted schematically in Fig. 6. The objects belong to the following classes:

CMmiView: The view in the active window where the result of rendering is displayed

CMmiDoc: The object associated with the currently active bitstream

Executive: The ‘‘master’’ object of the platform-independent application engine

Presenter (OpenGLPresenter): The object controlling the presentation of visual and audio objects, with core functionality and library specific functionality separated in the base and derived classes

VisualRenderer (OpenGLVisualRenderer): The object in charge for rendering video and graphics

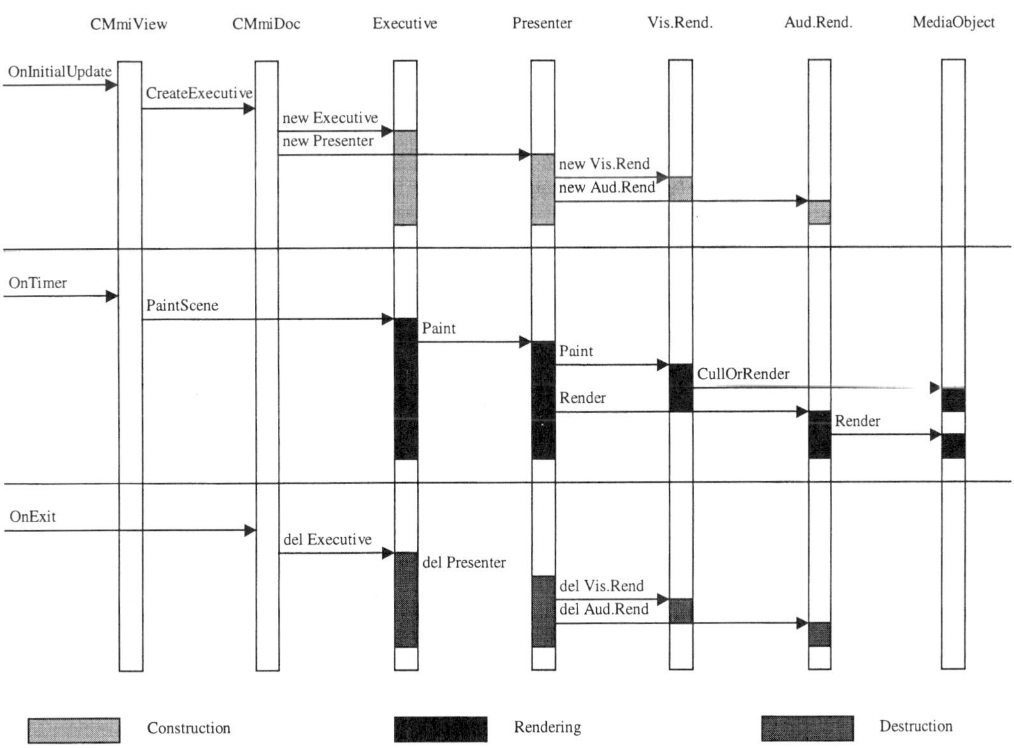

Figure 6 Time diagram of construction, rendering, and destruction of objects in the 3D player.

AudioRenderer (DirectSoundAudioRenderer): The object in charge for rendering audio
MediaObject (MediaObjectProxy): The single scene object, either video or audio

C. Sample Scene

This section described the "look and feel" of the 3D player with a sample scene, produced by Guillame (Ananda) Allys for France Telecom Research (CNET). The user interaction supports navigation in the scene by click and drag with the left mouse button and triggering of events on sensors with the right mouse button.

Figure 7 shows a screen shot of the 3D player with a scene representing a virtual room with three panels on the back wall, on which the same video sequence is mapped. The table at the center of the room has a blue 3D button (realized as a box with an associated sensor). When the user clicks on the 3D button, a TV set "falls" from the ceiling and another video sequence starts being decoded and mapped on the TV screen.

The sample scene provides a typical demonstration of how VRML objects (providing the 3D virtual environment) can be integrated with streaming media (the sequences on the panels and on the TV screen). Another significant feature is the possibility of synchronizing the streaming of information related to the scene (the scene description at the beginning of the multiplexed stream and the possible updates of the scene) together with streaming media and with user-generated events that modify the scene and/or trigger the start of other streaming media. It should be noted that user interaction with the scene has different effects according to the application scenario.

In a "push" system, in which MPEG-4 data is continuously streamed from a source and the user interaction at the receiver side cannot be fed back the transmitter side, the effect of adding other subscenes or media is necessarily synchronized with the currently available streaming information. The only way to provide an interactive feel is to stream data continuously from the transmitter side, eventually using a "carousel" technique for providing random access points at relatively short time distances.

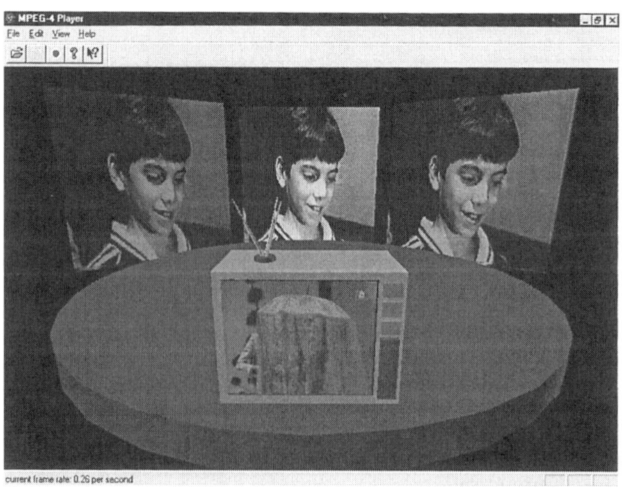

Figure 7 Screen shot of the 3D player (K. A. Øygard) with the scene "Salle.mpeg4" (A. Allys).

In a ''pull'' system, user interaction is fed back to the transmitter side and actually controls the insertion of a subscene or of additional media streams. This is the case of the sample scene presented here, where data is read from local storage.

VI. THE DMIF STACK

A. DMIF Architecture

The IM1 software platform adopts the DMIF model and implements its key aspects. Further value-added modules are also integrated within the IM1 platform but are not considered an integral part of it and are thus subject to different availability conditions. A comprehensive scheme of the modules integrated in the IM1 platform is shown in Fig. 8, where the light gray area defines the borders of a terminal, the dark gray area defines the borders of a process, and each box corresponds to a separate software module (library or application).

The DMIF filter implementation is the key element to support the DMIF model. The model assumes that the application provides a URL to indicate the desired service and that the DMIF filter inspects the URL in order to identify the correct DMIF instance in charge of providing that service. Further flows through the DAI are then (conceptually) directly managed by the identified DMIF instance.

The IM1 implementation of the DMIF filter defines the DAI as a set of methods of two classes, one to be implemented in an object in the DMIF filter, the other to be implemented in a sink in the application, for callbacks.

The correct DMIF instance is identified with the help of the Windows registry. All available DMIF instances are registered in a well-known branch of the registry; whenever

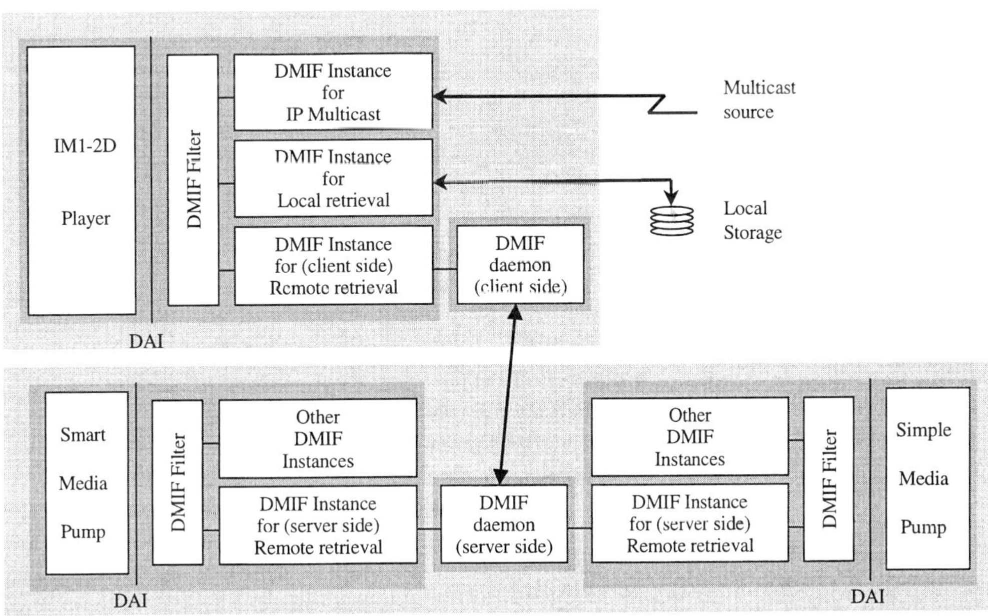

Figure 8 Software modules in the DMIF implementation integrated in IM1.

a ServiceAttach call is issued by the application, the DMIF filter looks at the registry and loads one DLL at a time, tests it against the URL identifying the requested service, and decides whether to unload or use it. Each DLL represents a DMIF instance implementation and provides a particular method to test whether or not it is able to access a service as identified by the URL. In this way, new DMIF instances can be added at run time in a system, or just updated, and be activated seamlessly through appropriate URLs.

The interface between the DMIF filter and the various DMIF instances is realized as well by means of a couple of class definitions, one for each direction of the flow. In particular, the interface defines a base class from which all DMIF instances inherit their own specific derived class. The derived class characterizes the DMIF instance, but the DMIF filter invokes only the methods defined in the base class. This allows the DMIF filter to control DMIF instances that are not known in advance and thus ensures one key feature of the DMIF architecture, that is, to allow full independence of an application from the delivery technology used.

As a side note, Winsock2 does not provide the same level of independence: applications using Winsock2 are aware of the protocol stack they select and cannot be used without changes to work on top of protocol stacks not supported initially.

B. DMIF Instances in IM1

Three DMIF instances have been integrated in the IM1 platform, with the goal of validating the key concept of the DMIF model, that is, the ability to hide delivery technologies as well as operating scenarios from the application making use of the DMIF-application interface. Thus, an instance for each scenario (local retrieval, broadcast, and remote retrieval) has been developed and successfully integrated.

1. The Local Retrieval Scenario

A DMIF instance for local retrieval scenarios has been developed to read from a file in ''Trivial File Format.'' This instance has been widely used for tests and demos, but will soon be replaced by a DMIF instance that reads files in ''MPEG-4 File Format.'' This instance is implemented as a single DLL.

2. The Broadcast Scenario

A DMIF instance has been integrated in IM1 to work with Internet Protocol (IP) multicast. This instance does not represent any ''official'' use of MPEG-4 over IP multicast, which is still under study, and has adopted a number of solutions that will probably differ from the still-to-come standardized mechanism. In particular, this instance obtains the stream map table and the initial object descriptor for a particular service through the use of a Web server (via Hyper Text Transport Protocol, HTTP), whereas the current direction is to extend the session description protocol (SDP) to provide that information in a suitable way. The implemented solution does not support dynamic addition of elementary streams within the same service session. This instance is implemented as a single DLL but in its current form also assumes the presence of a Web server and of well-formatted files to represent the stream map table and the initial object descriptor.

3. The Remote Retrieval Scenario

The retrieval scenario implementation is by far the most complex, as it also includes a server part and a connection management entity.

With reference to Figure 8, the implementation integrated in IM1 is made of the following modules:

DMIF instance for (client-side) remote retrieval (DLL loaded in the IM1 player process space)
DMIF daemon for connection management (client side)
DMIF daemon for connection management (server side)
DMIF instance for (server-side) remote retrieval (DLL loaded in one application server process space)
DMIF filter for the server side
Server applications (a simple media pump, for basic play/stop commands, and a smart media pump, for additional stream controls, e.g., pause, fast Forward)

This implementation provides hooks for quality of service management. Aspects that belong to the QoS management module are the monitoring of the objective QoS delivered (especially when operating over the Internet: in this case implementations are in progress making use of RTP/RTCP) and the decision on how to aggregate elementary streams into flexmux streams. A number of activities are in progress, in contexts other than IM1, to implement meaningful strategies for aggregating elementary streams, especially for IP networks.

Activity is also in progress to unify the client and server interfaces, as well as the client and server DLLs and daemons, thus allowing the exploitation of the IM1 software for applications based on MPEG-4 but other than a browser (e.g., conferencing applications, where a terminal acts as both a receiver and a transmitter).

VII. SUMMARY

In this chapter, team members of the IM1 project presented the software writing aspect of the MPEG-4 standard development process. It contains a brief and simplified description of some parts of the player developed by the group. We have tried to encompass the most important issues, but obviously, because of shortage of space, many aspects could not be covered. Engineers who want to benefit from the availability of the IM1 source code in the public domain, as part of the standard specifications, would probably need to read longer official documents.

MPEG-4 is in FDIS (Final Draft International Standard) stage as this chapter is written. This means we cannot give references to official ISO publications, because these will be published only when the specifications reach IS (International Standard) state. This also implies that by the time you read this the following references might have been superseded by newer versions. Moreover, as happens often, software development runs ahead of proper documentation, and the IM1 project was not an exception. Many development details came as result of internal discussions, and therefore the documents cited here might be outdated compared with the status of the software or even this chapter. Having said that, it is still useful to have pointers to a few MPEG documents.

The IM1 software is an implementation of the MPEG-4 Systems specifications [1]. The DMIF module implements MPEG-4 Delivery [2]. The player software and its basic documentation are also available, rather outdated [3]. The BIFS encoder and Multiplex tools, as well as the Decoder, IPMP, and DMIF APIs, are all packaged and described in one document [4].

REFERENCES

1. ISO/IEC JTC1/SC29/WG11 N2501. Text for ISO/IEC FDIS 14496-1 Systems, December 1998.
2. ISO/IEC JTC1/SC29/WG11 N2506. Text for ISO/IEC FDIS 14496-6 Delivery, January 1999.
3. ISO/IEC JTC1/SC29/WG11 N2805. Text for ISO/IEC FDIS 14496-5 Reference Software, July 1999.
4. ISO/IEC JTC1/SC29/WG11 N2523. IMI Tools, October 1998.

17
Multimedia Transport in ATM Networks

Daniel J. Reininger and Dipankar Raychaudhuri
C&C Research Laboratories, NEC USA, Inc., Princeton, New Jersey

I. INTRODUCTION

Emerging multimedia content on the Internet and digital video applications, such as video distribution, videoconferencing, remote visualization, and surveillance, are imposing ever more demanding performance goals for network service providers. The traffic profile of these multimedia applications is highly nonstationary and require large, heterogeneous, and dynamic network bandwidth allocation. In addition, wireless mobile computing technology (e.g., wireless broadband networks and embedded computing and communication devices) currently under active development will require network service providers to support a wider range of services, coverage areas, and terminals with various types of quality of service (QoS) requirements, while maintaining reasonably high network capacity utilization.

Efforts to standardize efficient mechanisms for multimedia transport have been under way for some time at networking standard bodies. For asynchronous transfer mode (ATM) technology, the framework for providing QoS is described in the ATM Forum's traffic management specification [1]. This framework can support static QoS requirements of real-time traffic based on per-stream bandwidth reservation set during connection setup phase. Realizing that multimedia traffic is highly nonstationary and that many multimedia applications can adapt to sporadic drops in bandwidth availability, we have developed a flexible and efficient framework for QoS control in ATM networks, called soft-QoS with VBR^+. This framework complements the traditional traffic management model for real-time variable bit rate (VBR) traffic supported by ATM. It allows source-initiated and network-initiated bandwidth renegotiation based on the concept of a connection's softness profiles. The softness profile represents the sensitivity of application performance to the connection's network bandwidth allocation. As a result, the network can achieve high utilization and maintain satisfactory application performance even during network congestion.

Because only transport-level parameters are renegotiated, the framework can be applied independently of the network layer. Internet Protocol (IP) networks and ATM networks can meet the soft-QoS requirements of an application via different mechanisms. For example, whereas Internet can use the Resource Reservation Protocol (RSVP) for bandwidth reservation [2], ATM uses Q.2963 [3] signaling. Therefore, the soft-QoS model can be used on both Internet and ATM broadband networks provided access network elements implement per-flow or per-Virtual Circuit (VC) bandwidth reservation mechanisms.

The chapter has eight sections. Section II describes the system model for computer-based multimedia applications over ATM. Because video is a critical component of multimedia applications, Section III outlines issues important to realize flexible, robust, and efficient video transport over ATM networks. Section IV summarizes the existing ATM QoS model. Section V describes the use of dynamic bandwidth allocation for multimedia traffic. Section VI introduces the soft-QoS model for multimedia applications. Section VII explains how dynamic bandwidth allocation and the soft-QoS concept can be combined to provide a dynamic framework for soft-QoS control within the ATM network infrastructure. Finally, Section VIII concludes the chapter with a summary and a brief discussion of how soft-QoS can be supported in future broadband networks that will integrate ATM and IP QoS mechanisms.

II. SYSTEM MODEL

A system model for multimedia transport in ATM networks is shown in Figure 1. The system consists of a client terminal that connects to a multimedia server via a broadband ATM network. Each of these elements can be viewed as having a separated path for data flow and control. Modules along the data flow operate continuously and at high speed, and the control modules operate sporadically, monitoring the underlying data flow and coordinating the allocation of system resources with other controllers along the connection's data flow path. These resource controllers are distributed objects at the server, the network, and the client that communicate via a binding software architecture.

At the client's terminal the data flow goes through a buffer and decoding and display modules. The buffer stores the network's packets until the decoder is ready to process them. In general, the decoder has an error concealment front end that detects packet losses and conceals them, utilizing information from previously decoded frames. Once a video frame is decoded, it is passed to the display module for presentation. The capabilities of these modules vary considerably depending on the terminal's architecture. Some terminals have hardware support for decoding and display, whereas others use a software-based

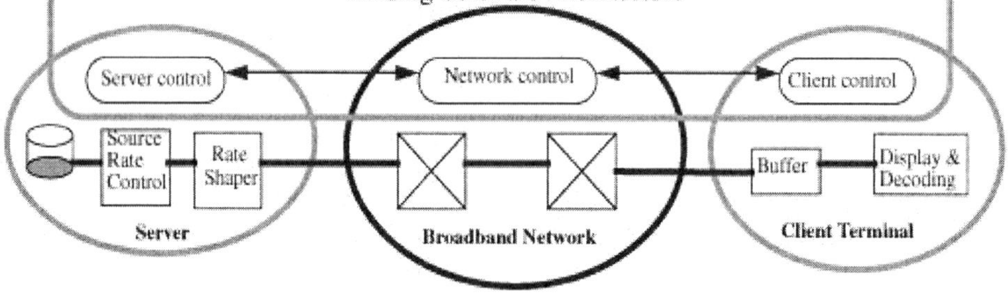

Figure 1 System model for multimedia transport on ATM networks.

approach on general-purpose computer processors. The client control takes into consideration the terminal's capabilities when specifying the service requirements to the server and the requirements of the application. Once the terminal and the application requirements are defined, the client control includes them in a software object, called a "QoS contract" [4]. The client control sends the QoS contract to the server via the binding software architecture.

Among the server's data flow modules are the source rate control and the rate shaper. The source rate control generates scalable multiquality video by adjusting the video bitstream's rate to match the available bandwidth. The rate shaper, at the interface between the server and the network, shapes the bitstream to ensure that the stream rate fits within the traffic descriptor allocated by the network.

The control plane in Figure 1 coordinates the use of system resources. The server control attempts to maintain the client's contracted QoS via bandwidth renegotiation with the network control. The network control is responsible for connection admission and bandwidth allocation at the network's switches. Within the VBR^+ traffic class, the control dynamically receives requests for bandwidth renegotiation from servers and allocates bandwidth along the connections' path based on available resources. If at a given time the control algorithm determines that a connection's bandwidth request cannot be fully granted, it will reallocate bandwidth using a novel QoS-based optimization criterion. This approach significantly differs from that of the admission and allocation models currently in practice in two main aspects. First, in current models requests are rejected if they cannot be fully granted; second, current allocation strategies strive only to achieve bandwidth fairness, which can lead to large performance variance among users. In the proposed framework, the applications' softness profiles make it possible to perform a bandwidth allocation sensitive to users' satisfaction.

III. VIDEO TRANSPORT OVER ATM NETWORKS

Because video is a critical component of multimedia content, in this section we outline technical approaches necessary to realize flexible, robust, and efficient video transport over ATM networks. The overall system design problem is discussed for a general distributed multimedia computing scenario and key video-related design issues are identified. A more detailed view of video transport issues in ATM networks is given in Refs. 5 and 6.

Core technologies for broadband networking and digital video have matured significantly during the past few years. MPEG standards [7] for video compression are being used successfully in several important application scenarios. Asynchronous transfer mode products following the ATM Forum [8] specifications are being used to transport digital TV and broadband multimedia applications.

However, because of the wide range of possible application scenarios, no uniform approach for video transport in broadband networks exists. Each application scenario [e.g., TV broadcasting, videoconferencing, video on demand (VOD), multimedia computing] will tend to have different design criteria, as summarized in Table 1. For example, the de facto MPEG-over-ATM solution for the TV broadcasting and VOD scenarios [9] calls for constant bit rate (CBR) MPEG with MPEG-2 systems [10] on the codec side, with CBR–ATM service via an existing ATM adaptation layer (AAL) protocol (such as AAL5) on the network side. Although this method may be acceptable for broadcasting and certain VOD applications, further work is required to identify a more flexible and efficient video transport framework for multimedia computing over broadband networks.

Table 1 Application Scenarios and Key Design Criteria for Video on Broadband Networks

Application Scenario	Key design criteria
TV–HDTV broadcasting	High picture quality
	High cell loss resilience
	Low receiver cost
Videoconferencing	Low interactive delay
	High bandwidth efficiency
	Low codec cost
Video on demand (VOD)	High bandwidth efficiency
	Service flexibility
	Low decoder cost
Distributed multimedia computing	Low interactive delay
	Service flexibility and scalability
	Software integration

These are some of the important design objectives:

1. Video quality versus bit rate characteristics necessary for cost-effective applications
2. High overall network efficiency for integrated services including video
3. Low end-to-end application delay, particularly for interactive applications
4. Hardware and software throughput appropriate for full-quality and/or multistream video
5. Integration of video into software framework, including quality-of-service support
6. Scalable system capable of working with various bandwidth–QoS, CPU, speeds, displays
7. Graceful degradation during network congestion or processing overload

Effective system design for network video requires prior consideration of each of these objectives when selecting subsystem-level technical approaches.

IV. ATM QoS MODEL

In this section we briefly summarize the existing ATM QoS model. Five service categories have been defined under ATM [11]. These categories are differentiated according to whether they support constant or variable rate traffic and real-time or non–real-time constraints. The service parameters include a characterization of the traffic and a reservation specification in the form of QoS parameters. Also, traffic is policed to ensure that it conforms to the traffic characterization and rules are specified for how to treat nonconforming traffic. ATM provides the ability to tag nonconforming cells and specify whether tagged cells are policed (and dropped) or provided with best-effort service.

Under UNI 4.0, the service categories are constant bit rate (CBR), real-time variable bit rate (rt-VBR), non–real-time variable bit rate (nrt-VBR), unspecified bit rate (UBR), and available bit rate (ABR). The definition of these services can be found in Ref. 11. Table 2 summarizes the traffic descriptor and QoS parameters relevant to each category within traffic management specification version 4.0 [1]. Here, the traffic parameters are

Table 2 ATM Traffic and QoS Parameters

	ATM service category				
Attribute	CBR	Rt-VBR	Nrt-VBR	UBR	ABR
Traffic parameters					
PCR and CDVT	Yes	Yes	Yes	Yes	Yes
SCR and MBS	N/A	Yes	Yes	N/A	N/A
MCR	N/A	N/A	N/A	N/A	Yes
QoS Parameters					
CDV	Yes	Yes	No	No	No
Maximum CTD	Yes	Yes	No	No	No
CLR	Yes	Yes	Yes	No	No

peak cell rate (PCR), cell delay variation tolerance (CDVT), sustainable cell rate (SCR), maximum burst size (MBS), and minimum cell rate (MCR). The QoS parameters characterize the network-level performance in terms of cell loss ratio (CLR), maximum cell transfer delay (max CTD), and cell delay variation (CDV). The definition of these parameters can be found in Ref. 1.

Functions related to the implementation of QoS in ATM networks are usage parameter control (UPC) and connection admission control (CAC). In essence, the UPC function (implemented at the network edge) ensures that the traffic generated over a connection conforms to the declared traffic parameters. Excess traffic may be dropped or carried on a best-effort basis (i.e., QoS guarantees do not apply). The CAC function is implemented by each switch in an ATM network to determine whether the QoS requirements of a connection can be satisfied with the available resources. Finally, ATM connections can be either point to point or point to multipoint. In the former case, the connection is bidirectional, with separate traffic and QoS parameters for each direction; in the latter case it is unidirectional. In this chapter, we consider only point-to-point connections for simplicity.

V. DYNAMIC BANDWIDTH ALLOCATION

The bit rates of multimedia applications vary significantly among sessions and within a session due to user interactivity and traffic characteristics. Contributing factors include the presence of heterogeneous media (e.g., video, audio, and images), compression schemes (e.g., MPEG, JPEG), presentation quality requirements (e.g., quantization, display size), and session interactivity (e.g., image scaling, VCR-like control). Consider, for example, a multimedia application using several media components or media objects (such as a multiwindow multimedia user interface or future MPEG-4 encoded video). These applications allow users to vary the relative importance of presence (IoP) of a given media object to match the various viewing priorities. In this case, there is a strong dependence of user–application interaction on the bandwidth requirements of individual media components. In addition, multimedia applications may require changing the level of detail (LoD) of media components dynamically during a session. For example, some applications may require more detailed spatial resolution, whereas others may trade off spatial resolution for higher display rate. When a user enlarges a video window during an interactive session rather than using the terminal processing resources to enlarge the low-resolution frames locally, it may be preferable to request a video bitstream with larger image resolution from the server. However, when the network is congested and/or the application migrates to a mobile personal terminal with wireless connectivity, the server has to reduce the stream's rate to comply with the network's limitations. In these scenarios, matching the target quality of applications, processing limitations of terminals, and capacity limitations of networks requires video sources able to scale their quality dynamically. A suitable network service for these applications should support bandwidth renegotiation to achieve high network utilization and simultaneously maintain acceptable performance.

An example of video traffic generated as the user changes the required LoD is shown in Figure 2. As shown, a suitable network service needs to provide bandwidth on demand. However, in current broadband networks bandwidth allocation is done at connection setup. For example, the variable bit rate (VBR) service class proposed for broadband networks requires connections to specify a usage parameter control (UPC) consisting of peak rate, burst length, and sustained rate [12,13]. The UPC is to be declared at call setup time and remains effective for the duration of the call. As the video VBR process is nonstationary due to changes in long-term scene activity and interactivity, it is generally not possible to find a single UPC that results in uniform video quality for the entire duration of the session without significantly overdimensioning the initially declared UPC. A new service class, called VBR$^+$, has been proposed [14,15] to solve the practical difficulties experienced in maintaining QoS and achieving significant statistical multiplexing gains for video and multimedia traffic. This class provides all the functionalities of the traditional VBR service class, with the added capability of dynamic bandwidth renegotiation between the server and the network during a session.

As network architectures evolve, some degree of programmability will be available to implement more effective dynamic resource allocation strategies. Initial support may be deployed on the access network, where most statistical multiplexing gain can be exploited. There, the combination of specialized media processing gateways with control–protocol function servers can be used to support VBR$^+$ transport.

Other contributions also emphasize the need for dynamic bandwidth reservation in ATM networks. In Ref. 16 a fast buffer reservation protocol is proposed as a means of

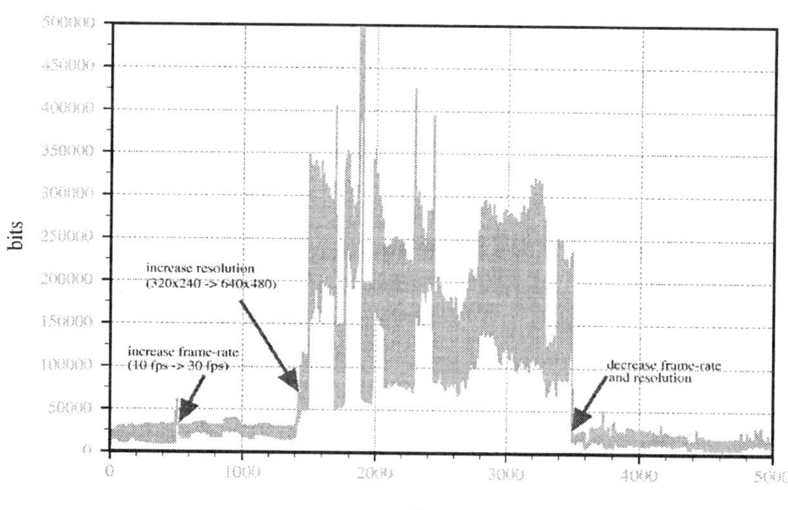

Figure 2 Video bit rate changes on an interactive multimedia session.

handling sources with large peak-to-link rate ratio. However, video sources mostly need bandwidth renegotiation to accommodate medium- to long-term burstiness caused by user interactivity and variable scene content. Short-term burstiness can be described sufficiently well by the parameters in the traffic descriptor, avoiding the signaling and processing overhead of fast reservations. A dynamic bandwidth allocation scheme that uses integer multiples of a basic reservation unit is proposed and analyzed in Ref. 17. The paper shows that the basic reservation unit has a significant impact on resource usage in terms of network bandwidth and processing requirements, but the impact of the scheme on QoS was not addressed.

The effect of bandwidth renegotiation on network utilization and QoS was addressed using VBR MPEG video in Ref. 18 and using CBR traffic in Ref. 19. The results show the impact of renegotiation blocking on delay and loss. More recently, Ref. 20 presented similar results based on analysis and simulations.

VI. SOFT-QoS MODEL FOR MULTIMEDIA

The notion of quality of service in distributed multimedia systems is associated with the provision of adequate resources for *acceptable* application performance. Distributed multimedia applications have a wide range of QoS requirements. For example, the network capacity needed to maintain acceptable *application-level* QoS depends on user performance requirements and the robustness of the application to bandwidth outage. Thus, the network requirements of multimedia applications significantly change depending on user requirements, session interactivity, and application softness. Existing network services for distributed multimedia either do not support the QoS requirements of applications (best-effort service) or provide only network-level QoS measured in terms of traffic engineering parameters. These parameters do not convey application-specific needs. As a result, the network does not consider the sensitivity of application performance to bandwidth allocation. There is an architectural gap between the provision of network-level QoS and the actual QoS requirements of applications. This gap causes distributed multimedia applications to use network bandwidth inefficiently, leading to poor end-to-end system performance.

Multimedia applications have a wide range of bandwidth requirements, but most can gracefully adapt to sporadic network congestion while still providing acceptable performance. This graceful adaptation can be quantified by a *softness profile* [21]. Figure 3 shows the characteristics of a softness profile. The softness profile is a function defined on the scales of two parameters: satisfaction index and bandwidth ratio. The satisfaction index is based on the subjective mean opinion score (MOS), graded from 1 to 5; a minimum satisfaction divides the scale into two operational regions: the acceptable satisfaction region and the low-satisfaction region. The bandwidth ratio is defined by dividing the current bandwidth allocated by the network to the bandwidth requested to maintain the desired application performance. Thus, the bandwidth ratio is graded from 0 to 1; a value of 1 means that the allocated bandwidth is sufficient to achieve fully the desired application performance. The point indicated as B is called the critical bandwidth ratio because it is the value that results in a minimum acceptable satisfaction. As shown in Figure 3, the softness profile is approximated by a piecewise linear "S-shaped" function, consisting of three linear segments. The slope of each linear segment represents the rate at which the application's performance degrades (satisfaction index decreases) when the network

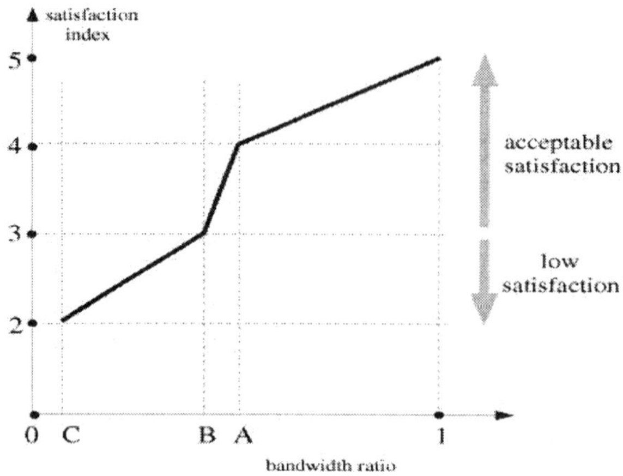

Figure 3 Softness profile.

allocates only a portion of the requested bandwidth: the steeper the slope, the ''harder'' the corresponding profile.

Applications can define a softness profile that best represents their needs. For example, the softness profile for digital compressed video is based on the nonlinear relationship between coding bit rate and quality, and the satisfaction index is correlated with the user perception of quality [22,23]. Although video-on-demand (VoD) applications may, in general, tolerate bit rate regulations within a small dynamic range, applications such as surveillance or teleconference may have a larger dynamic range for bit rate control. Other multimedia applications may allow a larger range of bit rate control by resolution scaling [4]. In these examples, VoD applications are matched to a harder profile than the other, more adaptive multimedia applications. Users on wireless mobile terminals may select a softer profile for an application in order to reduce the connection's cost, whereas a harder profile may be selected when the application is used on a wired desktop terminal.

VII. DYNAMIC FRAMEWORK FOR SOFT-QoS CONTROL

A QoS control framework that allows dynamic renegotiation of transport-level parameters is key to supporting application-level QoS. The framework should provide the ability to specify required bandwidth and application softness dynamically. The softness profile allows an efficient match of application requirements to network resource availability. With the knowledge of the softness profile, network elements can perform soft-QoS control that leads to QoS-fair allocation of resources among contending applications when congestion arises.* In this section we provide an overview of selected issues on the framework's implementation. More details and experimental results can be found in the cited references.

* Sporadic congestion is inevitable for a highly utilized broadband network carrying multimedia traffic, as this traffic is highly bursty and nonstationary.

A. Soft-QoS Software Model

Figure 4 shows the system and application programming interfaces (APIs) for dynamic QoS control with bandwidth renegotiation. The API between the adaptive application and the QoS control module is dynamic; i.e., its parameters can be modified during the session. For example, the Heidi soft-QoS API [4] allows applications to control dynamically high-level, application-specific, parameters for individual media components in a multimedia application. The QoS control module associates a channel ID and media type with each media component. When the media type is video, typical control functionality provided includes VCR commands (play/pause/ff/rew/stop), detail (quantization and resolution), and frame rate control. In addition, each media component can specify a *softness profile* and/or a *priority index* to characterize the dependence between media quality and network bandwidth. Through the same API, the QoS control module reports the *session cost* to the application and issues *alerts* when it cannot maintain the contracted QoS.

The QoS control module provides access to network services via a standard network API such as the Winsock 2 API in Microsoft Windows [24]. This API allows dynamic changing of the network QoS parameters such as the traffic descriptor as well as loss and delay requirements. In this framework, a new transport service provider implements the VBR$^+$ service to track the bit rate requirements of the bitstream. As shown in Figure 4, the VBR$^+$ service provides "bandwidth on demand" required for multimedia services.

B. ATM Signaling Extensions for Soft-QoS Control

ATM user network interface (UNI) signaling extensions are used to support dynamic bandwidth management. Figure 5 shows the messages, with appropriate information elements (IEs), used for dynamic bandwidth management.

These extensions are based on ITU-T recommendations for ATM traffic parameter modification while the connection is active [3]. Although these procedures are not finalized

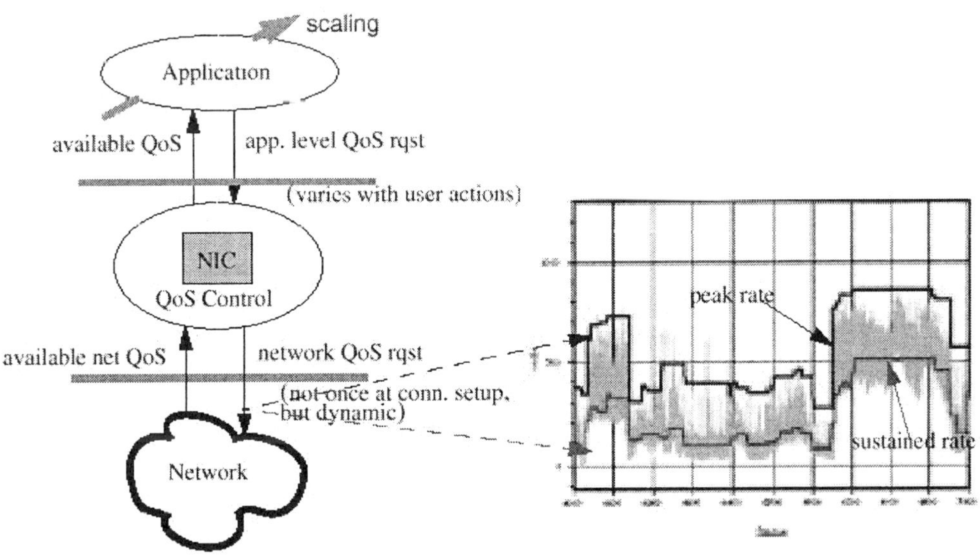

Figure 4 Dynamic framework for QoS control.

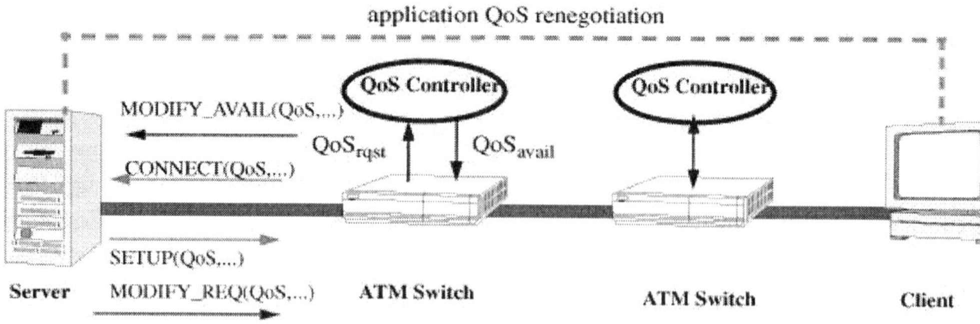

Figure 5 Dynamic bandwidth management within the ATM model.

at the time of this writing, an overview of the current state of the recommendation is given next.

The IETF Integrated Services Architecture (ISA) could support dynamic bandwidth management through RSVP [2]. RSVP allows senders to modify their traffic characterization parameters, defined by a *Sender TSpec*, and cause the receiver to modify its reservation request. As shown in Figure 6, the server uses the PATH message to request a bandwidth reservation along the path to the client. Each QoS controller on the IP routers along the path modifies the PATH message to reflect its resource availability. When the PATH message is received by the client, it places the resulting QoS and bandwidth availability on the RESV message and sends it upstream to the server.

1. *ITU-T Q.2963 Signaling*

ITU-T Q.2963 allows all three ATM traffic parameters, PCR, SCR and MBS, to be modified during a call. All traffic parameters must be increased or decreased; it is not possible to increase a subset of the parameters while decreasing others. Traffic parameter modification is applicable only to point-to-point connections and may be requested only by the terminal that initiated the connection while in the active state.

The following messages are added to the UNI:

MODIFY REQUEST is sent by the *connection owner* to request modification of the traffic descriptor. The bandwidth request is carried in the ATM *traffic descriptor* IE.

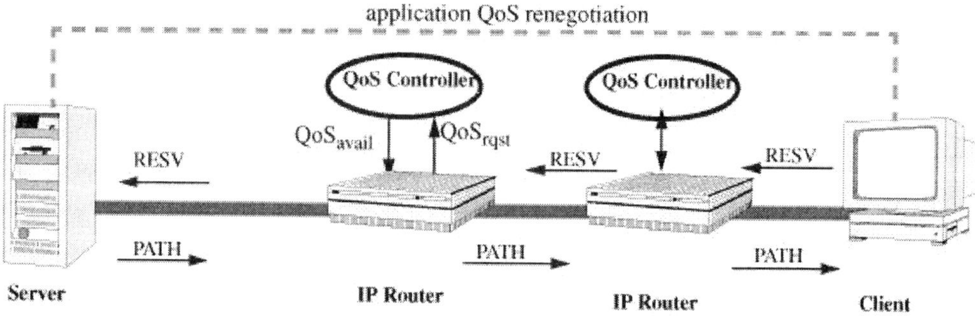

Figure 6 Dynamic bandwidth management within the Internet model.

MODIFY ACKNOWLEDGE is sent by the *called user* or *network element* to indicate that the modify request is accepted. The *broadband report type* IE is included in the message when the called user requires confirmation on the success of modification.

CONNECTION AVAILABLE is an optional message issued by the *connection owner* to confirm the connection modification. The need for explicit confirmation is indicated in the *modification confirmation* field in the MODIFY ACKNOWLEDGE broadband report IE.

MODIFY REJECT is sent by the *called user* or *network element* to indicate that the MODIFY REQUEST message is rejected. The cause of the rejection is given in the *cause* IE.

Figures 7 and 8 illustrate the use of these messages in typical signaling scenarios.

2. Q.2963 Extensions for Soft-QoS

In addition, the soft-QoS control framework requires the following extensions to the Q.2963 signaling mechanisms just explained:

A new message called BANDWIDTH CHANGE INDICATION (BCI) supports *network-initiated* and *called user–initiated* renegotiation. The BCI message can be issued by any network element along the connection path or by the called user. It signals the connection owner the need to modify its traffic descriptor. The BCI's ATM traffic descriptor IE carries the new parameters for the connection owner. When the connection owner receives a BCI message, it must issue a MODIFY REQUEST message. The BCI message specifies a new traffic descriptor to be requested in the MODIFY REQUEST. Timers are set when issuing the BCI message and cleared when the corresponding MODIFY REQUEST message is received; if the trimmer expires, the terminal and/or network element can modify the traffic policers to use the ATM traffic descriptor issued in the BCI message. Figure 9 illustrates the use of BCI for called user–initiated modification.

The *softness profile* and a *minimum satisfaction level* are two new IEs added to the MODIFY REQUEST message to specify the soft-QoS level.

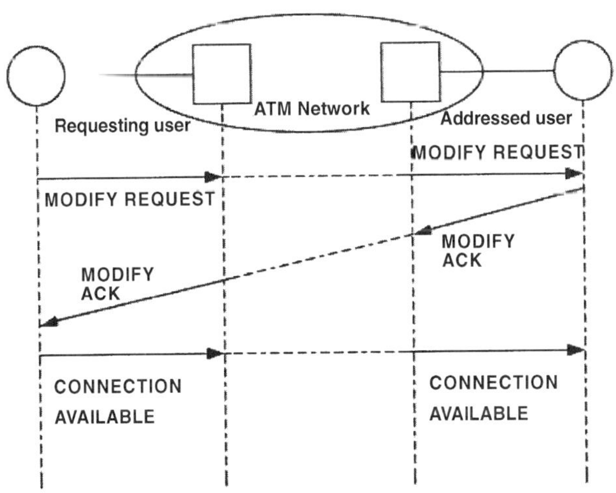

Figure 7 VBR$^+$ UNI extensions for soft-QoS: successful Q2963 modification of ATM traffic parameters with (optional) confirmation.

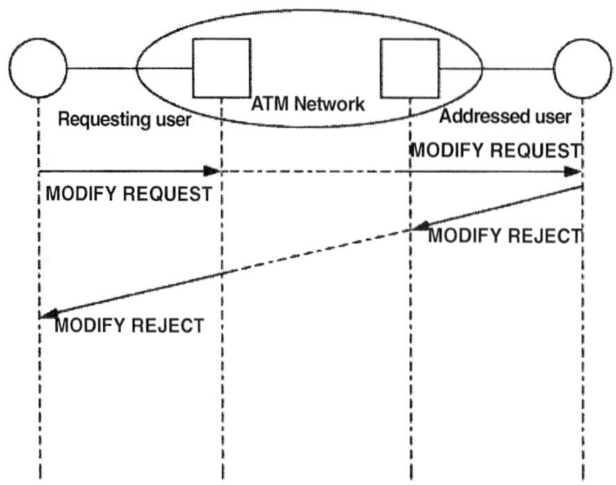

Figure 8 VBR+ UNI extensions for soft-QoS: rejection of modification by addressed user and/or network element.

Finally, an additional IE, called *available bandwidth fraction (ABF)*, is added to the MODIFY REJECT message. ABF is defined as the ratio of the available to requested value on a given traffic descriptor parameter. This results in ABF-PCR, ABF-SCR, and ABF-MBS for the peak, sustained, and maximum burst size, respectively. The connection owner may recompute the requested ATM traffic descriptor using the ABF information and reissue a suitable MODIFY REQUEST message.

Two new call states are added to the UNI4.0 protocol state machine to support modification. An entity enters the *modify request* state when it issues a MODIFY REQUEST or BCI message to the other side of the interface. An entity enters the *modify received* state when it receives a MODIFY REQUEST or BCI message from the other side of the interface.

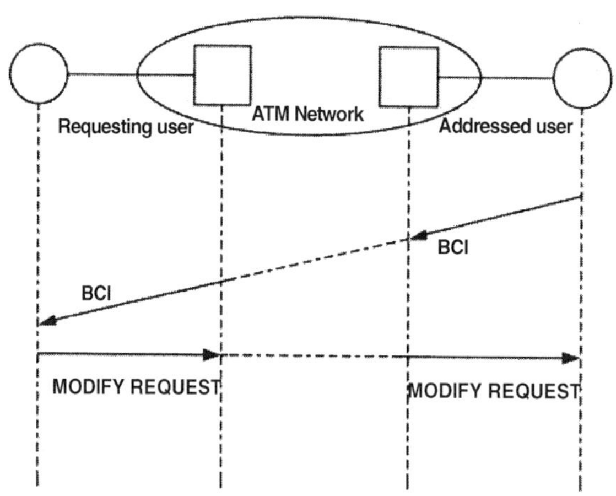

Figure 9 VBR+ UNI extensions for soft-QoS: modification by ATM traffic parameters by addressed user and/or network element.

VIII. CONCLUDING REMARKS

The provision of bandwidth on demand with strict quality-of-service guarantees is a fundamental property of ATM networks that makes them especially suitable for carrying real-time multimedia traffic. Realizing that multimedia traffic is highly nonstationary and that many multimedia applications can adapt to sporadic drops in bandwidth availability, we have developed a flexible and efficient framework for QoS control in ATM networks called soft-QoS with VBR$^+$. This framework complements the traditional traffic management model for ATM. It allows source-initiated and network-initiated bandwidth renegotiation based on the concept of a connection's softness profiles. The softness profile represents the sensitivity of application performance to the connection's network bandwidth allocation. Based on the relative softness of connections, QoS-fair allocation of resources is possible during network congestion. Traffic management mechanisms and signaling extensions used for dynamic bandwidth reservation within the access network are outlined.

Statistical multiplexing of VBR$^+$ connections within the backbone network allows effective aggregation and capacity engineering. Several statistical aggregation mechanisms (such as those based on equivalent bandwidth computation) can be used to dimension appropriately the capacity for backbone transport. Scalability is thus possible because per-connection explicit bandwidth renegotiation is not needed within the backbone. Bandwidth renegotiation frequency is high only within the access nodes, where the expected number of high-bit-rate multimedia connections is not too large. Experimental results show that the soft-QoS framework suits the traffic demand of real-time multimedia applications while providing high network resource utilization. Also, the computational load of soft-QoS control and bandwidth renegotiation is within the capabilities of switch controllers based on general-purpose PC processors.

In addition, the flexibility of soft-QoS control allows applications to specify a wide range of QoS expectations, from best effort to strict guarantees. Because only transport-level parameters are renegotiated, the framework can be applied independently of the network layer. IP networks and ATM networks can meet the soft-QoS requirements of an application via different mechanisms (e.g., RSVP [2] or Q.2963 [10] signaling). Therefore, the soft-QoS model can be used on both Internet and ATM broadband networks provided access network elements implement per-flow or per-VC bandwidth reservation mechanisms. Thus, the flexibility of the soft-QoS framework can be used to bridge IP and ATM QoS models that will certainly coexist in future network architectures. As these network technologies evolve, soft-QoS can be used as a mechanism for multimedia application users to specify QoS expectations independent of the underlying network implementation. At the same time, the soft-QoS specification can be used independently by various network technologies to ensure that users' end-to-end service expectations are met.

REFERENCES

1. The ATM Forum Technical Committee. Traffic management specification version 4.0. AF-95-0013R11, March 1996.
2. P White. RSVP and integrated services in the Internet: A tutorial. IEEE Commun Mag, May 1997.

3. ITU-T. ATM TraAEc descriptor modification by the connection owner. ITU-T Q.2963.2, September 1997.
4. M Ott, G Michelitsch, D Reininger, G Welling. An architecture for adaptive QoS and its application to multimedia systems design. Comput Commun 21:334–349 1998.
5. D Raychaudhuri, D Reininger, R Siracusa. Video transport in ATM networks: A systems view. Multimedia Syst J 4(6):305–315, 1996.
6. D Raychaudhuri, D Reininger, R Siracusa. Video transport in ATM networks. In: B Furht, ed. Handbook of Multimedia Computing. Boca Raton, FL: CRC Press, 1998.
7. International Organisation for Standardisation. Generic coding of moving pictures and associated audio. ISO/IEC/JTC1/SC29/WG11, March 1993.
8. The ATM Forum. ATM user–network interface (UNI) signalling specification. ATM Forum/95-1434R11, February 1996.
9. Technical Working Group ATM Forum. SAA audio-visual multimedia service (AMS) implementation agreement. ATM Forum/95-0012, 1995.
10. International Organisation for Standardisation. MPEG-2 Systems working draft. ISO/IEC/JTC1/SC29/WG-11-N0501, July 1993.
11. The ATM Forum Technical Committee. ATM user–network signalling specification, version 4.0, AF-95-1434R9, January 1996.
12. ITU-T. TraAEc control and congestion control in B-ISDN. Recomendation I.371, June 1992.
13. ITU-T. IVS base-line document. Report of Work Party 2/13, March 1994.
14. D Reininger, G Ramamurthy, D Raychaudhuri. VBR+, an enhanced service class for multimedia traAEc. ATM Forum/94-0353, May 1994.
15. D Reininger, G Ramamurthy, D Raychaudhuri. VBR MPEG Video coding with dynamic bandwidth renegotiation. Proceedings, 1995 IEEE International Conference on Communications, Seattle, June 1995, pp 1773–1777.
16. J Turner. Bandwidth management in ATM networks using Fast Buffer Reservation. IEEE Network Mag September 1992.
17. Z Tsai, W Wang, J Chang. A dynamic bandwidth allocation scheme for ATM networks. Proceedings Twelfth Annual International Phoenix Conference on Computers and Communications, March 1993, pp 289–295.
18. H Zhang, E Knightly. RED-VBR: A new approach to support VBR video in packet-switching networks. Proceedings IEEE Workshop on Network and Operating System Support for Digital Audio and Video (NOSSDAV'95), April 1995, pp. 275–286.
19. M Grossglauser, S Keshav, D Tse. A simple and efficient service for multiple time-scale traffic. Proceedings SIGCOMM'95, September 1995, pp 219–230.
20. M De Marco, V Trecordi. Bandwidth renegotiation in ATM for high-speed computer communications. Proceedings 1995 IEEE Global Telecommunication Conference, GLOBECOM'95, November 1995, pp 393–398.
21. D Reininger, R Izmailov. Soft quality-of-service with VBR+ video. Proceedings of Eight International Workshop on Packet Video (AVSPN97), Aberdeen, Scotland, September 1997.
22. JG Lourens, HH Malleson, CC Theron. Optimization of bit-rates, for digitally compressed television services as a function of acceptable picture quality and picture complexity. Proceedings IEE Colloquium on Digitally Compressed TV by Satellite, 1995.
23. E Nakasu, K Aoi, R Yajima, Y Kanatsugu, K Kubota. A statistical analysis of MPEG-2 picture quality for television broadcasting. SMPTE J November: 702–711, 1996.
24. Microsoft Corporation. Windows Quality of Service Technology. White paper available on line at http://www.microsoft.com/ntserver/
25. International Organisation for Standardisation. Generic coding of moving pictures and associated audio. ISO/IEC/JTC1/SC29/WG11, March 1993.
26. D Reininger, D Raychaudhuri, J Hui. Dynamic bandwidth allocation for VBR video over ATM networks. IEEE J Selected Areas Commun 14:1076–1086, August 1996.

18

Delivery and Control of MPEG-4 Content Over IP Networks

Andrea Basso and Mehmet Reha Civanlar
AT&T Labs, Red Bank, New Jersey

Vahe Balabanian
Nortel Networks, Nepean, Ontario, Canada

I. INTRODUCTION

The ubiquity of the Internet and the continuous increase in desktop computing power together with the availability of relatively inexpensive multimedia codecs have made multimedia content generation a readily available functionality on desktop computers. The new generation of multimedia tools and applications allow the user to manipulate, compose, edit, and transmit multimedia content over packet networks.

MPEG-4 is an ISO/IEC standard being developed by MPEG (Moving Picture Experts Group), the committee that also developed the Emmy Award–winning standards known as MPEG-1 and MPEG-2. MPEG-4 is building on the proven success of three fields: digital television, interactive graphics applications (synthetic content), and the World Wide Web (distribution of and access to content). It will provide the standardized technological framework enabling the integration of the production, distribution, and content access paradigms of the three fields.

The standard defines tools with which to represent individual *audiovisual objects*, both natural and synthetic (ranging from arbitrarily shaped natural video objects to sprites and face and body animations). These objects are encoded separately into their own elementary streams. In addition, *scene description* information is provided separately, defining the spatiotemporal location of these objects in the final scene to be presented to the user. This also includes support for user interaction. The scene description uses a tree-based structure, following the VRML (Virtual Reality Modeling Language) design. In contrast to VRML, scene descriptions can be dynamically updated. *Object descriptors* are used to associate scene description components that relate to digital video and audio with the actual elementary streams that contain the corresponding coded data. All these components are encoded separately and transmitted to the receiver. The receiving terminal then has the responsibility for composing the individual objects together for presentation and also managing user interaction.

MPEG-4 provides a broad framework for creating, distributing, and accessing digital audiovisual content with several advantages. For one, it leverages existing digital video content by allowing the use of MPEG-1 and MPEG-2. Furthermore, because of its powerful framework, it enables richer development of new digital video applications and services. Because of its flexibility, it becomes a tool in the hands of application developers rather than addressing a single, monolithic, vertical application. The inherent complexity of MPEG-4, as well as the fact that it breaks away from the pure stream-based structure of prior audiovisual representation schemes, creates a different operating environment for the entire chain of systems that utilize it, ranging from servers to playback clients.

This chapter describes the current joint efforts of the International Engineering Task Force (IETF) and MPEG communities in developing a mapping for MPEG-4 into the IETF Real Time Protocol (RTP) and its companion Real Time Control Protocol (RTCP) through the ISO/IEC MPEG Delivery Multimedia Integration Framework (DMIF). This would allow the use of RTP tools to monitor MPEG-4 delivery performance, the use of RTP mixers to combine MPEG-4 streams received from multiple end systems into a set of consolidated streams for multicasting, and the use of RTP translators for streaming MPEG-4 through firewalls. Furthermore, in this chapter a potential interaction between MPEG-4 DMIF and IETF real-time streaming protocol (RTSP) for media control will be discussed.

The chapter is organized as follows. Section II briefly presents the MPEG-4 end-system architecture. Section III discusses DMIF. Section IV overviews RTP and RTCP. Section V analyzes possible alternatives for the mapping of MPEG-4 on RTP. Section VI discusses a potential interface of RTSP with DMIF for media control purposes.

II. OVERVIEW OF MPEG-4 END-SYSTEM ARCHITECTURE

Figure 1 presents the general layered architecture of MPEG-4 terminals. The compression layer processes individual audiovisual media streams. The MPEG-4 compression schemes are defined in the ISO/IEC specifications 14496-2 [1] and 14496-3 [2]. The compression schemes in MPEG-4 achieve efficient encoding over a bandwidth ranging from several kilobits to many megabits per second. The audiovisual content compressed by this layer is organized into elementary streams (ESs). The MPEG-4 standard specifies MPEG-4–compliant streams. Within the constraint of this compliance the compression layer is unaware of a specific delivery technology, but it can be made to react to the characteristics of a particular delivery layer such as the path Maximum Transmission Unit (MTU) or loss characteristics.

Also, some compressors can be designed to be delivery specific for implementation efficiency. In such cases the compressor may work in a nonoptimal fashion with delivery technologies that are different from the one it is specifically designed to operate with.

The hierarchical relations, location, and properties of ESs in a presentation are described by a dynamic set of object descriptors (ODs). Each OD groups one or more ES descriptors referring to a single content item (audiovisual object). Hence, multiple alternative or hierarchical representations of each content item are possible.

The ODs are themselves conveyed through one or more ESs. A complete set of ODs can be seen as an MPEG-4 resource or session description at a stream level. The

```
media aware          +-----------------------------------------+
delivery unaware     |            COMPRESSION LAYER            |
14496-2 Visual       |streams from as low as Kbps to multi-Mbps|
14496-3 Audio        +-----------------------------------------+  Elementary
                                                                  Stream
=================================================================Interface
                                                                  (ESI)
                     +-----------------------------------------+
media and            |               SYNC LAYER                |
delivery unaware     | manages elementary streams, their syn-  |
14496-1 Systems      | chronization and hierarchical relations |
                     +-----------------------------------------+  DMIF
                                                                  Application
=================================================================Interface
                                                                  (DAI)
                     +-----------------------------------------+
delivery aware       |             DELIVERY LAYER              |
media  unaware       |provides transparent access to and delivery|
14496-6 DMIF         | of content irrespective of delivery     |
                     |                technology               |
                     +-----------------------------------------+
```

Figure 1 General MPEG-4 terminal architecture.

resource description may itself be hierarchical; i.e., an ES conveying an OD may describe other ESs conveying other ODs.

The session description is accompanied by a dynamic scene description, Binary Format for Scene (BIFS), again conveyed through one or more ESs. At this level, content is identified in terms of audiovisual objects. The spatiotemporal location of each object is defined by BIFS. The audiovisual content of the objects that are synthetic and static is also described by BIFS. Natural and animated synthetic objects may refer to an OD that points to one or more ESs that carry the coded representation of the object or its animation data.

By conveying the session (or resource) description as well as the scene (or content composition) description through their own ESs, it is made possible to change portions of the content composition and the number and properties of media streams that carry the audiovisual content separately and dynamically at well-known instants in time.

One or more initial scene description streams and the corresponding OD streams have to be pointed to by an initial object descriptor (IOD). The IOD needs to be made available to the receivers through some out-of-band means that are not defined in this document.

A homogeneous encapsulation of ESs carrying media or control (ODs, BIFS) data is defined by the sync layer (SL), which primarily provides the synchronization between streams. The compression layer organizes the ESs in access units (AUs), the smallest elements that can be attributed individual time stamps. Integer or fractional AUs are then encapsulated in SL packets. All consecutive data from one stream is called an SL-packetized stream at this layer. The interface between the compression layer and the SL is called the elementary stream interface (ESI). The ESI is informative.

The delivery layer in MPEG-4 consists of the Delivery Multimedia Integration Framework defined in ISO/IEC 14496-6 [3]. This layer is media unaware but delivery technology aware. It provides transparent access to and delivery of content irrespective of the technologies used. The interface between the SL and DMIF is called the DMIF Application Interface (DAI). It offers content location–independent procedures for establishing MPEG-4 sessions and access to transport channels. The specification of this payload format is considered as a part of the MPEG-4 delivery layer.

A. MPEG-4 Elementary Stream Data Packetization

The ESs from the encoders are fed into the SL with indications of AU boundaries, random access points, the desired composition time, and the current time. The sync layer fragments the ESs into SL packets, each containing a header that encodes information conveyed through the ESI. If the AU is larger than an SL packet, subsequent packets containing remaining parts of the AU are generated with subset headers until the complete AU is packetized. The syntax of the sync layer is not fixed and can be adapted to the needs of the stream to be transported. This includes the possibility to select the presence or absence of individual syntax elements as well as configuration of their length in bits. The configuration for each individual stream is conveyed in an SL ConfigDescriptor, which is an integral part of the ES descriptor for this stream.

III. DMIF in MPEG-4

The DMIF provides an interface and procedures for the delivery of MPEG-4 streams, which satisfy the following MPEG-4 requirements:

- Playing of streams transparent to their source locations whether on interactive remote end systems reached over networks, broadcast as a set of streams, or stored on file systems
- Improved delivery setup time and quality of service (QoS) through efficient packing of streams
- Improved critical stream availability in case of drop in transport capacity
- Define stream QoS without knowing delivery technology for subsequent mapping to specific delivery technologies
- Permit user action to reduce overload on network capacity
- Log the critical network resources used in a session for subsequent cost recovery through revenue generation
- Allow transparent concatenation of delivery networks for operation over a mix of access and backbone network transport technologies

In achieving these requirements, DMIF assumes that standards are in place and aims at integrating them within an integration framework. In time, DMIF will integrate all the relevant standards in its framework using the rules set for each standard.

A. The DMIF Model

DMIF as an integration framework uses a uniform procedure at the DAI interface to access the MPEG-4 content irrespective of whether the content is broadcast, stored on a local file, or obtained through interaction with a remote end system. The specific instance of interest in this chapter is the interaction with a remote end system. For this case DMIF uses an internal (informative) DMIF–Network Interface (DNI) to map the controls obtained from the application through DAI into the various signaling appropriate to the various networks. All networks are considered to be outside the scope of DMIF.

IV. STREAMING OVER IP NETWORKS

A. Real-Time Protocol—RTP

RTP [4] provides the basic functionality needed for carrying real-time data over packet networks. It does not offer mechanisms for reliable data delivery or protocol-specific flow and congestion controls such as the ones offered by TCP. RTP relies on other protocol layers, capable of managing and controlling network resources, to provide on-time delivery and the framing service. The services provided by RTP include payload type identification, sequence numbering, time stamping, and delivery monitoring. RTP typically runs on top of the User Datagram Protocol (UDP). The RTP specification is articulated into two parts. The first one is generic, where the basic roles, protocol services, and message formats are defined. The information relative to this generic part is carried in the *RTP fixed header*, shown in Fig. 2. The second part is media specific and offers auxiliary services to improve the quality of delivery of the specific media.

The RTP payload type identification service, together with the multiplexing services supported by the underlying protocols, provides the necessary infrastructure to multiplex a large variety of information. As an example, multicast transmission of several MPEG streams multiplexed together with other information (i.e., information relevant to the MPEG stream content) can easily be identified via the payload type and handled accordingly. The 7-bit payload type (PT) is an unsigned number in the range 0–127 that identifies the media carried in the payload. Payload types can be static or dynamic. A static payload type is defined a priori according to RFC1890 [10]; a dynamic one is specified by means external to RTP (i.e., via a Session Description Protocol, SDP). In the current specification all the payload types in the range 96–127 should be dynamic. As an example, MPEG audio and MPEG2 video payload types are static and have the values 14 and 33, respectively. The payload type of a given RTP stream can be changed on the fly during a session. This feature is particularly important in a scenario in which a sender, in response to a large set of RTCP receiver reports indicating severe packet losses, decides to change the current

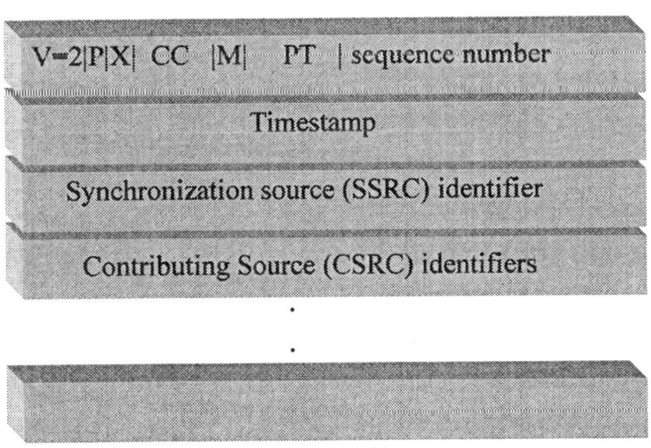

Figure 2 The RTP fixed header $(3 + n)$ *32 bits in length, where n is the number of CSRCs.

media encoder for another one that operates at a lower bit rate (i.e., from MPEG1 to H.263), which are identified with two different payload types.

The marker (M) bit semantics are dependent on the payload type and are intended to signal the occurrence of significant events such as media boundaries (i.e., video or audio frames) to be marked in the packet stream.

The sequence number field is 16 bits in length. RTP sequence numbers are incremented by one for each RTP packet transmitted. Their main use is to support packet loss detection. Note that RTP does not define any mechanisms for recovering from packet losses. Such mechanisms typically depend on the payload type and on the system implementation.

The RTP time stamping service provides support for synchronization of different media originating from a single source or several sources. RTP time stamps can also be used for measuring the packet arrival jitter. As shown in Fig. 1, each RTP packet carries a 32-bit time stamp that reflects the sampling instant of the first byte of the data contained in the packet payload. The interpretation and use of the RTP time stamps are payload dependent. For video streams, the time stamps depend on the frame rate. Furthermore, if a frame was transmitted as several RTP packets, these packets would all bear the same time stamp. Note that RTP depends on the Network Timing Protocol (NTP) [6] to relate the RTP time stamps to the wall clock time and to synchronize clocks on different networked hosts.

The synchronization source identifier (SSRC) field is 32 bits long. It identifies the source of the RTP stream. The SSRC is a number that the source assigns randomly when the RTP stream is started. Typically, each stream in an RTP session has a distinct SSRC. Note that because the SSRC of a given RTP source may change in time as a result of an SSRC conflict discovery and resolution or an application reinitialization, the receivers keep track of the participants of a session via the canonical name (see next section on RTCP) and not the SSRC.

Finally, the RTP fixed header carries a list of a maximum of 15 contributing sources identifiers (CSRC) for the payload carried in the packet. Such a list is typically inserted by RTP mixers or translators [4]. If no RTP translator or mixer is involved, such a list is empty.

B. Real-Time Control Protocol—RTCP

RTCP is designed to handle the delivery monitoring service of RTP. It has several functions: provide feedback on the quality of the data distribution; provide the timing information; establish an identification, i.e., a canonical name (CNAME) for each participant; scale the control packet transmission with the number of participants; and finally provide minimal session control information.

RTCP can be effectively used in both unicast and multicast scenarios. In unicast scenarios it can provide media delivery feedback and support for some form of rate control as well as media synchronization. RTCP is particularly effective in multicast scenarios. Both the RTP and RTCP streams of a given RTP session use the same multicast address with distinct port numbers. RTCP packets are sent periodically from all participants of a session to all other participants.

Five RTCP packet types are defined: source description packets; sender reports; receiver reports; BYE packets, which indicate the end of participation to the session; and APP packets, which carry information on application-specific functions. RTCP packets are stackable; i.e., receiver reception reports, sender reports, and source descriptors can be concatenated into a single compound RTCP packet.

1. Source Description Packets

RTCP carries a persistent identifier for a given RTP source called canonical name (CNAME). Such an identifier is needed because the SSRC of a given RTP source may change in time due to SSRC conflict discovery or application reinitialization. Receivers keep track of the participants of a session via the CNAME and not the SSRC. The CNAME as well as other information describing the source (such as e-mail address of the sender, the sender's name, and the application that generates the RTP stream and the SSRC of the associated RTP stream) is part of the source description packets.

2. Receiver and Sender Reports

RTCP compound packets may contain sender and/or receiver reports that carry statistics concerning the media delivery. For each RTP stream, each receiver generates a reception status that is transmitted periodically as RTCP receiver report packets. Some of the most important fields of an RTCP receiver report packet are the following:

The SSRC of the RTP stream for which the reception report is being generated.

The fraction of packets lost within the RTP stream. Each receiver calculates the number of RTP packets lost divided by the number of RTP packets sent as part of the stream. A sender may decide, in order to improve the reception rate, to switch to a lower encoding rate when, for example, a large population of the participants in the session signal that they are receiving only a small fraction of the sender's transmitted packets.

The last sequence number received relative to this RTP stream.

The packet interarrival jitter.

The senders create and transmit RTCP sender report packets for each RTP stream. These are the most important fields of such a packet type:

The SSRC of the RTP stream

The RTP time stamp and the corresponding wall clock time of the most recently generated RTP packet in the stream

The number of RTP packets and bytes sent

3. Media Synchronization

The RTCP sender report packets associate the sampling clock of the RTP time stamps with the NTP wall clock time. Such association is used for intra- and intermedia synchronization for sources whose NTP time stamps are synchronized. In addition, the NTP time stamps can be used in combination with the time stamps returned in reception reports from other receivers to measure round-trip propagation to those receivers.

The wall clock time is obtained via the Network Time Protocol [6]. The 64-bit NTP time stamp indicates the wall clock (absolute) time when the RTCP packet was sent. The RTP time stamp corresponds to the time of the NTP time stamp but in the units and with the random offset of the RTP time stamps in the RTP data packets (RTP time stamps start with a random offset to prevent plain-text attacks on encryption.)

4. RTCP Bandwidth Utilization

For a given RTP session the amount of RTCP traffic grows linearly with the number of receivers. Thus the RTCP bandwidth may be considerable when the group of participants is large. To obviate such problems, the RTCP rate for each participant is modified as a function of the number of participants in the session. The protocol keeps track of the total number of participants in the session (this is possible because every RTCP packet is received by all the other participants in the session) and limits its traffic to 5% of the session bandwidth. For example, in a scenario with a single sender and multiple receivers, RTCP assigns 75% of the 5% of the session bandwidth to the receivers and the remaining 25% to the sender. The bandwidth devoted to the receivers is equally shared among them. A participant (a sender or receiver) determines the RTCP packet transmission period by dynamically calculating the average RTCP packet size (across the entire session) and dividing the average RTCP packet size by its allocated rate. For a detailed discussion of the BYE and APP packets and their role the reader is referred to Ref. 4.

V. STREAMING OF MPEG-4 CONTENT OVER RTP

The RTP specification is articulated into two parts. The first one is generic, where the basic roles, protocol services, and message formats are defined (as illustrated in Sec. IV.A) and the second one is media specific, offering auxiliary services to improve the quality of delivery of the specific media. In the case of MPEG-4 streams, the media-specific information is carried in the media-specific RTP header called *RTP MPEG-4 specific header*. This header follows the *fixed RTP header* of Fig. 1.

Several services provided by RTP are beneficial for transport of MPEG-4–encoded data over the Internet. This section discusses an RTP [4] payload format for transporting MPEG-4–encoded data streams. The benefits of using RTP for MPEG-4 data stream transport include these:

- Ability to synchronize MPEG-4 streams with other RTP payloads
- Monitoring MPEG-4 delivery performance through RTCP
- Combining MPEG-4 and other real-time data streams received from multiple end systems into a set of consolidated streams through RTP mixers
- Converting data types through the use of RTP translators

A. Analysis of the Alternatives for Carrying MPEG-4 Over IP

Considering that the MPEG-4 SL defines several transport-related functions such as timing and sequence numbering, the transport of MPEG-4 content directly encapsulated in UDP packets seems to be the most straightforward alternative for carrying MPEG-4 data over IP. One group of problems with this approach, however, stems from the monolithic architecture of MPEG-4. No other multimedia data stream (including those carried with RTP) can be synchronized with MPEG-4 data carried directly over UDP. Furthermore, the dynamic scene and session control concepts cannot be extended to non–MPEG-4 data. Even if the coordination with non–MPEG-4 data is overlooked, carrying MPEG-4 data over UDP has the following additional shortcomings:

Mechanisms need to be defined to protect sensitive parts of MPEG-4 data. Some of these (such as forward error correction (FEC)) are already defined for RTP.
There is no defined technique for synchronizing MPEG-4 streams from different servers in the variable delay environment of the Internet.
MPEG-4 streams originating from two servers may collide (their sources may become unresolvable at the destination) in a multicast session.
An MPEG-4 back channel needs to be defined for quality feedback similar to that provided by RTCP.
RTP mixers and translators cannot be used.
The back-channel problem may be alleviated by developing a reception reporting protocol such as RTCP. Such an effort may benefit from RTCP design knowledge but needs extensions.

We will now analyze some of the potential alternatives for the transport of MPEG-4 content over IP. The alternative of having the RTP header followed by full MPEG-4 headers may be implemented by using the send time or the composition time coming from the reference clock as the RTP time stamp. In this way, no new feedback protocol needs to be defined for MPEG-4's back channel, but RTCP may not be sufficient for MPEG-4's feedback requirements, which are still in the definition stage. Also, because of the duplication of header information, such as the sequence numbers and time stamps, this alternative causes unnecessary increases in the overhead. Furthermore, scene description or dynamic session control cannot be extended to non–MPEG-4 streams.

The alternative of encapsulating MPEG-4 ESs over RTP by means of individual payload types is the most suitable alternative for coordination with the existing Internet multimedia transport techniques and does not use MPEG-4 systems at all. Its complete implementation requires definition of potentially many payload types and might lead to constructing new session and scene description mechanisms. Considering the amount of work involved, which essentially reconstructs MPEG-4 systems, this may be a long-term alternative only if no other solution can be found. The inefficiency of the approach described earlier can be fixed by using a reduced SL header that does not carry duplicate information following the RTP header.

B. Proposal for an RTP Payload Format for MPEG-4

Based on the preceding analysis, a reasonable compromise is to map the MPEG-4 SL packets onto RTP packets, such that the common pieces of the headers reside in the RTP header, which is followed by an optional reduced SL header providing the MPEG-4 specific information. The RTP payload consists of a single SL packet, including an SL packet header without the sequenceNumber and compositionTimeStamp fields. Use of all other fields in the SL packet headers that the RTP header does not duplicate (including the decodingTimeStamp) is optional. Packets should be sent in the decoding order. If the resulting, smaller, SL packet header consumes a noninteger number of bytes, zero padding bits must be inserted to byte-align the SL packet payload. The size of the SL packets should be adjusted so that the resulting RTP packet is not larger than the path-MTU. To handle larger packets, this payload format relies on lower layers for fragmentation, which may not be dependable. The semantics of the RTP packet fields is as follows:

Payload type (PT): Set to a value identified for MPEG-4 data.
Marker (M) bit: Set to one to mark the last fragment (or only fragment) of an AU.

Extension (X) bit: Defined by the RTP profile used.

Sequence number: Derived from the sequenceNumber field of the SL packet by adding a constant random offset. If the sequenceNumber is less than 16 bits long, the MSBs must initially be filled with a random value that is incremented by one each time the sequenceNumber value of the SL packet returns to zero. If the value sequenceNumber = 0 is encountered in multiple consecutive SL packets, indicating a deliberate duplication of the SL packet, the sequence number should be incremented by one for each of these packets after the first one. In implementations in which full SL packets are generated first, the sequenceNumber must be removed from the SL packet header by bit-shifting the subsequent header elements toward the beginning of the header. When unpacking the RTP packet, this process can be reversed with the knowledge of the SLConfigDescriptor. For using this payload format, MPEG-4 implementations that do not produce the full SL packet in the first place but rather produce the RTP header and stripped-down (perhaps null) SL header directly are preferable. If no sequenceNumber field is configured for this stream (no sequenceNumber field present in the SL packet header), then the RTP packetizer must generate its own sequence numbers.

Time stamp: Set to the value in the compositionTimeStamp field of the SL packet, if present. If compositionTimeStamp is less than 32 bits length, the MSBs of time stamp must be set to zero. Although it is available from the SL configuration data, the resolution of the time stamp may need to be conveyed explicitly through some out-of-band means to be used by network elements that are not MPEG-4 aware. If compositionTimeStamp is more than 32 bits long, this payload format cannot be used. In case compositionTimeStamp is not present in the current SL packet, but has been present in a previous SL packet, the same value must be taken again as the time stamp.

If compositionTimeStamp is never present in SL packets for this stream, the RTP packetizer should convey a reading of a local clock at the time the RTP packet is created. As with the handling of the sequence numbers in implementations that generate full SL packets, the compositionTimeStamp, if present, must then be removed from the SL packet header by bit-shifting the subsequent header elements toward the beginning of the SL packet header. When unpacking the RTP packet this process can be reversed with the knowledge of the SLConfigDescriptor and by evaluating the compositionTimeStampFlag.

Time stamps are recommended to start at a random value for security reasons [4, Sec. 5.1]. SSRC: Set as described in RFC1889 [4]. A mapping between the ES identifiers (ESIDs) and SSRCs should be provided through out-of-band means. CC and CSRC fields are used as described in RFC 1889. RTCP should be used as defined in RFC 1889.

C. Multiplexing

Because a typical MPEG-4 session may involve a large number of objects, which may be as many as a few hundred, transporting each ES as an individual RTP session may not always be practical. Allocating and controlling hundreds of multicast destination addresses for each MPEG-4 session may pose insurmountable session administration problems. The input–output processing overhead at the end points will also be extremely high.

In addition, low-delay transmission of low-bit-rate data streams, e.g., facial animation parameters, results in extremely high header overheads. To solve these problems, MPEG-4 data transport requires a multiplexing scheme that allows selective bundling of several ESs. This is beyond the scope of the payload format discussed here. MPEG-4's Flexmux multiplexing scheme may be used for this purpose by defining an additional RTP payload format for ''multiplexed MPEG-4 streams.'' On the other hand, considering that many other payload types may have similar needs, a better approach may be to develop a generic RTP multiplexing scheme usable for MPEG-4 data. The generic multiplexing scheme reported in Ref. 7 is a candidate for this approach.

For MPEG-4 applications, the multiplexing technique needs to address the following requirements:

The ESs multiplexed in one stream can change frequently during a session. Consequently, the coding type, individual packet size, and temporal relationships between the multiplexed data units must be handled dynamically.

The multiplexing scheme should have a mechanism to determine the ES identifier (ES_ID) for each of the multiplexed packets. The ES_ID is not a part of the SL header.

In general, an SL packet does not contain information about its size. The multiplexing scheme should be able to delineate the multiplexed packets, whose lengths may vary from a few bytes to close to the path-MTU.

D. Security Considerations

RTP packets using the payload format defined in this specification are subject to the security considerations discussed in the RTP specification [4]. This implies that confidentiality of the media streams is achieved by encryption. Because the data compression used with this payload format is applied end to end, encryption may be performed on the compressed data so that there is no conflict between the two operations. This payload type does not exhibit any significant nonuniformity in the receiver side computational complexity for packet processing to cause a potential denial-of-service threat.

VI. MEDIA CONTROL OF MPEG-4 CONTENT

Remote interactivity is becoming an important issue in the current advanced multimedia research as well as in the current developments of standards such as MPEG-4. A networked multimedia application may require support to react to a user action such as addition of a multimedia stream or its removal or to handle some modification of a complex scene that changes in time.

With the development of a large variety of multimedia end stations, from smart cellular phones to autoPCs and personal digital assistants (PDAs) to thin terminals, the support for client heterogeneity is becoming more and more important. The host (client or server) can no longer be considered as having guaranteed and time-invariant available resources. The presence of a media control channel allows renegotiating in time the available network and processing resources.

In the following sections we will discuss some of the most important requirements for remote interactivity and host resource handling, the nature of the associated messages, and the potential protocols that can handle such messages efficiently in the current Internet framework.

A media control framework should allow a multimedia client and one or more servers to exchange different types of control messages. Control messages can also be exchanged among different clients. This requires several components as shown in Fig. 3.

1. A description of a stored or live presentation
2. A set of protocols that can provide proper services for the back-channel message delivery
3. A set of protocols that can allocate resources for the involved hosts and networks

Such components can be implemented, as we will see, with existing protocols or with extensions of existing protocols.

A. Presentation Description

A client wants to have a media server play back a given presentation that includes high-quality video and audio of the speaker, the transparencies of the presentation, and finally the multimedia clips that the speaker wants to show as demos. The client will need to refer to a description of the presentation that expresses the temporal and static properties of the presentation itself. Such a *presentation description* contains information about the *media* involved in the presentation (e.g., media encoding schemes, language information, content description), about the *location of the media* (the media streams may be located on different media servers), the *network* (i.e., network addresses and ports), and *transport modalities* (unicast, multicast). A presentation description should provide multiple description instances of the same presentation, i.e., a list of servers that carry the same media streams but with different encodings. By means of the presentation description the client can specify a given combination of media streams and media servers that fit its presentation capabilities, the location of such, and how the media streams need to be transported. The

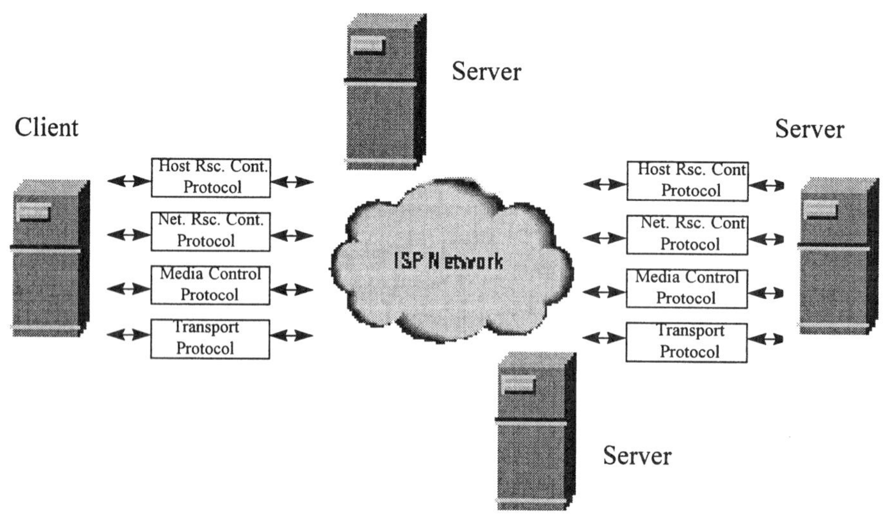

Figure 3 Media control framework.

presentation description may reside on a separate server (Title Server) and can be encoded with different schemes (Fig. 4).

For example, in the case of synthetic scene presentations the client should be able to select a presentation that is tailored to its capabilities. The presentation description for synthetic audio and graphics has to detail several properties describing the complexity of the scene. Useful descriptors include the number of polygons of the scene, size of texture maps, preferred render speed in frames per second, color resolution, audio coder and bandwidth, audio channels, spatial audio requirements (panning, azimuth, full 3D), and text-to-speech synthesizer. Based on this description, the client selects a presentation for which he is able to allocate the resources.

Note that according to this ''client centric'' model the client is the ''orchestrator'' of the presentation and the server is ''participating'' in the amount and with the modalities specified by the client. The analysis of the presentation description as well as the protocol for its delivery is outside the scope of this chapter.

B. Client and Server State Representations

The stream delivery and its control are in general performed via separate protocols and different transport modalities. Furthermore, the media stream may be modified by different sequential control commands. It is thus necessary that the server keep a state of the stream(s) status for each client it is serving. In a typical situation the server needs to keep track of a given media stream that is ready to be delivered (i.e., the media stream setup has been successfully carried out), if it is playing or if the server is performing some other task such as recording the media stream. The client on the other side has to keep trace of all the participating streams. The client and server state representations should take into account command precedence and preconditions; for example, the client should always issue a setup command to initialize the media streams before the request to play them.

Furthermore, it should be possible to address individual media streams as well as logical regrouped streams (e.g., a video and its corresponding audio) so that a single media control command can be sent to modify them simultaneously. For a more detailed analysis

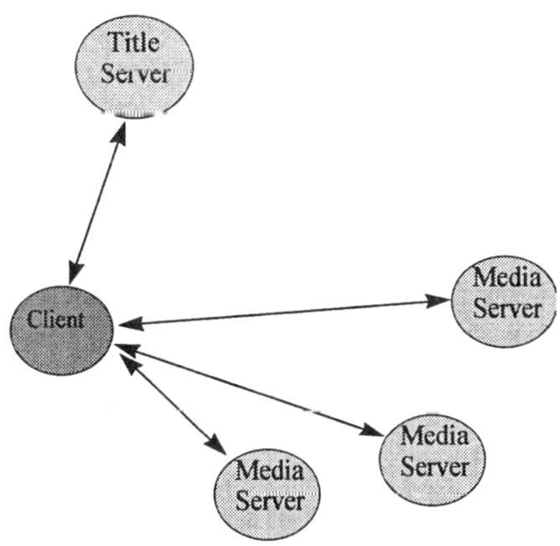

Figure 4 Title and media servers.

and examples of server and client state machines see Ref. 8. Note that while server state representation is important, it is also important that the computation of the status is very light.

C. Basic Media Control Messages

A networked multimedia system should have access to control messages ranging from remote VCR functions such as stop, play, fast forward, and fast reverse to messages generated in response to user actions to modify the presentation of a given object stream, to add an object, or to remove it to more complex interactive control schemes that allow real-time control of some of the client–server features such as the media encoders and decoder and network devices. The basic control functionality relates to presentation and stream setup; play, stop, pause teardown of the multimedia streams; and recording.

The SETUP process requires the capability of the client to request from a server the description of the presentation or of a server to ask from a client a description of a presentation that needs to be recorded. Messages such as the IETF's Real Time Streaming Protocol (RTSP) DESCRIBE and ANNOUNCE [8] have been designed with this in mind.

Note that the setup process involves a network and host resource reservation process. On the client side, enough processing resources should be reserved for the multimedia presentation. Similarly, the involved servers should reserve enough cycles to stream the requested media. Scenarios without resource reservation can also be possible, when, for example, client and servers are overdesigned with respect to the processing power requested. On the network the proper services and connection modalities need to be selected.

The PLAY command tells the server to start sending the data via the transport mechanism that has been specified in the setup process. It is important that the command precedence is respected; that is, a client should never issue a play command until the setup process has been completed and acknowledged successfully. On the other hand, a server should generate an error message if such a state is illegal in its representation of the associated stream. The play command implies a play time, i.e., the temporal extension of the media stream that should be delivered. Thus, proper support for time representation is needed. Finally, a stream that reaches the end of the playtime should move to the PAUSED state.

Note that the PLAY command can be associated with a given stream of a logically regrouped set of streams. The client, in reaction to a PLAY for a given set of media streams, has the responsibility to issue the PLAY command to all the media servers involved. Once the PLAY command is sent from the client, the server positions the stream to the prespecified position and starts to stream the data to the client.

A PAUSE command tells the server to stop temporarily a given task such as PLAY or RECORD. The associated streams are thus temporarily halted. The critical aspect of the PAUSE command is to ensure that the proper stream synchronization is kept once the stream playback is resumed or once recording restarts. Note that the PAUSE command implies that the server keeps the allocated resources for the task for a given time period still allocated. The resources can be released after a given time-out interval expires. Finally, the PAUSE command can refer to a PAUSE time in the future. The server should make sure that such a time in the future is inside the playback time interval.

Finally, the command TEARDOWN stops the stream delivery and frees the associated resources.

The RTSP [8] is an application-level protocol that provides an extensible framework to enable controlled delivery of real-time data, such as audio and video. Sources of data can include both live data and stored content. It can use several transport protocols such as UDP, multicast UDP, and TCP and it is designed to work with established protocols such as RTP and HyperText Transport Protocol (HTTP). RTSP provides support for bidirectional (client to server and server to client) communications and supports the state-based representation. It allows retrieval of media from a given server, invitation of a media server to a conference, and addition of media to an existing presentation. RTSP supports Web security mechanisms; it is transport independent and multiserver capable. Furthermore, it provides frame-level timing accuracy through the use of Society of Motion Picture and Television Engineers (SMPTE) time stamps to allow remote digital editing.

RTSP allows negotiation of the transport method of the continuous media as well as negotiation of the capabilities of the server; the client can interrogate the server capabilities via DESCRIBE messages.

D. The Use of RTSP Within the MPEG-4 Architecture

From the DMIF perspective, RTSP is an application alongside MPEG-4 Systems. Figure 5 shows the configuration between a sending and a receiving DMIF terminal with RTSP. At the receiver the RTSP client and server interact with MPEG-4 Systems. The RTSP client and server control the streams through the DAI by an RTSP–DMIF interface. This interface is kept simple by limiting it to field mapping between the RTSP fields and the DAI primitive parameters.

The RTSP client–server interactions are used to control the MPEG-4 elementary streams. The RTSP messages flow in both the downstream (sender to receiver) and upstream (receiver to sender) directions.

E. MPEG-4 RTSP DMIF Message Sequences

RTSP is used as the primary application agent for both setting up and control of the MPEG-4 elementary streams. Two stages are considered. In the first stage an MPEG-4 scene(s) is received from a peer MPEG-4 terminal, and in the second stage channels for selected elementary streams from a scene(s) are established and the elementary streams flowing over those channels are controlled.

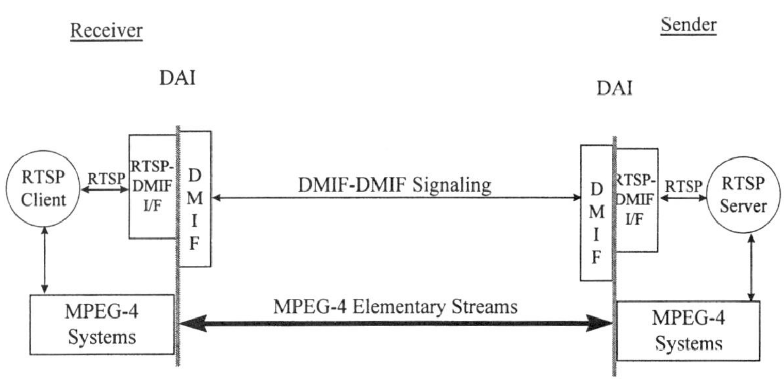

Figure 5 RTSP within the MPEG-4 architecture.

1. Receiving MPEG-4 Scene

In this stage the scene and the corresponding object descriptor streams are received from the sender. Figure 6 shows the message flow sequence. The RTSP messages and the corresponding DAI primitives are shown at both the receiver and the sender locations. The DMIF layer and the DMIF signaling are completely hidden and are not shown for simplicity.

A precondition for this message sequence is that the receiver possesses the initial object descriptor and has an active network session nsId and an active service session ssId. These are obtained through DA_ServiceAttach with the service identified through a DMIF URL. The method of obtaining the DMIF URL is not defined and could be through IETF SDP or any other means.

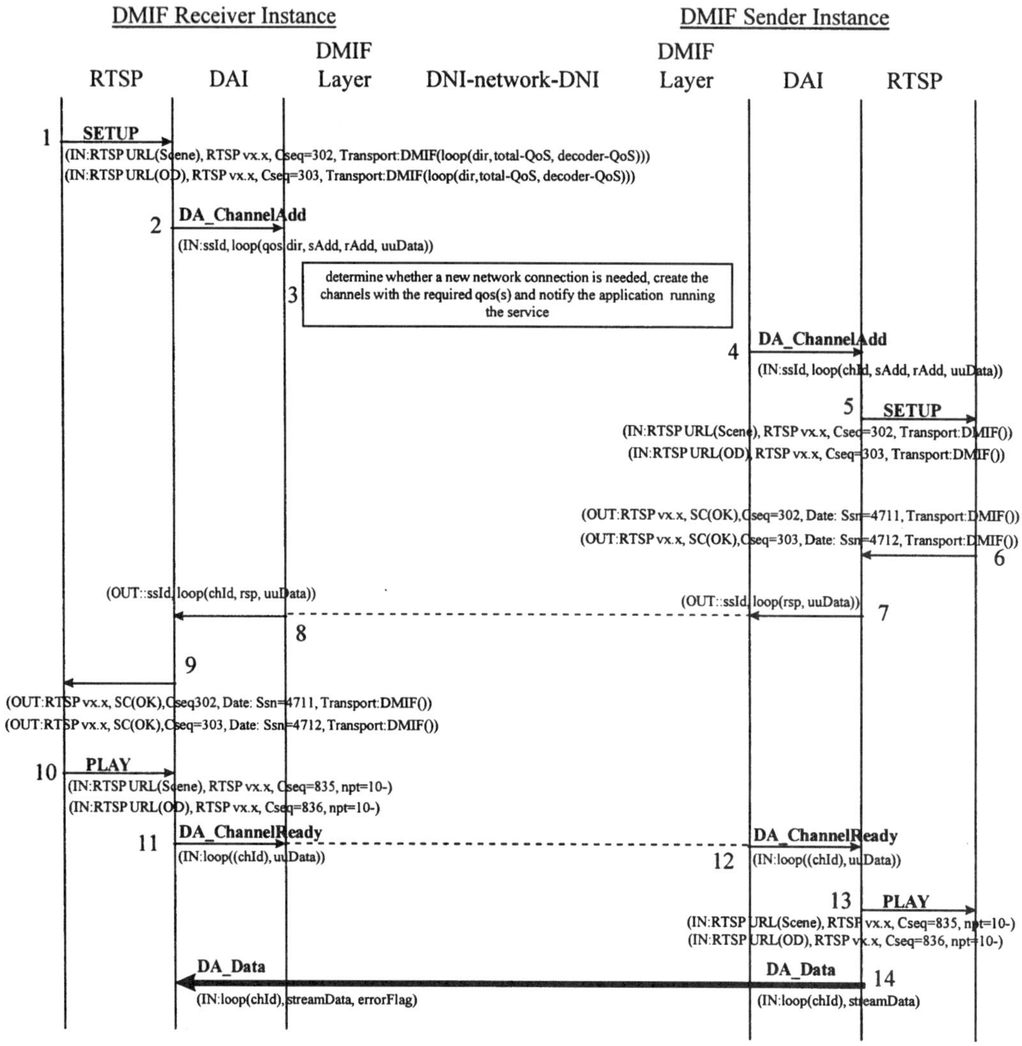

Figure 6 Message sequence with RTSP and DMIF for receiving an MPEG-4 scene.

Step 1 Using the scene and object descriptor URLs from the initial object descriptor, RTSP forms an RTSP URL and initiates setup of both streams. As transport, it uses DMIF and identifies the media-based QoS of each stream from the CD elementary stream descriptors of the scene and the OD streams from the initial object descriptor.

Step 2 The DA_ChannelAdd is prepared from the RTSP SETUP in the following manner: DMIF URL is used in the sAddr, indicating that the source location is derived from the RTSP URL. In the case of IP networks, the DMIF URL is the same as the RTSP. But in case a tunneling connection needs to be established to ISP gateways before reaching the source, the DMIF URL defines an additional address to get to the ISP gateway first before proceeding to establish a connection to the source location. In the case of RTSP, the rAddr, which is optional, is not used. The DMIF() in the RTSP transport header field is used to extract the QoS descriptors. The RTSP vx.x, Cseq are included in the uuData().

Step 3 The DMIF URL is used to identify the port of the source; the QoS is used in establishing the channel for the corresponding stream.

Step 4 Once the channels with the requested transport QoSs are established, the DA_AddChannel is generated at the sender's DAI. It contains the locally assigned chIds, which are channel handlers used to send or receive data.

Step 5 At this point the RTSP SETUP message is regenerated at the sender from the DA-AddChannel, except that the Transport DMIF() now carries the local chIds. These in turn are passed to MPEG-4 Systems by the RTSP server.

Step 6 If the SETUP is accepted, RTSP responds with a status code OK. The DMIF() is left empty.

Step 7 The DA_AddChannel response is generated using the parameters of the RTSP response from step 6.

Step 8 When the DA_AddChannel response is received, the appropriate chId is added for each channel over which an EsId will flow.

Step 9 The RTSP response is generated from the DA_AddChannel in step 8 with a status code OK. The DMIF() now carries the local chIds for the EsIds. These in turn are passed to MPEG-4 Systems by the RTSP client.

Step 10 The receiver now sends the RTSP PLAY command for both the scene and the corresponding OD streams. The DSM-CC IS Normal Play Time (NPT) is used to indicate the starting point. In the case of a live event, the NTP constant "now" is used.

Step 11 The RTSP play parameters are used to generate DA_ChannelReady. ChId indicates the local channel handle on which a particular scene or OD stream is to be received. The RTSP PLAY parameters are carried in uuData.

Step 12 DA_ChannelReady carries the local chIds at the sender for the scene or the OD streams. These are used to obtain the scene and OD URLs required by the RTSP PLAY message in step 13.

Step 13 The RTSP PLAY is used to request the media server or the encoder to begin sending the scene and OD stream data and also monitor for their type 1 back channel if available using the local channel handles chIds.

Step 14 The scene and OD streams and their back channels are sent and received at both terminals. If the request is to start a scene and an OD at NTP greater than value 0, then the first scene and OD AUs must contain the aggregate objects and not the increments.

2. Playing the Streams

When a receiver gets the first scene and the corresponding OD AUs, it decodes them at the DecodingTimeStamp before they are composited. From the decoded information, the desired elementary streams are selected and matching decoders are verified and instantiated. Proper computing resources are reserved to achieved the desired media-based QoS. Corresponding channels are also established with reserved QoS resources. All of these must be carried out in a timely manner so that the elementary stream composition units are available for rendering at the time needed by the scene. From this point on for live events, the increments in the scene are monitored for new object items and the timely process started for the stream composition units to be present at the time of their rendering. In case of stored programs, any scene changes caused by playback control are monitored and again the same process is followed. In addition, in the case of stored programs, the end user may choose an object and try to control its playback while not touching the rest of the scene. It may subsequently request the complete scene to progress in a normal manner. Finally, it is possible for an end user to replay a scene from the start and apply or correct the selections memorized from a prior viewing of the MPEG-4 presentation. Figure 7 provides the sequence of messages for control using RTSP that allows all the variations described. A precondition for this message sequence is that the receiver possesses the first scene and OD AUs. It contains the necessary decoders installed and ready for instantiation.

Steps 1 through 9 These are the same as those used to receive the scene and OD streams except that in this case the URLs and the QoSs are obtained from the scene and OD AUs as opposed to using the initial object descriptor.

Step 10 The receiver now sends RTSP PLAY command for both the ESId1 and ESId2 streams. The NPT is used to indicate the starting point. In the case of a live event, the NTP constant ''now'' is used. In the case of stored programs, the NTP time must be indicated relative to the scene (required new NPT field).

Steps 11 through 14 This message sequence is the same as the one shown in Fig. 6.

A scene can be advanced or rewound at any time. All objects still present in the scene will be advanced or rewound along with the scene. New object items selected from the current scene require both RTSP SETUP and PLAY to be sent for their streams. This will also be the case if new streams from the persisting objects are selected. Any object that disappears from the scene will require the RTSP TEARDOWN to be sent for their respective streams.

3. Pausing the Streams

The message sequence in Fig. 8 provides the use of RTSP PAUSE and PLAY with DMIF. Either a scene or individual streams can be paused. A paused scene can be followed by play of individual streams. Let us assume the normal play of a scene with selected streams is active.

Step 1 End user pauses the EsId1; an RTSP PAUSE is issued with the RTSP session of the EsId1. An NPT could be added based on the scene NPT at the instant the end user paused the stream for more accurate positioning of the pausing location of the EsId1 at the sender.

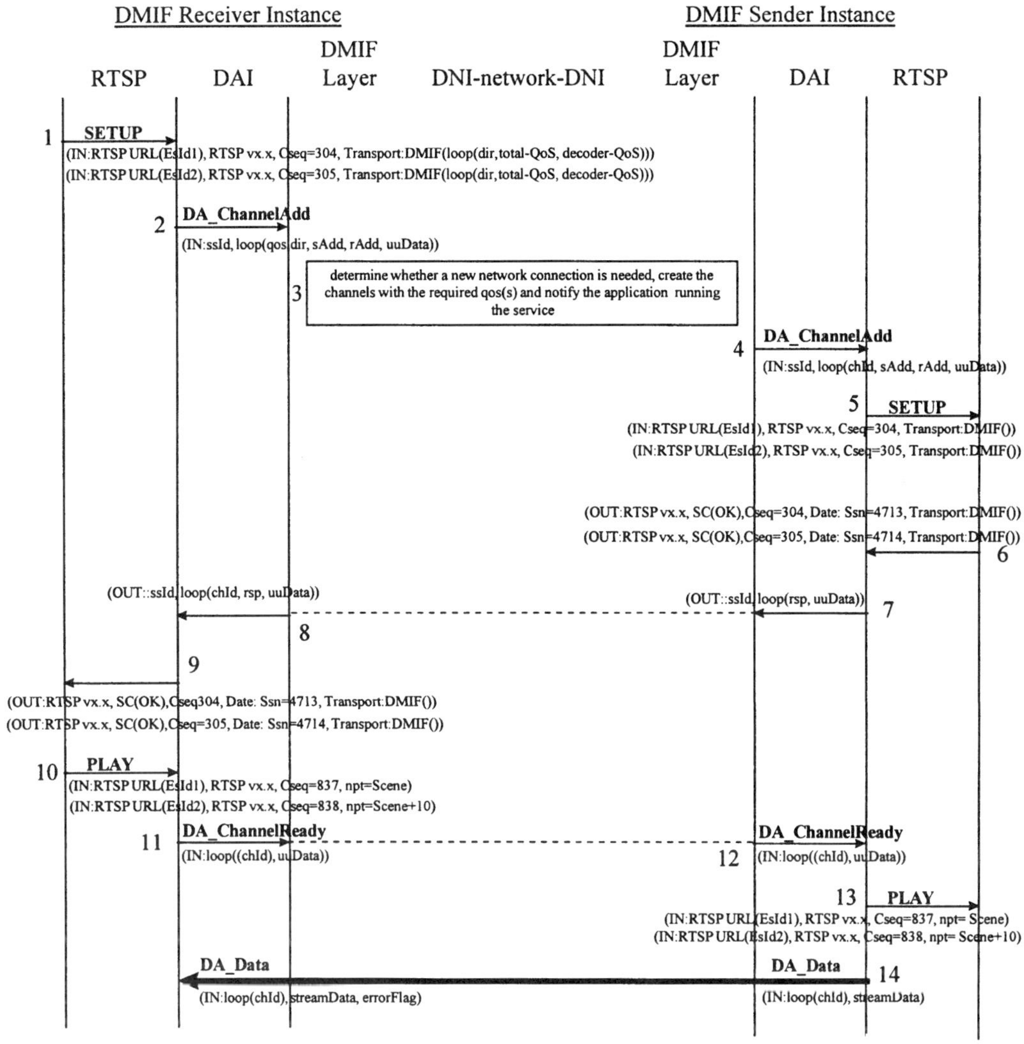

Figure 7 Message sequence with RTSP for MPEG-4 stream controls.

Step 2 DA_ChannelPause is generated. From the EsId1 the chId is derived. The RTSP PAUSE parameters except URL are carried in uuData. At this point it is possible to reuse the transport capacity not being used by the stream to carry lower priority streams through the use of DMIF–DMIF signaling, or if a new stream with high priority could not be established because of lack of sufficient bandwidth.

Step 3 DA_ChannelPause is received at the sender with a replaced sender local chId. The latter is used to identify EsId1.

Step 4 The RTSP PAUSE message is regenerated at the sender from the DA_ChannelPause.

Step 5 The EsId1 stream is paused at the instant the RTSP PAUSE command is

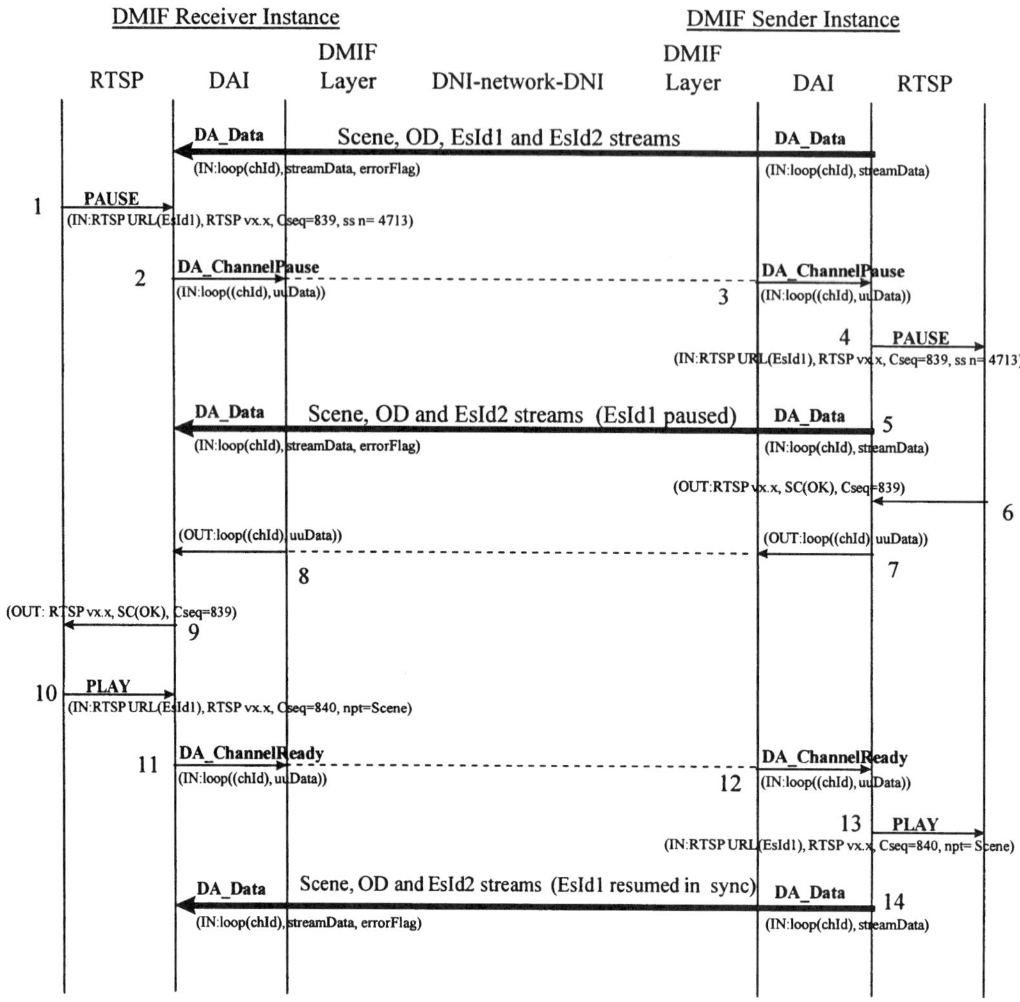

Figure 8 Message sequence with RTSP for MPEG-4 pause and play stream controls.

received and becomes in effect or is paused at the location specified by the NTP.

Step 6 The RTSP PAUSE response is received with a status code OK.

Step 7 The DA_ChannelPause response is generated using the parameters of the RTSP PAUSE response from step 6. At this point the transmission of the EsId ceases at the DAI. The channel has been reused in step 2.

Step 8 The DA_ChannelPause response is received with the local chId.

Step 9 The RTSP response is generated from the DA_ChannelPause in step 8 with a status code OK.

Steps 10 through 14 These remain the same as in Fig. 7.

4. *Tearing Down the Streams*

The message sequence in Fig. 9 covers the use of the RTSP TEARDOWN with DMIF. While a scene is active any individual stream can be torn down. Either a scene or individual

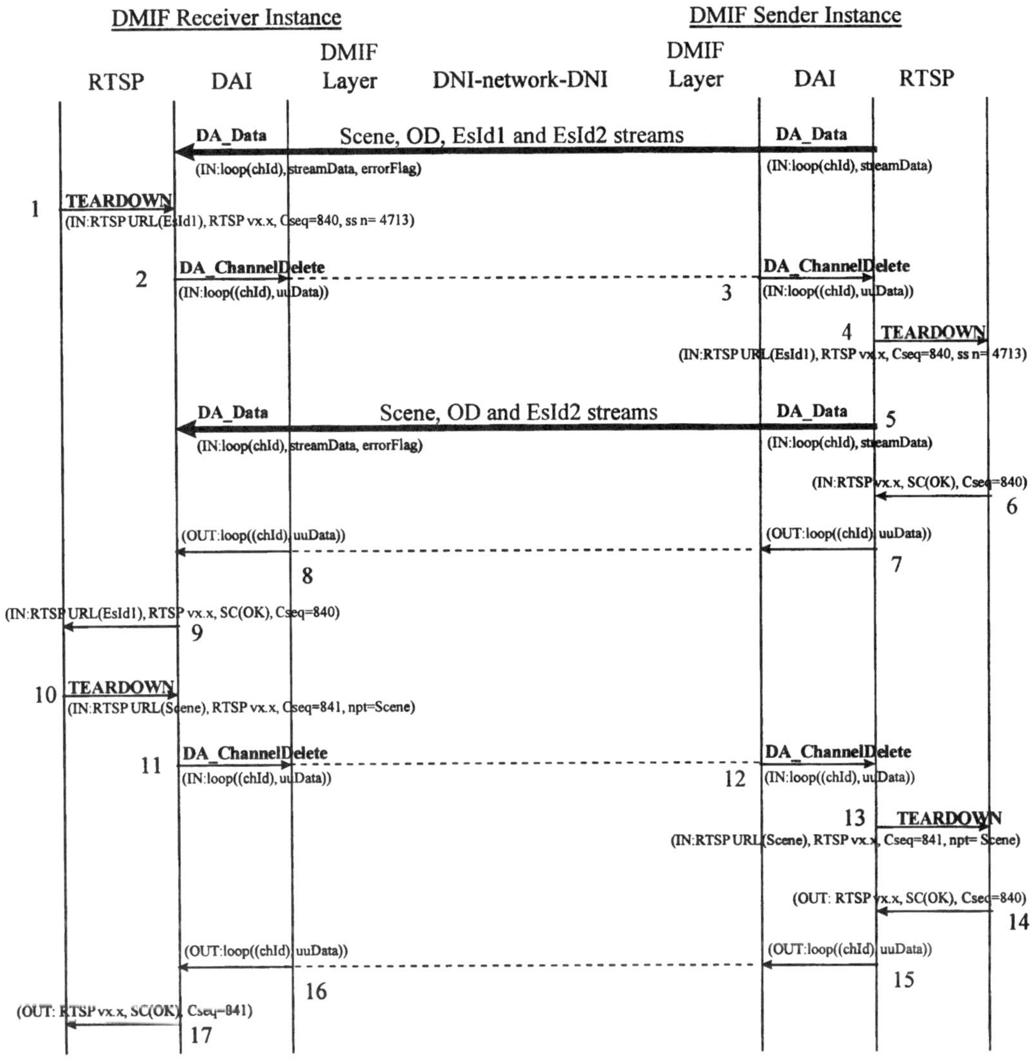

Figure 9 Message sequence with RTSP for MPEG-4 teardown stream controls.

streams can be torn down. If a scene is torn down it must be followed by a DA_ServiceDetach so that all the streams associated with that scene are torn down. If a torn-down scene is a master scene, all the embedded scenes are torn down.

Let us assume the normal play of a scene with selected streams is active.

Step 1 End user deselects the EsId1, and an RTSP TEARDOWN is issued with the RTSP session of the EsId1.

Step 2 DA_ChannelDelete is generated. From the EsIdl the chId is derived. The RTSP PAUSE parameters except URL are carried in uuData. At this point the channel is deleted by DMIF. This will result in the channel capacity being reused immediately or after a caching time-out. In the ultimate case in which channels are no longer being carried on the connection, DMIF will release the connection.

Step 3 DA_ChannelDelete is received at the sender with a replaced sender local chId. The latter is used to identify EsId1.

Step 4 The RTSP TEARDOWN message is regenerated at the sender from the DA_ChannelDelete.

Step 5 The EsId1 stream ceases at the DAI the instant the RTSP TEARDOWN command is received. The DMIF channel, however, was deleted in step 2.

Step 6 The RTSP TEARDOWN response is received with a status code OK.

Step 7 The DA_ChannelDelete response is generated using the parameters of the RTSP TEARDOWN response from step 6.

Step 8 The DA_ChannelDelete response is received with the receiver local chId.

Step 9 The RTSP TEARDOWN response is generated from the DA_ChannelPause in step 8 with a status code OK.

Step 10 through 17 These repeat steps 1 through 9, but in this case the scene stream as opposed to ESId1 is being deleted. In the semantics of RTSP, all streams of the presentation cease when the presentation is torn down. But this is not possible because the DMIF–RTSP interface element does not distinguish between a presentation RTSP URL and stream RTSP URL in order to send DA_ServiceDetach as opposed to DA_ChannelDelete in the case of RTSP URL for a presentation. The application is to send a DA_ServiceDetach that corresponds to the DA_ServiceAttach sent earlier in order to obtain the initial object descriptor as a precondition to activate RTSP.

F. Timing and Clock Considerations

The control of the scene playback is based on the NPT defined in MPEG DSM-CC IS [9], which indicates a current position within the stream relative to the beginning of the scene stream. There are two related NTP representations: AppNTP is used for human consumption and transport NTP is used to coordinate the position between the sender and receiver. AppNTP can be displayed like a VCR counter to the end user.

The NPT relates to an event. This is described in MPEG-2 Systems IS as consisting of a collection of streams with a common time base and with an associated start time and an associated end time. In MPEG-4, however, the event is embodied in the scene stream with all the other audio video object (AVO) elementary streams being controlled relative to the scene. Consequently, the NPT in MPEG-4 should refer only to the scene.

The relationship between the NPT and the scene OTB (and the derived OTB based on the STB) is defined by an NPT_Reference and the corresponding STC_Reference in the NPT Reference descriptor. This descriptor needs to be carried in the scene's first AU received at the receiver after each OTB time discontinuity at splicing points caused by editing a scene stream. Splicing may not be present if the scene is recreated from original stream contents.

The NTP Endpoint descriptor provides the startNPT and the stopNPT. These define the starting and ending points of a scene stream. This is also required in the scenes's first AU. The startNPT provides the value of the NPT at the startCompositionTimeStamp of the scene conveyed in the AL-PDU header in its first AU.

In normal play all the elementary streams selected from the scene will play congruently at the MediaTimeSensor.elapsedTime relative to the Composition Time Stamp of the scene AU. This time is converted by the receiver to the media time line of the related elementary stream, relative to its startCompositionTimeStamp. This is used to render the composition unit of this specific elementary stream.

In case an end user selects to advance a specific elementary stream at a rate faster or slower than a scene, a separate AppNTP may be shown for the elementary stream. In order to keep the relative position of the stream with respect to the scene, the MediaTimeSensor.elapsedTime may be incremented (or decremented to a maximum of Scene AU STC−startCompositionTimeStamp of the elementary stream) in order to render the successive composition units of the elementary stream.

As the MPEG-4 elementary streams are stored with the OTB at their encoding time, any jumps in the stream as a result of their control will result in discontinuities of the OTB. In order to alleviate this situation, the wallClockTime Stamp must be transmitted in the first AU of the stream after the OTB discontinuity.

The semantics of the RTSP TEARDOWN of a presentation is affected in the sense that with MPEG-4 it will not immediately tear down the complete presentation. For this an MPEG-4 application must rely on DA_ServiceDetach. It is noted that in Section VI.E.4 as a precondition before the RTSP is activated, the application sends a DA_ServiceAttach in order to obtain the initial object descriptor for the scene. The location of the initial object descriptor could be advertised in an SDP with a specific DMIF URL to be used in the DA_ServiceAttach.

MPEG-4 Systems needs to carry the NTP and NTP Endpoint descriptors soon after any scene OTB discontinuities. Any discontinuity of the OTB of an elementary stream must contain the wallClockTimeStamp in its first AU after the discontinuity.

DMIF needs to add DA_ChannelPause descriptor at the DAI. All the channel-related primitives and their callbacks must contain uuData to carry the field from the RTSP commands not carried explicitly in the primitives.

The present NPT event semantics based on MPEG-2 are to be changed from all streams with the same time base and with the same end and start times to the event consisting of only the MPEG-4 scene (all streams belonging to the scene are dealt with relative to the scene). A new NPT field is required to indicate the congruence of a stream with the scene in order to indicate a relative advance or rewinding of the stream with respect to the scene.

REFERENCES

1. ISO/IEC 14496-2 FDIS MPEG-4 Visual, November 1998.
2. ISO/IEC 14496-3 FDIS MPEG-4 Audio, November 1998.
3. ISO/IEC 14496-6 FDIS Delivery Multimedia Integration Framework, November 1998.
4. H Schulzrinne, S Casner, R Frederick, V Jacobson. RTP: a transport protocol for real time applications RFC 1889, Internet Engineering Task Force, January 1996.
5. ISO/IEC 14496-1 FDIS MPEG-4 Systems, November 1998.
6. D Miffs. Network time protocol version 3, RFC 1305, March 1992.
7. M Handley. GeRM: Generic RTP multiplexing, work in progress, November 1998.
8. H Schulzrinne, A Rao, R Lanphier. Real time streaming protocol, 1999.
9. DSM-CC IS ISO/IEC 13818-6.
10. H Schulzrinne. RTP profile for audio and video conferences with minimal control, RFC 1890, January 1998.

19
Multimedia Over Wireless

Hayder Radha, Chiu Yeung Ngo, Takashi Sato, and Mahesh Balakrishnan
Philips Research, Briarcliff Manor, New York

I. INTRODUCTION

Wireless is one of the fastest growing segments of the telecommunication and consumer electronics industries. The explosion of technological alternatives and the success of the second-generation digital cellular systems (e.g., Global System for Mobile, GSM, and Personal Digital Cellular, PDC) have established wireless communications as indispensable in modern life. With the emerging standards of IEEE 802.11 [1] in the United States and HIPERLAN (High-Performance Radio Local Area Network) [2,3] in Europe, a line of new wireless products will soon appear in the market. Because of the low-cost and low-consumption characteristics of emerging wireless products targeted at low to medium bit rate services, these products are expected to play an important role in wireless communications in the next few years. For example, the creation of the ''Bluetooth'' consortium [4] early in 1998 by industry leaders could revolutionize wireless connectivity for short-range voice communication and Internet access. Meanwhile, as people are becoming more aware of the benefits of wireless communications (e.g., flexibility, portability, and no wire necessary for installation), demand will begin to appear for a more advanced form of services, namely multimedia services.

Wireless multimedia communications (e.g., audiovisual telephony and videoconferencing) require medium to high bit rate channels (64 kbps to 2 Mbps per user). Therefore, for these applications, it will be necessary to have broadband wireless networks that support bit rates in excess of 2 Mbps per radio channel, where each radio channel could be shared by multiple users or sessions. In addition, these sessions have to be provided with some quality-of-service (QoS) guarantees over their respective, error-prone wireless connections. In order to achieve these goals, one has to address the following key issues: (1) how to increase the capacity of wireless channels, (2) how to provide QoS in a cost-effective way, and (3) how to combat the wireless channel impairments.

These questions and related technical issues have been addressed in numerous research papers, including several excellent overview papers in the field. For example, Li and Qiu [5] give an overview of personal communication systems (PCSs) that can provide timely exchange of multimedia information with anyone, anywhere, at any time, and at low cost through portable handsets. They also provide an excellent overview of various multiple-access schemes. Hanzo [6] describes various wireless multimedia concepts, sampling and coding theory, cellular concept, multiple access, modulation, and channel coding techniques. Hanzo also provides an extensive list of references. Mikkonen et al. [7] give an excellent overview of various emerging wireless broadband networks in Europe and discuss the frequency spectrum issue in details. Pahlavan et al. [2] give an overview of the status of wideband wireless local access technologies and a comparison of IEEE 802.11 and HIPERLAN. Correia and Prasad [3] give an overview of wireless broadband communications by addressing some of the applications and services that are foreseen, as well as some of the technical challenges that need to be solved.

In addition to the continuing interest in wireless audiovisual communication applications just mentioned, a great deal of interest has emerged in higher end wireless multimedia services (e.g., video and audio distribution of entertainment content to or within the home). However, current wireless networks, which are primarily low-bit-rate narrowband systems targeted for voice or data, are inadequate for supporting audiovisual communication applications (e.g., medium-bit-rate video telephony) or high-end multimedia services. To address this problem, two major trends have emerged in industry and research. First, numerous research and standardization activities have focused on the development of source and channel coding schemes for the transmission of audiovisual information over error-prone wireless networks. Second, a great deal of interest has emerged in the development of technologies, prototype systems, and standards for a reliable wireless broadband communication system (WBCS) with data rates exceeding 2 Mbps and perhaps up to 155 Mbps. In this chapter, we focus on these two areas of wireless multimedia: (1) source and channel coding for audiovisual transmission and (2) evolving WBCS technologies and standards. As an example of a very important WBCS application area, we also provide an overview of wireless in-home networking technologies and standards.

The remainder of this chapter is organized as follows. Section II scans recent overviews published in the area of wireless multimedia. Section III provides an overview of broadband wireless systems and technologies and highlights some of the key technical issues for the delivery of multimedia traffic over such networks. Section IV provides an overview of audiovisual and channel coding solutions for multimedia-over-wireless applications. Section V describes the latest trends in in-home networking technologies and standards and addresses how emerging wireless solutions fit in this critical application area.

II. OVERVIEW OF WIRELESS MULTIMEDIA

In this section, we provide a brief summary of recent overview papers on wireless multimedia communications. We identify some of the primary issues, problems, and solutions that have been addressed in these papers and point out the areas that have not received much attention.

Prior review papers on wireless multimedia have concentrated primarily on mobile communications [6]. Because of the growing interest in mobile multimedia communications, a series of conferences has been organized on that topic [8]. The main technical areas that have received considerable attention from researchers in this community are wireless networks, protocols for QoS and media access control, source coding, channel coding, modulation, power management, and mobility. There have also been papers on complete systems such as cellular telephone systems, as well as on subcomponent systems such as transceivers.

In the rest of this section, we will briefly touch upon the main subtopics that have received attention in prior wireless multimedia review papers.

A. Networks and Protocols

A good overview of the evolution of the cellular telephone system is given in Hanzo [6]. Currently, work is ongoing to develop third-generation cellular systems called universal mobile telecommunications systems (UMTSs). Other network technologies that are relevant to wireless systems include asynchronous transfer mode (ATM) and Internet Protocol (IP). An overview of the challenges that need to be addressed to make wireless ATM a reality is covered in Ayanoglu et al. [9]. Because ATM was originally designed to operate over very low error rate channels (e.g., a bit error rate of 10^{-10}), it has been questioned whether it will work at all at the higher error rates that can be expected in wireless systems. Some of these questions have begun to be answered.

In the Internet domain there are various proposals and Internet Engineering Task Force (IETF) drafts referred to by the combined name Mobile Internet Protocol [10]. The vast majority of them have concentrated on issues related to mobility as opposed to supporting wireless connections. It needs to be emphasized that wireless and mobility do not always go hand in hand. There are applications (surround sound speakers in the home, for example) that benefit from being untethered (wireless) but do not necessarily need to be mobile. The IETF working group on Mobile IP has examined the routing problem in the context of mobility and has adopted a twin mechanism to address it. There are other problems that are more related to wireless—for instance, how the congestion-based Transport Control Protocol (TCP) back-off approach will work in the case of packet loss due to channel errors (i.e., packet losses that are not due to router congestion, as is the case with standard IP networks). There are preliminary solutions to some of these issues.

Most multimedia services tend to be real-time in nature; i.e., the data being transported need to get to the destination by a certain time in order to be useful. This implies the need to develop techniques for call admission, bandwidth allocation, and the handling of real-time variable rate streams. These are problems, however, that apply to wired networks as well and are not, therefore, unique to wireless multimedia communication systems.

Other wireless networks have been specified or are still evolving. These include 802.11 [1] and Bluetooth [4] for use in corporate environments as well as possibly in the home, in addition to Wireless 1394 and HomeRF, which are predominantly for the home, as discussed in Section V. These networks are designed to support multimedia, but the carriage of multimedia over these networks has not yet been studied.

The two major protocol-related problems in wireless multimedia concern medium access and quality of service. Wireless systems are inherently multiple medium access in nature and, therefore, there needs to be a reliable medium access control (MAC) layer that also supports QoS. A good overview of many of the current approaches to designing a MAC layer is provided in Goodman and Raychaudhuri [8].

B. Source Coding

Audio, video, and graphics need to be compressed before transport over a bandwidth-constrained wireless channel. Given the emphasis on mobile wireless systems in the past, the media element that has received the most attention in the context of wireless multimedia is speech [6]. This is only natural because the most widely deployed wireless multimedia system today is cellular telephony, which is a fairly limited bandwidth system. There has also been a great deal of interest in wireless video [6,8] given the increased bandwidth capabilities of UMTS. The two video compression standards that are most relevant to these systems are MPEG-4 [11] and H.263 [12], both of which have already been evaluated for use in GSM systems. Because of the unreliable nature of wireless networks, it has become important to build source coding schemes that are robust to channel errors. Scalable compression schemes that offer graceful degradation with loss of data have become popular. There is, however, still a fair amount of work to be done to get scalable schemes that work satisfactorily. A somewhat mature approach that was first researched in the late 1970s, multiple description coding [13], is again receiving attention as a possible solution for wireless multimedia. Multiple description coding provides multiple coded versions of the same source that can be transported over different channels (or with different levels of error protection). The source can be fully recovered when one gets all of the descriptions. However, access to just one of the descriptions enables partial recovery of the source.

Audio and graphics are two source elements that have not received extensive research in the context of wireless systems. There has, however, been some work on handwriting coding [8]. Delivery of high-quality audio over wireless networks is an application that will in some cases arrive before video, particularly in the home. It is therefore ironic that video has received so much more attention than audio. This is again due to the fact that the wireless networks that have received the most attention so far are cellular telephony systems and wireless wide area networks (WANs) for corporate environments.

Even with scalable and multiple description-based source coding schemes, there will still be lost data on wireless systems. Error recovery and concealment at the receiver is therefore an important topic and has received some attention [14–16], again primarily for video. These error concealment techniques rely to a large extent on knowing the underlying source compression technique and exploiting some of the tools that are used therein. For example, many concealment techniques attempt to leverage the fact that video compression schemes use motion compensation and that motion vector information is available in the data stream. Error concealment techniques for audio have not received much attention. Solutions in this area seem to be difficult to come by.

New trends in source coding for wireless communications are covered in more detail in Section IV.

C. Channel Coding

An overview of channel coding techniques that are relevant to wireless multimedia is given in Goodman and Raychaudhuri [8]. Convolutional codes and block codes have been used successfully to tackle burst channel errors. Punctured codes have been used to provide unequal bit error protection within the data stream. This enables the more important bits within the source to be error protected at a higher level than the less significant bits.

Significant gains have been made in channel coding through the development of turbocodes. Turbocoding involves encoding the sequence twice using recursive systematic convolutional codes with an interleaving stage between the two coding stages in order to introduce randomness into the bitstream and reduce the correlation between the two encoded streams.

Shannon's theory states that it is theoretically possible to build an optimal communication system that is a combination of a source coder and a channel coder that are treated as separate entities. This ignores the complexity of the algorithms as well as coding delays. However, as a result of Shannon's fundamental work, source and channel coders have historically been optimized separately. Joint source–channel coding, in which source and channel coders are optimized in tandem, has immense applicability in wireless multimedia communication systems. Although many excellent research works have been reported in the literature, no comprehensive overview of joint source–channel coding can be found in overview papers on wireless multimedia. In Section IV, we provide an overview of research on joint source–channel coding that is relevant to video wireless communications.

D. Mobility

Most of the wireless systems today also support mobility. The velocity associated with mobility has been one of the key parameters that affect system design. For this reason, many of the approaches to solving channel-related problems associated with mobility have been developed for specific classes of mobile systems—pedestrian (velocity of a few meters/sec), vehicular (velocities of about 100 km/sec), and high-speed (velocities of hundreds of kilometers/sec). Examples of high-speed mobile systems are those deployed on the European high-speed train network (i.e., the TGV train network). In cellular systems, handoff from one cell to another is an issue that needs to be addressed and requires some fundamental rethinking for connection-oriented systems such as ATM [17]. Mobility also affects routing and addressing, which has received a significant amount of attention from the Mobile IP community [10]. Mobile middleware [8] attempts to offer high-level applications an interface that allows a common view of the underlying network, which may be heterogeneous in addition to hiding aspects of mobility from the application.

Tracking of mobile systems is also an important issue for many mobile multimedia applications. For example, to support a ''follow-me'' application within a building, the system has to know where the user is within the building and be able to track the user as he or she moves within the building. There is ongoing study of how to support location identification on many mobile telephone systems in order to accommodate emergency services such as ''911'' in the United States.

III. WIRELESS BROADBAND COMMUNICATION SYSTEM (WBCS) FOR MULTIMEDIA

Depending on its applications, there are two distinct approaches to the development of WBCS: wireless local area network (WLAN) and mobile broadband system (MBS). Whereas WLAN targets mainly indoor applications (e.g., wireless office and home networking), MBS is intended for a future generation of cellular systems that can provide full mobility to Broadband Integrated Services Digital Network (B-ISDN) users. Currently, both approaches are under intensive study by various standardization bodies. MBS is taking place both in the UMTS initiative in the European Telecommunication Standards Institute (ETSI) and in the International Mobile Telecommunications (IMT-2000) initiative in the International Telecommunications Union (ITU). WLAN is considered mainly in four organizations: ATM Forum, Internet Engineering Task Force (IETF), IEEE 802.11, and Broadband Radio Access Network (BRAN) in ETSI [18]. Whereas ATM Forum, ITU, and IETF focus on the core network standards, IEEE 802.11 and BRAN define the wireless access network standards.

Although the core network dilemma is still going strong between IP and ATM for broadband multimedia services, almost all of the WBCS technology demonstrations are based on ATM technology. This is not surprising because high-speed ATM technologies have being developed for many years now for wired networks. Moreover, ATM as a broadband infrastructure has been designed for multimedia communications to accommodate a variation of data rates, QoS requirements, and connection and connectionless paradigms. It is quite natural to assume a combination of wireless and ATM-based services at the consumer end of a wired network. A list of these hybrid technology demonstrations can be found in References 7, 19, and 20, including NEC's WATMnet, Bell Labs' BAHAMA, and various European Union–funded projects such as the Magic WAND (Wireless ATM Network Demonstrator), AWACS (Advanced Wireless ATM Communications Systems), SAMBA (System for Advanced Mobile Broadband Applications), and MEDIAN (Wireless Broadband CPN/LAN for Professional and Residential Multimedia Applications). Except for AWACS and SAMBA, which are targeted for both indoor and outdoor usage, all others are for indoor WLAN applications. As pointed out in Chelouche et al. [21], a cost-effective WLAN design can easily be obtained by exploiting its characteristics (e.g., license free, low operating frequency, and flexibility in architecture design such as ad hoc networking [22]).

In order to deliver multimedia traffic over broadband wireless networks, we need to have sufficient bandwidth and be able to support service-specific QoS requirements concerning delay, delay variation, and packet loss on a per-connection basis. That means we have to deal with all problems associated with time-varying channels and QoS assurance over the time-varying error-prone ''air interface.''* In the following, we address these problems hierarchically using a layered approach and discuss various multimedia-related issues in detail.

A. Physical Layer Issues

The radio physical layer is essentially the soul of any wireless network. Ideally, one wants to find a radio physical layer technology that is spectrum efficient (bits per hertz), minimizes the radio overhead (e.g., radio receiver–transmitter turnaround time, synchronization, and training sequences), and is robust in both indoor and outdoor environments. However, because of various channel impairments, it is very hard to get an optimal radio physical layer.

1. *Spectrum Consideration* [7,23]

The wireless broadband air interface will demand a relatively large frequency band to support bit rates in excess of 2 Mbps. This type of allocation is hard to find below 3 GHz and the availability of bandwidth becomes easier on higher frequencies, but at the cost of more complex and expensive techniques. Because at higher frequencies the path loss is greater, line-of-sight operation becomes important and wall penetration becomes a challenge for WLANs. Various spectra are considered for WBCS, as follows:

5 GHz band: In Europe, HIPERLAN will use 5.15–5.35 GHz and 5.47–5.725 GHz. In the United States, a SUPERNet band at 5.15–5.3 GHz and 5.725–5.825

* ''Air interface'' is commonly used jargon for a wireless connection.

GHz with 300 MHz bandwidth was proposed by the Federal Communications Commission (FCC) for unlicensed National Information Infrastructure (UNII) devices (wireless LANs, ATM) up to 20 Mbps. The 5.2 GHz band overlaps with HIPERLAN, and 5.7 GHz falls into the industrial, scientific, and medical (ISM) band in Europe.

17 GHz band: In Europe, 17.1 to 17.3 GHz is allocated for HIPERLINK system deployment. HIPERLINK is also covered under the ETSI BRAN standardization activity.

40 GHz band: In Europe, 39.5–40.5 GHz and 42.5–43.5 GHz will be used as bands for mobile broadband systems.

60 GHz band: In Europe, 62–63 GHz paired with 65–66 GHz is provisionally allocated for MBSs. In addition, the band at 59–62 GHz may be used by WLANs. In the United States, the 60-GHz band is considered for high-speed applications and is supported by an etiquette, so the frequencies 59–64 GHz are for general unlicensed applications.

Note that 5- and 60-GHz spectrum allocations are geographically widely available for MBS or WLAN. This makes them commercially important. BRAN's initial targets are the 5-GHz band that is available in Europe, the United States, and Japan. To ensure compatibility among the produced standards, close cooperation among the ATM Forum, IEEE 802.11, Japan's ARIB (Association of Radio Industries and Business), and BRAN must be established.

2. *Channel Characteristics* [3,5,21,23]

In the wireless environment, the transmitted radio signal is subject to various time-varying impairments that arise from inherent user mobility and unavoidable changes related to the movement of surrounding environment. This results in fading and shadowing effects. Another problem is the presence of multipath propagation, leading to fading (flat or frequency-selective) and time delay spread, which give rise to intersymbol interference (ISI) that can strongly increase the bit error rate (BER). One way to overcome fading-induced impairments is to use antenna diversity techniques. This is a useful concept for capacity enhancement. The use of adaptive antennas can be exploited to steer the beam. The combination of antenna diversity and equalization has the potential to offer significant performance and capacity gains. Another way to improve spectrum efficiency is via power control.

3. *Modulation and Channel Coding Schemes* [6,7]

Limited spectrum drives the need to have modulation and access schemes that are spectrally efficient and have a strong resistance to cochannel interference. In GSM and Digital Enhanced Cordless Telecommunication* (DECT), constant envelope partial response Gaussian Minimum Shift Keying (GMSK) is employed. Its main advantage is that it ignores any fading-induced amplitude fluctuation present in the received signal and hence facilitates the utilization of power-efficient nonlinear class C amplification [24]. In WBCS

* The phrases Digital European Cordless Telephone [6] and Digital European Cordless Telecommunication [26] have also been used to refer to the DECT standard. However, based on the official Web site for DECT (http://www.etsi.fr/dect/dect.htm), the abbreviation stands for Digital Enhanced Cordless Telecommunication.

there are two promising solutions, namely (1) traditional single-carrier modulation with powerful equalization and/or directional sectored antenna to combat the multipath fading and (2) multicarrier techniques such as OFDM. However, because WBCS will probably employ pico- and microcells with low transmitted power and low signal dispersion, the deployment of bandwidth-efficient multicarrier modulation will become more desirable. Although OFDM is more complex and costly, its great ability to combat frequency-selective fading allows it to be adopted in BRAN and the next version (i.e., the 5-GHz version) of IEEE 802.11 (up to 54 Mbps). OFDM is also a serious candidate for terrestrial systems both in Europe (where it is part of the DAB and DVB-T standard) and in the United States.

Research shows that trellis-coded modulation (TCM) provides a higher coding gain than consecutive forward error correction (FEC) coding and modulation over an AWGN channel. Significant advances have been made toward the Shannonian performance predications with the introduction of the turbocodes [25].

B. Data Link Control Layer Issues

The wireless transmission medium is a shared radio environment. Therefore, coordinated scheduling of transmission by a central access point (e.g., a central controller-based architecture) can be used to maximize throughput. The major issues are to define a flexible air interface and efficient error control and traffic scheduling algorithms. Data link control (DLC) is the core for multiplexing services with varying QoS demands. Although the specific scheduling algorithms should not be standardized, the DLC control framework and interface parameters should be harmonized. A generic wireless DLC will consist of a flexible packet access interface, delay- and delay variation–oriented scheduling for multimedia traffic, and error control per service requirement. In general, QoS control is taken care of by core network signaling.

Ideally, one wants to find an appropriate DLC protocol that can (1) maintain some availability of the shared wireless resource using a multiple access technique and (2) at the same time provide added protection on the transmitted packets using error control to mitigate packet loss related to the fading of the medium.

1. *Multiple Access Scheme* [5,7,21,23,27]

Multiple access (MA) protocol is required to minimize or eliminate the chance of collision of different information bursts transmitted from different users. Ideally, the desired MA scheme should be insensitive to the channel impairments. The services provided to the user must satisfy certain quality requirements no matter how bad the channel is. In addition, a good MA protocol can improve system capacity and lower the system cost. Hence, the MA scheme is a very important design issue for efficient and fair utilization of the available system resources. In order to provide multimedia services under the limited bandwidth constraint, a sophisticated MA scheme is crucial to cope with various traffic characteristics: voice is delay sensitive but relatively loss insensitive, data are loss sensitive but delay insensitive, and video data rate is generally much higher than either voice or ordinary data and is also delay sensitive. These traffic characteristics have to be considered in the integration of the MA protocol design. It is not surprising that there is no MA scheme optimized for every kind of traffic. Designing an integrated MA protocol is very challenging.

In wireless communication, the multiple access channel can be shared by a large number of users via an MA scheme. There are three types of MA schemes: Frequency

Division Multiple Access (FDMA), Time Division Multiple Access (TDMA), and Code Division Multiple Access (CDMA). FDMA assigns a unique frequency band to each user, TDMA assigns access in time slots, and CDMA assigns a unique code using a spread spectrum technique. FDMA, although suitable for the first-generation cellular systems (AMPS, TACS), lacks the attributes needed for high-capacity multimedia wireless systems. Between TDMA and CDMA, each has its own set of advantages. As experts are still debating their relative merits, the decision has not been finalized. Various combinations of these three basic sharing methods are also possible. According to the amount of coordination needed in the resource assignment, MA schemes can be categorized into three types, namely random access (e.g., ALOHA and CSMA/CD, CDMA), fixed assignment (FDMA and TDMA), and demand assignment [Dynamic TDMA (D-TDMA), Packet Reservation Multiple Access (PRMA), Resource Auction Multiple Access (RAMA)]. The duplex access scheme can be either FDD (Frequency Division Duplex) or TDD (Time Division Duplex). The TDD systems have the advantage of flexible bandwidth allocation between uplink and downlink and may therefore be suitable for applications in which asymmetrical transmission is needed.

2. *Error Control*

Whereas the DLC layer is used to enhance the transport capability of the physical layer, the Logic Link Control (LLC) layer is used to improve error performance. Error control is typically achieved by coding and/or retransmission. A trade-off between coding and retransmission has to be optimized for the efficient transmission of data over the air interface. For error control, FEC and automatic Repeat Request (ARQ) are very effective in improving QoS parameters.

C. Higher Layer Issues

1. *Channel Allocation* [5]

Channel allocation is an important issue in radio resource management. It involves allocating the radio resources systemwide to achieve the highest spectrum efficiency. There are two basic forms of channel allocation, namely fixed channel allocation (FCA) and dynamic channel allocation (DCA). An overview of these allocation schemes and their variants can be found in Li and Qiu [5].

2. *Handoff* [5,13,28]

This is under the umbrella of mobility management and involves all the issues associated with the mobility feature of wireless communications, such as location registration, update, paging, handoff, and call routing. This requires tracking the call to maintain connection and provide intercell handoff when needed. In order to maintain the desired QoS after handoff, a scheme called QoS-controlled handoff [28] may be required.

D. Multimedia-Related Issues

Application awareness and adaptive streaming are two essential concepts in QoS provisioning in the multimedia wireless environment [27]. In the following, we discuss how these concepts can be utilized to ensure QoS in continuous multimedia streams via dynamic bandwidth control, flow control, and admission control aspects [29].

1. *Dynamic bandwidth control* ensures seamless end-to-end delivery of multimedia traffic over both wireless and wired portions of a communication path. The realization of end-to-end QoS control and the exploitation of scalable flows can be achieved through resource binding and provision of a set of QoS-aware adaptive algorithms, such as adaptive network services, and adaptive and active transport services. The latter incorporate a QoS-based application programming interface (API) and a full range of transport algorithms to support the delivery of continuous media over wireless networks.
2. *Flow control* is information rate control exerted by a receiving system or network node (whether mobile or not) on the source sending the information. Control is typically exerted when a receiver is in danger of overflowing. This type of control cannot be exerted on real-time traffic such as video, which must be delivered with specified deadlines on time delay and delay jitter. It can be exerted only on non–real-time traffic with relatively loose delay bounds.
3. *Wireless call admission control* requires an estimate of effective bandwidth over the air and a resource (bandwidth or capacity) allocation strategy for different categories of users.

E. Other Considerations

1. *Hardware–software radio*: A software radio is a flexible radio in which many of the functions traditionally implemented in hardware are implemented in software and therefore easily reconfigurable. This is a useful concept in enabling radio systems to offer a wide range of data services. It will provide a common platform to support the multiple air interface standards that exist in the world today.
2. *Power consumption*: Low-power-consumption large-scale integration (LSI) or monolithic-microwave integrated circuit (IC) combined with high-density packaging technology is required to realize compact and lightweight personal terminals.
3. *Security services*: Cryptographic systems are necessary to provide adequate privacy and authentication for wireless communications.

IV. AUDIOVISUAL SOLUTIONS FOR WIRELESS COMMUNICATIONS

There has been a great deal of standardization and research effort in the area of audiovisual coding for wireless communications. Many excellent papers and reviews have been published covering different aspects of audiovisual coding for wireless communications (examples of recent publications include References 6, 26, 30, 31, and 32). In general, previous reviews have focused on wireless speech coding schemes and error-resilient video standards. Meanwhile, a great deal of research has been conducted in the area of joint source–channel coding for error-prone networks. In this section, we provide a very brief overview of speech coding schemes being used and proposed for wireless telephony applications. Then we focus our attention on providing a more comprehensive overview of recent image and video coding techniques for wireless multimedia.

In the area of wireless audio, the focus has been on the development of speech coding solutions for cordless systems, cellular telephony services, and emerging personal communication services (PCSs) [6,26]. A comprehensive overview of wireless speech coding solutions is given in Reference 26. Here we provide a short summary of the different speech coding methods covered in the literature. For more details, the reader is referred to the cited references.

Many of the speech coding solutions proposed for wireless applications have their roots in earlier wireline speech coding standards [26,33]. The International Telecommunications Union–Telecommunications sector (ITU-T) developed a series of speech coding standards for wireline telephony. These standards include ITU-T G.726, G.728, G.729, and G.723.1. The G.726 and G.728 standards provide "toll quality" speech at bit rates of 32 and 16 kbps, respectively. G.729 and G.723.1 are designed for low-bit-rate telephony applications at 8 kbps (G.729) or lower rates (G.723.1). The G.723.1 standard, which supports dual-rate coding modes at 6.3 and 5.3 kbps, has been adapted by ITU-T standards on audiovisual telephony terminals for low-bit-rate applications such as H.323 and H.324.

Cordless systems and many of the PCS systems developed in Europe (DECT), North America (e.g., PACS, personal access communications services), and Japan (PHS, personal handy-phone system) are based on the ITU-T G.726 speech coder. This is due, among other factors, to the low complexity of the G.726 Adaptive Differential Pulse Code Modulation (ADPCM) coder, which translates into low power for cordless and PCS devices.

For a comprehensive overview of speech coding solutions for digital cellular services (e.g., GSM), the reader is referred to Reference 26. In the following we shift our attention to visual coding for wireless applications.

A. Image and Video Coding for Wireless Applications

Because of the time-varying impairments that characterize wireless communication channels, special attention has to be paid when coding any multimedia signal to be delivered over such networks. In particular, compressed image or video signals can experience severe degradation if transmitted over error-prone channels. This is due mainly to (1) the usage of variable length codes (VLCs) in the compressed bitstream and (2) the deployment of prediction-based coding needed for eliminating both spatial and temporal redundancies in the original signal. Channel errors affecting a VLC could result in a loss of synchronization at the decoder. For example, after a VLC error occurs, the decoder may lose track of the image region to which the received compressed data pertains. This phenomenon usually creates major degradation in significant portions of the picture experiencing the error. An example is shown in Figure 1. The picture in this example is divided into equal regions of pixels where each region is referred to as a group of blocks (GOB). The example assumes that a synchronization code is used in the bitstream at the beginning of every GOB. Therefore, an error inside the compressed bitstream could damage the remainder of the GOB being decoded until synchronization is achieved at the beginning of the next GOB.

When interpicture prediction-based coding is employed, major degradation in a video sequence can be observed if a reference frame (e.g., an intra-coded picture) experiences any errors. In this case, when a corrupted picture is used to predict another picture, the error will propagate in the video sequence until the affected area is refreshed by trans-

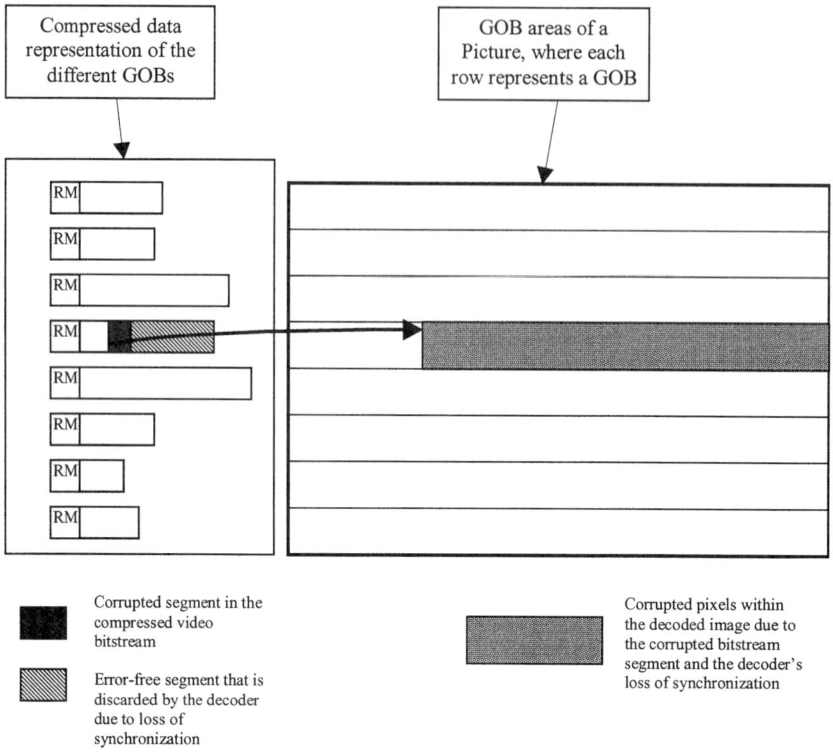

Figure 1 An example illustrating the impact of an error in the compressed data on the region of a picture in the pixel domain. RM is a resynchronization marker used at the beginning of each group of blocks (GOB). This enables the decoder to achieve synchronization at the next GOB after an error.

mitting intra-coded blocks or a whole new intra-coded picture. An example of an error propagation scenario in a video sequence is shown in Figure 2.

As mentioned in Section II, error concealment algorithms have been developed for image and video signals that experience channel impairments [14–16]. One advantage of error concealment methods is that, in general, they do not require any overhead data (i.e., maintaining good coding efficiency). However, extra complexity is needed at the receiver to implement these approaches. In Reference 14, a good overview of error concealment methods is provided. In Reference 15 and 16 new nonlinear error concealment filtering techniques are introduced. These nonlinear methods provide promising results and are more robust than linear filtering techniques. However, the nonlinear filters are usually more complex than their linear counterparts.

Because of these issues, recent video coding standards support error-resilient tools for protecting the compressed video signal against wireless network impairments. In addition, a great deal of research has been directed toward joint source–channel coding schemes designed to minimize the impact of channel errors on compressed video and image signals. In the following two subsections, we provide an overview of both video standards for wireless applications and new trends in joint source–channel coding optimized for error-prone channels.

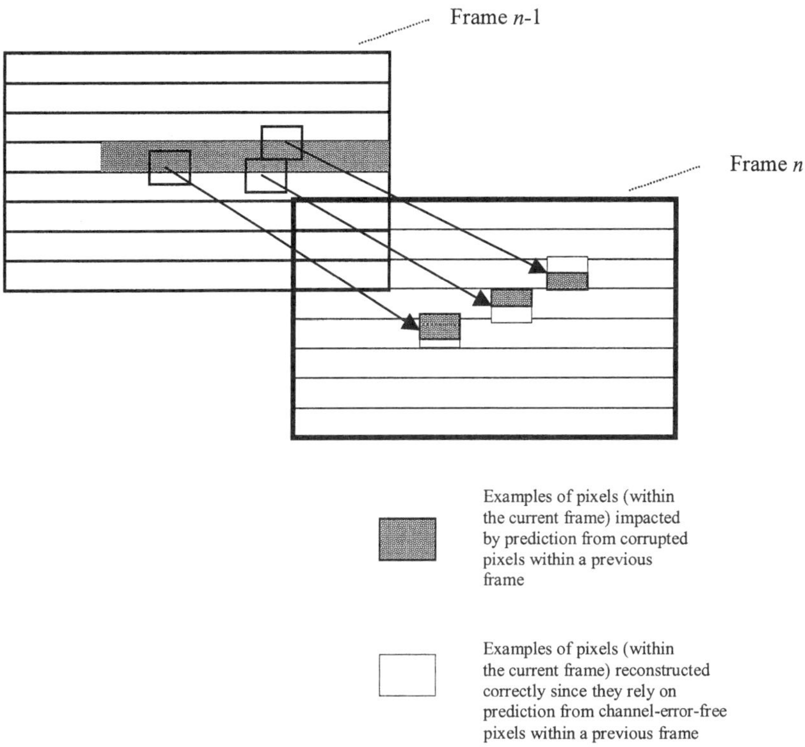

Figure 2 An example of error propagation due to interframe prediction coding.

1. Error-Resilient Video Coding Standards

Both the ITU-T H.263 and ISO MPEG-4 video standards include error-resilient tools suitable for multimedia wireless communications. These tools and related robust video coding mechanisms are described in Farber and Girod [32] and Steinbach et al. [34] for H.263 and in Talluri [31] for the ISO MPEG-4 video standard. Resynchronization markers and reversible VLCs (RVLCs) are among the error-resilient schemes used in both MPEG-4 and the H.263 standard.*

As mentioned before, resynchronization markers enable the decoder to regain synchronization after a VLC error. In general, when a channel error occurs, the closer these markers are in the bitstream, the smaller the affected region. In MPEG-4, the encoder has the option of (1) not using any GOB resynchronization markers (e.g., in error-free situations), (2) inserting these markers at the beginning of every GOB, or (3) distributing these markers in the bitstream between integer numbers of macroblocks. In the last case, the encoder can distribute the resynchronization markers evenly in the bitstream (e.g., every 256, 512, or 1024 bits depending on the bit rate). The baseline H.263 standard supports options (1) and (2) only.

* The reversible VLC feature is supported in an extension of H.263 [32].

Reversible VLCs [35] enable the receiver to decode the bitstream in a backward manner starting from the next resynchronization marker after an error. Therefore, for any bitstream segment located between two consecutive synchronization markers, RVLCs could assist the decoder in isolating, and consequently discarding, the region of the bitstream segment experiencing one or more errors as shown in Figure 3. The other error-

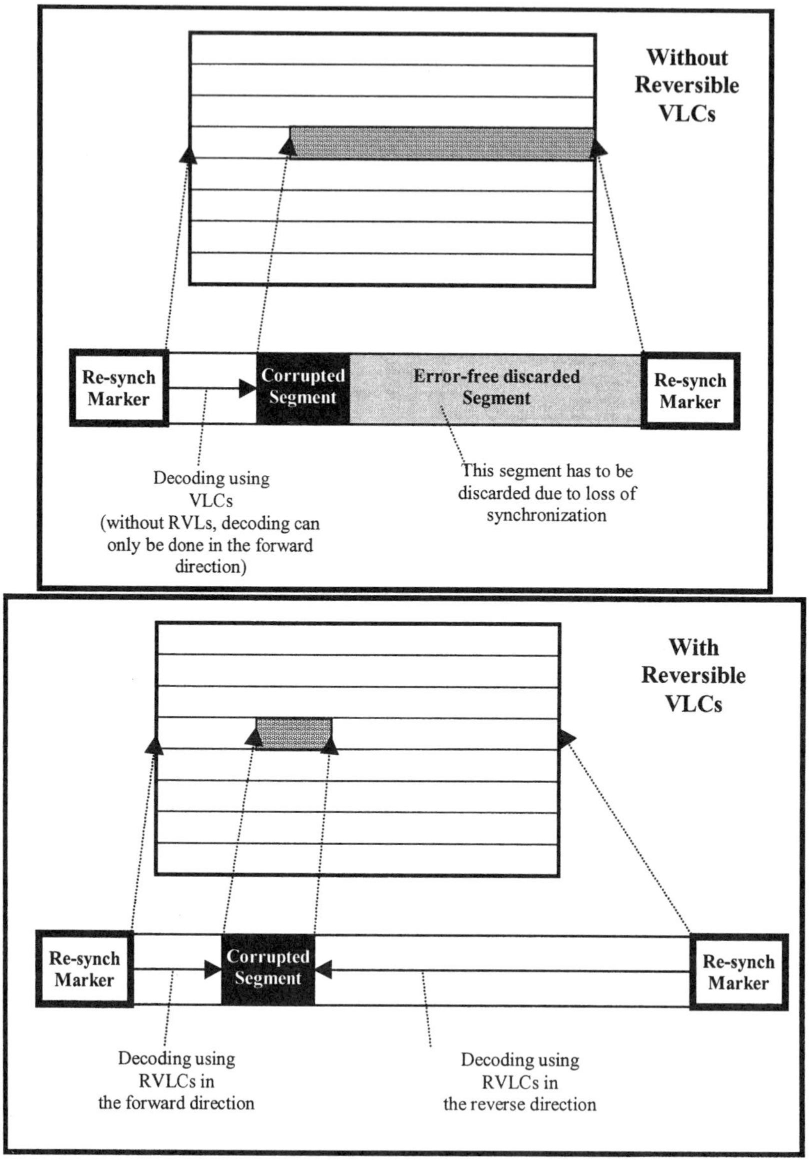

Figure 3 Example illustrating the benefit of using reversible variable length codes. The data segment for the case with RVLCs is shown wider (i.e., more bits) than the case without RVLCs because some coding efficiency is lost when using RVLCs.

free regions of the same bitstream segment can then be used by the decoder to generate viable pixels.

In general, adding any robustness to the compressed video signal in the form of resynchronization bits or using RVLCs reduces the coding efficiency. In order to maintain a good balance between coding efficiency and error resilience, other mechanisms have also been proposed in conjunction with standard-compliant bitstreams (see, for example, Refs. 32, 36, and 37). In Reference 32, an error-tracking scheme is described for basic* H.263 encoders. Under this scheme, the encoder receives feedback messages from the decoder via the ITU-T H.245 control protocol [38]. These messages inform the encoder about the temporal and spatial locations of any erroneous GOBs detected by the decoder. The encoder uses this information to track the affected regions in subsequent frames and consequently use intra coding to compress these affected regions. This adaptive refresh technique stops the propagation of errors in the video sequence while relying on the more efficient inter-coding mode during error-free periods.

In Reference 36 another adaptive intra-coding scheme is proposed that does not require a feedback channel. In this case, the encoder computes an error sensitivity metric for each macroblock. This metric is accumulated over subsequent coded pictures. When the accumulated metric of a given macroblock indicates a large degree of vulnerability to channel errors, then that macroblock is intra coded. In addition, macroblocks with small error sensitivity measurements are updated using a raster-based forced update scheme. This ensures that all macroblocks within the video sequence are eventually refreshed.

In Reference 37, a transcoding scheme is used to increase the robustness of an H.263 signal transmitted over a combined wired–wireless network. The transcoding mechanism, which is employed at the boundary between the wired and wireless segments of the combined network, is designed to improve the resilience of the video signal to errors while minimizing loss of coding efficiency.

2. *Joint Source–Channel Coding*

In joint source–channel (JSC) coding, the time-varying characteristics of an error-prone channel are taken into consideration when designing the source and channel coders of a wireless system. Much research centers around reducing the impact of channel errors and noise on the transmitted video by controlling the number of bits used by the source and/or channel coder (see, for example, Refs. 39–49).

Figure 4 shows a generic model of a visual source–channel coder. The source signal could be either a still image or a video sequence. For video sequences, the source signal could be either an original picture (for intra-coded frame) or a residual signal representing the difference between an original picture and a prediction of that picture (e.g., for motion-compensated prediction).

The source signal usually undergoes an orthogonal transform that provides clustering of high-energy coefficients in a compact manner. The discrete cosine transform (DCT) is an example of such a transform. In addition to being the transform most widely used by well-established and recent international, standards (JPEG, MPEG-1, MPEG-2, MPEG-4, H.261, H.263), the DCT has been proposed in the context of optimized JSC coding

* "Basic H.263" is a reference to the original H.263 standard, which does not support such error-resilient tools as RVLCs or data partitioning.

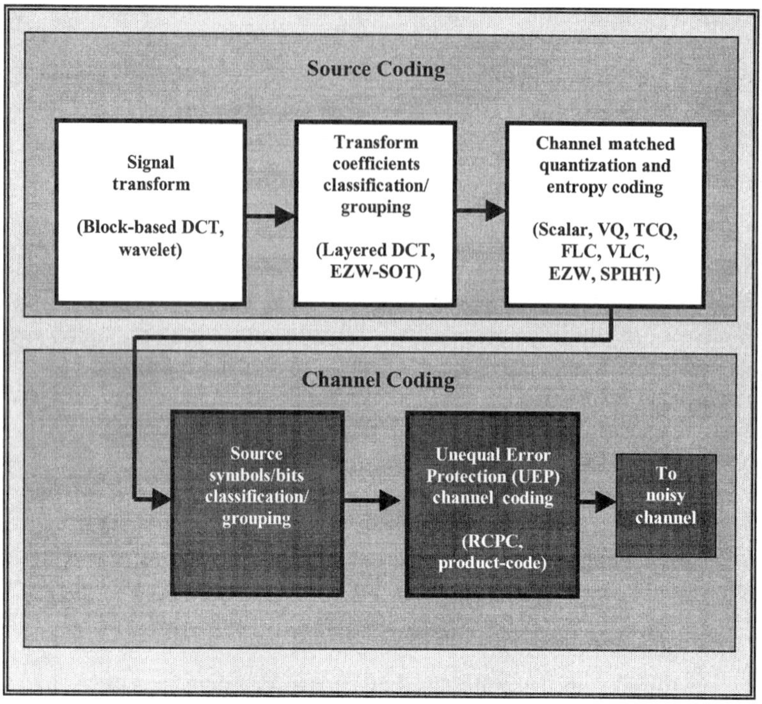

Figure 4 Generic model of a joint source–channel coding system for a noisy wireless network.

[e.g., 39,41,50,51]. Meanwhile, wavelet transform and subband-based visual coding have also been proposed for many JSC solutions [42–47]. Moreover, wavelet transform–based image compression has been adapted by MPEG-4 as the basis for the still-image texture coding tool [11] and is the lead candidate for the JPEG-2000 image-coding standardization activity. Wireless and mobile applications are among key target application areas for both of these standards.

In the remainder of this section, we describe in more detail the other components of the JSC coding system generic model.

a. Classification and Grouping of the Transform Coefficients. The second stage of the generic JSC model shown in Figure 4 is the classification and grouping of the transform coefficients. This stage is needed for more efficient and robust quantization and coding (source and/or channel) of the coefficients.

For example, in Regunathan et al. [39] each DCT block is classified into a mode m_j, $j = 1, 2, \ldots M$. This is used to assign a different scalar quantizer and corresponding bit allocation map for each DCT block. The classification mode m_j is transmitted to the receiver using a 1/3 error correction code (i.e., two parity bits are used for each data bit). This ensures that the classification mode information is reliably delivered even with severe channel error conditions. Initially, the blocks are classified based on their AC energy, and then a channel-optimized based algorithm is used to adjust the classification mode m_j used for the blocks.

In Li and Chen [41] a different type of classification and grouping is employed. Although a block-based DCT transform is used, the DCT coefficients are grouped on the basis of their frequencies into B "subband" images (subsources). Therefore, for $N \times N$ DCT blocks, there are N^2 subsources. Each subsource contains all coefficients with the same frequency (e.g., the DC coefficients). This enables the sender to allocate different amounts of bits for the different subsources.

Subband coding provides a natural grouping of the transform coefficients. For example, in References 42 and 47 3D subband video coding is used, and each subband is allocated a different amount of source and channel coding bits depending on some JSC distortion measure.

Most of the wavelet transform–based visual coding solutions proposed for error-prone channels [e.g., 43–46] are based on grouping the wavelet coefficients using Embedded Zero-tree Wavelet (EZW) [52,53]. All of the cited references [43–46] and many others employ an improved variation of the EZW algorithm known as SPIHT (Set Partitioning in Hierarchical Trees) [53]. For a good understanding of how the SPIHT algorithm works, it is important to discuss the EZW method briefly. Here we give a very brief overview of the original EZW algorithm before describing the remaining components of the JSC coding system of Figure 4.

The EZW algorithm groups the wavelet coefficients by partitioning them into sets of hierarchical *spatial orientation trees*. An example of a spatial orientation tree is shown in Figure 5. In the original EZW algorithm [52], each tree is rooted at the highest level (most coarse subband) of the multilayer wavelet transform. If there are m layers of sub-

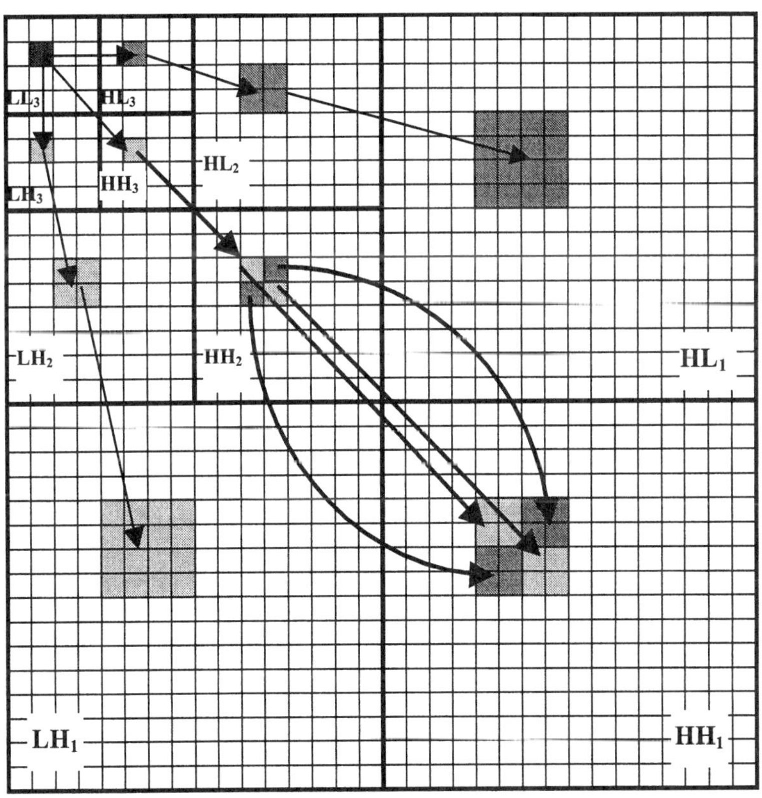

Figure 5 Example of the hierarchical, spatial orientation trees of the zero-tree algorithm.

bands in the hierarchical wavelet transform representation of the image, then the roots of the trees are in the LL_m subband of the hierarchy as shown in Figure 5. If the number of coefficients in subband LL_m is N_m, then there are N_m spatial orientation trees representing the wavelet transform of the image.

In EZW, coding efficiency is achieved on the basis of the hypothesis of ''decaying spectrum'': the energies of the wavelet coefficients are expected to decay in the direction from the root of a spatial orientation tree toward its descendants. Consequently, if the wavelet coefficient c_n of a node n is found insignificant (relative to some threshold $T_k = 2^k$), then it is highly probable that all descendants $D(n)$ of the node n are also insignificant (relative to the same threshold T_k). If the root of a tree and all of its descendants are insignificant, then this tree is referred to as a zero tree (ZTR). If a node n is insignificant (i.e., $|c_n| < T_k$) but one (or more) of its descendants is (are) significant, then this scenario represents a violation of the ''decaying spectrum'' hypothesis. Such a node is referred to as an isolated zero tree (IZT). In the original EZW algorithm, a significant coefficient c_n (i.e., $|c_n| > T_k$) is coded either positive (POS) or negative (NEG) depending on the sign of the coefficient. Therefore, if $S(n, T_k)$ represents the *significance symbol* used for coding a node n relative to a threshold $T_k = 2^k$, then

$$S(n, T_k) = \begin{cases} \text{ZTR} & \text{if } |c_n| < T_k \text{ and } \max\limits_{m \in D(n)} (|c_m|) < T_k \\ \text{IZT} & \text{if } |c_n| < T_k \text{ and } \max\limits_{m \in D(n)} (|c_m|) \geq T_k \\ \text{POS} & \text{if } |c_n| \geq T_k \text{ and } c_n > 0 \\ \text{NEG} & \text{if } |c_n| \geq T_k \text{ and } c_n < 0 \end{cases} \tag{1}$$

The coding procedure used for coding the significance of the wavelet coefficients is referred to as *significance map* coding. All of the different variations of the EZW algorithm, in general, and the SPIHT-based solutions, in particular, are based on grouping the wavelet coefficients using the spatial orientation trees (SOTs) framework [11,52–54]. However, these solutions differ in the way the significance map is coded as described in the following.

b. Quantization and Entropy Coding of the Transform Coefficients. The third stage of the generic JSC model of Figure 4 is the quantization and entropy coding of the classified transform coefficients. Fixed length entropy coding (i.e., using fixed length codes FLCs), variable length coding (i.e., using VLCs), hybrid fixed–variable length coding, or arithmetic entropy-based coding mechanisms are normally used in conjunction with some type of quantization. In addition to scalar quantization, vector quantization (VQ) [55] and trellis-coded quantization (TCQ) [56] are among the popular techniques proposed for wireless video.

For example, in Regunathan et al. [39] quasi–fixed length codes are generated by employing different allocation maps for the different (preclassified) DCT blocks (as explained before). A predetermined FLC is used for each coefficient (within each DCT block) based on a bit allocation map. However, the length of the codes could change from one block to another based on the block classification m_j, $j = 1, 2, \ldots M$.

In Li and Chen [41], fixed rate uniform threshold TCQ (UTTCQ) quantization and coding are employed for the ''subband'' DCT subsources described earlier. One of the advantages of the TCQ quantization is its relative insensitivity to channel errors. [If N is the number of trellis states, then one bit error affects, at most, $1 + \log_2(N)$ outputs.]

In Cheung and Zakhor [42], the 3D spatiotemporal subbands are successively quantized to generate a layered fine-granular scalable video. Entropy coding is achieved using a conditional arithmetic coder for each quantization layer. On the other hand, in the 3D subband coding proposed in Srinivasan et al. [47], full-search vector quantization is used for coding the subband coefficients except for the low-low (LL) subband, where a scalar quantizer is used.

Vector quantization has also been proposed using an end-to-end optimization framework for transmission of video over noisy channels [e.g., 48,57]. In Jafarkhani and Farvardin [48], a channel-matched (CM) hierarchical table-lookup VQ (CM-HTVQ) is employed based on the channel-optimized VQ (COVQ) paradigm proposed in Farvardin [57]. In addition to providing robustness against channel noise, the CM-HTVQ approach benefits from the simplicity of table-lookup encoding and decoding. Therefore, although the performance of the CM-HTVQ is slightly inferior to COVQ, its encoder complexity is significantly less than that of the COVQ encoder.

Regarding wavelet-based methods, as mentioned earlier many improvements in the original EZW algorithm have been proposed, most notably the SPIHT algorithm [53]. The SPIHT variation of EZW represents the basis for the majority of the wavelet-based, error-resilient visual coding works proposed recently in the literature [43–46]. Moreover, different variations of the EZW approach (including SPIHT) are being considered for transmission of video over nonguaranteed QoS networks (e.g., the Internet and certain wireless networks) [54].

With the SPIHT algorithm [53], a different method (from the original EZW) is used for coding the significance map. The algorithm simply uses only two symbols: 1 (significant) and 0 (insignificant). In this case, the insignificance of a node* refers to either the insignificance of its own coefficient or the insignificance of its descendants. Therefore, if $\aleph$ represents a set of nodes, the symbols used for coding the significance map under the SPIHT algorithm can be expressed as

$$S(\aleph, T_k) = \begin{cases} 1 & \text{if } \max_{m \in \aleph}(|c_m|) \geq T_k \\ 0 & \text{otherwise} \end{cases} \tag{2}$$

In this case, $\aleph$ could be a single node (i.e., $\aleph = \{n\}$) or it could be a set of nodes (e.g., $\aleph = \{D(n)\}$ = the set of descendants of a node n). (For more details of the EZW and SPIHT algorithms and how they differ in their coding approaches to the spatial orientation trees of the wavelet coefficients, the reader is referred to References 52–54).

Because no positive or negative symbols are used in the preceding equation, under SPIHT the encoder has to send the sign of each significant coefficient. Therefore, effectively, the SPIHT algorithm uses a variable-length code for the significance map symbols: 1 bit if the coefficient is insignificant and 2 bits (positive or negative) if the coefficient is significant. The VLC coding scheme of the SPIHT algorithm is used in References 43, 44, and 46. This VLC-based coding, of course, could create a major loss-of-synchronization event at the decoder if an error corrupts the significance map (in particular, if the error hits an ''insignificance'' symbol). Therefore, in Man et al. [45] a fixed-length coding scheme is used. In this case, in addition to using 2 bits for coding a significant coefficient (positive or negative), a 2-bit code is used for an ''insignificant'' wavelet coefficient. The

* Here ''node'' represents a wavelet coefficient in a spatial orientation tree as described before.

FLC in Man et al. [45] is used only when transmitting the significance symbol for a coefficient (i.e., the case in which $\aleph$ is a single node: $\aleph = \{n\}$). When coding the significance of a set of coefficients (e.g., $\aleph = \{D(n)\}$ = the set of descendants of a node n), the VLC coding scheme is used.

Another way to protect VLCs and other entropy-coded symbols is to use robust channel-coding schemes. This aspect of the JSC coding system model is described next.

c. Channel Coding for Video Transmission over Noisy Channels. After the source encoder generates its symbols, the channel coder provides the necessary protection to these symbols prior to their transmission over error-prone networks. One popular approach that has been used extensively is the unequal error protection (UEP) paradigm. The UEP paradigm enables the channel coder to use different levels of protection depending on the channel condition (''channel-driven'' UEP). In addition, under UEP, the channel coder can provide different levels of protection for the different source symbols depending on their importance (''source-driven'' UEP). The importance of a source symbol can be measured on the basis of the amount of distortion introduced when that symbol is corrupted by a channel error.

For any source-driven UEP approach, there is a need for a classification process for the source symbols, as shown Figure 4. For example, in Regunathan et al. [39] each bit of the quantized DCT coefficients is assigned a ''sensitivity measure'' S based on the amount of distortion caused by inverting that bit due to a channel error. In Li and Chen [41], a simple layered approach is used to classify the different bits of the UTTCQ coder. The higher layer in which the bits are located (within the trellis), the more important they are. All bits in the same layer are treated in the same way (in terms of importance) and therefore are grouped into one data block. Then different levels of protection are used for the different data blocks.

For the output symbols of 3D subband coders [e.g., 42, 47] and wavelet-based EZW encoders, the importance of the symbols naturally defined as already mentioned. This, in general, enables the employment of an UEP approach without the need for the extra stage of source symbol classification.

Once the source symbols are classified, the desired UEP channel-coding mechanism can be used. One way to implement this is to have multiple channel encoder–decoder pairs, running in parallel, where each pair provides a different level of protection depending on the importance of the source symbols or the current channel condition. This approach, however, is impractical, especially if the number of protection levels needed is high. Rate-compatible punctured convolutional (RCPC) codes were invented by Hagenauer [58] to provide a viable and practical channel-coding solution for the UEP paradigm [59]. RCPC codes are generated by a single (channel) encoder–decoder structure that prevents the transmission of certain code bits (i.e., ''puncturing'' the code) at the output of a convolutional coder. A ''puncturing table'' is used to control the rate at the output of the convolutional encoder, and Viterbi-based channel decoding can be employed at the receiver.

Because of its flexibility and low complexity, the RCPC channel coder has been extremely popular for wireless video transmission. Consequently, it has been employed in virtually all of the references cited [39,41–47] for source-driven, channel-driven, or both source and channel UEP.

Because of the error sensitivity of symbols resulting from an EZW-SPIHT source encoder, different RCPC-based error protection mechanisms have been proposed [43–46]. For example, in order to minimize the impact of errors on the significance map symbols (which could result in a total loss of synchronization at the decoder), in Xiong [45] and Man [46] source-driven RCPC channel coders are used to provide higher protection of these symbols. Meanwhile, the same RCPC coder is used to protect the wavelet coefficient quantized value (''refinement'' bits) using lower level protection than that used for the significance map symbols.

Moreover, cyclic redundancy code (CRC) is employed in Sherwood and Zeger [43] and Xiong et al. [46] prior to the RCPC channel coder. The CRC provides extra error detection capability that could be used to enhance the Viterbi-based decoding algorithm at the receiver. For example, in Xiong et al. [46] if a CRC check indicates an error in a received block, the decoder tries to find the path with the next lowest metric in the trellis. However, if the decoder fails to decode the received block within a predetermined depth of the trellis, then a negative acknowledgement (NACK) scheme is used to request the retransmission of that block from the transmitter.

The RCPC channel coder has also been used in conjunction with systematic product channel coders. In Sherwood and Zeger [44], an outer CRC coder is used for protecting each packet of information, while a Reed-Solomom (RS) coder is used across packets in a product coder configuration. Using the combination of a row-wise CRC error detection and a column-wise RS error correction structure doubles the correcting capability of the RS code (when compared with no CRC code). Each row of the two-dimensional product code structure is fed to an RCPC coder for a source-driven UEP.

V. WIRED AND WIRELESS IN-HOME DIGITAL NETWORKS

Various wired and wireless communication and interface technologies are being proposed to build in-home digital networks. However, no single technology is versatile enough to accommodate a wide range of devices, applications, and cost requirements in a home. Applications in a home include video, audio, voice, data, and device control. They come with different bandwidth and timing requirements. In general, real-time video, audio, and voice need to be transferred isochronously with guaranteed bandwidth and latency. Others can be transferred asynchronously. Asynchronous schemes do not guarantee bandwidth or latency but usually employ more rigid data transfer mechanisms such as an acknowledge-and-retry protocol. The bandwidth requirements vary significantly from application to application. Some applications such as uncompressed video may require a bandwidth in excess of 100 megabits per second (Mbps), whereas others such as device controls may require less than a few kilobits per second (kbps). Table 1 shows the bandwidth requirements for some of the applications commonly found in a home. The cost requirements may vary significantly as well. High-end audio and video equipment can easily absorb the additional cost of a high-performance digital interface. On the other hand, home appliances such as thermostats, lighting control devices, and garage openers have very limited budget for an additional interface cost.

A few radio frequency (RF) [4,60], power line [61], or phone line [62] technologies were proposed to address an in-home digital network. These technologies have advantages over other options because no new wiring will be required to enable communications among devices. However, their applications are limited to mainly voice and data communications because of their limited bandwidth. To enable real-time video and audio applications, a network technology that supports high-speed isochronous data transfer is needed.

Table 1 Bandwidth Requirements for Common Applications in a Home

Application	Required payload bandwidth
Digital television (DTV) over cable	38.8 Mbps
Digital video (DV)	28.8 Mbps
DSS	23.6 Mbps
DTV	19.4 Mbps
D-VHS (max.)	13.8 Mbps
DSS single program (e.g.)	4.8 Mbps
CD audio	1.4 Mbps
Dolby AC-3 audio (max.)	640 kbps
MPEG audio (e.g.)	256 kbps
Voice (e.g.)	64 kbps
Internet audio (e.g.)	16 kbps

There is at least one such technology currently available, namely, IEEE 1394 [63]. In fact, IEEE 1394 has been chosen as the backbone home network technology by the VESA (Video Electronics Standard Association) Home Network Committee.

Digital islands or clusters are naturally forming around computers and digital TV receivers in a home, as digital interface technologies such as USB [64] and IEEE 1394 start being incorporated in computers, peripherals, and consumer electronics devices. Therefore, it is important to accommodate these existing clusters when building an in-home digital network. To take advantage of these existing clusters, technologies that connect or bridge those clusters together will be needed. One such technology is the long-distance IEEE 1394b technology [65], which extends the reach of a 1394 cluster to over 50 m, using unshielded twisted pair (UTP) or optical fiber cables. Another such technology is the IEEE 1394.1 bus-bridge technology [66], which connects multiple 1394-bus clusters together without creating the problem of instability and/or bandwidth exhaustion often associated with a large 1394-bus cluster.

RF technologies were also discussed in the context of the IEEE 1394.1 bridge group in the past as a means to replace 1394 cables and/or an internal structure of a bridge. Such technologies can be called wireless 1394 technologies. A 1394.1-compliant bridge device that employs an RF technology as its internal fabric is called a 1394 wireless bridge. A 1394 wireless bridge will be a preferred solution to connect a 1394 cluster in one room to another cluster in another room for those who wish to avoid wiring the house.

It is also important to consider how digital services that arrive in a home will connect to an in-home digital network. There will be multiple digital services arriving at a home via satellite dish, RF antenna, cable, telephone line, ISDN line, and so on. Therefore, technologies will be needed to connect the in-home digital network to the outside world via individual network-interface modules. These technologies are called gateway technologies.

One can foresee that an in-home digital network will be heterogeneous, comprising multiple technologies including network, interface, bridge, and gateway technologies. The transfer media in a home may vary as well. They will include coax, power line, unshielded twisted pair, plastic optical fiber (POF), glass optical fiber (GOF), radio frequency, infrared

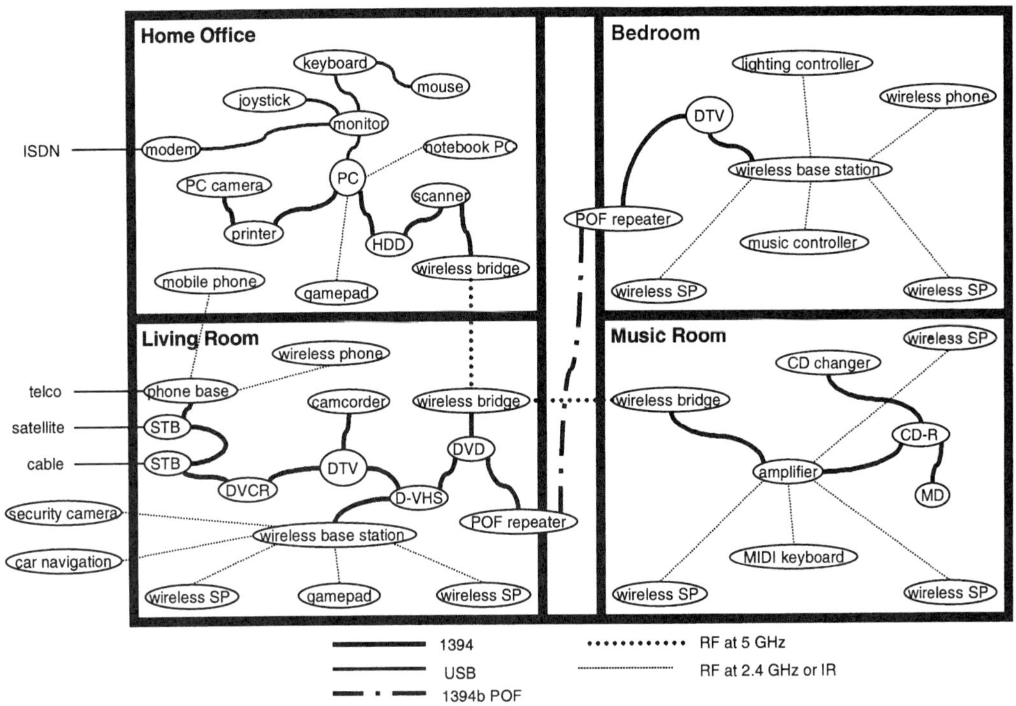

Figure 6 Example of a wireless–wired in-home network.

(IR), and standard interface cables such as USB and 1394 cables. Figure 6 shows an example of a heterogeneous in-home digital network. As can be seen from the figure, both wired and wireless technologies are required. Figure 7 shows various in-home digital network technologies that have been proposed or are being discussed. Figure 8 shows the capabilities of each of these technologies in terms of its bandwidth and transfer distance.

Most of the in-home digital wireless standards have roots in their counterpart wired standards. The following section provides an overview of key wired in-home digital interconnect standards. Then, in Section B, a more comprehensive overview of wireless in-home network technologies and standards is given.

A. Wired Digital Interconnect Solutions

1. USB (Universal Serial Bus)

USB [64] was developed in the computer industry as a peripheral interface for medium to low-speed peripherals, such as keyboards, mice, joysticks, modems, speakers, floppy-disk drives, and scanners. Virtually every computer now comes with at least one USB port.

USB supports plug and play, a maximum of 127 nodes, a distance of 5 m per hop, and a data rate of up to 12 Mbps. It supports isochronous data transfer as well as asynchronous data transfer. Therefore, it can be used for real-time audio and possibly low-speed video applications in addition to data and control applications.

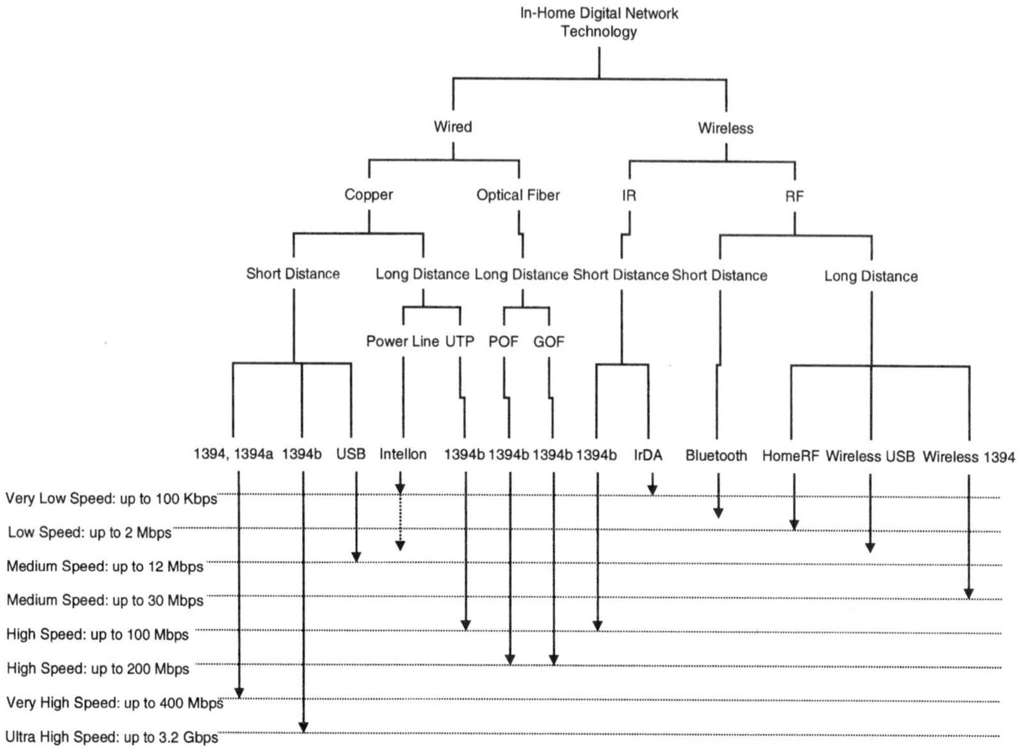

Figure 7 Wireless and wired in-home digital networking technologies.

Because USB requires a master controller, which is usually a computer, and does not support peer-to-peer communications, it has not been incorporated in consumer electronics devices yet. However, it is feasible that consumer electronics devices such as digital set-top boxes and home theater controllers will incorporate USB to communicate with their local peripherals such as keyboards and mice.

Although the reach of USB connections can be extended by use of USB hubs, a wired USB cluster will probably remain inside a room. USB is considered a local cluster technology rather than a home network technology.

2. IEEE 1394 (High-Performance Serial Bus)

IEEE 1394 high-performance serial bus [63] technology has already become a standard feature of certain consumer electronics devices and is expected to be pervasive among a variety of digital devices including digital TV receivers, digital VCRs, and digital set-top boxes. Furthermore, IEEE 1394 is becoming a standard feature of computers and high-speed computer peripherals.

IEEE 1394 is often simply called 1394, but it is also called FireWire in the computer industry and i.LINK in the consumer electronics industry. IEEE 1394 was born in the computer industry but raised in the consumer electronics industry during its infancy. The technology was then raised to maturity and fully supported by both industries.

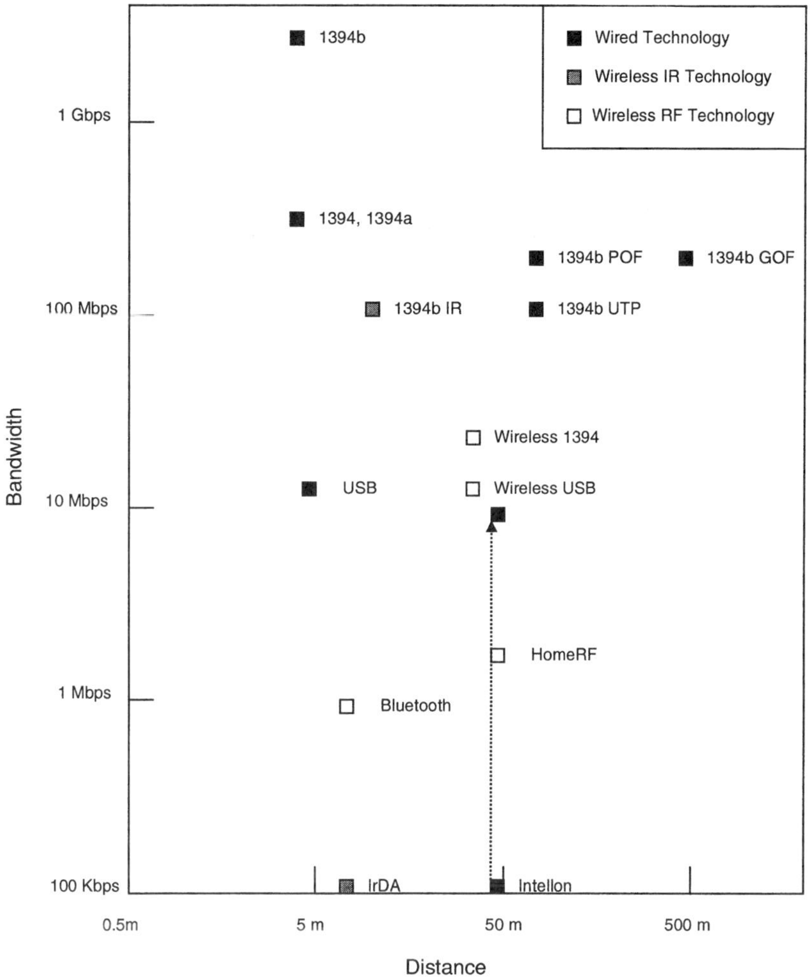

Figure 8 Bandwidth and distance constraints for wireless and wired in-home network technologies.

IEEE 1394 supports plug and play, a maximum of 63 nodes on a single bus cluster, a distance of 4.5 m per hop, and a data rate of up to 400 Mbps. It supports isochronous data transfer as well as asynchronous data transfer. It is suitable for a variety of applications, including high-speed video, audio, and data.

What distinguishes IEEE 1394 from USB most is its performance, namely its support of high-speed real-time applications, and also its distributed control model. USB would not be used by consumer electronics devices as a network technology even if it could support high-speed real-time applications. Consumer electronics devices need a peer-to-peer control model, not a computer–peripherals or master–slaves model as employed by USB. Thanks to its high performance and its control model, IEEE 1394 is the only technology currently available that can connect computers, peripherals, and consumer electronics devices all at once and enable "digital convergence."

Although the current IEEE 1394 standard was published in 1995 as IEEE Std 1394-1995, extensions and enhancements of 1394 are still under way. IEEE 1394a [67] is an ambiguity-free, more efficient, and enhanced version of 1394. IEEE 1394b [65] defines both long-distance (over 50 m up to a few hundred meters) and high-speed (up to 3.2 Gbps) versions of 1394. The 1394b standard addresses several transport media, including plastic optical fiber, glass optical fiber, and unshielded twisted pair. An infrared technology has also been demonstrated as its transport medium.

The IEEE 1394.1 [66] group is developing the 1394 bus bridge technology that connects multiple 1394 bus clusters together to form a larger 1394 network. One of the drawbacks of the 1394 technology is that every node on a bus shares the limited bandwidth with all other nodes on the same bus. For instance, one DV (digital video) stream from a digital camcorder occupies 36% of the total available isochronous bandwidth on the bus. Therefore, no more than two simultaneous DV streams can coexist on the entire bus, regardless of the speed capabilities of other nodes, the number of the nodes on the bus, or the size of the bus. However, if bridges are used to connect multiple buses together, each bus segment can maintain its own local bandwidth. Data will be transferred across a bridge only on a need-to-be-transferred basis. Therefore, adding bridges will increase the aggregate bandwidth of the network.

Another advantage of adding bridges to the network is that they can confine bus reset within a local bus, thus keeping the instability of the network to a minimum. A drawback of the 1394 technology is that whenever a device is added to or removed from a bus or a device is powered on or off, at least one bus reset will be generated. When a bus reset occurs, the node ID of each device may change; thus certain applications try to find out how the node IDs of other nodes have been changed, while other applications try to reallocate isochronous resources. This would create a storm of asynchronous traffic after each bus reset. In addition, if there is at least one 1394-1995 device on the bus, a bus reset will last over 160 μsec. Consequently, some packets will be lost and isochronous streams will be disrupted. Furthermore, if for any reason a node fails to receive all the self-ID packets correctly during a bus reset, it may generate another bus reset to start the process all over again. This could lead to a storm of bus resets under certain conditions. Therefore, the bigger the bus, the higher the probability that bus resets will be generated, and thus the worse the potential instability of the bus. The use of bridges can minimize this instability.

Other groups are defining new specifications over 1394. For instance, the Internet Engineering Task Force (IETF) is defining the specifications of IP over 1394 [68], which will enable IP applications in a 1394 network. This will be an enabling technology to build a local area network in a home. IP over 1394 will be used among computers as well as IP-capable consumer electronics devices in a home.

Another group, the Copy Protection Technology Working Group (CPTWG), is defining copy protection (CP) over 1394. This will be an essential technology to exchange digital contents between devices in a home. Without this technology, the content providers or Hollywood would not allow digital contents to be transferred over an in-home digital network.

There are already established specifications and protocols over 1394 as well. IEC 61883 [69] defines a way to transfer and manage real-time data streams over 1394. It also defines a basic mechanism to exchange device control information over 1394. The AV/C-CTSs (Audio Video/Control Command Transaction Sets) have already been defined for several consumer electronics and audio devices. Additional command sets are being defined under the 1394 Trade Association [70] as a new group of devices appears in the market. Most of the currently available 1394-enabled consumer electronics devices employ IEC 61883 and AV/C-CTS.

As we have seen, IEEE 1394 is the most versatile and capable in-home digital network technology currently available. It supports a variety of physical media, protocols, and applications. It will be even more versatile when wireless technologies are added, as described in the following Wireless 1394 section. It is expected that IEEE 1394 will be one of the most important in-home digital network technologies, if not the dominant technology.

B. Wireless in-Home Interconnect Solutions

1. HomeRF

The HomeRF technology [60] addresses the area of a wireless voice and data network in the home, using the Shared Wireless Access Protocol (SWAP) technology. The SWAP technology was derived from both existing DECT and IEEE 802.11 (wireless LAN) technologies. It supports both voice and data in the unlicensed ISM (industry, science, medical) band at 2.4 GHz. Isochronous traffic such as voice is carried by TDMA (Time Division Multiple Access) services. Asynchronous traffic such as data is carried by CSMA/CA (Carrier Sense Multiple Access/Collision Avoidance) services. The system supports a maximum of 127 nodes, a total bandwidth of up to 2 Mbps, and a range of up to 50 m.

The SWAP network can be ad hoc or managed by a special node called the connection point, which is required to enable isochronous services. In the absence of the connection point, the network control is equally distributed among all nodes, and only asynchronous services can take place.

The network can contain a mixture of the following four node types:

1. Connection point, which controls the isochronous (TDMA) and asynchronous (CSMA/CA) services.
2. I (isochronous) node or voice terminal, which uses the isochronous (TDMA) services to communicate with the connection point.
3. A (asynchronous) node or data node, which uses the asynchronous (CSMA/CA) services to communicate with the connection point and other A nodes.
4. I & A node, which is a combination of I and A node.

The HomeRF technology can cover the area of voice, low-speed data, and control applications in a home. A way to connect this HomeRF network to existing USB and 1394 clusters will be needed to enable a multimedia in-home digital network.

2. Bluetooth

The Bluetooth technology [4], which addresses the area of short-range wireless voice and data communications anywhere in the world, enables the wireless connectivity of personal and business mobile devices. Although its primary application is to connect devices such as cellular phones, PDAs (personal data assistants), and notebook computers, the Bluetooth technology may directly compete with the HomeRF technology in a home. It uses the same ISM band at 2.4 GHz and supports both voice and data at a gross data rate of up to 1 Mbps.

The Bluetooth technology is designed to be compact, low power, low cost, and robust against interference and fading, especially for mobile applications. It employs a fast frequency hopping scheme for the robustness and a time-division duplex scheme for full-duplex transmission. Its nominal link range is up to 10 meters (m). However, the range can be extended to more than 100 m if the transmission power is increased.

3. *IrDA Control*

The IrDA Control technology [71] addresses in-room data communications between multiple hosts and peripherals, using a bidirectional infrared technology. It employs TDMA and Packet Reservation Multiple Access (PRMA) protocols. Host devices include computers, consumer electronics devices, and home appliances. Peripherals include mice, keyboards, gamepads, PDAs, and remote control units. The IrDA Control technology supports up to 75 kbps and a distance of 10 m.

4. *Wireless USB*

It will be desired to replace some of the USB cables with wireless links for portable and room-to-room applications. USB's relatively low data rate will make it feasible to replace a cable with a wireless link, using either an IR or RF technology. A RF technology in the 2.4 GHz band will be a good candidate for the wireless USB because of its available bandwidth of around 10 Mbps. A wireless USB technology based on an RF technology would expand the reach of a USB cluster beyond room boundaries, which will make USB a step closer to a home network technology.

5. *Wireless 1394*

Because 1394 clusters will naturally form in separate rooms in a home, technologies that can connect those clusters will be needed to enable an in-home digital network. However, wiring the house will not be an option for many people because of the cost and hassle associated with wiring an existing house. What will be needed are technologies that can connect these clusters in a wireless manner. At least two such wireless technologies have been proposed or discussed: one will be called a 1394 wireless repeater and the other a 1394 wireless bridge.

A 1394 wireless repeater is similar to ordinary 1394 repeaters or 1394b long-distance repeaters in its design, but at least one of the 1394 ports is replaced by a wireless port or transceiver. It repeats the signals among the wired or wireless ports according to the physical layer specifications defined in 1394, 1394a, or 1394b. A pair of these 1394 wireless repeaters will be used to merge two 1394 clusters into a single larger 1394 bus or network.

In order to replace a 1394 cable with a wireless link, the wireless link must meet the requirements for the physical layer and the cable defined in the 1394 standard. This means that the wireless link must support at least S100 speed or 98.304 Mbps. In addition, it must guarantee a sufficiently low error rate and low data latency. Because of these requirements, candidates for the wireless link for a wireless repeater are practically limited to high-speed infrared technologies for consumer use. NEC demonstrated their 1394 IR repeaters [72] at Comdex '97 and on other occasions.

The main drawback of IR technologies is that infrared light cannot penetrate a wall. Therefore, communications will be confined within a room. Furthermore, at a high speed such as the one required for 1394 wireless repeaters, a line of sight needs to be secured, and the diversities of the two repeaters need to be precisely aligned. This will limit possible applications and installation sites for the 1394 IR repeaters. There will be no possibility for mobile use, either.

A RF technology has advantages over IR technologies because of its ability to penetrate walls and its ease of installation. It can be used for mobile applications as well. However, RF technologies will not be used for 1394 wireless repeaters due to their relatively limited bandwidth. RF technologies are best suited for 1394 wireless bridges as described in the following.

A 1394 wireless bridge is a 1394 bus bridge as defined in the IEEE 1394.1 standard but uses a wireless technology as its internal fabric. Unlike repeaters, a bridge will connect one bus cluster to another without merging the two into a single bus. It isolates local bus resets and local traffic from global traffic, thus adding extra stability and extra aggregate bandwidth to the network. Each bus stays separate in terms of bus-reset propagation and bandwidth usage. The bus bridge technology will be essential to build a large 1394 network.

A 1394 bridge can naturally be divided into two parts, each of which is called a half-bridge. Each half-bridge contains a portal, which is itself a 1394 node on a 1394 bus. The so-called internal fabric connects and transfers data between these two half-bridges. As a bridge keeps local traffic local, there will be no need for the internal fabric to be capable of transferring the entire traffic from one bus to another. In fact, a bridge can specify the data transfer capability of its internal fabric.

Because of the softer restrictions on the bandwidth and data transfer latency, a low-bandwidth link such as an RF link can be used as the internal fabric of a bridge, without breaking its compliance to the 1394.1 specifications. Each wireless half-bridge device will consist of a 1394.1-compliant portal part and a wireless transceiver part. Because radio signals can penetrate walls, one of these half-bridges can be placed in one room and the other in another room without ever wiring the house. Therefore, a 1394 RF wireless bridge will be a preferred solution to enable inter-1394-cluster communications between rooms for those who wish to avoid wiring the house.

A few RF technologies can be used as the internal fabric of a 1394 wireless bridge. A wireless link for the internal fabric must meet certain requirements. For instance, it must support isochronous data transfer and clock synchronization to support real-time applications. Candidates for the internal fabric include the IEEE 802.11a and ETSI BRAN technologies in a 5-GHz band. A data rate of 20–25 Mbps will be achieved using these technologies. Their MAC layer may need to be modified to accommodate these isochronous requirements.

A proposal to the 1394.1 group [73] shows a way of connecting more than two bus segments together at the same time, rather than using a wireless link as a point-to-point internal fabric of a two-portal bridge. It proposes a bridge device that consists of a real half-bridge and a virtual half-bridge. The virtual half-bridge can be implemented solely in software. The virtual bridges in a network communicate with each other over a wireless network. However, they conceal the properties of the wireless network from the rest of the 1394 network by behaving as if they are real half-bridges connected on a real 1394 bus. This solution provides a more intuitive and efficient way to integrate a wireless technology into a home network. Sato [73] also shows that the same principle can be used for connecting individual wireless nodes to a 1394 network.

Long before wiring a new or existing house becomes a reasonable option for the general public, natural 1394 clusters will start forming around a computer in the home office and around a digital TV in the living room. It is foreseeable that the first need for room-to-room communications will be satisfied by the 1394-wireless-bridge technology, which will be less expensive and much easier to install than the wiring alternatives as shown in Figure 6.

VI. CONCLUDING REMARKS

In this chapter, we focused on two important and emerging areas of wireless multimedia: (1) source and channel coding for audiovisual transmission over wireless networks and (2) evolving wireless broadband communication system (WBCS) technologies and standards.

In Section II, we gave a brief overview of other summary papers published recently covering different aspects of wireless multimedia. There are, however, many other issues that are relevant to wireless multimedia systems that our overview did not touch upon. These include power utilization management in untethered systems and its impact on the computational complexity of the algorithms that can be used, as well as antenna design to compensate for fading channels and moving objects in shorter distance communication links in indoor environments.

In Section III, we addressed some specific constraints on wireless multimedia in general and WBCS systems in particular. These constraints include access protocol efficiency, channel propagation effects, and limited available frequency bands. In order to overcome these constraints, one needs to choose the best possible compromises in cost and performance on the physical, the data link control (DLC), and even higher layers. Of course, an efficient modulation scheme with strong coding and smart antenna techniques in the physical layer is always an important step toward a successful WBCS.

Although various studies of WBCS are being conducted, it will take a while before we can realize ''true'' wireless multimedia. For the time being, medium-bit-rate wireless LANs (WLANs) such as Bluetooth, IEEE 802.11, and HIPERLAN will play an essential role in the current indoor applications.

In Section IV, we provided an overview of recent image and video coding techniques for wireless multimedia. Because of the sensitivity of compressed image and video bitstreams to channel errors, recent video coding standards support error-resilient tools for protecting the compressed signal against wireless network impairments. As we highlighted in Section IV, a great deal of research has been directed toward joint source–channel coding schemes designed for minimizing the impact of channel errors on compressed video and image signals.

One promising area in source coding that can be suitable for wireless multimedia applications is the Multiple Description Coding (MDC) paradigm. There has been new interest in this area for the transmission of image and video coding over error-prone channels. In fact, a whole new session in a recent IEEE conference on image processing (ICIP98) was dedicated to MDC algorithms and methods.

Finally, and as an example of a very important WBCS application area, we provided in Section V an overview of wireless in-home networking technologies and standards. We also showed how wired and wireless interconnect solutions can coexist in a heterogeneous in-home network.

ACKNOWLEDGMENTS

The authors would like to thank Alan Cavallerano and Narciso Tan from Philips Research for their excellent reviews of an original version of the manuscript. The authors would also like to thank Atul Puri from AT&T Labs for his feedback and encouragement during the preparation of the manuscript.

REFERENCES

1. C Andren. IEEE 802.11 wireless LAN: Can we use it for multimedia? IEEE Multimedia April–June:84–89, 1998.
2. K Pahlavan et al. Wideband local access: Wireless LAN and wireless ATM. IEEE Commun Mag. November:34–40, 1997.
3. L Correia, R Prasad. An overview of wireless broadband communications. IEEE Commun Mag. January:28–33, 1997.
4. Bluetooth, http://www.bluetooth.com.
5. VOK Li, XX Qiu. Personal communication systems (PCS). Proc IEEE 83:1210–1243, 1995.
6. L Hanzo. Bandwidth efficient wireless multimedia communications. Proc IEEE 87:1342–1382, 1998.
7. J Mikkonen et al. Emerging wireless broadband networks. IEEE Commun Mag. February: 112–117, 1998.
8. DJ Goodman, D Raychaudhuri, eds. Mobile Multimedia Communications. New York: Plenum, 1997.
9. E Ayanoglu, KY Eng. MK Karol. Wireless ATM: Limits, challenges, and proposals. IEEE Personal Commun Mag 3:18–34, August 1996.
10. C Perkins. Mobile IP: Design Principles and Practice. Reading MA: Addison-Wesley Longman, 1998.
11. ISO/IEC 14496-2. Information technology—Coding of audio-visual objects: Visual. Committee Draft, ISO/IEC JTC1/SC29/WG11, N2202, March 1998.
12. ITU-T Recommendation H.263. Video coding for low bitrate communication, 1996.
13. JK Wolf, A Wyner, J Ziv. Source coding for multiple descriptions. Bell Syst Tech J 59:1417–1426, 1980.
14. Y Wang, Q-F Zhu. Error control and concealment for video communication: A review. Proc IEEE 86:974–997, 1998.
15. HR Rabiee, RL Kashyap, H Radha. Error concealment of encoded still image and video streams with multidirectional recursive nonlinear filters. IEEE 3rd International Conference on Image Processing, ICIP'96, Lausanne, Switzerland, September 16–19, 1996.
16. HR Rabiee, H Radha, RL Kashyap. Nonlinear error concealment of compressed video streams over ATM networks. Submitted to IEEE Trans C&S Video Tech (under revision).
17. A Acampora, M Nagashineh. An architecture and methodology for mobile executed handoffs in cellular ATM networks. IEEE JSAC 12(8):1365–1375, 1994.
18. ETSI/BRAN, http://www.etsi.fr/bran/bran.htm
19. D Raychaudhuri, et al. WATMnet: A prototype wireless ATM system for multimedia personal communication. IEEE J Selected Areas Commun. January:83–95, 1997.
20. KY Eng. BAHAMA: A broadband ad-hoc wireless ATM local-area network. Proceedings IEEE ICC, 1995, p. 1216.
21. M Chelouche, et al. Digital wireless broadband corporate and private networks: RNET concepts and applications. IEEE Commun Mag. January:42–51, 1997.
22. Y Du, et al. Adhoc wireless LAN based upon cellular ATM air interface. Submitted to IEEE JSAC.
23. M Shafi, et al. Wireless communications in the twenty-first century: A perspective. Proc IEEE 85:1622–1637, 1997.
24. D Fink, D Christiansen, eds. Electronic Engineers' Handbook, 3rd ed. New York: McGraw-Hill, 1989.
25. C Berrou, A Glavieux, P Thitimajshima. Near Shannon limit error-correcting coding and decoding: Turbo-codes Proceedings ICC 1993, pp 1064–1070.
26. M Budagavi, JD Gibson. Speech coding in mobile radio communications. Proc IEEE 87(7): 1998.
27. M Naghshineh, M Willebeek-LeMair. End-to-end QoS provisioning multimedia wireless/mo-

bile networks using an adaptive framework. IEEE Commun Mag. November:72–81, 1997.
28. A Campbell. QoS-aware middleware for mobile multimedia communications. Multimedia Tools Appl July:67–82, 1998.
29. M Schwartz. Network management and control issues in multimedia wireless networks. IEEE Personal Commun June: 8–16, 1995.
30. P Bahl, B Girod. Wireless video. IEEE Commun Mag 36(6):1998.
31. R Talluri. Error resilient video coding in the ISO MPEG-4 standard. IEEE Commun Mag 36(6):1998.
32. N Farber, B Girod. Extensions of ITU-T recommendation H.324 for error-resilient video transmission. IEEE Commun Mag 36(6):1998.
33. RV Cox, MH Sherif. Standardization and characterization of G.729. IEEE Commun Mag 35(9):1997.
34. E Steinbach, N Farber, B Girod. Standard compatible extensions of H.263 for robust video transmission over mobile environments. IEEE Trans Circuits Syst Video Technol 7(6):1997.
35. G Wen, J Villansenor. A class of reversible variable length codes for robust image and video coding. Proceedings of IEEE ICIP '97, October 1997, pp 65–68.
36. JY Liao, JD Villasenor. Adaptive update for video coding over noisy channels. Proceedings of IEEE ICIP '97, September 1997, pp 763–766.
37. G Reyes, et al. Video transcoding for resilience in wireless channels. Proceedings of IEEE ICIP, October 1998, pp 338–342.
38. ITU-T Recommendation H.245. Control protocol for multimedia communication, 1996.
39. SL Regunathan, K Rose S Gadkari. Multimode image coding for noisy channels. Proceedings of IEEE Data Compression Conference, March 1997, pp 82–90.
40. AA Alatan, JW Woods. Joint utilization of fixed and variable-length codes for improving synchronization immunity for image transmission. Proceedings of IEEE ICIP '98, October 1998, pp 319–323.
41. H Li, CW Chen. Joint source and channel optimized block TCQ with layered transmission and RCPC. Proceedings of IEEE ICIP, October 1998, pp 644–648.
42. G Cheung, A Zakhor. Joint source/channel coding of scalable video over noisy channels. Proceedings of IEEE ICIP '96, September 1996, pp 767–770.
43. PG Sherwood, K Zeger. Progressive image coding for noisy channels. IEEE Signal Process Lett 4(7):189–191, 1997.
44. PG Sherwood, K Zeger. Error protection for progressive image transmission over memoryless and fading channels. Proceedings of IEEE ICIP '98, October 1998, pp 324–328.
45. H Man, F Kossentini, JT Smith. Robust EZW image coding for noisy channels. IEEE Signal Process Lett 4(8):227–229, 1997.
46. Z Xiong, BJ Kim, WA Pearlman. Progressive video coding for noisy channels. Proceedings of IEEE ICIP '98, October 1998, pp 334–337.
47. M Srinivasan, R Chellepa, P Burlia. Adaptive source channel subband video coding for wireless channels. Proceedings IEEE Multimedia Signal Processing Workshop, June 1997, pp 407–412.
48. H Jafarkhani, N Farvardin. Channel-matched hierarchical table-lookup vector quantization for transmission of video over wireless networks. Proceedings of IEEE ICIP '96, September 1996, pp 755–758.
49. DW Redmill, NG Kingsbury. The EREC: An error-resilient technique for coding variable-length blocks of data. IEEE Trans Image Process 5(4):565–574, 1996.
50. R Chandramouli, N Ranganathan, SJ Ramadoss. Empirical channel matched design and UEP for robust image transmission. Proceedings of IEEE Data Compression Conference, May 1998.
51. R Chandramouli, N Ranganathan, SJ Ramadoss. Joint optimization of quantization and on-line channel estimation for low bit-rate video transmission. Proceedings of IEEE ICIP '98, October 1998, pp 649–653.

52. JM Shapiro. Embedded image coding using zerotrees of wavelets coefficients. IEEE Trans Signal Process 41:3445–3462, 1993.
53. A Said, W Pearlman: A new fast and efficient image codec based on set partitioning in hierarchical trees. IEEE Trans Circuits Syst Video Technol 6:243–250, 1996.
54. H Radha, Y Chen, K Parthasarathy, R Cohen. Scalable Internet video using MPEG-4. Signal Processing: Image Communication 15, October 1999, pp. 95–126.
55. A Gersho, RM Gray. Vector Quantization and Signal Compression. Boston: Kluwer Academic Publishers, 1992.
56. MW Marcellin, TR Fischer. Trellis coded quantization of memoryless and Gauss–Markov sources. IEEE Trans Commun 38(1):82–93, 1990.
57. N Farvardin, V Vaishampayan. On the performance and complexity of channel-optimized vector quantizers. IEEE Trans on Inform Theory 36(1):1990.
58. J Hagenauer. Rate-compatible punctured convolutional codes (RCPC codes) and their applications. IEEE Trans Commun 36(4):389–400, 1988.
59. LHC Lee. New rate-compatible punctured convolutional codes for Viterbi decoding. IEEE Trans Commun 42(12):3037–3079, 1994.
60. HomeRF, http://www.homerf.org.
61. Intellon, http://www.intellon.com.
62. http://www.homepna.org.
63. IEEE Std 1394-1995. Standard for a high performance serial bus.
64. Universal Serial Bus Specification, Revision 1.0, January 15, 1996.
65. IEEE Project P1394b Draft 0.87, Draft standard for a high performance serial bus (Supplement), September 28, 1999.
66. IEEE Project P1394.1, Draft standard for high performance serial bus bridges, Draft 0.04, February 7, 1999.
67. IEEE Project P1394a Draft 0.14, Draft standard for a high performance serial bus (Supplement), March 15, 1998.
68. IETF, Ipv4 over IEEE 1394 Internet Draft, draft-ietf-ip1394-ipv4-19.txt, September 1999.
69. ISO/IEC 61883:1998.
70. 1394 Trade Association, http://www.1394ta.org.
71. IrDA Control Specification, Final Revision 1.0, June 30, 1998.
72. NEC 1394 IR Repeater, http://www.nec.co.jp/english/today/newsrel/9710/2102.html.
73. T Sato. 1394 wireless bridge with virtual bus. IEEE P1394.1 WG, Doc. BR029R00, June 9, 1998.

20
Multimedia Search and Retrieval

Shih-Fu Chang
Columbia University, New York, New York

Qian Huang, Atul Puri, and Behzad Shahraray
AT&T Labs, Red Bank, New Jersey

Thomas Huang
University of Illinois at Urbana-Champaign, Urbana, Illinois

I. INTRODUCTION

Multimedia search and retrieval has become an active research field thanks to the increasing demand that accompanies many new practical applications. The applications include large-scale multimedia search engines on the Web, media asset management systems in corporations, audiovisual broadcast servers, and personal media servers for consumers. Diverse requirements derived from these applications impose great challenges and incentives for research in this field.

Application requirements and user needs often depend on the context and the application scenarios. Professional users may want to find a specific piece of content (e.g., an image) from a large collection within a tight deadline, and leisure users may want to browse the clip art catalog to get a reasonable selection. Online users may want to filter through a massive amount of information to receive information pertaining to their interests only, whereas offline users may want to get informative summaries of selected content from a large repository.

With the increasing interest from researchers and application developers, there have been several major publications and conferences dedicated to survey of important advances and open issues in this area. Given the dynamic nature of applications and research in this broad area, any survey paper is also subject to the risk of being incomplete or obsolete. With this perspective, we focus this chapter on several major emerging trends in research, as well as standards related to the general field of multimedia search and retrieval. Our goal is to present some representative approaches and discuss issues that require further fundamental studies.

Section II addresses a promising direction in integrating multimedia features in extracting the syntactic and semantic structures in video. It introduces some domain-specific techniques (e.g., news) combining analysis of audio, video, and text information in analyzing content at multiple levels. Section III focuses on a complementary direction in which visual objects and their features are analyzed and indexed in a comprehensive way. These approaches result in search tools that allow users to manipulate visual content directly to form multimedia queries. Section IV shows a new direction, incorporating knowledge from machine learning and interactive systems, to break the barriers of decoding semantics from multimedia content. Two complementary approaches are presented: the probabilistic graphic model and the semantic template. Section V covers an important trend in the multimedia content description standard, MPEG-7, and its impact on several applications such as an interoperable metasearch environment.

II. VIDEO SEGMENTATION, INDEXING, AND BROWSING

As discussed in the previous section, different methods are suitable for different contexts when people access large collections of information. Video content has unique characteristics that further affect the role of each access method. For example, the sequential browsing method may not be suitable for long video sequences. In this case, methods using content summaries, such as those based on the table of contents (ToC), are very useful in providing quick access to structured video content.

Different approaches have been used to analyze the structures of video. One technique is to do it manually, as in the cases of books (ToC) or broadcast news (closed captions) delivered by major American national broadcast news companies. Because manual generation of an index is very labor intensive and, thus, expensive, most sources of digital data in practice are still delivered without metainformation about content structures. Therefore, a desirable alternative is to develop automatic or semiautomatic techniques for extracting semantic structure from the linear data of video. The particular challenges we have to address include identification of the semantic structure embedded in multiple media and discovery of the relationships among the structures across time and space (so that higher levels of categorization can be derived to facilitate further automated generation of a concise index table).

Typically, to address this problem, a hierarchy with multiple layers of abstractions is needed. To generate this hierarchy, data processing in two directions has to be performed: first, hierarchically *segmenting* the given data into smaller retrievable data units, and second, hierarchically *grouping* different units into larger, yet meaningful, categories. In this section, we focus on issues in segmenting multimedia news broadcast data into retrievable units that are directly related to what users perceive as meaningful. The basic units after segmentation can be indexed and browsed with efficient algorithms and tools. The levels of abstraction include commercials, news stories, news introductions, and news summaries of the day. A particular focus is the development of a solution that effectively integrates the cues from video, audio, and text.

Much research has concentrated on segmenting video streams into ''shots'' using low-level visual features [1–3]. With such segmentation, the retrievable units are low-level structures such as clips of video represented by key frames. Although such systems provide significant reduction of data redundancy, better methods of organizing multimedia data are needed in order to support true content-based multimedia information archiving, search, and retrieval capabilities. The insufficiency associated with low-level visual feature–based approaches is multifold. First, these low-level structures do not correspond in a direct and convenient way to the underlying semantic structure of the content, making it difficult for users to browse. Second, the large number of low-level structures generated from the segmentation provide little improvement in browsing efficiency compared with linear search. When users interact with an information retrieval system, they expect the system to provide a condensed, concise characterization of the content available. The system should offer a mechanism for users to construct clear, unambiguous queries and return requested information that is long enough to be informative and as short as possible to avoid irrelevant information [4]. Thus, multimedia processing systems need to understand, to a certain extent, the content embedded in the multimedia data so that they can recover the semantic structure originally intended to be delivered. The conventional ''scene cut''–based systems cannot meet such requirements although they are able to facilitate retrieval in a limited sense. Therefore, more attention has been directed to automatically recovering semantically meaningful structures from multimedia data [5–13].

A. Integrated Semantic Segmentation Using Multimedia Cues

One direction in semantic-level segmentation focuses on detecting an isolated, predefined set of meaningful events. For example, Chang et al. [14] attempted to extract ''touchdown'' events from a televised football game video by combining visual and audio data. Another approach generates a partition over the continuous data stream into distinct events, so that each segment corresponds to a meaningful event such as a news story [4–12,15–17]. In other words, the aim is to recover the overall semantic structure of the data that reflects the original intention of the information creator (such structure is lost when the data are being recorded on the linear media). Some work along this line has made use of the information from a single medium (visual or text) [5–7,15–17]. With visual cues only, it is extremely difficult to recover the true semantic structure. With text information only, because of the typical use of fixed processing windows, it has been shown that obtaining a precise semantic boundary is difficult and the accuracy of segmentation varies with the window size used [16,17]. Other work integrated the cues from different media to achieve story segmentation in broadcast news [4,8,10–13]. By combining cues, the story boundaries can be identified more precisely.

Some existing approaches rely on fixed textual phrases or exploit the known format of closed captions. In the work of Merlino and Maybury [8–10], particular phrases (e.g., ''still to come on the news. . .'') are exploited as cues for story boundaries. Shahraray and Gibbon [18,19] use the closed-caption mode information to segment commercials from stories. In Hauptmann's work [4], particular structures of closed captions (markers for different events) are used to segment both commercials and stories. Other techniques are aimed at developing an approach that can be used in more general settings, i.e., solutions that do not make assumptions about certain cue phrases, the availability of closed captions, or even the particular format of closed captions [20]. Acoustic, visual, and textual signals are utilized at different stages of processing to obtain story segmentation [11,18]. Audio features are used in separating speech and commercials; anchorperson segments are detected on the basis of speaker recognition techniques; story-level segmentation is achieved using text analysis that determines how blocks of text should be merged to form news stories, individual story introductions, and an overall news summary of the day. When different semantic units are extracted, they are aligned in time with the media data so that proper representations of these events can be constructed across different media in such a way that the audiovisual presentation can convey the semantics effectively [5,6,11,12].

In this section, we review an integrated solution for automated content structuring for broadcast news programs. We use this example to illustrate the development of tools for retrieving information from broadcast news programs in a semantically meaningful way at different levels of abstraction.

A typical national news program consists of news and commercials. News consists of several headline stories, each of which is usually introduced and summarized by the anchor prior to and following the detailed reports by correspondents, quotations, and interviews of newsmakers.

Commercials are usually found between different news stories. With this observation, we try to recover this content hierarchy by utilizing cues from different media whenever it is appropriate. Figure 1 shows the hierarchy we intend to recover. In this hierarchy, the lowest level contains the continuous multimedia data stream (audio, video, text). At the next level, we separate news from commercials. The news is then segmented into the anchorperson's speech and the speech of others. The intention of this step is to use the recognized anchor's identity to hypothesize a set of story boundaries that consequently partition the continuous text into adjacent blocks of text. Higher levels of semantic units can then be extracted by grouping the text blocks into news stories and news introductions. In turn, each news story can consist of either the story by itself or the story augmented by the anchorperson's introduction. Detailed semantic organization at the story level is shown in Figure 2.

The remaining content in a news program after the commercials are removed consists of news segments that are extracted using the detected anchor's speech [11,12]. Figure 2 illustrates what each news segment can be classified into and further can be merged with others to form different semantics. Using duration information, each segment is initially

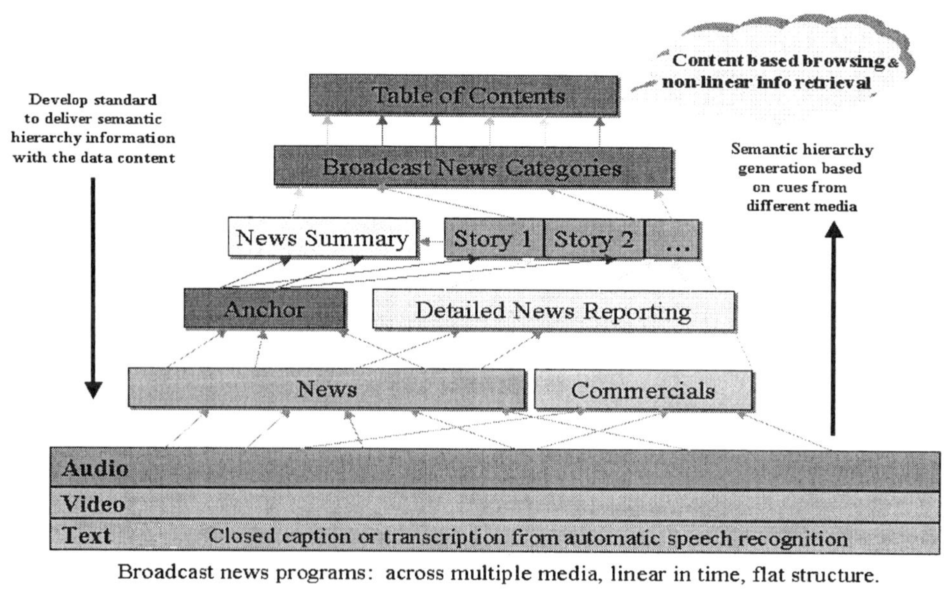

Figure 1 Content hierarchy of broadcast news programs.

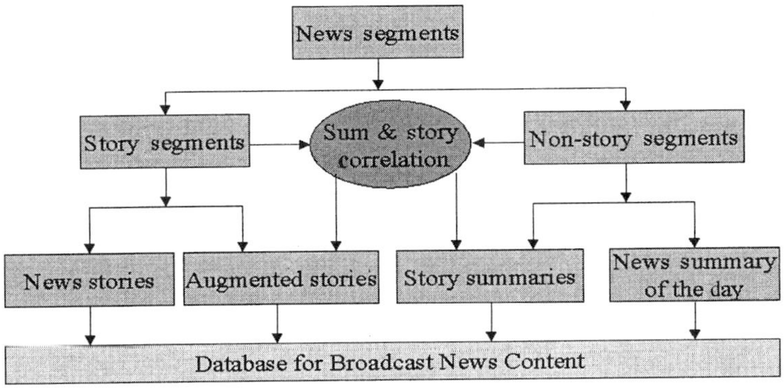

Figure 2 Relationship among the semantic structures at the story level.

classified as either the story body (having longer duration) or news introduction or non-story segments (having shorter duration). Then text analysis verifies and refines the story boundaries, extracts the introduction segment associated with each story (by exploiting the word correlation between stories and anchor's segment), and then form various types of semantics such as augmented news stories (by concatenating a highly correlated anchor speech segment to a story body) or news summary of the day (by identifying and concatenating a small set of anchor speech segments that covers each and every news story once) [12].

The news data are segmented into multiple layers in a hierarchy to meet different needs. For instance, some users may want to retrieve a story directly, others may want to listen to the news summary of the day in order to decide which story sounds interesting before making further choices, and yet others (e.g., a user employed in the advertising sector) may have a totally different need to monitor commercials of competitors in order to come up with a competing commercial. The segmentation mechanism partitions the broadcast data in different ways so that direct indices to the events of different interests can be automatically established.

B. Representations and Browsing Tools

When semantic structures are recovered and indexed, efficient tools are needed to present the extracted semantics in a form that is compact, concise, easy to understand, and at the same time visually pleasing. Now, we discuss the representation issue at three levels for broadcast news data: (1) how to present the *semantic structure* of the broadcast news to the users, (2) how to represent particular *semantics* based on the content of a news story, and (3) how to form the representation for *news summary* of the day.

A commonly used presentation for semantic structure is in the form of a table of contents. In addition, in order to give users a sense of time, we also use a streamline representation for the semantic structure. Figure 3 shows one presentation for the semantic structure of a news program. On the left side of the screen, different semantics are categorized in the form of a table of contents (commercials, news, and individual news stories, etc.). It is in a familiar hierarchical fashion that indexes directly into the time-stamped media data. Each item listed is color coded by an icon or a button. To play back a particular item, a user simply clicks on the button for the desired item in this hierarchical table. At

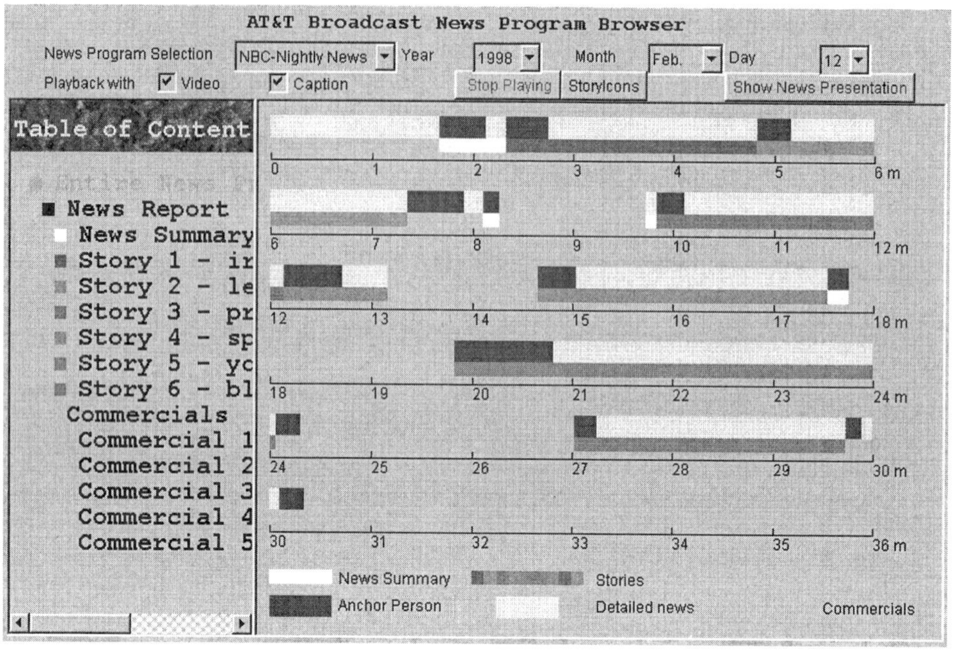

Figure 3 Representation for extracted semantic structures.

the right of this interface is the streamline representation, where the time line runs from left to right and top to bottom. The time line has two layers of categorization. The first layer is event based (anchor's speech, others' speech, and commercials) and the second layer is semantic based (stories, news introduction, and news summary of the day). Each distinct section is marked by a different color and the overall color codes correspond to the color codes used in the table of contents. Obviously, the content categorized in this representation is aligned with time.

To represent each news story, two forms are considered. One is static (StoryIcon) and the other is dynamic (multimedia streaming). To form the static presentation of a news story, we automatically construct the representation for a story that is most relevant to the content of the underlying story. Textual and visual information is combined. Figure 4 gives an example of this static representation for the story about the damage that El Niño caused in California. In Figure 4, the ToC remains on the left so that users can switch to a different selection at any time and the right portion is the static presentation of the story that is currently being chosen (story 3 in this example). In this static presentation, there are three parts: the upper left corner lists a set of keywords chosen automatically from the segmented story text based on the words' importance evaluated by their term frequency/inverse document frequency (TF/IDF) scores, the right column displays the transcription of the story (which can be scrolled if users want to read the text), and the center part is the visual presentation of the story consisting of a number of key frame images chosen automatically from the video in a content-sensitive manner [12]. Notice that the keywords for each story are also listed next to each story item in the ToC so that users can get a feeling about the content of the story before they decide which story to browse.

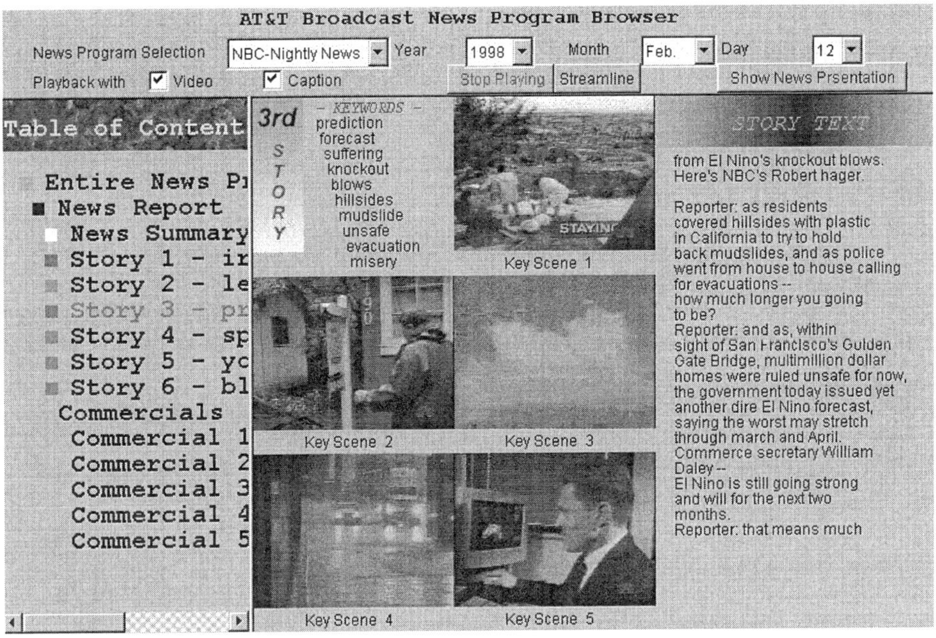

Figure 4 Visual representation for stories about El Niño.

From this example, we can see that the static representation is compact, semantically revealing, and visually informative with respect to the content of the story. It provides a more detailed review of the content than simply the keywords (listed in the ToC). On the basis of such review, a user may decide to listen to the story by invoking the dynamic presentation of the story. The user can either click on any of the images in the static view to listen to that particular section of the story or simply click on the button of that story in the ToC to play back the entire story. In both cases, a player window will pop up that will play back the synchronized audio, video, and text of the chosen story with precise timing information [11,12]. Such playback is timed from the beginning of the story and ends when the story ends. Compared with linear browsing or low-level scene cut–based browsing, this system allows much more effective *content-based* nonlinear information retrieval.

Finally, we construct the representation for the news summary of the day. It is composed of K images, where K is the number of headline stories detected on a particular day. The K images are chosen so that they are the most important in each story, using the same criterion in selecting story representation [11,12]. Figure 5 gives the visual presentation for the news summary of the day for the NBC Nightly News on February 12, 1998. From this presentation, a user can see immediately that there are six headline stories on that particular day. Below the representative image for each story, the list of keywords from that story is displayed in a rotating and flickering fashion (the interval of flickering is about 1 sec for each word) so that users can get a sense of the story from the keywords. The six headlines stories covered on that particular day are about (1) Russians tipping Iraq about weapon inspection, (2) Clinton's scandal, (3) El Niño in California, (4) controversy about whether the secret service should testify, (5) high suicide rate in an Indian

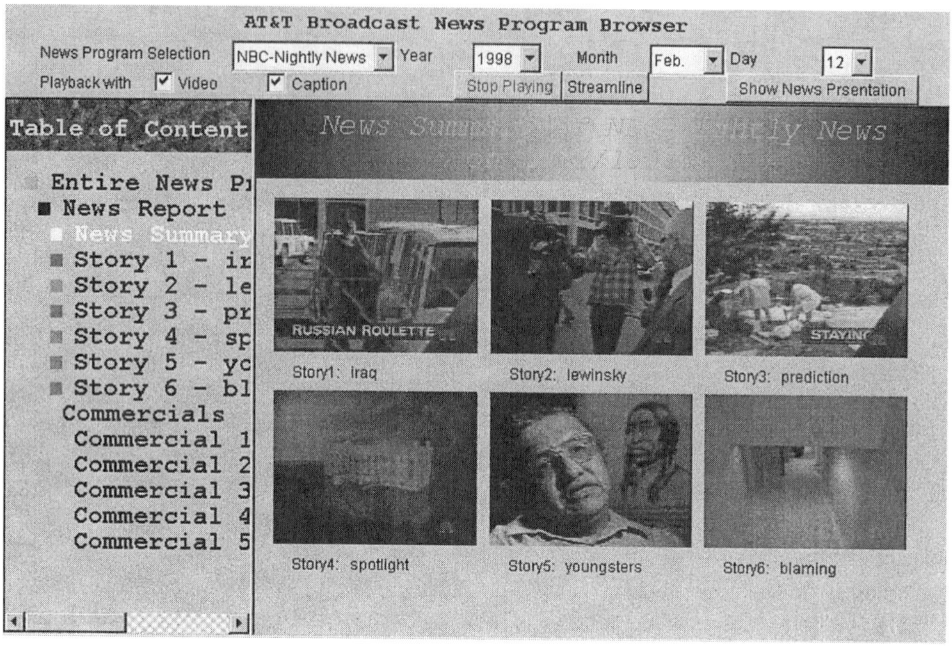

Figure 5 Representation for news summary of the day.

village, and (6) why taxpayers should pay for empty government buildings. From these examples, the effectiveness of this storytelling visual representation for the news summary is evident.

III. OBJECT-BASED SPATIOTEMPORAL VISUAL SEARCH AND FILTERING

An active research direction complementary to the preceding one using semantic-level structuring is the one that directly exploits low-level objects and their associated features in images or videos. An intuitive and popular approach is to segment and provide efficient indexes to salient objects in the images. Such segmentation processes can be implemented using automatic or semiautomatic tools [21,22]. Examples of salient objects may correspond to meaningful real-world objects such as houses, cars, and people or low-level image regions with uniform features such as color, texture, or shape.

Several notable image–video search engines have been developed using this approach, namely searching images or videos by example, by features, or by sketches. Searching for images by examples or templates is probably the most classical method of image search, especially in the domains of remote sensing and manufacturing. From an interactive graphic interface, users select an image of interest, highlight image regions, and specify the criteria needed to match the selected template. The matching criteria may be based on intensity correlation or feature similarity between the template image and the target images.

In feature-based visual query, users may ask the computer to find similar images according to specified features such as color, texture, shape, motion, and spatiotemporal structures of image regions [23–26]. Some systems also provide advanced graphic tools for users to draw visual sketches to describe the images or videos they have in mind [21,26,27]. Users are also allowed to specify different weightings for different features. Figure 6 shows query examples using the object color and the motion trail to find a video clip of a downhill skier and using color and motions of two objects to find football players.

A. Object-Based Video Segmentation and Feature Extraction

To demonstrate the processes involved in developing the preceding object-based video search tools, we review the process of developing the VideoQ search system [21] in this subsection.

Video sequences are first decomposed into separate shots. A video shot has a consistent background scene, although the foreground objects may change dynamically (they may even occlude each other, disappear, or reappear). Video shot separation is achieved by scene change detection. Scene change may include abrupt scene change, transitional changes (e.g., dissolve, fade in or out, and wipe). Once the video is separated into basic segments (i.e., video shots), salient video regions and video objects are extracted. Video object is the fundamental level of indexing in VideoQ. Primitive regions are segmented according to color, texture, edge, or motion measures. As these regions are tracked over time, temporal attributes such as trajectory, motion pattern, and life span are indexed. These low-level regions may also be used to develop a higher level of indexing that includes links to conceptual abstractions of video objects.

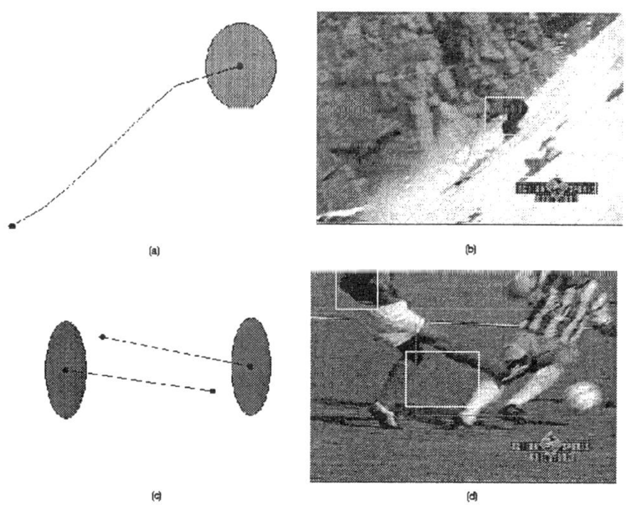

Figure 6 Object-oriented search using example image objects, features, and their relationships. By matching the object features and their spatiotemporal relationships [e.g., query sketches shown in (a) and (c)], video clips containing most similar objects and scenes [e.g., (b) downhill skiing and (d) soccer players] are returned. (Video courtesy of Actions, Sports, and Adventure, Inc.)

Based on the object-based representation model, a visual feature library (VFL) has also been developed. The VFL includes a rich set of visual features automatically extracted from the video objects. For example, color may include single color, average color, representative color, color histogram, and color pairs [28–30]. Texture may include wavelet-domain textures, texture histogram, Tamura texture, and Laws filter–based texture [31–34]. Shape may include geometric invariants, moments of different orders, polynomial approximation, spline approximation, and algebraic invariants. Motion may include trajectories of the centroid of each object and the affine models of each object over time (with camera motion compensation). The concept of VFL is to capture distinctive features that can be used to distinguish salient video objects efficiently and effectively. The final selection of features and their specific representations should depend on the resource (computing and storage) available and the application requirements (e.g., geometric invariance in object matching).

In measuring the similarity between video objects, various distance functions for measuring the feature similarity could be used, such as the Euclidean distance, the Mahalalobis distance, the quadratic distance with cross-element correlation, and the L1 distance. Usually it is difficult to find the optimal distance function. Instead, it is more beneficial to develop a flexible framework that can support various search methods, feature models, and matching metrics.

The preceding visual paradigm using the object-based representation and the VFL opens up a great opportunity for image–video search. In VisualSEEk [26], we integrated image search on the basis of visual features and spatial constraints. A generalized technique of 2D strings was used to provide a framework for searching for and comparing images by the spatial arrangement of automatically segmented feature regions. Multiple regions can be queried by either their absolute or relative locations. As shown in Figure 7, the overall query strategy consists of ''joining'' the queries based on the individual regions in the query image. Each region in the query image is used to query the entire database of regions. The resulting matched regions are then combined by the join operation, which consists of intersecting the results of region matches. The join operation identifies the candidate images that contain matches to all the query regions and then evaluates the relative spatial locations to determine the best matched image that satisfies the constraints of relative region placement. For video, more sophisticated spatiotemporal relationships can be verified at this stage to find video clips including rich spatiotemporal activities.

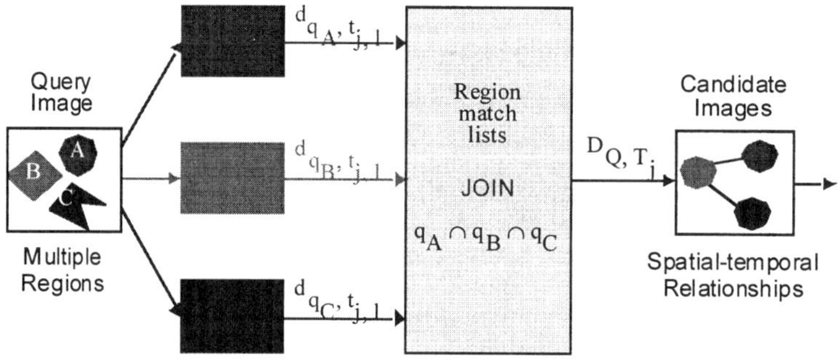

Figure 7 Query processing architecture supporting multiple regions and verification of spatiotemporal relationships among regions.

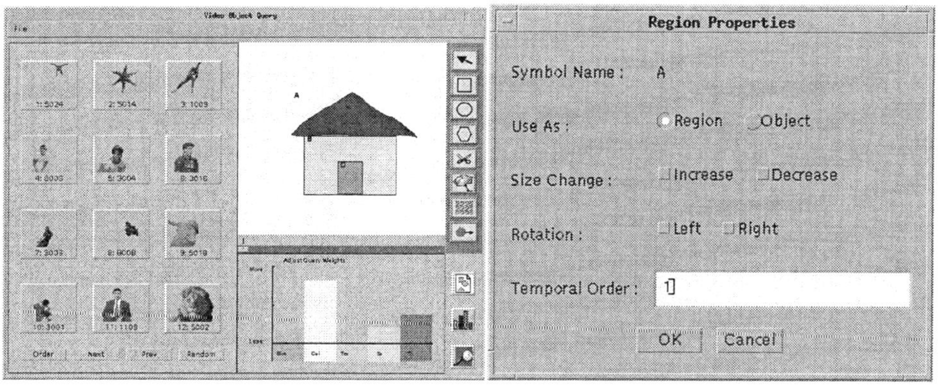

Figure 8 Search interface for the AMOS semantic object search engine. Objects are at the semantic level (e.g., people, animals, cars) rather than the region-level objects used in VideoQ.

This video object search system achieves fully automatic object segmentation and feature extraction at a low level. However, the results of object segmentation remain at a low level (e.g., image regions with uniform features). These usually do not correspond well to the real-world physical objects. One way to solve this problem is to use some input from users or content providers. For example, in the new MPEG-4 standard [35], video scenes consist of separate video objects that may be created, compressed, and transmitted separately. The video objects in MPEG-4 usually correspond to the semantic entities (e.g., people, car, and house). Region segmentation and feature extraction can be applied to analyze these objects and obtain efficient indexes for search and retrieval. Some systems combine some initial user interaction and automatic image region tracking to extract semantic video objects in a semiautomatic process [36]. Figure 8 shows the interface of a search engine for MPEG-4 semantic video objects [37]. Note that in addition to the single-level spatiotemporal search functions available in VideoQ, the semantic object search engine indexes spatiotemporal relationships at multiple levels (i.e., regions and objects) and thus allows more flexible search.

IV. SEMANTIC-LEVEL CONTENT CLASSIFICATION AND FILTERING

The spatiotemporal search tools just described provide powerful capabilities for searching videos or images at a low level. In many situations, users prefer to use simple and direct methods. For example, users may just want to browse through the content categories in which each image or video is classified into one or several meaningful classes. In other cases, we have found that the combination of feature-based similarity search tools and the subject navigation utilities achieves the most successful search results.

Recently, more research interest has emerged in developing automatic classification or filtering algorithms for mapping images or videos to meaningful classes, e.g., indoor, outdoor, or people [38–41]. In this section, we discuss two complementary approaches for identifying semantic concepts in multimedia data.

A. Content Modeling Using Probabilistic Graphic Models

The first approach uses probabilistic graphic models to identify events, objects, and sites in multimedia streams by computing probabilities such as *P*(underwater AND shark | segment of multimedia data). The basic idea is to estimate the parameters and structure of a model from a set of labeled multimedia training data.

A *multiject* (multimedia object) has a semantic label and summarizes the time sequences of low-level features of multiple modalities in the form of a probability *P*(semantic label | multimedia sequence). Most multijects fall into one of the three categories *sites*, *objects*, and *events*. Whereas some multijects are supported mainly by video (e.g., shark), others are supported mainly by audio (e.g., interior of traveling train) and still others are strongly supported by both audio and video (e.g., explosion).

The *lifetime* of a multiject is the duration of multimedia input that is used to determine its probability. In general, some multijects (e.g., family quarrel) will live longer than others (e.g., gunshot) and the multiject lives can overlap. For simplicity, we could break the multimedia into shots and within each shot fix the lifetimes of all multijects to the shot duration. Given the multiject probabilities, this leads to a simpler static inference problem within each shot. Although this approximation ignores event sequences within a shot (e.g., gunshot followed by family quarrel), it leaves plenty of room for useful inferences, because directors often use different shots to highlight changes in action and plot.

We model each modality in a multiject with a hidden Markov model (HMM) [41]. We investigate what combinations of input features and HMM structure give sufficiently accurate models. For example, we found that in the case of the explosion multiject, a color histogram of the input frames and a three-state HMM give a reasonably accurate video model [42]. In contrast, a video model for a bird may require object detection and tracking.

Each HMM in a multiject summarizes the time sequence for its corresponding modality. In order to summarize modalities, it is necessary to identify likely correspondences between events in the different modalities. For example, Figure 9a and b show the posterior probabilities that an explosion occurred in a movie clip at or before time t under an audio and a video HMM that were each trained on examples of explosions. In this case, the sound of the explosion begins roughly 0.15 sec (eight audio frames at 50 Hz or five video frames at 30 Hz) later than the video of the explosion. In other cases, we found that the audio and video events were more synchronized. To detect explosions accurately, the explosion multiject should be invariant to small time differences between the audio and video events. However, if the multiject is overly tolerant of the time difference, it may falsely detect nonexplosions. For example, a shot consisting of a pan from a sunrise to a waterfall will have a flash of bright red followed, after a large delay, by a thundering sound with plenty of white noise. Although the audio and video features separately match an explosion, the large time delay indicates that it is not an explosion.

We have used two methods [42,43] for summarizing the modalities in a multiject. In the first method, each HMM models an event in a different modality and the times at which the event occurs in the different modalities are loosely tied together by a kernel (e.g., Gaussian). An example of a pair of such event-coupled HMMs is shown in Figure 9c, where t^A and t^V are the times at which the explosion event begins in the audio and video. In Reference 42, we discussed an efficient algorithm for computing the probability

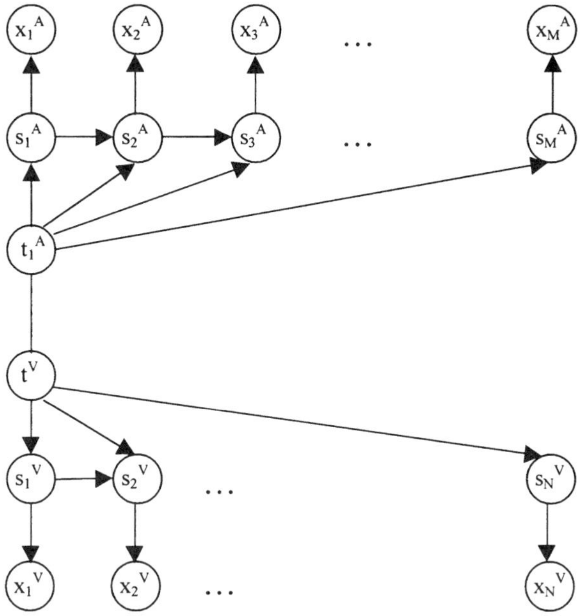

Figure 9 (a and b) The posterior probabilities that an explosion occurred in a movie clip at or before time *t* under an audio and a video HMM. (c) A graphic model for a multiject (multimedia object) that couples together the two HMMs.

of the multiject, e.g., P(explosion | video sequence, audio sequence). In the second method [43], a high-level HMM is trained on the states of the HMMs for the different modalities. Using a fast, greedy, bottom-up algorithm, the probability of the multiject can be approximated quite accurately. Note that the preceding model and structure can be greatly expanded to determine a multiject representation that is simple enough for learning and inference while being powerful enough to answer useful queries.

Suppose we are interested in automatically identifying movie clips of exotic birds from a movie library. We could compute the probability of the bird multiject for each multimedia shot in the library and then rank the shots according to these probabilities. However, by computing the probabilities of other related multijects, we may be able to derive additional support for the bird multiject. For example, an active waterfall multiject provides evidence of an exotic location where we are more likely to find an exotic bird. As a contrasting example, an active underwater multiject decreases the support for a bird being present in the shot.

We use the *multinet* (multiject network) as a way to represent higher level probabilistic dependences between multijects. Figure 10 shows an example of a multinet. In general, all multijects derive probabilistic support from the observed multimedia data (directed edges) as described in the previous section. The multijects are further interconnected to form a graphical probability model. Initially, we will investigate Boltzmann machine models [44] that associate a real-valued weight with each undirected edge in the graph. The weight indicates to what degree two multijects are correlated a priori (before the data are observed). In Figure 10, plus signs indicate that the multijects are correlated a priori whereas minus signs indicate that the multijects are anticorrelated.

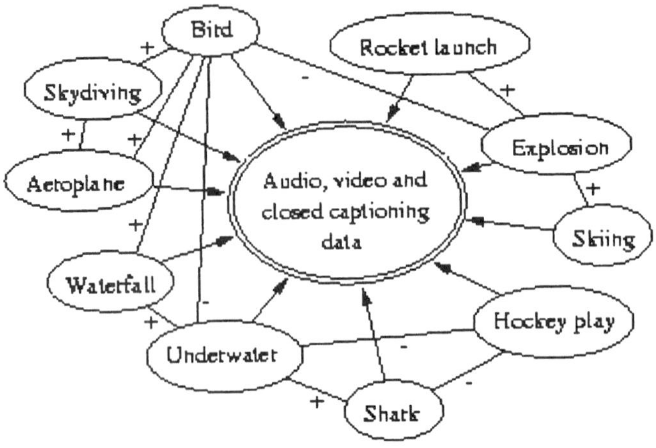

Figure 10 A multinet (multimedia network) probabilistically links multijects (multimedia objects) to the data and also describes the probabilistic relationships between the multijects. Plus signs indicate the multijects are correlated a priori (before the data are observed); minus signs indicate the multijects are anticorrelated.

Returning to the bird example, a plus sign on the connection between the bird multiject and the waterfall multiject indicates that the two multijects are somewhat likely to be present simultaneously. The minus sign on the connection between the bird multiject and the underwater multiject indicates that the two multijects are unlikely to be present simultaneously. The graphical formulation highlights interesting second-order effects. For example, an active waterfall multiject supports the underwater multiject, but these two multijects have opposite effects on the bird multiject.

In general, exact inference such as computing $P(\text{bird} \mid \text{multimedia data})$ in a richly connected graphical model is intractable. The second-order effects just described imply that many different combinations of multijects need to be considered to find likely combinations. However, there has been promising work in applying approximate inference techniques to such intractable networks in areas including pattern classification, unsupervised learning, data compression, and digital communication [45]. In addition, approaches using Markov chain Monte Carlo methods [44] and variational techniques [46] are promising.

B. Indexing Multimedia with Semantic Templates

In this subsection, we discuss a different semantic-level video indexing technique, semantic templates (STs) [47]. Semantic templates associate a set of exemplar queries with each semantic. Each query in the template has been chosen because it has been successful at retrieving the concept. The idea is that because a single successful query rarely completely represents the information that the user seeks, it is better to *cover* the concept using a set of successful queries. Semantic templates can be defined over indexable media of any type. Here we focus on video, semantic visual templates (SVTs). Figure 11 shows example SVTs for the ''high jumper'' concept and the ''sunsets'' concept.

The goal of the ST is similar to that of multijects, to detect and recognize video objects, sites, or events at the semantic level. However, the approach is based on different but synergistic principles. The generation of a semantic template involves no labeled

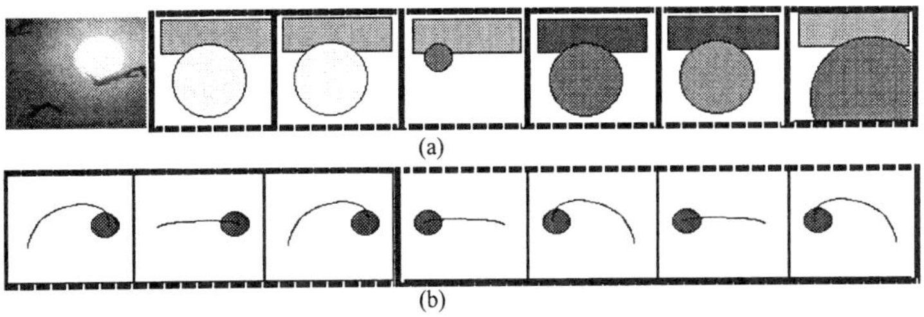

Figure 11 A semantic-level search paradigm using semantic visual templates. Image icons shown are subsets of optimal templates for each concept [(a) ''sunset'' (b) ''high jumper''] generated through the two-way interactive system.

ground truth data. What we do require is that some positive examples of the concept be present in the database so that they can be used in the interactive process when users interact with the system to generate the optimal set of STs. Development of STs utilizes the following unique principles.

Two-way learning: The template generation system emphasizes the two-way learning between the human and the machine. Because the human being is the final arbiter of the ''correctness'' of the concept, it is essential to keep the user in the template generation loop. The user defines the video templates for a specific concept with the concept in mind. Using the returned results and relevance feedback, the user and the system converge on a small set of queries that *best* match (i.e., provide maximal recall of) the user's concept.

Intuitive models: Semantic templates are intuitive, understandable models for semantic concepts in the videos. The final sets of SVTs can be easily viewed by the user. Users can have direct access and make manipulation to any template in the library.

Synthesizing new concepts: Different STs can be graphically combined to synthesize more complex templates. For example, templates for *high jumpers* and *crowds* can be combined to form a new template for ''high jumpers in front of a crowd.'' The audio templates for ''crowd'' and ''ocean sounds'' and the visual template for ''beach'' can be combined to form the ''crowds at a beach'' template.

The template framework is an extended model of the object-oriented video search engine described in Section III. The video object database consists of video objects and their features extracted in the object segmentation process. A visual template may consist of two types of concept definitions: *object icons* and *example scenes/objects*. The object icons are animated sketches such as the ones used in VideoQ. In VideoQ, the features associated with each object and their spatial and temporal relationships are important. The example scenes or objects are represented by the feature vectors extracted from these scenes/objects. Typical examples of feature vectors that could be part of a template are histograms, texture information, and structural information (i.e., more or less the global characteristics of the example scenes). The choice between an icon-based realization and an example-based realization depends on the semantic that we wish to represent. For example, a ''sunset'' can be very well represented by using a couple of objects, whereas a waterfall or a crowd is better represented using example scenes characterized by a global feature set. Hence, each template contains multiple icons and example scenes/objects to represent the idea. The elements of the set can overlap in their coverage. The goal is to come up with a minimal template set with maximal coverage.

Each icon for the concept comprises multiple objects that are associated with a set of visual attributes. The relevance of each attribute and each object to the concept is also specified using a context specification questionnaire. For example, for the concept ''sunsets,'' color and spatial structures of the objects (sun and sky) are more relevant. The object ''sun'' may be nonmandatory because some sunset videos may not have the sun visible. For the concept ''high jumper,'' the motion attribute of the foreground object (mandatory) and the texture attribute of the background object (nonmandatory) are more relevant than other attributes.

Development and application of semantic templates require the following components:

Generation: This is used to generate STs for each semantic concept. We will describe an interactive learning system in which users can interactively define their customized STs for a specific concept.

Metric: This is used to measure the ''fitness'' of each ST in modeling the concept associated with the video shot or the video object. The fitness measure can be modeled by the spatiotemporal similarity between the ST and the video.

Applications: An important challenge in applications is to develop a library of semantic concepts that can be used to facilitate video query at the semantic level. We will describe the proposed approaches to achieving such systems later.

Automatic generation of the STs is a hard problem. Hence we use a two-way interaction between the user and system in order to generate the templates (shown in Figure 12). In our method, given the initial query scenario and using relevance feedback, the system converges on a small set of icons (exemplar queries for both audio and video) that gives

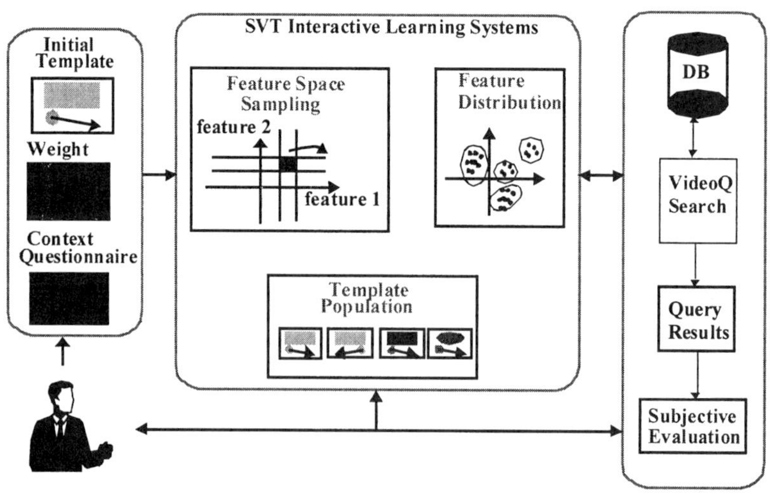

Figure 12 System architecture for semantic visual template.

us maximum recall. We now explain the mechanisms for generation of semantic visual templates.

The user comes to the system and sketches out the concept for which he wishes to generate a template. The sketch consists of several objects with spatial and temporal constraints. The user can also specify whether or not the object is mandatory. Each object is composed of several features. The user also assigns relevance weights to each feature of each object. This is the initial query scenario that the user provides to the system.

The initial query can also be viewed as a point in a high-dimensional feature space. Clearly, we can also map all videos in the database in this feature space. Now, in order to generate the possible icon set automatically, we need to make jumps in each of the features for each object. Before we do so, we must determine the jump step size, i.e., quantize the space. This we do with the help of the weight that the user has input along with the initial query. This weight can be thought of as the user's belief in the relevance of the feature with respect to the object to which it is attached. Hence, a low weight gives rise to coarse quantization of the feature and vice versa.

Because the total number of icons possible using this technique increases very rapidly, we do not allow for joint variation of the features. For each feature in each object, the user picks a plausible set for that feature. The system then performs a join operation on the set of features associated with the object. The user then picks the joins that are most likely to represent variations of the object. This results in a candidate icon list.

In a multiple-object case, we do an additional join with respect to the candidate lists for each object. Now, as before, the user picks the plausible scenarios. After we have generated a list of plausible scenarios, we query the system using the icons the user has picked. Using relevance feedback on the returned results (the user labels the returned results as positive or negative), we then determine the icons that provide us with maximum recall.

We now discuss a detailed example showing the generation mechanism for creating the semantic visual template for slalom skiers.

We begin the procedure by answering the context questionnaire shown in Figure 13a. We label the semantic visual template "slalom." We specify that the query is object based and will be composed of two objects. Then (Fig. 13b), we sketch the query. The large, white background object is the ski slope and the smaller foreground object is the skier with its characteristic zigzag motion trail.

We assign maximum relevance weights to all the features associated with the background and skier. We also specify that the features belonging to the background will remain static while those of the skier can vary during template generation. Then the system automatically generates a set of test icons, and we select plausible feature variations in the skier's color and motion trajectory.

A set of potential icons including both the background and the foreground skier are shown in Figure 13c. The user then chooses a candidate set to query the system. The 20 closest video shots are retrieved for each query. The user provides relevance feedback, which guides the system to a small set of exemplar icons associated with slalom skiers.

As mentioned earlier, the framework of the semantic templates can be applied to multiple medium modalities including audio and video. Once we have the audio and the

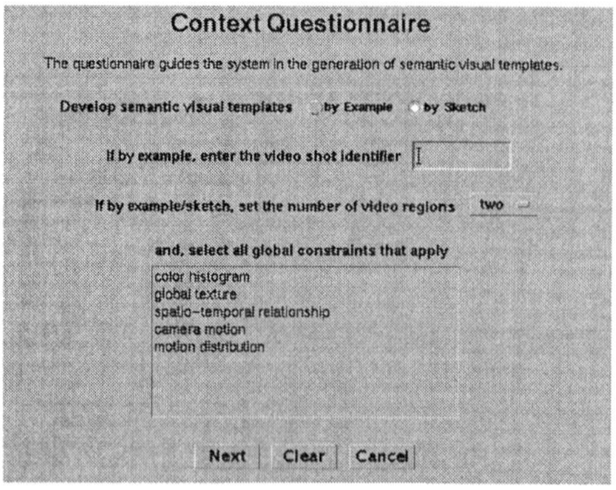

(a)

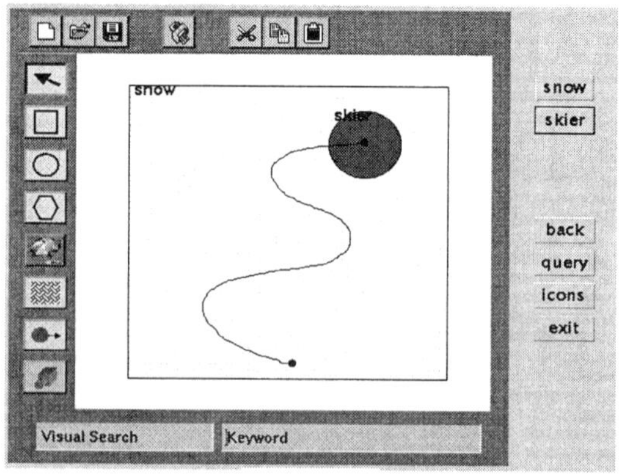

(b)

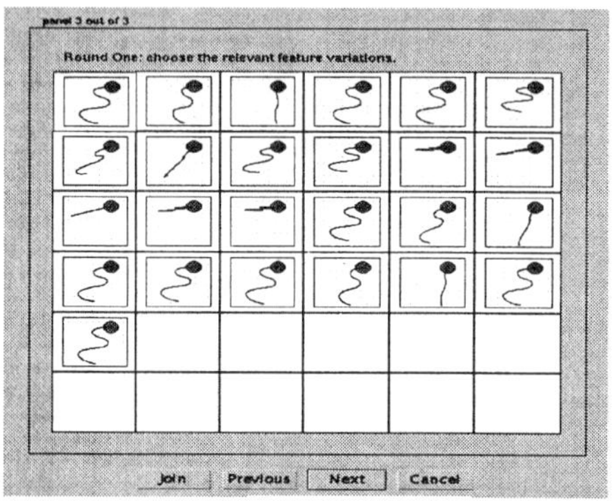

(c)

Figure 13 Example of semantic template development—Slalom. (a) Questionnaire for users to define the objects and features in a template; (b) initial graphic definition of the template; and (c) candidate icons generated by the system.

visual templates for different concepts such as ''skiing'' and ''sunsets,'' the user can interact with the system at the concept level. The user can compose a new multimedia concept that is built using these templates. For example if the user wanted to retrieve a group of people playing beach volleyball, he would use the visual templates of beach volleyball and beach sounds to generate a new query: {{Video: Beach volleyball, Beach}, {Audio: Beach Sounds}}. Then, the system would search for each template and return a result based on the user's search criterion. For example, he may indicate that he needs only some of the templates to be matched or that he needs all templates to be matched. Also, once we have a collection of audio and video templates for a list of semantics, we use these templates to match with the new videos and thereby generate a list of potential audio and video semantics (thereby generating a semantic index) that are associated with the video clip. Early results on querying using SVTs indicate that the concept works well in practice. For example, in the case of sunsets, the original query over a large heterogeneous database yielded only 10% recall. Using eight icons for the template, we boosted this result to 50%.

V. INTEROPERABLE CONTENT DESCRIPTION SCHEMES AND METASEARCH ENGINES

Techniques discussed earlier contribute to the state of the art in multimedia search and retrieval. Efficient tools and systems have been developed at different levels, including the physical level (e.g., image object matching) and the semantic level (e.g., news video content structure and semantic labeling). These tools can be optimized for the maximum power in specialized application domains. However, in many cases customized techniques may be used by different service providers for specialized content collection. Interoperability among the content indexes and the search functions becomes a critical issue. How do we develop a transparent search and retrieval gateway to hide the proprietary indexes and search methods and for users to access content in heterogeneous content sources?

A. Content-Describing Schemes and MPEG-7

To describe various types of multimedia information, the emerging MPEG-7 standard [48] has the objective of specifying a standard set of descriptors as well as description schemes (DSs) for the structure of descriptors and their relationships. This description (i.e., the combination of descriptors and description schemes) will be associated with the content itself to allow fast and efficient searching for material of a user's interest. MPEG-7 will also standardize a language to specify description schemes, (i.e., a description definition language, DDL), and the schemes for encoding the descriptions of multimedia content.

In this section, we briefly describe a candidate interoperable content description scheme [49] and some related research on image metasearch engines [50]. The motives of using these content description schemes for multimedia can be explained with the following scenarios.

Distributed processing: The self-describing schemes will provide the ability to interchange descriptions of audiovisual material independently of any platform, any vendor, and any application. The self-describing schemes will enable the distributed processing of multimedia content. This standard for interoperable content descriptions will mean that data from a variety of sources can be easily plugged into a variety of distributed applications such as multimedia processors, editors, retrieval systems, and filtering agents.

Content exchange: A second scenario that will greatly benefit from an interoperable content description is the exchange of multimedia content among heterogeneous audiovisual databases. The content descriptions will provide the means to express, exchange, translate, and reuse existing descriptions of audiovisual material.

Customized views: Finally, multimedia players and viewers compliant with the multimedia description standard will provide the users with innovative capabilities such as multiple views of the data configured by the user. The user could change the display's configuration without requiring the data to be downloaded again in a different format from the content broadcaster.

To ensure maximum interoperability and flexibility, our description schemes use the eXtensible Markup Language (XML), developed by the World Wide Web consortium (W3C) [51]. Here we briefly discuss the benefits of using XML and its relationship with other languages such as SGML. SGML (Standard Generalized Markup Language, ISO 8879) is a standard language for defining and using document formats. SGML allows documents to be self-describing; i.e. they describe their own grammar by specifying the tag set used in the document and the structural relationships that those tags represent. However, full SGML contains many optional features that are not needed for Web applications and has proved to be too complex to current vendors of Web browsers.

The W3C has created an SGML Working Group to build a set of specifications to make it easy and straightforward to use the beneficial features of SGML on the Web [52]. This subset, called XML, retains the key SGML advantages in a language that is designed to be vastly easier to learn, use, and implement than full SGML. A major advantage of using XML is that it allows the descriptions to be self-describing in the sense that they combine the description and the structure of the description in the same format and document. XML also provides the capability to import external document type definitions (DTDs) into the image description scheme DTD in a highly modular and extensible way.

The candidate description scheme consists of several basic components: *object*, *object hierarchy*, *entity relation graph*, and *feature structure*. Each description includes a set of objects. The objects can be organized in one or more object hierarchies. The relationships between objects can be expressed in one or more entity relation graphs.

An object element represents a region of the image for which some features are available. It can also represent a moving image region in a video sequence. There are two different types of objects: *physical* and *logical* objects. Physical objects usually correspond to continuous regions of the image with some descriptors in common (semantics, features, etc.). Logical objects are groupings of objects (which may not occupy a continuous area in the image) based on some high-level semantic relationships (e.g., all faces in the image). The object element comprises the concepts of group of objects, objects, and regions in the visual literature. The set of all objects identified in an image is included within the object set element.

The description scheme also includes one or more object hierarchies to organize the object elements in the object set element. Each object hierarchy consists of a tree of object node elements. Each object node points to an object. The objects in an image can be organized by their location in the image or by their semantic relationships. These two ways to group objects generate two types of hierarchies: physical and logical hierarchies. A physical hierarchy describes the physical location of the objects in the image. On the other hand, a logical hierarchy organizes the objects based on a higher level semantics, such as information in categories of *who*, *what object*, *what action*, *where*, *when*, *why*.

In addition to the object hierarchy, an *entity–relationship* model is used to describe general relationships among object elements. Examples include spatial relationships, temporal relationships, and semantic-level relationships (e.g., A is shaking hands with B).

Each object includes one or more associated features. Each object can accommodate any number of features in a modular and extensible way. The features of an object are grouped together according to the following categories: *visual*, *semantic*, and *media*. Multiple abstraction levels of features can be defined. Each object may be associated with objects in other modalities through modality transcoding.

Examples of visual features include color, texture, shape, location, and motion. These features can be extracted or assigned automatically or manually. Semantic features include annotations and semantic-level description in different categories (people, location, action, event, time, etc.). Media features describe information such as compression format, bit rate, and file location.

Each feature of an object has one or more associated descriptors. Each feature can accommodate any number of descriptors in a modular and extensible way. External descriptors may also be included through linking to an external DTD. For a given descriptor, the description scheme also provides a link to external extraction code and similarity matching code.

The unified description scheme may be applied to image, video, and combinations of multimedia streams in a coherent way. In the case of multimedia, a multimedia stream is represented as a set of multimedia objects that include objects from the composing media streams or other multimedia objects. Multimedia objects are organized in object hierarchies. Relationships among two or more multimedia objects that cannot be expressed in a tree structure can be described using multimedia entity relation graphs. The tree structures can be efficiently indexed and traversed, while the entity relation graphs can model more general relationships.

Details of the description schemes just mentioned can be found in References 53–55.

B. Multimedia Metasearch Engines

The preceding self-describing schemes are intuitive, flexible, and efficient. We have started to develop an MPEG-7 testbed to demonstrate the feasibility of our self-describing schemes. In our testbed, we are using the self-describing schemes for descriptions of images and videos that are generated by a wide variety of image–video indexing systems. In this section, we will discuss the impact of the MPEG-7 standard on a very interesting research topic, image metasearch engines.

Metasearch engines act as gateways linking users automatically and transparently to multiple search engines. Most of the current metasearch engines work with text. Our earlier work on a metasearch engine, MetaSEEk [50], explores the issues involved in querying large, distributed, online visual information systems. MetaSEEk is designed to select intelligently and interface with multiple online image search engines by ranking their performance for different classes of user queries. The overall architecture of MetaSEEk is shown in Figure. 14. The three main components of the system are standard for metasearch engines; they are the *query dispatcher*, the *query translator*, and the *display interface*. The procedure for each search is as follows:

Upon receiving a query, the dispatcher selects the target search engines to be queried by consulting the performance database at the MetaSEEk site. This database contains performance scores of past query successes and failures for each supported search engine. The query dispatcher selects only search engines that provide capabilities compatible with the user's query (e.g., visual features and/or keywords).

The query translators then translate the user query to suitable scripts conforming to the interfaces of the selected search engines.

Finally, the display component uses the performance scores to merge the results from each search engine and displays them to the user.

MetaSEEk evaluates the quality of the results returned by each search engine based on the user's feedback. This information is used to update the performance database. The operation of MetaSEEk is very restricted by the interface limitations of current search engines. For example, most existing systems can support only query by example, query

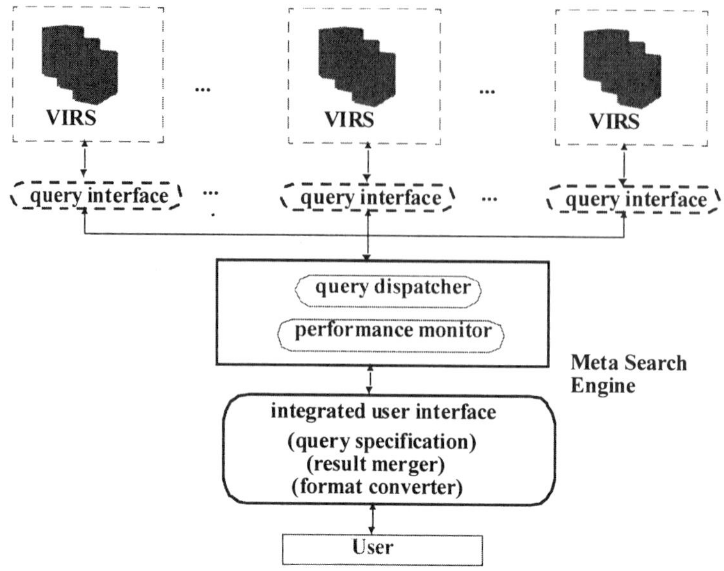

Figure 14 System architecture of metasearch engines such as MetaSEEk.

by sketch, and keyword search. Results usually are just a flat list of images (with similarity scores, in some cases).

As discussed in the previous section, we envision a major transformation in multimedia search engines thanks to the development of the MPEG-7 standard. Future systems will accept not only queries by example and by sketch but also queries by MPEG-7 multimedia descriptions. Users will be able to submit desirable multimedia content (specified by MPEG-7 descriptions) as the query input to search engines. In return, search engines will work on a best effort basis to provide the best search results. Search engines unfamiliar with some descriptors in the query multimedia description may just ignore those descriptors. Others may try to translate them to local descriptors. Furthermore, queries will result in a list of matched multimedia data as well as their MPEG-7 descriptions. Each search engine will also make available the description scheme of its content and maybe even proprietary code.

We envision the connection between the individual search engines and the metasearch engine to be a path for MPEG-7 streams, which will enhance the performance of metasearch engines. In particular, the ability of the proposed description schemes to download programs dynamically for feature extraction and similarity matching by using linking or code embedding will open the door to improved metasearching capabilities. Metasearch engines will use the description schemes of each target search engine to learn about the content and the capabilities of each search engine. This knowledge also enables meaningful queries to the repository, proper decisions to select optimal search engines, efficient ways to merge results from different repositories, and intelligent display of the search results from heterogeneous sources.

VI. DISCUSSION

Multimedia search and retrieval involves multiple disciplines, including image processing, computer vision, database, information retrieval, and user interfaces. The content types contained in multimedia data can be very diverse and dynamic. This chapter focuses on the multimedia content structuring and searching at different levels. It also addresses the interoperable representation for content description. The impact of the emerging standard, MPEG-7, is also discussed from the perspective of developing metasearch systems.

Many other important research issues are involved in developing a successful multimedia search and retrieval system. We briefly discuss several notable ones here. First, user preference and relevance feedback are very important and have been used to improve the search system performance. Many systems have taken into account user relevance feedback to adapt the query features and retrieval models during the iterated search process [56–59].

Second, content-based visual query poses a challenge because of the rich variety and high dimensionality of features used. Most systems use techniques related to prefiltering to eliminate unlikely candidates in the initial stage and to compute the distance of sophisticated features on a reduced set of images [60]. A general discussion of issues related to high-dimensional indexing for multimedia content can be found in Reference 61.

Investigation of search and retrieval for other types of multimedia content has also become increasingly active. Emerging search engines include those for music, audio clips, synthetic content, and images in special domains (e.g., medical and remote sensing). Wold et al. [62] developed a search engine that matches similarities between audio clips based on the feature vectors extracted from both the time and spectral domains. Paquet and Rioux [63] presented a content-based search engine for 3D VRML data. Image search engines specialized for remote sensing applications have been developed [33,34] with focus on texture-based search tools.

Finally, a very challenging task in multimedia search and retrieval is performance evaluation. The uncertainty of user need, the difficulty in obtaining the ground truth, and the lack of a standard benchmark content set have been the main barriers to developing effective mechanisms for performance evaluation. To address this problem, MPEG-7 has incorporated the evaluation process as a mandatory part of the standard development process [64].

REFERENCES

1. H Zhang, A Kankanhalli, S Smoliar. Automatic partitioning of full-motion video. A Guided Tour of Multimedia Systems and Applications. IEEE Computer Society Press, 1995.
2. JS Boreczky, LA Rowe. Comparison of video shot boundary detection techniques. Proceedings of SPIE Conference, on Storage and Retrieval for Still Image and Video Databases IV, SPIE vol 2670, San Jose, February 1996, pp 170–179.
3. B Shahraray. Scene change detection and content-based sampling of video sequences. In: RJ Safranek, AA Rodriquez, eds. Digital Video Compression: Algorithms and Technologies. 1995.
4. A Hauptmann, M Witbrock. Story segmentation and detection of commercials in broadcast news video. Proceedings of Advances in Digital Libraries Conference, Santa Barbara, April 1998.

5. M Yeung, B-L Yeo. Time-constrained clustering for segmentation of video into story units. Proceedings of International Conference on Pattern Recognition, Vienna, Austria, August 1996, pp 375–380.
6. M Yeung, B-L Yeo, B Liu. Extracting story units from long programs for video browsing and navigation. Proceedings of International Conference on Multimedia Computing and Systems, June 1996.
7. Y Rui, TS Huang, S Mehrotra. Constructing table-of-content for videos. ACM Multimedia Syst 1998.
8. M Maybury, M Merlino, J Rayson. Segmentation, content extraction and visualization of broadcast news video using multistream analysis. Proceedings of ACM Multimedia Conference, Boston, 1996.
9. I Mani, D House, D Maybury, M Green. Towards content-based browsing of broadcast news video. Intelligent Multimedia Information Retrieval, 1997.
10. A Merlino, D Morey, D Maybury. Broadcast news navigation using story segmentation. Proceedings of ACM Multimedia, November 1997.
11. Q Huang, Z Liu, A Rosenberg, D Gibbon, B Shahraray. Automated generation of news content hierarchy by integrating audio, video, and text information. 1999 IEEE International Conference on Acoustics, Speech, and Signal Processing, Phoenix, AZ, March 15–19, 1999.
12. Q Huang, Z Liu, A Rosenberg. Automated semantic structure reconstruction and representation generation for broadcast news. SPIE: Electronic Imaging: Storage and Retrieval of Image and Video Databases, San Jose, January 1999, pp 50–62.
13. A Hanjalic, RL Lagendijk, J Biemond. Semi-automatic analysis, indexing and classification system based on topics preselection. SPIE: Electronic Imaging: Storage and Retrieval of Image and Video Databases, San Jose, CA, January 1999, pp 86–97.
14. YL Chang, W Zeng, I Kamel, R Alonso. Integrated image and speech analysis for content-based video indexing. Proceedings of Multimedia, September 1996, pp 306–313.
15. J Nam, AH Tewfik. Combined audio and visual streams analysis for video sequence segmentation. Proceedings of International Conference on Acoustic, Speech, and Signal Processing, vol 4, 1997, pp 2665–2668.
16. MA Hearst. Multi-paragraph segmentation of expository text. The 32nd Annual Meeting of the Association for Computational Linguistics, New Mexico, June 1994.
17. MG Brown, J Foote, GJF Jones, KS Jones, SJ Young. Automatic content-based retrieval of broadcast news. Proceedings of ACM Multimedia Conference, San Francisco, 1995, pp 35–42.
18. B Shahraray, D Gibbon. Efficient archiving and content-based retrieval of video information on the Web. AAAI Symposium on Intelligent Integration and Use of Text, Image, Video, and Audio Corpora, Stanford, CA, March 1997, pp 133–136.
19. B Shahraray, D Gibbon. Automated authoring of hypermedia documents of video programs. Proceedings Third International Conference on Multimedia (ACM Multimedia '95), San Francisco, November 1995.
20. Q Huang, Z Liu, A Rosenberg. Automated semantic structure reconstruction and representation generation for broadcast news. Proceedings SPIE, Storage and Retrieval for Image and Video Databases VII, San Jose, CA, January 1999.
21. S-F Chang, W Chen, HJ Meng, H Sundaram, D Zhong. VideoQ—An automatic content-based video search system using visual cues. ACM Multimedia 1997, Seattle, November 1997 (demo:http://www.ctr.columbia.edu/videoq).
22. D Zhong, S-F Chang. AMOS—An active system for MPEG-4 semantic object segmentation. IEEE International Conference on Image Processing, Chicago, October 1998.
23. S-F Chang, A Eleftheriadis, R McClintock. Next-generation content representation, creation and searching for new media applications in education. Proc IEEE 86:884–904, 1998.
24. M Flickner, H Sawhney, W Niblack, J Ashley, Q Huang, B Dom, M Gorkani, J Hafner, D

Lee, D Petkovic, D Steele, P Yanker. Query by image and video content: The QBIC system. IEEE Comput Mag 28(9):23–32, 1995.
25. A Pentland, RW Picard, S Sclaroff. Photobook: Tools for content-based manipulation of image databases. Proceedings Storage and Retrieval for Image and Video Databases II, vol 2185, SPIE, Bellingham, WA, 1994, pp 34–47.
26. JR Chang, S-F Chang. VisualSEEk: A fully automated content-based image query system. ACM Multimedia Conference, Boston, November 1996 (demo:http://www.ctr.columbia.edu/VisualSEEk).
27. CE Jacobs, A Finkelstein, DH Salesin. Fast multiresolution image querying. ACM SIGRAPH, August 1995, pp 277–286.
28. A Nagasaka, Y Tanaka. Automatic video indexing and full-video search for object appearances, Visual Database Systems II, 1992.
29. JR Smith, S-F Chang. Tools and techniques for color image retrieval. SPIE Conference on Storage and Retrieval for Image and Video Database, San Jose, CA, February 1996.
30. M Swain, D Ballard. Color indexing. Int J Comput Vision 1:11–32, 1991.
31. T Chang, C-CJ Kuo. Texture analysis and classification with tree-structured wavelet transform. IEEE Trans Image Process 2(4):1993.
32. JR Smith, S-F Chang. Transform features for texture classification and discrimination in large image databases. Proceedings, IEEE First International Conference on Image Processing, Austin, TX, November 1994.
33. L Bergman, V Castelli, C-S Li. Progressive content-based retrieval from satellite image archives. CNRI Digital Library Magazine (on line), October 1997 (http://www.dlib.org).
34. WY Ma, BS Manjunath. Texture features for browsing and retrieval of image data. IEEE Trans Pattern Anal Machine Intell 18:837–842, 1996.
35. ISO/IEC JTC1/SC29/WG11 N1909, MPEG-4 Version 1 Overview, October 1997.
36. D Zhong, S-F Chang. AMOS—An active system for MPEG-4 semantic object segmentation. IEEE International Conference on Image Processing, Chicago, October 1998.
37. D Zhong, S-F Chang. An integrated system for content-based video object segmentation and retrieval. Part II: Content-based searching of video objects. Submitted to IEEE Trans Circuits Syst Video Technol.
38. M Szummer, R Picard. Indoor-outdoor image classification. IEEE International Workshop on Content-Based Access of Image and Video Databases CAIVD '98, Bombay, India, January 1998.
39. A Vailaya, A Jain, HJ Zhang. ON image classification: City vs. landscape. IEEE Workshop on Content-Based Access of Image and Video Libraries, Santa Barbara, CA, June 1998.
40. D Forsyth, M Fleck. Body plans. IEEE Conference on Computer Vision and Pattern Recognition, Puerto Rico, June 1997.
41. A Jaimes, S-F Chang. Model based image classification for content-based retrieval. SPIE Conference on Storage and Retrieval for Image and Video Database, San Jose, CA, January 1999.
42. TT Kristjansson, BJ Frey, TS Huang. Event-coupled hidden Markov models. Submitted to Adv Neural Inform Process Syst.
43. MR Naphade, TT Kristjansson, BJ Frey, TS Huang. Probabilistic multimedia objects (multijects): A novel approach to video indexing and retrieval in multimedia systems. IEEE International Conference on Image Processing, Chicago, October 1998.
44. GE Hinton, TJ Sejnowski. Learning and relearning in Boltzmann machines. In: DE Rumelhart, and JL McClelland, eds. Parallel Distributed Processing: Explorations in the Microstructure of Cognition. vol I, Cambridge, MA: MIT Press, 1986, pp 282–317.
45. BJ Frey. Graphic Models for Machine Learning and Digital Communication. Cambridge, MA: MIT Press, 1998.
46. MI Jordab, Z Ghahramani, TS Jaakkola, LK Saul. An introduction to variational methods for

graphic models. In: MI Jordan, ed. Learning and Inference in Graphic Models. Norwell, MA: Kluwer Academic Publishers, 1998.

47. S-F Chang, W Chen, H Sundaram. Semantic visual templates—Linking visual features to semantics. IEEE International Conference on Image Processing, Chicago, October 1998.
48. MPEG-7's Call for Proposals for Multimedia Content Description Interface, http://drago.cselt.it/mpeg/public/mpeg-7_cfp.htm.
49. S Paek, AB Benitez, S-F Chang. Self-describing schemes for interoperable MPEG-7 multimedia content descriptions. SPIE.VCIP'99, Visual Communication and Image Processing, San Jose, CA, January 1999.
50. AB Benitez, M Beigi, S-F Chang. Using relevance feedback in content-based image metasearch. IEEE Internet Comput Mag 2(4):59–69, 1998 (demo: http://www.ctr.columbia.edu/MetaSEEk).
51. World Wide Web Consortium's (W3C) XML web site http://www.w3.org/XML.
52. World Wide Web Consortium's (W3C) XML Web site http://www.w3.org/XML.
53. S Paek, AB Benitez, S-F Chang, C-S Li, JR Smith, LD Bergman, A Puri. Proposal for MPEG-7 image description scheme. ISO/IEC JTC1/SC29/WG11 MPEG98/P480 MPEG document, Lancaster, UK, February 1999.
54. S Paek, AB Benitez, S-F Chang, A Eleftheriadis, A Puri, Q Huang, C-S Li, JR Smith, LD Bergman. Proposal for MPEG-7 video description scheme. ISO/IEC JTC1/SC29/WG11 MPEG98/P481 MPEG document, Lancaster, UK, February 1999.
55. Q Huang, A Puri, AB Benitez, S Paek, S-F Chang. Proposal for MPEG-7 integration description scheme for multimedia content. ISO/IEC JTC1/SC29/WG11 MPEG98/P477 MPEG document, Lancaster, UK, February 1999.
56. J Huang, SR Kumar, M Mitra. Combining supervised learning with color correlograms for content based image retrieval. ACM Multimedia '97, Seattle, November 1997.
57. TP Minka, R Picard. Interactive learning using a "society of models." MIT Media Lab Perceptual Computing Section Technical Report 349.
58. Y Rui, T Huang, S Mehrotra, M Ortega. A relevance feedback architecture for content-based multimedia information retrieval systems. CVPR'97 Workshop on Content-Based Image and Video Library Access, June 1997.
59. IJ Cox, ML Miller, SM Omohundro, PY Yianilos. The PicHunter Bayesian multimedia retrieval system. ADL '96, Forum, Washington, DC, May 1996.
60. J Hafner, HS Sawhney, W Equitz, M Flickner, W Niblack. Efficient color histogram indexing for quadratic form distance functions. IEEE Trans PAMI, July 1995.
61. C Faloutsos. Searching Multimedia Databases by Content. Boston: Kluwer Academic Publishers, 1997.
62. E Wold, T Blum, D Keislar, J Wheaton. Content-based classification, search, and retrieval of audio. IEEE Multimedia Mag 3(3):27–36, 1996.
63. E Paquet, M Rioux. A content-based search engine for VRML databases. Proceedings of the 1998 Computer Society Conference on Computer Vision and Pattern Recognition (CVPR), Santa Barbara, CA, June 1998, pp 541–546.
64. MPEG Requirements Group, MPEG-7 Evaluation Procedure. Doc. ISO/MPEG N2463, MPEG Atlantic City Meeting, October 1998.
65. JR Bach, C Fuller, A Gupta, A Hampapur, B Horowitz, R Humphrey, RC Jain, C Shu. Virage image search engine: An open framework for image management. Symposium on Electronic Imaging: Science and Technology—Storage & Retrieval for Image and Video Databases IV, IS&T/SPIE, February 1996.
66. J Meng, S-F Chang. CVEPS: A compressed video editing and parsing system. ACM Multimedia Conference, Boston, November 1996. (demo: http://www.ctr.columbia.edu/webclip).
67. L Rabiner, BH Juang. Fundamentals of Speech Recognition. Ch. 5, Englewood Cliffs, NJ: Prentice Hall, 1993, chap 5.

21

Image Retrieval in Digital Libraries

Bangalore S. Manjunath and Yining Deng
University of California at Santa Barbara, Santa Barbara, California

David A. Forsyth, Chad Carson, Sergey Ioffe, Serge J. Belongie, and Jitendra Malik
University of California at Berkeley, Berkeley, California

Wei-Ying Ma
Hewlett-Packard Laboratories, Palo Alto, California

The enabling technologies of multimedia content search and retrieval are crucial for the development of digital libraries. The past few years have seen a significant growth in multimedia content as new, sophisticated, yet inexpensive devices are introduced that make it easier than ever before to create multimedia documents. Many of the science and engineering journals have gone on line toward electronic publication. Geographic information systems and medical imaging are rich in image and video content. New content description techniques that go beyond keyword-based methods are needed for searching for information in such heterogeneous data. Recent efforts in standardizing such descriptions for multimedia data in the MPEG community clearly underscore this point.

This chapter focuses on image search and retrieval issues in the context of the two University of California (UC) Digital Library projects. The UC Berkeley's California Environmental Digital Information system includes many different types of environmental data [1]. The goal of the Alexandria Digital Library project at UC Santa Barbara [2] is to establish an electronic library of spatially indexed information. Image data from the two projects include a large number of ground photographs, aerial imagery, and satellite pictures.

The UC Berkeley (http://elib.cs.berkeley.edu) and UC Santa Barbara (http://www.alexandria.ucsb.edu) projects are two of the six projects sponsored in 1994 by the National Science Foundation (NSF), the Department of Defense's Advanced Research Projects Agency (ARPA), and the National Aeronautics and Space Administration (NASA) as part of the interagency Digital Library Initiative (DLI). The University of Illinois at Urbana-Champaign (UIUC) (http://dli.grainger.uiuc.edu) project's goal is to develop Web technology for effective searching of technical documents on the Internet. The University of Michigan's project focuses on an agent-based architecture for digital library services (http://www.si.umich.edu/UMDL/), whereas the Stanford University project's main goal was interoperability (http://diglib.stanford.edu). The Carnegie Mellon Informedia project (http://www.informedia.cs.cmu.edu) emphasizes multimedia databases, integrating textural, audio, and visual descriptions to annotate multimedia data automatically.

I. IMAGES IN DIGITAL LIBRARIES

There is an extensive literature on retrieving images from large collections using features computed from the whole image, including color histograms, texture measures, and shape measures [3–13]. However, in the most comprehensive study of usage practices (a paper by Enser [14] surveying the use of the Hulton Deutsch collection), there is a clear user preference for searching these collections on image semantics. An ideal search tool would be a quite general recognition system that could be adapted quickly and easily to the types of objects sought by a user. Building such a tool requires a much more sophisticated understanding of the process of recognition than currently exists. Toward this objective, we emphasize image representations that are local—region- or object-based image descriptors and their spatial organization, for search and retrieval.

This chapter presents recent research results on querying image databases—from using simple image descriptors such as texture to more sophisticated representations that involve spatial relationships, for finding *objects* in images. Section II introduces a *texture thesaurus* that demonstrates the power of image texture descriptors in characterizing a variety of salient image structures in aerial pictures. In Section III an online image retrieval system called NeTra (which means *eyes* in the Sanskrit language) is described. NeTra uses region color and shape descriptors, in addition to texture, to annotate a collection of 2500 images from a Corel photo CD database. An automated image segmentation scheme and an efficient color indexing structure are the key components of NeTra. Section IV describes the *blobworld*, which further demonstrates the substantial advantages to structuring systems around *regions*—image patches of coherent color and texture that could correspond to objects. Most important, the results presented by such a system are easy for a user to understand because the representation is relatively transparent. This means that refining the query and following new paths is easy and clean; when individual query results are understandable, it is simpler to produce a stream of queries that satisfies the user's need. Finally, as Section V demonstrates, some kinds of objects can be found quickly and quite accurately by reasoning about spatial relationships between regions. In particular, animals and people look like assemblies of cylinders—which we call *body plans*—where the relationships between the cylinders are constrained by the kinematics of the animal. We show how to take advantage of these constraints to assemble groups of regions that indicate the presence of a person or of a horse.

II. A TEXTURE THESAURUS

In recent years image texture has emerged as an important visual primitive to search and browse through large collections of similar-looking patterns. An image can be considered as a mosaic of textures, and texture features associated with the regions can be used to index the image data. For instance, a user browsing an aerial image database may want to identify all parking lots in the image collection. A parking lot with cars parked at regular intervals is an excellent example of a textured pattern when viewed from a distance, such as in an air photo. Similarly, agricultural areas and vegetation patches are other examples of textures commonly found in aerial and satellite imagery. Examples of queries that could be supported in this context include ''Retrieve all Landsat images of Santa Barbara that have less than 20% cloud cover'' and ''Find a vegetation patch that looks like this region.''

A. Texture Features

In [28] we proposed the use of Gabor-filtered outputs of an image for texture analysis and provided a comprehensive evaluation and comparison with other multiresolution texture features using the Brodatz texture database [15]. Our experimental results indicate that the Gabor texture features provide very good pattern retrieval accuracy. This scheme has been used to annotate large aerial photographs (typically 5K × 5K pixels) in our current collection by performing the feature extraction on nonoverlapping subimages (of dimension 64 × 64 pixels). A subimage pattern is processed through a bank of Gabor filters at different scales and orientations. The mean and standard deviation of the filtered outputs are then used to construct a feature vector to represent that pattern.

What distinguishes image search from traditional texture classification is the need to retrieve more than just the best match. Typically, the search results are rank ordered on the basis of the distance to the query pattern, so the comparison in the feature space should preserve visual similarities. This is an important criterion but is difficult to achieve. In Ma and Manjunath [16] a hybrid neural network algorithm to learn pattern similarity in the Gabor texture feature space is proposed. The neural network uses training data containing similarity information (provided by the users) to partition the feature space into visually similar texture pattern clusters. An evaluation of this learning scheme on the Brodatz database indicates that significantly better retrieval performance can be achieved [17]. This partitioning of the feature space into visually similar subclusters is used in constructing a *visual thesaurus*, and this is discussed later.

B. Image Segmentation

Automated image segmentation is clearly a significant bottleneck in enhancing retrieval performance. We again emphasize the fact that it is important to localize the image feature information. Region- or object-based search is more natural and intuitive than search using the whole image information. With automated segmentation as the primary objective, we have developed a robust segmentation scheme, called EdgeFlow, that has yielded very promising results on a diverse collection of a few thousand images [18].

The EdgeFlow scheme utilizes a simple predictive coding model to identify and integrate changes in visual cues such as color and texture at each pixel location. As a result of this computation, a flow vector that points in the direction of the closest image boundary is constructed at each pixel location in the image. The magnitude of this edge flow vector is the *total edge energy* at that location.

The distribution of these edge flow vectors on an image grid forms a *flow field*, which is allowed to propagate. At each pixel location, the flow is in the estimated direction of a boundary element. A boundary location is characterized by flows in opposing directions toward it. On a discrete grid, the flow typically takes few iterations to converge.

The flow direction gives the direction with the most information change in the image feature space. Because any of the image attributes such as color, texture, or their combination can be used to define the edge flow, this provides a simple framework for integrating

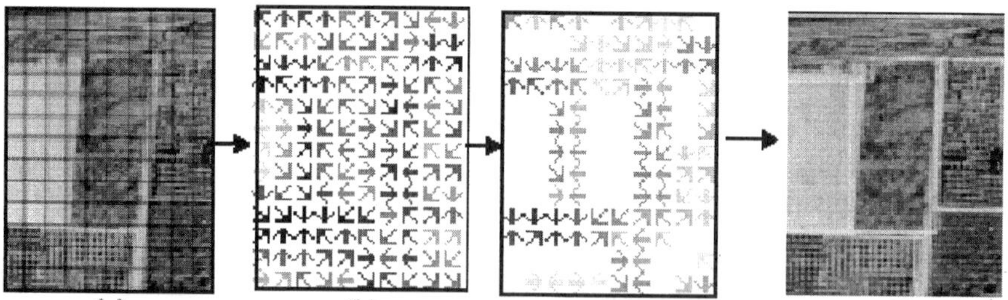

Figure 1 Image segmentation based on local texture flow. The texture flow is indicated by the arrows in (b) whose directions point to the texture boundaries, with darker arrows representing stronger texture gradient. After a few iterations the propagation reaches a stable state (c). The final image segmentation is shown in (d).

diverse visual cues for boundary detection. Figure 1 shows different stages of the segmentation algorithm on a small region in an aerial photograph. The flow vectors are constructed in the texture feature space. For computational reasons, the image texture is computed in blocks of 64 $\times$ 64 pixels. The initial texture flow vectors so computed are shown in Figure 1b and, after convergence, in Figure 1c. The final detected boundaries are shown in Figure 1d.

C. A Texture Thesaurus for Aerial Photographs

A texture thesaurus can be visualized as an image counterpart of the traditional thesaurus for text search. It contains a collection of code words that represent visually similar clusters in the feature space. A subset of the air photos is used as training data for the hybrid neural network algorithm to create the first level of an indexing tree [16]. Within each subspace, a hierarchical vector quantization technique is used to partition the space further into many smaller clusters. The centroids of these clusters are used to form the code words in the texture thesaurus, and the training image patterns of these centroids were used as icons to visualize the corresponding code words. In the current implementation, the texture thesaurus is organized as a two-level indexing tree. Figure 2 shows some examples of the visual code words in the texture thesaurus designed for air photos. Associations of the code words can be made to semantic concepts as well. This is being investigated in a related project [19].

When an air photo is ingested into the database, texture features are extracted using 64 $\times$ 64 subpatterns. These texture features are used in the EdgeFlow method for a coarse segmentation of the image and the region texture features are computed. These features are then used to compare with the code words in the thesaurus. Once the best match is identified, a two-way link between the image region and the corresponding code word is created and stored as the image metadata. During query time, the feature vector of the query pattern is used to search for the best matching code word, and by tracing back the links to it, all similar patterns in the database can be retrieved (Fig. 3). Some examples of retrievals are shown in Figure 4 and Figure 5. An online Web demonstration of this texture-based retrieval can be found at http://vivaldi.ece.ucsb.edu/AirPhoto.

Our long-term goal is to construct a visual thesaurus for images/video where the thesaurus code words are created at various levels of visual hierarchy by grouping primi-

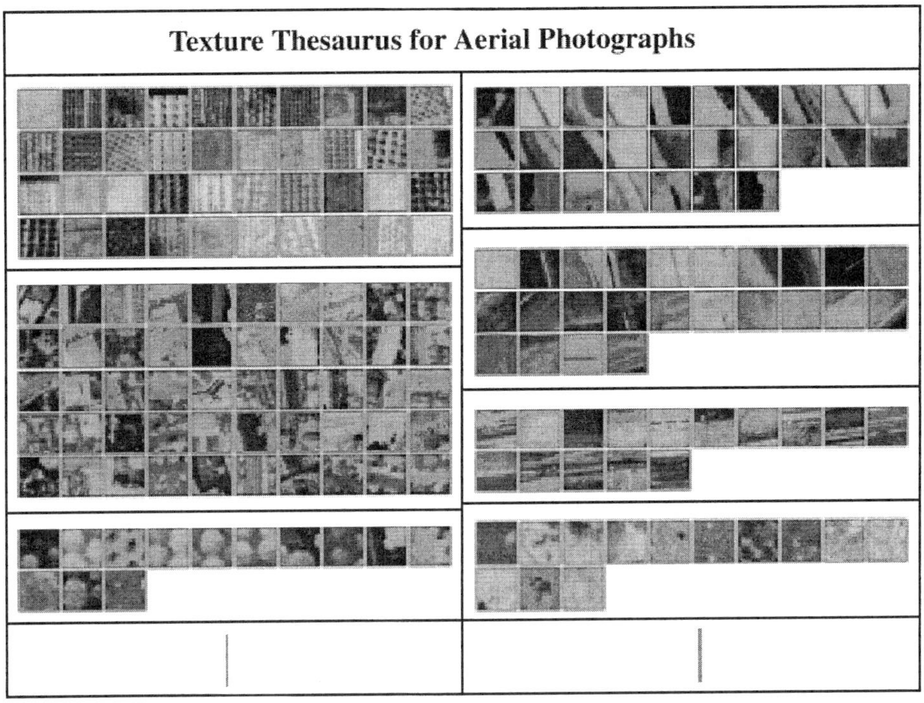

Figure 2 Examples of the code words obtained for the aerial photographs. The patterns inside each block belong to the same class.

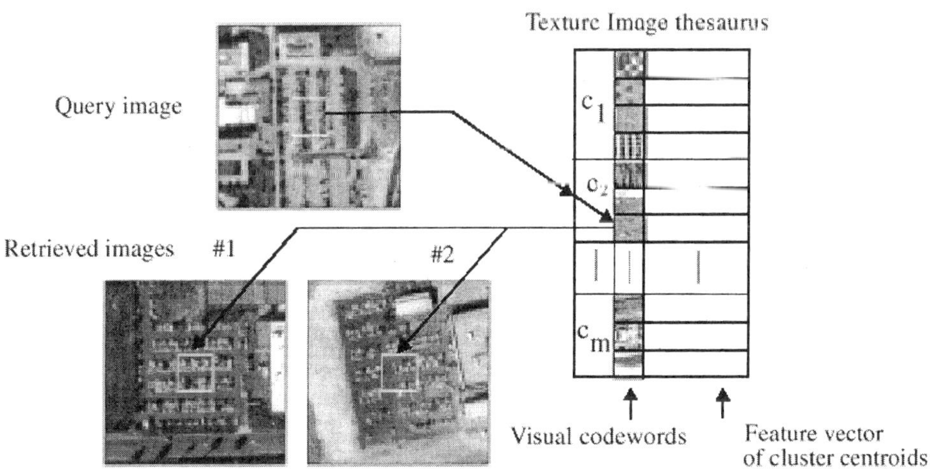

Figure 3 A texture image thesaurus for content-based image indexing. The image tiles shown here contain parking lots.

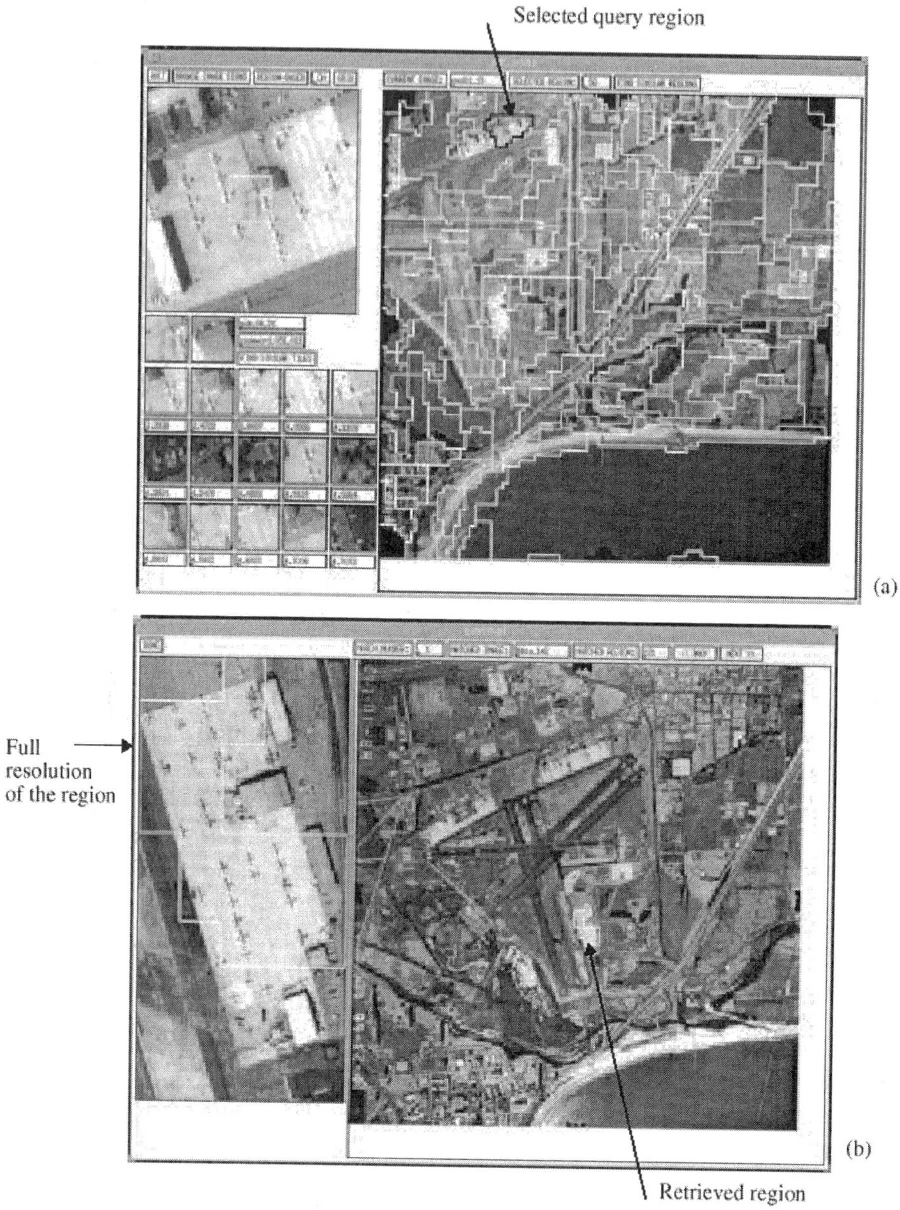

Figure 4 An example of the region-based search. (a) The downsampled version of a segmented large air photo (on the right) and higher resolution picture of the selected airport region (on the left). Some of the retrievals based on the 64×64 tiles are shown as well (left bottom). (b) A region-based retrieval result using the airport region shown in (a). Both the query and retrieval patterns are from the airport areas.

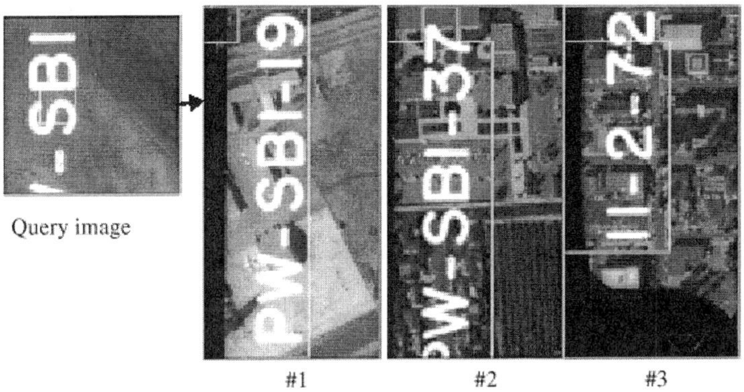

Figure 5 Another retrieval example. The query is from a region containing an image identification number.

tives such as texture, color, shape, and motion. For complex structured patterns, these code words take the form of a labeled graph, with the nodes in the graph representing primitive image attributes and the links the part relationships.

III. NeTra—SEARCHING IMAGES USING LOCALIZED DESCRIPTORS

NeTra [8] is an online image retrieval system that, in addition to the texture descriptors, uses region color and shape to annotate a large collection of natural images. At the time of ingestion into the database, images are segmented into homogeneous regions using the EdgeFlow method described earlier. Texture, color, and shape descriptors for each of these regions are then computed and indexed. This representation allows users to compose such queries as ''retrieve images that contain regions that have the color of object A, texture of object B, shape of object C and lie in the upper one-third of the image,'' where the individual objects could be regions belonging to different images. A Web demonstration of NeTra is available at http://vivaldi.ece.ucsb.edu/NeTra.

As in the texture thesaurus for air photos, NeTra uses the Gabor texture features for representing region texture. For shape, NeTra uses well-known representations such as the curvature function, the centroid distance and Fourier descriptors of the contour [8]. In the following, we explain the color feature and indexing used in NeTra.

A. Color Descriptors and Color Indexing

Color features are extracted from the segmented image regions. First, colors in each region are quantized into a small number of representative colors using a perceptual quantization algorithm [20]. The percentage of the quantized colors in each image region is then computed. Each quantized color and its corresponding percentage form a pair of attributes for that color for a given region. The color descriptor F is defined to be the set of such attribute pairs:

$$F = \{\{c_i, p_i\}, i = 1, \ldots, N\} \tag{1}$$

where N is the total number of quantized colors in the image region, c_i is a 3-D color vector, p_i is its percentage, and $\sum_i p_i = 1$. Colors are represented in the perceptual uniform CIE LUV space.

The distance between two color features F_1 and F_2 is measured by

$$D^2(F_1, F_2) = \sum_{i=1}^{N_1} p_{1i}^2 + \sum_{i=1}^{N_2} p_{2j}^2 - \sum_{i=1}^{N_1} \sum_{i=1}^{N_2} 2a_{1i,2j} p_{1i} p_{2j} \quad (2)$$

where $a_{k,l}$ is the similarity coefficient between colors c_k and c_l,

$$a_{k,l} = \begin{cases} 1 - d_{k,l}/d_{\max} & d_{k,l} \leq T_d \\ 0 & d_{k,l} > T_d \end{cases} \quad (3)$$

where $d_{k,l}$ is the distance between color c_k and c_l,

$$d_{k,l} = \|c_k - c_l\| \quad (4)$$

T_d is the maximum distance for two colors to be considered similar and is determined experimentally, $d_{\max} = \alpha T_d$, and α is set to 1.2 in the experiments. The preceding distance measure can be shown to be equivalent to the quadratic color histogram distance measure [21].

1. Color Indexing

A lattice structure is used to index the colors in the 3D space. In the current implementation of NeTra, the lattice points are uniformly distributed in 3D. Each color in each of the regions is assigned to its nearest lattice point. As explained in Deng and Manjunath [22], nearest neighbor and range searches can be efficiently performed using this indexing method. Insertions and deletions are quite straightforward as well.

Figure 6 illustrates this indexing structure in the 2D space, where the desired search radius r is the query search range and the actual search radius R is the minimum search distance for lattice points such that the desired sphere of radius r is covered. Let ρ denote the minimum radius of a sphere that can cover a *Voronoi* cell, as shown in Figure 6, and R is given by:

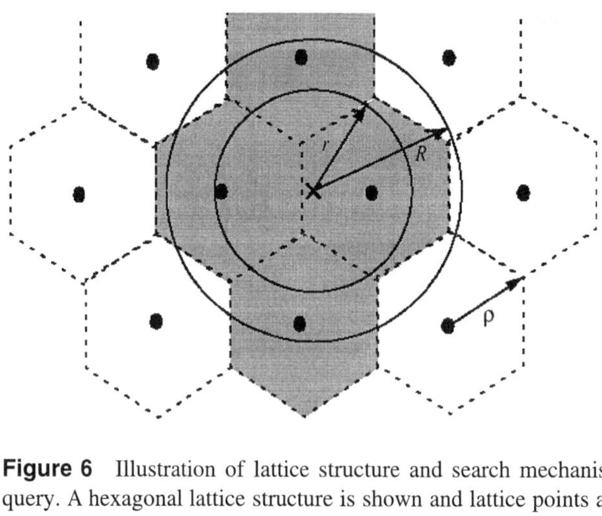

Figure 6 Illustration of lattice structure and search mechanism in 2D plane. Point ''x'' is the query. A hexagonal lattice structure is shown and lattice points are marked. r and R are the desired and actual search radius, respectively; ρ is the minimum radius of a sphere that can cover a Voronoi cell.

$$R = r + \rho \tag{5}$$

Because an indexing node contains all the entries in its Voronoi cell, part of the search space could be unnecessary. In Figure 6 it can be seen that the actual search space includes all the shaded areas. For a given query range r, the value of ρ in the lattice design is important to the retrieval performance. A small value of ρ means that the actual search space is only slightly larger than the desired search space, and therefore most of the indexing nodes accessed are relevant. However, there is a trade-off because the number of indexing nodes is increased and the indexing itself becomes less efficient. Also, the retrievals are not as fully sorted and this will affect the performance in the later joining process. Regardless of the design, however, the search complexity is low. The number of indexing nodes that need to be accessed per query color is on the order of $O(R^3/\rho^3)$ and does not depend on the database size.

The D_3^* lattice is chosen for indexing because of its optimal properties [23]:

1. *Accuracy*: It gives the minimal total mean squared quantization errors among all the lattice structures in the 3D space.
2. *Efficiency*: It has the most optimal covering of the 3D space; i.e., the volume of the smallest sphere that covers the Voronoi cell divided by the volume of the cell is also minimum.

The structure of the D_3^* lattice is quite simple. The basic lattice consists of the points (x, y, z) where x, y, and z are all even or all odd integers. These points can be scaled and shifted to have desired lattice point intervals and locations.

By using the proposed indexing scheme, insertions and deletions of database entries are straightforward, which allows the database to be dynamic.* Because the positions of the lattice points are known, only a few calculations are needed to find the nearest lattice point for any color to be indexed. For a point c, the nearest lattice point is the closer one of the nearest even lattice point and the nearest odd one, i.e.,

$$q_c = \arg\min(\|q - c\|) \quad \text{and} \quad q = \{q_e, q_o\} \tag{6}$$

where

$$q_e = \text{round}\left(\frac{c}{L}\right) L, \qquad q_o = \text{round}\left(\frac{c - L/2}{L}\right) L + \frac{L}{2} \tag{7}$$

where L is the scale factor or the increment step along the coordinate. For example, $L = 2$ for the basic lattice. In Eqs. (6) and (7), the origin is assumed to be a lattice point. After the indexing node is found, the region label of the new entry is then compared with the list of sorted region labels in the node and is inserted in the right order.

It is to be noted that the lattice structure can have more than one layer of representation. Each Voronoi cell can be further divided into a set of subcells, resulting in a hierarchical lattice tree structure [24]. This will be useful for refining the retrieval results in the case in which the colors in the database have a very unbalanced distribution.

---

* In the database literature, an index structure is dynamic if it supports insertions and deletions to the database without reconstructing the index afresh; i.e., the cost of insertion or deletion is less than the cost of reconstruction.

B. Query Processing

Color, texture, and shape descriptors of each region are indexed separately. For texture and shape features, a data structure similar to the SS-tree [25] is generated. A modified *k*-means clustering algorithm is used to construct the indexing tree and the balancing of the tree is achieved by imposing a constraint on the minimum number of nodes in each cluster. If the clustering in any iteration results in clusters with fewer than the minimum, such clusters are deleted and their members are assigned to other needy clusters. The cluster centers are then recomputed and used as the initial condition for the next iteration. The process is repeated until no cluster is smaller than the specified threshold or the number of iterations exceeds a given number.

For a query consisting of more than one image feature, the intersection of the results of search using individual features are computed and then sorted based on a weighted similarity measure. The current implementation of NeTra uses an implicit ordering of the image features to prune the search space. The first feature that the user specifies is used to narrow down the space, within which a more detailed search is performed to similarity order the retrieval results.

C. Experiments

The online database consists of 2500 images from a Corel photo CD database. There are 25 categories of images, and after segmentation more than 26,000 regions are obtained. Table 1 summarizes some of the experimental data for color indexing.

For the color features, a performance evaluation is done as follows. Because the proposed color descriptor can be considered as a variation of the well-known color histogram, comparisons are made with a 1024-bin color histogram. A 1024-dimensional color histogram feature vector is extracted from each image region in the database. The dimension is set high to achieve the best possible results. The histogram features of the query set are compared with all the histogram features in the database through exhaustive searches to obtain the top 100 retrievals. Note that, in contrast, the color descriptor used in NeTra uses (on the average) 14 numbers per region. (Note that each color is represented by four numbers: three numbers for the three color components and one number to represent the percentage. On the average, there are 3.5 colors per region or, equivalently, 14 numbers per region.) An alternative way to use only 14 numbers per feature is to perform SVD on quadratic matrix A as suggested in [20] to obtain a 14D transformed color histogram vector. Again, exhaustive searches in the database are performed to find the top 100 retrievals by using this approach.

Before the evaluation, subjective testing was done to determine the relevant matches of the query image regions in the database. The retrievals from both the histogram and the new approaches are marked by five subjects to decide whether they are indeed visually similar in color. Those marked by at least three out of the five subjects are considered

Table 1 Experimental Data

Average number of colors per region	3.5
Total number of indexing nodes	1553
Average number of nodes accessed per query feature	134.9
Average execution time (CPU plus I/O) per query	0.16 s

the relevant matches and the others are considered false matches. Because it is impractical to go through the entire database to find all the relevant matches for the queries, the union of relevant retrievals from the two methods is used as approximate ''ground truth'' to evaluate the retrieval accuracy, which is measured by precision and recall, defined as [26].

$$\text{Precision}(K) = C_K/K \quad \text{and} \quad \text{Recall}(K) = C_K/M \tag{8}$$

where K is the number of retrievals considered, C_K is the number of relevant matches among all the K retrievals, and M is the total number of relevant matches in the database obtained through the subjective testing.

The average precision and recall curves are plotted in Figure 7. It can be seen from the table and the figures that the new method achieves good results in terms of retrieval accuracy compared with the histogram method. Its performance is close to that of the high-dimensional histogram method and is much better than that of the SVD approach. Figure 8 shows two example retrievals of the query set. One query is the dark blue and white colored snowcap on the mountain. The other is the mixed red, yellow, and green foliage in autumn. The top six retrievals in both examples show good match of colors.

An example of color plus texture retrieval is shown in Figure 9. It is much more difficult to provide meaningful evaluation criteria or even to design a methodology for combined features. Perhaps, as the next two sections detail, this could be addressed at a semantic level in terms of finding objects in the scene.

IV. BLOBWORLD—RETRIEVAL BY REGIONS

Object recognition will not be comprehensively solved in the foreseeable future. However, understanding what users want to find means that we can build systems that are more effective. *BlobWorld* provides a new framework for image retrieval based on segmentation into regions and querying using properties of these regions [27,28]. These regions generally correspond to objects or parts of objects. The segmentation algorithm is fully automatic; there is no parameter tuning or hand pruning of regions. An online system for retrieval in a collection of 10,000 Corel images using this approach is available at http://clib.cs.berkeley.edu/photos/blobworld.

We present results indicating that querying for distinctive objects such as tigers, zebras, and cheetahs using Blobworld produces higher precision than does querying using color and texture histograms of the entire image. In addition, Blobworld false positives are easier to understand because the matching regions are highlighted. This presentation of results means that interpreting and refining the query are more productive with the Blobworld system than with systems that use low-level features from the entire image.

A. Blobworld's Representation

The Blobworld representation is related to the notion of photographic or artistic scene composition. Blobworld is distinct from color-layout matching as in query by image and video content (QBIC) [3] in that it is designed to find objects or parts of objects; each image is treated as an ensemble of a few ''blobs'' representing image regions that are roughly homogeneous with respect to color and texture. A blob is described by its color distribution and mean texture descriptors. Figure 10 illustrates the stages in creating Blobworld. Details of the segmentation algorithm may be found in Belongie et al. [27].

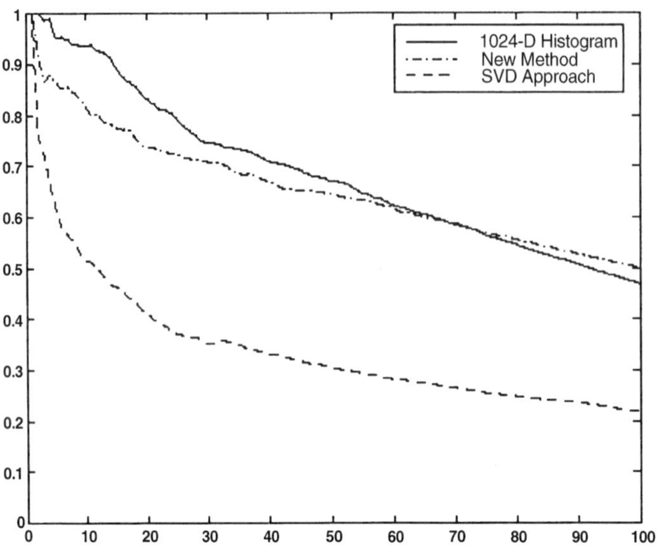

(a)

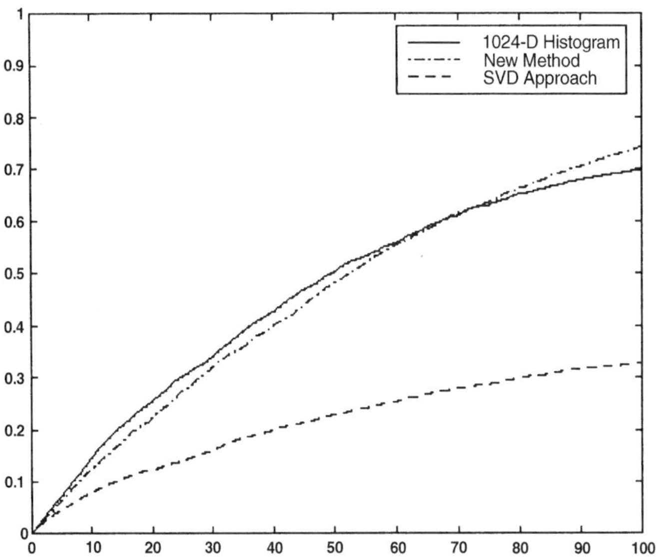

(b)

Figure 7 (a) Average precision versus number of retrievals. (b) Average recall versus number of retrievals.

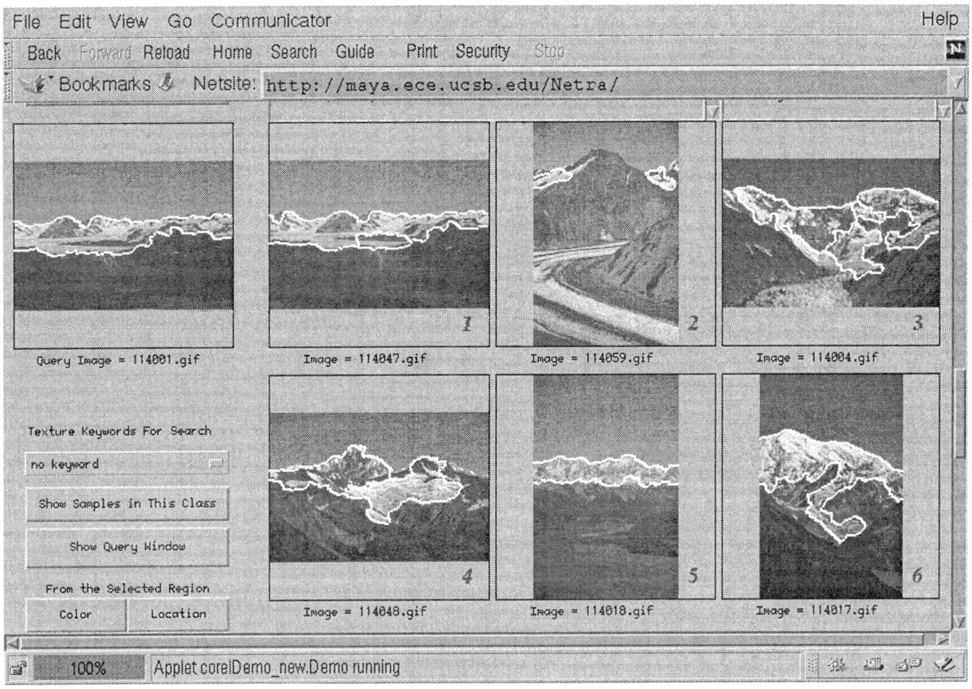

a

b

Figure 8 Two example of region-based image search using the proposed color indexing method. (a) The query is the snowcap on thc mountain. (b) The autumn foliage is the query. The top six retrievals are shown.

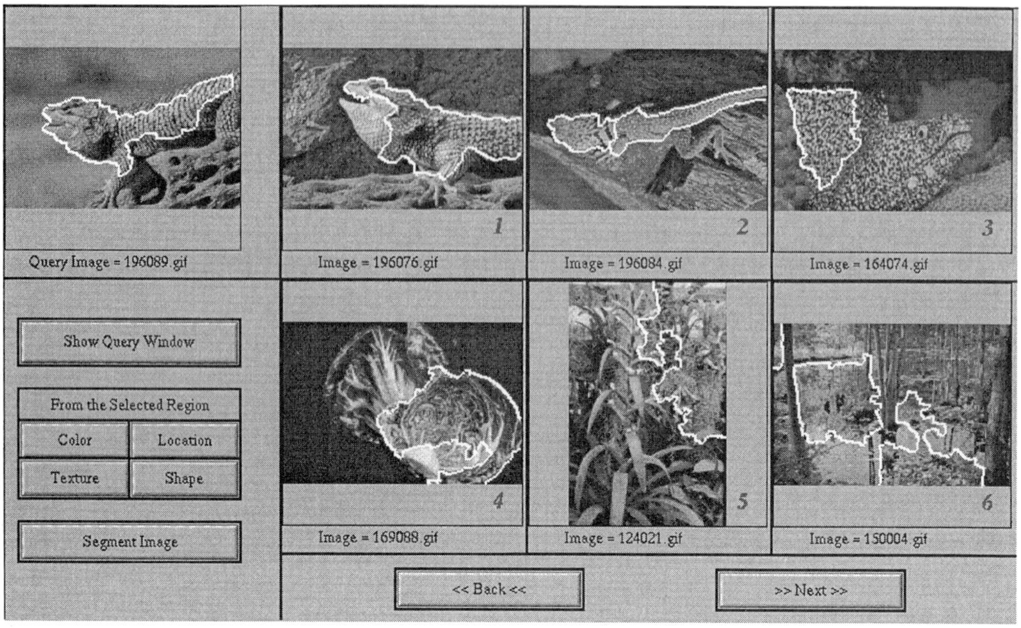

Figure 9 An example of region-based image retrieval using color and texture.

B. Grouping Pixels into Regions

Each pixel is assigned a vector consisting of color, texture, and position features. The three color features are the coordinates in the L*a*b* color space [29]; we smooth these features to avoid oversegmentation arising from local color variations due to texture. The three texture features are contrast, anisotropy, and polarity, extracted at a scale that is selected automatically [27]. The position features are simply the position of the pixel; including the position generally decreases oversegmentation and leads to smoother regions.

We model the distribution of pixels in the 8D space using mixtures of two to five Gaussians. We use the expectation-maximization algorithm [30] to fit the mixture of Gaussians model to the data. To choose the number of Gaussians that best suits the natural number of groups present in the image, we apply the minimum description length (MDL) principle [31,32]. Once a model is selected, we perform spatial grouping of connected pixels belonging to the same color–texture cluster. Figure 11 shows segmentations of a few sample images.

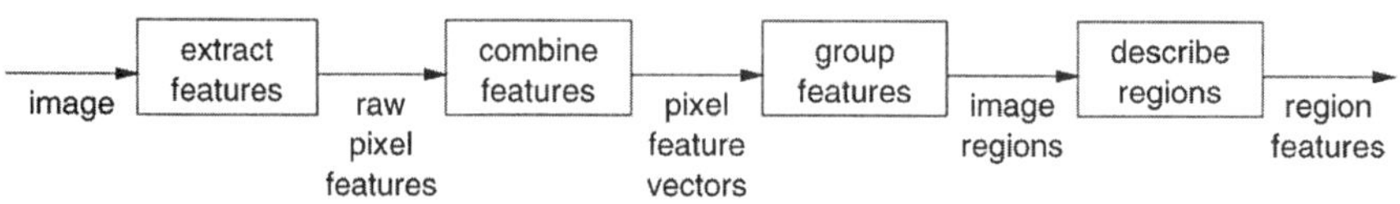

Figure 10 The stages of Blobworld processing: from pixels to region descriptions.

Figure 11 Results from a tiger query. The user selected the tiger blob and grass blob. Matching regions in each returned image are shown in red.

C. Describing the Regions

We store the color histogram over pixels *within each region*. This histogram is based on bins with width 20 in each dimension of L*a*b* space. This spacing yields 5 bins in the L* dimension and 10 bins in each of the a* and b* dimensions, for a total of 500 bins. However, not all of these bins are valid; the gamut corresponding to $0 \leq R, G, B \leq 1$ contains only 218 bins that can be filled.

To match the color of two regions, we use the quadratic distance between their histograms x and y [21]:

$$d_{\text{hist}}(x, y) = (x - y)^{\mathrm{T}} A (x - y) \tag{9}$$

where $A = [a_{ij}]$ is a symmetric matrix of weights between 0 and 1 representing the similarity between bins i and j based on the distance between the bin centers; neighboring bins have a weight of 0.5. This distance measure allows us to give a high score to two regions with similar colors even if the colors fall in different histogram bins.

For each blob we also store the mean texture contrast and anisotropy. The distance between two texture descriptors is defined as the Euclidean distance between their respective values of contrast and (anisotropy × contrast). (Anisotropy is modulated by contrast because it is meaningless in areas of low contrast.) We do not include polarity in the region description because it is generally large only along edges; it would not help us distinguish among different kinds of regions.

D. Querying in Blobworld

The user composes a query by submitting an image in order to see its Blobworld representation, selecting the relevant blobs to match, and specifying the relative importance of the blob features. We define an ''atomic query'' as one that specifies a particular blob to match (e.g., ''like-blob-1''). A ''compound query'' is defined as either an atomic query or a conjunction of compound queries (''like-blob-1 and like-blob-2''). The score μ_i for each atomic query with feature vector v_i is calculated as follows:

1. For each blob b_j in the database image (with feature vector v_j):
   (a) Find the distance between v_i and v_j: $d_{ij} = (v_i - v_j)^T \Sigma(v_i - v_j)$.
   (b) Measure the similarity between b_i and b_j using $\mu_{ij} = \exp(-d_{ij}/2)$. This score is 1 if the blobs are identical in all relevant features; it decreases as the match becomes less perfect.
2. Take $\mu_i = \max j \, \{\mu_{ij}\}$

The matrix Σ is block diagonal. The block corresponding to the texture features is an identity matrix, weighted by the texture weight set by the user. The block corresponding to the color features is the A matrix used in finding the quadratic distance, weighted by the color weight set by the user.

The compound query score for the database image is calculated using fuzzy logic operations [33], and the user may specify a weight for each atomic query.

We then rank the images according to overall score and return the best matches, indicating for each image which set of blobs provided the highest score; this information helps the user refine the query. After reviewing the query results, the user may change the weighting of the blob features or may specify new blobs to match and then issue a new query. Figure 11 shows the results of a sample query.

E. Experiments

We expected that Blobworld querying would perform well in cases in which a distinctive object is central to the query. In order to test this hypothesis, we performed 50 queries using both Blobworld and global color and texture histograms.

We selected 10 object categories: airplanes, black bears, brown bears, cheetahs, eagles, elephants, horses, polar bears, tigers, and zebras. There were 30 to 200 examples of each category among the 10,000 images.

We compared the Blobworld results to a ranking algorithm that used the global color and texture histograms of the same 10,000 images. The color histograms used the same 218 bins as Blobworld, along with the same quadratic distance. For texture histo-

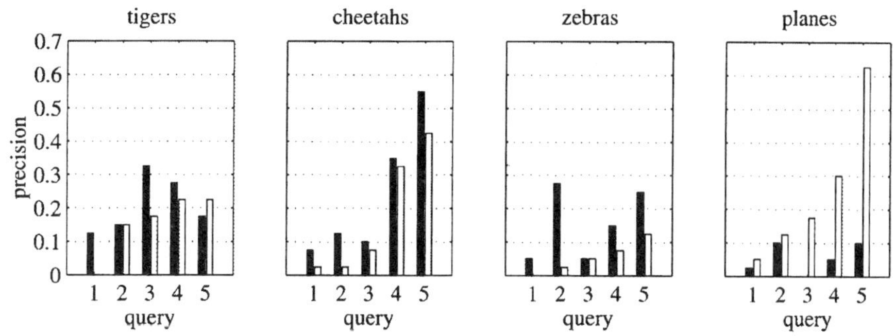

Figure 12 Precision of Blobworld (black) and global histogram (white) queries in the top 40 retrieved images for each of five queries in four example categories. The Blobworld queries used two blobs. Blobworld performs better on queries for distinctive *objects* such as tigers, cheetahs, and zebras, whereas global histograms perform better on queries for distinctive *scenes* such as airplane scenes. (Chance would yield precision ranging from 0.003 for zebras to 0.02 for airplanes.)

grams, we discretized the two texture features into 21 bins each. In this global image case, we found that color carried most of the useful information; varying the texture weight made little difference to the query results.

For each category we tested queries using one blob, two blobs, and global histograms. In each case we performed a few test queries to select the weights for color and texture and for each blob. We then queried using five new images. Note that we used the same weights for each image in a category. We also were not allowed to choose whether to use one or two query blobs, which penalizes the Blobworld results; for example, in some bear images a background blob was salient to the category, whereas in others there was not a meaningful background blob. In Figure 12 we plot the precision (fraction correct) of the top 40 matches for each query in the tiger, cheetah, zebra, and airplane categories; the differences between Blobworld and global histograms show up most clearly in these categories. Figure 13 illustrates how the precision drops as more images are retrieved.

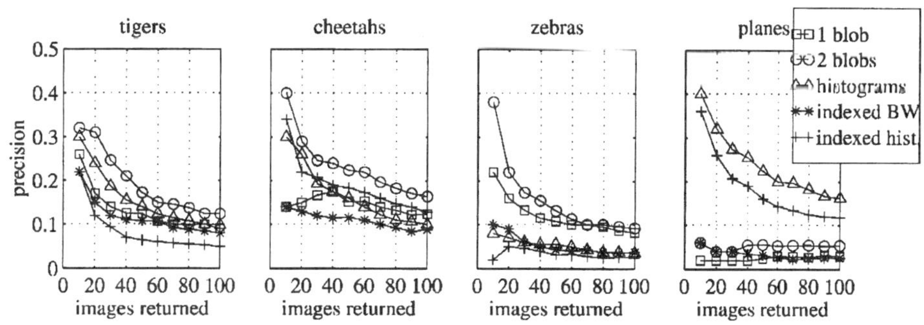

Figure 13 Average precision versus number of retrieved images for several query types. The index generally tracks the corresponding full query, except for the zebra case, in which the indexed Blobworld precision is much lower than the full Blobworld precision because we index only color, not texture.

The results indicate that the categories fall into three groups:

Distinctive objects: The color and texture of cheetahs, tigers, and zebras are quite distinctive, and the Blobworld result quality was clearly better than the global histogram result quality.

Distinctive scenes: For most of the airplane images the entire scene is distinctive (a small gray object and large amounts of blue), but the airplane region itself has quite a common color and texture. Histograms did better than Blobworld in this category. (We expect that adding blob size information to the query would yield better Blobworld results, as then we could specify ''large sky blob'' to avoid matching the small regions of sky found in thousands of images in the database.)

Other: The two methods perform comparably on the other six categories. Blobs with the same color and texture as these objects are common in the database, but the overall scene (a general outdoor scene) is also common, so neither Blobworld nor global histograms have an advantage, given that we used only color and texture. However, histograms can be taken no further, but Blobworld has much room left for improvement. For example, the shapes of elephants, bears, and horses (and airplanes) are quite distinctive.

These results support our hypothesis that Blobworld yields good results when querying for distinctive objects. Blobworld queries for a range of distinctive objects provide high precision and understandable results because Blobworld is based on finding coherent image regions. Furthermore, Blobworld queries can be indexed to provide fast retrieval while maintaining precision.

V. BODY PLANS—FINDING PEOPLE AND ANIMALS BY GROUPING

People and many animals can be thought of as assemblies of cylinders, each with characteristic color and texture. If one adopts this view, then a natural strategy for finding them is as follows:

Use masked images for regions of appropriate color and texture.

Find regions of appropriate color and texture that look roughly cylindrical (nearly straight boundaries with nearly parallel sides).

Form and test increasingly large assemblies against a sequence of predicates.

Although this approach does not work for clothed people, because of the wide variation in the color and texture of clothing, it does work for people who are unclad. An automatic test for whether an image contains a person not wearing clothing has obvious applications; many people wish to avoid (or to find) such pictures.

We have experimented with using these relational representations to find unclad people and to find horses (Fig. 14). A sequence of classifiers group image components that could correspond to appropriate body segments. The stages mean the process is efficient: it incrementally checks increasingly large groups, so that not all groups of four segments are presented to the final classifier.

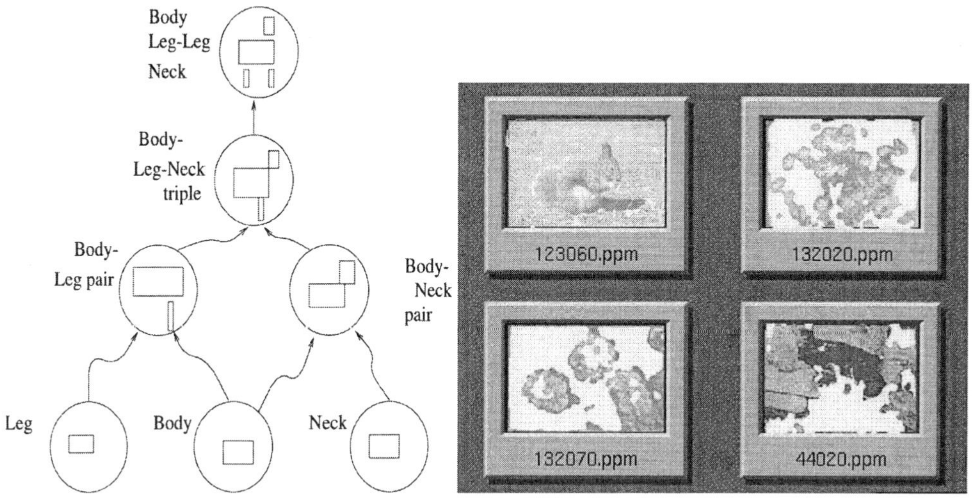

Figure 14 (Left) The body plan used for horses. Each circle represents a classifier, with an icon indicating the appearance of the assembly. An arrow indicates that the classifier at the arrowhead uses segments passed by the classifier at the tail. The topology was given in advance. The classifiers were then trained using image data from a total of 38 images of horses. (Right) Typical images with large quantities of hidelike pixels (white pixels are not hidelike; others are hidelike) that are classified as *not* containing horses, because there is no geometric configuration present. Although the test of color and texture is helpful, the geometric test is important, too, as the results in Fig. 15 suggest.

A. Finding People

Our first example is a program that identifies pictures containing people wearing little or no clothing. This program has been tested on a usually large and unusually diverse set of images; on a test collection of 565 images known to contain lightly clad people and 4289 control images with widely varying content, one tuning of the program marked 241 test images and 182 control images. The performance of various different tunings is indicated in Figure 15; more detailed information appears in Refs. 34 and 35. The recall is comparable to full-text document recall [36–38] (which is surprisingly good for so abstract an object recognition query), and the rate of false positives is satisfactorily low. In this case, the representation was entirely built by hand. Each test in the sequence of tests used is defined by intuition and guesswork.

B. Learning to Find Horses

The second example used a representation whose combinatorial structure—the order in which tests were applied—was built by hand but the tests were learned from data. This program identified pictures containing horses and is described in greater detail in Ref. 39. Tests used 100 images containing horses and 1086 control images with widely varying content (Fig. 16). The performance of various different configurations is shown in Figure 15. For version ''F,'' if one estimates performance omitting images used in training and images for which the segment finding process fails, the recall is 15%—i.e., about 15% of the images containing horses are marked—and control images are marked at the rate

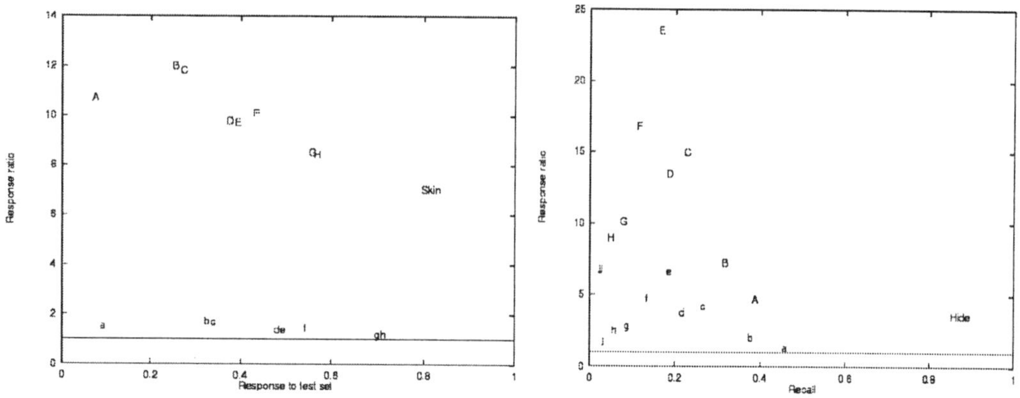

Figure 15 The *response ratio* (percent incoming test images marked/percent incoming control images marked), plotted against the percentage of test images marked, for various configurations of the two finding programs. Data for the nude human finder appear on the left, for the horse finder on the right. Capital letters indicate the performance of the complete system of skin–hide filter and geometric grouper, and lowercase letters indicate the performance of the geometric grouper alone. The label ''skin'' (resp. ''hide'') indicates the selectivity of using skin (resp. hide) alone as a criterion. For the human finder, the parameter varied is the type of group required to declare a human is present—the trend is that more complex groups display higher selectivity and lower recall. For the horse finder, the parameter being varied affects resistance to finding groups in very large collections of segments.

of approximately 0.65%. In our test collection, this translates to 11 images of horses marked and 4 control images marked.*

Finding using body plans has been shown to be quite effective for quite special cases in quite general scenes. It is relatively insensitive to changes in aspect [39]. It is quite robust to the relatively poor segmentations that our criteria offer, because it is quite effective in dealing with nuisance segments—in the horse tests, the average number of four-segment groups was 2,500,000, which is an average of 40 segments per image. Nonetheless, the process described is crude: it is too dependent on color and texture criteria for early segmentation, the learning process is absent (humans) or extremely simple (horses), and there is one recognizer per class.

C. Learning to Assemble People

The recognition processes we have described have a strong component of correspondence; in particular, we are pruning a set of correspondences between image segments and body segment labels by testing for kinematic plausibility. The search for acceptable correspondences can be made efficient by using *projected classifiers*, which prune labelings using the properties of smaller sublabelings, as in Grimson and Lozano [40] who use manually determined bounds and do not learn the tests. Given a classifier C that is a function of a

* These figures are *not* 15 and 7, because of the omission of training images and images where the segment finder failed in estimating performance.

Figure 16 Images of horses recovered from a total of 100 images of horses and 1086 control images, using the system described in the text. Notice that although the recall is low, there are few false positives. The process is robust to aspect—frontal views, three-quarter views, and lateral views are found. There is insufficient precision in the representation to tell a brown quadruped (the deer) from a horse; this is a strength, because once the rough body configuration is known, it is easier to know where to look for, say, the horns.

set of features whose values depend on segments with labels in the set $L = \{l_1, \ldots, l_m\}$, the projected classifier $C_{l_1,\ldots,l_k}$ is a function of all the features that depend only on the segments with labels $L' = \{l_1 \ldots l_k\}$. In particular, $C_{l_1,\ldots,l_k}(L') > 0$ if there is some extension L of L' such that $C(L) > 0$. The converse need not be true: the feature values required to bring a projected point inside the positive volume of C may not be realized with any labeling of the current set of segments 1, . . . , N. For a projected classifier to be useful, it must be easy to compute the projection, and it must be effective in rejecting labelings at an early stage. These are strong requirements that are not satisfied by most good classifiers; for example, in our experience a support vector machine with a positive definite quadratic kernel projects easily but typically yields unrestrictive projected classifiers.

We have been using an axis-aligned bounding box, with bounds learned from a collection of positive labellings, for a good first separation and then using a boosted version of a weak classifier that splits the feature space on a single feature value (as in Ref. 5). This yields a classifier that projects particularly well and allows clean and efficient algorithms for computing projected classifiers and expanding sets of labels (see Ref. 40).

The segment finder may find either one or two segments for each limb, depending on whether it is bent or straight; because the pruning is so effective, we can allow segments to be broken into two equal halves lengthwise, both of which are tested.

1. Results

The training set included 79 images without people, selected randomly from the Corel database, and 274 images each with a single person on a uniform background. The images with people have been scanned from books of human models [41]. All segments in the test images were reported; in the control images, only segments whose interior corresponded to human skin in color and texture were reported. Control images, both for the training and for the test set, were chosen so that all had at least 30% of their pixels similar to human skin in color and texture. This gives a more realistic test of the system performance by excluding regions that are obviously not human and reduces the number of segments in the control images to the same order of magnitude as those in the test images.

The models are all wearing either swimsuits or no clothes, otherwise segment finding fails; it is an open problem to segment people wearing loose clothing. There is a wide variation in the poses of the training examples, although all body segments are visible. The sets of segments corresponding to people were then hand labeled. Of the 274 images with people, segments for each body part were found in 193 images. The remaining 81 resulted in incomplete configurations, which could still be used for computing the bounding box used to obtain a first separation. Because we assume that if a configuration looks like a person then its mirror image does too, we double the number of body configurations by flipping each one about a vertical axis. The bounding box is then computed from the resulting 548 points in the feature space, without looking at the images without people.

The boosted classifier was trained to separate two classes: the points corresponding to body configurations and 60,727 points that did not correspond to people but lay in the bounding box, obtained by using the bounding box classifier to build labelings incrementally for the images with no people. We added 1178 synthetic positive configurations obtained by randomly selecting each limb and the torso from one of the 386 real images of body configurations (which were rotated and scaled so the torso positions were the same in all of them) to give an effect of joining limbs and torsos from different images, rather as in childrens' flip books. Remarkably, the boosted classifier classified each of the real data points correctly but misclassified 976 out of the 1178 synthetic configurations as negative; the synthetic examples were unexpectedly more similar to the negative examples than the real examples were.

The test data set was separate from the training set and included 120 images with a person on a uniform background and varying numbers of control images, shown in Table 2. We report results for two classifiers, one using 567 features and the other using a subset

Table 2 Number of Images of People and Without People Processed by the Classifiers with 367 and 567 Features and False-Negative (Images with a Person Where No Body Configuration Was Found) and False-Positive (Images with No People Where a Person Was Detected) Rates

Features	Number of test images	Number of control images	False negatives (%)	False positives (%)
367	120	28	37	4
567	120	86	49	10

of 367 of those features. Table 2 also shows the false-positive and false-negative rates achieved for each of the two classifiers. By marking 51% of test images and only 10% of control images, the classifier using 567 features compares extremely favorably with that of Ref. 34, which marked 54% of test images and 38% of control images using hand-tuned tests to form groups of four segments. In 55 of the 59 images where there was a false negative, a segment corresponding to a body part was missed by the segment finder, meaning that the overall system performance significantly understates the classifier performance.

VI. CONCLUSIONS

Knowing what users want makes it possible to build better retrieval systems. Local image descriptors of color and texture are quite powerful in representing characteristics of *objects*, as evidenced by the texture thesaurus for air photos and the online demonstrations of Netra and Blobworld. For special cases, we can reason about the spatial relationships between particular kinds of region and use this approach to find people and horses. This offers a strategy for incorporating the user preference to find objects into practical systems in a flexible way: we offer users the options of accepting a small range of specially precomposed queries for special cases that find particular objects, or of querying on the basis of spatial relationships between various special types of region, or of querying on regions.

Key to the success of such region-based representation is automated segmentation. Netra and Blobworld utilize two distinctly different segmentation schemes, but both place strong emphasis on automating the process with no parameter tuning. As we move toward the standardization of MPEG-4 and MPEG-7, such tools for automated spatiotemporal segmentation will increasingly play a critical role in the respective applications.

ACKNOWLEDGMENTS

This work was supported in part by NSF grants IRI-9704785 and IRI-9411330.

REFERENCES

1. R Wilensky. Toward work-centered digital information services. IEEE Comput Mag May: 37–45, 1996.

2. TR Smith. A digital library of geographically referenced materials. IEEE Comput M 54–60, 1996.
3. M Flickner, H Sawhney, W Niblack, J Ashley. Query by image and video content: The {Q system. IEEE Comput 28:23–32, September 1995.
4. D Forsyth, J Malik, R Wilensky. Searching for digital pictures. Sci Am 276:72–77, June 199
5. Y Freund, RE Schapire. Experiments with a new boosting algorithm. Machine Learning 13: 1996.
6. A Gupta, R Jain. Visual information retrieval. Commun ACM 40:70–79, May 1997.
7. P Kelly, M Cannon, D Hush. Query by image example: The {CANDID} approach. SPIE Proceedings Storage and Retrieval for Image and Video Databases, 1995, pp 238–248.
8. WY Ma, BS Manjunath. NeTra: A toolbox for navigating large image databases. ACM Multimedia Systems, Vol. 7, No. 3, pp 184–198, May 1999.
9. V Ogle, M Stonebraker. Chabot: Retrieval from a relational database of images. IEEE Comput 28:40–48, September 1995.
10. A Pentland, R Picard, S Sclaroff. Photobook: Content-based manipulation of image databases. Int J Comput Vision 18:233–254, 1996.
11. JR Smith, SF Chang. VisualSEEK: A fully automated content-based image query system. ACM Multimedia 96, Boston, November 20, 1996.
12. JR Bach, C Fuller, A Gupta, A Hampapur, B Horowitz, R Humphrey, RC Jain, C Shu. Virage image search engine: An open framework for image management. Proceedings of SPIE, Storage and Retrieval for Image and Video Databases IV, San Jose, CA, February 1996, pp 76–87.
13. BS Manjunath, WY Ma. Texture features for browsing and retrieval of image data. IEEE Trans PAMI 18(8):837–842, 1996.
14. P Enser. Query analysis in a visual information retrieval context. J Doc Text Manage 1:25–52, 1993.
15. P Brodatz. Textures: A Photographic Album for Artists & Designers. New York: Dover Publications, 1966.
16. WY Ma, BS Manjunath. A texture thesaurus for browsing large aerial photographs. J Am Soc Inform Sci 49:633–648, 1998.
17. WY Ma, BS Manjunath. Texture features and learning similarity. Proceedings of IEEE International Conference on Computer Vision and Pattern Recognition, San Francisco, June 1996, pp 425–430.
18. WY Ma, BS Manjunath. EdgeFlow: A framework for boundary detection and image segmentation. Proceedings of IEEE CVPR'97, 1997, pp 744–749.
19. H Chen. Semantic interoperability for geographic information systems, http://ai.bpa.arizona.edu/gis.
20. Y Deng, C Kenney, MS Moore, BS Manjunath. Peer-group filtering and perceptual color quantization. Submitted to ISCAS-99 (ECE TR #98-25, UCSB), 1998
21. J Hafner, HS Sawhney, W Equitz, M Flickner, W Niblack. Efficient color histogram indexing for quadratic form distance functions. IEEE Trans Pattern Anal Machine Intell 17:729–736, 1995.
22. Y Deng, BS Manjunath. An efficient color indexing scheme for region based image retrieval. Proceedings of ICASSP '99, 1999, pp 3017–3020.
23. JH Conway, NJA Sloane. Sphere Packings, Lattices and Groups. New York: Springer-Verlag, 1993.
24. D Mukherjee, SK Mitra. Vector set-partitioning with successive refinement Voronoi lattice VQ for embedded wavelet image coding. Proceedings of ICIP, 1998.
25. DA White, R Jain. Similarity indexing with the SS-tree. Proceedings of International Conference on Data Engineering, 1996, pp 516–523.
26. JR Smith. Image retrieval evaluation. Proceedings of IEEE Workshop on Content-Based Access of Image and Video Libraries, 1998, pp 112–113.

27. S Belongie, C Carson, H Greenspan, J Malik. Color- and texture-based image segmentation using EM and its application to content-based image retrieval. Proceedings, International Conference on Computer Vision, 1998.
28. C Carson, S Belongie, H Greenspan, J Malik. Region-based image querying. Proceedings IEEE Workshop on the Content-Based Access of Image and Video Libraries, 1997.
29. G Wyszecki, W Stiles. Science: Concepts and Methods, Quantitative Data and Formulae, 2nd ed. New York: Wiley, 1982.
30. A Dempster, N Laird, D Rubin. Maximum likelihood from incomplete data via the {EM} algorithm. J R Stat Soc Ser B 39:1–38, 1997.
31. J Rissanen. Modeling by shortest data description. Automatica 14:465–471, 1978.
32. J Rissanen. Stochastic complexity in statistical inquiry. World Scientific, 1989.
33. JS Jang, and C-T Sun, Mizutani E. Neuro-Fuzzy and Soft Computing. Englewood Cliffs. NJ: Prentice Hall, 1997.
34. MM Fleck, DA Forsyth, C Bregler. Finding naked people. European Conference on Computer Vision 1996, vol II, 1996, pp 592–602.
35. DA Forsyth, MM Fleck. Identifying nude pictures. IEEE Workshop on Applications of Computer Vision, 1996, pp 103–108.
36. DC Blair. Stairs redux: Thoughts on the stairs evaluation, ten years after. J Am Soc Inform Sci 47(1):4–22, 1996.
37. DC Blair, ME Maron. An evaluation of retrieval effectiveness for a full text document retrieval system. Commun ACM 28:289–299, 1985.
38. G Salton. Another look at automatic text retrieval systems. Commun ACM 29:649–657, 1986.
39. DA Forsyth, MM Fleck. Body plans. IEEE Conference on Computer Vision and Pattern Recognition, 1997.
40. WEL Grimson and T Lozano. Localizing overlapping party by searching the interpretation tree. IEEE Trans. on Pattern Analysis and Machine Intelligence 9(4):469–482, 1987.
41. Unknown. Pose file, vol 1–7. Books Nippan, 1993–1996. A collection of photographs of human models, annotated in Japanese.

22
MPEG-7: Status and Directions

Fernando Pereira
Instituto Superior Técnico/Instituto de Telecomunicações, Lisbon, Portugal

Rob H. Koenen
KPN Research, Leidschendam, The Netherlands

I. CONTEXT AND MOTIVATIONS

The significant advances of the past decade in the area of audiovisual technology, in terms of both algorithms and products, made digital audiovisual information accessible to everybody, not only in consumption but increasingly also in terms of production. Digital still cameras directly storing in JPEG format have hit the mass market. Together with the first digital video cameras directly recording in MPEG-1 format, e.g., Hitachi's MPEG-1A, the world's first tapeless camcorder, this represents a major step for the acceptance, in the consumer market, of digital audiovisual acquisition technology. This step transforms everyone into a potential content producer, capable of creating content that can be easily distributed and published using the Internet.

But if it is now easier and easier to acquire, process, and distribute audiovisual content, it must also be equally easy to access the available information, because incommensurable amounts of digital audiovisual information are being generated all over the world every day. In fact, there is no point in making available audiovisual information that can be found only by chance. Unfortunately, identifying the desired information by retrieving or filtering is only becoming more difficult, because acquisition and processing technologies progress and so does the ability of consumers to use these new technologies and functionalities. In this context, a growing number of people are making a business of digital audiovisual information, notably its acquisition, processing, and creation, to make it finally available to potential consumers. For them, there is an easy rule: the easier it is for their consumers to identify the right information by searching, retrieving, accessing, and filtering, the higher its value and consequently the revenues. This conclusion applies to very different types of environments, ranging from broadcasting (let people find and tune into *your* broadcast channel) to database searching (e.g., multimedia directory services).

Although there is an ancient saying, ''Seek and ye shall find,'' those familiar with text-based search engines such as those available on the Internet know that this is not a very accurate statement. However, the amount of ''noise'' that people are willing to accept from these tools in return for valuable information shows that a real demand exists. These text-based search engines can work only with textually indexed content, which is still mostly created by manual annotation. It is, however, hard to imagine that all the audiovisual content currently available and yet to become available will be manually indexed using textual descriptions. Indeed, it is expected that automatic, or at least semiautomatic, content description will be important in the near future. This may, for instance, be achieved through the automatic use of transcripts, captions, embedded text, and speech recognition. Moreover, some searching and filtering needs cannot be well expressed by a query that uses only text. For example, if you are looking for a video with blue sky on the top, emerald water on the bottom and some fast motion with the corresponding sound, e.g., racing boats, text may not be the best medium to express this need. This type of query also requires that the content description be based not just on text but also on adequate audiovisual features. Although automatic description is basically still limited to low-level features, such as color, motion, and shape for pictures and rhythm for audio, intensive research is in progress to reach higher levels of abstraction, at least when some structuring information is available a priori, typically context dependent. It is expected that non–text-based features will be essential in providing content consumers richer and more powerful content identification engines able to deal with audiovisual information at a higher semantic level. The concept of ''identification engine'' encompasses here both the searching (pull) and filtering (push) paradigms, as both target the identification of the right content.

The desire to retrieve audiovisual content efficiently and the difficulty of doing so directly lead to the need for a powerful solution for quickly and efficiently identifying (searching, filtering, etc.) various types of audiovisual content of interest to the user, also using non–text-based technologies. In July 1996, at the Tampere (Finland) meeting, MPEG [1] recognized this important need, stating its intention to provide a solution in the form of a ''generally agreed-upon framework for the description of audiovisual content.'' To this end, MPEG initiated a new work item, formally called Multimedia Content Description Interface, better known as MPEG-7. MPEG-7 will specify a standard way to describe various types of audiovisual information, including still pictures, video, speech, audio, graphics, 3D models, and synthetic audio, irrespective of its representation format, e.g., analog or digital, and storage support, e.g., paper, film, or tape.

MPEG, bearing a major responsibility for the widespread use of digital audiovisual content in both professional and consumer markets by setting the MPEG-1, -2, and -4 standards, took on the task of ''putting some order in the audiovisual content chaos.'' That order should come from the tools that allow efficiently identifying, locating, and using the digital content. Participants in the development of MPEG-7 represent broadcasters, equipment and software manufacturers, digital content creators and managers, telecommunication service providers, publishers and intellectual property rights managers, and university researchers.

MPEG-7 represents an even bigger challenge for the MPEG committee than MPEG-4, because many of the required technical skills and backgrounds were not present within the group when the decision was taken to start the work. This means that MPEG needed to transform itself again, this time into a world-class group of experts in audiovisual description technology. The goal of MPEG-7 also means that MPEG entered a technological area where the idea of a ''standard'' is not yet familiar and boundaries of what to standardize and what not are often unclear. These boundaries will become clearer only with time and work. As in the best MPEG tradition, however, MPEG-7 should again be like a contract, in the sense of a ''process by which individuals recognize the advantage of doing certain things in an agreed way and codify that agreement in a contract. In a contract, we compromise between what we want to do and what others want to do. Though it constrains our freedom, we usually enter into such an agreement because the perceived advantages exceed the perceived disadvantages'' [2].

In MPEG-7 the perceived advantages are the ease of fast and efficient identification of audiovisual content that is of interest to the user by

Allowing the same indexed material to be identified by many identification (e.g., search and filtering) engines
Allowing the same identification engine to identify indexed material from many different sources

As with the other MPEG standards, the major advantages of MPEG-7 are this increased ''interoperability,'' the prospect of offering lower cost products through the creation of mass markets, and the possibility of making new, standards-based services ''explode'' in terms of number of users. This agreement should stimulate both content providers and users and simplify the entire content consumption process, giving the user the tools to easily ''surf on the seas of audiovisual information''—also known as the pull model—or ''filter floods of audiovisual information''—also known as the push model. Of course, the standard needs to be technically sound, because otherwise proprietary solutions will prevail, hampering interoperability. But the problem is not confined to ''human end users.'' Also, more and more automated systems operate on (digital) audiovisual data, trying to extract or manipulate information for some specific purpose. Examples are a surveillance system monitoring a highway, or an intelligent web server supplying content to a wide range of access devices and networks. Both the human end users and the automated systems need information about the content in order to be able to take decisions with respect to that content. This information is not normally readily present in the encoded representation of the content. Either it is absent (for instance information about recording time and place) or, when it is available, it may require time and computationally expensive decoding and processing of the digital data (for example, in a ''low-level'' search for sports material, to detect whether a dominant green color is present.)

The immense number of books, journals, special issues, conferences, and workshops on topics such as audiovisual content description, feature extraction, indexing and retrieval, digital libraries, content based querying, and electronic program guides, etc. and also products in the market show that technology is currently available to address the need for audiovisual content description. Moreover, these topics are among the most intensively researched areas in multimedia groups all around the world at present. Matching the needs and the technologies in audiovisual content description is thus the new task of MPEG.

This chapter addresses the status and directions of MPEG-7 as expressed by the most recent versions of the MPEG-7 reference documents: MPEG-7: Context and objectives and technical road map [3], Applications for MPEG-7 [4], MPEG-7 requirements [5], and MPEG-7 evaluation process [6].

II. OBJECTIVES

MPEG-7, like the other members of the MPEG family, will be a standard representation of audiovisual information satisfying a set of well-defined requirements. In this case, the requirements relate to the identification of audiovisual content. Audiovisual information includes still pictures, video, speech, audio, graphics, 3D models, and synthetic audio, although priorities among these data types may be expressed by the companies represented in MPEG. Because the emphasis is on audiovisual content but textual information is of great importance in the description of audiovisual information, the standard will not specify new description tools for text but will rather consider existing solutions for describing text documents and support them as appropriate [5]. In fact, quite a few standardized descriptions for text exist, such as HTML, SGML, and RDF. A clear example of where text comes in handy is in giving names of people and places. Moreover, MPEG-7 will allow linking audiovisual data descriptions to any associated data, notably textual data. Care must be taken, however, that the usefulness of the descriptions is as independent of the language as possible.

MPEG-7 will be generic and not especially tuned to any specific application. This means that the applications addressed can use content available in storage, on line and off line, or streamed, e.g., broadcast and Internet streaming. MPEG-7 will support applications operating in both real-time and non–real-time environments. In this context, a real-time environment means that the description information is associated with the content while that content is being captured.

MPEG-7 descriptions will often be useful stand-alone descriptions, e.g., if only a summary of the audiovisual information is needed. More often, however, they will be used to locate and retrieve the same audiovisual content represented in a format suitable for reproducing the content: digitally coded or even analog. In fact, MPEG-7 data are mainly intended for content *identification* purposes, whereas other representation formats, such as MPEG-2 and MPEG-4, are mainly intended for content *reproduction* purposes, although the boundaries may be not so sharp. This means that they fulfill different requirements. MPEG-7 descriptions may be physically colocated with the ''reproduction data,'' in the same data stream or in the same storage system. The descriptions may also live somewhere else on the globe. When the various audiovisual representation formats are not colocated, mechanisms linking them are needed. These links should be able to work in both directions: from the ''description data'' to the ''reproduction data'' and vice versa.

In principle, MPEG-7 descriptions will not depend on the way the content is available, either on the reproduction format or on the form of storage. Video information could, for instance, be available as MPEG-4, -2, or -1, JPEG, or any other coded form—or not even be coded at all: it is entirely possible to generate an MPEG-7 description for an analog movie or for a picture that is printed on paper. Undeniably, however, there is a special relationship between MPEG-7 and MPEG-4, as MPEG-7 will be grounded on an object-based data model, which also underpins MPEG-4 [7]. Like MPEG-4, MPEG-7 can describe the world as a composition of audiovisual objects with spatial and temporal behavior, allowing object-based audiovisual descriptions. As a consequence, each object in an MPEG-4 scene can have a description (stream) associated with it; this description can be accessed independently. It is interesting to note that MPEG-4, already foreseeing this close relationship, specifies how to attach an MPEG-7 description stream to an MPEG-4 object by using MPEG-4's ''object descriptor'' tool. MPEG-4 also supports a simple text-based description of MPEG-4 objects by using the ''object content information (OCI)'' tool [8]. The object-based approach does not exclude more conventional data models, such as frame-based video, but it does allow additional functionalities, such as

Object-based querying and identification: Users may not only identify complete scenes but also individual objects that fulfill their needs.

Object-based access and manipulation: The object descriptions may be independently accessed, processed, and reused.

Object-based granularity: Different objects may be described with different levels of detail, semantic abstraction, etc.

Targeting a wide range of application environments, MPEG-7 will offer different levels of discrimination by allowing several levels of granularity in its descriptions, along axes such as time, space, and accuracy. Because descriptive features must be meaningful in the context of an application, the descriptions can differ between user domains and applications. This implies that the same material can be described in various ways, using different features, tuned to the area of application. Moreover, features with different levels of abstraction will be considered. It will thus be the task of the content description generator to choose the right features and corresponding granularity. A very important conclusion can be drawn now: there is no single right way to describe audiovisual data! All descriptions may be equally valid from their own usage point of view. The strength of MPEG-7 is that these descriptions will all be based on the same description tools, syntax, and semantics, increasing interoperability.

Because a standard is always a constraint of freedom, it is important to make it as minimally constraining as possible. To MPEG this means that a standard must offer the maximum of advantages by specifying the minimum necessary, allowing competition and evolution of technology in the so-called nonnormative areas. This implies that just the audiovisual description itself will be standardized and not the extraction, encoding, or any part of the consumption process (see Fig. 1). Although good analysis and identification tools will be essential for a successful MPEG-7 application, their standardization is not required for interoperability. In the same way, the specification of motion estimation and rate control is not essential for MPEG-1 and MPEG-2 applications, and the specification of segmentation is not essential for MPEG-4 applications. Nor will the description production engine (the ''MPEG-7 encoder'') be specified, but only the syntax and semantics of the description tools and the corresponding decoding process. Following the principle of specifying the minimum for maximum usability, MPEG will concentrate on standardizing the tools to express the audiovisual description. The development of audiovisual content analysis tools—automatic or semiautomatic—as well as of the tools that will use the MPEG-7 descriptions—search engines and filters—will be a task for the industries that will build and sell MPEG-7 enabled products [9]. This strategy ensures that good use can be made of the continuous improvements in the relevant technical areas. The consequence is that new automatic analysis tools can always be used, even after the standard is finalized, and that it is possible to rely on competition to obtain ever better results. In fact, it will be

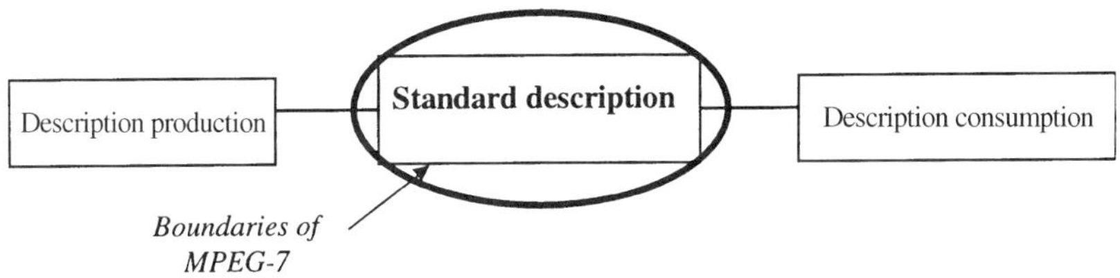

Figure 1 Boundaries of the normative parts of the MPEG-7 standard.

the very nonnormative tools that products will use to distinguish them, which only reinforces their importance.

Content may typically be described using so-called low-level features, such as color and shape for images and pitch for speech, as well as through more high-level features such as genre classification, and rating. For example, textual data, e.g., annotations, are essential to express concepts such as names, places, and dates. It should be clear that the burden of creating descriptions, thereby introducing more or less interpretation and ''meaning,'' is on the description generator and hence outside the boundaries of the standard. Whereas more conceptual descriptions are still mainly based on manual or semiautomatic processes that create textual output, low-level descriptions rely on automatic analysis tools and use other types (nontextual) of audiovisual features. The identification engines must provide the adequate interface and querying tools to allow mixing low- and high-level querying data to get the best results in each of the application domains. Modalities such as text-based only, subject navigation, interactive browsing, visual navigation and summarization, and search by example, as well as using features and sketches [10], will be supported by MPEG-7 compliant descriptions. Not only the type of tasks to perform but also the application, the environment it is used in, and the available descriptions will determine the most adequate identification modality to use. Again, this will be a competition area outside MPEG-7's jurisdiction.

Although low-level features are easier to extract, the truth is that most (nonprofessional) consumers would like to express their queries as much as possible at the semantic level. This discrepancy raises an important issue in audiovisual content description: where is the semantic mapping between signal processing–based, low-level features and image understanding–based, high-level semantic features done? Because the mapping from low level to high level is intrinsically context dependent, there are basically two major approaches:

1. *Semantic mapping before description*: The description is high level, context dependent, according to the mapping used (and corresponding subjective criteria). Only high-level querying is possible because low-level description information is not available. The description process limits the set of possible queries that can be dealt with. This approach may not be feasible for real-time applications.
2. *Semantic mapping after description*: The description is low level, not context dependent, leaving the mapping from low level to high level to the identification engine. This approach allows different mapping solutions in different application domains, avoiding querying limitations imposed by the description process. Moreover, low-level and high-level querying may coexist.

The first solution puts the semantic mapping in the description generator, which limits the genericity of the description; the second solution moves the mapping to the identification engine, maximizing the number of queries that can be answered because many more semantic mapping criteria can be used. As it is possible to design a description framework that combines low-level and high-level features in every single description (and thus combines the two approaches), there is no need to choose one and only one of the two approaches. This mapping dichotomy, however, is central to MPEG-7, notably for the selection of the features to be expressed in a standardized way, and constitutes one of the major differences between MPEG-7 and other emerging audiovisual description solutions. In fact, according to this dichotomy, description genericity requires low-level, non–context-dependent features, pointing toward the harmonious integration of low-level, signal processing–based features and high-level, audiovisual understanding–based features.

It is widely recognized that audiovisual content description strongly depends on the application domain. As it is impossible to have MPEG-7 specifically addressing every single application—from the most relevant to the more specific—it is essential that MPEG-7 is an open standard. Thus it must be possible to extend it in a normative way to address description needs, and thus application domains, that cannot be fully addressed by the core description tools. This extensibility must give MPEG-7 the power to address as many applications as possible in a standard way, even if only the most important applications drive its development. Building new description tools (possibly based on the standard ones) requires a description language. This language is the description tool that should allow MPEG-7 to keep growing, by both answering to new needs and integrating newly developed description tools.

In conclusion, in comparison with other available or emerging solutions for multimedia description, MPEG-7 can be characterized by:

a. Its genericity: the capability to describe content from many application environments;
b. Its object-based data model: it is capable of independently describing individual objects within a scene (MPEG-4 or any other content);
c. The integration of low-level and high-level features/descriptors into a single architecture, allowing to combine the power of both types of descriptors; and
d. Its extensibility, provided by the Description Definition Language which allows MPEG-7 to keep growing, to be extended to new application areas, to answer to newly emerging needs, and to integrate novel description tools.

III. NORMATIVE TOOLS AND TERMINOLOGY

As established before, it is essential that the standard define only the minimum set of tools essential for interworking. Following this principle, MPEG chose to standardize the following elements [5]:

A set of *descriptors*
A set of *description schemes*
A (set of) *coding scheme*(s) for the descriptions
A language to specify description schemes and (possibly) descriptors called *Description Definition Language* (DDL)

These are the normative elements of the standard. ''Normative'' means that if these elements are implemented, they must be implemented according to the standardized specification. Typically, these are the elements whose normative specification is essential to guarantee interoperability. Feature extraction, similarity measures, and search engines are also relevant but will not be standardized. Due to their relevance for the MPEG-7 process, the MPEG-7 requirements document [5] defines the following terms:

Data: Data is audio-visual information that will be described using MPEG-7, regardless of storage, coding, display, transmission medium, or technology. This definition is intended to be sufficiently broad to encompass video, film, music, speech, text, and any other medium. Examples are an MPEG-4 stream, a video tape, a CD containing music, sound or speech, a picture printed on paper, and an interactive multimedia installation on the Web.

Feature: A Feature is a distinctive characteristic of the data, which signifies something to somebody. Some examples are the color of an image, pitch of a speech segment, rhythm of an audio segment, camera motion in a video, style of a video, the title of a movie, the actors in a movie, etc.

Descriptor (D): A Descriptor (D) is a representation of a Feature. A Descriptor defines the syntax and the semantics of the Feature representation. It is possible to have several descriptors representing a single feature, e.g., to address different relevant requirements. Possible descriptors are a color histogram, the average of the frequency components, a motion field, the text of the title, etc.

Descriptor Value: A Descriptor Value is an instantiation of a Descriptor for a given data set (or subset thereof). Descriptor values are combined via the mechanism of a description scheme to form a description. Examples are a value to express the duration of a movie, and a set of characters to give the name of a song.

Description Scheme (DS): A Description Scheme (DS) specifies the structure and semantics of the relationships between its components, which may be both Descriptors and Description Schemes. A DS provides a solution to model and describe audiovisual content in terms of structure and semantics. Examples are a movie, temporally structured as scenes and shots, including some textual descriptors at the scene level, and color, motion and some audio descriptors at the shot level.

Description: A Description consists of a DS (structure) and the set of Descriptor Values (instantiations) that describe the Data. A description contains or refers to a fully or partially instantiated DS.

Coded Description: A Coded Description is a Description that has been encoded to fulfil relevant requirements such as compression efficiency, error, resilience, random access, etc.

Description Definition Language (DDL): The DDL is a language that allows the creation of new Description Schemes and, possibly, Descriptors. It also allows the extension and modification of existing Description Schemes.

These concepts and tools will play a major role both in the development of the MPEG-7 standard and in the implementation of MPEG-7 enabled applications.

IV. STANDARDIZATION PROCESS

The technological landscape has changed considerably in the transition from analog to digital, and it is essential that standard makers acknowledge this by adapting the way in which they operate. Standards should offer interoperability across regions, services, and applications and no longer follow a ''system-driven approach'' by which the value of a standard is limited to a specific, vertically integrated, system. Responding to this change, MPEG has adopted a toolkit approach: a standard provides a minimum set of relevant tools, which can be assembled according to specific industry needs, to provide the maximum interoperability at minimum complexity and cost [2].

The success of MPEG standards is largely based on this toolkit approach, bounded by the ''one functionality, one tool'' principle. This translates to a standard development process with the following major steps:

1. Identify relevant applications using input from MPEG members.
2. Identify the functionalities needed by the preceding applications.
3. Describe the requirements following from the functionalities in such a way that common requirements can be identified for different applications.
4. Identify which requirements are common across the areas of interest and which are not common but still relevant.
5. Specify tools that support the preceding requirements in three phases:
    a. A public call for proposals is issued, asking all interested parties to submit technology that is relevant to fulfill the identified requirements.
    b. The proposals are evaluated in a well-defined, adequate, and fair evaluation process, which is published with the call itself. The process can entail, e.g., subjective testing, objective comparison, and evaluation by experts.
    c. As a result of the evaluation, the technology best addressing the requirements is selected. This is the start of a collaborative process to draft and improve the standard. The collaboration includes the definition and improvement of a working model, which embodies early versions of the standard and can include nonnormative parts. The working model evolves by comparing different alternative tools with those already in the working model, the so-called core experiments (CEs).
6. Verify that the tools developed can be used to assemble the target systems and provide the desired functionalities. This is done by means of verification tests. For MPEG-1 through 4, these tests were subjective evaluations of the decoded quality. For MPEG-7, they will have to assess efficiency in identifying the right content described using MPEG-7.

This process is not rigid: some steps may be taken more than once and iterations are sometimes needed. The time schedule is, however, always closely observed by MPEG. Although all decisions are taken by consensus, the process keeps a high pace, allowing MPEG to provide timely technical solutions.

For MPEG-7, this process translates to the work plan presented in Table 1. After

Table 1 MPEG-7 Time Schedule

October 16, 1998	Call for proposals Final version of the MPEG-7 Proposal Package Description.
December 1, 1998	Preregistration of proposals
February 1, 1999	Proposals due
February 15–19, 1999	Evaluation of proposals (in an ad hoc group meeting held in Lancaster, UK)
March 1999	First version of MPEG-7 experimentation model
December 1999	Working draft stage (WD)
October 2000	Committee draft stage (CD)
March 2001	Final committee draft stage (FCD) after ballot with comments
July 2001	Final draft international standard stage (FDIS) after ballot with comments
September 2001	International Standard after yes or no ballot

defining the requirements (an ongoing process, as new requirements may arise), an open call for proposals was issued in October 1998 [11]. The call asked for relevant technology fitting the MPEG-7 requirements [12]. After the evaluation of the technology that was received [6], choices were made and the collaborative phase started with the most promising tools. In the course of developing the standard, additional calls may be issued when not enough technology is available within MPEG to meet the requirements, but there must be indications that the technology does indeed exist.

Two working tools play a major role in the collaborative development phase that follows the initial competitive phase: the experimentation model (XM) and core experiments (CEs) [13].

In MPEG-1 the (video) working model was called the simulation model (SM), in MPEG-2 the (video) working model was called the test model (TM), and in MPEG-4 the (video, audio, SNHC, and systems) working models were called verification models (VMs). In MPEG-7 the working model is called experimentation model (XM), intending an alphabetical pun. It is important to realize that neither the experimentation model nor any of the core experiments will end up in the standard itself, as these are just working tools to ease the development phase. The ''collaboration after competition'' approach followed by MPEG is one of its major strengths because it allows concentration of the effort of many research teams throughout the world on the further improvement of technology that was already demonstrated to be (among) the best.

According to the MPEG-7 proposal package description (PPD) [13], the two MPEG major development tools are described as follows.

A. Experimentation Model (XM)

The MPEG-7 experimentation model is a complete description framework such that an experiment performed by multiple independent parties will produce essentially identical results. The XM enables checking the relative performance of different tools, as well as improving the performance of selected tools [14].

The XM will be built after screening the proposals answering the call for proposals. The XM is not the best proposal but a combination of the best tools, independent of the proposal to which they belong. The XM will have components for evaluating and improving the DDL, DSs, Ds, and coding tools. The XM typically includes normative and nonnormative tools to create the ''common framework'' that allows performing adequate evaluation and comparison of tools targeting the continuous improvement of the technology included in the XM. After the first XM is established, new tools can be brought to MPEG-7 and will be evaluated inside the XM following a core experiment procedure. The XM evolves through versions as core experiments verify the inclusion of new techniques or prove that included techniques should be substituted. At each XM version, only the best performing tools are part of the XM. If any part of a proposal is selected for inclusion in the XM, the proposer must provide the corresponding source code for integration into the XM software in the conditions specified by MPEG.

B. Core Experiments

The improvement of the XM will start with a first set of core experiments, which will be defined at the conclusion of the evaluation of proposals. The core experiments process allows multiple, independent, directly comparable experiments to be performed to determine whether or not a proposed tool has merit. Proposed tools may target the substitution of a tool in the XM or the direct inclusion in the XM to provide a new relevant functionality. Improvements and additions to the XM will be based on the results of core experiments.

A core experiment has to be completely and uniquely defined so that the results are unambiguous. In addition to the specification of the tool to be evaluated, a core experiment specifies the conditions to be used, again so that results can be compared. A core experiment is proposed by one or more MPEG experts and it is accepted by consensus, provided that two or more independent experts agree to perform the experiment. Proposers whose tools are accepted for inclusion in the XM must provide the corresponding source code for the XM software in the conditions specified by MPEG.

V. APPLICATIONS AND REQUIREMENTS

As already mentioned, MPEG-7 requirements are application driven. The relevant applications are all those that should be enabled by the MPEG-7 vision and associated tools. Addressing new applications has the same priority as improving the power of existing ones. There are many application domains that should benefit from the MPEG-7 standard, and no application list drawn up today can be exhaustive. According to the MPEG-7 applications document [4], MPEG-7 targets at least the following application domains:

Education	Surveillance
Journalism	Biomedical applications
Tourist information	Shopping
Cultural services	Architecture, real estate, interior design
Entertainment	Social
Investigation services, forensics	Film, video, and radio archives
Geographical information systems	Audiovisual content production
Remote sensing	Multimedia portals

The MPEG-7 applications document includes examples of both improved existing applications and new ones and organizes the example applications into three sets, as follows [4]:

Pull applications: Applications mainly following a pull paradigm, notably storage and retrieval of audiovisual databases, delivery of pictures and video for professional media production, commercial musical applications, sound effects libraries, historical speech database, movie scene retrieval by memorable auditory events, and registration and retrieval of trademarks.

Push applications: Applications mainly following a push paradigm, notably user agent–driven media selection and filtering, personalized television services, intelligent multimedia presentations, and information access facilities for people with special needs.

Specialized professional applications: Applications that are particularly related to a specific professional environment, notably teleshopping, biomedical, remote sensing, educational, and surveillance applications.

This is a living list, which will be augmented in the future, intended to give the industry—clients of the MPEG work—some hints about the application domains addressed. If MPEG-7 enables new and ''unforeseen'' applications to emerge, this will show the strength of the toolkit approach. For each application listed, the MPEG-7 applications document gives a description of the application, the corresponding requirements, and a list of relevant work and references, including a generic architecture of MPEG-7 enabled systems.

Figure 2 shows a generic architecture of MPEG-7 enabled systems, notably for searching and filtering applications. After the generation of the MPEG-7 description using the multimedia (MM) content, whatever its format, the description is encoded using an MPEG-7 encoder engine and stored or transmitted. The MPEG-7 coded description is then consumed in a searching or filtering framework by a user, who may be human or a machine and who may finally request the access to the corresponding multimedia content.

In order to develop useful tools for the MPEG-7 toolkit, functionality requirements have been extracted from the identified applications. The MPEG-7 requirements [5] are currently divided into descriptor and description scheme requirements, DDL requirements, and system requirements. Whenever applicable, visual and audio requirements are considered separately. The requirements apply, in principle, to both real-time and non–real-time as well as to offline and streamed applications. They should be meaningful to as many applications as possible. Just to get a taste of the current MPEG-7 requirements, which are still under development and refinement, a list of the most relevant requirements is presented in the following:

A. Descriptor and Description Scheme Requirements

It is expected that the descriptors and description schemes to be standardized will constitute the core of MPEG-7 to which the DDL will add some extensibility capabilities. The descriptor and description scheme requirements are organized in three groups [5]:

1. General requirements
   Types of features: MPEG-7 shall support descriptions using various types of features such as annotations, *N*-dimensional spatiotemporal features (e.g., the duration of a music segment), objective features (e.g., number of beds in a hotel, shape of an object, audio pitch), subjective features (features subject to different interpretations such as how nice, happy, or fat someone is, topic, style), production features (e.g., date of data acquisition, producer, director, performers, roles,

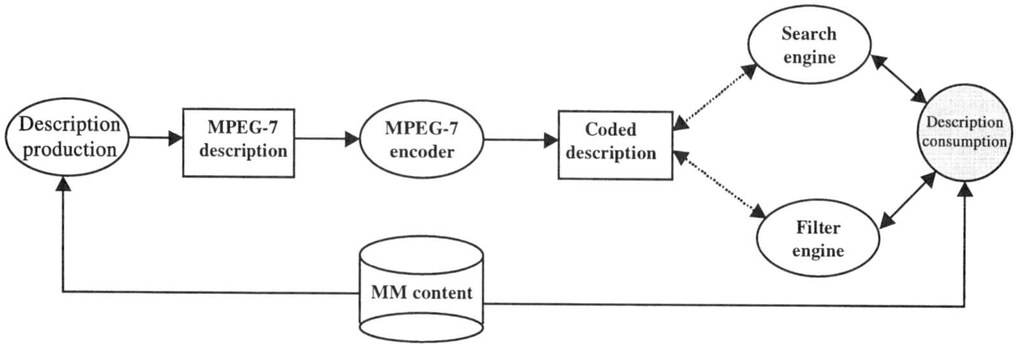

Figure 2 Generic architecture of MPEG-7 enabled systems.

production company, production history), composition information (e.g., how the scene is composed, editing information), and concepts (e.g., events, activity).
Abstraction levels for multimedia material: MPEG-7 shall support the means to describe multimedia data hierarchically according to different abstraction levels to represent efficiently user's information needs at different levels.
Cross-modality: MPEG-7 shall support audio, visual, or other descriptors that allow queries based on visual descriptions to retrieve audio data and vice versa (e.g., using an excerpt of Pavarotti's voice as the query to retrieve video clips where Pavarotti is singing or video clips where Pavarotti is present).
Multiple descriptions: MPEG-7 shall support the ability to handle multiple descriptions of the same material at several stages of its production process as well as descriptions that apply to multiple copies of the same material.
Description scheme relationships: MPEG-7 description schemes shall express relationships between descriptors to allow the use of the descriptors in more than one description scheme. The capability to encode equivalence relationships between descriptors in different description schemes shall also be supported.
Descriptor priorities: MPEG-7 shall support the prioritization of descriptors in order that queries may be processed more efficiently. The priorities may denote some sort of level of confidence, reliability, etc.
Descriptor hierarchy: MPEG-7 shall support the hierarchical representation of different descriptors in order that queries may be processed more efficiently in successive levels where N-level descriptors complement $(N - 1)$-level descriptors.
Descriptor scalability: MPEG-7 shall support scalable descriptors in order that queries may be processed more efficiently in successive layers where N-layer description data are an enhancement or refinement of $(N - 1)$-layer description data (e.g., texture and shape scalability in MPEG-4).
Description of temporal range: MPEG-7 shall support the association of descriptors to different temporal ranges, both hierarchically (descriptors are associated with the whole data or a temporal subset of it) and as sequentially (descriptors are successively associated to successive time periods).
Direct data manipulation: MPEG-7 shall support descriptors that can act as handles referring directly to the data, to allow manipulation of the multimedia material.
Language of text-based descriptions: MPEG-7 text descriptors shall specify the language used. MPEG-7 text descriptors shall support all natural languages.
Translations in text descriptions: MPEG-7 text descriptions shall provide the means to contain several translations, and it shall be possible to convey the relation between the descriptions in different languages.

2. Functional requirements.
Content-based retrieval: MPEG-7 shall support the effective (''you get what you are looking for and not other stuff'') and efficient (''you get what you are looking for, quickly'') retrieval of multimedia data, based on their contents, whatever the semantic involved.
Similarity-base retrieval: MPEG-7 shall support descriptions making it possible to rank-order multimedia content by the degree of similarity with the query.
Associated information: MPEG-7 shall support the use of information associated to the multimedia data, such as text, to complement and improve data

retrieval (e.g., diagnostic medical images are retrieved not only in terms of image content but also in terms of other information associated with the images, such as text describing the diagnosis, treatment plan, etc.).

Streamed and stored descriptions: MPEG-7 shall support both streamed (synchronized with content) and nonstreamed data descriptions.

Distributed multimedia databases: MPEG-7 shall support the simultaneous and transparent retrieval of multimedia data in distributed databases.

Referencing analog data: MPEG-7 descriptions shall support the ability to reference and describe audiovisual objects and time references of analog format.

Interactive queries: MPEG-7 descriptions shall support mechanisms to allow interactive queries.

Linking: MPEG-7 shall support a mechanism allowing source data to be located in space and in time. MPEG-7 shall also support a mechanism to link to related information.

Prioritization of related information: MPEG-7 shall support a mechanism allowing the prioritization of the related information mentioned above.

Browsing: MPEG-7 shall support descriptions making it possible to preview information content in order to aid users to overcome their unfamiliarity with the structure and/or types of information or to clarify their undecided needs.

Associate relations: MPEG-7 shall support relations between components of a description.

Intellectual property information: MPEG-7 shall enable the inclusion (e.g., create the hooks) of copyright, licensing, and authentication information related to the data described and the corresponding descriptions. As copyright and licensing may change, suitable timing information and other relevant associated information shall also be considered.

3. Coding requirements.

   Description efficient representation: MPEG-7 shall support the efficient representation of data descriptions.

   Description extraction: MPEG-7 shall standardize descriptors and description schemes that are easily extractable from uncompressed and compressed data, notably according to the most widely used compression formats.

   Robustness to information errors and losses: MPEG-7 shall provide error resilience capabilities at the description layer that guarantee graceful decoding behavior.

Moreover, the following visual and audio specific requirements were identified:

Data visualization: MPEG-7 shall support a range of multimedia data descriptions with increasing capabilities in terms of visualization. This means that MPEG-7 data descriptions shall allow a more or less sketchy visualization of the indexed data.

Data sonification: MPEG-7 shall support a range of multimedia data descriptions with increasing capabilities in terms of sonification.

B. DDL Requirements

Although it is largely recognized that MPEG-7 shall have some degree of extensibility, it is as yet unclear how and to what extent this target will be achieved. For the moment, the following requirements express the boundaries of the MPEG-7 DDL:

Compositional capabilities: The DDL shall allow new DSs and Ds to be created and existing DSs to be modified or extended.

Unique identification: The DDL shall allow a unique identification of Ds and DSs. An example for unique identification is a namespace, that uniquely qualifies the element names and relationships and makes these names recognizable to avoid name collisions on elements that have the same name but are defined in different vocabularies.

Primitive data types: The DDL shall provide a set of primitive data types, e.g., text, integer, real, date, time/time index, version etc.

Composite data types: The DDL shall be able to describe composite data types such as histograms, graphs, RGB values, enumerated types, etc.

Multiple media types: The DDL shall provide a mechanism to relate Ds to data of multiple media types of inherent structure, particularly audio, video, audio-visual presentations, the interface to textual description, and any combinations of these.

Various types of DS instantiations: The DDL should allow various types of DS instantiations full, partial, full-mandatory, partial-mandatory.

Relationships within a DS and between DSs: The DDL shall be able to express spatial, temporal, structural, and conceptual relationships between the elements of a DS, and between DSs.

Relationship between description and data: The DDL shall supply a rich model for links and/or references between one or several descriptions and the described data.

Link to ontologies: the DDL shall supply a linking mechanism between a description and several ontologies.

Platform independence: The DDL shall be platform and application independent.

Grammar: The DDL shall follow an unambiguous and easily parsable grammar.

Validation of constraints: An MPEG-7 DDL parser shall be capable of validating the following elements: values of properties, structures, related classes, and values of properties of related classes.

IPMP mechanisms: The DDL shall provide a mechanism for the expression of intellectual property management and protection (IPMP) for description schemes and descriptors.

Human readability: The DDL shall allow that Ds and DSs can be read by humans.

C. Systems Requirements

Although at this early stage of the MPEG-7 evolution the boundaries of MPEG-7 Systems are not at all sharp, it is becoming clear that there are a few requirements that necessarily fall in the area of what has been traditionally known in MPEG as Systems. This area encompasses the tools that are above audio and visual tools, such as multiplexing and synchronization tools. The very initial MPEG-7 systems requirements include [5]:

Multiplexing of descriptions: MPEG-7 shall make possible to: a) embed multiple MPEG-7 descriptions into a single data stream: b) embed MPEG-7 description(s) into a single data stream together with associated content.

Flexible access to partial descriptions at the systems level: MPEG-7 shall allow the selection of partial descriptions at the systems level without the need to decode the full description.

Temporal synchronization of content with descriptions: MPEG-7 shall allow the temporal association of descriptions with content (AV objects) that can vary over time.

Synchronization of multiple descriptions over different physical locations: MPEG-7 shall provide a mechanism to keep consistent multiple descriptions of a content, if they exist.

Physical location of content with associated descriptions: MPEG-7 shall allow the association descriptions with content (AV objects) that can vary in physical location.

Transmission mechanisms for MPEG-7 streams: MPEG-7 shall allow the transmission of MPEG-7 descriptions using a variety of transmission protocols.

File format: MPEG-7 shall define a file format supporting flexible user-defined access to parts of the description(s) that are contained. Linking mechanisms with other files and described content should also be provided.

Robustness to information errors and losses: MPEG-7 shall provide error resilience mechanisms that guarantee graceful behavior of the MPEG-7 system in the case of transmission errors. The precise error conditions to withstand are to be identified.

Quality of Service (QoS): MPEG-7 shall provide a mechanism for defining Quality of Service (QoS) for the transmission of MPEG-7 description streams.

IPMP mechanisms: MPEG-7 shall define intellectual property management and protection (IPMP) mechanisms for the protection of MPEG-7 descriptions.

The requirements drive the MPEG-7 development process by determining criteria for the evaluation of technical proposals and for the core experiments, setting the objectives for the later verification of the standard, and defining possible conformance points.

VI. EVALUATION PROCESS

After an initial period dedicated to the specification of objectives and the identification of applications and requirements, MPEG-7 issued in October 1998 a call for proposals [11] to gather the best available technology. MPEG called for the following technology:

For the normative part of the standard (normative tools):

Descriptors
Description schemes
Description Definition Language (DDL)
Coding methods for compact representation of descriptions
Systems tools addressing the MPEG-7 Systems requirements [5]

The proposals for descriptors, description schemes, DDL, and coding schemes were evaluated by MPEG experts in February 1999 in Lancaster (UK), following the procedures defined in the MPEG-7 Evaluation document [6,12,15,16]. The proposals addressing MPEG-7 Systems tools were only reviewed and summarized, because the MPEG-7 Systems requirements were not yet completely stable.

For the development of the standard, to be used in the XM (nonnormative tools):

Feature extraction methods
Searching methods
Evaluation and validation techniques

As for the MPEG-7 Systems tools, nonnormative tools were reviewed and summarized in order to gather relevant information for the building of the first MPEG-7 XM.

As in past MPEG calls for proposals, proposals were assessed to identify the most promising technology to be further developed collaboratively. It was soon clear that the task of evaluating proposals for descriptors, description schemes, and the DDL was not a simple one, notably because of the many variables that determine their performance. For example, the retrieval performance of a certain descriptor can depend on the described feature, the extraction algorithm, the coding, and the matching criteria applied. It also varies with the type of query, such as query by example or any other type of query. Because the extraction algorithm and the matching criteria are not to be standardized within MPEG-7, evaluating the performance of a descriptor just using normative standard elements is basically impossible. Similar reasoning applies for description schemes and the DDL. These difficulties were confirmed by the absolute lack of recognized methodologies for the evaluation of audiovisual description tools.

To overcome this situation, it was important to remember the objective of the evaluation process: to allow the start of the collaborative phase. This means that evaluation is not the end but rather the start of the process, bearing in mind that there is time for further developments and clarifications. In fact, the evaluation process has produced not only several top-performing tools but also a set of less evident cases to be examined in well-defined core experiments.

To deal with the constraints and conditions mentioned before, the MPEG-7 evaluation process for descriptors, description schemes, DDL, and coding schemes included two default evaluation phases as follows:

1. *Evaluation by experts of proposal documentation*: MPEG experts assessed the merits of the proposals against evaluation criteria and requirements using the proposal documentation. This documentation included a summary and a detailed structured description of the proposal, as well as the answers to a questionnaire proposed by MPEG [6]. The MPEG-7 evaluation document [6] lists the relevant set of criteria used in the evaluation of the various types of proposals.
2. *Presentation and demonstration by proposers*: Most proposers made a presentation to MPEG evaluation experts demonstrating the value of the proposed tool and highlighting its range of use. Some demonstrations also provided evidence for the claims.

Besides these two default evaluation phases to be used for all types of proposals, some additional phases were performed for some descriptors and the DDL. For the descriptors allowing similarity-based retrieval, a third evaluation phase was performed:

3. *Similarity-based retrieval*: The proposed descriptors were used to retrieve, e.g., pictures from the available set of pictures. To this end, proposers had indexed the content relevant for their proposal (e.g., pictures for shape descriptors) and brought a running similarity-based retrieval program to the MPEG-7 evaluation meeting. Using relevant input selected by the MPEG experts, proposers had to find similar items from the relevant content set, using their descriptor and pro-

gram. Experts following their own perception of similarity and dissimilarity for the relevant feature judged the ranking provided by the retrieval program.

Moreover, for the DDL a third evaluation phase was also used:

3. *DDL core experiments*: Because it is essential that the DDL is at least able to express the selected description schemes, some core experiments for the most promising DDL proposals after the first two evaluation phases were defined. These core experiments were to be performed after the two first default evaluation phases and their definition took into account the evaluation results for the descriptors and description schemes.

A special set of multimedia content was provided to the proposers for use in the evaluation process and is still used in the collaborative phase [17]. The set consists of 32 compact discs with sounds, pictures, and moving video. This content has been made available to MPEG under the licensing conditions defined in Reference 18.

Although very simple methodologies were used for the evaluation of the audiovisual description tools in the MPEG-7 competitive phase, it is expected that new, more powerful methodologies will be developed during the collaborative phase in the context of the core experiments. As in the past, e.g., for MPEG-4 with the Double Stimulus Continuous Quality Evaluation (DSCQE) method, new evaluation methodologies are one of the most interesting side results of the MPEG standardization processes.

VII. CONCLUSION

MPEG-1 and MPEG-2 have been successful standards that have enabled widely adopted commercial products, such as digital audio broadcasting, digital television, the DVD, and many video-on-demand trials and services. MPEG-4 sets the first intrinsically digital audiovisual representation standard in the sense that audiovisual data are modeled as a composition of objects, both natural and synthetic, with which the user may interact. Following these projects, MPEG has decided to address the problem of quickly and efficiently identifying various types of multimedia material of interest to the user. For this, a standardized way of describing audiovisual information will be developed.

It looks as though MPEG, a major factor responsible for the widespread use of digital audiovisual content in both professional and consumer markets by setting the MPEG-1, -2, and -4 standards, took on the task of ''putting some order in the audiovisual content chaos.'' Without doubt, a big challenge!

ACKNOWLEDGMENTS

The research is in part funded by the European Commission under the ACTS DICEMAN project. The authors thank all the MPEG-7 members for the interesting and fruitful discussions in meetings and by e-mail, which substantially enriched their technical knowledge. As parts of this chapter were written on the basis of the official MPEG documents, the MPEG community is in a way a coauthor. Finally, F. Pereira acknowledges the Portuguese framework PRAXIS XXI for his participation in MPEG meetings under the project ''Normalização de métodos avançados de representação de vídeo.''

REFERENCES

1. MPEG home page, http://www.cselt.it/mpeg/.
2. L Chiariglione. The challenge of multimedia standardization. IEEE Multimedia 4(2):79–83, 1997.
3. MPEG Requirements Group. MPEG-7: Context, objectives, technical roadmap. Doc. ISO/MPEG N2729 MPEG Seoul Meeting, 1999.
4. MPEG Requirements Group. Applications for MPEG-7. Doc. ISO/MPEG N2728. MPEG Seoul Meeting, 1999.
5. MPEG Requirements Group. MPEG-7 requirements. Doc. ISO/MPEG N2727. MPEG Seoul Meeting, 1999.
6. MPEG Requirements Group. MPEG-7 evaluation process. Doc. ISO/MPEG N2463. MPEG Atlantic City Meeting, 1998.
7. R Koenen, F Pereira, L Chiariglione. MPEG-4: Context and objectives. Image Commun 9(4): 295–304, 1997.
8. MPEG Systems Group. Text of ISO/IEC FDIS 14496-1: Systems. Doc. ISO/MPEG N2501. Atlantic City MPEG Meeting, 1998.
9. P Correia, F Pereira. The role of analysis for content-based representation, retrieval, and interaction. Signal Process J Special issue on Video Sequence Segmentation for Content-Based Processing and Manipulation. 66(2):125–142, 1998.
10. S Chang, A Eleftheriadis, R McClintock. Next-generation content representation, creation, and searching for new-media applications in education. Proc IEEE 86:884–904, 1998.
11. MPEG Requirements Group. MPEG-7 call for proposals. Doc. ISO:MPEG N2469. MPEG Atlantic City Meeting, 1998.
12. MPEG Requirements Group. MPEG-7 list of proposal pre-registrations. Doc. ISO/MPEG N2567. MPEG Rome Meeting, 1998.
13. MPEG Requirements Group. MPEG-7 Proposal Package Description (PPD). Doc. ISO/MPEG N2464. MPEG Atlantic City Meeting, 1998.
14. MPEG Requirements Group. MPEG-7 eXperimentation Model (XM). Doc. ISO/MPEG N2571. MPEG Rome Meeting, 1998.
15. MPEG Requirements Group. MPEG-7 proposal evaluation logistics guide. Doc. ISO/MPEG N2569. MPEG Rome Meeting, 1998.
16. MPEG Requirements Group. MPEG-7 evaluation results. Doc. ISO/MPEG N2730. MPEG Seoul Meeting, 1999.
17. MPEG Requirements Group. Description of MPEG-7 Content Set. DOC. ISO/MPEG N2467, MPEG Atlantic City Meeting, 1998.
18. MPEG Requirements Group. Licensing agreement for MPEG-7 content set. Doc. ISO/MPEG N2466. MPEG Atlantic City Meeting, 1998.

Index